PROGRESS IN POLYAMINE RESEARCH

Novel Biochemical, Pharmacological, and Clinical Aspects

ADVANCES IN EXPERIMENTAL MEDICINE AND BIOLOGY

Recent Volumes in this Series

Volume 243
EICOSANOIDS, APOLIPOPROTEINS, LIPOPROTEIN PARTICLES, AND ATHEROSCLEROSIS
Edited by Claude L. Malmendier and Petar Alaupovic

Volume 244
THE EXPANDING ROLE OF FOLATES AND FLUOROPYRIMIDINES IN CANCER CHEMOTHERAPY
Edited by Youcef Rustum and John J. McGuire

Volume 245
MECHANISMS OF PHYSICAL AND EMOTIONAL STRESS
Edited by George P. Chrousos, D. Lynn Loriaux, and Philip W. Gold

Volume 246
PREDIABETES
Edited by Rafael A. Camerini-Davalos and Harold S. Cole

Volume 247A
KININS V, Part A
Edited by Keishi Abe, Hiroshi Moriya, and Setsuro Fujii

Volume 247B
KININS V, Part B
Edited by Keishi Abe, Hiroshi Moriya, and Setsuro Fujii

Volume 248
OXYGEN TRANSPORT TO TISSUE XI
Edited by Karel Rakusan, George P. Biro, Thomas K. Goldstick, and Zdenek Turek

Volume 249
MINERAL ABSORPTION IN THE MONOGASTRIC GI TRACT:
Chemical, Nutritional, and Physiological Aspects
Edited by Frederick R. Dintzis and Joseph A. Laszlo

Volume 250
PROGRESS IN POLYAMINE RESEARCH:
Novel Biochemical, Pharmacological, and Clinical Aspects
Edited by Vincenzo Zappia and Anthony E. Pegg

A Continuation Order Plan is available for this series. A continuation order will bring delivery of each new volume immediately upon publication. Volumes are billed only upon actual shipment. For further information please contact the publisher.

PROGRESS IN POLYAMINE RESEARCH

Novel Biochemical, Pharmacological, and Clinical Aspects

Edited by

Vincenzo Zappia

Institute of Biochemistry of Macromolecules
University of Naples
Naples, Italy

and

Anthony E. Pegg

The Milton S. Hershey Medical Center
The Pennsylvania State University
Hershey, Pennsylvania

PLENUM PRESS • NEW YORK AND LONDON

Library of Congress Cataloging in Publication Data

International Symposium on Polyamines in Biochemical and Clinical Research (1988: Sorrento, Italy)
Progress in polyamine research: novel biochemical, pharmacological, and clinical aspects / edited by Vincenzo Zappia and Anthony E. Pegg.
p. cm. — (Advances in experimental medicine and biology; v. 250)
"Proceedings of the International Symposium on Polyamines in Biochemical and Clinical Research, held June 13–17, 1988, in Sorrento, Naples, Italy" — T.p. verso.
Includes bibliographies and index.
ISBN 978-1-4684-5639-4 ISBN 978-1-4684-5637-0 (eBook)
DOI 10.1007/978-1-4684-5637-0
1. Polyamines — Congresses. I. Zappia, Vincenzo. II. Pegg, Anthony E. III. Title.
[DNLM: 1. Polyamines — congresses. W1AD559 v. 250 / QU 60 I609p 1988]
QP801.P638I58 1988
574.19′24 — dc19
DNLM/DLC 89-3664
for Library of Congress CIP

Proceedings of the International Symposium on Polyamines in Biochemical and Clinical Research, held June 13–17, 1988, in Sorrento, Naples, Italy

Softcover reprint of the hardcover 1st edition 1988
A Division of Plenum Publishing Corporation
233 Spring Street, New York, N.Y. 10013

PREFACE

This book contains the scientific contributions presented at an International Symposium held in Sorrento, Italy, in June 1988 under the auspices of the University of Naples, the Italian Society of Biochemistry, and the National Research Council.

The modern history of polyamines dates back to 1958 when the Tabors and Rosenthal first described the outlines of their biological synthesis. From then on, and particularly in the last ten years, a veritable explosion of Literature, characterized by thousands of papers, has witnessed the interest of the scientific community toward these molecules. Perhaps the old statement that "polyamines are molecules in search of a function" is no longer true today. A large number of effects exerted by these simple molecules are well known, and in many cases the mechanisms underlying these effects have been elucidated.

The first section of the volume is entirely devoted to the enzymology and molecular biology of ornithine decarboxylase. Since its discovery by Gale more than forty years ago, this can be considered among the most widely studied enzymes in biology, and one of the most complex models in enzyme regulation. The mechanism of control of the enzyme activity at the transcriptional, post-transcriptional and post-translational levels, as well as the fine regulation by antizyme, are discussed in detail.

The second group of contributions deals with AdoMet decarboxylase, propylamine transferase, polyamine oxidase and the other enzymes related to polyamine interconversion and regeneration.

A separate section concerns the biological roles of MTA and the mechanism of its cellular disposal. In the context of the history of science, it is interesting to recall that the biogenesis of this thioether from methionine was demonstrated by F. Schlenk even before the discovery of its precursor AdoMet by Cantoni.

The number of regulatory roles exerted by MTA on several enzyme systems and the lack of MTA phosphorylase observed, so far, solely in some malignant cells represent fascinating new areas of investigation with interesting perspectives both from the diagnostic and pharmacological points of view. In this respect, the hypothesis formulated by Guy Williams-Ashman some time ago, that in certain cells the primary function of spermine and spermidine synthases may well be to produce MTA rather than polyamines, seems less provocative today.

A central problem in the cell physiology of polyamines is the mechanism underlying their effects on cell growth and differentiation. Although the molecular basis of such phenomena is still obscure, a few universally accepted statements can be made: cellular levels of polyamines are elevated during growth, enzymatic inhibitors of their biosynthesis exert antiproliferative effects, and mutants lacking polyamine biosynthetic ezymes are endowed with low growth rates. As evidenced by the various contributions, the innovative approaches and models offered by molecular biology provide the right tools for elucidating this complex field.

The analysis of the interactions with acidic macromolecules represents another valid approach to the comprehension of the mechanism(s) of action of polyamines, although these studies are characterized by intrinsic difficulties due to the weak energy of the bonds involved. Significant attention has been devoted to the stoichiometry of binding with nucleic acids, as well as to its biological significance.

Another group of papers is related to the interactions of polyamines with acidic phospholipids in cell and mitochondrial membranes: their functional roles seem to be a consequence of this binding, particularly in membrane fusion as well as protein translocation.

The covalent binding of polyamines in some post-translational modifications of proteins, with special emphasis on transglutaminases and hypusine biosynthesis, is the subject of another chapter in this volume.

The well-documented biological effects exerted by a number of inhibitors of polyamine biosynthesis make this class of compounds good candidates for antineoplastic therapy. Moreover, notable success in therapy has been achieved in several parasitic protozoal diseases; the efficiency in the clinical treatment of trypanosomiasis by difluoromethylornithine promises to revolutionize the therapeutic protocols of this endemic disease, for which an estimated 45 million people are at risk of infection. A collection of papers discusses the role of polyamines in the regulation of the cell cycle of several protozoa, including Plasmodium and Leishmania. In particular, new data on the metabolic peculiarities of the microorganisms with respect to glutathione and trypanothione metabolism are reported.

We hope that the present volume will serve both as a reference point for those just entering the field, and will inspire new directions and approaches for on-going research programs.

The Editors would like to express their gratitude to the authors of the papers and to the Plenum Press for having made possible the rapid publication of this volume.

Vincenzo Zappia
Anthony E. Pegg

ACKNOWLEDGEMENTS

The Editors wish to express their sincere appreciation to the University of Naples, the Italian Society of Biochemistry and the National Research Council for sponsoring the Symposium.

They also gratefully acknowledge the support received from the following Italian Institutions: Ministero della Sanità, Ministero della Pubblica Istruzione, Regione Campania, Provincia di Napoli, Ente Provinciale del Turismo di Napoli, Azienda Autonoma di Soggiorno, Cura e Turismo di Napoli, Azienda Autonoma di Soggiorno di Sorrento, Beckman Analytical S.r.l., Kontron Instruments S.p.A. and Varian S.p.A. In particular, the cooperation of ARFACID, (Italian Association for Pure and Applied Research on Cancer, Ageing and Degenerative Diseases) was essential to this undertaking.

The Editors are indebted to the Organizing Committee for the successful outcome of the Symposium in Sorrento, and in particular to Maria Cartenì-Farina for her collaboration in reviewing the manuscripts.

They also thank Patricia Reynolds, who acted as Editorial Assistant, for her efficient performance in overseeing and editing this volume.

CONTENTS

ORNITHINE DECARBOXYLASE : MOLECULAR GENETICS, POST-SYNTHETIC MODIFICATIONS AND ANTIZYME

ENZYMES OF POLYAMINE METABOLISM AND RECYCLING

METABOLISM OF 5'-METHYLTHIOADENOSINE IN NORMAL AND TRANSFORMED CELLS

POLYAMINES IN CELL GROWTH AND DIFFERENTIATION

POST-SYNTHETIC MODIFICATIONS OF PROTEINS INVOLVING POLYAMINES

INTERACTIONS OF POLYAMINES WITH MACROMOLECULES

POLYAMINES IN PLANTS

POLYAMINES IN MICROORGANISMS

INHIBITORS OF POLYAMINE BIOSYNTHESIS IN THERAPY

ORNITHINE DECARBOXYLASE : MOLECULAR GENETICS, POST-SYNTHETIC MODIFICATIONS AND ANTIZYME

ANDROGEN REGULATION OF ORNITHINE DECARBOXYLASE AND S-ADENOSYLMETHIONINE DECARBOXYLASE GENE EXPRESSION

Olli A. Jänne, Anne Crozat, Mervi Julkunen, Noreen J. Hickok, Leonard Eisenberg, and Evie Melanitou

The Population Council and The Rockefeller University
1230 York Avenue, New York, New York 10021

INTRODUCTION

The action of androgenic steroids in their target tissues occurs in a receptor-mediated fashion similar to that of other steroid hormones (1-3). Over the last several years, a number of gene products regulated by androgens have been characterized and their induction kinetics in vivo elucidated. The best defined genes fall, with regard to the tissues of their expression, into three main categories: (i) α_{2u}-globulin and the major urinary protein genes are regulated by androgens in rodent liver; (ii) prostatein and seminal vesicle basic protein genes are controlled by androgens in rat accessory sex organs; and (iii) ornithine decarboxylase, β-glucuronidase, RP2, and kidney androgen-regulated protein genes exhibit androgen regulation in murine kidney.

There are some salient differences in the expression of these genes, including the multi-hormonal control of rodent liver genes (4,5), the relatively high abundancy of the gene products in rat accessory sex organs (6,7), the genetically regulated androgen responsiveness of β-glucuronidase and ornithine decarboxylase genes, and the induction kinetics of the genes (8). The different genes are, however, regulated by androgen in a similar manner in that increased accumulation of the respective mRNAs is always involved. Furthermore, at least in the case of murine renal genes, the induction is dependent on functional androgen receptors, as no change [except for some increase in kidney androgen-regulated protein mRNA accumulation (9)] in their steady-state mRNA levels is elicited by androgens in testicular feminized (Tfm/Y) mice (8,10). An intriguing feature of androgen-regulated mRNA accumulation in murine kidney is the timing and sensitivity of the response, which appear to be unique for each individual gene. We have recently reviewed these comparative aspects in the action of androgens on mouse renal cells (8). In this chapter, we describe characteristics and androgen regulation of ornithine decarboxylase (ODC) and S-adenosylmethionine decarboxylase (AdoMetDC) enzymes in rodent kidney and accessory sex organs.

ORNITHINE DECARBOXYLASE PROTEIN

ODC protein has been purified to apparent homogeneity from a variety of tissues, including the murine kidney. The enzyme protein has a subunit molecular weight of approximately 50,000 under denaturing conditions, but apprears to be a dimeric molecule in its active form (11-13). However, dimerization may not be required for the catalytic activity, since the individual 50,000-dalton subunits bind covalently the active-site directed suicidal inhibitor, 2-difluoromethylornithine (13). On two-dimensional gel electrophoresis, the purified enzyme protein exhibits some charge heterogeneity (13), but the biochemical reason for this characteristic is currently not known.

In view of the crucial importance of ODC activity for the growth and survival of eukaryotic cells, it is not very surprising that the amino acid sequence of the enzyme is well conserved among different species. Comparison of the amino acid sequences deduced from murine and human ODC cDNAs indicated that the enzymes of both species comprise 461 residues, with predicted molecular weights of 51,156 and 51,172 for the human and murine enzyme, respectively (14,15). The overall identity of the two amino acid sequences is about 90%, with 44 differences among their 461 residues. Of these changes in the sequence, 57% (25 of 44) are conservative substitutions. The distribution of amino acid differences between the two enzymes is uneven in that the amino-terminal and more hydrophobic half (residues 1-220) has only three nonconservative changes, and there are two stretches of 61 (residues 13-74) and 38 (residues 137-175) amino acids without any change. The most hydrophilic region (residues 290-340) is flanked with identical sequences of 42 and 33 amino acids in the two enzymes. In view of these similarities, it is tempting to speculate that the hydrophobic amino-terminal part and the hydrophilic region comprising residues 290-340 of the protein play an important role in the enzyme's catalytic function, such as binding of pyridoxal 5'-phosphate and L-ornithine. The carboxy-terminal part of the amino acid sequence is less well conserved and contains more than half of the nonconservative amino acid substitutions. We and other investigators (16-18) have previously shown that cleavage of a 2,000- to 3,000-dalton peptide from the murine ODC does not abolish its catalytic activity. On the basis of the amino acid data, one can predict that this peptide is derived from the nonconserved region in the carboxyl terminus of the enzyme protein. This notion is supported by studies in which truncated murine ODC cDNA sequences were expressed in *Escherichia coli*; deletion of 38 carboxy-terminal amino acids from the enzyme protein changed its catalytic activity only very little (19).

A salient feature of ODC protein is its very fast turnover rate; the half-life of the enzyme is about 15 min in mouse kidney (13,20). The murine and human enzyme proteins contain two regions rich in proline, glutamic acid, serine, and threonine (so-called PEST sequences), which seem to be typical of eukaryotic proteins exhibiting short half-lives (21). Interestingly, androgen administration at pharmacological doses altered the turnover rate of renal ODC by prolonging the half-life from 10-15 min to 100-150 min (13). The mechanisms responsible for this phenomenon remain to be elucidated. It may not have, however, any major physiological importance, since the half-life of the protein in intact male mice appears to be only 10-15 min (13).

Induction of ODC activity by virtually all stimuli is fast and short-lived. This may be due to a number of reasons, such as a short duration of action of the inducer itself, rapid turnover rate of ODC protein and/or ODC mRNA, or co-ordinate induction of molecules that facilitate inactivation of ODC. In addition to these possibilities, recent studies have revealed that putrescine and other polyamines inhibit the translation of ODC mRNA, thereby controlling their own synthesis at this level (22,23). These latter findings also explain, at least in part, why androgen induction of ODC activity in murine renal cells is much longer-lived than in other tissues; in contrast to most other experimental systems, putrescine and other polyamines are excreted and do not accumulate in renal cells in concentrations commensurate with the extent of ODC activation (24). In keeping with this postulate, Sertich and Pegg (25) have shown that administration of exogenous putrescine to androgen-treated mice leads to an acute and dramatic reduction of renal ODC activity without affecting appreciably the ODC mRNA levels in the kidney.

ODC mRNA SPECIES

Size Heterogeneity of ODC mRNAs

Mouse kidney contains at least two different ODC mRNA species of about 2.2 and 2.7 kb in size (10). DNA sequencing of ODC cDNA clones and the ODC gene has indicated that the size difference of these two mRNA species is due to a longer 3'-nontranslated region in the 2.7-kb ODC mRNA (15,26-28). The longer mRNA contains two polyadenylation/termination signals (AAUAAA); the second one is 422 nucleotides downstream from the first one, resulting in a 3'-noncoding region of 748 nucleotides in this mRNA as opposed to a 329-nucleotide long 3'-noncoding sequence in the 2.2-kb ODC mRNA (26). No appreciable heterogeneity exists at the 5'-ends of the messages (28,29). However, more than the above-mentioned two ODC mRNA species may be present in murine tissues. For example, we have identified and partially sequenced an ODC cDNA clone from a mouse kidney cDNA library, which has the poly(A)-tail added 5 nucleotides downstream from that in the 2.2-kb mRNA. That this represents an authentic mRNA sequence is supported by the presence of a corresponding ODC pseudogene in the mouse genome. In contrast to rodents, human tissues express only one ODC mRNA species that is slightly longer than the 2.2-kb mRNA in murine cells. The human genome contains at least two ODC gene loci (14,30) that were mapped to chromosomes 2 and 7. In keeping with the presence of a single mRNA species, our recent isolation and sequencing of the expressed human chromosomal ODC gene has indicated that its 3'-end contains only one polyadenylation/termination signal.

Androgen Regulation of ODC mRNA Accumulation

There is a clear sexual dimorphism in ODC activity, immunoreactive enzyme protein concentration, and ODC mRNA content in renal tissue of most mouse strains so far studied, indicating that physiological androgen concentrations regulate the expression of the ODC gene(s) in the kidney (10,13,24,31). The biological significance of this phenomenon is unclear at present. It is of interest to note, however, that the ODC gene has maintained its androgen regulation throughout the evolution of the *Mus* species as opposed to most other androgen-induced renal gene products (F.G. Berger, personal communication). In contrast to the wild-type animals, androgen-insensitive (Tfm/Y) mice show no change in ODC gene expression during prolonged testosterone administration, indicating that androgen receptors are required for this regulation (10,31).

Administration of testosterone at pharmacological doses increases the accumulation of the two ODC mRNA species in a co-ordinate fashion with relatively rapid kinetics. After a single dose of androgen, the induction of ODC mRNA parallels closely to that of the immunoreactive enzyme protein, with the initial increase in the mRNA content being detected between 2 and 6 h after steroid administration (10,13). When cycloheximide is given with testosterone, no induction of the enzyme protein occurs. These data thus indicate that the androgenic induction of ODC activity takes place throughnew enzyme protein synthesis, rather than *via* activation of existing inactive forms of the enzyme (31). Long-term androgen treatment with testosterone-releasing implants elicits changes in ODC activity, immunoreactive protein concentration, and ODC mRNA accumulation in a fashion similar to those in acute, single-dose studies. In all these experiments, the changes in ODC activity were always proportional to those in the enzyme protein content, indicating that post-translational modifications of the protein play a minor, if any, role in androgen regulation of ODC activity in murine kidney (13). Implantation of female mice for 7 days with Silastic rods releasing 10, 20, 40, 80, and 200 µg testosterone per day resulted in a dose-dependent increase in renal ODC activity, with the maximal enzyme protein concentration being 200- to 500-fold higher than that in intact female animals. Under these conditions, the extent of induction of ODC activity was directly proportional to the nuclear androgen receptor concentration (J. Bertolini, C.W. Bardin and O.A. Jänne, unpublished observations).

These data again emphasize the importance of androgen receptors as regulators of ODC gene expression.

Administration of androgens at pharmacological doses for several days increases ODC mRNA accumulation 10- to 50-fold over the values in intact females or castrated male mice (10,26). Therefore, it seems that the increases in the enzyme activity and the immunoreactive ODC concentration (200- to 500-fold) are relatively greater than those in the mRNA accumulation. This notion is in agreement with the findings that prolonged androgen treatment decreases the rate of ODC turnover (13,20) and that the rate of enzyme protein synthesis is increased about 25-fold in murine kidney by androgen administration (17). Despite the dual mechanism leading to increased ODC protein content (increased synthesis and decreased degradation), induction of the enzyme activity by androgens is always accompanied by enhanced mRNA accumulation. These results along with those on nuclear androgen receptors do not, however, prove that androgens stimulate directly the rate of transciption of the renal ODC gene(s) [see below].

In addition to murine renal cells, ODC mRNA accumulation is androgen-regulated also in rat kidney (Fig. 1). The two mRNA species expressed in rat tissues are co-ordinately induced by testosterone administration. The resposes in murine and rat kidneys have, however, some differences: first, the relative changes in ODC mRNA accumulation are more extensive in the murine kidney; second, the induction of ODC mRNA accumulation appears to be transient in the rat renal cells despite the continuous presence of testosterone-releasing implants; and third, ODC mRNA accumulation seems to peak earlier in the rat than mouse kidney. The reason for these differences is not known at present; however, one possibility is that ODC induction in the rat kidney occurs in cell types other than the epithelial cells of proximal tubules, which are the targets for androgen regulation in the murine kidney.

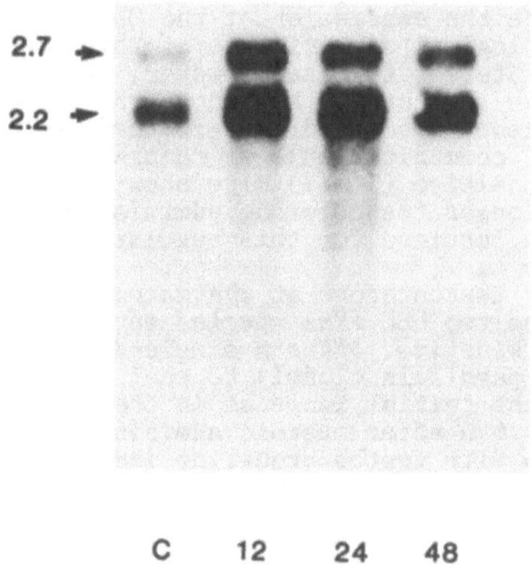

Fig. 1. Changes in ODC mRNA concentration in rat kidney after androgen administration. Sixty-day-old male rats were castrated 4 days prior to implantation with testosterone-releasing rods (400 µg steroid/day) for 12, 24 and 48 h. C = castrated control animals. Poly(A)-containing RNA was isolated from renal tissue, fractionated on 1% agarose gel containing 2.2 M formaldehyde, transferred to nitrocellulose, and hybridized to a 1.7-kb long [^{32}P]ODC cRNA. After treatment with single-strand specific RNase, the filter was exposed to Kodak XAR film for 5 h.

Recent studies have suggested that accumulation of ODC mRNAs in murine kidney would occur mainly through mechanisms not requiring an increase in the rate of ODC gene transcription. As illustrated in Fig. 2, administration of testosterone to female mice fails to elicit a significant increase in the rate of transcription of the ODC gene, as measured by nuclear run-on assays. In these experiments, single-stranded RNA sequences covering the entire protein coding sequence and most of the 5'- and 3'-nontranslated regions of the ODC mRNA were immobilized to nitrocellulose and used to hybridize with [^{32}P]UTP-labeled nuclear RNA. Antisense and sense RNA sequences were immobilized separately to assure that transcription of a specific strand was measured by the RNA-RNA hybridization. The data in Fig. 2 indicate that both strands of the ODC gene are actively transcribed in renal cell nuclei, the transcription of the opposite strand being somewhat less active than that coding for the message strand. In view of the presence of multiple ODC gene sequences in the murine genome, it is possible that the two hybridizable RNAs do not originate from the same transcription unit. In a similar study but using double-stranded DNA sequences as hybridization probes for three different renal gene products (ODC, kidney androgen-regulated protein, and RP2 mRNAs), Berger et al. (32) showed that androgens had little or no effect on the synthesis of these mRNAs, despite their 10- to 20-fold increases in steady-state levels. Taken together, these data suggest that induction of ODC mRNA accumulation by androgens is not fully accounted for by stimulation of gene transcription but must be generated predominantly at the level of mRNA processing and/or turnover. This notion seems to apply also to androgen-regulated mRNAs in accessory sex organs, such as those encoded in rat prostate by the prostatein genes (33).

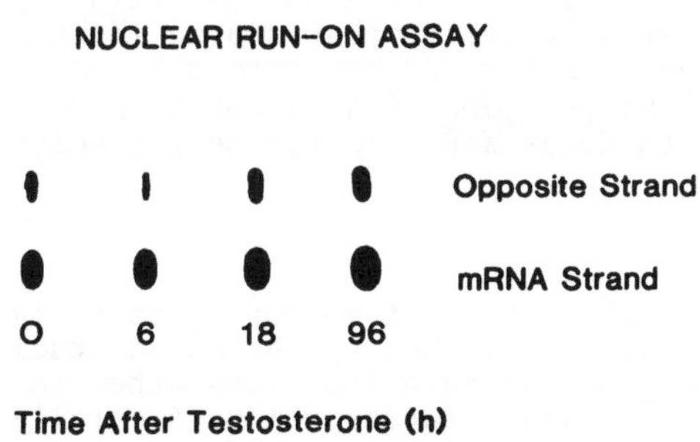

Fig. 2. Effect of androgen administration on the rate of transcription of the renal ODC gene(s). Nuclei were isolated from kidneys of intact female mice (0), female mice treated with single 5-mg doses of testosterone for 6 and 18 h, and females implanted with testosterone-releasing rods (200 µg steroid/day) for 96 h. Nuclear RNA synthesis in the presence of [^{32}P]UTP was carried out at 25 C for 20 min, nuclear RNA isolated and hybridized to nitrocellulose-bound antisense and sense ODC RNAs. After treatment with single-strand specific RNase, the filter was exposed to Kodak XAR film.

Androgen-induced accumulation of ODC mRNA species can be seen not only in cytoplasmic but also in nuclear RNA (Fig. 3). We were unable to detect any preferential accumulation of ODC mRNA precursors in nuclei of intact or androgen-treated female animals, suggesting that processing of the primary transcripts of the ODC gene is neither rate-limiting nor an androgen-regulated event. The RNA samples shown in Fig. 3 are polyadenylated; however, simultaneous Northern blotting of nuclear and cytoplasmic RNAs that failed to bind to oligo(dT)-cellulose did not reveal marked changes in specifically hybridizable RNA species. Comparison of the relative quantities of nuclear and cytoplasmic ODC mRNA species indicated that there are about 50 times more ODC mRNA sequences in the cytoplasmic than nuclear compartment; this ratio is not markedly different between the intact and androgen-treated female animals.

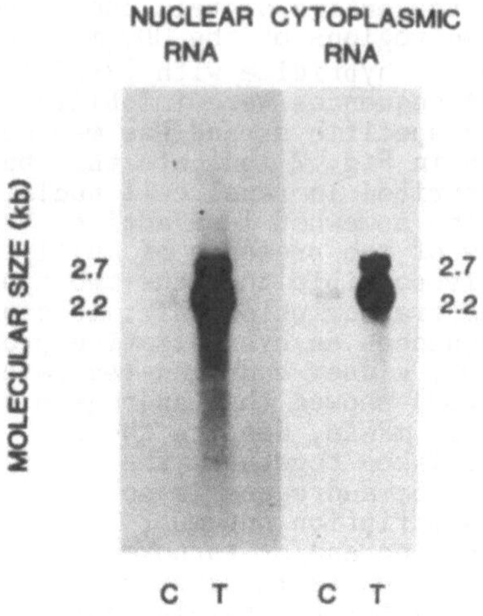

Fig. 3. Androgen regulation of nuclear and cytoplasmic ODC mRNA species. RNA was isolated from kidneys of intact female mice (C) and females treated with testosterone-releasing implants (200 µg steroid/day) for 96 h (T), enriched for poly(A)-containing RNA by oligo(dT)-cellulose chromatography, separated by electrophoresis on 1% agarose gel containing 2.2 M formaldehyde, and transferred to nitrocellulose. Hybridization was performed using a 1.7-kb [^{32}P]ODC cRNA. After treatment with RNase, the filters were exposed to Kodak XAR film for 24 h (nuclear RNA) or 3 h (cytoplasmic RNA).

If the major regulatory site of androgen action is at the level beyond transcription, as illustrated by the above studies, it suggests that the androgen receptor may have functions other than being only a trans-acting regulator of gene transcription. A corollary to this notion is that gene-specific responses in mouse kidney may reflect a direct interaction of the androgen receptor with induced mRNAs or processes responsible for their degradation/stabilization, which seem to take place already in the nuclei (Fig. 3). Alternatively, androgen action may involve a receptor-mediated production of a trans-acting factor in renal cells during the hormone response, which, in turn, brings about its action through posttranscriptional mechanisms. It should be kept in mind, however, that most of the current data originate from investigations in vivo and that no direct studies on the regulatory regions of the renal androgen-responsive genes have been carried out.

Genetic Variation in Androgen Regulation of Renal ODC

A number of tissue functions and/or specific gene products are expressed in inbred strains of mice in a fashion dependent on the genetic background of the animals. Among the specific, hormone-regulated genes are those coding for renal β-glucuronidase and RP2 mRNAs, submaxillary gland renin mRNAs, and hepatic major urinary protein mRNAs. These examples have revealed that several mechanisms may account for the genetic variation, such as polymorphism in the structural gene, differential expression of two non-allelic structural genes, and the presence of cis-acting regulatory locus in close proximity to the structural gene. Recent studies in our laboratory have demonstrated that the renal ODC gene(s) can be included in the list of the genes, the control of which is regulated by the genetic background of the animals.

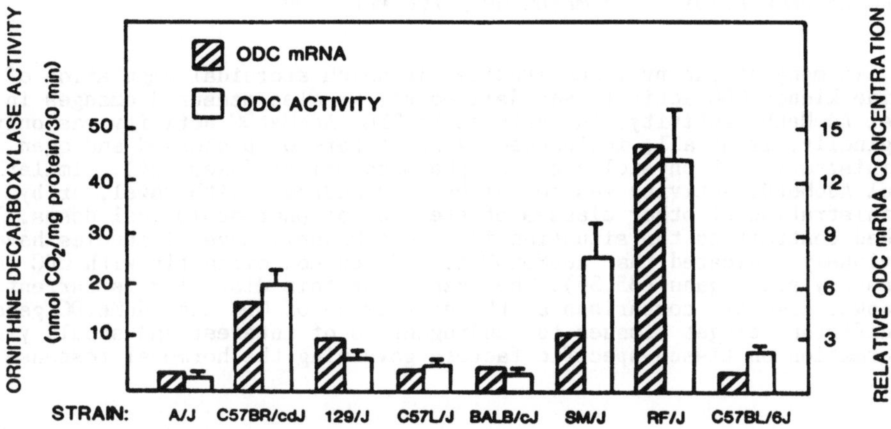

Fig. 4. Renal ODC activity and ODC mRNA concentration in male mice of eight inbred strains. The mRNA concentrations are expressed relative to those of females in a given strain. There was no strain-dependent variation in the enzyme activity or ODC mRNA content of females among the different strains. Data adapted from Melanitou et al. (34).

Renal ODC activity and ODC mRNA concentration among eight inbred strains of mice (A/J, C57BR/cdJ, 129/J, C57L/J, BALB/cJ, SM/J, RF/J, and C57BL/6J) do not exhibit significant differences in intact females; however, there is a marked strain-dependent variation in intact males (34). Male animals of three strains (RF/J, SM/J, and C57BR/cdJ) have 5- to 20-fold higher enzyme activities than other strains, and similar albeit less marked differences are present in the renal ODC mRNA concentrations (Fig. 4). As mentioned above, two ODC mRNA species (2.2 and 2.7 kb) are constitutively expressed and androgen-regulated in murine kidney. The relative amounts of these mRNAs varied among the eight strains studied, with the 2.7-kb mRNA representing 3.3%-19.2% and 2.9%-23.9% of total ODC mRNA content in kidneys of female and male animals, respectively. This variation appears to be unrelated to the genetic regulation that governs the extent of androgen responsiveness of the ODC gene(s) (34). In addition to the three high-responsive strains, there are two strains of mice (A/J and C57BL/6J) with a low male-to-female ratio in the basal ODC activity and a blunted increase in the enzyme activity and mRNA concentration after treatment with superphysiological doses of testosterone. In contrast to their high enzyme activity and ODC mRNA content in the males, C57BR/cdJ, SM/J and RF/J strains did not exhibit a greater response to exogenous androgen than other strains except A/J and C57BL/6J (34). We interprete these data to indicate that the major genetic regulation of ODC gene expression is at the level of androgen sensitivity, as illustrated by the strain-dependent variation in the response to endogenous testosterone in intact males. By contrast, six out of eight strains had very similar maximal responses, which may not be controlled by the same mechanisms as the androgen sensitivity.

The factors responsible for the genetic control of the ODC gene(s) in murine kidney are not known at present. The strain-dependent variation in the males cannot be explained by differences in serum androgen concentrations, since there is no correlation between renal ODC activity and serum testosterone level in individual male mice (34). Similarly, differences in nuclear androgen receptors among the strains do not relate to the renal ODC concentration or ODC mRNA content. These data are taken to mean that it is the renal ODC gene, or its regulator, whose androgen responsiveness is controlled by the genetic background of the mice, rather than genetic regulation of testicular androgen production or nuclear androgen receptor occupancy.

ANDROGEN REGULATION OF AdoMetDC mRNA ACCUMULATION

In many of our previous studies, in which steroidal regulation of murine kidney ODC activity was delineated, we also measured changes in renal AdoMetDC activity. In contrast to ODC, AdoMetDC activity was only marginally, if at all, influenced by short-term or prolonged androgen administration at physiological or pharmacological doses (24). Similarly, renal AdoMetDC activity was not altered by androgen withdrawal, or by administration of other classes of steroids at pharmacological doses. In marked contrast to the situation in rodent kidney, several studies have previously indicated that AdoMetDC is induced concomitantly with ODC in accessory sex organs (35,36). The reason for this disparity is currently unknown; however, comparison of the expression of ODC and AdoMetDC genes in different target tissues for androgens is of interest and should yield information on tissue-specific factors governing the hormonal response.

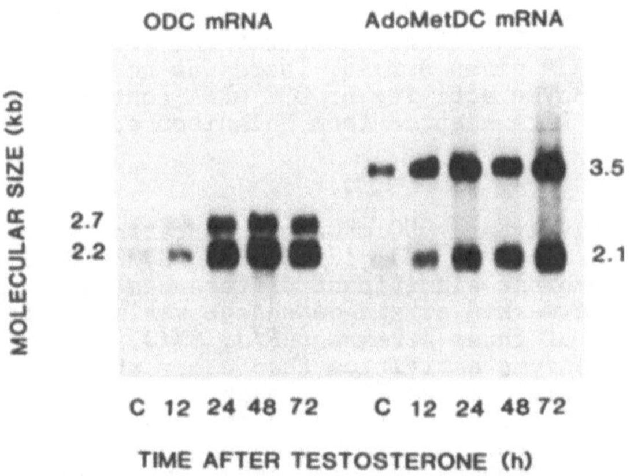

Fig. 5. Androgen regulation of ODC and AdoMetDC mRNA accumulation in rat ventral prostate. Sixty-day-old male rats were castrated 4 days prior to implantation with testosterone-releasing rods (400 μg steroid/day) for the indicated time periods; control animals (C) were mock-implanted. Poly(A)-containing RNA was isolated from prostate, fractionated by electrophoresis on 1% agarose gels containing 2.2 M formaldehyde and transferred to sheets of nitrocellulose. Hybridization probes were a 1.7-kb murine [^{32}P]ODC cRNA and 1.6-kb human [^{32}P]AdoMetDC cRNA. After hybridization and washings, the filters were treated with single-strand specific RNase and exposed to autoradiography for 24 h (ODC mRNA) and 12 h (AdoMetDC mRNA).

Recent cloning and sequencing of cDNAs for rat and human AdoMetDC has shown that this enzyme, similar to ODC, is highly conserved among different species (37). The protein is encoded by two mRNA species with molecular sizes of about 2.1 and 3.5 kb, which appear to be formed through differential utilization of two polyadenylation/termination signals. Using labeled human AdoMetDC cRNA as the hybridization probe, only minor changes in the concentration of AdoMetDC mRNA were seen in mouse or rat kidney after androgen administration (not shown). By contrast, the two AdoMetDC mRNAs are induced co-ordinately with the ODC mRNA species in rat ventral prostate (Fig. 5), rat seminal vesicle, and in murine accessory sex organs. The changes in ODC and AdoMetDC mRNA levels after testosterone administration are similar in their timing and maximal response, and correlate fairly well with the changes measured in the respective enzyme activities.

CONCLUDING REMARKS

Androgenic regulation of ODC in murine kidney has a number of characteristics that render this system an attractive model to study steroid hormone action. First, there are several levels of regulation by androgens, including mRNA accumulation, enzyme protein synthesis and enzyme protein degradation. Second, androgen induction of ODC mRNA accumulation seems to involve mainly stabilization of the mRNA as opposed to enhancement of the rate of ODC gene transcription. Third, there are multiple copies of the ODC gene in the murine genome. Although we have tacitly assumed that androgen regulation of ODC does not involve activation of previously silent genes, this question has not been answered by direct experiments. Fourth, androgen sensitivity of ODC induction is dependent on the genetic background of the animals, as is the relative abundancy of the 2.2- and 2.7-kb ODC mRNA species. Finally, comparison of the mechanisms regulating the expression of ODC and AdoMetDC genes in rodent kidney and accessory sex organs provides an attractive experimental system to investigate tissue-specific factors governing the dissimilar effects of androgens on these two genes in different tissues.

ACKNOWLEDGEMENT

This work was supported by the National Institutes of Health (Grant Nos. HD-13541 and DK-37692).

REFERENCES

1. C. W. Bardin and J. F. Catterall, Testosterone, a major determinant of extragenital sexual dimorphism, Science 211:1285 (1981).

2. J. E. Griffin, M. Leishin, and J. D. Wilson, Androgen resistance syndromes, Am. J. Physiol. 243:E81 (1982).

3. K. K. Kontula, O. A. Jänne, and C. W. Bardin, Intracellular hormone receptor defects and disease, in: "Receptors IV", P.M. Conn, ed., Academic Press, Orlando, pp. 37-74 (1986).

4. A. B. Kulkarni, R.M. Gubito, and P. Feigelson, Developmental and hormonal regulation of α_{2u}-globulin gene transcription, Proc. Natl. Acad. Sci. USA 82:2579 (1985).

5. N. J. Kuhn, M. Woodworth-Gutai, K. W. Gross, and W. A. Held, Subfamilies of the mouse urinary protein (MUP) multi-gene family: Sequence analysis of cDNA clones and differential regulation in the liver, Nucleic Acids Res. 12:6073 (1984).

6. M. G. Parker, R. White, and J. G. Williams, Cloning and characterization of androgen-dependent mRNA from rat ventral prostate, J. Biol. Chem. 255:6996 (1980).

7. M. K. Kistler, R. E. Taylor, Jr., J. C. Kandala, and W. S. Kistler, Isolation of recombinant plasmids containing structural gene sequences for rat seminal vesicle secretory proteins IV and V, Biochem. Biophys. Res. Commun. 99:1161 (1981).

8. J. F. Catterall, K. K. Kontula, C. S. Watson, P. J. Seppänen, B. Funkenstein, E. Melanitou, N. J. Hickok, C. W. Bardin, and O. A. Jänne,

Regulation of gene expression by androgens in murine kidney, Recent Prog. Horm. Res. 42:71 (1986).

9. C. S. Watson, D. Salomon, and J. F. Catterall, Structure and expression of androgen-regulated genes in mouse kidney, Ann. N.Y. Acad. Sci. 438:101 (1984).

10. K. K. Kontula, T. K. Torkkeli, C. W. Bardin, and O. A. Jänne, Androgen induction of ornithine decarboxylase mRNA in mouse kidney as studied by complementary DNA, Proc. Natl. Acad. Sci. USA 81:731 (1984).

11. A. E. Pegg and P. P. McCann, Polyamine metabolism and function, Am. J. Physiol. 243:C212 (1982).

12. J. E. Seely, H. Pösö, and A. E. Pegg, Purification of ornithine decarboxylase from kidneys of androgen treated mice , Biochemistry 21:3394 (1982).

13. V. V. Isomaa, A. E. I. Pajunen, C. W. Bardin, and O. A. Jänne, Ornithine decarboxylase in mouse kidney. Purification, characterization and radioimmunological determination of the enzyme protein , J. Biol. Chem. 258:6735 (1983).

14. N. J. Hickok, P. J. Seppänen, G. L. Gunsalus, and O. A. Jänne, Complete amino acid sequence of human ornithine decarboxylase deduced from complementary DNA, DNA 6:179 (1987).

15. M. Gupta and P. Coffino, Mouse ornithine decarboxylase. Complete amino acid sequence deduced from cDNA, J. Biol. Chem. 260:2941 (1985).

16. O. A. Jänne, K. K. Kontula, V. V. Isomaa, T. K. Torkkeli, and C. W. Bardin, Androgen-receptor dependent regulation of ornithine decarboxylase gene expression in mouse kidney, in: "Steroid Hormone Receptors: Structure and Function", H. Eriksson and J.-A. Gustafsson, eds., Elsevier Science Publishers BV, Amsterdam, pp. 461-476 (1983).

17. L. Persson, J. E. Seely, and A. E. Pegg, Investigation of structure and rate of synthesis of ornithine decarboxylase protein in mouse kidney, Biochemistry 23:3777 (1984).

18. A. Pulkka, T. Taskinen, H. Aaltonen, J. Ramberg, and A. E. I. Pajunen, Studies on the degradation of ornithine decarboxylase by the immunoblotting technique, Biochem. Int. 11:845 (1985).

19. M. Macrae and P. Coffino, Complementation of a polyamine-deficient Escherichia coli mutant by expression of mouse ornithine decarboxylase, Mol. Cell. Biol. 7:564 (1987).

20. J. E. Seely and A. E. Pegg, Changes in mouse kidney ornithine decarboxylase activity are brought about by changes in the amount of enzyme protein as measured by radioimmunoassay, J. Biol. Chem. 258:2496 (1983).

21. S. Rogers, R. Wells, and M. Rechsteiner, Amino acid sequences common to rapidly degraded proteins: The PEST hypothesis, Science 234:364 (1986).

22. C. Kahana and D. Nathans, Translational regulation of mammalian ornithine decarboxylase by polyamines, J. Biol. Chem. 260:15390 (1985).

23. E. Hölttä and P. Pohjanpelto, Control of ornithine decarboxylase in Chinese hamster ovary cells by polyamines. Translational inhibition of synthesis and acceleration of degradation of the enzyme by putrescine, spermidine, and spermine, J. Biol. Chem. 261:9502 (12986).

24. A. E. I. Pajunen, V. V. Isomaa, O. A. Jänne, and C. W. Bardin, Androgenic regulation of ornithine decarboxylase activity and its

relationship to changes in cytosol and nuclear androgen receptors, J. Biol. Chem. 257:8190 (1982).

25. G. J. Sertich and A. E. Pegg, Polyamine administration reduces ornithine decarboxylase activity without affecting the mRNA content, Biochem. Biophys. Res. Commun. 143:424 (1987).

26. N. J. Hickok, P. J. Seppänen, K. K. Kontula, P. A. Jänne, C. W. Bardin, and O. A. Jänne, Two ornithine decarboxylase mRNA species in mouse kidney arise from size heterogeneity at their 3' termini, Proc. Natl. Acad. Sci. USA 83:594 (1986).

27. C. Kahana and D. Nathans, Nucleotide sequence of murine ornithine decarboxylase mRNA, Proc. Natl. Acad. Sci. USA 82:1673 (1985).

28. A. Katz and C. Kahana, Isolation and characterization of the mouse ornithine decarboxylase gene, J. Biol. Chem. 263:7604 (1988).

29. M. Brabant, L. McConlogue, T. van Daalen Wetters, and P. Coffino, Mouse ornithine decarboxylase gene: Cloning, structure, and expression, Proc. Natl. Acad. Sci. USA 85:2200 (1988).

30. R. Winqvist, T. P. Mäkelä, P. Seppänen, O. A. Jänne, L. Alhonen-Hongisto, J. Jänne, K.-H. Grzeschik, and K. Alitalo, Human ornithine decarboxylase sequences map to chromosome regions 2pter->p23 and 7cen->qter but are not co-amplified with the NMYC oncogene, Cytogenet. Cell. Genet. 42:133 (1986).

31. O. A. Jänne, K. K. Kontula, V. V. Isomaa, and C. W. Bardin, Ornithine decarboxylase mRNA in mouse kidney: A low abundancy gene product regulated by androgens with rapid kinetics, Ann. N.Y. Acad. Sci. 438:72 (1984).

32. F. G. Berger, D. Loose, H. Meisner, and G. Watson, Androgen induction of messenger RNA concentrations in mouse kidney is posttranscriptional, Biochemistry 25:1170 (1986).

33. M. J. Page and M. G. Parker, Effect of androgen on the transcription of rat prostatic binding protein genes, Mol. Cell. Endocrinol. 27:343 (1982).

34. E. Melanitou, D. A. Cohn, C. W. Bardin, and O. A. Jänne, Genetic variation in androgen regulation of ornithine decarboxylase gene expression in inbred strains of mice, Mol. Endocrinol. 1:266 (1987).

35. A. E. Pegg, D. H. Lockwood, and H. G. Williams-Ashman, Concentrations of putrescine and polyamines and their enzymic synthesis during androgen-induced prostatic growth, Biochem. J. 117:17 (1970).

36. K. Käpyaho, A. Kallio, and J. Jänne, Differential effects of 2-difluoromethylornithine and methylglyoxal bis(guanylhydrazone) on the testosterone-induced growth of ventral prostate and seminal vesicles of castrated rats, Biochem. J. 219:811 (1984).

37. A. Pajunen, A. Crozat, O. A. Jänne, R. Ihalainen, P. H. Laitinen, B. Stanley, R. Madhubala, and A. E. Pegg, Structure and regulation of mammalian S-adenosylmethionine decarboxylase, J. Biol. Chem., in press (1988).

REGULATION OF POLYAMINE BIOSYNTHESIS IN *ESCHERICHIA COLI* BY THE ACIDIC ANTIZYME AND THE RIBOSOMAL PROTEINS S20 AND L34

C. A. Panagiotidis, S. C. Huang, S. A. Tsirka*,
D. A. Kyriakidis* and E. S. Canellakis

Department of Pharmacology, Yale University
School of Medicine, New Haven, Connecticut 06510, U.S.A.
and *Laboratory of Biochemistry, Faculty of Chemistry
University of Thessaloniki, Thessaloniki, Greece

Two roads diverged into a wood and I-
I took the one less travelled by,
and that has made all the difference.

Robert Frost

I personally appreciate the friendly and scientific support that I have received from Shin Ichi Hayashi, Claudio Caldarera, Margaret Brosnan, John Mitchell, Paivi Laitinen and Flavio Flamigni. These investigators and others have made the journey on the less travelled path, the post-translational control of polyamine biosynthesis by antizyme in eukaryotic cells, a very pleasant experience. In the course of this meeting we shall learn the latest on the molecular aspects of this control mechanism in eukaryotic cells.

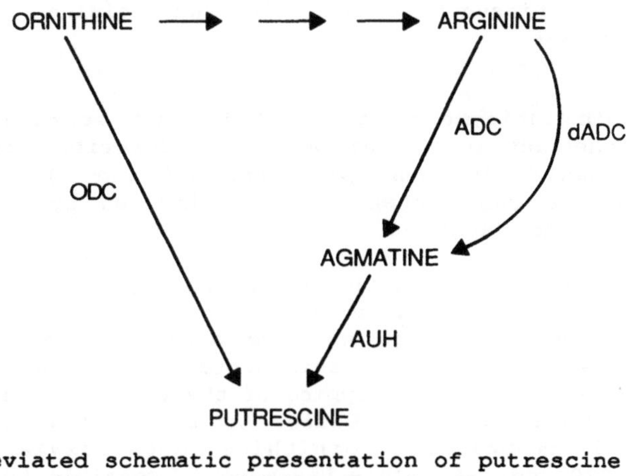

Fig. 1. Abbreviated schematic presentation of putrescine biosynthesis in *E. coli*. ODC = ornithine decarboxylase; ADC = arginine decarboxylase; dADC = biodegradative arginine decarboxylase AUH = agmatine ureohydrolase.

Lately, we have been devoting much of our time to the study of the comparable control mechanism in *Escherichia coli*. *E.coli* has always impressed me with the insurance policy it maintains to guarantee that polyamines will be available. As shown by Morris and Pardee[1], it has two sets of almost every enzyme leading to the synthesis of polyamines. In the first set are the biosynthetic enzymes for polyamine biosynthesis and in the second set are the biodegradative enzymes. This is a very exceptional insurance policy, when one considers that *E.coli* has only one set of glycolytic enzymes to provide the energy it requires for growth.

Figure 1 shows that the second set of polyamine synthesizing enzymes, the biodegradative enzymes, can be called into service when the polyamine biosynthetic enzymes have been deleted[2]. In this particular case the biodegradative arginine decarboxylase converts arginine to agmatine which is then converted to putrescine[3]. These *E. coli* deletion mutants can survive with minute amounts of polyamines compared to controls. Such alternative pathways make the study of polyamine control mechanisms in *E.coli* very interesting, and emphasize the importance of polyamine synthesis to these organisms.

We had earlier indicated that polyamines induce the synthesis of three proteins which inhibit *E.coli* ornithine decarboxylase[4,5]. These proteins and their properties are outlined in Table 1. One of these is the acidic antizyme. I shall be emphasizing the "acidic" to contrast it with the two "basic" ribosomal proteins, the S20 of the small subunit and the L34 of the large subunit[6].

TABLE 1. Properties of acidic antizyme, S20 and L34

Polyamines induce:			
	Acidic antizyme: MW= 50-60 KD, pI approx.=3		
	S20 : MW= 9,553 D	basic r-protein	pI>12
	L34 : MW= 5,381 D	basic r-protein	pI>12

Common Properties: Nucleic acid binding proteins (Az and S20, probably also L34); present in ribosome fraction; inhibit the biosynthetic ornithine and arginine decarboxylase.

The acidic antizyme is the protein that we considered to be analogous to the antizyme that we first described in eukaryotic cells[7,8]. The basic ribosomal proteins, S20 and L34 have antizyme properties in the sense that they are induced by polyamines and inhibit ornithine decarboxylase.

As we kept working with the acidic antizyme it became apparent that it degraded rapidly upon purification. We do not know why this degradation occurs. Although we have overcome these experimental difficulties, the instability of this protein has greatly delayed our progress in the sequencing and mapping of the acidic antizyme gene. We have been able to prepare antibodies to it; we have also prepared antibodies and probes for ornithine decarboxylase, arginine decarboxylase, S20 and L34. This has made it possible to evaluate the changes that occur in these various components and to define how they may participate in the regulation of polyamine biosynthesis.

That the acidic antizyme inhibits both ornithine decarboxylase and arginine decarboxylase is shown in Figure 2A. This figure shows that if we add increasing amounts of arginine decarboxylase to antizyme-inhibited ornithine decarboxylase, the antizyme-inhibited ornithine decarboxylase is reactivated. Figure 2A also indicates that arginine decarboxylase can only relieve about 50% of the inhibition of ornithine decarboxylase by antizyme, suggesting that there is also a very tight association of ornithine decarboxylase with antizyme that cannot be competed for by arginine decarboxylase. The existence of such a tight association of ornithine decarboxylase with antizyme is also shown by the titration of antizyme-inhibited arginine decarboxylase by ornithine decarboxylase as shown in Figure 2B. Here, we find that ornithine decarboxylase reverses the inhibition of arginine decarboxylase by antizyme by more than 80%.

The simplest interpretation of these results is, of course, that the excess arginine decarboxylase binds the antizyme and makes less of it available for ornithine decarboxylase. The activity of ornithine decarboxylase is correspondingly increased. A similar set of results is obtained if we repeat these experiments using the basic ribosomal proteins S20 and L34 instead of acidic antizyme.

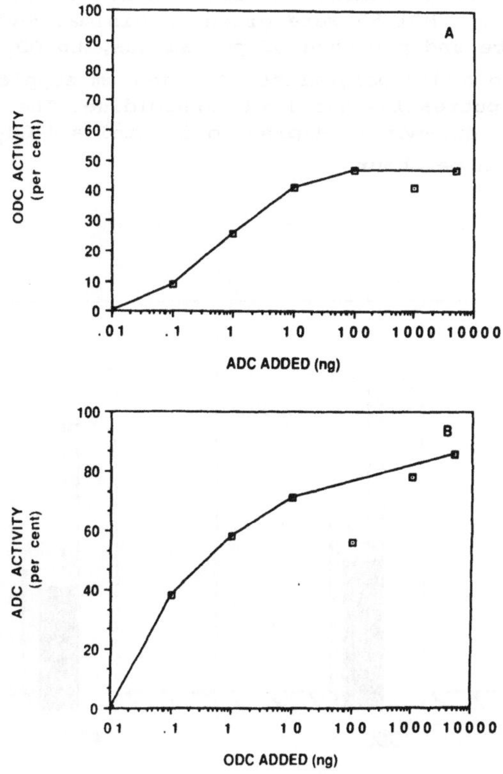

Fig. 2. Reactivation of antizyme-inhibited ODC and ADC by ADC and ODC, respectively. Highly purified preparations (1 ng, >99 pure) of ornithine (A) or arginine decarboxylase (B) were titrated with acidic antizyme (ODC,70% inhibited; ADC, 32% inhibited). To each preparation were added the indicated amounts of purified ornithine decarboxylase (B) or arginine decarboxylase (A), and the degree of reactivation of the originally inhibited enzyme was determined.

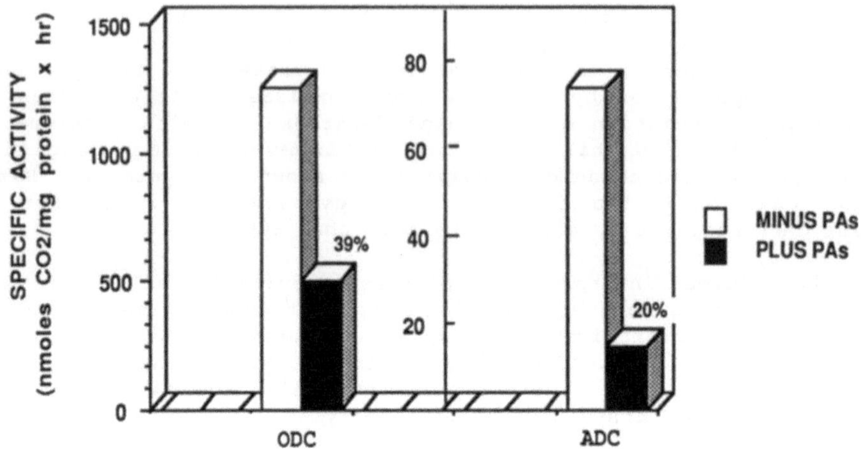

Fig. 3. Effect of polyamines on ODC and ADC activities. *E.coli* MG1655 were grown in minimal medium 56, in the absence and presence of polyamines, to OD_{600}= 0.5. - = no added polyamines; + = medium supplemented with 1 mM putrescine and 1 mM spermidine. The specific activity of the enzymes is expressed in nmoles $^{14}CO_2$ per mg total protein per hour.

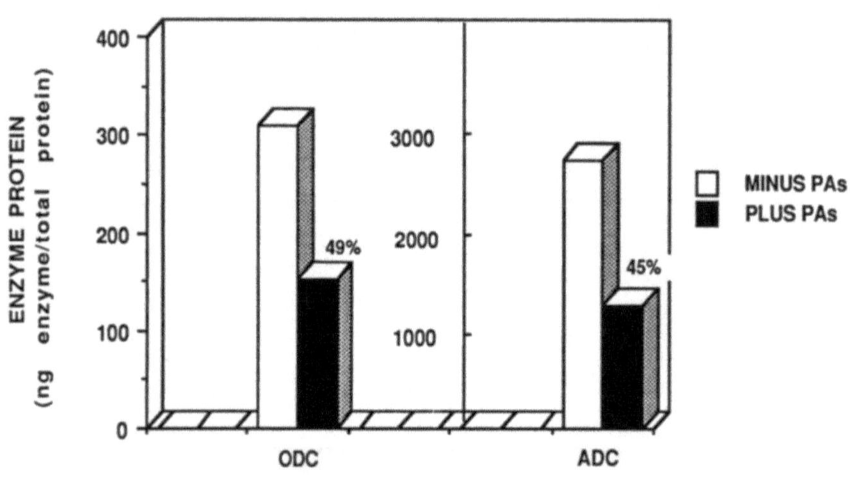

Fig. 4. Effect of polyamines on the ODC and ADC protein levels. *E.coli* MG1655 were grown as described in Fig.3. The amounts of ODC and ADC proteins were quantitated by scanning Western immunoblots with a soft-laser scanning densitometer.

When *E. coli* is grown in a polyamine-supplemented medium, in agreement with the Tabors' results[9] we find that the specific activity of both ornithine decarboxylase and arginine decarboxylase falls (Fig. 3). In addition, the amount of ornithine decarboxylase and arginine decarboxylase protein in the cells drops (Fig. 4), suggesting that the synthesis of these enzymes decreases in response to the presence of polyamines in the medium.

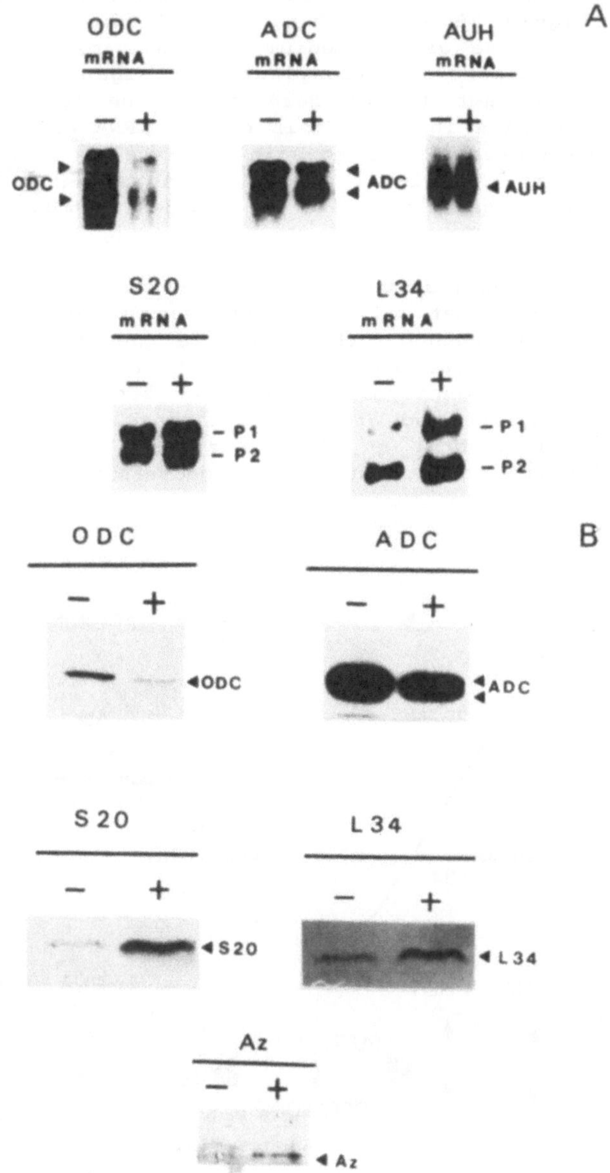

Fig. 5. Effect of polyamines on the protein and mRNA content of ODC, ADC, AUH, S20, L34 and acidic antizyme in *E. coli*. *E. coli* was grown as described in the legend of Fig. 3.
A. Northern blots of ODC, ADC, S20, L34 and AUH mRNAs.
B. Wester blots of ODC, ADC, S20, L34 and acidic antizyme.
- = no added polyamines, + = medium supplemented with 1 mM putrescine and 1 mM spermidine.

Figure 5A shows that when *E. coli* is grown in a medium supplemented with 1 mM putrescine and 1 mM spermidine, the levels of the mRNAs of ornithine decarboxylase and arginine decarboxylase are decreased. Our results indicate that polyamines do not affect the level of the mRNA of agmatine ureohydrolase (AUH). This result conforms with Boyle's finding[10] that polyamines do not affect the activity of agmatine ureohydrolaseIn contrast, the presence of polyamines increases the levels of the mRNAs of S20 and of L34. These results are substantiated when we examine the protein levels of these components. Figure 5B shows that, as indicated by immunoblots, supplementation of the growth medium with polyamines decreases the level of the ornithine decarboxylase and arginine decarboxylase proteins; we have not as yet quantitated the levels of the AUH protein. The increase in the levels of the mRNA of S20 and L34 is accompanied with a parallel increase in the levels of the S20 and L34 ribosomal proteins. The level of the acidic antizyme protein also increases. We still have to define the changes in mRNA of the acidic antizyme.

The preceding results are presented in summary form in Figure 6. This figure indicates that polyamines exert opposing effects on these various cellular components; they decrease the levels of ornithine decarboxylase and arginine decarboxylase but raise the levels of S20 and L34 and of acidic antizyme. Since the latter are inhibitors of ornithine decarboxylase and arginine decarboxylase, they can exert post-translational control on these enzymes. These two sets of opposing effects act in concert towards decreasing the synthesis of polyamines.

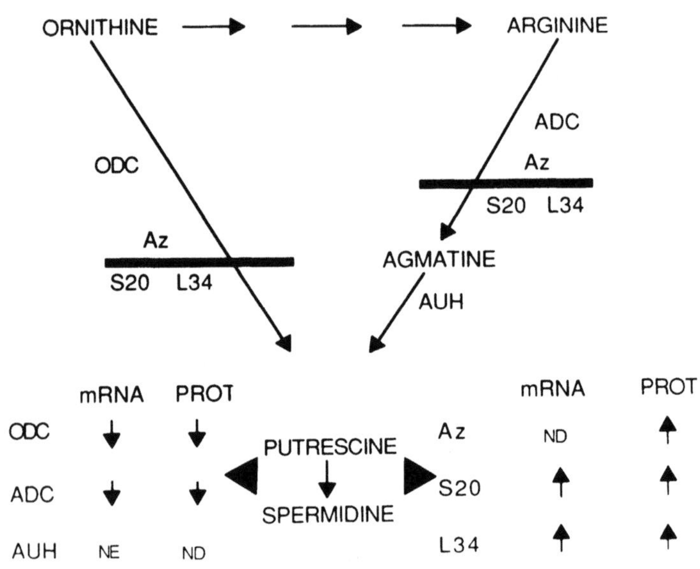

Fig. 6. Schematic presentation of the polyamine biosynthetic pathway in *E.coli* indicating the known sites of inhibition of ornithine decarboxylase and arginine decarboxylases by the antizyme, and the ribosomal proteins S20 and L34, as well as the effects of polyamines on the mRNA and protein levels of the various components. Az= antizyme; NE= no effect; ND= not determined; for the other abbreviations see Fig. 1. ↑= increase; ↓ = decrease.

The question has been raised by Kashiwagi and Igarashi[11] whether the inhibition of ornithine decarboxylase and of arginine decarboxylase by the S20 and L34 ribosomal proteins has any physiological significance. It has been suggested that perhaps the inhibition by S20 and L34 is an artifact due to the nonspecific inhibition of ornithine decarboxylase by basic proteins.

We have investigated this question in three different ways. We reasoned that if these two ribosomal proteins inhibit ornithine decarboxylase and arginine decarboxylase, then overexpression of these enzymes should affect the levels of the ribosomal proteins, and vice versa, if we overexpress the S20 and L34 ribosomal proteins then this overexpression should affect the levels of the two decarboxylases.

TABLE 2. Effect of overexpression of ornithine decarboxylase on the levels of S20 and L34 mRNA and protein

S20		L34	
mRNA	protein	mRNA	protein
Per cent increase over control			
110	60	20	100

The levels of S20 and L34 mRNA and protein were determined in *E. coli* MG1655 transformed with plasmid pODC-1 and compared to the respective mRNA and protein levels of *E. coli* transformed with the parent plasmid pBR322. The transformation with pBR322 did not affect the amounts of the proteins and of the mRNAs of interest.

We have overexpressed ornithine decarboxylase, by transforming *E. coli* with the pODC-1 plasmid[12] and examined the effects on the mRNA and protein levels of S20 and L34. As shown in Table 2, overexpression of ornithine decarboxylase causes a 2.1-fold increase in the levels of the S20 mRNA and a 1.6-fold increase of its protein. The level of the ribosomal protein L34 is also increased by 2-fold when ornithine decarboxylase is overexpressed, although its mRNA level is only slightly increased.

This result conforms with our expectations, in the sense that the cellular response to increased ornithine decarboxylase is to increase the levels of its inhibitors, so as to maintain constant polyamine levels.

The other approach we used, shown in Table 3, was to overexpress the ribosomal proteins S20 and L34, by transforming *E. coli* with the pS20 (pGP2) plasmid[13] or with the pL34 plasmid and to determine their effects on ornithine decarboxylase activity and arginine decarboxylase activity. The finding that overexpression of S20 and L34 decreases the activity of ornithine decarboxylase and arginine decarboxylase (Table 3), substantiates the notion that these ribosomal proteins act as inhibitors of these enzymes.

TABLE 3. Effect of overexpression of S20 and L34 ribosomal proteins on ornithine and arginine decarboxylase activity

Overexpressed r-protein	ODC specific activity	ADC specific activity
	Per cent inhibition over control	
S20	20	44
L34	37	37

The specific activities of ODC and ADC were determined in *E.coli* MG1655 transformed with pS20 (pGP2)[13] or with pL34. Plasmid pL34 was constructed by subcloning the EcoR 1 fragment of plasmid pRB1A[14], containing the L34 structural gene, in pBR322. They are compared with the specific activities of the respective enzymes of *E. coli* transformed with the parent plasmid pBR322. The transformation with pBR322 did not affect the specific activities of these enzymes. ODC and ADC specific activities are expressed in nmoles $^{14}CO_2$ per mg protein per hour.

We have obtained plasmids for various basic ribosomal proteins of both the large and small subunits [plasmids for S10, L4, L2, L23, L22 and S19 were obtained from Lindahl[15] and for S15 from Takata[16]]. By transforming *E. coli* with these plasmids, we have overexpressed each of these ribosomal proteins and we have determined their effects on ornithine and arginine decarboxylase activity. In this way we were able to compare the effects of overexpression of these ribosomal proteins to the effects of overexpression of S20 and L34 (Table 4). The pI values of the individual ribosomal proteins, as determined by Kaltschmidt[17], are included in this table.

The results presented in Table 4 indicate that overexpression of seven ribosomal proteins, the S10, L4, L2, L23, L22, S19 and S15 does not inhibit the activity of ornithine decarboxylase, nor does overexpression of S10 or L4 affect arginine decarboxylase. Overexpression of the relatively neutral ribosomal protein S10 has a very similar effect on ornithine decarboxylase as does the very basic ribosomal protein S15. We do not know what significance to attach to the 12-15% inhibition of arginine decarboxylase that occurs with L2 & L23 and L22 & S19, but this is a considerably lower inhibition than that exerted by S20 and L34.

The third approach that we used was to determine whether the mRNA levels of these various ribosomal proteins were altered when the cells were grown in polyamine-supplemented media. Figure 7 shows the results of RNA dot blot experiments with the RNA of *E. coli* grown in the absence and presence of polyamines. In accord with the previous results shown in Figure 5A, growth in polyamine-supplemented medium increases the levels of the mRNAs of S20 and L34. However, polyamine supplementation does not alter the levels of the mRNAs of the ribosomal proteins S10, L4, L2, L23, L22, S19 and S15, nor does it alter the level of the mRNA of AUH.

TABLE 4: Effect of overexpression of the ribosomal proteins S10, L4, L2, L23, L22, S19, S15 and of S20 and L34 on ODC and ADC activity.

Overexpressed r- protein	ODC specific activity	ADC specific activity	pI
	Per cent change		
S10	+10	+10	7.9
L4	- 3	+ 5	7.6
L2 & L23	+ 5	-15	>12 & 9.6
L22 & S19	+ 1	-12	11.5 & >12
S15	+20	- 9	>12
S20	-20	-44	>12
L34	-37	-37	>12

The specific activities of ODC and ADC were determined in *E. coli* transformed with multicopy plasmids carrying the genes of ribosomal proteins S10, L4, L2 & L23, L22 & S19, S15, S20 and L34. These values were compared with the respective enzyme specific activities of the parent untransformed cells.

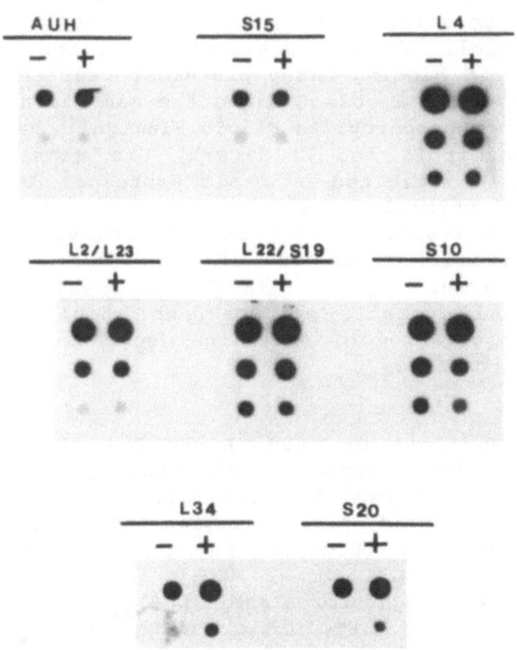

Fig. 7. Dot blot analysis of the mRNAs of AUH and of the ribosomal proteins S10, L4, L2 & L23, L22 & S19, S15, S20 and L34. *E.coli* MG1655 were grown in minimal medium 56 to OD_{600}= 0.5, in the absence and presence of polyamines. - = no added polyamines; + = medium supplemented with 1 mM putrescine and 1 mM spermidine.

To summarize: We have shown that supplementation of the growth medium of *Escherichia coli* with polyamines, lowers the mRNA and protein levels of ornithine decarboxylase and arginine decarboxylase. However, it raises the mRNA and protein levels of the basic S20 and L34 ribosomal proteins as well as the protein levels of the acidic antizyme. Such polyamine supplementation of the growth medium does not affect the mRNA levels of seven other ribosomal proteins of the small and large subunits, five of which are basic proteins, nor that of agmatine ureohydrolase. We have also shown that, although overexpression of the ribosomal proteins S20 and L34 in *E.coli* decreases the activities of both ornithine decarboxylase and arginine decarboxylase, overexpression of the seven ribosomal proteins referred to above does not affect ornithine decarboxylase and arginine decarboxylase activities. So far, our studies do not indicate that there is an obvious correlation between the isoelectric points of ribosomal proteins and ornithine and arginine decarboxylase, if the experiments are performed *in vivo*. Our experiments do not exclude the possibility that other ribosomal proteins may affect the enzymes associated with polyamine biosynthesis, but they suggest that at least the ribosomal proteins S20 and L34 and the acidic antizyme participate in a unique manner in the regulation of polyamine biosynthesis.

Closing remarks: In closing I would like to make two suggestions. One is that there is enough justification to carefully examine whether in mammalian cells, ornithine decarboxylase is also affected by and possibly affects descrete ribosomal proteins. Such a study may provide us with some insight on the interaction of ODC with the ribosomes and its relationship to protein synthesis.

The other suggestion is based upon a number of correlates. The ribosomal proteins constitute the bulk of the nucleic acid binding basic proteins of *E.coli*, whereas in animal cells, the histones are the prominent nucleic acid binding basic proteins. Despite the structural dissimilarity between the *E.coli* ODC and the mammalian ODC, these two enzymes share a common property. As Flavio Flamigni[18] has shown, along with Guarnieri, Marmiroli and Caldarera, the mammalian ornithine decarboxylase is also inhibited by basic proteins. Is this a chance phenomenon or is it physiologically significant? Is there some some specificity to these reactions? Are certain individual histones the primary physiological effectors of the mammalian ornithine decarboxylase? In order to understand the physiological role of polyamines in mammalian cells perhaps we should also look more carefully at the association of ornithine decarboxylase with various components of the chromosomes.

Acknowledgements. We thank Dr. S.M. Boyle for plasmids pODC-1 and pKA5, Dr. L. Lindahl for plasmids pLL36, pLL133, pLL153 and pLL161 carrying the genes for ribosomal proteins S10, L4, L2 & L23, L22 & S19, and Dr. R. Takata for plasmid A127S1 carrying the ribosomal protein S15 gene. We also thank Dr. G. Stoffler for the antibodies against the ribosomal proteins S20 and L34, Dr. G. Mackie for plasmid pGP2 carrying the S20 gene and Dr. A. Wright for plasmid pRB1A carrying the L34 gene. This work was supported by U.S. Public Health Service grant CA-26546 (to E.S.C.) and by U.S.P.H.S. Career Award 5 KO6 GM03070 (to E.S.C.).

REFERENCES

1. D. R. Morris and A. B. Pardee, Multiple pathways of putrescine biosynthesis in *Escherichia coli*, J. Biol. Chem. 24:3129 (1966).

2. H. Tabor, E. W. Hafner and C. W. Tabor, Construction of an *Escherichia coli* strain unable to synthesize putrescine, spermidine or cadaverine: characterization of two genes controlling lysine decarboxylase, J. Bacteriol. 144:952 (1980).
3. C. A. Panagiotidis, S. Blackburn, K. B. Low and E. S. Canellakis, Biosynthesis of polyamines in ornithine decarboxylase, arginine decarboxylase, and agmatine ureohydrolase deletion mutants of *Escherichia coli* strain K-12, Proc. Natl. Acad. Sci. U.S.A. 84:4423 (1987).
4. D. A. Kyriakidis, J. S. Heller and E. S. Canellakis, Modulation of ornithine decarboxylase activity in *Escherichia coli* by positive and negative effectors, Proc. Natl. Acad. Sci. U.S.A. 75:4699 (1978).
5. D. A. Kyriakidis, J. S. Heller and E. S. Canellakis, Purification of ornithine decarboxylase antizymes (*Escherichia coli*) Meth. Enzymol. 94:193 (1983).
6. C. A. Panagiotidis and E. S. Canellakis, Comparison of the basic *Escherichia coli* antizyme 1 and antizyme 2 with the ribosomal proteins S20/L26 and L34, J. Biol. Chem. 259:15025 (1984).
7. W. F. Fong, J. S. Heller and E. S. Canellakis, The appearance of an ornithine decarboxylase inhibitory protein upon the addition of putrescine to cell cultures, Biochim. Biophys. Acta 428:456 (1976).
8. J. S. Heller, W. F. Fong and E. S. Canellakis, Induction of a protein inhibitor to ornithine decarboxylase by the end products of its reaction, Proc. Natl. Acad. Sci. U.S.A. 73:1858 (1976).
9. H. Tabor and C. W. Tabor, Formation fo 1,4-diaminobutane and of spermidine by an ornithine auxotroph of *Escherichia coli* grown on limiting ornithine or arginine, J. Biol. Chem. 244:2286 (1969).
10. C. Satishchandran and S. M. Boyle, Antagonistic transcriptional regulation of the putrescine biosynthetic enzyme agmatine ureohydrolase by cyclic AMP and agmatine in *Escherichia coli*, J. Bacteriol. 157:552 (1984).
11. K. Kashiwagi and K. Igarashi, Nonspecific inhibition of *Escherichia coli* ornithine decarboxylase by various ribosomal proteins: detection of a new ribosomal protein possessing strong antizyme activity, Biochim. Biophys. Acta 911:180 (1987).
12. S. M. Boyle, G. D. Markham, E. W. Hafner, J. M. Wright, H. Tabor and C. W. Tabor, Expression of the cloned genes encoding the putrescine biosynthetic enzymes and methionine adenosyltransferase of *Escherichia coli* (*speA*, *speB*, *speC* and *metK*), Gene 30:129 (1984).
13. G. A. Mackie, Nucleotide sequence of the gene for ribosomal protein S20 and its flanking regions, J. Biol. Chem. 256:8177 (1981).
14. R. E. Braun, K. O'Day and A. Wright, Autoregulation of the DNA replication gene *dnaA* in *E. coli* K-12, Cell 40:159 (1985).
15. J. M. Zengel, D. Mueckl and L. Lindahl, Protein L4 of the *E. coli* ribosome regulates an eleven gene r protein operon, Cell 21:523 (1980).
16. R. Takata, T. Mukai, M. Aoyagi and K. Hori, Nucleotide sequence of the gene for *Escherichia coli* ribosomal protein S15 (*rpsO*), Mol. Gen. Genet. 197:225 (1985).
17. E. Kaltschmidt, Ribosomal proteins. XIV. Isoelectric points of ribosomal proteins of *E.coli* as determined by two-dimensional polyacrylamide gel electrophoresis, Anal. Biochem. 43:25 (1971).

18. F. Flamigni, C. Guarnieri, S. Marmiroli and C.M. Caldarera, Inhibition of rat heart ornithine decarboxylase by basic polypeptides, Biochem. J. 229:807 (1985).

STUDIES ON ORNITHINE DECARBOXYLASE ANTIZYME

S. Hayashi, Y. Murakami, S. Matsufuji, M. Nishiyama,
R. Kanamoto and T. Kameji

Department of Nutrition
The Jikei University School of Medicine
Tokyo 105, Japan

Tissue levels of ornithine decarboxylase (ODC) are under negative feedback regulation by polyamines. Thus, exogenously added polyamines cause a rapid decrease of ODC activity. Recent studies in various laboratories have indicated that the rapid decay of ODC activity caused by polyamines is due both to suppression of ODC synthesis (1-3) and to acceleration of ODC degradation (1,2,4) and that a new protein synthesis is needed for the latter effect (2). Antizyme was discovered in 1976 by Canellakis and his collaborators as a unique protein inhibitor of ODC induced by polyamines (5,6). Since then, antizyme has been shown to be present in various tissues and cells as a free form and/or a complex with ODC (7). Previous studies in this and other laboratories have led to a hypothesis that antizyme not only inhibits ODC activity but also accelerates its degradation in a recycling manner (8-10). The strongest evidence for that hypothesis was the existence of a good correlation between ODC decay rate and antizyme/ODC ratio in HTC cells (9). In the present paper, we report the results of our recent studies that support the above hypothesis. We also present some of our results on the cloning of cDNAs for rat liver antizyme.

EXISTENCE AND ROLE OF MOUSE KIDNEY ANTIZYME

If antizyme plays an essential role in rapid ODC decay, then it should be present in all tissues and cells in which ODC turns over rapidly. It has been reported by several groups of investigators, however, that antizyme was not detected in mouse kidney (11-13), raising a question on the physiological role of antizyme. We therefore examined the presence or absence of antizyme in mouse kidney (14).

Female mice of ICR strain which had been starved overnight were injected intraperitoneally with 2 mmol each of purescine/kg body wt., three times with 1 h intervals, and killed 1 h after the last injection. Crude kidney extracts were found to contain an ODC inhibitor which was identified as antizyme from the following evidence. (i) It inhibits ODC activity in a stoichiometric and time-independent manner. (ii) The inhibition is reversed by antizyme inhibitor (15). (iii) It has an aparent molecular weight of 26000, as judged by gel filtration analysis. (iv) It is inducible with polyamines. (v) It cross-reacts with seven monoclonal antibodies to rat liver antizyme.

Small but significant amounts of ODC-antizyme complex were also found to be present in the kidney of untreated female mice. Thus, upon gel-filtration analysis in the presence of 1M NaCl, where ODC-antizyme complex can be dissociated and the components can be resolved, both ODC and antizyme activities were eluted from the column at positions approximately corresponding to MW 50000 and Mw 27000, respectively (Fig. 1). The amount of ODC-antizyme complex in these kidney extracts was about 25 percent of total ODC, as determined with antizyme inhibitor, which strongly binds to antizyme and therefore replaces ODC in ODC-antizyme complex (15). The presence of ODC-antizyme complex in the kidney of untreated mice indicates that antizyme is a physiological cellular component in the kidney and is presumably synthesized in response to endogenous polyamines.

Several reasons are conceivable for the previous failures to detect antizyme in mouse kidney. (i) An extremely high renal ODC activity of male mice may have prevented an appearance of free antizyme by neutralizing it. (ii) A substantial capacity of mouse kidney particulates to bind antizyme might have have prevented an apppearance of antizyme in cytosol (unpublished result). (iii) Administered polyamines might have been rapidly excreted into urine (16).

Mouse kidney ODC is known to be dependent on androgens (17,18). In order to study the role of mouse kidney antizyme in ODC degradation, we first examined the change in antizyme/ODC ratio in the kidney of female mice

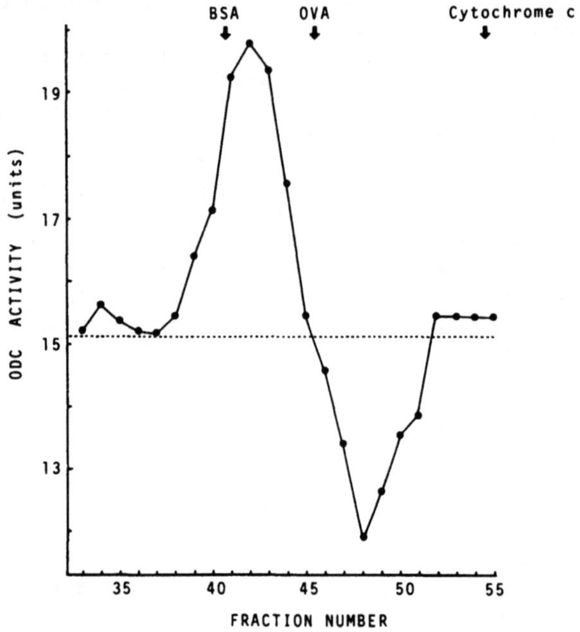

Fig. 1. Gel-filtration chromatography showing the presence of ODC-antizyme complex in the kidney of untreated female mice. A DEAE-cellulose fraction of pooled kidney extracts, containing 6.2 units of ODC activity, was chromatographed on TSKgel G3000SW (7.5 x 600mm) in the presence of 1M NaCl. An aliquot of each fraction was assayed for ODC activity in the presence of 15.1 units of partially purified ODC. Collectively 28.6 units of ODC and 16.1 units of antizyme were recovered. BSA, bovine serum albumin (MW 67500); OVA, ovalbumin (MW 43000): cytochrome c (MW 12400).

during a testosterone treatment (14), which is known to cause a 4- to 10-fold increase in the half-life of renal ODC (17,18). As shown in Table 1, renal content of ODC-antizyme complex was increased three-fold by the treatment. However, because of a large increase in ODC activity, antizyme/ODC ratio (the proportion of ODC present as a complex with antizyme) was decreased to one fiftieth. It is quite understandable that such a decrease in antizyme/ODC ratio was responsible for the stabilization of ODC.

It is also known that renal ODC activity of male mouse is much higher than that of female mouse (19). We examined sex differences in the decay rate of ODC and antizyme/ODC ratio in the kidneys of untreated mice (14). As shown in Table 2, renal ODC activity of males was about 100-times as high as that of females, whereas renal amount of ODC-antizyme complex was only 3-fold higher in males. Therefore, antizyme/ODC ratio was about 25-fold higher in females than in males. Concomitantly, half-life of renal ODC activity measured upon administration of cycloheximide was about one-fifth in females compared with that in males. This suggested that there is a correlation between ODC decay rate and antizyme/ODC ratio in mouse kidney. Fig. 2 shows that such a correlation was in fact present in mouse kidney, in good agreement with the previously reported correlation observed in HTC cells. These results strongly suggest that antizyme plays an essential role in ODC degradation in mouse kidney, too. We have also confirmed that pretreatment of testosterone-injected mice with putrescine increased the antizyme/ODC ratio and accelerated the rate of ODC decay in the presence of cycloheximide.

In the mean time, the decay rate of renal ODC-antizyme complex was also about 4-fold higher in females than in males (Table 2). This could totally account for the 3-fold larger amount of the complex in the males compared with females, and therefore synthesis rate of renal antizyme was presumed to be about the same between male and female mice. In agreement with this, sex

Table 1. Effects of testosterone on ODC activity, the amount of ODC-antizyme complex and antizyme/ODC ratio in the kidney of female mouse

Testosterone treatment[a]	ODC activity (units/g kidney)	ODC-antizyme complex[b] (units/g kidney)	Antizyme/ODC ratio
None	96	30	0.24
1 day	1760	19	0.01
2 days	2018	86	0.04
9 days	21132	93	0.004
11 days	15479	93	0.006

[a]Female ICR mice were given a single subcutaneous injection of testosterone enanthate (25 mg/mouse).
[b]The amount of ODC-antizyme complex was determined by the competitive method of Murakami et al. (8) after removal of free ODC by an affinity column.
Results were average of two to four mice or of two experiments with pooled extracts from four mice.
Modified from Murakami et al. (14)

Table 2. Comparisons of ODC activity, amount of antizyme/ODC complex, and their half-lives between male and female mice

	Males	Females
ODC activity*	4639 ± 2119	37.0 ± 11.8
ODC-antizyme complex*	73.6	24.2
Antizyme/ODC ratio	0.016	0.40
Half-life of ODC activity	38 min	8 min
Half-life of the complex	74 min	19 min

*units/g kidney

Half-lives were measured from values obtained before and 20 min after an injection of cycloheximide (50mg/kg body wt.). Results on ODC activity are means ± SD for five mice. Assay method for the amount of ODC-antizyme complex was described in Murakami et al. (14). Modified from Murakami et al. (14).

difference in renal polyamine levels was surprisingly small (Fig. 3), in spite of the large sex difference in renal ODC activity, as previously reported (19). Thus, both spermidine and spermine levels were about the same between male and female mice and only putrescine level was about twice as high in males as in females. These results were compatible with the concept that antizyme synthesis is regulated by endogenous polyamine levels under physiological conditions.

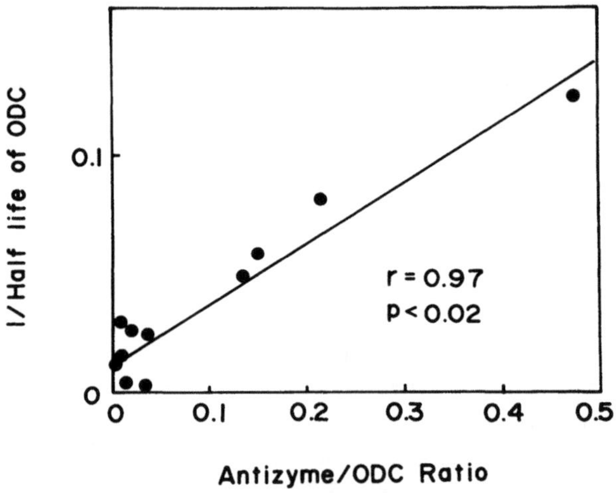

Fig. 2. Correlation between the reciprocal of half-life od ODC activity and antizyme/ODC ratio in mouse kidney. The results were obtained from male and female mice with and without testosterone treatment.

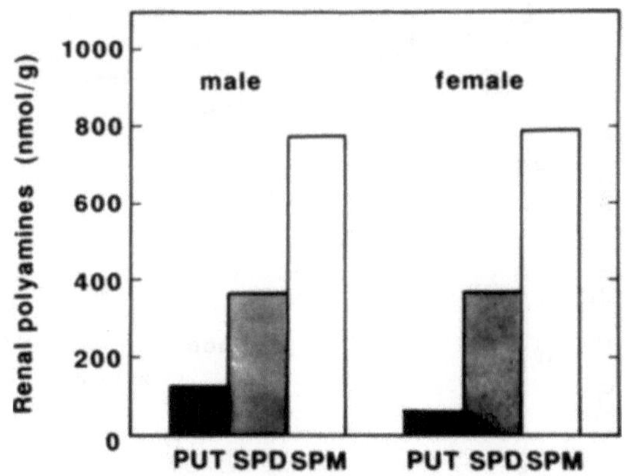

Fig. 3. Renal polyamine contents of male and female mice. Two percent perchloric acid extract of a kidney from each mice was analyzed for polyamines by high-performance liquid chromatography. Results are means ± SD for 9 mice. Modified from Murakami et al. (14).

ANTIZYME DEFICIENCY IS THE MECHANISM UNDERLYING ODC STABILIZATION CAUSED BY INHIBITORS OF POLYAMINE SYNTHESIS

If cellular polyamines regulate the rate of ODC degradation via antizyme, inhibitors of polyamine synthesis are expected to bring about ODC stabilization since they lower cellular polyamine levels. In fact, ODC is known to be stabilized not only by ODC inhibitors, DL-alpha-hydrazino-ornithine (20), DL-alpha-hydrazino-delta-aminovaleric acid (21) and alpha-methylornithine (22,23), but also by a spermidine synthase inhibitor, cyclohexylamine (24), and by S-adenosylmethionine decarboxylase inhibitors, methylglyoxalbis(guanylhydrazone) and its analogs (24-26). Persson et al. has reported, however, that ODC was not stabilized when Ehrlich ascites cells were treated with alpha-difluoromethylornithine (DFMO), an enzyme-activated irreversible inhibitor of ODC (27). We therefore examined this problem in HTC cells (28).

High-density cells were preincubated for 20 h and then incubated on fresh medium. ODC activity usually increased after change of medium, reaching a peak at 4 h under such a condition. DFMO (5mM) was added to half of cells throughout these incubations. When examined at 4 h after change of medium, the rate of ODC decay was not significantly affected by DFMO. When examined at 6 h after change of medium, however, the half-life of ODC decay in the presence of cycloheximide was increased more than three-fold by the DFMO treatment (Fig. 4, Table 3). At that time point, DFMO-treated cells had a 7-fold higher amount of total ODC, most of which being DFMO-inactivated form, compared with control cells. The amount of ODC-antizyme complex was one-third and antizyme/ODC ratio was one-twentieth in the DFMO-treated cells compared with control. Cellular putrescine and spermidine levels of the treated cells were shown to be 15% and 83% of the control, respectively, whereas spermine level was 127% of the control. In conclusion, ODC in HTC cells could be stabilized by DFMO under an

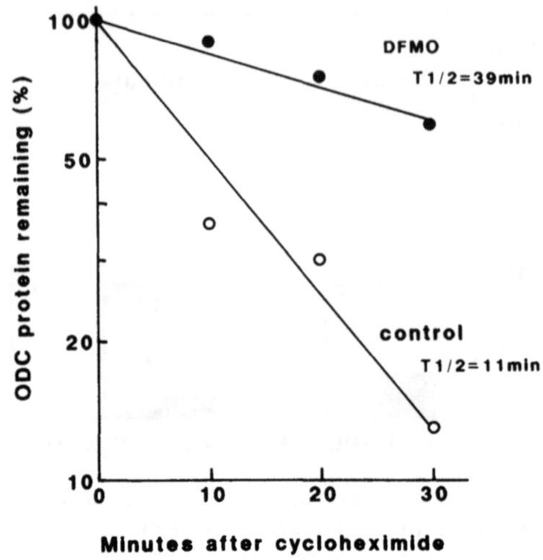

Fig. 4. Effect of DFMO treatment on the decay rate of ODC protein in HTC cells. Cells were treated with 5mM DFMO as described in the text. Cycloheximide (50 microgram/ml) was added 6 h after change of medium. Total amount of ODC protein was measured by an enzyme immunosorbent assay (Nishiyama et al, presented in this symposium). Initial amounts of ODC protein were 0.9 ng/10^6 cells for the control and 6.8 ng/106 cells for DFMO-treated cells. Modified from Murakami et al. (28).

Table 3. Effects of DFMO treatment on half-life of total ODC and antizyme/ODC ratio in HTC cells

Cells	Total ODC	ODC-antizyme complex	Antizyme/ODC ratio	Half-life of total ODC (min)
	(units/mg protein)			
Control	2.7	1.4	0.52	11
DFMO	19.2	0.52	0.027	38

Experimental condition was the same as for Fig. 4. Total ODC for control cells was the sum of free ODC and ODC-antizyme complex and that for DFMO cells was the sum of free ODC, DFMO-ODC, ODC-antizyme complex and DFMO-ODC/antizyme complex. Each component was assayed after separation by affinity chromatography. DFMO-ODC was determined by its ability to reactivate ODC-antizyme complex. Half-life of total ODC was calculated from the amounts of total ODC before and 20 min after cycloheximide addition. Modified from Murakami et al. (28).

appropriate condition and such a stabilization was accompanied by a marked decrease in antizyme/ODC ratio and by substantial decreases in putrescine and spermine levels.

Why did not DFMO stabilize ODC of HTC cells at 4 h after change of medium? Similarly, why did not DFMO stabilize ODC of Ehrlich ascites cells in the study of Persson et al.? When HTC cells are cultured in a fresh medium, cellular amount of ODC-antizyme complex increases only after the peak of ODC activity and accumulates during ODC-declining phase. Therefore, antizyme/ODC ratio at first decreases from an initial high value to the lowest value at around the ODC peak and then markedly increases during ODC-declining phase. Concomitantly, ODC decay rate at first decreases from initial high value to the lowest value at around ODC peak and then increases again during ODC-declining phase. It is easily understandable that DFMO hardly exhibited ODC-stabilizing effect at around ODC peak (4 h after change of medium) when the enzyme was already quite stable. Similar interpretation is possible for the result of Persson et al., in which DFMO effect on ODC decay rate was examined in the middle of ODC-increasing phase (27). Meanwhile, the time-dependency of the ODC-stabilizing effect by DFMO excludes the possibility that the stabilization is the direct result of the binding between the enzyme and the inhibitor. These results therefore supports the concept that polyamine synthesis inhibitors stabilize ODC by lowering cellular polyamine levels and consequently decreasing antizyme/ODC ratio.

Mitchell et al. reported that ODC activity in HTC cells was stabilized by cyclohexylamine, a spermidine synthase inhibitor (24). We examined the effect of this inhibitor on cellular amount of ODC-antizyme complex in order to confirm whether antizyme/ODC ratio was decreased by the inhibitor. For this purpose we used HMO_A cells, an ODC-stabilized variant cell line of HTC, because the content of ODC-antizyme complex is very high in this cell line (8) and therefore fluctuation of the complex can be determined accurately. As shown in Table 4, cyclohexylamine treatment caused more than two-fold

Table 4. Effects of cyclohexylamine (CHA) on ODC activity, amount of ODC-antizyme complex and polyamine contents in HMO_A cells.

Cells	ODC	ODC-antizyme complex	PUT	SPD	SPM
	(units/mg protein)		(nmol/mg protein)		
Control	19.4 ±2.4	14.2 ±0.6	12.1 ±1.3	16.5 ±1.2	10.3 ±2.3
CHA	41.9 ±4.9	4.1 ±2.2	27.4 ±1.1	0.46 ±0.02	16.6 ±2.6

Half-confluent cells were cultured in a fresh medium for 20 h with or without 1 mM cyclohexylamine. Data are means ± SD for cell extracts from triplicate cultures. PUT, Putrescine; SPD, spermine; SPM, spermine. Modified Murakami et al. (28).

accumulation of ODC and conversely decreased the amount of the complex to one third of control cells. As a result the treatment caused a marked decrease in antizyme/ODC ratio. This supports the generality of the correlation between ODC decay rate and antizyme/ODC ratio.

Another object of this study was to determine which polyamine induces antizyme. This is a difficult problem to solve in normal cells, in which polyamines are freely interchangeable. As shown in Table 4, the spermidine synthase inhibitor caused an almost complete depletion of cellular spermidine and concomitant increases of putrescine and spermine. Taking into account the marked decrease in the amount of ODC-antizyme complex under the condition, this result indicates that spermidine is the most important polyamine for the induction of antizyme. When added exogenously to the cyclohexylamine-treated cells, however, both putrescine and spermine were also effective to increase the amount of ODC-antizyme complex, although higher concentrations were needed compared with spermidine. Putrescine is hardly converted to spermidine in the presence of cyclohexylamine. Analysis of cellular polyamine levels revealed that cellular spermidine concentration did not increase upon addition of either putrescine or spermine, suggesting that the latter polyamines induced antizyme by themselves. These results therefore strongly suggest that while spermidine was the most effective in the induction of antizyme, putrescine and spermine were also effective at higher concentrations.

CLONING OF cDNA FOR RAT LIVER ANTIZYME

In order to elucidate the structure and gene expression of antizyme, we have tried to obtain cDNA for rat liver antizyme. For that purpose, rat liver cDNA library constructed in λgt11 expression vector was screened for antizyme cDNA with biotinized HZ-3H1 monoclonal antibody to rat liver antizyme as a probe. Eleven positive clones were selected out of 1.3 million plaques. The specificity of the cDNA was confirmed by its expression in E. coli-λgt11 system as a fusion protein with β-galactosidase, as indicated by the following evidence: (i) An ODC-inhibitory activity was induced in the λgt11 recombinant lysogen by a Lac-inducer. (ii) The ODC inhibitory activity was exhibited in a time-independent and stoichiometric manner, and reversed by an escess amount of antizyme inhibitor partially purified from rat liver. (iii) The ODC inhibitor was reactive both with monoclonal antibodies to rat liver antizyme and with anti-β-galactosidase serum upon immunoblotting analysis and immunoaffinity chroimatography.

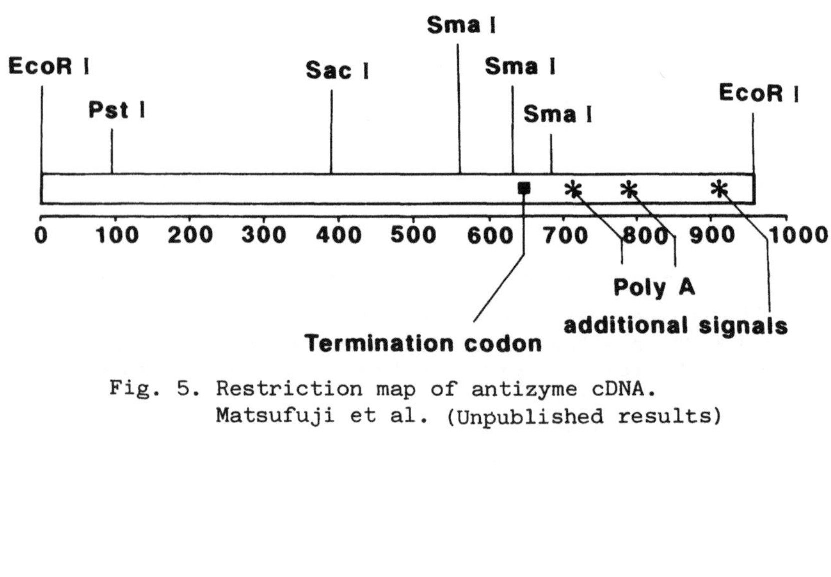

Fig. 5. Restriction map of antizyme cDNA.
Matsufuji et al. (Unpublished results)

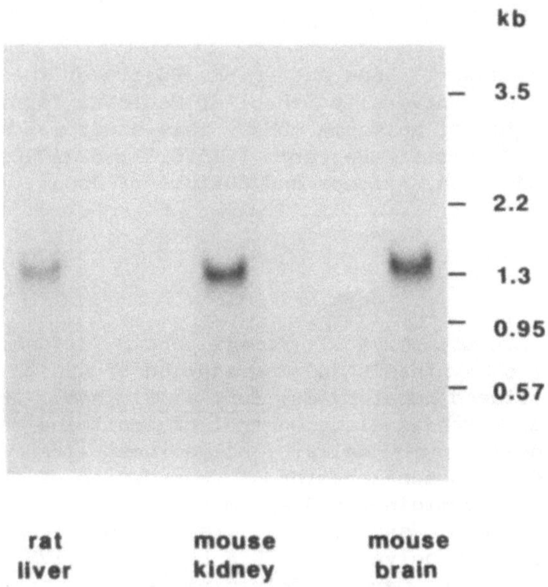

Fig. 6. Northern blot analysis of antizyme mRNA.
Matsufuji et al. (Unpublished results)

Fig. 5 shows a restriction map of the longest cDNA for rat liver antizyme. The cDNA had an open reading frame of 640 nucleotides and a 3'-noncoding region of 310 nucleotides, which contained three sites for poly(A) addition, and a poly(A) segment at 3' end. The cDNA lacked a part of coding region including the initiation codon and 5'-noncoding region.

A Northern blot analysis of antizyme mRNA was carried out using cDNA as a probe. As shown in Fig. 6, a single mRNA band of the same size was detected in total cellular RNA fraction of rat liver as well as of mouse kidney and mouse brain. From the size of the mRNA (ca 1300 nucleotides) and the sizes of 3'-noncoding region (310 nucleotides) and the partly defective coding region (640 nucleotides) of our cDNA, the length of the full coding region was presumed to be between 640 and 1000 nucleotids. Therefore, the molecular weight of native antizyme should be between 23000 and 36000.

The fact that mouse kidney contained a species of RNA of similar size that cross-hybridized to cDNA for rat liver antizyme offers further evidence for the presence of antizyme in mouse kidney. It is interesting to note that a similar cross-hybridizing RNA was also present in mouse brain. We have recently confirmed that mouse brain contains antizyme from several lines of evidence, including kinetic characteristics of ODC inhibition, reversibility of the inhibition by antizyme inhibitor and its apparent molecular weight (26000) (30). The mouse brain antizyme was clearly different from rat liver or mouse kidney antizymes, since it did not cross-react with any of the seven monoclonal antibodies to rat liver antizyme, in contrast to mouse kidney antizyme which cross-reacted with all of the antibodies to various extents. Apparently, cross-hybridization of mRNA is more sensitive than immunocross-reactivity of protein in detecting their structural resemblance between different tissues and species. In any case, it appears that there are tissue-specific antizymes in the mouse and rats. Studies are under way in our laboratory on the comparative biochemistry of antizyme. These studies might give us a clue on the evolutionary history of this unique regulatory protein.

ACKNOWLEDGEMENTS

We thank Professor T. Ohno and Dr. K. Fujita of the Department of Microbiology of Jikei University School of Medicine for their valuable advice on the cloning of antizyme cDNA. This study was supported in part by Grant-in-Aids for Scientific Research (61570131, 62790281 and 62010071) from the Ministry of Eduction, Science and Culture of Japan.

REFERENCES

1. R. Kanamoto, K. Utsunomiya, T. Kameji, and S. Hayashi, Effects of putrescine on synthesis and degradation of ornithine decarboxylase in primary cultured hepatocytes, Eur. J. Biochem., 154:539 (1986)
2. E. Hölttä and P. Pohjanpelto, Control of ornithine decarboxylase in Chinese hamster ovary cells by polyamines. Translational inhibition of synthesis and acceleration of degradation of the enzyme by putrescine, spermidine, and spermine, J. Biol. Chem., 261:9502 (1986)
3. L. Persson, I. Holm., and O. Heby, Translational regulation of ornithine decarboxylase by polyamines, FEBS Lett., 205:175 (1986)
4. J. R. Glass and E. W. Gerner, Polyamine-mediated turnover of ornithine decarboxylase in Chinese hamster ovary cells, Biochem. J., 236:
5. W. F. Fong, J. S. Heller, and E. S. Canellakis, The appearance of an ornithine decarboxylase inhibitory protein upon addition of putrescine to cell cultures, Biochim. Biophys. Acta, 428:456 (1876)
6. J. S. Heller, W. F. Fong, and E. S. Canellakis, Induction of a protein inhibitor to ornithine decarboxylase by the end products of its reaction, Proc. Natl. Acad. Sci. USA, 73:1858 (1876)
7. S. Hayashi and E. S. Canellakis, Ornithine decarboxylase antizymes, in "Ornithine Decarboxylase: Biology, Enzymology, and Molecular Genetics," S. Hayashi, ed., Pergamon Press, Elmsford, N.Y., (In the press)
8. Y. Murakami, K. Fujita, T. Kameji, and S. Hayashi, Accumulation of ornithine decarboxylase-antizyme complex in HMO_A cells, Biochem. J., 225:689 (1985)
9. Y. Murakami and S. Hayashi, Role of antizyme in degradation of ornithine decarboxylase in HTC cells, Biochem. J., 226:893 (1985)
10. J. E. Seely and A. E. Pegg, Effect of 1,3-diaminopropane on ornithine decarboxylase enzyme protein in thioacetamide-treated rat liver, Biochem. J., 216:701 (1983)
11. L. Persson and E. Rosengren, Inhibition of ornithine decarboxylase activity in mouse kidney by administration of diamines, Acta Chem. Scand., 33:537 (1979)
12. J. E. Seely and A. E. Pegg, Changes in mouse kidney ornithine decarboxylase activity are brought about by changes in the amount of enzyme protein as measured by radioimmunoassay, J. Biol. Chem., 258:2496 (1983)
13. P. H. Laitinen, R.-L. Huhtinen, O. A. Hietala, and A. E. I. Pajunen, Ornithine decarboxylase activity in brain regulated by a specific macromolecule, the antizyme, J. Neurochem., 44:1885 (1985)
14. Y. Murakami, M. Marumo, and S. Hayashi, Ornithine decarboxylase antizyme in kidneys of male and female mice, Biochem. J., (In the press)
15. K. Fujita, Y. Murakami, and S. Hayashi, A macromolecular inhibitor of the antizyme to ornithine decarboxylase, Biochem. J., 204:647 (1982)
16. L. Persson, Putrescine turnover in kidneys of testosterone-treated mice, Biochim. Biophys. Acta,, 674:204 (1981)
17. J. E. Seely, H. Pösö, and A. E. Pegg, Effect of androgens on turnover of ornithine decarboxylase in mouse kidney. Studies using labeling of the enzyme by reaction with (^{14}C)alpha-difluoromethylornithine, J. Biol. Chem., 257:7549 (1982)

18. V. V. Isomaa, A. E. I. Pajunen, C. W. Bardin, and O. A. Jänne, Ornithine decarboxylase in mouse kidney. Purification, characterization, and radioimmunological determination of the enzyme protein, J. Biol. Chem., 258:6735 (1983)
19. S. Henningsson and E. Rosengren, Biosynthesis of histamine and putrescine in mice during post-natal development and its hormone dependence, J. Physiol., 245:467 (1975)
20. S. I. Harik, M. D. Hollenberg, and S. H. Snyder, Ornithine decarboxylase turnover slowed by alpha-hydrazinoornithine, Mol. Pharmacol., 10:41 (1974)
21. H. Inoue, Y. Kato, M. Takigawa, K. Adachi, and Y. Takeda, Effect of DL-alpha-hydrazino-delta-aminovaleric acid, an inhibitor of ornithine decarboxylase, on polyamine metabolism in isoproterenolstimulated mouse parotid glands, J. Biochem., 77:879 (1975)
22. P. P. McCann, C. Tardif, M. C. Duchesne, and P. S. Mamont, Effect of alpha-methylornithine on ornithine decarboxylase activity of rat hepatoma cells in culture, Biochem. Biophys. Res. Commun., 76:893 (1977)
23. P. P. McCann, Regulation of ornithine decarboxylase in eukaryotes, in "Polyamines in Biomedical Research," G. M. Gaugas, ed., John Wiley and Sons, New York (1980)
24. J. L. A. Mitchell, D. W. Maha,. P. P. McCann, and P. Qasba, Dicyclohexylamine effects on HTC cell polyamine content and ornithine decarboxylase activity, Biochim. Biophys. Acta, 840:309 (1985)
25. E. Karvonen and H. Pösö, Stabilization of ornithine decarboxylase and N^1-spermidine acetyltransferase in rat liver by methylglyoxal-bis(guanylhydrazone), Biochim. Biophys. Acta, 791:239 (1984)
26. P. Nikula, L. Alhonen-Hongisto, and J. Jänne, Effects of bis(guanylhydrazone)s on the activity and expression of ornithine decarboxylase, Biochem. J., 231:213 (1985)
27. L. Persson, S. M. Oredsson, S. Anehus, and O. Heby, Ornithine decarboxylase inhibitors increase the cellular content of the enzyme: Implications for translational regulation, Biochem. Biophys. Res. Commun., 131:239 (1985)
28. Y. Murakami, M. Nishiyama, and S. Hayashi, Involvement of antizyme in stabilization of ornithine decarboxylase caused by inhibitors of polyamine synthesis, (Submitted to Eur. J. Biochem.)
29. A. E. Pegg, Recent advances in the biochemistry of polyamines in eukaryotes, Biochem. J., 234:249 (1986)
30. H. Onoue, S. Matsufuji, M. Nishiyama, Y. Murakami, and S. Hayashi, Changes in ornithine decarboxylase and antizyme activities in developing mouse brain, Biochem. J., 250:797 (1988)

ORNITHINE DECARBOXYLASE ANTIZYME IN MAMMALIAN TISSUES

Margaret E. Brosnan and Yu-Wan Hu

Department of Biochemistry
Memorial University of Newfoundland
St. John's, Newfoundland

INTRODUCTION

Antizyme was first described by Canellakis and coworkers (1) as a protein inhibitor of ornithine decarboxylase which is induced in response to an elevated level of putrescine in cells in culture or in liver of rats treated with putrescine. It was characterized as protease-sensitive, nuclease-insensitive and heat-labile, while the inhibition is concentration-dependent, time-independent and non-competitive with respect to ornithine (1).

Kitani and Fujisawa (2) purified antizyme from rat liver to homogeneity. They reported that it has an approximate molecular mass of 19-22 kDa and bound to ornithine decarboxylase with high affinity (Ka = 1.4 x 10^{11} M^{-1}). Thus antizyme appeared to behave as a typical modifier protein which inhibits its target enzyme, like the protein phosphatase inhibitors described by Cohen (3). It has even been described as a subunit of ornithine decarboxylase (4), like the cAMP-binding subunit of protein kinase.

The name "antizyme" was chosen by Canellakis and his colleagues (1) to indicate that it could be induced by the end products of the enzyme reaction, namely the polyamines. Since cycloheximide prevents the appearance of antizyme in response to diamines (5), and also prevents the diamine-induced inhibition of ornithine decarboxylase (6), it is generally thought that antizyme must be newly synthesized and that it plays a role in the inhibition of ornithine decarboxylase. Several reports, however, have indicated that antizyme could also be present in organelles as well as in cytoplasm in liver (7), HMO_A cells (8) and kidney (9).

In the earlier studies, it was only possible to measure free antizyme by its ability to inhibit added ornithine decarboxylase (5), so the apparent rate of appearance of antizyme depended upon the level of pre-existing ornithine decarboxylase in a tissue (10). More recently, it has been reported (11) that inactivated ornithine decarboxylase (for example, by the covalent attachment of difluoromethyl ornithine, an enzyme-activated irreversible inhibitor of ornithine decarboxylase) can compete with and displace native ornithine decarboxylase from the antizyme-enzyme complex. Thus it is possible to determine the presence of antizyme-enzyme complex in the absence of free antizyme by recording the increase in ornithine decarboxylase activity upon addition of DFMO-inactivated ornithine decarboxylase to tissue extract (11).

WHERE IS ANTIZYME FOUND?

Although treatment of cells or animals with diamines provides a useful means of producing sufficient free antizyme to study, it is not a normal situation and may well obscure the actual physiological response. Thus it is important to consider normal situations in which antizyme is present. We have detected free antizyme in the mammary gland of 24 hour-fasted lactating rats (12), a physiological situation in which ornithine decarboxylase activity is very low (13), and polyamine levels are elevated (12). Hayashi and coworkers have measured free antizyme in liver of starved rats (14) and RK13 kidney cells (9). Other tissues in which antizyme has been detected without application of exogenous polyamines include brains of young mice (15) and lymphoid tissues (thymus and spleen) of dexamethasone-treated rats (16). Antizyme complexed with ornithine decarboxylase has been detected in mammary gland of 6 hour-fasted lactating rats (17), HMO_A cells (11), brain of normal young mice (18, 19) and in RK 13 kidney cells (9). Thus antizyme can be detected in a number of tissues under physiological conditions. Biochemical detection of antizyme, however, requires excess of antizyme over ornithine decarboxylase in order to assay free antizyme, and there should be little or no free antizyme if antizyme-enzyme complex is to be assayed.

Hayashi and Matsufuji obtained seven monoclonal antibodies against antizyme purified from rat liver (19). These antibodies were capable of precipitating antizyme from liver (19) and kidney (Matsufuji, personal communication), but not from brain under the conditions studied (19). We have used these antibodies, generously provided by Drs. Matsufuji and Hayashi, to study antizyme using immunochemical techniques. We were initially interested in determining how similar antizyme from mammary gland was to that from liver. The two antizymes seem to show the same characteristics (17), but antizyme from mammary gland is about twice as large as that from liver (17). In order to determine whether these antizymes are similar or different proteins, we have compared the immunological behaviour of antizyme from mammary gland with that from liver.

For antibody neutralization of antizyme, assays contained 3 units antizyme, 500 μg bovine serum albumin and 10 μg mouse anti-antizyme IgG in a total volume of 610 μl Buffer B (50 mM Tris-Cl, pH 7.5, 50 mM KCl, 0.1 mM EDTA, 5 mM 2-mercaptoethanol). Goat anti-mouse IgG immobilized on agarose (10 μl packed gel) was added and samples were incubated overnight at 4^oC to allow complete interaction of antibody and antizyme. Samples were centrifuged to remove the pelleted agarose-IgG-IgG-antizyme complex and the supernatant was assayed for antizyme activity, as described (17).

Using the optimum ratio of antibody to antizyme (ie, that giving maximal precipitation), antibody inactivated approximately 37% of liver antizyme, but only 15% of mammary gland antizyme. Control mouse IgG did not precipitate any of either antizyme. Thus antibody to antizyme from rat liver was able to precipitate antizyme from rat mammary gland, but not as efficiently as it did antizyme from rat liver.

The antizyme-antibody complex was visualized using a system of diffusion in one dimension, as described by Oudin (20). One ml 0.5%

Table 1. Immunoprecipitation of antizyme.

Source of antizyme	% antizyme precipitated
Liver	36.8 ± 6.1^{a}
Mammary gland	15.3 ± 3.1

[a] Mean ± SD (4)

(w/v) agar in Buffer B containing 2 units of antizyme was pipetted into a test tube (10 x 75 mm) and allowed to harden. One-half ml 0.4% agar in Buffer B containing 20 μg antibody was carefully layered on top. Tubes were incubated at room temperature for 3 to 7 days to allow development of a visible precipitin line. Antibody from both tissues gave a clear precipitin line, indicating antigen-antibody complex formation. The interaction of antizyme from mammary gland with antibody raised against liver antizyme is shown in Figure 1. The precipitin line in the tube containing liver antizyme was indistinguishable from this one.

Therefore, we have shown that antizyme from mammary gland can form a complex with antibody to liver antizyme, although perhaps not quite as well as antizyme from liver itself.

Antibodies can be used to detect immunoreactive antizyme under conditions in which antizyme activity is not measurable. Crude cytosol preparations from fed or fasted rats were subjected to polyacrylamide gel electrophoresis in the presence of sodium dodecyl sulphate (21). Proteins were then transferred to nitrocellulose membranes and probed with antibody to liver antizyme. A second antibody labelled with peroxidase was then overlaid and peroxidase localized by reaction with o-dianisidine. Using this technique, we were able to show the presence of immunoreactive antizyme protein in a number of tissues from either fed or fasted rats (22). In some cases, free antizyme could be detected biochemically, in other cases antizyme complexed to ornithine decarboxylase was detected (+/- in Table 2) and in others no antizyme could be detected biochemically but immunoreactive antizyme protein was clearly present in the cytosol. The results are summarized qualitatively in Table 2.

There have been reports of antizyme in compartments of the cell in addition to cytoplasm (7-9). The antizyme we observe in cytosol might be released on homogenization of tissue (as suggested for brain 18, 19) or it might be present naturally in the cytosol at all times, but just be detected biochemically when the ornithine decarboxylase activity is very low.

WHAT ROLE DOES ANTIZYME PLAY?

No one really knows what antizyme is or does so we are free to speculate. Antizyme was originally described as a polyamine-inducible protein which specifically inhibits ornithine decarboxylase (1).

Table 2. Qualitative summary of antizyme assays in four tissues of fed or fasted rats.

TISSUE	ANTIZYME DETECTED	
	BIOCHEMICALLY	IMMUNOLOGICALLY
LIVER-FED	+/-[a]	+
FASTED	+[b]	+
MAMMARY GLAND-FED	-[c]	+
FASTED	+	+
PANCREAS-FED	+/-	+
FASTED	+	+
KIDNEY-FED	-	+
FASTED	-	+

[a] antizyme detected complexed to ornithine decarboxylase
[b] free antizyme readily measured
[c] no antizyme detected; high ornithine decarboxylase activity.

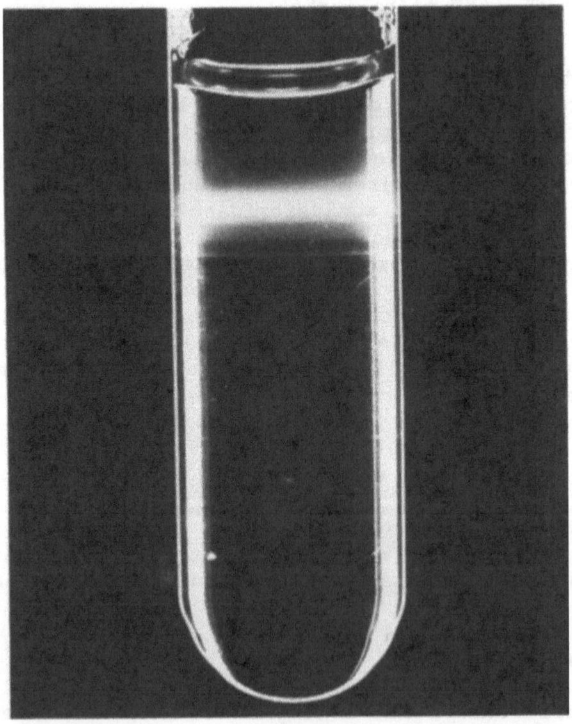

Fig. 1. Antizyme interaction with antibody, as visualized by a simple diffusion system.

Certainly the presence of putrescine does cause ornithine decarboxylase activity to fall markedly and free antizyme is easily assayed. Certainly, also, antizyme does inhibit ornithine decarboxylase specifically and antizyme binds to the enzyme with high affinity (2). But is antizyme specifically involved in the regulation of ornithine decarboxylase in vivo or is it an unrelated protein which "finds" and binds ornithine decarboxylase only as the tissue is being homogenized?

Putrescine is known to cause a rapid disappearance of ornithine decarboxylase protein, which in turn would permit antizyme activity to be detected. If this interpretation option should happen to be the correct one, then release of antizyme could cause considerable error in decarboxylation assays for ornithine decarboxylase, especially in those tissues with low activity. Such a suggestion has been made by Laitinen et. al. (18) for brain, and we also think that it is a problem in pancreas.

If, on the other hand, antizyme is specific for polyamines and ornithine decarboxylase, one could imagine a number of potential roles.

Antizyme may be a modifier protein, as originally proposed (1). The family of modifier proteins is growing rapidly and contains a number of proteins which complex with, and inhibit, their target enzymes, like the inhibitors of the protein phosphatases (3). In this scenario, antizyme could hold ornithine decarboxylase in an inactive form until it is needed, at which time it could be released as active enzyme. Alternatively, antizyme could target ornithine decarboxylase for proteolytic degradation (akin to ubiquitin) or in some way enhance its rate of turnover (6, 11).

Antizyme has been reported to be present in ribonucleoprotein particles in liver (7). It is known that ornithine decarboxylase is subject to translational control (23) but the mechanism is not yet known. Is it possible that antizyme could inhibit the translation, as well as the activity, of ornithine decarboxylase?

Some modifier proteins have recently been shown to serve as "locators" to place their enzymes in the correct position in the cell. For example, protein phosphatase in muscle is held on the glycogen particle by a specific protein (24) and thus is correctly positioned to dephosphorylate enzymes involved in glycogen metabolism. Could antizyme normally be associated with a particular subcellular organelle or membrane fraction and hold ornithine decarboxylase in a particular geographical location which might be important for its function? Although free ornithine decarboxylase is found in the cytosol after homogenization (25), some studies using cytochemistry have placed the immunoreactive protein and/or DFMO-labelled enzyme in the nucleus of some cells (26, 27). Thus ornithine decarboxylase released from antizyme could give rise to a marked and rapid increase in the local polyamine concentration.

A new role recently suggested for a modifier protein is that of organizer of an enzyme complex. Specifically, Roy and Cannon (28) have suggested that a 60 kDa binding protein could be involved in some way in the assembly of the ribulose bisphosphate carboxylase complex in plant cells? Could ornithine decarboxylase be part of a multienzyme complex, containing the polyamine synthetic arginase and perhaps other enzymes of polyamine synthesis?

Whatever the final role, obviously there are a number of questions still to be answered about antizyme.

ACKNOWLEDGEMENTS

Work from the authors' laboratory was supported by a grant from the Medical Research Council of Canada. Monoclonal antibodies to rat liver antizyme were provided by Dr. S. Matsufuji and Dr. S. Hayashi, Jikei University, Tokyo, Japan. The authors wish to thank Mrs. Donna Osborne for preparation of the manuscript.

REFERENCES

1. Heller, J.S., Fong, W.F. and Canellakis, E.S. Induction of a protein inhibitor to ornithine decarboxylase by the end products of its reaction. Proc. Natl. Acad. Sci (USA) 73, 1858 (1976).
2. Kitani, T. and Fujisawa, H. Purification and some properties of a protein inhibitor (antizyme) of ornithine decarboxylase from rat liver. J. Biol. Chem. 259, 10036 (1984).
3. Cohen, P. The role of protein phosphorylation in the hormonal control of enzyme activity. Eur. J. Biochem. 151, 439 (1985).
4. Small, C. and Traut, T. Antizymes:inhibitor proteins or regulatory subunits? Trends Biochem. Sci. 9, 49 (1984).
5. Fong, W.F., Heller, J.S. and Canellakis, E.S. The appearance of an ornithine decarboxylase inhibitory protein upon the addition of putrescine to cell cultures. Biochim. Biophys. Acta 428, 456 (1976).
6. Seely, J.E. and Pegg, A.E. Effect of 1,3 diamino propane on ornithine decarboxylase enzyme protein in thioacetamide-treated rat liver. Biochem. J. 216, 701 (1983).
7. Heller, J.S., Kyriakidis, D., Fong, W.F. and Canellakis, E.S. Ornithine decarboxylase antizyme is a normal component of uninduced H-35 cells and rat liver. Eur. J. Biochem. 81, 545 (1977).
8. McCann, P.P., Tardif, C., Hornsperger, J.-M. and Bohlen, P. Two distinct mechanisms for ornithine decarboxylase regulation by polyamines in rat hepatoma cells, J. Cell. Physiol. 99, 183 (1979).
9. Murakami, Y., Marumo, M. and Hayashi, S. Existence of antizyme and ornithine decarboxylase-antizyme complex in RK 13 kidney cells by FEBS Letts. 199, 49 (1986).
10. Heller, J.S. and Canellakis, E.S. Cellular control of ornithine decarboxylase activity by its antizyme. J. Cell. Physiol. 107, 209 (1981).
11. Murakami, Y., Fujita, K., Kameji, T. and Hayashi, S. Accumulation of ornithine decarboxylase-antizyme complex in HMO_A cells. Biochem. J. 225, 689 (1985).
12. Brosnan, M.E., Farrell, R., Wilansky, H. and Williamson, D.H. Effect of starvation and refeeding on polyamine concentrations and ornithine decarboxylase antizyme in mammary gland of lactating rats. Biochem. J. 212, 149 (1983).
13. Brosnan, M.E., Ilic, V. and Williamson, D.H. Regulation of the activity of ornithine decarboxylase and S-adenosylmethionine decarboxylases in mammary gland and liver of lactating rats. Effects of starvation, prolactin and insulin deficiency. Biochem. J. 202, 693 (1982).
14. Fujita, K., Matsufuji, S., Murakami, Y. and Hayashi, S. Antizyme to ornithine decarboxylase is present in the liver of starved rats Biochem. J. 218, 557 (1984).
15. Laitinen, P.H., Huhtinen, R.-L., Hietala, O.A. and Pajunen, A.E.I. Ornithine decarboxylase activity in brain regulated by a specific macromolecule, the antizyme. J. Neurochem. 44, 1885 (1985).

16. Bishop, P.B., Young, J., Peng. T. and Richards, J.F. An inhibitor of ornithine decarboxylase in the thymus and spleen of dexamethasone-treated rats. Biochem. J. 226, 105 (1985).
17. Hu, Y.-W. and Brosnan, M.E. Properties of ornithine decarboxylase-antizyme from mammary gland of lactating rats. Arch. Biochem. Biophys. 254, 637 (1987).
18. Laitinen, P.H., Hietala, O.A., Pulkka, A.E. and Pajunen, A.E.I. Purification of mouse brain ornithine decarboxylase reveals its presence as an inactive complex with antizyme. Biochem. J. 236, 613 (1986).
19. Onoue, H., Matsufuji, S., Nishiyama, M., Murakami, Y. and Hayashi, S. Changes in ornithine decarboxylase and antizyme activities in developing mouse brain. Biochem. J. 250, 797 (1988).
20. Oudin, J. Qualitative analysis of antigen-antibody reactions in gels. In: "Methods in Immunology and Immunochemistry", vol. 3, C.A. Williams and M.W. Chase, eds, p 118. Academic Press, New York (1971).
21. Laemmli, U.K. Cleavage of structural proteins during the assembly of the head of bacteriophage T4, Nature 227, 680 (1970).
22. Hu, Y.-W. and Brosnan, M.E. Manuscript in preparation.
23. Kahana, C. and Nathans, D. Translational regulation of mammalian ornithine decarboxylase by polyamines. J. Biol. Chem. 260, 15390 (1985).
24. Stralfors, P., Hiraga, A. and Cohen, P. The protein phosphatases involved in cellular regulation. Purification and characterisation of the glycogen-bound form of protein phosphatase-1 from rabbit skeletal muscle. Eur. J. Biochem. 149, 295 (1985).
25. Murphy, B.J. and Brosnan, M.E. Subcellular localization of ornithine decarboxylase in liver of control and growth-hormone-treated rats. Biochem. J. 157, 33 (1976).
26. Greenfield, A.R.L., Taffet, S.M. and Haddox, M.K. Immunocytochemical localization of ornithine decarboxylase in cultured murine cells. Cell Tissue Res 243, 33 (1986).
27. Emanuelsson, H. and Heby, O. Nuclear ornithine decarboxylase. Electron microscope autoradiographic identification of the active enzyme using α-[5-^{3}H] difluoromethylornithine. Eur. J. Cell Biol. 30, 288 (1983).
28. Roy, H. and Cannon, S. Ribulose bisphosphate carboxylase assembly: what is the role of the large subunit binding protein? Trends Biochem. Sci. 13, 163 (1988).

PHOSPHORYLATION BY CASEIN KINASE-2 AND REVERSIBLE ALTERATION OF THIOL GROUPS: MECHANISMS OF CONTROL OF ORNITHINE DECARBOXYLASE?

Flavio Flamigni, Flavio Meggio*, Sandra Marmiroli,
Carlo Guarnieri, Lorenzo A. Pinna* and
Claudio M. Caldarera

Dipartimento di Biochimica, University of Bologna
and *Dipartimento di Chimica Biologica, University
of Padova, Italy

INTRODUCTION

The regulation of ornithine decarboxylase (ODC) in mammalian cells appears to be complex and not completely clarified[1,2]. This enzyme is characterized by a very short half-life, but the mechanism of ODC degradation is as yet unknown[3]. Because of the rapid turnover, changes in the rate of synthesis are promptly expressed as fluctuations of ODC activity[1,4]. However, a post-translational control of ODC has been proposed by several Authors[1], and include interaction with a specific inhibitor, antizyme[5], and interconversion between distinct charged species[6]. These mechanisms could account for the occurence of inactive ODC forms or play a function in the process of ODC decay, although their precise role remains to be defined. The present paper focuses on two additional covalent modifications of ODC molecule, which could cooperate at the control of ODC activity and decay: the reversible oxidation of sulphydrylic groups and the phosphorylation by a particular class of protein kinases, called casein kinase-2 (CK-2).

EXPERIMENTAL

Phosphorylation of rat heart ODC. ODC was purified from rat heart as detailed elsewhere[7]. Purification of CK-2 from rat liver cytosol and *in vitro* phosphorylation of ODC were carried out essentially as previously described[8]. In particular, for time curve experiments, the phosphorylation mixture contained 50 mM Tris-HCl, pH 7.5, 12 mM $MgCl_2$, 10 μM [$^{32}\gamma$ P]ATP, 2 mM DTT and 10 μM pyridoxalphosphate. For the kinetic experiments, ATP concentration was 30 μM. After stopping the reaction, ODC was isolated by gel electrophoresis and evidenced by autoradiography as previously described[8]. The ODC radiolabeled band was then excissed from the gel and counted for radioactivity.

This paper is dedicated to Profs. M.F. Zuretti and G. Mezzetti, members of the Italian Group for the study on Polyamine Research, who prematurely deceased in 1988.

Cell culture and ODC assay. Friend erythroleukemia cells (clone 745) were grown in RPM-I (Sigma) supplemented with 10% foetal calf serum. In order to induce ODC, stationary phase cells were seeded at a density of 3×10^5 cell/ml in fresh medium plus 10% serum. At the time indicated, aliquots of cells were harvested and washed with phosphate - buffered saline. The cell pellet was extracted with 0.3 ml of 50 mM Tris-HCl, pH 7.2, 0.1 mM EDTA, 50 μM pyridoxal phosphate, 5 mM dithiothreitol (DTT) and 0.5% triton X-100 for 2 min. The cell lysate was centrifuged and the supernatant assayed for ODC activity[7] by adding 0.1 μCi of L-[1-^{14}C] ornithine (final concn. of L-ornithine: 0.44 mM) and for protein concentration[9]. In the experiments with thiol oxidizing agents (Table 3), DTT was omitted from the ODC extraction buffer and final concn. of L-ornithine was 37 μM.

RESULTS AND DISCUSSION

Phosphorylation of ODC by Casein Kinase-2

Casein Kinase-2 (CK-2) is a type of multifunctional and ubiquitous protein kinase, which can phosphorylate serine or threonine residues using GTP as well as ATP as phosphoryl donors[10]. Although the present knowledge about the mechanism of regulation and the physiological role of CK-2 is limited, some aspects of CK-2 appear very interesting in view of a possible relation to ODC. CK-2 activity can be affected by polyamines[11], enhanced CK-2 activity has been found in several growth conditions[12-15], and CK-2 is known to phosphorylate key proteins of transcription and translation[10, 16]. We have previously shown that ODC purified from rat heart can be readily phosphorylated by CK-2 *in vitro*[8]. More recently, this finding has been confirmed by other Authors, using different sources of ODC or CK-2[17, 18]. Particularly interesting is the report by Donato *et al.*[18], who provided indirect evidence that such a phosphorylation occurs *in vivo*. Interestingly, Kahana and Nathans[19] reported the presence of alkali-labile phosphate in immunoprecipitated ODC.

Table 1. Effect of polycations on phosphorylation rate of rat heart ODC by CK-2

Addition	Concn.	Phosphorylation rate (cpm)
None	-	380
Polylysine	0.1 mg/ml	1348
Polyarginine	0.1 mg/ml	854
Polyornithine	0.1 mg/ml	1561
Histones*	0.2 mg/ml	710
Protamine	0.2 mg/ml	896
Arg_4 Ala Gly Arg_4	1 mg/ml	877
Spermine	1 mM	370
Putrescine	1 mM	351

*Type II AS from Sigma.

In the present paper we report some features of the phosphorylation reaction, which indicate that ODC is a very good potential substrate for CK-2. Table 1 shows that the phosphorylation rate of ODC is increased by various basic polypeptides, which are known to stimulate CK-2 activity [20], but not by polyamines. The phosphorylation reation is fast and is accelerated by polylysine (Fig.1). In fact the reaction was essentially completed after 5 min of incubation in the presence of polylysine. Fig.1 also shows that the incorporation of phosphate is relevant approaching 1 mole per mole of ODC. The Km of CK-2 for rat heart ODC is very low (Fig.2). Polylysine further decreased the Km value without affecting Vmax. The low Km for ODC may be related to the very low amounts of ODC in mammalian tissues or cells. Phosphorylation by CK-2 is known to require a cluster of acidic residues C terminal to the phosphorylatable aminoacid [21]. In this respect serine 303 in the aminoacid sequence of rat ODC represents an ideal target being adjacent to five acidic residues [22]. We have provided evidence that this serine may be indeed the main if not the only phosphorylatable aminoacid for CK-2 [23]. This phosphorylation site is conserved in all the mammalian species of which the ODC primary structure is published [19,22,24-26]. ODC from a lower eukaryote, Trypanosoma brucei, presents a core region of homology to the mammalian enzyme, but lacks, this sequence [27]. It is very interesting that, unlike the mammalian enzyme, ODC from T.brucei is very stable according to Phillips et al. [27], and does not go a detectable turnover. Indeed, the phosphorylation site is contained within one of the two "PEST regions" present in mammalian ODC [28]. These regions rich in proline, glutamic acid, serine and threonine, are typically present in rapidly degraded proteins. Furthermore, Rogers et al. [28] observed that PEST regions of many proteins contain potential sites for phosphorylation by CK-2 and proposed this phosphorylation as

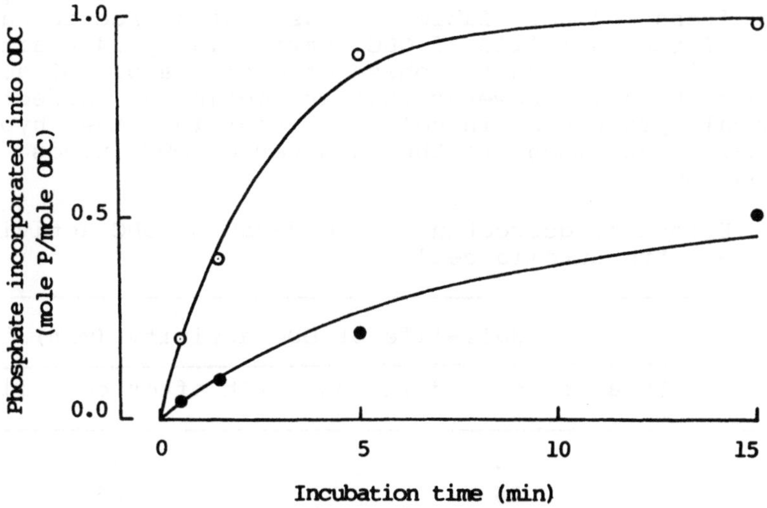

Fig.1. Time curve of the phosphorylation reaction of ODC by CK-2. ODC purified from rat heart was incubated with CK-2 for different times in the absence (●) or in the presence (○) of polylysine (0.1 mg/ml). Polylysine (180 residues on average) was from Sigma.

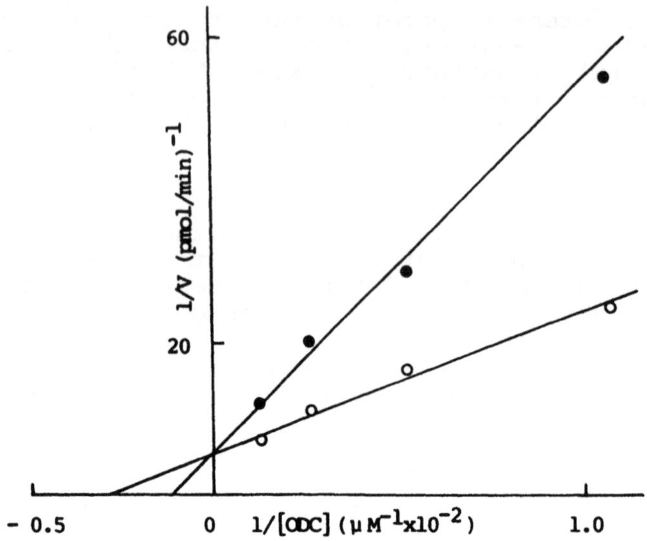

Fig.2. Lineweaver-Burk plots of the phosphorylation of rat heart ODC by CK-2 as a function of ODC concentration. Km for ODC was 83 nM in the absence (●) and 34 nM in the presence (○) of polylysine (0.1 mg/ml). Vmax was 0.18 pmol/min in each case.

one of the possible mechanisms to account for the rapid turnover. In order to test this hypothesis we have studied the effect of quercetin, a flavanoid that has been shown to be a powerful CK-2 inhibitor[29], on the half-life of ODC in Friend erythroleukemia cells. Addition of quercetin to these cells resulted in a rapid and essentially complete inhibition of casin kinase activity, which was relieved after 1 or 2 hours (not shown). ODC activity was transiently induced in erythroleukemia cells following dilution of stationary phase cells in fresh medium. Table 2 shows that quercetin caused an increase of the half-life of ODC, particularly 24 h after cell dilution, i.e. during the phase of decrease of ODC activity-It should be noted however that quercetin may affect other biochemical processes in cells[30]; therefore the hypothesis that CK-2 might modulate the turnover of ODC needs further investigation.

Table 2. Effect of quercetin on the decay of ODC activity in erythroleukemia cells.

Agent	Half-life of ODC activity (min)	
	5h after cell dilution	24h after cell dilution
Control	59	37
Quercetin	76	66

ODC was induced by dilution of the cells in fresh medium. The half-life of ODC activity was measured following cycloeximide treatment (0.2 mM). Quercetin (0.1 mM) was added 15 min before cycloeximide.

Reversible Oxidation of "Essential" Thiol Groups

It is well known that ODC contains SH group(s) essential for its activity[31-37]. In fact DTT or other synthetic dithiols are usually added to preparative and assay buffer to preserve ODC activity. Removal of DTT from ODC buffers results in a loss of ODC stability: in this conditions, the catalytic activity is progressively lost when the enzyme is maintained in presence of a O_2 containing atmosphere; adding of microsomal fractions[32], L-ornithine[33], oxygen radicals[34], iron[35], disulfides[36] and various SH reagents[36] accelerate the inactivation. This loss of activity is considered reversible since addition of DTT usually results in a noticeable recovery of activity. The oxidation of a thiol group in proximity of the active site[33,36] with formation of protein or mixed disulfides may be responsible of this inactivation. We have recently[37] tried to answer the question of which physiological thiol(s) can substitute DTT in maintaining _in vivo_ ODC in the reduced, active status. We have found that non diffusible factors present in rat liver cytosol could re-activate oxidized ODC provided that NADPH or reduced glutathione (GSH) was added. It should be noted that NADPH by itself was ineffective and GSH had only a limited effect. More recently[38], the NADPH- and GSH-dependent reducing systems were separated into three components. The chromatographic and catalytic properties of these factors suggest that the NADPH-dependent factors may be identified as thioredoxin and thioredoxin reductase, whereas the GSH-dependent factor may be identified as thiol transferase. In Fig. 3 it is depicted a scheme showing that oxidized ODC may be reduced to the active form by the thiol transferase or the thioredoxin system. Both thioredoxin and thiol transferase are small dithiol proteins (Mr approx. 12000), which have been proposed to exert a thiol-redox control of proteins by acting as effective thiol-disulfide oxidoreductases[39-41]. _In vivo_, the thioredoxin system, GSH and thiol transferase may protect the SH groups of ODC and maintain the enzyme in the active status. Thus, they counteract the oxidizing action of oxygen radicals[34], oxidized glutathione[36] or microsomal fraction[32,35]. The thiol redox status of ODC may depend on the balance between these opposite effectors. So an excess of oxidizing agents or an impairment of the reducing systems may lead to oxidation of the enzyme.

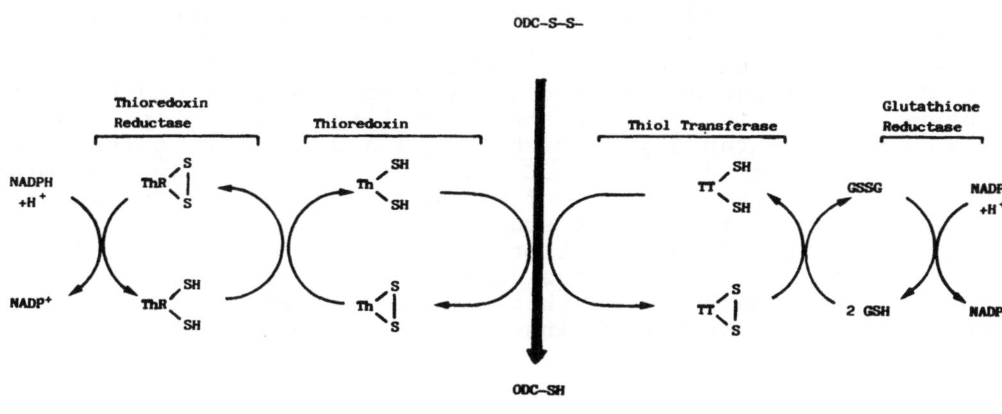

Fig. 3. Reduction of ODC disulfide by thiol transferase and thioredoxin systems.

Table 3. Effect of thiol oxidizing agents on ODC activity in erythroleukemia cells.

Agent	concn. (mM)	ODC activity (% of control)	
		-DTT	+DTT
Control	-	100	100
tBHP	0.16	59	66
BCNU	0.10	54	62
Vinblastine	0.10	80	89
Azelaic acid	0.50	36	37
$NaAsO_2$	0.20	32	37

tBHP, BCNU and $NaAsO_2$ were added 0.5 h, vinblastine 1h, and azelaic acid 5h before harvesting cells. ODC activity was measured in the absence or in the presence of 1 mM DTT.

In an attempt to investigate the physiological role of essential thiol groups of ODC, we have incubated erythroleukemia cells with agents known to alter, by different mechanisms, the thiol redox status in cells: t-BHP (t-butylhydroperoxide) decreases GSH/GSSG as a substrate of glutathione reductase[42,43], BCNU (1,3-bis-2-chloroethyl-1-nitrosurea) decreases GSH/GSSG by inhibiting glutathione reductase[42], vinblastine decreases GSH/GSSG possibly by binding to microtubules[43,44], azelaic acid inhibits thioredoxin reductase[45] and sodium arsenite inhibit dithiol enzymes[46] including those of the thioredoxin and thiol transferase systems. As shown in Table 3, all these compounds were able to inhibit ODC activity induced in these cells. However, adding of DTT to the assay mixture resulted only in a limited recovery of ODC activity. This indicates that in these conditions most of the activity is irreversibly lost. Inhibition of ODC synthesis or enhanced ODC degradation could also be hypothisized.Actually, oxidation of SH groups of ODC has been proposed as a step preceding degradation of the enzyme[32,35,47]. An altered form of ODC, with a longer half-life *in vivo* and a stronger resistance to inactivation by microsomal fractions, has been described[47].However, more recently, Zuretti *et al.*[48], comparing ODC from thioacetamide-treated rat liver and from regenerating liver, found that the enzyme exhibiting the greater stability *in vitro* appears to turn over more rapidly *in vivo*. Using a different approach, we have studied the decay of ODC activity following cycloeximide treatment in the presence of various thiol oxidizing agents. The results indicate that stabilization of the enzyme rather than accelerated degradation was the conseguence of the altered cell thiol status (not shown). It remains to be investigated whether some thiol-dependent factor, involved in ODC turnover might become inactivated in these conditions.

Acknowledgements. We thank Ms.Giovanna Grandi for her careful secretarial support.

REFERENCES

1. C.W.Tabor and H.Tabor,Polyamines, Ann.Rev.Biochem. 53: 749 (1984).
2. A.E.Pegg, Recent advances in the biochemistry of polyamines in eukaryotes, Biochem. J. 234:249 (1986).
3. J.R.Glass and E.W.Gerner, Spermidine mediates degradation of ornithine decarboxylase by a non-lysosomal,ubiquitin-independent mechanism, J.Cell.Physiol. 130:133 (1987).
4. D.H.Russell,Ornithine decarboxylase,a key regulatory enzyme in normal and neoplastic growth, Drug Metab.Rev 16:1 (1985).
5. E.S.Canellakis,D.A.Kyriakidis,C.A.Rinehart Jr., S.C.Huang, C.Panagiotidis and W.F.Fong, Regulation of polyamine biosynthesis by antizyme and some recent development relating the induction of polyamine biosynthesis to cell growth, Bioscience Reports, 5:189 (1985).
6. J.L.A.Mitchell,P.Qasba,R.E.Stofko and M.A.Fransen,Ornithine decarboxylase modification and polyamine-stimulated enzyme inactivation in HTC cells,Biochem.J., 228:297 (1985).
7. F.Flamigni,C.Guarnieri,C.Stefanelli, F.Meggio, L.A.Pinna and C.M.Caldarera,Ornithine decarboxylase in the early phase of cardiac hypertrophy induced by isoproterenol,in: "Recent Progress in Polyamine Research", L.Selmeci, M.E.Brosnan,N.Seiler, eds., Akademiai Kiado', Budapest (1985).
8. F.Meggio, F.Flamigni, C.M.Caldarera, C.Guarnieri and L.A.Pinna, Phosphorylation of rat heart ornithine decarboxylase by type-2 casein kinase,Biochem.Biophys.Res. Commun, 122:997 (1984).
9. M.M.Bradford,A rapid and sensitive method for the quantitation of microgram quantities of protein utilizing the principle of protein-dye binding,Anal.Biochem., 72:248 (1976).
10. G.M.Hathaway and J.A.Traugh,Casein Kinase II,Meth.Enzymol, 99:317 (1983).
11. C.Cochet, G.Job, F.Pirollet and E.M.Chambaz, Cyclic nucleotide independent casein kinase (G type) in bovine adrenal cortex, Biochim.Biophys.Acta,658:191 (1981).
12. H.R.Schneider,G.H.Reichert and O.G.Issinger,Enhanced casein kinase II activity during mouse embryogenesis,Eur.J.Biochem., 161:733 (1986).
13. K.Prowald,H.Fisher and O.G.Issinger,Enhanced casein kinase II activity in human tumour cell cultures,FEBS Lett.,176: 479 (1984).
14. A.M.Brunati, D.Soggioro, L.Chieco-Bianchi and L.A.Pinna, Altered protein kinase activities of lymphoid cells transformed by Abelson and Moloney Leukemia viruses,FEBS Lett., 206:59 (1986).
15. J.Sommercorn,J.A.Mulligan,F.J.Lozeman and E.G.Krebs, Activation of casein kinase II in responce to insulin and to epidermal growth factor,Proc.Natl.Acad.Sci USA,84:8834 (1987).
16. P.Ackerman, V.C.Glover and N.Osheroff, Phosphorylation of DNA topoisomerase II by casein kinase II: modulation of eukaryotic topoisomerase II activity in vitro,Proc.Natl. Acad.Sci USA, 82: 3164 (1985).
17. U.R.Tipnis and M.K.Haddox,Casein kinase II mediated phosphorylation of ornithine decarboxylase,J.Cell.Biol., 101:356a (1985).

18. N.J.Donato,C.F.Ware and C.V.Byus,A rat monoclonal antibody which interacts with mammalian ornithine decarboxylase at an epitope involved in phosphorylation,Biochim.Biophys.Acta, 884:370 (1986).
19. C.Kahana and D.Nathans,Nucleotide sequence of murine ornithine decarboxylase in RNA,Proc.Natl.Acad.Sci. USA,82: 1673 (1985).
20. F.Meggio, A.M.Brunati and L.A.Pinna,Autophosphorylation of type 2 casein kinase TS at both its α- and β -subunits.Influence of different effectors, FEBS Lett., 160:203 (1983)
21. L.A.Pinna,P.Agostinis and S.Ferrari,Selectivity of protein kinases and protein phosphatases:a comparative analysis, Adv.Prot.Phosphatases, III:327 (1986).
22. H.J.Van Kranen,L.Van de Zande,C.F.Van Kreijl, A.Bishop and B.Wieringa, Cloning and nucleotide sequence of rat ornithine decarboxylase cDNA, Gene, 60:145 (1987).
23. F.Meggio,F.Flamigni, C.Guarnieri and L.A.Pinna,Location of the phosphorylation site for casein kinase-2 within the aminoacid sequence of ornithine decarboxylase, Biochim. Biophys.Acta, 929:114 (1987).
24. M.Gupta and P.Coffino, Mouse ornithine decarboxylase.Complete aminoacid sequence deduced from cDNA,J.Biol.Chem., 260:2941 (1985).
25. N.J.Hickock,P.J.Seppanen, G.L.Gunsalus and O.A.Janne, Complete aminoacid sequence of human ornithine decarboxylase deduced from complementary DNA, DNA, 6:179 (1987).
26. P.R.Srinivasan, P.N.Tonin, E.J.Wensing and W.H.Lewis, The gene for ornithine decarboxylase is co-amplified in hydroxyurea-resistant hamster cells,J.Cell.Biol.,262: 1287 (1987).
27. M.A.Phillips,P.Coffino and C.C.Wang,Cloning and sequencing of the ornithine decarboxylase gene from Trypanosoma brucei. Implications for enzyme turnover and selective difluoromethylornithine inhibition.J.Biol.Chem.,262:8721 (1987).
28. S.Rogers, R.Wells and M.Rechsteiner, Aminoacid sequences common to rapidly degraded proteins:the PEST hypothesis, Science, 234:364 (1986).
29. C.Cochet,J.J.Feige,F.Pinollet,M.Keramidis and E.M.Chambaz, Selective inhibition of a cyclic nucleotide independent protein kinase (G type casein kinase) by quercetin and related polyphenols,Biochem.Pharmacol.,31:1357 (1982).
30. Y.Graziani and R.Chayoth, Regulation of cyclic AMP level and synthesis of DNA, RNA and protein by quercetin in Ehrich ascites tumor cells, Biochem.Pharmacol, 28:397 (1979).
31. J.Janne and H.G.Williams-Ashman,On the purification of L-Ornithine decarboxylase from rat prostate and effects of thiol compounds on the enzyme,J.Biol.Chem,246:1725 (1971)
32. M.F.Zuretti and E.Gravela,Studies on the mechanisms of ornithine decarboxylase in vitro inactivation,Biochim. Biophys.Acta, 742:269 (1983).
33. C.Danzin and L.Persson,L-ornithine-induced inactivation of mammalian ornithine decarboxylase in vitro,Eur.J.Biochem., 166:45 (1987).
34. C.Guarnieri,A.Lugaresi,F.Flamigni,C.Muscari and C.M.Caldarera, Effect of oxygen radicals and hyperoxia on rat heart ornithine decarboxylase activity,Biochim.Biophys. Acta, 718:157 (1982).
35. Y.Murakami, T.Kameji, S.Hayashi, Cysteine-dependent inactivation of hepatic ornithine decarboxylase,Biochem.J.,

217:573 (1984).
36. J.L.A.Mitchell,Post-translational controls of ornithine decarboxylase activity, Adv.Polyamine Res., 3:15 (1981).
37. F.Flamigni,C.Guarnieri and C.M.Caldarera,Rat liver cytosol contains NADPH- and GSH-dependent factors able to restore ornithine decarboxylase inactivated by removal of thiol reducing agents, Biochem.J., 250:53 (1988).
38. F.Flamigni, S.Marmiroli, C.M.Caldarera and C.Guarnieri, Involvement of thiol transferase and thioredoxin-dependent systems in the protection of "essential" thiol groups of ornithine decarboxylase, (submitted for publication).
39. K.Axelsson, S.Eriksson and B.Mannervik, Purification and characterization of cytoplasmic thiol transferase (Glutathione:disulfide oxidoreductase) from rat liver, Biochemistry, 17:2978 (1978).
40. A.Holmgren,Thioredoxin,Ann.Rev.Biochem., 54:237 (1985).
41. D.M.Ziegler,Role of reversible oxidation-reduction of enzyme thiols-disulfides in metabolic regulation.Ann.Rev. Biochem., 54:305 (1985).
42. C.E.Olson, A.H.Soll and N.Kaplowitz, Modulating effect of thiol-disulfide status on ^{14}C aminopyrine accumulation in the isolated parietal cell,J.Biol.Chem.,260:8020 (1985).
43. K.Sato,H.Mimura,K.Wakai,N.Tomari,T.Tsushiwa and K.Shizume, Modulating effect of glutathione disulfide on thyroxine-5'-deiodination by rat hepatocytes in primary culture: effect of glucose, Endocrinology, 113:878 (1983).
44. W.T.Beck,Increase by vinblastine of oxidized glutathione in cultured mammalian cells, Biochem.Pharmacol.,29:2333 (1980).
45. K.U.Shallreuter and J.M.Wood, Azelaic acid as a competitive inhibitor of thioredoxin reductase in human melanoma cells. Cancer Lett., 36:297 (1987).
46. F.C.Knowles and A.A.Benson, The biochemistry of arsenic, Trends Biochem.Sci., 9:178 (1983).
47. E.Karvonen,L.C.Andersson and H.Poso,A human neuroblastoma cell line with a stable ornithine decarboxylase in vivo and in vitro,Biochem.Biophys.Res.Commun.,126:96 (1985).
48. M.F.Zuretti, O.Brossa, P.Gili and E.Gravela,Ornithine decarboxylase properties: is there a role for a microsome-bound inactivating activity, Cell.Biochem.Funct., 6: 107 (1988).

MODIFICATIONS OF ORNITHINE DECARBOXYLASE INDUCED BY PHOSPHATASES

J.L.A. Mitchell, M.F. Hicks, H.J. Chen and J.A. Hoff

Department of Biological Sciences
Northern Illinois University
DeKalb, IL 60115 U.S.A.

INTRODUCTION

The sensitivity with which a tissue regulates polyamine biosynthesis during normal growth, development and hormone response has long been appreciated, but not well understood. In particular, the activity of the initial enzyme in this pathway, ornithine decarboxylase (ODC), appears to be subject to many cellular regulatory signals. Through recent advances in cDNA probes for ODC mRNA, and the use of radioactively labeled DFMO to tag ODC, along with the isolation and use of monospecific antibodies, great progress has been made in our understanding of the control of ODC synthesis at the levels of transcription and translation. However, this enzyme protein is also noted for its unusual instability (T1/2 can be less than 20 min) in mammalian tissues, and therefore the factors influencing this instability must also be considered if we are to gain complete understanding of the observed rapid modulation of cellular ODC activity. Unfortunately very little is known about how a cell may rapidly target and specifically degrade such a protein. This process must be of great significance, as this extreme enzyme instability is not common among cellular proteins, yet it is a feature shared by many proto-oncogene products.

The rate at which ODC is inactivated and degraded in a cell appears to be subject to physiological controls. Stimulation of ODC activity by hormones[1-5] and selected (even non-metabolizable) amino acids[6,7] is generally associated with prolongation of the ODC half life. Tumor cell lines have been noted to demonstrate abnormally stable ODC protein[7] either due to an alteration in the enzyme itself[8-10] or to possible defects in the mechanism responsible for ODC degradation.[11] Conversely, the immediate and subsequent products of this enzyme (the polyamines) actively stimulate the degradation of ODC.[12-18] Mitchell et al.[12] and others[18,19] have shown ODC half lives of greater than 3 hr in mammalian cells inhibited in spermidine uptake or production. Upon addition of exogenous polyamine, and an obligatory period of protein synthesis, this enzyme demonstrates a half life shorter than 10 min.[13] These experiments suggest that polyamines induce a specific protein that is responsible for rapid ODC degradation. Such a protein, ODC-antizyme, was initially described by Fong, Heller and Canellakis.[20]

ODC-antizyme is a small (MW approx. 26,500) protein whose cellular level increases in response to elevations in polyamine content and is temporally associated with rapid ODC inactivation.[21-26] Highly purified preparations of this antizyme have been shown to bind to ODC with an equilibrium constant as high as 1.4×10^{11} M^{-1}.[27] While complexed to this regulatory protein, ODC appears to be catalytically inactive, and this complex has been demonstrated to exist transiently during ODC inactivation.[12,23,26,28,29] It has been suggested that this antizyme acts catalytically to initiate ODC degradation,[12,19] however there has been no evidence to indicate the mechanism of this catalytic action.

Alternatively, polyamines may stimulate the rapid inactivation of ODC by inducing a particular conformational change in the protein that targets it for a non-specific protein degradation mechanism. The absence of partially degraded intermediates during rapid in vivo inactivation of ^{35}S or DFMO-labeled ODC suggests that such targeting, and not proteolytic degradation, is rate limiting. According to Hershko and Ciechanover[30] the probable mechanism for specific degradation of an unstable cellular protein is an ATP-dependent, nonlysosomal proteolytic activity that is initiated by the formation of a ubiquitin-protein complex. Glass and Gerner,[18] however, demonstrated that spermidine-induced ODC degradation does not involve lysosomal activity, yet it is also independent of ubiquitin. Recently several models have emerged suggesting that exposure and or modification of specific sequences on select proteins could induce their rapid, specific degradation. Rogers et al.[31] have described an amino acid region (PEST), common to very rapidly degraded eukaryotic proteins, that appears to mark these proteins for instability. Interestingly, mammalian ODC contains two of these sequences, one very near to its presumed active site and overlapping the putative casein kinase 2 target site. Another very suggestive model for rapid protein turnover is that recently proposed by Bachmair et al.[32] in which the N-terminal residue of a protein can determine stability. This model would suggest that a structural alteration or subunit dissociation, could expose the N-terminal end and promote rapid enzyme degradation.[33]

Over the past several years our laboratory has examined several spermidine-induced alterations in ODC charge and conformation in an effort to correlate these with the enzyme's interaction with antizyme, and its inactivation and degradation. In this paper we review some of the progress being made in understanding this aspect of the control of ODC activity.

MULTIPLE CHARGE SPECIES OF MAMMALIAN ODC

In homogenates of many tissues, ODC can be chromatographically separated into multiple peaks of activity, suggesting either isoenzymes or possible post-translational enzyme modification(s). There are isolated reports of multiple ODC forms distinguished by thiol-sepharose columns,[34] differential heat sensitivity,[35] molecular weight[36] and enzyme kinetics [37,38], but the greatest interest has been in the multiple ionic states of active ODC that are found upon DEAE anion-exchange chromatography.[12,14,39-42] The relative proportions of the various charge forms of the enzyme have been correlated with hormone induction[41] and ODC activity modulation,[42] yet these forms appear to be identical in their enzyme catalytic properties.[42] Interestingly, one of these forms (ODC-I) does decline much more rapidly during enzyme activity decay following cycloheximide addition, strongly suggesting that these enzyme states are differentially sensitive to the rapid ODC inactivation mechanism.[14]

Analysis by two-dimensional gel electrophoresis has also indicated the presence, in some tissues,[43] of distinct forms of ODC that differ with respect to their isoelectric point[1,15,43] yet are identical in their molecular weight. We have recently shown that the two ODC forms separated as native enzyme monomers on anion exchange columns will also remain distinct when denatured with 8 M urea and separated by anion exchange chromatography.[39] In that report it was further demonstrated that these individual peaks of activity are identical to the distinct isoelectric forms observed in both rat liver cell and mouse kidney tissues. What, however, is the relationship between, and physiological function of, these ODC charge isoforms?

In studies using ODC derived from rat liver cells (HTC) in culture, it was found that the more acidic enzyme form (ODC-II) readily converts to ODC-I in the presence of spermidine and unidentified factors present in the crude cell homogenate[14]. Thus the appearance of distinct ODC charge states must be due to some post-translational modification of the enzyme. More recently we have correlated this ODC-II to I conversion with the presence of phosphatase activity in these homogenates. Accordingly phosphatase inhibitors such as 10 uM levamisole and 1.0 mM tartrate were found to partially

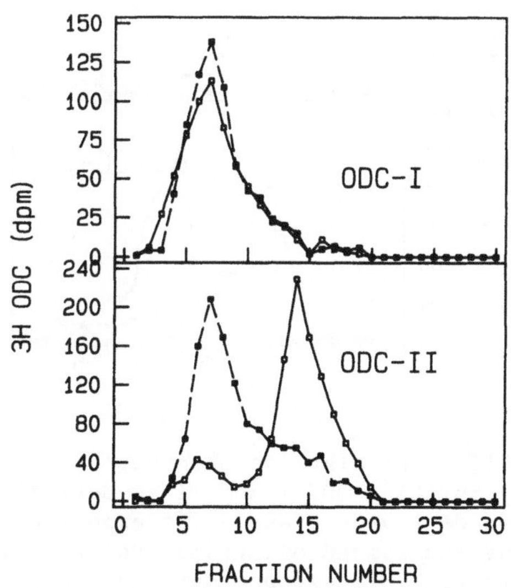

Fig. 1. Alkaline phosphatase alteration of partially purified native ODC-I and II. ODC enzyme was labeled by reaction with ^3H-DFMO and partially purified by ammonium sulfate precipitation, gel filtration chromatography and Mono Q anion exchange chromatography, as previously described.[39] Samples of peak I and II enzyme from the Mono Q column were individually incubated with purified calf intestine alkaline phosphatase (7.2 U/ml) for 15 min at 37°C. The tritium labeled ODC was subsequently chromatographed again on the Mono Q column. For each form of the enzyme the position of elution of the enzyme protein is noted before (□) and after (■) the phosphatase treatment.

inhibit this form conversion, and orthovanadate (1 mM) completely inhibited this transformation.

We also have observed that the addition of either alkaline or acid phosphatases purified from several different sources all were capable of inducing the charge modification associated with the conversion of ODC-II to I. Phosphatase-induced conversion is observed in both mouse and rat liver ODC, in crude and highly purified fractions, and even in the absence of the obligate coenzyme, pyridoxal 5'-phosphate (PLP). Fig. 1 illustrates the effect of alkaline phosphatase upon partially purified ODC-I and ODC-II. In this experiment each of the isoforms had been labeled by inactivation with ^{3}H-DFMO to permit us to follow alterations in the ODC protein apart from any

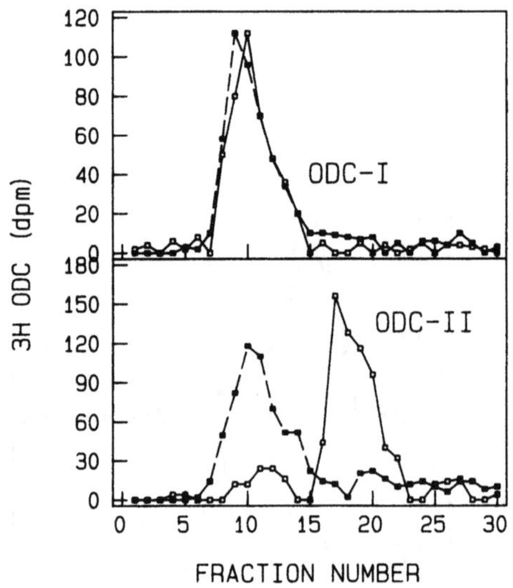

Fig. 2. Effect of alkaline phosphatase on partially purified ODC-I and II as determined in their urea denatured state. The experimental conditions and legend notations are exactly the same as in Fig. 1, except that the ODC protein is denatured with 8M urea before Mono Q chromatography, which is also in the presence of 8 M urea. The fraction numbers of the eluted enzyme peaks are different than in Fig. 1 because of a slightly different salt gradient used to elute the enzyme in the presence of urea.[39]

changes in enzyme activity. ODC-I was essentially unaffected by this phosphatase treatment whereas ODC-II was almost completely converted to the ODC-I charge state. In order to diminish the possibility that the alteration in charge state might be due to a non-covalently associated molecule, or to a conformational change in the ODC protein, the phosphatase treatments were repeated on similar samples of purified ODC (Fig. 2). In this experiment the enzyme samples were subsequently denatured and separated by ion exchange chromatography in the presence of 8 M urea. From such experiments it is clear that there is a distinct charge difference between the ODC isoforms that can be eliminated by simple phosphatase action.

ASSOCIATION OF PHOSPHATE WITH ODC *in vivo*

We are increasingly aware of the role of reversible protein phosphorylation in the regulation of activity, cellular location and stability of a wide variety of proteins. Very little, however, is known about the role such phosphorylation(s) may play in the control of ODC activity. Atmar and Kuehn[44] initially described a polyamine-stimulated nuclear kinase (presumably N II) that phosphorylated Physarum nuclear ODC and rendered it inactive. Mammalian ODC, however, was found not to be phosphorylated by protein kinase N II.[45] The amino acid sequence of mouse ODC derived from cDNA clones does show a potential substrate site for casein kinase 2 at Ser_{303}. Meggio et al.[46,47] showed that ODC can be phosphorylated *in vitro* by casein kinase 2. The same group[47,48] has further demonstrated that this phosphorylation may be at the Ser_{303} and that this phosphorylation may occur in intact rat heart tissue. Evidence has also emerged from Haddox's laboratory[49,50] suggesting that phosphorylation of ODC by casein kinase 2 may occur in RAW264 macrophage-like cells in culture. In that system the phosphorylated form of the enzyme separates from the non-phosphorylated enzyme on isoelectric focusing gels. More recently, Donato et al.[51] have used a specific rat monoclonal antibody to indicate a particular site on ODC that is phosphorylated with casein kinase 2. Using this immunological approach they have presented evidence to suggest that native mouse kidney ODC appears in both the phosphorylated and non-phosphorylated forms. In all of these cases, however, the addition of phosphate does not appear to alter the activity of ODC.

Since our results on the conversion of ODC-II to I suggested that a phosphate group was being removed from the ODC protein, it was of interest to ascertain the location and function of the phosphate involved. Our initial hypothesis was that it was the same as that probed by Donato et al.[51] using their specific monoclonal antibody. If one of the two ODC charge states, specifically ODC-II, contained a phosphate in the casein kinase-2 site that serves as the epitope for their monoclonal antibody, then that antibody should be unable to recognize it. To test this we titrated their antibody against partially purified preparations of ODC-I and II isolated from HTC cells. As evidenced by loss of ODC activity, there was no difference in the ability of this monoclonal antibody to recognize these two enzyme forms. Thus, the phosphorylation site monitored by Donato et al.[51] does not appear to be involved in the interconversion that we have been observing between the two active forms of ODC.

Although several groups have reported the phosphorylation of purified ODC by casein kinase-2,[46,47,51] evidence for phosphorylation of this enzyme *in vivo* has actually been indirect[51], or not well substantiated.[46,50] Since our previous results would be consistent with a phosphorylation of at least ODC-II, we attempted to demonstrate such phosphorylation by inducing ODC-II formation in HTC cultures maintained for 4h in ^{32}P-labeling media. The cell pellets were homogenized in the presence of phosphatase inhibitors, including 1mM orthovanadate, which we have shown to prevent ODC-II conversion to ODC-I. These homogenates were reacted with mono-specific, polyclonal antibody[52] and subsequently precipitated with goat anti-rabbit IgG with or without insoluble protein A. After appropriate washing the pellets were chromatographed using SDS-PAGE, and the gels autoradiographed. These experiments have been unsuccessful, so far, in demonstrating ^{32}P covalently linked specifically to ODC; however, the existence of such a phosphorylation has not been disproved. On the other hand, as shown in the following experiments, the phosphate group purportedly associated specifically with ODC-II need not be covalently linked.

Contrary to the SDS-PAGE results, analysis of the antibody-precipitated ODC from the above ^{32}P-labeling experiments, using Mono Q chromatography in the presence of 8 M urea, indicated that an easily measurable quantity of ^{32}P was firmly and specifically bound to ODC. Almost all of the label still associated with the antibody pellet after extensive washing was found to bind to the Mono Q anion exchange column in the presence of urea, and to

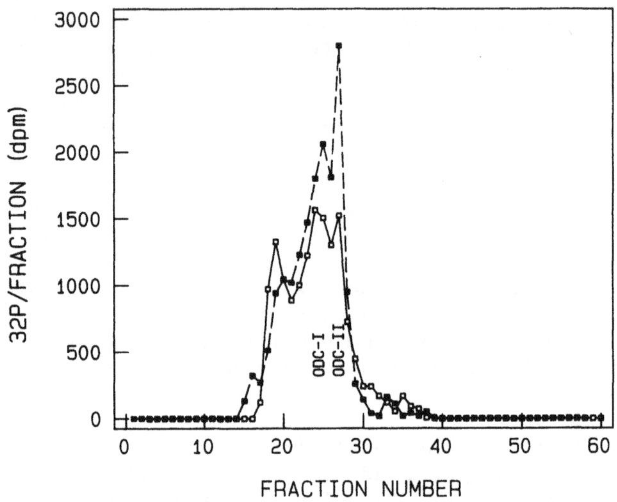

Fig. 3. Loss of ^{32}P from the ODC peak during incubation of crude homogenate. A 50 ml culture of HTC cells (approx. 1.0×10^6 cells/ml) was labeled with 0.25 mCi ^{32}P for 4h during enzyme induction in phosphate-depleted MEM media. After washing several times in phosphate-buffered saline, a pellet of half of the cells was homogenized in the presence of phosphatase inhibitors, including 1mM orthovanadate (■). The second pellet was homogenized without the inhibitors and incubated at 37°C for 10 min before cooling to ice temperature and addition of the phosphatase inhibitors (□). The crude supernatants were reacted with goat anti-rabbit IgG and insoluble protein A, and the resultant precipitates removed. The supernatants were subsequently precipitated with anti-ODC antibody[52], followed by the goat anti-rabbit IgG and protein A. These pellets were then washed extensively and finally the proteins were denatured by incubation in 8 M urea, 10 min at 37°C. The soluble material was then applied to the Mono Q column and eluted with a 0 to 0.5 M NaCl gradient. This is a much steeper gradient than used in Fig. 2, and all eluted material is shown up to the 0.5 M salt level. ^3H-ODC was added before antibody precipitation to serve as internal markers of the elution of ODC-I and II.

elute in the salt gradient fractions containing ^{3}H-DFMO-ODC (Fig. 3). This label appears to be specifically associated with ODC as it was not found in these same fractions upon chromatography of antibody pellets from ^{32}P-labeled cells in which ODC had been deinduced by the addition of 1 mM spermidine 90 min before harvest. 8 M urea is a strong protein denaturant and is generally used to prevent protein-ligand and protein-protein interactions in chromatography and isoelectric focusing. It was therefore somewhat surprising to observe this ODC-associated phosphate when our previous results indicated that it was not covalently bound. These experiments suggest that some unidentified phosphate-containing molecule remains associated with the ODC protein even in the presence of mild, non-ionic detergent washes and denaturation by 8 M urea.

The putative ODC-associated, phosphate-containing molecule appears to be related to the previously described alterations in ODC charge state. Incubation of crude cell homogenates in the absence of vanadate resulted in loss of ODC-II and increase in ODC-I, followed by a general decrease in ODC activity. As shown in Fig. 3, similar incubation of the ^{32}P-labeled cell homogenates resulted in a net loss in the ODC-associated label, with a noticeable shift from ODC-II to ODC-I. Interestingly, this incubation of crude cell homogenate also produced an additional peak, heretofore unobserved, that is even less acidic than ODC-I. ODC form conversion and inactivation in crude homogenates is accelerated by the presence of

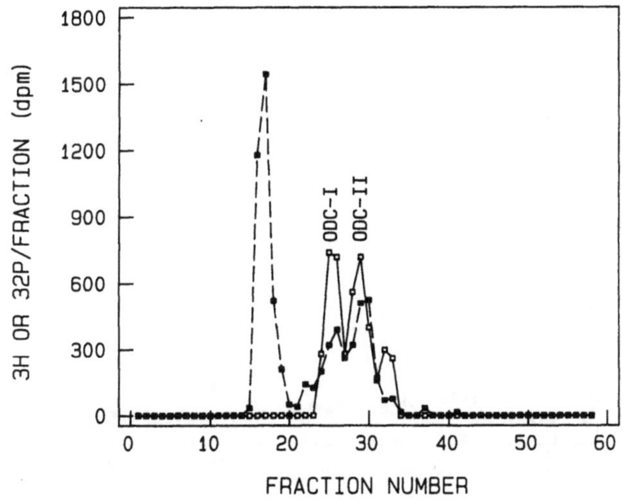

Fig. 4. Alteration in ^{32}P associated with ODC during incubation of crude cell homogenates with spermidine. A ^{32}P-labeled pellet of HTC cells was prepared exactly as described in Fig. 3, however in this case the crude homogenate was incubated without the phosphatase inhibitors and in the presence of 1.0 mM spermidine for 10 min, before the antibody precipitation (■). The internal markers of ^{3}H-ODC are also indicated (□). An improvement in the gradient program allowed a better resolution of the peaks as compared with Fig. 3.

spermidine. As seen in Fig. 4, spermidine promoted the loss of label from both ODC-I and II while increasing, somewhat, the label in the less acidic peak. Alkaline phosphatase, added to the crude homogenate, also promoted the shifting of the ^{32}P from the ODC-I and II location to the less acidic peak.

We postulated previously that conversion of ODC-II to I involved removal of a phosphate group, a process that could be mimicked by either alkaline or acid phosphatases. Now it appears that mono-specific ODC antibody can recognize three different charge species of ODC and all three are associated with phosphate-containing molecules that remain bound to ODC under the denaturing conditions used in isoelectric focusing. This opens the unsettling possibility that the distinct charge species of ODC that have been shown to have distinct isoelectric points upon isoelectric focusing in 8 M urea actually may not be due to covalent changes in the protein structure. The charge differences, for example, might be due to specific, tightly bound phospholipids. Indeed, there is evidence of specific phospholipid interaction with ODC and neutral phospholipids have been found to stimulate, and acidic phospholipids to inhibit, ODC activity.[53] Alternatively, GTP has recently been shown to have striking effects on ODC activity,[54] and could be found to be tightly bound to this protein. We are currently attempting to discover the identity of these phosphate-containing, ODC-associated molecules in an effort to ascertain their relevance to the observed ODC charge states, and to the activity and stability of this enzyme.

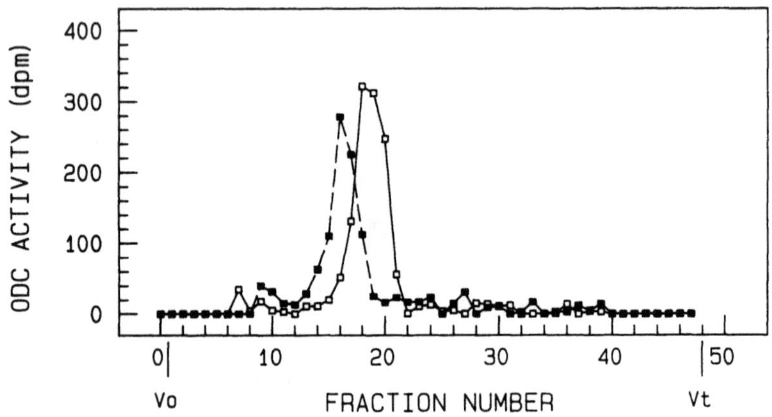

Fig. 5. Effect of spermidine on ODC conformation. ODC was isolated from HTC cell homogenates by ammonium sulfate precipitation and gel filtration chromatography. Samples of this enzyme preparation were placed in MOPS buffer (pH 7.2) containing 1 mM DTT, 50 uM PLP, 1 mM ornithine, 0.1 M NaCl either with (□) or without (■) 1 mM spermidine. The enzyme was then chromatographed on a Superose 12 high performance gel filtration column in the same buffer and the resultant 0.3 ml fractions were assayed for ODC activity as indicated previously.[39] Fraction 19 corresponds to Mw of approx. 50,000 while fraction 16 represents about 100,000.

Table 1. Antizyme binding to monomeric vs. dimeric ODC[a]

	Units of ODC Activity: Without Antizyme	Units of ODC Activity: With (2.3 units) Antizyme	ODC inactivated: units	ODC inactivated: control
Monomer ODC - alone	2.595	0.35	2.245	87.5%
- with Ab precipitation of antizyme	2.335	0.21	2.125	91.0%
Dimer ODC - alone	2.535	2.315	0.22	8.7%
- with Ab precipitation of antizyme	2.36	2.165	0.195	8.3%

[a]Antizyme was pepared as described in Fig. 7. ODC was prepared from the HTC cells and partially purified as described previously. This preparation was dissociated into ODC monomers by gel filtration chromatography in the presence of MOPS column buffer (20 mM, pH 7.2) containing 1.0 mM DTT, 0.5 mM EDTA, 0.25 M NaCl, 1.0 uM PLP. Aliquots containing approx. 2.5 units ODC were either left in this buffer (monomers) or the buffer was exchanged by G-25 chromatography to contain 1.0 mM DTT, 0.1 M NaCl, 50 uM PLP, 0.5 uM $MgCl_2$ and 0.5 mM ornithine (dimer buffer). Samples of each of these conditions were analyzed by Superose 12 gel filtration chromatography in the respective buffers, and it was found that about 10 % of the enzyme failed to return to the dimer form under these conditions. These samples were incubated with 2.3 units of antizyme at 4°C for 10 min and then the buffer of the monomer ODC was adjusted to be equivalent to the dimer sample. Small samples were removed for ODC assay before and after the incubation with antizyme, and the difference in activity attributed to antizyme binding. Identical aliquots of the enzyme before and after antizyme incubation were also precipitated with monoclonal antibody HZ1B3(given to us by S. Matsufuji) to remove any enzyme-antizyme complex and any free antizyme before sampling for ODC activity. This last procedure was to eliminate the possibility that either free or bound antizyme would interfere directly with the assay of the ODC activity.

SPERMIDINE-INDUCED ALTERATIONS IN ODC

ODC-II to I conversion, and the degradation of this enzyme, both are strongly stimulated by either spermidine or spermine. Perhaps these polyamines exert their effect by altering the physical conformation of the ODC molecule, making it more susceptible to modifying enzymes. The polyamines, for example, may influence the state of polymerization of this enzyme. It is generally accepted that ODC exists as a homodimer within a cell, and Solano et al.[55] suggest that the monomeric unit itself is inactive. Kitani and Fujisawa[56] described the dissociation of the ODC dimer by salts, while others[55] suggest that the substrate, ornithine, is critical in promoting dimer formation. We have extended these observations to show that the coenzyme, PLP, is essential for dimer formation from isolated monomers[39]. More recently we have examined the effect putrescine, polyamines and several other factors have on the state of ODC polymerization. Of special interest, spermidine and spermine were both found to facilitate the dissociation of mammalian ODC dimers into monomers, even in buffers containing 50 uM PLP, 0.5 mM ornithine and less than 0.1 M NaCl (Fig. 5). These data suggest that, although the cellular milieu would normally allow the ODC protein to remain almost exclusively in the active dimeric conformation, the elevation of free polyamines may shift this dynamic equilibrium to favor the appearance of free monomers. Although this may directly affect the total enzymatic activity only slightly, it could greatly stimulate the interaction of ODC with a regulatory molecule, if that molecule were specific for the monomeric state of this enzyme.

ODC-antizyme, a protein purportedly involved in the regulation of ODC, does appear to selectively interact with the ODC monomer. In the experiments reported in Table 1, we reacted 2.3 units of antizyme with partially purified preparations of HTC cell ODC that were maintained under dimer or monomer conditions. The actual polymeric states of the enzyme preparations were determined by Superose 12 gel filtration analysis. After this incubation the buffer conditions were adjusted to induce ODC dimerization in both reaction tubes and the remaining enzyme activity was quantitated. In a second experimental protocol the antizyme-ODC complex, and any free antizyme, were removed from the reaction tubes before the mixtures were adjusted to induce ODC dimers prior to the enzyme assays. The results were essentially the same for both procedures; antizyme binds almost exclusively to the monomer form of ODC. The small amount of enzyme inactivation noted in the dimer preparation may actually have been due to our inability to induce 100 % ODC dimer formation before the antizyme was added. It should also be noted that once the enzyme-antizyme complex was formed it was not reversed by subsequent attempts to induce enzyme dimers.

The specificity of antizyme for the ODC monomer has implications that have not been fully appreciated in regards to other studies concerning the regulation of ODC activity. For example, many investigators utilize radioactively labeled DFMO to specifically mark the ODC protein for studies

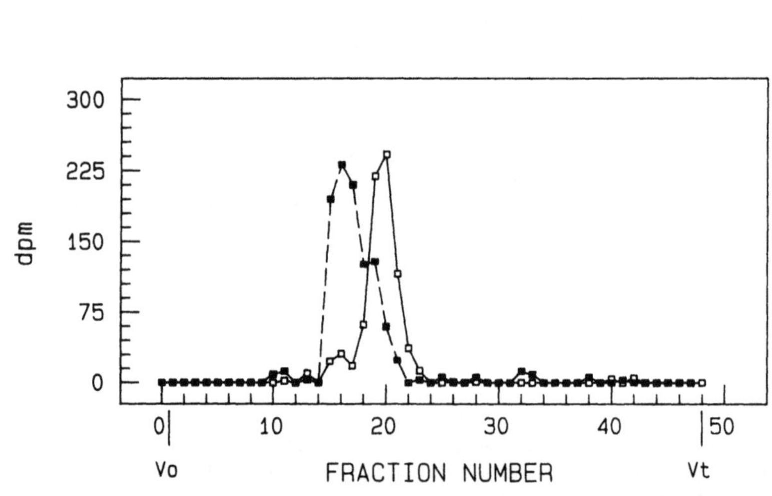

Fig. 6. Comparison of molecular size of native and DFMO inactivated ODC. Enzyme was prepared and chromatographed essentially as in Fig. 5 without spermidine added. In this experiment the enzyme was partially inactivated with ^{3}H-DFMO and the unbound DFMO removed by gel filtration chromatography before application to the Superose 12 column. Both the ^{3}H (□) and the ODC activity (■) of each fraction were analyzed.

in vivo and *in vitro*. Yet, as shown by the experiment in Fig. 6, labeling of ODC with DFMO strongly promotes the transition from the ODC dimeric to monomeric state. This transition may greatly affect the general disposition of such labeled enzyme relative to the unlabeled ODC that the investigator had intended to observe. Specifically, ODC inactivated with DFMO may be expected to interact more readily with antizyme. That this differential binding does indeed occur is demonstrated by the type of experiment illustrated in Fig. 7. In this study ODC-I was isolated, partially purified and reacted with ^{3}H-DFMO to produce a sample containing both active and ^{3}H-DFMO-bound ODC (panel A). Antizyme was also isolated and chromatographed separately on the Mono Q column, with a peak emerging later than the ODC-I (Fig. 7, panel A). The antizyme and enzyme preparations were mixed under conditions that normally maintain ODC in its active, dimer ODC conformation, and subsequently chromatographed on the Mono Q as before. As seen in Fig. 7 panel B, approximately half of the DFMO-labeled enzyme shifted to a later-eluting peak characteristic of the ODC-antizyme complex, whereas there did not appear to be any loss in the activity of the unlabeled, active ODC-I. This selective binding of antizyme to the DFMO-labeled enzyme was also observed in experiments comparing DFMO-labeled ODC-II against active ODC-I, and in the reciprocal titration.

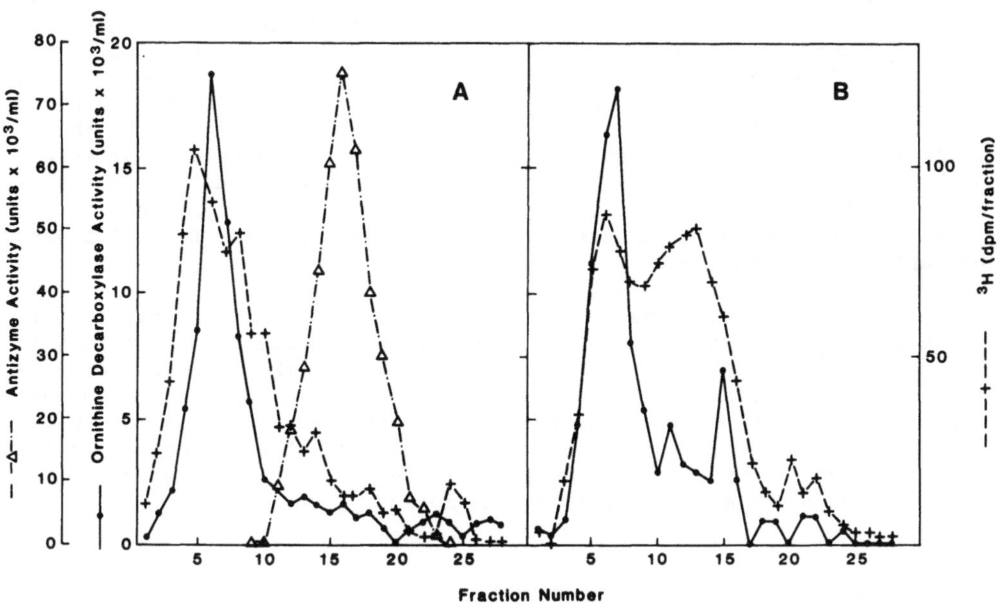

Fig. 7. Relative affinity of native and DFMO-inactivated ODC for antizyme. ODC-I was labeled by reaction with ^{3}H-DFMO, isolated and partially purified as described in Fig. 1. Antizyme was prepared from HTC cells treated with 5 mM spermidine for 5 h and isolated by ammonium sulfate fractionation and gel filtration chromatography. As shown in panel A, antizyme activity (Δ) elutes from the Mono Q column at a higher salt concentration than ODC. In panel B, the partially purified antizyme was combined with an excess of a mixture of active and ^{3}H-DFMO-labeled ODC-I, and subsequently chromatographed on the Mono Q column and eluted with the same salt gradient as in panel A. The DFMO-labeled ODC-I bound to antizyme can be seen as the peak at fraction number 13, intermediate between the ODC-I and antizyme elution peaks. The activity of the ODC-I peak (+) did not decrease, and there was no remaining antizyme activity.

Although the appearance of the antizyme molecule has been temporally associated with ODC degradation in intact tissues, its physiological function has still not been determined. Hayashi and coworkers[28] have suggested that it acts like ubiquitin in tagging this protein to indicate that it is to be rapidly degraded by some common proteolytic mechanism. Since the above data suggest that antizyme binds to, and maintains, ODC in the monomer state, it follows that normally hidden regions of the enzyme will be exposed in the enzyme-antizyme complex. These may include the N-terminal region,[32,33] one or another of the PEST regions,[31] or even a critical sulfhydral group as discussed by Danzin and Persson[57] and Flamingi et al.[58]. According to this scenario, exposure of such determinant protein regions would initiate the degradation of this protein by existing cytoplasmic proteases.

Another model is also suggested by the experiments described above. Although antizyme has not been shown to irreversibly alter ODC by itself, it may serve as a recognition subunit for another enzyme that can modify this enzyme. For example, in mammalian cells there appear to be a rather limited number of protein phosphatases that act on a large number of protein substrates in a very specific and carefully controlled manner. It is thought that the phosphatases themselves are located in a complex that reversibly associates with a number of different recognition subunits that facilitate the binding of appropriate substrate phosphoproteins to the phosphatase complex. In this scheme antizyme would serve to scavenge monomer ODC units and promote their modification by another, unknown enzyme. Since degradation of ODC does seem to be preceded by removal of at least one phosphate group from the ODC protein itself or a closely associated cofactor, it may be that antizyme facilitates the recognition of ODC by a phosphatase. Consistent with this paradigm, the elevation of spermidine levels and the binding of DFMO both enhanced the recognition of this enzyme by antizyme, and they both promoted a change in the charge of the ODC protein similar to that induced by purified phosphatases.

In the models described above, it is suggested that antizyme facilitates ODC degradation by capturing dissociated ODC molecules, but this is not to imply that antizyme itself is the signal for this degradation. Certain cells[59] and tissues[29] have been shown to contain significant amounts of antizyme even when the ODC activity is fully induced. Perhaps the initiating signal actively controlling the stability of this enzyme is simply the relative level of free spermidine influencing the proportion of total ODC dissociated to the monomer state at any moment.

CONCLUSION

Elevation of spermidine concentration within the cell appears to initiate several different events that could potentially lead to the inactivation and degradation of ODC. Not only does spermidine induce the synthesis of antizyme, but it also promotes dissociation of the active ODC dimer into monomeric units that are the substrate for antizyme binding. The studies reviewed above also suggest that spermidine stimulates an alteration in the charge state of the ODC protein that appears to correlate with enzyme instability. Furthermore, this work suggests that the changes in charge properties of this enzyme may actually be associated with tight, but noncovalently bound small, phosphate-containing molecules. It remains to be seen exactly how these various responses are orchestrated to result in the finely-tuned, product feedback control of ODC stability.

ACKNOWLEDGMENTS

This work was supported in part by NIH Grants GM 33841 and BRSG RR 07176.

REFERENCES

1. V. V. Isomaa, A. E. I. Pajunen, C. W. Bardin and O. A. Janne, Ornithine decarboxylase in mouse kidney. Purification, characterization and radioimmunological determination of the enzyme protein, J. Biol. Chem. 258:6735-6740 (1983).
2. J. E. Seely, H. Poso and A. E. Pegg, Effect of androgens on turnover of ornithine decarboxylase in mouse kidney, J. Biol. Chem. 257:7549-7553 (1982).
3. B. L. M. Hogan, A. McIlhinney and S. Murden, Effect of growth conditions on the activity of ornithine decarboxylase in cultured hepatoma cells. Effect of serum and insulin, J. Cell. Physiol. 83:353-358 (1973).
4. C. A. Rinehart and E. S. Canellakis, Induction of ornithine decarboxylase activity by insulin and growth factors is mediated by amino acids, Proc. Nat. Acad. Sci. USA 82:4365-4368 (1985).
5. M. Costa, M. Meloni and M. K. Jones, Regulation of ornithine decarboxylase activity by amino acids, cyclic AMP and luteinizing hormone in cultured mammalian cells, Biochim. Biophys. Acta 608:398-408 (1980).
6. D. A. Sens, J. H. Levine and M. G. Buse, Stimulation of hepatic and renal ornithine decarboxylase activity by selected amino acids, Metabolism 32:787-792 (1983).
7. O. Black and B. K. Chang, Ornithine decarboxylase enzyme activity in human and hamster pancreatic tumor cell lines, Cancer Lett. 17:87-93 (1982).
8. H. Poso, E. Karvonen, H. Suomalainen and L. C. Andersson, A human neuroblastoma cell line with an altered ornithine decarboxylase, J. Biol. Chem. 259:12307-12310 (1984).
9. E. Karvonen, L. C. Andersson and H. Poso, A human neuroblastoma cell line with a stable ornithine decarboxylase in vivo and in vitro, Biochem. Biophys. Res. Commun. 126:96-102 (1985).
10. T. G. O'Brien, T. Madara, J. A. Pyle and M. Holmes, Ornithine decarboxylase from mouse epidermis and epidermal papillomas: Differences in enzymatic properties and structure, Proc. Nat. Acad. Sci. USA 83:9448-9452 (1986).
11. M. L. Pritchard, A. E. Pegg and L. S. Jefferson, Ornithine decarboxylase from hepatoma cells and a varient cell line in which the enzyme is more stable, J. Biol. Chem. 257:5892-5899 (1982).
12. J. L. A. Mitchell, P. Qasba, and D. W. Mahan, Alternative charge states of mammalian ornithine decarboxylase and their polyamine-induced inactivation, in: "Recent Progress in Polyamine Res," L. Selmeci, M. E. Brosnan and N. Seiler, eds., Akademiai Kiado, Budapest and VNU, (1985).
13. J. L. A. Mitchell, D. W. Mahan, P. P. McCann and P. Qasba, Dicyclohexylamine effects on HTC cell polyamine content and ornithine decarboxylase activity, Biochim. Biophys. Acta 840:309-316 (1985).
14. J. L. A. Mitchell, P. Qasba, R. E. Stofko and M. A. Franzen, Ornithine decarboxylase modification and polyamine-stimulated enzyme inactivation in HTC cells, Biochem. J. 228:297-308 (1985).
15. L. Persson, J. E. Seely and A. E. Pegg, Investigation of structure and rate of synthesis of ornithine decarboxylase protein in mouse kidney, Biochemistry 23:3777-3783 (1984).
16. R. H. Davis, G. N. Krasner, J. J. DiGangi and J. L. Ristow, Distinct

roles of putrescine and spermidine in the regulation of ornithine decarboxylase in Neurospora crassa, Proc. Nat. Acad. Sci. USA 82:4105-4109 (1985).

17. E. Holtta and P. Pohjanpelto, Control of ornithine decarboxylase in Chinese hamster ovary cells by polyamines, J. Biol. Chem. 261:9502-9508 (1986).
18. J. R. Glass and E. W. Gerner, Spermidine mediates degradation of ornithine decarboxylase by a non-lysosomal, ubiquitin-independent mechanism, J. Cell. Physiol. 130:133-141 (1987).
19. Y. Murakami and S. Hayashi, Role of antizyme in degradation of ornithine decarboxylase in HTC cells, Biochem. J. 226:893-896 (1985).
20. W. F. Fong, J. S. Heller and E. S. Canellakis, The appearance of an ornithine decarboxylase inhibitory protein upon the addition of putrescine to cell cultures, Biochim. Biophys. Acta 428:456-465 (1976).
21. J. S. Heller and E. S. Canellakis, Cellular control of ornithine decarboxylase activity by its antizyme, J. Cell. Phys. 107:209-217 (1981).
22. E. S. Canellakis, D. A. Kyriakidis, C. A. Rinehart, S. C. Huang, C. Panagiotidis and W. F. Fong, Regulation of polyamine biosynthesis by antizyme and some recent developments relating the induction of polyamine biosynthesis to cell growth, Bioscience Reports 5:189-204 (1985).
23. K. Fujita, Y. Murakami, T. Kamaji, S. Matsufuji, K. Utsunomiya, R. Kanamoto and S. Hayashi, Regulation of hepatic ornithine decarboxylase by antizyme and antizyme inhibitor, Adv. Polyamine Res. 4:683-691 (1983).
24. K. Fujita, S. Matsufuji, Y. Murakami and S. Hayashi, Antizyme to ornithine decarboxylase is present in the liver of starved rats, Biochem. J. 218:557-562 (1984).
25. P. H. Laitinen, R. L. Huhtinen, O. A. Hietala and A. E. I. Pajunen, Ornithine decarboxylase activity in brain regulated by a specific macromolecule, the antizyme, J. Neurochem. 44:1885-1891 (1985).
26. Y. Murakami, M. Marumo and S. Hayashi, Existence of antizyme and ornithine decarboxylase-antizyme complex in RK13 kidney cells, FEBS Lett. 199:49-52 (1986).
27. T. Kitani and H. Fujisawa, Purification and some properties of a protein inhibitor (antizyme) in ornithine decarboxylase from rat liver, J. Biol. Chem. 259:10036-10040 (1984).
28. S. I. Hayashi, T. Kameji, K. Fujita, Y. Murakami, R. Kanamoto, K. Utsunomiya, S. Matsufuji, M. Takiguchi, M. Mori and M. Tatibana, Molecular mechanism for the regulation of hepatic ornithine decarboxylase, Adv. Enz. Reg. 23:311-329 (1985).
29. P. H. Laitinen, Involvement of an "antizyme" in the inactivation of ornithine decarboxylase, J. Neurochem. 45:1303-1307 (1985).
30. A. Hershko and A. Ciechanover, Mechanisms of intracellular protein breakdown, Ann. Rev. Biochem. 51:335-364 (1982).
31. S. Rogers, R. Wells and M. Rechsteiner, Amino acid sequences common to rapidly degraded proteins: The PEST hypothesis, Science 234:364-367 (1986).
32. A. Bachmair, D. Finley and A. Varshavsky, In Vivo Half-Life of a Protein Is a Function of Its Amino-Terminal Residue, Science 234:179-234 (1986).
33. J. M. Thornton and B. L. Sibanda, Amino and carboxy-terminal regions in globular proteins, J. Mol. Biol. 167:443-460 (1983).
34. M. F. Obenrader and W. F. Prouty, Detection of multiple forms of rat liver ornithine decarboxylase, J. Biol. Chem. 252:2860-2865 (1976).
35. D. Loeb, P. W. Houben and L. P. Bullock, Two forms of ornithine decarboxylase activity in mouse kidney, Molec. Cell. Endocrin. 38:67-73 (1984).

36. T. Kitani and H. Fujisawa, Strain-specific occurrence of two ornithine decarboxylase species in mouse kidney, Biochem. Biophys. Res. Commun. 151:450-457 (1988).
37. C. Lau and T. A. Slotkin, Regulation of rat heart ornithine decarboxylase: change in affinity for ornithine evoked by neuronal, hormonal, and ontogenetic stimuli, Molec. Pharm. 16:504-512 (1979).
38. E. Gravela, M. F. Zuretti and O. Brossa, Studies on ornithine decarboxylase from liver, kidney and tumoral tissues during activity decay following cycloheximide administration, Cell Biol. Interl. Rep. 11:573-582 (1987).
39. J. L. A. Mitchell, M. D. Rynning, H. J. Chen and M. F. Hicks, Interrelation between the charge isoforms of mammalian ornithine decarboxylase, Arch. Biochem. Biophys. 260:585-594 (1988).
40. M. A. Pereira, R. E. Savage and C. Guion, Induction by chloroform of two forms of ornithine decarboxylase in rat liver, Biochem. Pharmacol. 32:2511-2514 (1983).
41. J. F. Richards, K. Lit, R. Fuca and C. Bourgeault, Multiple species of ornithine decarboxylase in rat tissues: Effects of dexamethasone, Biochem. Biophys. Res. Commun. 99:1461-1467 (1981).
42. J. L. A. Mitchell and G. K. Mitchell, Ornithine decarboxylase protein diversity and activity modulation in HTC cells, Biochem. Biophys. Res. Commun. 105:1189-1197 (1982).
43. J. E. Seely, L. Persson, G. J. Sertich and A. E. Pegg, Comparison of ornithine decarboxylase from rat liver, rat hepatoma and mouse kidney, Biochem. J. 226:577-586 (1985).
44. V. J. Atmar and G. D. Kuehn, Phosphorylation of ornithine decarboxylase by a polyamine-dependent protein kinase, Proc. Nat. Acad. Sci. USA 78:5518-5522 (1981).
45. J. E. Seely, D. A. Stetler, S. T. Jacob and A. E. Pegg, Absence of inactivation or phosphorylation of ornithine decarboxylase by nuclear protein kinase NII and of immunological cross-reactivity between RNA polymerase I and ornithine decarboxylase, Biochem. Biophys. Res. Commun. 120:219-225 (1984).
46. F. Meggio, F. Flamigni, C. M. Calderera, C. Guarnieri and L. A. Pinna, Phosphorylation of rat heart ornithine decarboxylase by type-2 casein kinase, Biochem. Biophys. Res. Commun. 122:997-1004 (1984).
47. F. Meggio, F. Flamigni, C. Guarnieri and A. L. Pinna, Location of the phosphorylation site for casein kinase-2 within the amino acid sequence of ornithine decarboxylase, Biochim. Biophys. Acta 929:114-116 (1987).
48. L. A. Pinna, F. Flamigni, F. Meggio, C. Guarnieri and C. M. Caldarera, Phosphorylation of ornithine decarboxylase by casein kinase-2: A signal for proteolytic degradation?, in: "Biomedical Studies of Natural Polyamines," C. M. Caldarera, C. Clo and C. Guarnieri, eds., Cooperativa Libraria Universitaria Editrice Bologna, Bologna, Italy (1986).
49. U. R. Tipnis and M. K. Haddox, Casein kinase II mediated phosphorylation of ornithine decarboxylase, J. Cell Biol. 101:356a (1985).
50. D. J. Reich, L. Worth and M. K. Haddox, Phosphorylation of ornithine decarboxylase in situ, J. Cell Biol. 101:246a (1985).
51. N. J. Donato, C. F. Ware and C. V. Byus, A rat monoclonal antibody which interacts with mammalian ornithine decarboxylase at an epitope involved in phosphorylation, Biochim. Biophys. Acta 884:370-382 (1986).
52. L. Persson, Antibodies to ornithine decarboxylase. Immunochemical cross-reactivity, Acta Chem. Scand. B 36:685-688 (1982).
53. T. Kitani and H. Fujisawa, The effect of phospholipids on the activity of ornithine decarboxylase from rat liver, FEBS Lett. 132:296-298 (1981).
54. T. G. O'Brien, O. Hietala, K. O'Donnell and M. Holmes, Activation of

mouse epidermal tumor ornithine decarboxylase by GTP: Evidence for different catalytic forms of the enzyme, Proc. Nat. Acad. Sci. USA 84:8927-8931 (1987).

55. F. Solano, R. Penafiel, M. E. Solano and J. A. Lazano, Equilibrium between active and inactive forms of rat liver ornithine decarboxylase mediated by L-ornithine and salts, FEBS Lett. 190:324-328 (1985).

56. T. Kitani and H. Fujisawa, Influence of salts on the activity and the subunit structure of ornithine decarboxylase from rat liver, Biochim. Biophys. Acta 784:164-167 (1984).

57. C. Danzin and L. Persson, L-Ornithine-induced inactivation of mammalian ornithine decarboxylase in vitro, Eur. J. Biochem. 166:45-48 (1987).

58. F. Flamigni, C. Guarnieri and C. M. Caldarera, Rat liver cytosol contains NADPH-and GSH-dependent factors able to restore ornithine decarboxylase inactivated by removal of thiol reducing agents, Biochem. J. 250:53-58 (1988).

59. Y. Murakami, K. Fujita, T. Kamaji and S. Hayashi, Accumulation of ornithine decarboxylase-antizyme complex in HMOA cells, Biochem. J. 225:689-697 (1985).

ENZYMES OF POLYAMINE METABOLISM AND RECYCLING

S-ADENOSYLMETHIONINE DECARBOXYLASE FROM ESCHERICHIA COLI AND FROM SACCHAROMYCES CEREVISIAE: CLONING AND OVEREXPRESSION OF THE GENES

K. Kashiwagi, S.K. Taneja, Q.-W. Xie, C.W. Tabor and H. Tabor

National Institute of Diabetes and Digestive and Kidney Diseases, National Institutes of Health
Bethesda, Maryland 20892, U.S.A.

INTRODUCTION

S-Adenosylmethionine decarboxylase is a key enzyme in the biosynthesis of spermidine and spermine in both prokaryotes and eukaryotes.[1] We first detected this enzyme in our initial studies on the biosynthesis of spermidine in *E. coli*[2,3]; subsequently S-adenosylmethionine decarboxylase has been found widely distributed in bacteria, fungi, plants, and animal cells.[1] The enzyme has been purified to homogeneity from *E. coli*,[4,5] *S. cerevisiae*,[6,7] liver,[8,9] lymphocytes,[11] and mammary gland.[12]

The most notable characteristic of S-adenosylmethionine decarboxylase is the presence in the enzyme protein of covalently-bound pyruvoyl groups that are essential for enzymatic activity. The presence of such pyruvoyl groups in S-adenosylmethionine decarboxylase was first shown in our studies on the *E. coli* enzyme with Wickner in 1970 [4] and in our studies on the yeast enzyme with Cohn in 1977*.[7] It is noteworthy that both the *E. coli* and the yeast enzymes contain covalently-bound pyruvoyl groups, even though the enzymes from these two sources differ considerably; i.e., they have markedly different molecular weights and have different requirements for Mg^{2+} and putrescine.[1] In addition, as shown by our current studies, the two enzymes have completely different nucleotide and amino acid sequences.

---

*Covalently-bound pyruvoyl groups have also been shown to be present in S-adenosylmethionine decarboxylase that was isolated from liver both in this laboratory[8] and by Pegg.[10]

Covalently-bound pyruvate was first described by Riley and Snell in histidine decarboxylase from Lactobacillus in 1968 and was shown to be required for enzymatic activity.[13] Subsequently, in addition to S-adenosylmethionine decarboxylase, covalently-bound pyruvate has also been described in proline reductase, α-aspartate decarboxylase, and phosphatidylserine decarboxylase.[14] The mechanism of the formation of the pyruvoyl group has been studied only in histidine decarboxylase; in these studies Snell and his colleagues showed that the enzyme is first formed as a proenzyme, which is then cleaved at a serine-serine peptide bond to form the pyruvoyl moiety. The cleavage reaction was most unusual since it did not need any additional enzyme, but resulted from an internal serinolysis of the serine-serine peptide bond.[14,15]

We have been interested in finding out the source of the pyruvoyl moiety that is present in S-adenosylmethionine decarboxylase in both Escherichia coli and Saccharomyces cerevisiae. In view of the differences in the characteristics of the two enzymes from the two sources, we wanted to see if the mechanism of the formation of the pyruvoyl group was the same in both cases, and whether this mechanism was similar to that described for the Lactobacillus histidine decarboxylase.[14,15]

S-Adenosylmethionine Decarboxylase from Escherichia coli

To study the formation of the pyruvoyl moiety in S-adenosylmethionine decarboxylase in E. coli, the structural gene for S-adenosylmethionine decarboxylase (speD) was placed in a plasmid, downstream to a T7 promoter in the T7 promoter-expression system of S. Tabor and C. C. Richardson.[16] ^{35}S-Pulse-labeling experiments showed that S-adenosylmethionine decarboxylase was first formed as a proenzyme (molecular weight = 30,400), which was then cleaved to form two polypeptides (molecular weights of 12,400 and 18,000).[17,18] Studies on the structure of purified S-adenosylmethionine decarboxylase showed that both of these polypeptides were present in the purified enzyme, and that the pyruvoyl group was covalently bound to the N-terminal end of the 18,000 M_r polypeptide.[18,19]

We had already cloned the speD gene, and by sequencing studies had already demonstrated that the gene for S-adenosylmethionine decarboxylase would encode a 30,000 M_r proenzyme.[18] By comparison of the amino acid sequence deduced from this nucleotide sequence with the N-terminal amino acid sequence of the subunits (as determined by Anton and Kutny[19]), we were able to identify the site at which the processing occurred as the Lys111-Ser112 peptide linkage.

In subsequent studies[20] we were able to show by site-specific mutagenesis that substitutions in the Lys111 position markedly affected the degree of processing and the total enzyme activity. Thus, when this lysine was replaced by proline or isoleucine, there was essentially no processing and only traces of enzymatic activity. Substitution by methionine, leucine, or threonine had much less effect.

This same T7 expression system could be used to prepare large amounts of the processed enzyme. Under optimal conditions, 5-10% of the cellular protein was found to be S-adenosylmethionine decarboxylase. The enzyme was easily purified to homogeneity by a three-step procedure involving streptomycin precipitation, Phenyl-Sepharose chromatography, and DEAE-HPLC (unpublished studies).

In other studies[21] we have shown that speD is part of an operon that also contains the gene for spermidine synthase (speE). Studies are also underway on the promoter and termination areas of this operon.[22]

The above results are summarized in Figure 1.

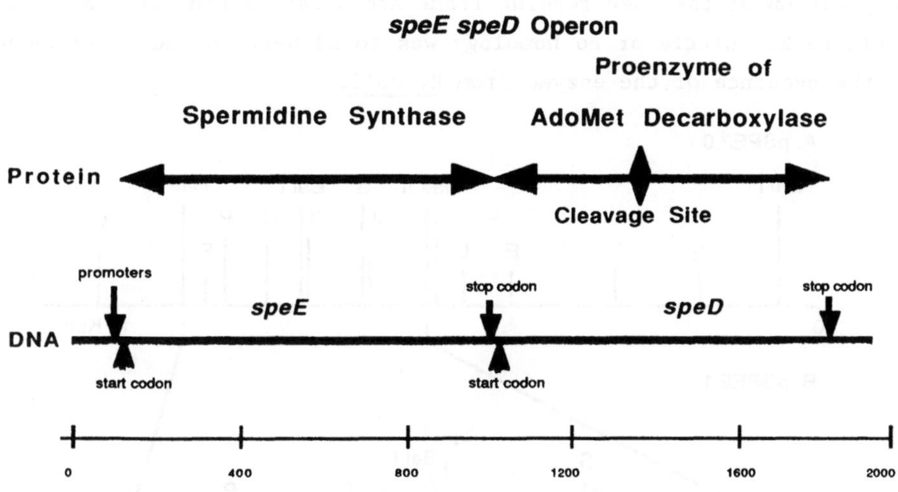

Spermidine synthase: Mol.Wt. 64,652 (2 subunits)

Adenosylmethionine Decarboxylase

Proenzyme Mol. Wt. 121,556 (4 subunits)

Processed enzyme: Mol. Wt. 121,560

4 subunits Mol. Wt. 12,442

4 subunits Mol. Wt. 17,948 (contains the N-terminal pyruvoyl groups)

Fig. 1. Summary of the speEspeD operon from references 18 and 21.

S-Adenosylmethionine decarboxylase has been purified to homogeneity from S. cerevisiae.[7] The enzyme from this source was shown to have a molecular weight of 84,000, and a subunit size of 41,000. The gene for this enzyme (SPE2) is closely linked to the arg1 gene[24], which was shown by Hilger and Mortimer[23] to be located on chromosome XV. S. cerevisiae strains (spe2) defective in this gene grow slowly in purified media, are not able to sporulate, and do not maintain the double-stranded RNA killer plasmid.[24,25] Addition of spermidine to the medium reversed these phenotypes.

In our current studies we have cloned the SPE2 gene into single- and multiple-copy shuttle vectors. A 10 kb nucleotide fragment was first cloned into the single-copy vector YCp50, using as an assay the ability of the hybrid plasmid to complement the growth defect of spe2 mutants. A 3500 bp fragment was then prepared by restriction endonuclease treatment, and sequenced[§]. This 3500 bp fragment was shown to contain an open reading frame encoding a 396 amino acid protein with a calculated molecular weight of 46,214. This open reading frame is large enough to encode the subunit size of 41,000 that we had reported for the purified enzyme. The position of the open reading frame and a restriction map is presented in Figure 2. Little or no homology was found between the yeast sequence and the sequence of the enzyme from E. coli.

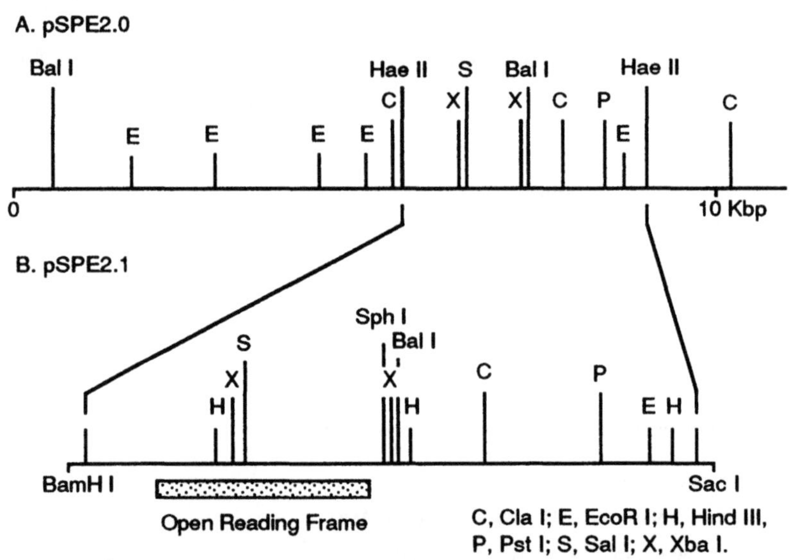

Fig. 2. Restriction map of the 10 kbp DNA insert from S. cerevisiae used for cloning of the SPE2 gene.

[§]The nucleotide sequence is being deposited in GenBank.

A 1700 bp insert, containing the above open reading frame, was placed into the multicopy vector (YEp351). When spe2 mutants were transformed by this plasmid, the activity of the S-adenosylmethionine decarboxylase in the extracts increased about 50 times over the wild type activity (Table 1). Consequently these extracts are very suitable for purification of the yeast adenosylmethionine decarboxylase.

Table 1. Increase in S-Adenosylmethionine Decarboxylase Activity in Yeast Strains Carrying a $SPE2^+$ Plasmid

Strain	$A_{550\ nm}$ [a]	Specific Activity [b]
72-1D	0.9	0
72-1D/pSPE2.1	0.3	9.0
	1.2	11.1
72-1D/pSPE2.2	0.3	151
	0.9	199
	1.9	132
AN33	0.3	4.1
(wild type)	0.9	2.3
	1.7	1.3

Yeast strains: 72-1D (α leu2-27 ura3 spe2-4), AN33 (α thr1 arg1).

[a] These cultures were grown in H medium to the $A_{550\ nm}$ value indicated.

[b] pmol/min/mg protein. Plasmid pSPE2.1 was a single-copy plasmid (YCp50) containing the 3.5 kb insert, and pSPE2.2 was a multi-copy plasmid (YEp351) containing the 1.7 kb insert.

Studies are now underway for the determination of the N-terminal amino acid sequence of each subunit before and after reductive amination. Comparison of these sequences with the deduced amino acid sequence of the proenzyme should define the site of cleavage.

REFERENCES

1. Review Articles:
C. W. Tabor and H. Tabor, Methionine adenosyltransferase (S-adenosylmethionine synthetase) and S-adenosylmethionine decarboxylase, Adv. Enzymol. Relat. Areas Mol. Biol., 56:251-282 (1984).
C. W. Tabor and H. Tabor, Polyamines, Annu. Rev. Biochem. 53:749-790 (1984).
A. E. Pegg, Recent advances in the biochemistry of polyamines in eukaryotes, Biochem. J. 234:249-262 (1986).
2. H. Tabor, S. M. Rosenthal, and C. W. Tabor, The biosynthesis of spermidine and spermine from putrescine and methionine, J. Biol. Chem. 233:907-914 (1958).

3. C. W. Tabor, Adenosylmethionine decarboxylase, Methods Enzymol. 5:756-760 (1962).
4. R. B. Wickner, C. W. Tabor, and H. Tabor, Purification of adenosylmethionine decarboxylase from Escherichia coli W: evidence for covalently bound pyruvate, J. Biol. Chem. 245:2132-2139 (1970).
5. G. D. Markham, C. W. Tabor, and H. Tabor, S-Adenosylmethionine decarboxylase of Escherichia coli: studies on the covalently-linked pyruvate required for activity, J. Biol. Chem. 257:12063-12068 (1982).
6. H. Pösö, R. Sinervirta, and J. Jänne, S-Adenosylmethionine decarboxylase from baker's yeast, Biochem. J. 151:67-73 (1975).
7. M. S. Cohn, C. W. Tabor, and H. Tabor, Identification of a pyruvoyl residue in S-adenosylmethionine decarboxylase from Saccharomyces cerevisiae, J. Biol. Chem. 252:8212-8216 (1977).
8. A. A. Demetriou, M. S. Cohn, C. W. Tabor, and H. Tabor, Identification of pyruvate in S-adenosylmethionine decarboxylase from rat liver, J. Biol. Chem. 253:1684-1686 (1978).
9. A. E. Pegg, Purification of rat liver S-adenosyl-L-methionine decarboxylase, Biochem. J. 141:581-583 (1974).
10. A. E. Pegg, Evidence for the presence of pyruvate in rat liver S-adenosylmethionine decarboxylase, FEBS Lett. 84:33-36 (1977).
11. C. E. Seyfried, O. E. Oleinik, J. L. Degen, K. Resing, and D. R. Morris, Purification, properties, and regulation of the level of bovine S-adenosylmethionine decarboxylase during lymphocyte mitogenesis, Biochim. Biophys. Acta 716:169-177 (1982).
12. T. Sakai, C. Hori, K. Kano, and T. Oka, Purification and characterization of S-adenosyl-L-methionine decarboxylase from mouse mammary gland and liver, Biochemistry 18:5541-5548 (1979).
13. W. D. Riley and E. E. Snell, Histidine decarboxylase of Lactobacillus 30a. IV. The presence of covalently bound pyruvate as the prosthetic group, Biochemistry 7:3520-3528 (1968).
14. Reviewed by P. A. Recsei and E. E. Snell, Pyruvoyl enzymes, Annu. Rev. Biochem. 53:357-358 (1984).
15. P. A. Recsei, Q. K. Huynh, and E. E. Snell, Conversion of prohistidine decarboxylase to histidine decarboxylase: peptide chain cleavage by nonhydrolytic serinolysis, Proc. Natl. Acad. Sci. USA 80:973-977 (1983).
16. S. Tabor and C. C. Richardson, DNA sequence analysis with a modified bacteriophage T7 DNA polymerase, Proc. Natl. Acad. Sci. USA 84:4767-4771 (1987).
17. C. W. Tabor, and H. Tabor, Adenosylmethionine decarboxylase of E. coli is synthesized from a proenzyme, Proceedings of the 13th International Congress of Biochemistry, Amsterdam, The Netherlands, Abstract No. FR-1 83, p. 180 (1985).
18. C. W. Tabor and H. Tabor, The speEspeD operon of Escherichia coli: formation and processing of a proenzyme form of S-adenosylmethionine decarboxylase, J. Biol. Chem. 262:16037-16040 (1987).
19. D. L. Anton and R. Kutny, Escherichia coli S-adenosylmethionine decarboxylase: subunit structure, reductive amination, and NH_2-terminal sequences, J. Biol. Chem. 262:2817-2822 (1987).
20. H. Tabor and C. W. Tabor, Effect of site-specific mutagenesis on the processing of the proenzyme of S-adenosylmethionine decarboxylase. FASEB J. 2, Abstract No. 1602, p. A572 (1988).
21. C. W. Tabor, H. Tabor, and Q.-W. Xie, Spermidine synthase of Escherichia coli: localization of the speE gene, Proc. Natl. Acad. Sci. USA 83:6040-6044 (1986).
22. Q.-W. Xie, C. W. Tabor, and H. Tabor, manuscript in preparation.
23. F. Hilger and R. K. Mortimer, Genetic mapping of arg1 and arg8 in Saccharomyces cerevisiae by trisomic analysis combined with interallelic complementation, J. Bacteriol. 141:270-274 (1980).

24. M. S. Cohn, C. W. Tabor, and H. Tabor, Isolation and characterization of *Saccharomyces cerevisiae* mutants deficient in *S*-adenosylmethionine decarboxylase, spermidine, and spermine, *J. Bacteriol.* 134:208-213 (1978).
25. M. S. Cohn, C. W. Tabor, H. Tabor, and R. B. Wickner, Spermidine or spermine requirement for killer double-stranded RNA plasmid replication in yeast, *J. Biol. Chem.* 253:5225-5227 (1978).

The reader is also referred to the article by Pegg et. al. in this volume for a summary of their work on the formation and processing of a proenzyme form of mammalian *S*-adenosylmethionine decarboxylase.

STRUCTURAL AND MECHANISTIC PROPERTIES OF E. COLI ADENOSYLMETHIONINE DECARBOXYLASE

David L. Anton and Rusty Kutny

Central Research and Development
E.I. du Pont de Nemours Co.
Wilmington, DE 19898

ABSTRACT

Adenosylmethionine decarboxylase catalyzes one of the first committed steps in polyamine biosynthesis. It is a member of a small class of decarboxylases that use a pyruvoyl prosthetic group rather than the more common pyridoxal cofactor. We have recently shown that AdoMet decarboxylase from E. coli is composed of stoichiometric amounts of two types of subunits; α (M_r = 19,000), and β (M_r = 14,000). The NH_2-terminal of the α subunit is blocked by the pyruvoyl group and can be sequenced only after reductive amination, which converts this to an alanine residue. The β subunit, on the other hand, has an unblocked NH_2-terminal and sequences normally. The molecular weight of the holoenzyme, estimated by gel filtration, is 136,000 suggesting that the enzyme is an $\alpha_4\beta_4$ octamer.

AdoMet decarboxylase undergoes a time dependent inactivation during turnover. The mechanism of this inactivation involves a transamination from the product, decarboxylated AdoMet, and the pyruvoyl group generating an NH_2-terminal alanine. The nascent product aldehyde then eliminates methylthioadenosine, resulting in the formation of acrolein, which covalently labels the α subunit. How this mechanism may explain AdoMet decarboxylase turned over, and how AdoMet decarboxylase inhibitors can affect its half life will be discussed.

INTRODUCTION

S-Adenosylmethionine (AdoMet) decarboxylase and ornithine decarboxylase catalyze the first two committed reactions in polyamine biosynthesis. Although both are amino acid decarboxylases, they differ in cofactor requirements. Ornithine decarboxylase requires pyridoxal phosphate whereas AdoMet decarboxylase uses the less common pyruvoyl prosthetic group. Several other pyruvoyl containing enzymes have been described including: histidine decarboxylase from several Gram-Positive bacteria[1], aspartate α-decarboxylase from E. coli[2], Phosphatidylserine decarboxylase from E. coli[3], proline reductase from Clostridia[4], and 4'-phosphopatothenoylcysteine decarboxylase from E. coli[5].

Much interest has focused on histidine decarboxylase which is the best characterized pyruvoyl containing enzyme. It is composed of two different subunits; α, β, with the pyruvoyl group covalently linked, by an amide bond, to the NH_2-terminal of the α subunit. The generation of the pyruvoyl group was shown in a series of elegant experiments by Snell and coworkers to be derived by the "serinolysis" of a larger proenzyme[1]. The mechanism (outlined in scheme 1) involves a N- to O-acyl transfer of the carboxyl group of residue 80 to the hydroxyl group of serine 81. Carboxylate is then eliminated generating the COOH-terminal of the β subunit and the formation of a dehydroalanyl residue at the NH_2-terminal of the α subunit. The dehydroalanyl residue then rearranges, forming an imine, which on hydrolysis gives the pyruvoyl moiety and ammonia. This reaction is autolytic, that is requires no additional catalyst, but is stimulated by monovalent cations[1].

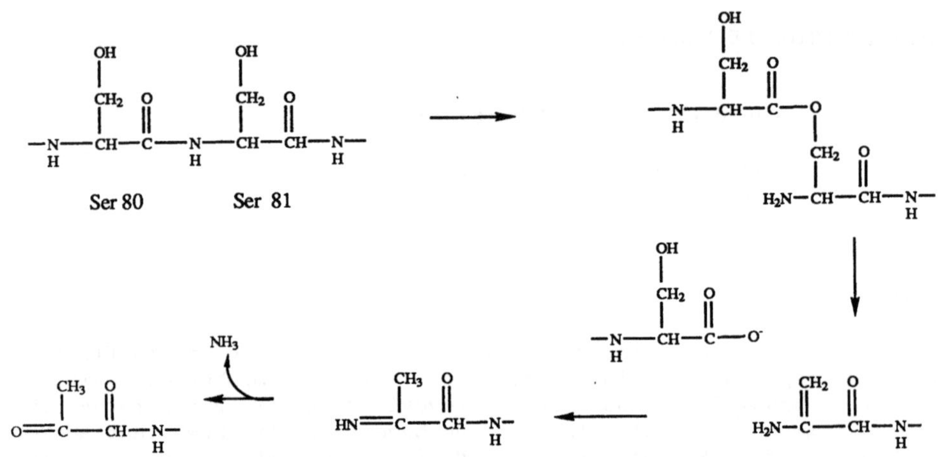

Scheme 1

The pyruvoyl group is thought to function in much the same way in these enzymes as pyridoxal phosphate does in the more common amino acid decarboxylases. A Schiff base is formed between the amino group of the substrate and the carbonyl of the pyruvoyl group. The pyruvoyl group then acts as an electron sink to facilitate the decarboxylation. The nascent product is protonated and the Schiff base hydrolyzed regenerating product (Scheme 2).

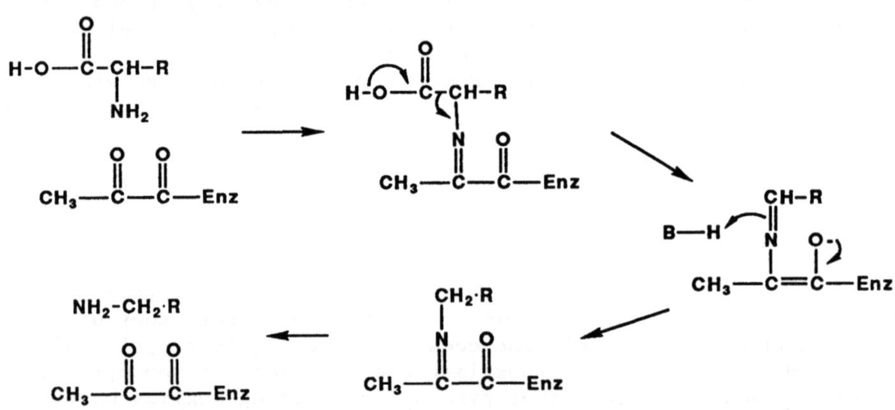

Scheme 2

All AdoMet decarboxylases yet studied are pyruvoyl containing enzymes. However they differ in their requirements, the E.coli enzyme has a absolute requirement for divalent metal ions (Mg^{++} being the best) whereas the enzyme from most other sources are stimulated by putrescine, but not Mg^{++}[6,7]. The E. coli enzyme is also less sensitive to the competitive inhibitor methylglyoxal bis(guanylhydrazone) than are the other enzymes[8].

STRUCTURAL ASPECTS OF ADOMET DECARBOXYLASE

Early reports for AdoMet decarboxylase suggested there was a single type of subunit with a molecular weight of 17,000 based on SDS-PAGE for the cloned enzyme (E. coli strain HT527)[8,9]. The AdoMet decarboxylase from the clone appeared to have a blocked NH_2 terminal. Native molecular weight was determined to be 108,000 which suggested the enzyme was a hexamer[8]. This suggested that the AdoMet decarboxylase differed from his decarboxylase in subunit composition, but might still have the pyruvoyl group linked to the NH_2-terminal. The rat prostate, and Saccharomyces enzymes are both reported to be homo-dimers; the Saccharomyces enzyme has a holoenzyme of 88,000 M_r with a subunit of 41,000 M_r, the molecular weight of the rat prostate enzyme is 68,000 and 31,00086.

We have been examining the structure of the pyruvoyl group in AdoMet decarboxylase from E. coli HT527 (which was a kind gift of Herbert and Celia Tabor). The enzyme was purified on MGBG Sepharose, by the procedure of Markham et al,[8] was found to contain two major bands (M_r = 14,000 & 19,000) on SDS-PAGE and a minor band in variable amounts (M_r = 35,000)[10].

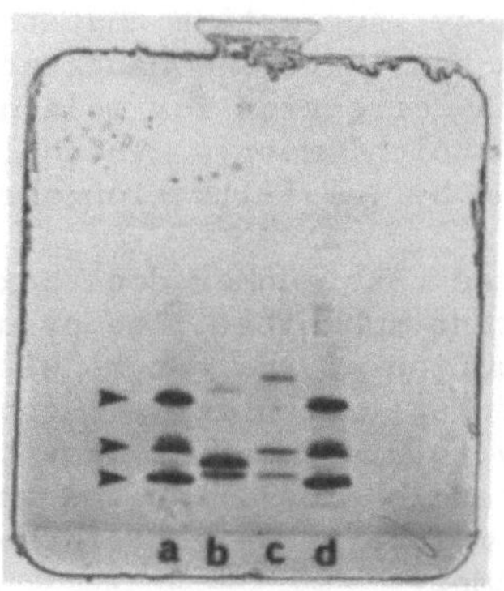

Fig. 1. SDS-PAGE. PhastGel (8-25%) Lane a and d contain carbonic anhydrase (M_r = 30,000), soybean trypsin inhibitor (M_r = 20,100), and cytochrome c (M_r = 13,600). Lane b, Adomet decarboxylase. Lane c, AdoMet decarboxylase after reduction and carboxymethylation. Reprinted from J. Biol. Chem.[10]

Since AdoMet decarboxylase had been reported to comprise a single type of subunit[8] additional chromatography columns were used to try to resolve these polypeptides including: DEAE-Sephadex, phenyl-Superose, Superose-12, Sephacryl S300, Mono Q, and Zorbax GF-250. In all cases the two polypeptides co-eluted.

NH_2-Terminal sequence analysis on the mixture, after reduction and carboxymethylation, resulted in a single major sequence, inconsistent with the report of a blocked NH_2-terminal for this enzyme[8]. After reduction and carboxymethylation the two major polypeptides were separable by reversed phase HPLC (Fig. 2). The peak that eluted first (I) from the column was identified as 14,000 M_r polypeptide and also had the NH_2-terminal sequence found for the mixture (Fig. 3). The other two major peaks in the chromatogram (II and III) both co-migrate with the 19,000 M_r polypeptide on SDS-PAGE, have identical amino acid composition, and blocked NH_2-terminals. These two peaks probably represent the same protein that differs by the hydrolysis of amide side chains or oxidation.

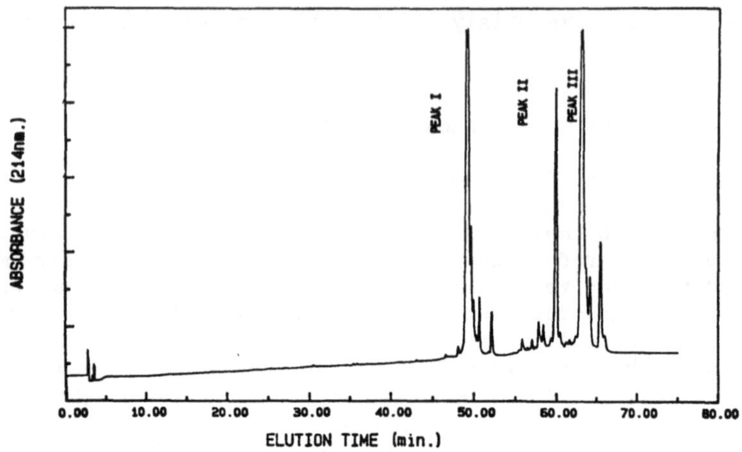

Fig. 2. Revered phase HPLC separation of subunits. Reduced and carboxymethylated AdoMet decarboxylase chromatographed on Vydac C_{18} column (4.8 X 250) eluted with a linear gradient from 9.9 % to 75% acetonitrile in 1% trifluoroacetic acid in water. Reprinted from J. Biol. Chem.[10]

Peptide maps of Peaks I and III (14,000 and 19,000 M_r) indicated no substantial sequence homology, suggesting that the smaller peptide is not derived, by cleavage, from the larger. Quantitation of the two proteins by absorbance at 214 nm and correction for molecular weight suggests that they are present in equal stoichiometry. The holoenzyme has a molecular weight of 136,000 estimated by gel filtration chromatography.

This data suggested that AdoMet decarboxylase like His decarboxylase was composed of two different subunits. The pyruvoyl group was located on the larger subunit by reduction of the enzyme with $NaCNBH_3$ in the presence of [Me-^{3}H] or [3,4-^{14}C] AdoMet, which forms a covalent bond between enzyme and substrate. By analogy to His decarboxylase the larger, pyruvoyl containing, subunit has been labeled α and the smaller β. The stoichiometry of the subunits and molecular weight of the holoenzyme suggest that the enzyme is an $\alpha_4\beta_4$ octamer.

Sequencing of the α subunit of His decarboxylase was accomplished by reductive amination of the pyruvoyl group, which converts it to a NH_2-terminal alanyl residue, with $NaCNBH_3$ and 2 M ammonium phosphate pH 7.5, which could then be sequenced by Edman degradation[1]. AdoMet decarboxylase proved to be recalcitrant to reductive amination under these conditions, and was ultimately reduced by extended reaction with ammonium acetate at pH 6.5. After reductive amination the α-subunit AdoMet decarboxylase gave the sequence shown in Fig. 3 with the first residue the expected Ala. This strongly suggest that the pyruvoyl group is linked through an amide linkage to the amino terminal. The difficultly in reduction of AdoMet decarboxylase might be due to the histidine adjacent to the pyruvoyl group(Fig. 3), which would affect the formation of the protonated Schiff base, which is the only form reduced by $NaCNBH_3$.

More recently Tabor and Tabor have described the formation of a 35,000 M_r proenzyme, which is subsequently cleaved to form 17,000 and 19,000 M_r proteins[11]. This may account for its variable quantitation. These workers have completed the oligonucleotide sequence of the gene (speD) for AdoMet decarboxylase and deduced the amino acid sequence[12]. They find the NH_2-terminal sequences described for the two subunits within the open reading frame for a larger protein with a molecular weight calculated for the derived amino acid sequence of 30,400. The sequence α Subunit:

Prv-His-Ile-Cys-Val-His-Thr-Tyr-Pro-Glu-Ser-His-Pro-Glu-Gly-Gly-Leu-Cys-Thr-Phe-Arg-Ala-Asp-Ile-Glu-Val-Ser-Thr-Cys-Gly-Val-Ile-

β Subunit

NH_2-Met-Lys-Lys-Leu-Lys-Leu-His-Gly-Phe-Asn-Asn-Leu-Thr-Lys-Ser-Leu-Ser-Phe-X-Ile-Tyr-

Fig. 3 NH_2-terminal Sequences of α and β subunits of AdoMet decarboxylase as determined by automated Edman degradation using a gas vapor sequencer (Applied Biosystems model 470) as previously described. (Prv determined as Ala see text).

for the β subunit is found at the NH_2-terminal of the reading frame. The sequence for the α subunit starts at residue 112 with serine as the precursor of the pyruvoyl group. The molecular weights determined from the amino acid sequences are 18,000, and 12,400 for α and β subunits respectively. The difference between these molecular weights and those determined by SDS-PAGE are considered to be due to the small size of the proteins[12].

SUBSTRATE INACTIVATION

As part of our interest in the mechanism of AdoMet decarboxylase, and pyruvoyl enzymes in general, we have investigated the time dependent lose in activity under turnover conditions[13]. As shown in Fig. 4 the lose of enzyme activity with time is linear on a semi-log plot versus time, indicating that the reaction is pseudo first order for enzyme. The rate of inactivation is dependent on the concentration of AdoMet in the reaction mixture and is saturable as shown by the re-plot of the rate of

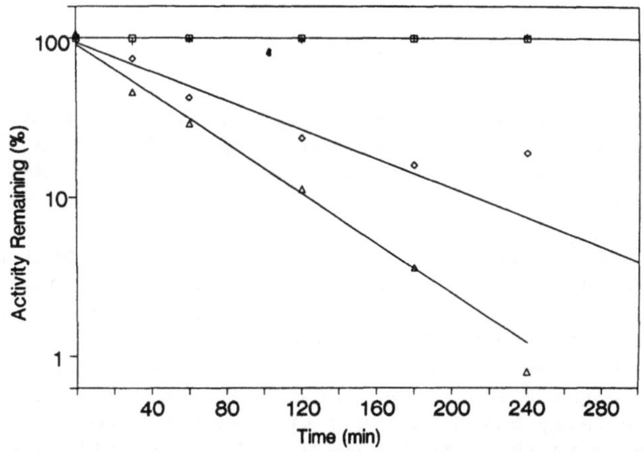

Fig. 4. Rate of Substrate Inactivation of AdoMet Decarboxylase. AdoMet decarboxylase (0.06 mg) was incubated in 25 mM potassium phosphate, 1 mM dithiothreitol, pH 7.5 with 0 (-+-), 0.2 (-◇-), or 1 mM AdoMet (-△-) and 10 mM $MgCl_2$, or 1 mM AdoMet (-□-) without $MgCl_2$ (total volume 100 μl) at 37°C for the indicated times. Aliquots were removed and the enzyme activity determined.

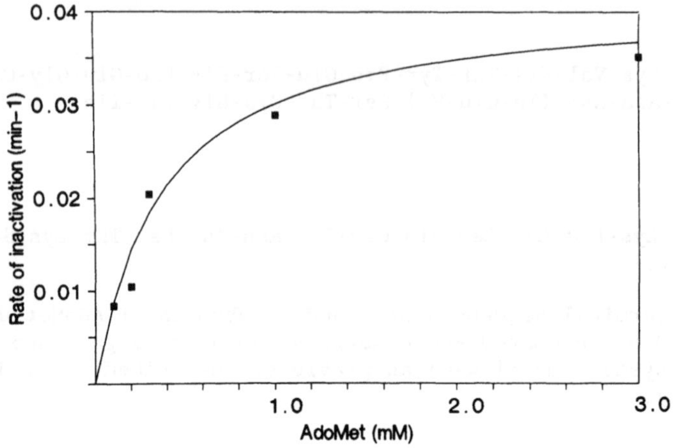

Fig. 5. Re-plot of the Rate of Inactivation of AdoMet decarboxylase vs. AdoMet concentration. Rate determined for the inactivation assuming pseudo first order reaction [$k = \ln(A_0/A)/t$] and plotted against AdoMet concentration.

inactivation versus concentration (Fig. 5). The inactivation is absolutely dependent on the addition of Mg^{++} and AdoMet. The rate of inactivation at 3 mM AdoMet is 0.035 min^{-1}, the turnover number for this enzyme is 230 min^{-1}, which results in a ratio of turnover to inactivation of 6600. The rate of half maximal rate of inactivation is reached at 300 μM AdoMet which is similar to the Km (200 μM) determined for this enzyme.

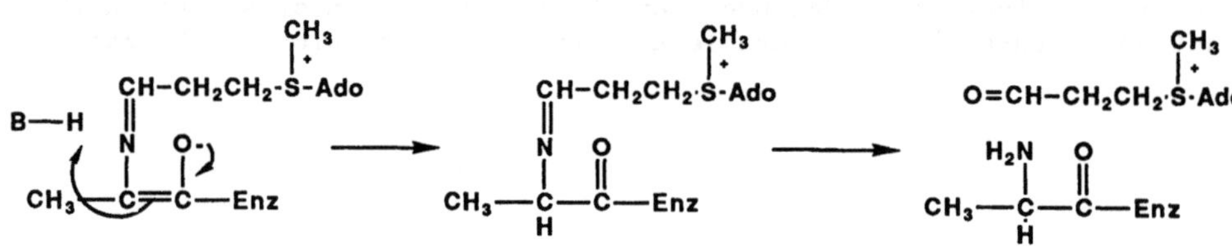

Scheme 3

The inactivation of AdoMet decarboxylase by 1 mM AdoMet is essentially completely prevented by including 1 mM MGBG in the incubation (data not shown). MGBG is a competitive inhibitor for AdoMet decarboxylase with a Ki of 20 μM[8].

Substrate inactivation is well known for pyridoxal phosphate containing decarboxylase[14]. The mechanism of inactivation in those cases involves transamination of the pyridoxal to a pyridoxamine moiety after decarboxylation of the substrate has occurred. A similar transamination mechanism has been shown for inactivation of AdoMet decarboxylase. By automated Edman degradation it was shown that the pyruvoyl group on the α subunit is essentially quantitatively transaminated to an Ala residue. The proposed mechanism is shown in Scheme 3. The main difference between catalysis (Scheme 2) and inactivation is that the intermediate iminium formed after decarboxylation is protonated on the pyruvoyl group rather than on the nascent product. Hydrolysis of the resulting Schiff base yields product aldehyde and an alanyl residue. Interestingly the histidine decarboxylase from Lactobacillus 30a does not undergo substrate inactivation even after extended incubation.

When AdoMet decarboxylase was inactivated in the presence of [methionyl-3,4-^{14}C] AdoMet, label was incorporated in to the protein in a stoichiometry of approximately 0.9 mole per α subunit. This radioactivity is not released by reduction and carboxymethylation. However when the inactivation of AdoMet decarboxylase was carried out in the presence of either [Me-^{3}H] AdoMet or [8-^{14}C] AdoMet only a small amount of radioactivity was associated with the reduced and carboxymethylated protein. This unexpected labeling with [methionyl-3,4-^{14}C] AdoMet but not [Me-^{3}H] AdoMet or [8-^{14}C] AdoMet suggests that there is elimination of methylthioadenosine from the aldehyde product (or from the nascent product before hydrolysis of the Schiff base) generating the Michael acceptor 2-propenal (Scheme 4), which can alkylate the enzyme.

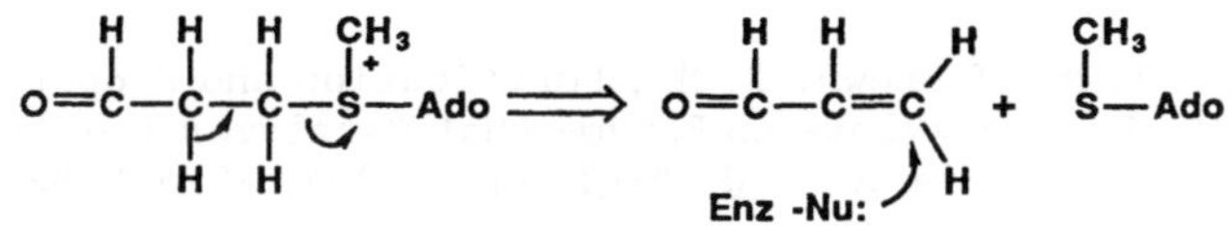

Scheme 4

Attempts to identify the nucleoside product of the inactivation by incubation with [8-^{14}C] AdoMet failed as a result of the non-enzymatic decomposition of the substrate, AdoMet, and reaction product, decarboxylated AdoMet[15]. Both compounds degrade to adenosine and methylthioadenosine at pH 7.5. Therefore, although both adenosine and methylthioadenosine, as well as decarboxylated AdoMet, were recovered from the reaction mixtures, the long reaction times required for complete inactivation made it difficult to be certain of the identity of the nucleoside product of the enzyme inactivation.

Less than 5% of the total radioactivity recovered is associated with the β subunit, while 96% is associated with the α subunit. No radioactivity was found in the residues of the first 20 cycles during automated Edman degradation of the α subunit from [methionyl-3,4-^{14}C] AdoMet inactivated AdoMet decarboxylase. The bulk of the radioactivity remained bound to the filter support; however, exact quantitation was not possible due to quenching. A single labeled polypeptide has been isolated after trypsin digestion of the reduced and carboxymethylated α subunit (Diaz, E. and Anton, D.L. unpublished observation) and the modification is being determined.

Rat liver AdoMet decarboxylase has also been shown to undergo a time-dependent inactivation in the presence of AdoMet, decarboxylated AdoMet, or analogs capable of forming Schiff's bases with the pyruvoyl group[16,17]. A similar transamination mechanism for substrate inactivation was proposed, although not demonstrated. In that case, inactivation with [Me-^{3}H] AdoMet led to no incorporation of radioactivity into the protein[16]. Since inactivations with AdoMet labeled elsewhere were not reported, it is not known whether an alkylation reaction such as we describe occurs in rat liver enzyme. The half time for inactivation of that enzyme is approximately 40 min in the presence of 1 mM AdoMet, slightly longer than what we have obtained for the E. coli enzyme at 1 mM AdoMet ($k = 0.03$ min^{-1}, $t_{1/2} = 23$ min).

The half life for AdoMet decarboxylase activity has been reported to be about 50 min in rat liver, just slightly shorter than the immunologically determined half life of 65 min for the AdoMet decarboxylase protein[18]. It is possible that the conversion of the pyruvoyl group to an alanine not only inactivates the enzyme but serves as a signal for AdoMet decarboxylase protein removal. Futher evidence for this hypothesis is suggested by the inhibition of inactivation by MGBG. Treatment of rats with this competitive inhibitor has been know to significantly increase the amount of AdoMet decarboxylase activity. This increase is due to an increased half life of the enzyme. It has been shown that MGBG prevents the inactivation of the E. coli enzyme, since MGBG is a better inhibitor of the mammalian AdoMet decarboxylases, it is likely that it also prevents the inactivation of this enzyme as will. This would lead to the observed increase in activity due to increased half life.

REFERENCES

1. Recsei, P. A., & Snell, E. E., Pyruvoyl Enzymes, Ann. Rev. Biochem. 53, 357-387 (1984).

2. Williamson, J. M., & Brown, G. M., Purification and Properties of the L-Aspartate alpha Decarboxylase an Enzyme that Catalyzes the Formation of Beta Alanine in Escherichia coli, J. Biol. Chem. **254**, 8074-8082 (1979).

3. Satre, M., & Kennedy, E. P., Identification of Bound Pyruvate Essential for the Activity of Phosphatidylserine Decarboxylase of Escherichia coli, J. Biol. Chem. **253**, 479-483 (1978).

4. Hodgins, D. S., & Abeles, R. H., The Presence of Covalently Bound Pyruvate in D-Proline Reductase and Its Participation in the Catalytic Process, J. Biol. Chem. **242**, 5158-5159 (1967).

5. Yang, H. and Abeles, R.H., Purification and Properties of Escherichia coli 4' Phosphopantothenoylcysteine Decarboxylase: Presence of Covalently Bound Pyruvate, Biochemistry **26**, 4076-4081 (1987).

6. Tabor, C.W., and Tabor, H. Methionine Adenosyltransferase (S-Adenosylmethionine Synthetase) and S-Adenosylmethionine Decarboxylase, Adv. Enzymol. Related Areas Molecular Biol. **56**, 251-282 (1984).

7. Bitonti, A.J., Dumont, J.A. and McCann, P.P. Characterization of Trypanosoma Brucei brucei S-Adenosyl-L-Methionine Decarboxylase and its Inhibition by Berenil, Pentamidine and Methylglyoxal bis(guanylhydrazone), Biochem. J. **237**, 685-690 (1986).

8. Markham, G. D., Tabor, C. W., & Tabor, H., S-Adenosylmethionine Decarboxylase of Escherichia coli: Studies on the Covalently Linked Pyruvate Required for Activity, J. Biol. Chem. **251**, 12063-12068 (1982).

9. Wickner, R.B., Tabor, C.W., and Tabor, H., Purification of Adenosylmethionine Decarboxylase from Escherichia coli W: Evidence for Covalently Bound Pyruvate, J. Biol. Chem. **245**, 2132-2139 (1970).

10. Anton, D.L. and Kutny, R., Escherichia coli S-Adenosylmethionine Decarboxylase: Subunit Structure, Reductive Amination and NH_2-Terminal Sequences, J. Biol. Chem. **262**, 2817-2822 (1987).

11. Tabor C.W., and Tabor, H. Adenosylmethionine Decarboxylase of E. coli is Synthesized from a Proenzyme, Proc. Int. Congr. Biochem. **13**, 686 (1985).

12. Tabor C.W. and Tabor, H., The SPE-E SPE-D Operon of Escherichia coli Formation and Processing of a Proenzyme form of S-Adenosylmethionine Decarboxylase. J. Biol. Chem. **262**, 16037-16040 (1987).

13. Anton, D.L. and Kutny, R., Mechanism of Substrate Inactivation of Escherichia coli S-Adenosylmethionine Decarboxylase, Biochemistry 26, 6444-6447 (1987).

14. O'Leary, M. H., & Herreid, R. M., Mechanism of Inactivation of Ornithine Decarboxylase (EC-4.1.1.17) by Alpha Methyl Ornithine, Biochemistry 17, 1010-1014 (1978).

15. Hoffman, J. L., Chromatographic Analysis of the Chiral and Covalent Instability of S-Adenosyl-L-Methionine, Biochemistry, 25, 4444-4449 (1986).

16. Pankaskie, M., & Abdel-Monem, M. M., Inhibitors of Polyamine Biosynthesis 8. Irreversible Inhibition of Mammalian S-Adenosyl-L_Methionine Decarboxylase (EC-4.1.1.50) by Substrate Analogs, J. Med. Chem. 23, 121-127 (1980).

17. Kolb, M., Danzin, C., Barth, J., & Claverie, N., Synthesis and Biochemical Properties of Chemically Stable Product Analogues of the Reaction Catalyzed by S-Adenosyl-L-Methionine Decarboxylase, J. Med. Chem. 25, 550-556 (1982).

18. Pegg, A.E., Investigation of the Turnover of Rat Liver S-Adenosylmethionine Decarboxylase Using a Specific Antibody, J. Biol. Chem. 254, 3249-3253 (1979).

S-ADENOSYLMETHIONINE DECARBOXYLASE FROM THE THERMOPHILIC ARCHAEBACTERIUM SULFOLOBUS SOLFATARICUS

Giovanna Cacciapuoti, Marina Porcelli, *Agata Gambacorta, Teresa Romano and Mario De Rosa

Department of Biochemistry of Macromolecules
University of Naples First Medical School
*Institute for the Chemistry of Molecules of Biological Interest, CNR, Naples, Italy

INTRODUCTION

It is well known that S-adenosylmethionine decarboxylase (AdoMet DC) plays a key role in the polyamine biosynthetic pathway by catalyzing the formation of S-adenosyl(5')-3-methylthiopropylamine, the donor of the propylamine moiety of polyamines (1-3).

AdoMet DC was first described in E. coli W and was then purified to homogeneity from E. coli, Saccharomyces cerevisiae and some mammalian tissues and to a lesser extent from a variety of prokaryotic and eukaryotic sources, including plants and the slime mold Physarum polycephalum (1,4-6). These enzymes differ in their subunit structures and cation requirement, but all contain a covalently linked pyruvate as prosthetic group (7,9).

E. coli AdoMet DC requires divalent metal cations for activity and is composed of two types of subunits: α, Mr 19000 and β, Mr 14000 present in equivalent amounts (10). They probably arise from a single peptide (11) with a mechanism similar to that reported for histidine decarboxylase, which involves a post-translational autolytic cleavage at a single serine-serine bond to generate the pyruvoyl group on the NH_2 terminus of the α-subunit (12). An $\alpha 4\beta 4$ octameric structure has been proposed for the oloenzyme that is characterized by a molecular weight of 136000 daltons (10).

Mammalian AdoMet DC, is strongly and specifically activated by putrescine (put) (13) and is composed of two subunits of 32000 daltons (3,4). It has been recently demonstrated that the enzyme is synthesized as a proenzyme having a molecular weight of 37000 daltons which is in turn cleaved to form the enzyme subunit of 32000 daltons in a putrescine-stimulated reaction (14).

AdoMet DC from Saccharomyces cerevisiae, like mammalian enzyme, requires putrescine for catalytic activity and is a dimeric protein of 88000 daltons (7).

At present no data are available in the Literature on AdoMet DC from archaebacteria.

ARCHAEBACTERIA AND POLYAMINE BIOGENETIC PATHWAY

A series of recent phylogenetic studies based on 16S ribosomal ribonucleic acid composition and other general biochemical features such as RNA polymerase, translation system, transfer RNA, 5S ribosomal RNA, cell wall and lipids, has revealed the existence of a new group of microorganisms (15). The small size and structural simplicity, the absence of a nuclear membrane and organelles, and low deoxyribonucleic acid content, assimilate archaebacteria to prokaryotes (16). In contrast, it has been proposed that archaebacteria are closely related to the ancestor eukaryotes since several properties such as the presence of histone-like proteins, the nature of their translation system and the sequence of their 5S RNA and that of ribosomal proteins, are considered to be typical of eukaryotes (17). On the basis of these properties, archaebacteria can be truly envisioned as a third line of evolution completely distinct from eukaryotic and eubacterial lines (18-19). They include three major phenotypes confined to a few ecological niches such as thermal habitat for extreme thermophiles, saturated brine for extreme halophiles and strict anaerobic environment for methanogens (16).

Archaebacteria are quite interesting microorganisms from the point of view of the early evolution of life; indeed, their name was given to underline the hypothesis that these organisms were the dominant inhabitants of the earth in early ancient times (15-19). In this respect, archaebacteria, as a distinct primary kingdom, are important as they give us some indications on the early events in the evolution of cells, thus contributing to a better understanding of the universal ancestor.

Polyamine pathway in archaebacteria has been extensively studied in Sulfolobus solfataricus, an extreme thermoacidophilic archaebacterium optimally growing at 87°C, at pH 3.5 (20). S-adenosylmethionine (AdoMet), a key intermediate in polyamine biosynthesis, has been detected at levels similar to those reported in other microorganisms and mammalian tissues, in spite of its thermal lability (21). Moreover, the role of the sulfonium compound as methyl donor to proteins has also been established (22).

The polyamine pool of S. solfataricus is characterized by two unusual polyamines, sym-nor-spermidine and sym-nor-spermine (Fig. 1). These polyamines, that have been also identified in a series of genetically different thermophilic eubacteria (23,24) and in some halophilic archaebacteria (25), represent about 50% of the polyamine pool of S. solfataricus (20). These polyamines are symmetrical polycations consisting of linear multiples of amino propyl groups, two in sym-nor-spermidine and three in sym-nor-spermine respectively. As demonstrated by in vivo studies with labeled precursors (20), an unusual reaction occurs, in which spermidine plays the role of precursor of the trimethylenic moiety. In fact, 1,3-diaminopropane, originated from spermidine breakdown, is rapidly propylaminated to form sym-nor-spermidine, which is in turn converted to sym-nor-spermine by an analogous reaction. According to this pathway, the entire molecular skeleton of the two symmetrical polycations derives from decarboxylated AdoMet, since also the propylamine moiety coming from spermidine is in turn originated from the sulfonium compound.

Several enzymes related to polyamine pathway in S. solfataricus, such as AdoMet synthetase (21), protein carboxyl-methyltransferase (22), propylamine transferase (26) and MTA phosphorylase (27) have been purified and characterized in order to elucidate the evolutionary and comparative bioche-mistry of polyamine biosynthesis and to make a more direct approach toward the problem of protein thermostability. These enzymes show an optimum temperature at 90°C, nearly the optimal growth temperature of the microorganism and are also highly thermostable. In fact, the enzymatic activity is not lost even after 1 hour preincubation

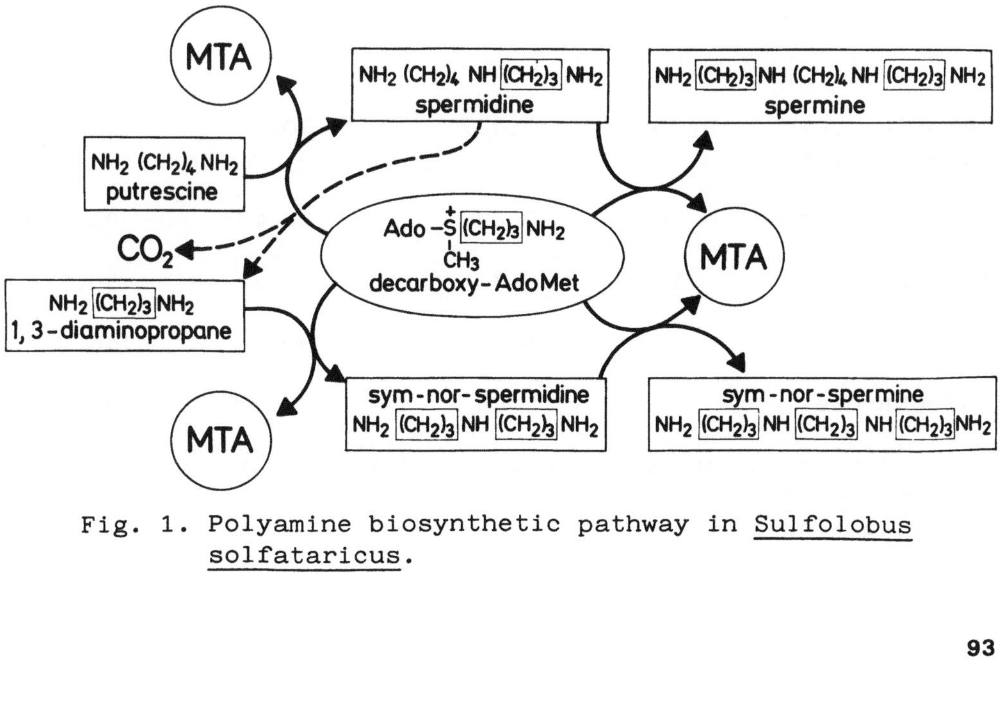

Fig. 1. Polyamine biosynthetic pathway in Sulfolobus solfataricus.

at 100°C for propylamine transferase (26) and MTA phosphorylase (27), or 90°C, in the presence of ATP, for AdoMet synthetase (21). It is interesting to note that some properties of these enzymes are typically eukaryotic such as the presence of isoenzymatic forms for AdoMet synthetase (21) or the phosphorolytic mechanism of action for MTA phosphorylase (27). In contrast, the unusual substrate specificity of propylamine transferase makes this enzyme completely different from its eukaryotic and eubacterial counterparts. In fact, the enzyme is able to recognize as substrates trimethylenic as well as tetramethylenic moieties (26).

PURIFICATION AND MOLECULAR PROPERTIES OF ADOMET DC FROM SULFOLOBUS SOLFATARICUS

AdoMet DC from S. solfataricus has been purified 1300-fold with 20% yield to a final specific activity of 12 nmoles of CO_2 formed per minute per mg of protein (Table 1). The homogeneity of the enzyme has been checked by SDS gel electrophoresis under native and denaturating conditions and isoelectric focusing.

Table 1. Purification of AdoMet DC from Sulfolobus solfataricus.

Step	Total protein (mg)	Specific activity (units*/mg)	Yield (%)	Purification (fold)
Crude extract§	29,200	0.009	100	1
Ammonium sulfate (30-65%)	9,407	0.027	90.16	3
DEAE cellulose	768	0.228	62.14	25.3
Hydroxylapatite	151	0.77	41.32	85.5
Sephadex G-75	14	4.02	30.20	446.6
Q-Sepharose	5	12.05	20.20	1338.0

*One unit is the amount of enzyme which forms 1 nmol of CO_2 per min at 70°C.
§A typical purification from 700 g of wet cells.

Table 2 reports some physico-chemical and kinetic properties of AdoMet DC from S. solfataricus. The affinity for AdoMet is similar to that of AdoMet DC from E. coli (6) and mammalian tissues (1). The calculated Km value, when compared with the cellular amount of the sulfonium compound (70-90 nmoles/g wet cells), indicates that AdoMet DC is almost saturated in physiological conditions. In this respect, it is a biochemical puzzle how the sulfonium compound, which is quite thermolabile, is stabilized in this organism optimally living at 87°C.

AdoMet DC from S. solfataricus shows an isoelectric point at pH 5.1 and in contrast with the enzyme from E. coli (6), does not require reducing agents for its catalytic activity, as demonstrated by the lack of any effect exerted by thiol inhibitors.

AdoMet DC is strongly inhibited by carbonyl group reagents, such as sodium borohydride, semicarbazide, phenylhydrazine and hydroxylamine suggesting that a carbonyl group is involved in the catalytic mechanism. At present there is no direct experimental evidence that either pyridoxal phosphate or pyruvate functions as coenzyme. Therefore, the observation that the spectrum of the purified enzyme contains no chromophore in the region of pyridoxal phosphate supports the view that a pyruvate, bound to the enzyme, serves as a prosthetic group.

The enzyme is competitively inhibited by methyl glyoxal bis(guanylhydrazone) (MGBG). The calculated Ki is much higher than that reported for eukaryotic enzyme (1,4), resembling, in this respect, the E. coli AdoMet DC (2). Furthermore, when MGBG is added to the culture medium, a marked increase (about 10-fold) in its specific activity is observable. This finding resembles the effect of MGBG on AdoMet DC levels observed in many normal and malignant tissues or in cultured cells (1).

Table 2. Physico-chemical and kinetic properties of AdoMet DC from Sulfolobus solfataricus

- Km for AdoMet = 96 μM
- No activation by reducing agents
- pI = 5.1
- Inhibition by carbonyl group reagents
- Competitive inhibition by MGBG (Ki = 33 μM)
- Increase of enzyme levels (10-fold) in cells grown in the presence of MGBG (16 mg/l)

This increase is accompanied by a prolongation of the half-life of the enzyme, suggesting the hypothesis that the _in vivo_ binding of this drug to the enzyme protein could result in a protection against a rapid proteolytic degradation.

The thermostability of AdoMet DC is notably high, even if it is still lower than that of some other characterized enzymes from _S. solfataricus_ (21,26-27). In fact, as shown in Fig. 2, only 20% of maximal activity is present after 1 hour of preincubation at 100°C. Furthermore, the enzyme is stable for 16 hours at 50°C and retains about 60°C of its maximal activity after 16 hours preincubation at 70°C.

The enzyme exhibits a maximum at 75°C (Fig. 3a) and is still active at temperatures below 40°C, unlike most enzymes from thermophilic organisms (28). The corresponding Arrhenius plot (Fig. 3b) is biphasic, indicating two different activation energy-dependent processes, occurring below and above 55°C. This result, observed also in other thermophilic proteins (28), can be interpreted as an adjustement of protein conformation in order to counterbalance the changes in thermic environment occurring as the temperature rises.

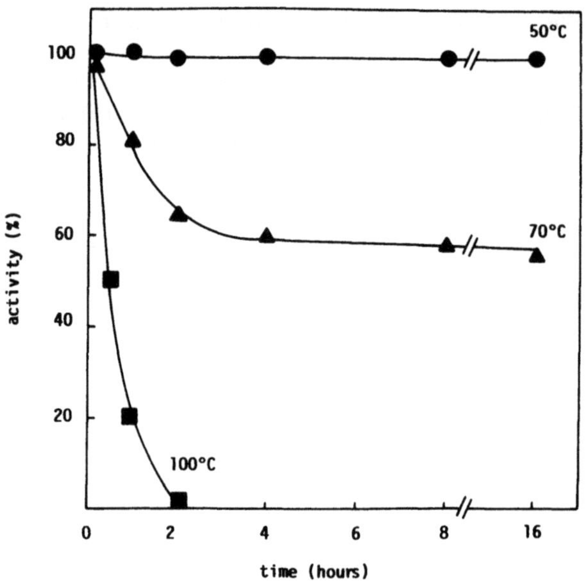

Fig. 2. Stability of AdoMet DC as a function of preincubation time.
The enzyme was preincubated at 50°, 70° and 100°C for the time indicated and an aliquot (20 µl) was withdrawn and assayed in the presence of 5 µmol of Tris/HCl, pH 7.5 and 20 nmoles of (carboxyl-^{14}C) AdoMet in a final volume of 0.2 ml. The enzyme assay was carried out at 70°C for 5 min according to (6) with minor modifications.

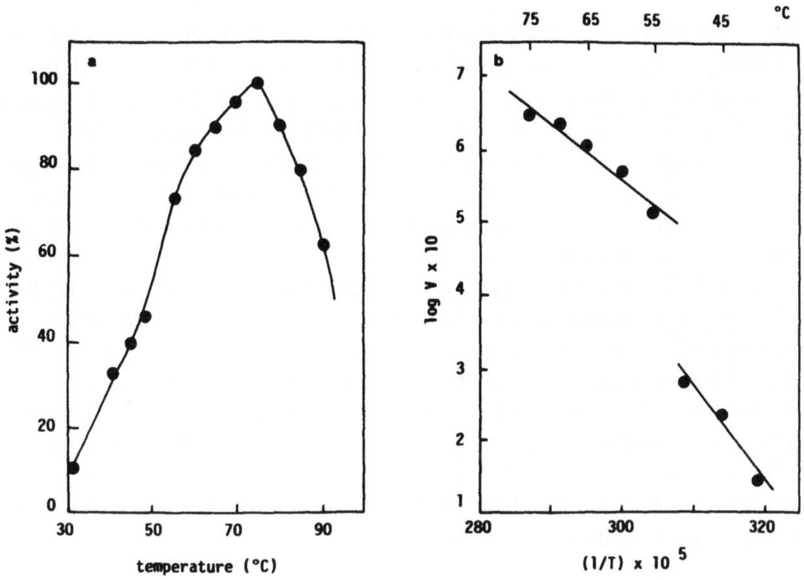

Fig. 3. (a) Effect of temperature on the reaction rate.
(b) Arrhenius plot.

A comparative analysis of AdoMet decarboxylases from the three evolutive lines: eukaryotes, prokaryotes and archaebacteria, is reported in Table 3.

Table 3. Comparative analysis of AdoMet decarboxylase from different evolutive lines.

Properties	Eukaryotes		Eubacteria	Archaebacteria
	rat liver	S.cerevisiae	E. coli	S.solfataricus
Cations requirement	put	put	Mg^{2+}	not activated
Native molecular weight	64000	88000	136000	32000
Subunit molecular weight	32000	41000	α = 19000 β = 14000	—
Protein structure	dimeric	dimeric	$\alpha_4 \beta_4$ octamer	monomeric

The most relevant differences are related to the cation requirements, the size and subunit composition. The enzyme from S. solfataricus does not require either putrescine or magnesium and has a more simple molecular organization than that of E. coli (10) and eukaryotic enzyme (14). In fact, it is a monomeric protein with a molecular weight of 32000 daltons, similar to the subunit molecular weight of AdoMet DC from mammalian tissues.

All together these findings indicate that the enzyme from S. solfataricus possesses molecular features distinct either from eukaryotic or prokaryotic counterparts, thus confirming the unique taxonomic position of the archaebacteria.

Studies on AdoMet DC from thermophilic eubacteria and mesophilic archaebacteria could contribute to clarifying if the differences in protein structure are related to the phylogenesis of archaebacteria or to the environmental pressure of high temperatures.

ACKNOWLEDGEMENTS

This work was supported by a grant from CNR research group Biological Macromolecules n. 87.01381.04

REFERENCES

1. H. G. Williams-Ashman and A.E. Pegg, Amino propyl group transfers, in: "Polyamines in Biology and Medicine" D.R. Morris and L.J. Marton, eds, Dekker, New York (1981).
2. C. W. Tabor and H. Tabor, Methionine adenosyltransferase (S-adenosylmethionine synthetase) and S-adenosylmethionine decarboxylase, Advances in Enzymol. 56:25 (1984).
3. A. E. Pegg, Recent advances in the biochemistry of polyamines in eukaryotes, Biochem. J. 234:249 (1986).
4. A. E. Pegg, S-adenosylmethionine decarboxylase. A brief review, Cell Biochem. Function 2:11 (1984).
5. C. W. Tabor and H. Tabor, Polyamines in microorganisms, Microbial Review 49:81 (1985).
6. G. D. Markham, C. W. Tabor and H. Tabor, S-adenosylmethionine decarboxylase of Escherichia coli, studies on the covalently linked pyruvate required for activity, J. Biol. Chem. 257:12063 (1982).
7. M. S. Cohn, C. W. Tabor and H. Tabor, Identification of pyruvoyl residue in S-adenosylmethionine decarboxylase from Saccharomyces cerevisiae, J. Biol. Chem. 252:8212 (1977).
8. A. E. Pegg, Evidence for the presence of a pyruvate in rat liver S-adenosylmethionine decarboxylase, FEBS Lett. 84:33 (1977).

9. R. B. Wickner, C. W. Tabor and H. Tabor, Purification of adenosylmethionine decarboxylase from E. coli W: evidence for covalently bound pyruvate, J. Biol. Chem. 245:2132 (1970).
10. D. L. Anton and R. Kutny, E. coli S-adenosylmethionine decarboxylase. Subunit structure, reductive amination and NH terminal sequence, J. Biol. Chem. 262:2817 (1987).
11. C. W. Tabor and H. Tabor, The speEspeD operon of E. coli formation and processing of a proenzyme form of S-adenosylmethionine decarboxylase, J. Biol. Chem. 262:16037 (1987).
12. P. A. Recsei, W. M. Moore and E. E. Snell, Pyruvoyl-dependent histidine decarboxylase from Clostridium perfringens and Lactobacillus buchneri, J. Biol. Chem. 258:439 (1983).
13. V. Zappia, M. Cartenì-Farina and G. Della Pietra, S-adenosylmethionine decarboxylase from human prostate, activation by putrescine, Biochem. J. 120:703 (1972).
14. A. E. Pegg, L. Wiest and A. Pajunen, Detection of proenzyme form of S-adenosylmethionine decarboxylase in extracts from rat prostate, Biochim. Biophys. Res. Commun. 150:788 (1988).
15. K. H. Schleifer and E. Stackenbrandt, Molecular systematics of prokaryotes, Annu. Rev. Microbial. 37:143 (1982).
16. C. R. Woese, L. J. Magrum and G. E. Fox, Archaebacteria J. Mol. Evol. 11:245 (1978).
17. L. M. Van Valen and V. C. Maiorana, The archaebacteria and eukaryotic origins, Nature (London) 287:248 (1980).
18. C. R. Woese and R. S. Wolfe, "The Bacteria," vol. 8 Academic Press Inc, New York (1985).
19. C. R. Woese, Archaebacteria and cellular origins: an overview, in: "Archaebacteria", O. Kandler, ed., Gustav Fisher-Verlag, Stuttgart (1982).
20. M. De Rosa, S. De Rosa, A. Gambacorta, M. Cartenì-Farina and V. Zappia, The biosynthetic pathway of new polyamines in Caldariella acidophila, Biochem. J. 176:1 (1978)
21. M. Porcelli, G. Cacciapuoti, M. Cartenì-Farina and A. Gambacorta, S-adenosylmethionine synthetase in the thermophilic archaebacterium Sulfolobus solafataricus: purification and characterization of two isoforms, Eur. J. Biochem. (1988), in press.
22. P. Galletti, M. De Rosa, A. Gambacorta, C. Manna, R. Festinese and V. Zappia, Protein methylation in Caldariella acidophila an extreme thermoacidophilic archaebacterium FEBS Lett. 124:62 (1981).
23. T. Oshima, Novel polyamines in Thermus thermophilus: isolation, identification, and chemical synthesis, in: "Methods in Enzymology", H. Tabor and C. W. Tabor eds, vol. 94, Academic Press, New York (1983)

24. M. De Rosa, S. De Rosa, A. Gambacorta, M. Cartenì-Farina and V. Zappia, Occurrence and characterization of new polyamines in the extreme thermophile Caldariella acidophila, Biochim. Biophys. Res. Commun. 69:253 (1976).
25. M. Cartenì-Farina, M. Porcelli, G. Cacciapuoti, M. De Rosa A. Gambacorta, W. D. Grant and H. N. M. Ross, Polyamines in halophilic archaebacteria, FEMS Micrbiol. Lett. 28:323 (1985).
26. G. Cacciapuoti, M. Porcelli, M. Cartenì-Farina, A. Gambacorta and V. Zappia, Purification and characterization of propylamine transferase from Sulfolobus solfataricus an extreme thermophilic archaebacterium,Eur. J. Biochem. 161:263 (1986).
27. M. Cartenì-Farina, A. Oliva, G. Romeo, G. Napolitano, M. De Rosa, A. Gambacorta and V. Zappia, 5'-Methylthioadenosine phosphorylase from Caldariella acidophila : purification and properties, Eur. J. Biochem. 101:317 (1979).
28. H. Zuber, Comparative studies of thermophilic and mesophilic enzymes: objectives, problems, results, in : "Biochemistry of Thermophily," S. N. Friedman, ed., Academic Press, New York (1978).

PROPERTIES OF HUMAN AND RODENT S-ADENOSYLMETHIONINE DECARBOXYLASE

Anthony E.Pegg [a], Bruce Stanley [a], Antti Pajunen [b], Anne Crozat [c] and Olli A. Jänne [c]

[a] Departments of Physiology and Pharmacology, Milton S. Hershey Medical Center, Pennsylvania State University, Hershey, Pennsylvania 17033, USA
[b] Department of Biochemistry, University of Oulu, Linnanmaa, SF-90570 Oulu, Finland
[c] The Population Council and the Rockefeller University, 1230 York Avenue New York, New York 10021, USA

INTRODUCTION

S-Adenosylmethionine decarboxylase (AdoMetDC) is a key enzyme in the biosynthesis of the polyamines spermidine and spermine. Its product, decarboxylated S-adenosylmethionine (dcAdoMet) is used as an aminopropyl donor by spermidine synthase and spermine synthase (Williams-Ashman and Pegg, 1981; Pegg and McCann, 1982; Tabor and Tabor, 1984a). Once decarboxylated by the action of AdoMetDC, S-adenosylmethionine (AdoMet) becomes committed to polyamine production since methyltransferases use dcAdoMet very poorly, if at all (Pegg, 1984,1986). In fact, the only known metabolic route for further metabolism of dcAdoMet apart for its use as an aminopropyltransferase substrate is its acetylation (Wagner et al., 1985; Pegg et al., 1986) The supply of dcAdoMet is normally regulated very tightly by the cellular polyamine content and this regulation is brought about by changes in the activity of AdoMetDC (Pegg, 1984). In this way, the cellular content of dcAdoMet is usually kept very low (about 1-3% of AdoMet content) as its synthesis is linked to the ability of the aminopropyltransferases to use it to form polyamines. Only when cellular polyamine metabolism is deranged by inhibition of the other enzymes in the polyamine biosynthetic pathway does the dcAdoMet content rise. Increases of several hundred fold occur when ornithine decarboxylase (ODC) activity is inhibited by drugs such as α-difluoromethylornithine (DFMO) and, only under these conditions, is the acetyl derivative of dcAdoMet formed in significant amounts (Pegg, 1986). The increased content of dcAdoMet is due to both an increase in the activity of AdoMetDC (Alhonen-Hongisto, 1980; Mamont et al., 1981; Pegg, 1984) and to the inability of the aminopropyltransferases to utilize the dcAdoMet formed by it because of the absence of putrescine and spermidine to serve as aminopropyl acceptors.

It is clear, therefore, that the regulation of AdoMetDC is a critical factor in the maintenance of normal cellular polyamine levels and that the abnormal accumulation of dcAdoMet and its acetylated derivative in cells treated with polyamine biosynthesis inhibitors may contribute to the effects of these compounds. A detailed knowledge on the properties and regulation of AdoMetDC is needed to understand the physiological regulation of polyamine content and to provide information on the optimal use of polyamine biosynthesis inhibitors. Furthermore, there is now convincing evidence that inhibitors of AdoMetDC can be used to deplete cells of spermidine and spermine and that such depletion leads to an inhibition of cell growth (Artamonova et al., 1986; Pegg, 1988; Pegg et al., 1988a). Potent and specific inhibitors of AdoMetDC have considerable potential as therapeutic agents but the use of inhibitors of this enzyme has been limited by the fact that the presently available compounds are either not strictly specific or are analogs of AdoMet of limited stability under biological conditions. The design of inhibitors of AdoMetDC will be facilitated by a greater knowledge of the structure and mechanism of the enzyme.

ENZYMOLOGY AND INHIBITION OF ADOMETDC

Early work on the properties and structure of AdoMetDC has been summarized in several reviews (Williams-Ashman and Pegg, 1981; Tabor and Tabor, 1984b; Pegg, 1984; White and Morris, 1988). The mammalian enzyme has been purified from several rat tissues including prostate, which is the richest source, mouse liver and mammary gland, and bovine liver. In all cases, the purifications have been achieved using methylglyoxal bis(guanylhydrazone) (MGBG) as a tool. This drug is a powerful competitive inhibitor of AdoMetDC but its administration leads to a substantial accumulation of the enzyme protein (see below) which enriches the starting material for purification. Furthermore, MGBG-bound to Sepharose retains a high affinity for the enzyme and can be used for affinity chromatography. Homogeneous preparations of the enzyme have been obtained using this methodology but only small amounts of protein have been obtained because AdoMetDC is a very minor fraction of the total protein in mammalian tissues. Even after MGBG pretreatment, AdoMetDC represents only about 1 part in 2000 of the soluble protein in the rat prostate (Shirahata and Pegg, 1986).

The purified AdoMetDC preparations show a single band of M.W. about 32,000 on electrophoresis under denaturing conditions and the native enzyme has a M.W. of about 70,000 (Pegg, 1984; White and Morris, 1988). This led to the supposition that the enzyme was a dimer of two identical sub-units of 32,000 but as described below, this conclusion is incorrect and a smaller sub-unit of M.W. of about 6,500 was missed in these experiments.

Mammalian AdoMetDC is activated by putrescine (Williams-Ashman and Pegg, 1981; Pegg and McCann, 1982). This activation provides an important regulatory control since it ensures that the enzyme is maximally active when there are high levels of putrescine and ensures that sufficient dcAdoMet is produced to serve as a co-substrate with putrescine for the spermidine synthase reaction. This activation therefore ensures the efficient conversion of putrescine into spermidine and, in most mammalian cells the content of spermidine is much higher than that of putrescine. Many microorganisms including *E. coli* have an AdoMetDC which is not stimulated by putrescine and in these cells, putrescine concentration usually exceeds that of spermidine. The mechanism of activation of AdoMetDC by putrescine is not fully understood but binding of putrescine leads to a decrease in the K_m for AdoMet. The activation is pH dependent, being greater at lower pH, and probably results from a change in the conformation of the enzyme on binding putrescine.

In the presence of saturating amounts of putrescine the K_m for AdoMet for the prostatic AdoMetDC is about 65 µM and the K_i for MGBG is about 0.3µM (Pegg and Jacobs, 1983). The most potent nucleoside inhibitor which acts as a competitive inhibitor is S-methyl-5'-(methylthio)adenosine [$AdoS^+(CH_3)_2$], which has a K_i of 2-6 µM (Kolb et al., 1982; Pegg and Jacobs, 1983) , but much more powerful inhibitory nucleosides which act as irreversible inhibitors have been synthesized recently. These include S-(5'-deoxy-5'-adenosyl)-methylthioethylhydroxylamine (Artamonova et al., 1986), 5'-deoxy-5'-[N-methyl-N-[2-(aminooxy)ethyl]]aminoadenosine [MAOEA] and 5'-deoxy-5'-[N-methyl-N-(3-hydrazino-propyl)]aminoadenosine [MHZPA] (Pegg et al., 1988a). Although a detailed kinetic analysis of these inhibitors has not been carried out, these compounds are clearly potent irreversible inactivators of the enzyme and more than 95% of the activity was lost within 15 min of exposure to 5 µM MAOEA or 0.5 µM MHZPA. Administration of either MAOEA and MHZPA to L1210 cells led to a large reduction in the production of dcAdoMet and of 5'-methylthioadenosine (Pegg et al., 1988a). These results indicate that the compounds do block AdoMetDC activity in the cell. Analysis of the polyamine content also confirmed this, since putrescine levels were greatly increased whilst spermidine and spermine content declined. These inhibitors also brought about an inhibition of cell growth which could be reversed by the addition of spermidine or spermine (Pegg et al., 1988a). These results indicate the potential for the use of AdoMetDC inhibitors but unfortunately it appears that these compounds have quite limited stability in cultured cells possibly because of their reaction with free pyruvate. Therefore, repeated addition is needed to maintain a sufficient level of inhibitor to block polyamine synthesis.

AdoMetDC has a covalently bound pyruvate group which serves as a prosthetic group by forming a Schiff's base with the substrate (Williams-Ashman and Pegg, 1981; Tabor and Tabor, 1984b; Pegg, 1984; White and Morris, 1988). This pyruvate is clearly bound to the larger 32,000 M.W. sub-unit since this subunit is labeled irreversibly by the addition of [^{35}S]dcAdoMet and Na cyanoborohydride to form a covalent bond between the product and the pyruvate (Shirahata et al., 1985). The stability of the pyruvate-protein bond indicates that it is probably contained in an amide linkage and, by analogy with bacterial enzymes known to contain a similar pyruvate co-factor (Recsei and Snell, 1984), it is likely that it is present at the amino terminus of the polypeptide and is formed from a serine residue during the processing of a proenzyme (see below).

There is some evidence for multiple forms of AdoMetDC. The enzyme purified from psoas muscle differed from that purified from rat liver in isoelectric point, affinity for MGBG, extent of activation by putrescine and K_m for AdoMet (Pösö and Pegg, 1982). Furthermore, the enzyme in skeletal muscle was more sensitive to repression by administration of spermidine than many other tissues (Pösö and Pegg, 1981). The molecular basis of these differences is not yet clear.

SYNTHESIS OF ADOMETDC IN VITRO

When prostatic mRNA was translated in a reticulocyte lysate system and the polypeptides corresponding to AdoMetDC were precipitated with a specific antiserum, it was found that the enzyme is synthesized as a precursor of M.W. about 38,000 (Shirahata and Pegg, 1986). The proenzyme was converted into the 32,000 sub-unit on continued incubation in the lysates. It is not known whether factors in the reticulocyte lysate were needed for the processing to take place or if the conversion was autocatalytic. However, a similar formation of the proenzyme and processing was seen when the mRNA was translated in a wheat germ system (unpublished observations). More recently, the *in vitro* synthesis of proenzyme and processed AdoMetDC has been carried out using reticulocyte lysates supplemented with a synthetic mRNA synthesized from a portion of a human cDNA which contains all of the coding sequence (Pajunen et al, 1988). [This fragment was obtained by digestion with *Pst* I and *Xba* I, inserted into the pT3/T7-19 vector and RNA synthesized by the action of T7 polymerase.] Using this system it has been possible to confirm that active AdoMetDC produced as the proenzyme is processed and also to observe the formation of the smaller polypeptide of M.W. about 6,500 in this processing. Both the 32,000 M.W. and the 6,500 M.W. polypeptides are precipitable by antiserum to AdoMetDC indicating that the smaller subunit does remain as part of the final enzyme (Pajunen et al, 1988).

EFFECT OF PUTRESCINE ON THE SYNTHESIS OF ADOMETDC

Addition of putrescine to the reticulocyte lysate synthesis system for AdoMetDC described above did not affect the rate of synthesis of the proenzyme but greatly increased the rate of conversion of the proenzyme into the enzyme subunits (Kameji and Pegg, 1987a). This provides another point at which the supply of active AdoMetDC is regulated by the concentration of putrescine. Proof that putrescine levels affect this conversion in vivo was obtained by Western blot analysis of the prostate polypeptides reacting with antibodies to AdoMetDC (Pegg et al., 1988b). The proenzyme could clearly be seen in this analysis but the proportion of the protein which was present as the proenzyme was increased from about 4% in control rats to about 25% in rats depleted of putrescine by treatment with DFMO. Conversely, the amount of proenzyme was reduced to almost undetectable levels by pretreatment with MGBG which increases putrescine content (Pegg et al., 1988b). The mechanism by which putrescine facilitates the processing is not understood but it may act by binding to the proenzyme and changing the conformation to a form in which the cleavage reaction is facilitated. If this is the case, the site of binding of putrescine could be the same as that which changes the K_m for AdoMet in the catalytic reaction.

EFFECT OF SPERMIDINE AND SPERMINE ON CONTENT OF ADOMETDC

Depletion of these polyamines by exposure to inhibitors of their biosynthesis leads to an increase in the amount of AdoMetDC protein . This rise is mediated both by an increase in AdoMetDC synthesis and by a decreased rate of degradation of the protein (Alhonen-Hongisto, 1980; Mamont et al., 1981; Pösö and Pegg, 1981; Shirahata and Pegg, 1986). Virtually nothing is known about the mechanism underlying the rapid degradation of AdoMetDC but it is clear that the protein is greatly stabilized by binding of MGBG (Williams-Ashman and Pegg, 1981; Shirahata and Pegg, 1985) and it is also stabilized by binding of the nucleoside inhibitors, $AdoS^{+}(CH_3)_2$, MAOEA or MHZPA (Pegg et al., 1987; Madhubala et al., 1988). It has been suggested that the substrate for degradation may be a spermidine (or spermine) protein complex and that the compounds listed above may reduce degradation by reducing the formation of this complex (Pösö and Pegg, 1981). This would explain the inverse relationship between the polyamine content and the half life of the protein but, at present, there is no direct experimental evidence for this hypothesis.

The increased synthesis of AdoMetDC when polyamines are reduced is produced by at least two factors. It has been demonstrated clearly that depletion of spermidine by DFMO and other agents leads to an increase in the content of AdoMetDC mRNA. This increase has been measured both by Northern and dot blot analysis using cDNA probes for AdoMetDC (Pajunen et al., 1988; Madhubala et al., 1988) and by translating the mRNA in a reticulocyte lysate and quantitating the formation of AdoMetDC protein (Shirahata and Pegg, 1986; Pegg et al., 1987) However, in some cases, and particularly the increase in response to inhibition of spermine synthesis (Pegg et al., 1987; Pajunen et al., 1988; Madhubala et al., 1988), it is clear that the change in mRNA content is not sufficient to account for the effect on AdoMetDC synthesis and it appears that there is also an effect on the translation of the mRNA. This hypothesis is firmly supported by experiments in which it was found that the synthesis of AdoMetDC in reticulocyte lysates was strongly inhibited by spermidine or spermine (Kameji and Pegg, 1987b). Inhibition was produced by concentrations of these polyamines which did not reduce the synthesis of other proteins including albumin. Thus, it appears that spermidine and spermine selectively inhibit the translation of the AdoMetDC mRNA. Whether this inhibition is mediated by binding of the polyamines to the mRNA (perhaps in the 5' non-translated region) and thereby changing the secondary structure, or involves the interaction of the polyamines with proteins or other factors needed for the translation requires to be established.

The content of AdoMetDC is also reduced by exogenous application of polyamines or certain polyamine analogs such as *sym*-norspermidine and N^1,N^{12}-bis(ethyl)spermine (Pösö and Pegg,1981; Porter and Bergeron, 1988). These analogs appear to act in the same way as the natural polyamines and in the case of N^1,N^{12}-bis(ethyl)spermine it has been shown that the translation of AdoMetDC mRNA is inhibited (unpublished observations).This analog may have potential as an antiproliferative agent since it causes a reduction in AdoMetDC and hence in polyamine content without itself substituting for the function of polyamines needed for cell growth (Porter and Bergeron, 1988)

REGULATION OF ADOMETDC ACTIVITY DURING CELL PROLIFERATION

Increases in AdoMetDC activity occur in a wide variety of systems in which growth is stimulated (reviewed by Pegg, 1984 and by White and Morris, 1988). These changes are due to increases in the amount of AdoMetDC protein and, in several cases, it has been shown that there is an enhanced rate of synthesis of the protein. This increase appears to be brought about both by a rise in the content of AdoMetDC mRNA and by an elevated rate of translation (Mach et al., 1986; Pegg et al., 1988c). The best studied experimental system is the activation of bovine lymphocytes with concanavalin A. In this system, there was a 4-fold rise in mRNA levels and a 10 fold increase in the rate of synthesis of the protein. The increase at the translational level was confirmed by demonstration of an increase in the average size of polysomes containing AdoMetDC mRNA (Mach et al., 1986; White and Morris, 1988). These results are therefore consistent with studies of the effects of changing polyamine content on the synthesis of AdoMetDC [described above] in showing that regulation occurs by means of changes in the content of the mRNA and in its translation.

PROPERTIES OF ADOMETDC MESSENGER RNA

In order to better understand the regulation of AdoMetDC, we have isolated cDNA clones for this enzyme from rat prostate and human fibroblasts (Pajunen et al., 1988). The inserts from these clones have been sequenced and used to probe the mRNA content from various cells and tissues. In most cases, the AdoMetDC mRNA detected in this way shows two sizes. The smaller mRNA has a size of about 2.1 kb in human and various rodent cells. The larger mRNA was about 3.4 kb in rat prostate and may be slightly larger (around 3.6 kb) in human cells (Pajunen et al., 1988). These results agree with those of Mach et al., (1986) who obtained a partial cDNA clone for bovine AdoMetDC and reported that the two mRNA species in bovine lymphocytes which were detected using this clone had sizes of 2.4 and 3.6 kb. It appears that the two different mRNA species correspond to the presence of multiple polyadenylation sites. Sequencing of the cDNAs corresponding to the shorter mRNA from human and rat cells indicates that the mRNA contains a long 3'- non-coding region of 557 nucleotides before the start of the poly(A) tail. The sequence AUUAAA serves a poly(A) addition signal (Pajunen et al., 1988). Sequencing of a cDNA corresponding to the longer mRNA of rat prostate is not yet complete but it is apparent that the 3' non-coding region consists of at least 1.8 kb and that the poly(A) signal is AAUAAA (unpublished observations).

The 3' non-coding regions of the smaller AdoMetDC mRNAs of rat and human show more than 85% sequence identity, which is only slightly less than the sequence identity in the coding sequence. Another interesting feature of the 3' sequence is that it contains an open reading frame which starts 9 bases after the termination codon for the AdoMetDC proenzyme. This open reading frame could code for a protein of 124 amino acids (human) or 125 amino acids (rat) and this polypeptide is also highly conserved between the two species but it is not known if the second open reading frame is actually translated *in vivo*.

The coding region of the AdoMetDC mRNA consists of 1002 nucleotides in the human and 999 nucleotides in the rat. The sequences are remarkably similar with only 52 base differences including the missing triplet (Pajunen et al., 1988).

The rat and human AdoMetDC mRNAs also contain a long 5' non-translated sequence. The presently available rat cDNA clones do not encompass this region but primer extension studies indicate that the mRNA contains about 320 nucleotides preceding the initiation codon and this is consistent with the length of the mRNA when the size of the coding and 3' non-translated regions are taken into account. The human cDNA clone sequence contains 248 nucleotides corresponding to the 5' non-translated sequence and thus provides the sequence of a substantial portion of this region even though it may not be quite full length (Pajunen et al., 1988). It is apparent therefore that AdoMetDC belongs to a very small class of mammalian proteins which have 5' leader sequences in their mRNAs of more than 200 nucleotides (Kozak, 1987). Such sequences may have important regulatory significance and it is noteworthy that AdoMetDC resembles ODC in belonging to this family. The similarity to ODC also extends to the 3' non-translated sequences since both enzymes also have long 3' non-translated sequences and have two forms of mRNA which result from multiple polyadenylation sites (Hickok et al., 1986).

Southern blot analysis indicates that there are multiple genes for AdoMetDC in rat and mouse cells (Pajunen et al., 1988). At present, it is not known whether more than one of these genes are active. Human cells appear to contain a more limited number of genes and sequences corresponding to the AdoMetDC were localized to chromosomes 6 and X (Radford et al., 1988).

AMINO ACID SEQUENCE OF ADOMETDC PROENZYME

At present, no direct amino acid sequence information is available for the rat and human AdoMetDC. However, Morris and colleagues were able to obtain some sequences from Lys-C digests of the bovine protein (Mach et al., 1986). These sequences are given in Figure 1 along with the complete amino acid sequences of the rat and the human AdoMetDC proenzymes as deduced from the open reading frame in the cDNAs (Pajunen et al., 1988). It is readily apparent that there is a remarkable similarity between the mammalian enzymes. All of the the sequences reported by Mach et al. (1986) from direct sequencing of the bovine enzyme, and some additional sequences deduced by them from the nucleotide sequence of partial cDNA, are present without change in the human sequence. It should be noted that this includes the sequence from amino acids 12 to 21. This sequence is so close to the amino terminus of the proenzyme that it must be present in the smaller subunit after cleavage and its presence in the digests of the purified bovine enzyme confirms the conclusion that the smaller polypeptide formed in the cleavage of the proenzyme remains as part of the enzyme. The human AdoMetDC proenzyme has 334 amino acids and a M.W. of 38,331. The rat counterpart contains only 333 residues and the sequence shows only 11 differences when compared to the human sequence. Even this underestimates the similarity between them since only 4 of the differences are non-conservative changes and the others include 6 conservative changes and a missing glutamine residue adjacent to the the carboxyl terminus.

```
Human   MEAAHFFEGTEKLLEVWFSRQQPDANQGSGNLRTIPRSEWDILLKDVQCS-   50
Rat     MEAAPFFEGTEKLLEVWFSRQQSDASQGSGNLRTIPRSEWDVLLKDVQCS-   50
Bovine            KLLEVWFSRQ

Human   IISVTKTDKQEAYVLSESSMFVSKRRFILKTCGTTLLLKALVPLLKLARD-  100
Rat     IISVTKTDKQEAYVLSESSMFVSKRRFILKTCGTTLLLKALVPLLKLARD-  100

Human   YSGFDSIQSFFYSRKNFMKPSHQGYPHRNFQEEIEFLNAIFPNGAGYCMG-  150
Rat     YSGFDSIQSFFYSRKNFMKPSHQGYPHRNFQEEIEFLNAIFPNGAAYCMG-  150

Human   RMNSDCWYLYTLDFPESRVISQPDQTLEILMSELDPAVMDQFYMKDGVTA-  200
Rat     RMNSDCWYLYTLDLPESRVINQPDQTLEILMSELDPAVMDQFYMKDGVTA-  200

Human   KDVTRESGIRDLIPGSVIDATMFNPCGYSMNGMKSDGTYWTIHITPEPEF-  250
Rat     KDVTRESGIRDLIPGSVIDATLFNPCGYSMNGMKSDGTYWTIHITPEPEF-  250
Bovine  KDVTRESGIRDLIPGSVIDATMFNPCGYSMNGMKSDGTYWTIHITPEPEF-

Human   SYVSFETNLSQTSYDDLIRKVVEVFKPGKFVTTLFVNQSSKCRTVLASPQ-  300
Rat     SYVSFETNLSQTSYDDLIRKVVEVFKPGKFVTTLFVNQSSKCRTVLSSPQ-  300
Bovine  SYVSFETNLSQTSYDD                          KC*TVL

Human   KIEGFKRLDCQSAMFNDYNFVFTSFAKKQQQQQS                   334
Rat     KIDGFKRLDCQSAMFNDYNFVFTSFAKKQQQQ-S                   333
Bovine  KIEGFK
```

Fig. 1. Complete amino acid sequences for the AdoMetDC proenzyme from human and rat and partial sequences for the bovine enzyme. The residues which differ are underlined. The rat and human sequences are derived from the cDNA sequence (Pajunen et al., 1988) and the bovine sequences are obtained from direct analysis of peptides obtained by digestion of the purified enzyme by Lys-C and by deduction from the nucleotide sequence of a partial cDNA clone (Mach et al., 1986). The lysine preceding the Lys-C peptides is also included in the sequences shown. The * indicates an unknown residue and the – a missing residue.

It is not yet known exactly where the cleavage site is in the proenzyme. It is highly probable that this site has a serine residue on the carboxyl side since the serine should then be converted to a pyruvate during the cleavage. This type of serinolysis reaction has been demonstrated for all of the bacterial enzymes which contain pyruvate in amide linkage (Recsei and Snell, 1984; Huynh and Snell, 1986; Tabor and Tabor, 1987; Dowhan and Li, 1988). The smaller sub-unit can be labeled with [^{35}S]cysteine when this is included in the reticulocyte lysate synthesis system so it must occur after residue 49. Based on the sizes of the sub-units as measured by polyacrylamide gel electrophoresis under denaturing conditions, the possible sites are those involving the serines at positions 50, 53, 66, 68 and 69. Experiments with site specific mutagenesis are under way to answer this question.

It is possible that AdoMetDC is subject to other forms of post-translational modification. No amino terminal amino acid could be detected when either the bovine enzyme (Mach et al., 1986) or the rat enzyme (unpublished observations) were subjected to Edman degradation procedures. The amino terminus of the large sub-unit is expected to be blocked by the pyruvate residue but the smaller sub-unit must be blocked by some other change. Also, the arginine residue at position 294 was present in the peptide sequenced by Mach et al. (1986) from the bovine enzyme as a residue which did not coincide with any of the normal amino acids but did not prevent the continued sequencing of this peptide. A modification of this arginine by methylation could account for this phenomenon.

The AdoMetDC proenzyme does contain a strong "PEST" sequence located at residues 243 to 269 which fall within the large subunit. It has been postulated that such sequences are signals for degradation in proteins with short half lives and similar sequences are found in an impressive number of these proteins including ODC (Rogers et al., 1986; Rechsteiner, 1988). Examination of the effects of modifications of this sequence on the turnover of AdoMetDC should permit a further test of this interesting hypothesis.

CONCLUSIONS AND FUTURE DIRECTIONS

AdoMetDC is a particularly interesting enzyme in a number of respects. These include: the regulation of its synthesis and degradation; the mechanism of formation of the pyruvate prosthetic group; the design of specific and potent inhibitors; and the mechanism of stimulation by putrescine. All of these aspects are now amenable to study using the available tools which include cDNAs, monospecific antisera and convenient systems in which the activity can be changed in response to regulatory controls. The limited availability of the purified enzyme which has restricted detailed structural studies can now be overcome by cloning a full length cDNA into a bacterial or yeast expression vector. Genomic clones, which are becoming available, could be used to study the transcription of the AdoMetDC mRNA and the possible effects of polyamines on this process. Site specific mutagenesis techniques can be used to study the processing of the the enzyme and the essential residues for activity and putrescine activation. Site specific mutagenesis and use of appropriate deletions and other constructions to alter the mRNA provide powerful tools to study both the regions of the mRNA which are important in the regulation of translation by polyamines and the features of the protein which result in the rapid turnover of AdoMetDC. The processing reaction by which the proenzyme is converted to the active subunits and the pyruvate group is generated may provide a particularly attractive step for the design of new inhibitors of AdoMetDC. Only a very limited number of pyruvoyl enzymes are known and it is likely that a highly specific inhibitor could be designed for this step when further details of the cleavage reaction become available.

ACKNOWLEDGEMENTS

This research was supported by grants CA 18138, HL-07223 and HD-13541 from the National Institutes of Health, Bethesda, MD, USA and by grants from the National Research Council for Natural Sciences, Academy of Finland and the Cancer Society of Finland.

REFERENCES

Alhonen-Hongisto, L., 1980, Regulation of S-adenosylmethionine decarboxylase by polyamines in Ehrlich ascites-carcinoma cells grown in culture, Biochem. J., 190: 747.

Artamonova, E. Y., Zavalova, L. L., Khomutov, R. M., and Khomutov, A. R., 1986, Irreversible inhibition of S-adenosylmethionine decarboxylase by hydroxylamine-containing analogues of decarboxylated S-adenosylmethionine, Biorg. Khim., 12: 206.

Dowhan, L., and Li, Q., 1988, Formation of the pyruvate residue of phosphatidylserine decarboxylase of E. coli, FASEB J., 46: Abst. 2349.

Hickok, N. J., Seppänen, P. J., Kontula, K. K., Jänne, P. A., Bardin, C. W., and Jänne, O. A., 1986, Two ornithine decarboxylase mRNA species in mouse kidney arise from size heterogeneity at their 3' termini, Proc. Natl. Acad. Sci. USA, 83: 594.
Huynh, Q. K., and Snell, E., 1986, Histidine decarboxylase from Lactobacillus 30a. Hydroxylamine clevages of the -seryl-seryl- bond at the activation site of prohistidine decarboxylase, J. Biol. Chem., 261: 1521.
Kameji, T., and Pegg, A. E., 1987a, Effect of putrescine on the synthesis of S-adenosylmethionine decarboxylase, Biochem. J., 243: 285.
Kameji, T., and Pegg, A. E., 1987b, Inhibition of translation of mRNAs for ornithine decarboxylase and S-adenosylmethionine decarboxylase by polyamines, J. Biol. Chem., 262: 2427.
Kolb, M., Danzin, C., Barth, J., and Calverie, N., 1982, Synthesis and biochemical properties of chemically stable product analogues of the reaction catalyzed by S-adenosylmethionine decarboxylase, J. Med. Chem., 25: 550.
Kozak, M., 1987, At least six nucleotides preceding the AUG initiator codon enhance translation in mammalian cells, J. Mol. Biol., 196: 947.
Mach, M., White, M. W., Neubauer, M., Degen, J. L., and Morris, D. R., 1986, Isolation of a cDNA clone encoding S-adenosylmethionine decarboxylase. Expression of the gene in mitogen-activated lymphocytes, J. Biol. Chem., 261: 11697.
Madhubala, R., Secrist III, J. A., and Pegg, A. E., 1988, Effect of inhibitors of S-adenosylmethionine decarboxylase on the contents of ornithine decarboxylase and S-adenosylmethionine decarboxylase in L1210 cells, Biochem. J., 254: 45.
Mamont, P., Joder-Ohlenbusch, A. M., Nussli, M., and Grove, J., 1981, Indirect evidence for a strict negative control of S-adenosylmethionine decarboxylase by spermidine in rat hepatoma cells, Biochem. J., 196: 411.
Pajunen, A., Crozat, A., Jänne, O. A., Ihalainen, R., Laitinen, P. H., Stanley, B., Madhubala, R., and Pegg, A. E., 1988, Structure and regulation of S-adenosylmethionine decarboxylase, J. Biol. Chem., in press.
Pegg, A. E., 1984, S-adenosylmethionine decarboxylase: a brief review, Cell Biochem. and Function, 2: 11.
Pegg, A. E., 1986, Recent advances in the biochemistry of polyamines in eukaryotes, Biochem. J., 234: 249.
Pegg, A. E., 1988, Polyamine metabolism and its importance in neoplastic growth and as a target for chemotherapy, Cancer Res., 48: 759.
Pegg, A. E., and Jacobs, G., 1983, Comparison of inhibitors of S-adenosylmethionine decarboxylase from different species, Biochem. J., 213: 495.
Pegg, A. E., and McCann, P. P., 1982, Polyamine metabolism and function, Am. J. Physiol., 243: C212.
Pegg, A. E., Wechter, R., Clark, R. S., Wiest, L., and Erwin, B. G., 1986, Acetylation of decarboxylated S-adenosylmethionine by mammalian cells, Biochemistry, 25: 379.
Pegg, A. E., Wechter, R., and Pajunen, A., 1987, Increase in S-adenosylmethionine decarboxylase activity in SV-3T3 cells treated with S-methyl-5'-methylthioadenosine, Biochem. J., 244: 49.
Pegg, A. E., Jones, D. B., and Secrist III, J. A., 1988a, Effect of inhibitors of S-adenosylmethionine decarboxylase on polyamine content and growth of L1210 cells, Biochemistry, 24: 1408.
Pegg, A. E., Wiest, L., and Pajunen, A., 1988b, Detection of proenzyme form of S-adenosylmethionine decarboxylase in extracts from rat prostate, Biochem. Biophys. Res. Commun, 150: 788.
Pegg, A. E., Kameji, T., Shirahata, A., Stanley, B., Madhubala, R., and Pajunen, A., 1988c, Regulation of mammalian S-adenosylmethionine decarboxylase, Advan. Enzyme Regul., 27: 43.
Porter, C. W., and Bergeron, R. J., 1988, Enzyme regulation as an approach to interference with polyamine biosynthesis- an alternative to enzyme inhibition, Advan. Enzyme Regul., 27: 57.
Pösö, H., and Pegg, A. E., 1981, Differences between tissues in response of S-adenosylmethionine decarboxylase to administration of polyamines, Biochem. J., 200: 629.
Pösö, H., and Pegg, A. E., 1982, Comparison of S-adenosylmethionine decarboxylases from rat liver and muscle, Biochemistry, 21: 3116.
Radford, D. M., Eddy, R., Haley, L., Henry, W. M., Pajunen, A., Pegg, A. E., and Shows, T. B., 1988, Gene sequences coding for S-adenosylmethionine decarboxylase

are present on chromosomes 6 and X and are not amplified in colon neoplasia, Cytogenet. Cell Genet., submitted.
Rechsteiner, M., 1988, Regulation of enzyme levels by proteolysis: the role of PEST regions, Advan. Enzyme Regul., 27: 135.
Recsei, P., and Snell, E., 1984, Pyruvoyl enzymes, Annu. Rev. Biochem.., 53: 357.
Rogers, S., Wells, S., and Rechsteiner, M., 1986, Amino acid sequences common to rapidly degraded protein: the PEST hypothesis, Science, 234: 364.
Shirahata, A., and Pegg, A. E., 1985, Regulation of S-adenosylmethionine decarboxylase activity in rat liver and prostate, J. Biol. Chem., 260: 9583.
Shirahata, A., and Pegg, A. E., 1986, Increased content of mRNA for a precursor of S-adenosylmethionine decarboxylase in rat prostate after treatment with 2-difluoromethionine, J. Biol. Chem., 261: 13833.
Shirahata, A., Christman, K., and Pegg, A. E., 1985, Quantitation of S-adenosylmethionine decarboxylase protein, Biochemistry, 24: 4417.
Tabor, C. W., and Tabor, C. H., 1984a, Polyamines, Annu. Rev. Biochem., 53: 749.
Tabor, C. W., and Tabor, H., 1984b, Methionine adenosyltransferase (S-adenosylmethionine synthetase) and S-adenosylmethionine decarboxylase, Advan. Enzymol. Related Areas Mol. Biol., 56: 251-282.
Tabor, C. W., and Tabor, H., 1987, The speEspeD operon of Escherichia coli. Formation and processing of a proenzyme form of S-adenosylmethionine decarboxylase, J. Biol. Chem., 262: 16037.
Wagner, J., Hirth, Y., Piriou, F., Zakett, D., Claverie, N., and Danzin, C., 1985, N-Acetyl decarboxylated S-adenosylmethionine, a new metabolite of decarboxylated S-àdenosylmethionine: isolation and characterization, Biochem. Biophys. Res. Commun., 133: 546-553.
White, M. W., and Morris, D. R., 1988, S-adenosylmethionine decarboxylase: genes and expression, In: "The Physiology of Polyamines," U. Bachrach, and Y. M. Heimer, Eds., CRC Press, Boca Raton, in press.
Williams-Ashman, H. G., and Pegg, A. E., 1981, Aminopropyl group transfers, In: "Polyamines in Biology and Medicine," D. R. Morris, and L. J. Marton, Eds., Marcel Dekker, New York, p. 407.

AMINOPROPYLTRANSFERASES: MECHANISTIC STUDIES AND THE SYNTHESIS OF SPECIFIC INHIBITORS

James K. Coward

Departments of Chemistry and Medicinal Chemistry
The University of Michigan
Ann Arbor, MI 48109, USA

We have studied the mechanism of enzyme-catalyzed alkyl transfer reactions by a variety of methods including non-enzymic model reactions, steady-state kinetics, and stereochemistry. In this brief paper, the results of our earlier work on catechol O-methyltransferase (COMT, E.C. 2.1.1.6) and more recent work on putrescine aminopropyltransferase (PAPT, E.C. 2.5.1.16, sometimes referred to as spermidine synthase) will be reviewed. These investigations have resulted in the demonstration that each of these alkyltransferases catalyze a reaction which proceeds *via* a ternary complex involving the enzyme and both nucleophilic and electrophilic substrates. In addition, the demonstration of general base catalysis in related model reactions, provides chemical precedent for the involvement of this type of catalysis by basic residues of the enzyme. The mechanistic conclusions have allowed us to design and synthesize several mechanism-based inhibitors which exhibit great potency and specificity for individual alkyltransferases[1].

The nature of the enzyme-substrate complex can be investigated by the use of kinetics and stereochemistry. We have studied COMT by both methods and have shown that the reaction proceeds *via* a single-displacement mechanism involving direct nucleophilic attack by the nucleophilic catecholamine on the electrophilic S-adenosylmethionine (AdoMet). This conclusion is supported by both steady-state kinetics[2] and stereochemistry[3] studies carried out with the enzyme isolated and partially purified from rat liver. In addition, the demonstration of general base catalysis in a series of intramolecular ring closure reactions has allowed us to propose a similar type of catalysis in the action of enzymes catalyzing alkyl transfer reactions[4]. Kinetics studies also revealed a strong inhibition of

COMT by one of the products, S-adenosylhomocysteine (AdoHcy)[2]. This has been the basis of a large effort in the synthesis of analogs of AdoHcy as direct inhibitors of methyltransferases and also the synthesis of purine nucleosides as inhibitors of S-adenosylhomocysteine hydrolase[5].

We have recently completed a similar series of investigations on the mechanism of enzyme-catalyzed aminopropyl transfer reactions. The study of the stereochemical course of the reaction catalyzed by *E. coli* PAPT required the synthesis of chirally deuteriated substrate, decarboxylated AdoMet (dcAdoMet), and the development of a ^1H NMR method for distinguishing between the two possible chiral spermidine products. The results obtained provide strong evidence that the PAPT-catalyzed reaction proceeds by a single-displacement mechanism much like that described above for COMT[6]. In the case of PAPT, the nucleophilic substrate is putrescine and the electrophilic substrate is dcAdoMet. A similar mechanism can be envisioned for the closely related enzyme, spermidine aminopropyl-transferase (SAPT, E.C. 2.1.1.22, otherwise known as spermine synthase). However, this enzyme is not found in prokaryotes and is not yet available in sufficient quantities from eukaryotes to enable us to carry out the type of NMR studies described for PAPT. It should be noted that with both enzymes discussed above (COMT and PAPT), the stereochemical investigations answered questions about the nature of the enzyme-substrate complex which were either ambiguous or in error based on steady-state kinetics studies.

As an extension of our earlier work on general base catalysis of the intramolecular cyclization of alcohols and phenols[4], we sought to investigate the possibility of general base catalysis in the intramolecular cyclization of amines[7]. The structures of the molecules synthesized for these studies are shown on the next page (Figure 1). In this work we have shown that the observation of general base catalysis in aqueous buffers is dependent on the pK_a of the nucleophile, the nucleofugality of the leaving group, and possibly Π-Π interactions between heterocycles such as imidazole and the aryl leaving group. Kinetics studies of the lyate reaction (not buffer-catalyzed) revealed that the reactivity of the intramolecular nucleophile ($\log k_{nuc}$) is linearly related to the pK_a of that nucleophile. The slope of the lines for the series of sulfonium salts (β_{nuc}=0.34) vs. the ammonium salts (β_{nuc}=0.49) showed the expected enhanced reactivity of the sulfonium salts such as AdoMet and dcAdoMet over the corresponding ammonium salts. While the data on general base catalysis obtained with the reactions of amines in the present work[7] are not as clear cut as those obtained in our previous work with alcohols and phenols[4], it is apparent that general catalysis of amine nucleophilic attack at sp^3 carbon can be observed in a limited number of cases.

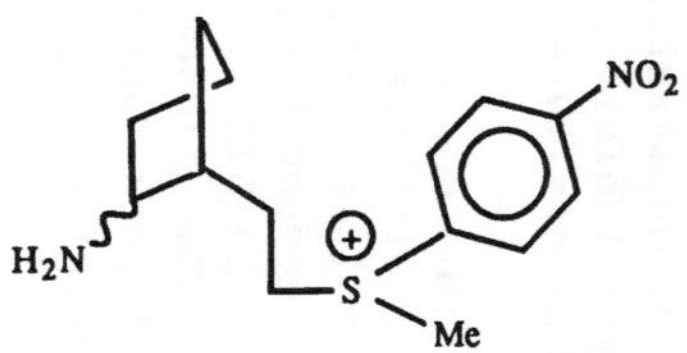

1

a, *cis*
b, *trans*

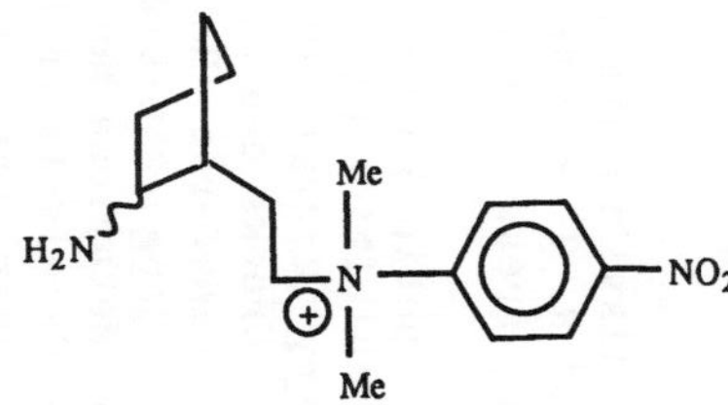

2

a, *cis*
b, *trans*

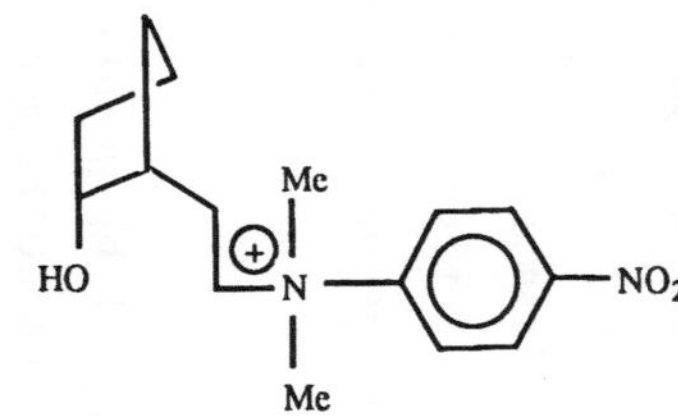

3

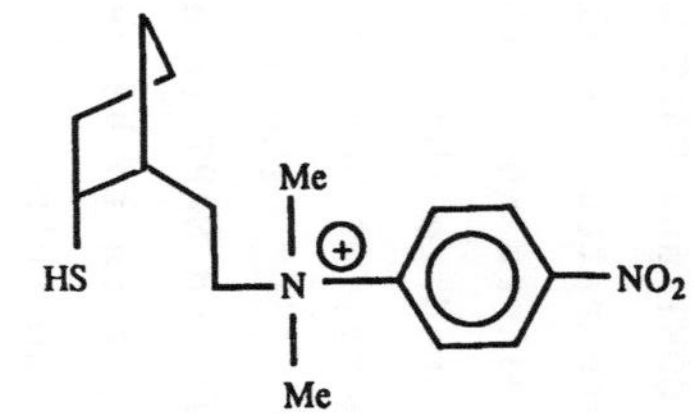

4

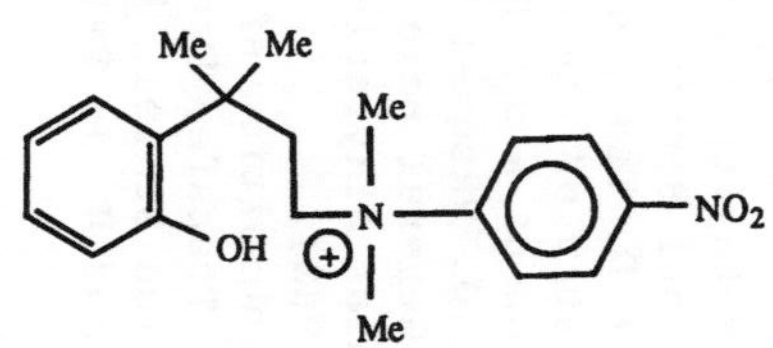

5

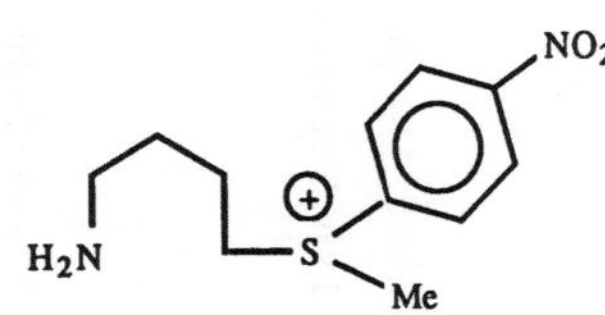

6

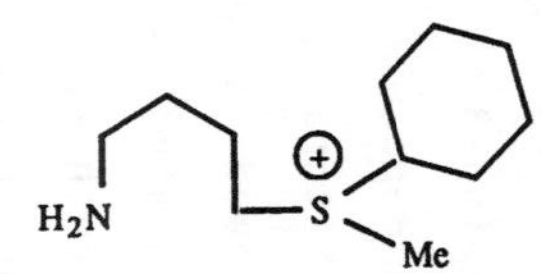

7

Figure 1. Structures of compounds studied in alkyltransferase model reactions.

The finding that the reactions catalyzed by COMT and PAPT both proceed *via* a ternary complex of enzyme and two substrates has allowed us to design and synthesize a series of specific inhibitors which incorporate both substrates in their structures. We have referred to these as "multisubstrate adduct inhibitors"[5]. In recent years, we have synthesized inhibitors of this type for PAPT[8], SAPT[9], and COMT[10,11]. The inhibitors of PAPT (S-adenosyl 1,8-diamino-3-thiooctane, AdoDATO) and SAPT (S-adenosyl 1,12-diamino-3-thio-9-azadodecane, AdoDATAD) have proven to be very specific and potent inhibitors of these enzymes in selected cell culture systems[12] and have been used in the study of polyamine biosynthesis *in vitro*[13]. The mechanistic studies described briefly in this paper have allowed the design and synthesis of specific and potent inhibitors of selected enzymes involved in the biosynthesis of polyamines. Their utility as specific drugs *in vitro* has been demonstrated. It remains to be seen if this high degree of selectivity between enzymes can be observed *in vivo*.

ACKNOWLEDGMENT

The research described in this paper was support by grants from the United States Public Health Service, NIH, GM30286, CA28097, and CA37606.

REFERENCES

1. J. K. Coward, Mechanism-Based Inhibitors of Alkyltransferases, in: "Biological Methylation and Drug Design," R. T. Borchardt, C. R. Creveling, and P. M. Ueland, eds., Humana Press, Clifton, New Jersey (1986).
2. J. K. Coward, E. P. Slisz, and F. Y-H. Wu, Kinetic Studies on Catechol O-Methyltransferase. Product Inhibition and the Nature of the Catechol Binding Site, Biochemistry 12:2291 (1973).
3. R. W. Woodard, M-D. Tsai, H. G. Floss, P. A. Crooks and J. K. Coward, Stereochemistry of the Catechol O-Methyltransferase Reaction: Demonstration of a S_N2 Reaction, J. Biol. Chem. 255:9124 (1980).
4. J. O. Knipe, P. J. Vasquez and J. K. Coward, Methylase Models: Studies on General-Base vs. Nucleophilic Catalysis in the Intramolecular Alkylation of Phenols, J. Amer. Chem. Soc. 104:3202 (1982).
5. J. K. Coward, Inhibition of Intracellular Methyl Transfer and Aminopropyl Transfer Reactions by Analogs of Substrates, Products and Transition States, in: "Design of Enzyme Inhibitors as Drugs," M. Sandler and H. J. Smith, eds., Oxford University Press, Oxford (1988).
6. G. R. Orr, D. W. Kullberg, G. Pontoni, P. C. Prabhakaran, S. J.Gould, and J. K. Coward, The Synthesis of Chirally Deuterated S-Adenosylmethylthiopropylamines and Spermidines. Elucidation of the Stereochemical Course of Putrescine Aminopropyltransferase (Spermidine Synthase), J. Amer. Chem. Soc. 110:5791 (1988).
7. R. J. Miller, A. Kuliopulos, and J. K. Coward, The Synthesis of p-Nitrothioanisoles and N-Methyl-p-Nitroanilines Containing Stereochemically Defined Neighboring Nucleophiles, J. Org. Chem. in preparation.

8. K- C. Tang, R. Mariuzza and J. K. Coward, The Synthesis and Evaluation of Some Stable Multisubstrate Adducts as Inhibitors of Spermidine Synthase, J. Med. Chem. 24:1277 (1981).
9. P. M. Woster, A. Y. Black, K. J. Duff, J. K. Coward, and A. E. Pegg, Synthesis and Biological Evaluation of S-Adenosyl-1,12-diamino-3-thio-9-azadodecane, A Multisubstrate Adduct Inhibitor of Spermine Synthase, J. Med. Chem. submitted.
10. G. L. Anderson, D. L. Bussolotti and J. K. Coward, The Synthesis and Evaluation of Some Stable Multisubstrate Adducts as Inhibitors of Catechol O-Methyltransferase, J. Med. Chem. 24:1271 (1981).
11. E. Yau and J. K. Coward, manuscript in preparation.
12. J. K. Coward and A. E. Pegg, Specific Multisubstrate Adduct Inhibitors of Aminopropyltransferases and Their Effect on Polyamine Biosynthesis in Cultured Cells, Adv. Enzyme Regul. 26:107 (1987).
13. A. E. Pegg, Polyamine Metabolism and Its Importance in Neoplastic Growth and as a Target for Chemotherapy, Cancer Res. 48:759 (1988).

APPROACHING THE STRUCTURES OF MAMMALIAN PROPYLAMINE TRANSFERASES AND THEIR GENES

Terho Eloranta, Olavi Kajander, Leila Kauppinen, Tapani Hyvönen, Annikka Linnala-Kankkunen, *Nisse Kalkkinen, Markku Kulomaa**, Leena Alhonen and Juhani Jänne

Department of Biochemistry, University of Kuopio, P.O.Box 6, SF-70211 Kuopio; *Department of Gene Technology, University of Helsinki, Valimotie 7, SF-00380 Helsinki and **Department of Biomedical Sciences, University of Tampere, P.O.Box 607, SF-33101 Tampere, Finland

INTRODUCTION

Propylamine transferases, spermidine synthase (EC 2.5.1.16) and spermine synthase (EC 2.5.1.22) catalyze the two final, irreversible steps in the conversion of arginine and methionine to spermidine and spermine, respectively. Both of the amino acid precursors are nutritionally essential for animals and are present in animal tissues at low, well-conserved concentrations. Arginine plays a central role in protein catabolism by being an essential intermediate in the conversion of toxic ammonia to excretable nontoxic urea. Methionine, on the other hand, is vitally important for protein synthesis by being the amino acid needed for peptide chain initiation and for the transmethylation reactions involved in the control of metabolism at several steps of transcriptional, translational and post-translational level. The fraction of the precursor amino acids consumed in polyamine synthesis varies a great deal depending on the cell type, the stage of differentiation and the proliferation rate. As to methionine consumption, the share of polyamine synthesis is less than 1 % in rat liver (Eloranta and Kajander, 1984) but may exceed 20 % in some cultured cells (Iizasa and Carson, 1985). Although detailed action mechanisms of polyamines in animal physiology are poorly known, their essentiality for cellular functions have been widely demonstrated (for references see Tabor and Tabor, 1984; Pegg, 1986).

Spermidine synthase has been purified to apparent homogeneity from a few mammalian tissues and its properties have been characterized (Samejima and Yamanoha, 1982; Yamanoha et al., 1984; Raina et al., 1984). Partial purification of the plant enzyme has also been reported (Hirasawa and Suzuki, 1983; Sindhu and Cohen, 1983). Bacterial enzyme was isolated and thoroughly characterized already 15 years ago (Bowman et al, 1973) and more recently it has been cloned and its full amino acid sequence has been deduced from the nucleotide sequence of the genome (Tabor and Tabor, 1987).

Spermine synthase has only been found in eukaryotic cells and was first isolated and characterized from bovine brain (Pajula et al., 1979). There is some indirect genetic evidence of the existance of spermine synthase in yeast (Cohn et al., 1980) but the production of spermine in some bacteria is likely to result from reactions catalyzed by spermidine synthase or other enzymes having poor substrate specificity (Tabor and Tabor, 1985).

Kinetic properties and substrate specificities of mammalian spermidine and spermine synthases have been well documented (Pegg et al., 1981; Samejima and Yamanoha, 1982; Pajula, 1983; Raina et al., 1984). Their distribution in rat tissues (Raina et al., 1976) as well as in human tissues and in some malignant human cell lines (Kajander et al., 1988) have been reported. Both enzymes are known to be dimers that do not need any cofactors for activity, have a long half-life and do not fluctuate to any great extent as a response to growth stimuli or treatment with drugs. The "constitutive" character of these enzymes have made polyamine researchers to underestimate their importance and, as a result from that, very little is known of their structural and regulatory details.

EXPERIMENTAL

Spermidine synthase was purified from bovine brain and spleen and from human spleen and spermine synthase from bovine brain and spleen and from human kidneys and placenta essentially according to the published methods (Samejima et al., 1983; Raina et al., 1983b) and following the procedure described in Fig. 1. Details of the purification, characterization and properties of the human enzymes will be published elsewhere (Kajander et al., unpublished). The enzyme activities were determined as previously described (Raina et al. 1976 and 1983a).

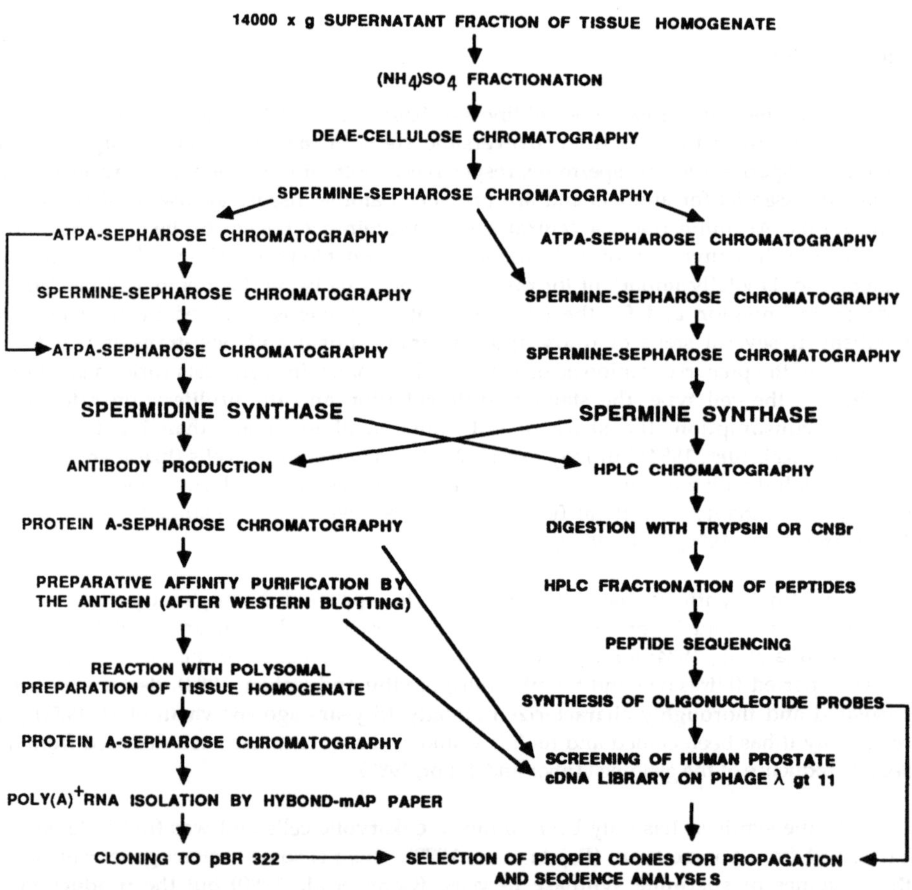

Fig. 1. Experimental procedures for solving the structures of spermidine and spermine synthases. ATPA; S-adenosyl-(5')-3'-thiopropylamine.

Polyclonal antibodies against human enzymes were prepared in mice and rabbits by conventional methods (Kajander et al., unpublished) and the antisera were further purified by protein A-Sepharose column chromatography as described by Lynch (1987). Antibodies against spermine synthase were also enriched by the preparative Western blotting procedure described by Olmsted (1981). The antibodies were then used to isolate spermine synthase - specific polysomes from bovine spleen and human placenta for mRNA isolation (Lynch, 1987; Aviv and Lederer, 1972) and cloning into the Eco RI site of pBR 322 (Kontula et al., 1984). Spermine synthase antibodies were also used to screen the human prostate cDNA library constructed in phage λ gt 11 (Gope et al., 1987).

Isolated spermidine and spermine synthases were subjected to high pressure liquid chromatography (HP 1090 Liquid Chromatograph, Hewlett Packard, Waldbronn, FRG) using C_4 and C_{18} reversed phase columns (Vydac) and elution with a linear gradient of 0.1 % trifluoroacetic acid to 68 % acetonitrile in 0.075 % trifluoroacetic acid over 66 min at a flow rate of 1 ml/min. Elution of protein was monitored at 218 nm, protein peaks were collected, lyofilized and treated with trypsin (Sigma) or CNBr (Eastman). Trypsin treatment was performed in 0.1 M $NaHCO_3$, pH 8.2, at 37°C for 4 hours using 3 µg of trypsin and 30-200 µg of enzyme protein in the total volume of 200 µl. The whole reaction mixture was then subjected to high pressure liquid chromatography as above, peptide fractions were collected, lyofilized, dissolved in H_2O and subjected to sequence analysis (Applied Biosystems 477 A Pulse Liquid Protein/Peptide Sequencer with 120 A Analyzer). Treatment with CNBr was performed overnight in 0.1 M HCl after which the mixture was evaporated to dryness under nitrogen flow, dissolved in H_2O with the aid of sonication and subjected to sequence analysis as above. Peptide sequences and the deduced gene sequences were compared with the data stored in the PIR data base of Georgetown university. A 20-mer oligonucleotide mixture, containing 48 chain possibilities, was deduced from a selected aminoacid sequence of human placental spermine synthase, synthesized and 5'-end labeled with polynucleotide kinase (Boehringer) and used as a probe in the screening the human prostate cDNA library constructed in phage λ gt 11. Similarly, a 23-mer oligonucleotide probe mixture, containing 128 chain possibilities, was constructed for bovine spleen spermidine synthase.

RESULTS AND DISCUSSION

Purification and antibody production

Spermidine synthase has been previously purified to apparent homogeneity from rat prostate (Samejima and Yamanoha, 1982), bovine brain (Raina et al., 1984) and rat and porcine liver (Yamanoha et al., 1984), whereas spermine synthase has only been purified from bovine brain (Pajula et al., 1979). In general, richest mammalian sources of spermidine synthase are rat pancreas, prostate and uterus (Raina et al., 1976), porcine and horse thymus (Pajula et al., unpublished) and human pancreas (Kajander et al., unpublished), whereas highest activities of spermine synthase can be found in porcine kidneys and bovine spleen, kidneys and thymus (Pajula et al., punpublished). Species and tissue distribution of these enzyme activities vary a great deal and the activity pattern and its age dependance reported for rat tissues (Raina et al., 1976) is not applicable to other mammals, e.g. rabbit, pig, bovine, sheep and horse (Pajula et. al., unpublished) or humans (Kajander et al., unpublished). Moreover, some tissues contain proteolytic activities that make them useless sources for isolation of propylamine transferases even though the activities in fresh tissue preparations might be promisingly high. While spermine synthase seems to be fairly stable in all studied mammalian and human tissue preparations, spermidine synthase is very labile in the pancreas preparations and rather unstable in the kidney and liver extracts of most mammalian species including human. Furthermore, while spermidine synthase activity is very stable in the homogenate of human spleen special care is needed to avoid a rapid loss of its activity in the preparations of bovine spleen.

For a simultaneous purification of both spermidine and spermine synthases, brain and thymus tissue of all mammalian species seems a good choice as a raw material. Since bovine tissues are easily available, we isolated both propylamine transferases initially from bovine brain and later also from bovine spleen. These preparations were apparently homogeneous proteins and good for the characterization of the kinetic and physical properties of the enzymes. However, all attempts to raise antibodies against them in mice failed. Apparently, the antigenic properties of the bovine enzymes were not strong enough to create significant response in rodents. We then turned to human tissues isolating spermidine synthase from frozen autopsy samples of spleens. To avoid contamination of the final preparation by spermine synthase, either gel filtration step (Samejima et al., 1983) or spermine-Sepharose step (Kajander et al., unpublished; see Fig. 1.) was found absolutely necessary. Spermidine synthase is not bound to spermine-Sepharose ,and can thus be easily released from any spermine synthase provided that the binding capacity of the column is not exceeded. For the latter reason, gel filtration can be considered a safer method to guarantee homogeneous final

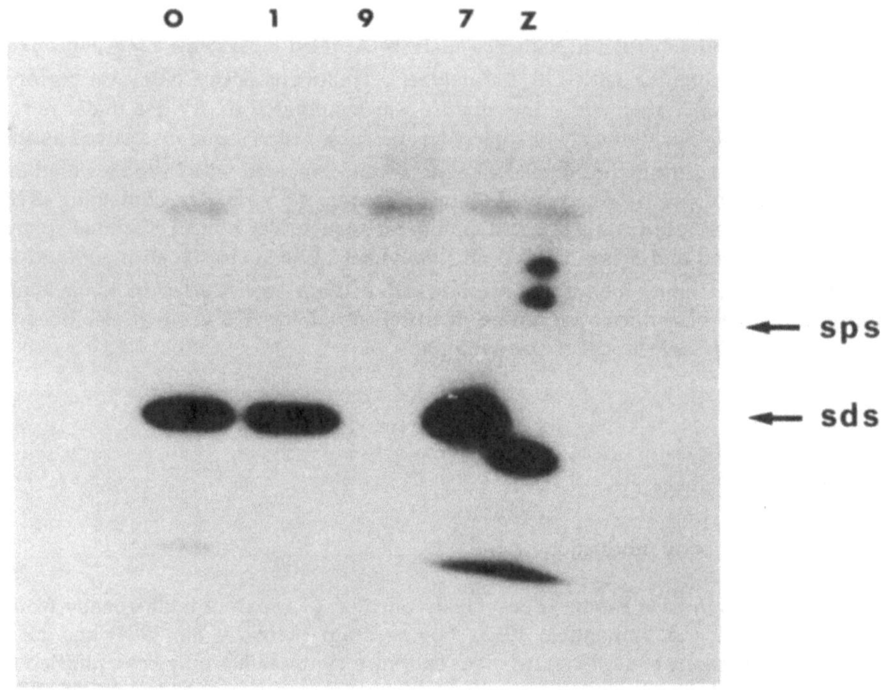

Fig. 2. Detection of spermidine synthase from bovine preparations by a mouse immune serum raised against the human spleen enzyme. Immunoblots of the enzyme preparations were prepared according to Billings et al. (1983), treated with antimouse IgG and I^{125}-labeled protein A, and developed by autoradiography. Slots 0 and 1 were loaded with partially purified mixtures of bovine spleen spermidine synthase (sds) and spermine synthase (sps), each containing 1 μg of protein. Slots 7 and 9 were loaded with 5-7-fold amounts of extensively purified spermidine and spermine synthases from the same source, respectively. Slot 7 was loaded with a 4-year old, totally autolyzed and inactivated preparation of bovine brain spermidine synthase.

preparation. We, however, got good results with adding the spermine-Sepharose step to the previously published purification procedure (Raina et al., 1984). Using this preparation, specific polyclonal antibodies were raised in mice without any difficulties and the antisera were satisfactory as such for detecting spermidine synthase in crude tissue preparations of all tested mammalian species. The antisera had no cross-reactivity with spermine synthase but sensitively recognized also the aggregates and some inactive degradation products of isolated bovine enzyme (Fig. 2) as well as the native dimeric forms of the enzyme (data not shown).

The purification procedure initially developed for the isolation of spermine synthase from brain tissue (Pajula et al., 1979) proved unsatisfactory in the case of human kidney and placenta. Both tissues, and especially the placenta, contain a variety of unspecified polyamine-binding proteins that are attached to spermine-Sepharose even more tightly than spermine synthase. Their total amount in the placenta greatly exceeds that of spermine synthase and interferes with the binding and elution of the latter enzyme at the affinity chromatographic step. Several successive spermine-Sepharose steps and careful adjustment of the chromatographic conditions are necessary to obtain apparently homogeneous protein (Kajander et al., unpublished). A better choice to ensure pure spermine synthase preparation is to use ATPA-Sepharose chromatography as an additional purification step (Fig. 1). Spermine synthase is even more firmly bound to ATPA-Sepharose than spermidine synthase, and it can not be recovered in equally high yields by any convenient concentration of decarboxylated adenosylmethionine in the elution buffer. The enzyme may partially become covalently linked to the affinity matrix, since it can not be totally eluted out of the column even by 1 M NaCl in 6 M urea.

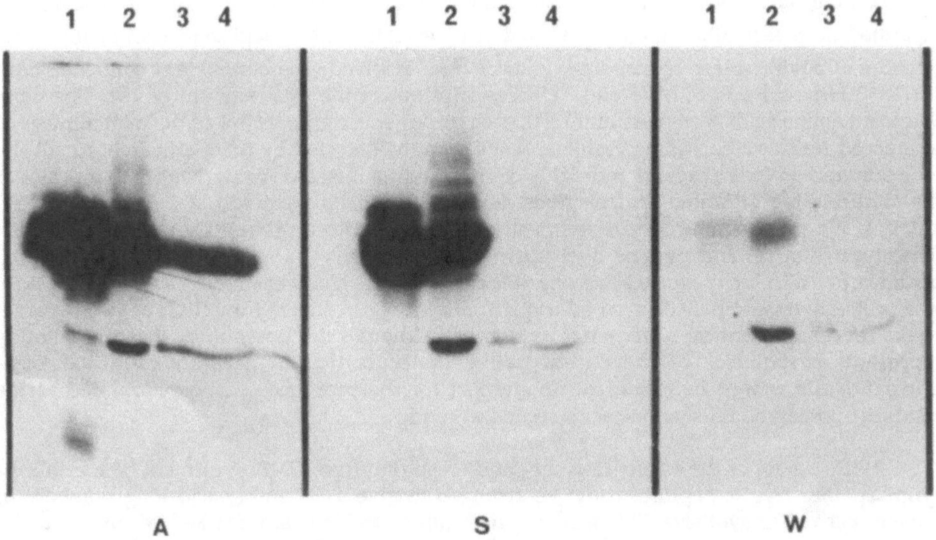

Fig. 3. Detection of spermine synthase by an immunoblotting technique. A; rabbit antiserum raised against the human kidney spermine synthase, S; the former preparation purified by protein A sepharose chromatography, W; preparation S further purified according to Olmsted (1981), 1; crude supernatant fraction of human placental homogenate (15 µg of protein), 2; the former preparation after purification by DEAE chromatography (7 µg of protein), 3; spermine synthase extensively purified from human placenta, 4; crude supernatant fraction of bovine spleen homogenate (36 µg of protein). For other details, see the legend for Fig. 2.

An apparently homogeneous preparation of human kidney spermine synthase, judged by SDS-PAGE with silver staining and also by HPLC chromatography, was used to raise polyclonal antibodies in rabbits. Although the response was good and the antisera revealed no cross-reactivity with human spermidine synthase, the antisera were apparently not specific to spermine synthase (Fig. 3). Significantly improved specificity was obtained by purifying the antiserum with the antigen attached to nitrocellulose by the Western blot technique. However, even this product still reacted with the unidentified proteins present in human tissue extracts. It thus appears that human spermine synthase shares common antigenic properties with some other human proteins. Apparently, such proteins are not dominating in bovine spleen (Fig. 3) enabling the use of the obtained antiserum for a fairly specific immunoprecipitation of spermine synthase in bovine tissue extracts. This was tested to isolate spermine synthase-specific polysomes from bovine spleen. A huge amount of chromosomal DNA released from spleen lymphocytes, however, made this approach unsucцesful.

Peptide mapping and amino acid sequence analysis

The purity of the propylamine transferase preparations used for tryptic digestion and amino acid sequence analysis was assured by subjecting them first to a reversed-phase HPLC. Both enzymes eluted from the C_4 and C_{18} columns as single peaks with clearly different retention times, spermine synthase requiring much higher hydrophobicity for elution than spermidine synthase. The elution profiles support the view that both enzymes are composed of identical, though distinct, subunits. Neither enzyme, whether purified from human or bovine tissues, gave any sequence data when applicated to the sequence analyser in amounts up to 4 nmol. This indicates that more than 98 % of the subunits of both enzymes have an N-terminal modification and do not contain any free amino groups at their N-terminal ends.

Taking advantage of the fact that many peptides overlapping in C_{18} HPLC could be separated from each other in C_4 HPLC, and vice versa, the amino acid sequences of 16 tryptic peptides of bovine spleen spermidine synthase were resolved. These sequences contained only 1 TRP, 3 HIS and 4 LYS, MET and THR residues out of the total amount of 152. The most abundant amino acids were VAL and LEU, each comprising roughly 10 % of the total number of sequenced residues. Excluding cysteine, that cannot be detected by the methods used, all the other amino acids were present in roughly equimolar amounts, each comprising 8-10 residues of the total number of amino acid residues detected. With the exception of a fairly high TYR amount, the present amino acid composition resemble those previously reported for HCl hydrolyzates of rat and porcine liver spermidine synthases (Yamanoha et al., 1984). In the peptides, one GLY-rich sequence and one SER-rich sequence could be found. One of the peptides was probably unspecificly cleaved by trypsin, since it was recovered in HPLC as two separate peaks revealing identical N-terminal sequences. Although the possibility that this peptide represents a sequence of the parent native molecule that is partially modified post-translationally cannot be excluded, no support for the presence of phosphorylated serine residues or other modified amino acids could be found.

With the aid of the sequence of a peptide resulting from CNBr-treatment of spermidine synthase, four tryptic peptides could be arranged to give a 33 amino acid long sequence of bovine spermidine synthase. This sequence does not resemble to any marked extent any of the known protein sequences stored in the PIR data base of the Georgetown University. No significant similarities were either found between the polynucleotide sequences stored in the PIR data base and all the possible template and antitemplate strands constructed for the solved portion of spermidine synthase. However, it must be emphasized that there are several regions in the bovine spleen spermidine synthase that reveal fairly good sequence homology with short segments of a variety of unrelated proteins. As an example, two identical sequence regions, one 5 amino acid residues and the other 6 amino acid residues in length, can be found in bovine spleen spermidine synthase and in the hypothetical 21K protein of Escherichia coli.

Somewhat surprisingly, there seems to be very little homology in the amino acid sequence or in the open-frame gene sequence of bovine spermidine synthase with its bacterial counterpart (Tabor et al., 1987). The inability of the polyclonal antibody raised against human spermidine synthase to recognize the bacterial enzyme also supports the view that mammalian and bacterial spermidine synthases are very different proteins. According to the best matching with the bacterial sequence, however, location of some of the tryptic peptides in the whole protein could be imagined (Fig. 4.). The longest resolved sequence of the bovine spleen enzyme seems to correspond to the C-terminal end of the bacterial one. Although mRNA sequences of bovine and bacterial enzymes are very different, there seems to be sufficient similarity in the properties of the unmatching amino acids to maintain portions of the protein molecules fairly identically folded. At least the hydrophilicity/hydrophobicity profiles of the middle and the C-terminal regions of the bacterial enzyme are very similar to the corresponding profiles of the bovine enzyme, as judged by the method of Hopp and Woods (1981). Whether this information can be used for localization of the centers involved in the reaction catalysis remains to be clarified.

The elution profile of the tryptic peptides of bovine spleen spermine synthase in C_{18} HPLC was totally different from that of spermidine synthase. All of the major peptide peaks of spermine synthase had retention times clearly separate from those of spermidine synthase. In the sequences of 24 tryptic peptides comprising the total of 247 amino acid residues, the most abundant were LEU (12,1 %), ASP (9,3 %), SER (8,5 %) and VAL (8,1 %). Only two residues of TRP, 4 residues of HIS and 6 residues of MET were found. Some of the resolved sequences are given in Table I. In general, the elution profiles of tryptic peptides of bovine and

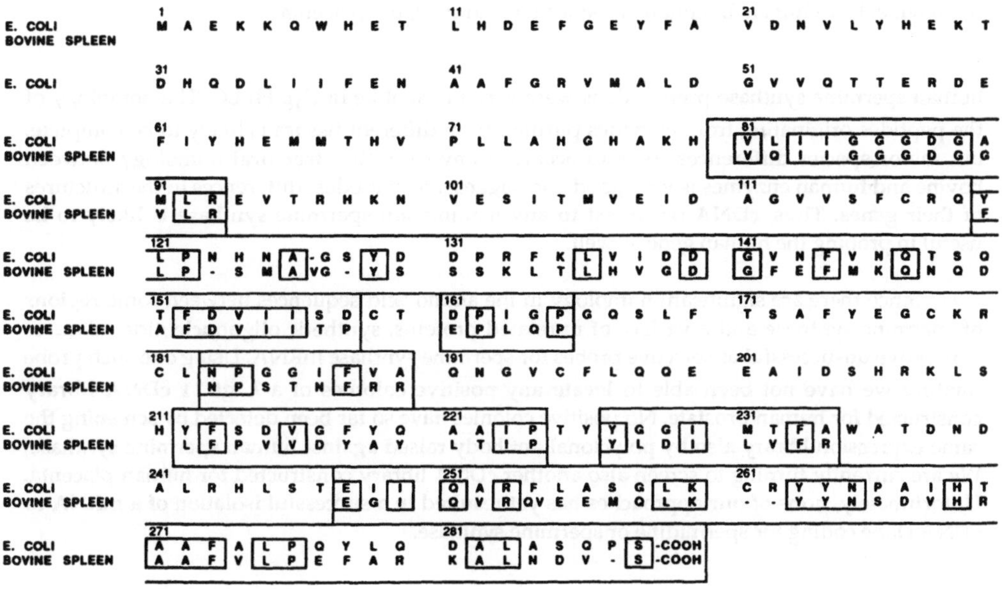

Fig. 4. Comparison and tentative localization of some resolved amino acid sequences of bovine spleen spermidine synthase with the known total sequence of the bacterial enzyme (Tabor and Tabor, 1987). The inner frames indicate locations of matching amino acids and the outer frames the regions of the bacterial enzyme that are compared with the resolved sequences of the bovine enzyme. The numbers refer to the location of amino acid residues in the sequence of the bacterial enzyme.

Table 1. Sequence comparison of some tryptic peptides of spermine synthase isolated from bovine and human tissues

Source	Sequence																				
	1	2	3	4	5	6	7	8	9	10	11	12	13	14	15	16	17	18	19	20	21
Bs, Hp:	A	D	G	E	T	I	L	K													
Bb, Bs, Hk, Hp:	Y	W	P	F	A	D	L	R													
Bp, Bs, Hp:	N	G	S	T	A	N	G	R													
Bp, Bs:	L	V	E	Y	D	I	D	K	V	V	Y	D	E	D	S	P	Y	Q			
Hk, Hp:	"	"	"	"	"	"	"	E	"	"	"	"	"	"	"	"	"	"	N	I	K
Hp:	G	L	Q	S	I	F	Q	E	Q	G	M	A	E	S	V	H	T	W	Q		
Bs:	"	"	"	?	"	"	"	"	"	"	"	T	"	"	"	"	"	"	"		
												Q									
Hp:	E	F	D	Y	V	I	N	D	L	T	A	V	P	I							
Bs:	?	"	?	"	"	"	"	?	"	"	"	"	"	"							

B = Bovine, H = Human, b = brain, s = spleen, k = kidney, p = placenta

human spermine synthase preparations were very much alike in C_{18} HPLC. The homology of the peptides originating from enzymes purified from different tissues is likely to be complete, but minor species differences seem to occur. In any case, the structural homology between bovine and human enzymes is very good and suggests only modest differences in the structures of their genes. Thus, cDNA produced to any mammalian spermine synthase is likely to be useful in probing the human gene as well.

Since there are significant homology in the amino acid sequences between some regions of spermine synthase and a variety of unrelated proteins, synthetic oligonucleotide mixtures may prove unsuccessful as selective probes for spermine synthase mRNA. Using one such probe mixture we have not been able to locate any positive colonies of a λ gt 11 cDNA library constructed for human prostate. No positive colonies have so far been detected by screening the same expression library also by polyclonal antibody raised against human spermine synthase. We are currently turning to screen also another cDNA library constructed for human placenta. Unfortunately, none of our approaches has yet resulted in a successful isolation of a mRNA or cDNA clone coding for spermidine or spermine synthase.

ACKNOWLEDGEMENTS

These studies are supported by the National Research Councils for Natural and Medical Sciences, Finland. The technical assistance of Miss Eija Korhonen is gratefully recognized.

REFERENCES

Aviv, H., and Lederer, P., 1972. Purification of biologically active globin messenger RNA by chromatography on oligothymidylic acid-cellulose, Proc. Natl. Acad. Sci USA, 69: 1408.

Billings, P.B., Hoch, S.O., White, P.J., Carson, D.A., and Vaugham, J.H., 1983, Antibodies to the Epstein-Barr virus nuclear antigen and to rheumatoid arthritis nuclear antigen identify the same polypeptide, Proc. Natl. Acad. Sci U.S.A., 80: 7104.

Bowman, W.H., Tabor, C.W., and Tabor, H., 1973, Spermidine biosynthesis. Purification and properties of propylamine transferase from Escherichia coli, J. Biol. Chem., 248: 2480.

Cohn, M.S., Tabor, C.W., and Tabor, H., 1980, Regulatory mutations affecting ornithine decarboxylase activity in Saccharomyces cerevisiae, J. Bacteriol., 142: 791.

Eloranta, T.O., and Kajander, E.O., 1984, Catabolism and lability of S-adenosyl-L-methionine in rat liver extracts, Biochem. J., 224: 137.

Gope, M.L., Keinänen, R.A., Kristo, P.A., Conneely, O.M., Beattie, W.G., Zaruchi-Schulz, T., O'Malley, B.W., and Kulomaa, M.S., 1987, Molecular cloning of the chicken avidin cDNA, Nucl. Acids Res., 15: 3595.

Hirasawa, E., and Suzuki, Y., 1983, Biosynthesis of spermidine in maize seedlings, Phytochem., 22: 103.

Hopp, T., and Woods, K.R., 1981, Prediction of protein antigenic determinants from amino acid sequences, Proc. Nat.. Acad. Sci U.S.A., 78: 3824.

Iizasa, T., and Carson, D.A., 1985, Differential regulation of polyamine synthesis and transmethylation reactions in methylthioadenosine phosphorylase deficient mammalian cells, Biochem. Biophys. Acta, 844: 280.

Kajander, E.O., Kauppinen, L.I., Pajula, R.-L., Kuronen, I., Eloranta, T.O., and Collan, Y., 1988, Polyamine synthesizing enzymes in human malignancies, in: "Proceedings of the 3rd International Congress on Hormones and Cancer", Pergamon Press, New York.

Kontula, K.K., Torkkeli, T.K., Bardin, C.W., and Jänne, O.A., 1984, Androgen induction of ornithine decarboxylase mRNA in mouse kidney as studied by complementary DNA, Proc. Natl. Acad. Sci USA, 81: 731.

Lynch, D.C., 1987, Use of antibodies to obtain specific polysomes, Methods Enzymol., 152: 248.

Olmsted, J.B., 1981, Affinity purification of antibodies from diazotized paper blots of heterogenous protein samples, J. Biol. Chem., 256: 11955.

Pajula, R.-L., 1983, Kinetic properties of spermine synthase from bovine brain, Biochem. J., 215: 669.

Pajula, R.-L., Raina, A., and Eloranta, T., 1979, Polyamine synthesis in mammalian tissues. Isolation and characterization of spermine synthase from bovine brain, Eur. J. Biochem., 101: 619.

Pegg, A.E., 1986, Recent advances in the biochemistry of polyamines in eukaryotes, Biochem. J., 234: 249.

Pegg, A.E., Shuttleworth, K., and Hibasami, H., 1981, Specificity of mammalian spermidine synthase and spermine synthase, Biochem. J., 197: 315.

Raina, A., Pajula, R.-L., and Eloranta, T., 1976, A rapid assay method for spermidine and spermine synthases. Distribution of polyamine synthesizing enzymes and methionine adenosyltransferase in rat tissues, FEBS Lett., 67: 252.

Raina, A., Eloranta, T., and Pajula, R.-L., 1983a, Assays for putrescine aminopropyltransferase (spermidine synthase) and spermidine aminopropyltransferase (spermine synthase), Methods Enzymol., 94: 257.

Raina, A., Pajula, R.-L., and Eloranta, T., 1983b, Purification of spermidine aminopropyltransferase (spermine synthase) from bovine brain, Methods Enzymol., 94: 276.

Raina, A., Hyvönen, T., Eloranta, T., Voutilainen, M., Samejima, K., and Yamanoha, B., 1984, Polyamine synthesis in mammalian tissues. Isolation and characterization of spermidine synthase from bovine brain, Biochem. J., 219: 991.

Samejima, K., and Yamanoha, B., 1982, Purification of spermidine synthase from rat ventral prostate by affinity chromatography on immobilized S-adenosyl-(5')-3-thiopropylamine, Arch. Biochem. Biophys., 216: 213.

Samejima, K., Raina, A., Yamanoha, B., and Eloranta, T., 1983, Purification of putrescine aminopropyltransferase (spermidine synthase) from eukaryotic tissues, Methods Enzymol., 94: 270.

Sindhu, R.K., and Cohen, S.S., 1983, Putrescine aminopropyltransferase (spermidine synthase) of chinese cabbage, Methods Enzymol., 94: 279.

Tabor, C.W., and Tabor, H., 1985,Polyamines in microorganisms, Microbiol. Rev., 49: 81.

Tabor, C.W., and Tabor, H., 1984, Polyamines, Annu. Rev. Biochem., 53: 749.

Tabor, C.W., and Tabor, H., 1987, The speEspeD operon of Escherichia coli. Formation and processing of a proenzyme form of S-adenosylmethionine decarboxylase, J. Biol. Chem., 262: 16037.

Yamanoha, B., Samejima, K., Nakajima, T., and Yasuhara, T., 1984, Differences between homogeneous spermidine synthases isolated from rat and pig liver, J. Biochem., 96: 1273.

POTENTIAL ROLES OF POLYAMINE INTERCONVERSION IN THE MAMMALIAN ORGANISM

Nikolaus Seiler

Merrell Dow Research Institute, 16, rue d'Ankara
67084 Strasbourg Cedex, France

INTRODUCTION

Presumably many cell types have an enzymatic machinery suitable for the degradation of spermine to spermidine, and of spermidine to putrescine, but systematic studies of polyamine catabolism have not been carried out. In the case of vertebrates it is now firmly established that N^1-acetylation is the first step in this degradative transformation of one polyamine into the other. This reaction may be catalyzed either by the basal or the induced form of acetylCoA:spermidine/spermine N^1-acetyltransferase (cSAT), both of which are cytosolic enzymes (Persson et al., 1985). In the following step the N^1-acetylpolyamines are oxidatively cleaved to form spermidine and putrescine, respectively (Seiler, 1981), an oxygen requiring reaction which is catalyzed by the flavine enzyme polyamine oxidase (PAO) (Hölttä, 1977).

The pools of putrescine that are supplied by decarboxylation of ornithine (Orn), and those pools which derive from the oxidative cleavage of N^1-acetylspermidine, are in equilibrium. At least, there is no evidence against this notion, as far as vertebrates are concerned, to which the following considerations are limited. Analogous considerations are also valid for spermidine. Therefore, it seems justified to formulate polyamine metabolism as a cyclic process (Fig. 1) (Seiler et al., 1985a). Direct evidence in favor of the coupling of synthetic and catabolic reactions to a cyclic process came from stud-

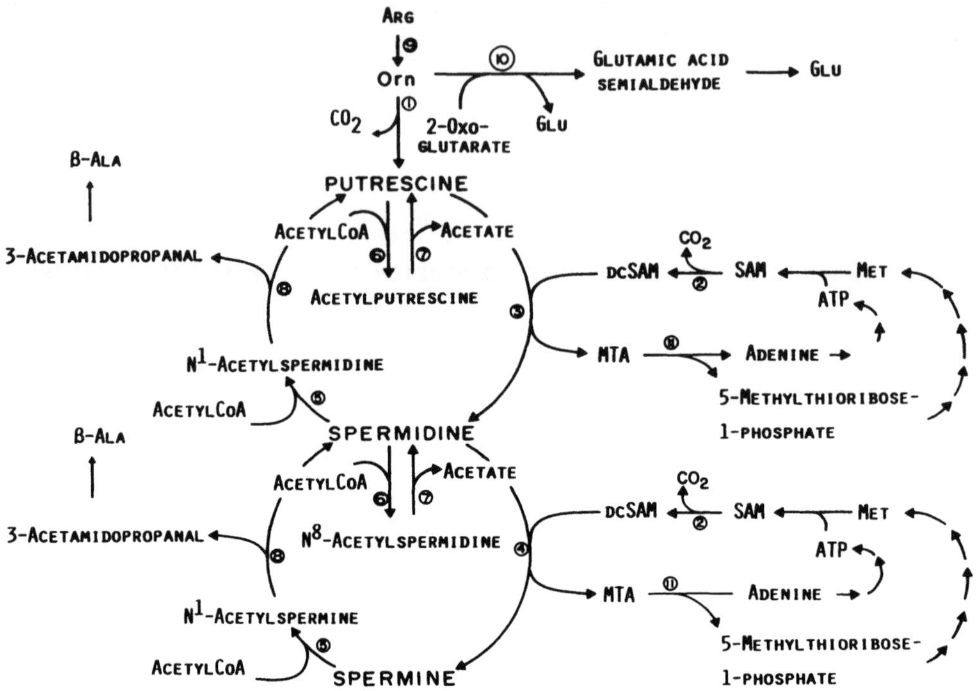

Fig. 1. The polyamine metabolic cycle.

(1) L-Ornithine decarboxylase (ODC) ; (2) S-Adenosyl-L-methionine decarboxylase (SAMDC) ; (3) Spermidine synthase ; (4) Spermine synthase ; (5) AcetylCoA:spermidine/spermine N^1-acetyltransferase (cytosolic) (cSAT) ; (6) AcetylCoA:spermidine N^8-acetyltransferase ; (7) N^8-Acetylspermidine deacetylase (cytosolic) ; (8) Polyamine oxidase (flavin dependent) (PAO) ; (9) Arginase ; (10) L-ornithine:2-oxoglutarate aminotransferase (OAT) ; (11) 5'-Deoxy-5'-methylthioadenosine phosphorylase.

Arg = L-arginine ; Orn = L-ornithine ; Glu = glutamic acid Met = methionine ; β-Ala = β-alanine ; SAM = S-adenosyl-L-methionine ; dcSAM = decarboxylated S-adenosyl-L-methionine (S-3-aminopropyl-5'-methylthioadenosine) ; MTA = 5'-deoxy-5'-methylthioadenosine.

ies of the metabolism of labelled polyamines in mouse and fish brains. They suggested a much longer biological half-life of the putrescine moiety of spermidine and spermine, than of the aminopropyl moiety, which derives from methionine (Seiler et al., 1979 ; Antrup and Seiler, 1980). Thus the reutilisation of the spermidine derived putrescine for de novo spermidine synthesis became evident. In the cyclic process, as formulated in Fig. 1, the methionine (Met)-derived decarboxylated S-adenosylmethionine (dcSAM) is used as source of the aminopropyl moieties, which are attached to putrescine and spermidine respectively. With each spermidine and spermine molecule, one molecule of 5'-deoxy-5'-methylthioadenosine (MTA) is formed. It can be salvaged to Met and ATP (Williams-Ashman, 1985). The aminopropyl moieties are removed in the catabolic branch of the polyamine cycle in the form of 3-acetamidopropanal, which can be transformed to β-alanine. This may undergo further metabolism, for example degradation by transamination or conjugation to form carnosine. In any case is the Met-derived aminopropyl moiety irreversibly eliminated from the polyamine metabolic cycle, whereas the putrescine moiety can be conserved.

The polyamine interconversion cycle appears to be a futile metabolic pathway because it transforms in an energy-requiring sequence Met into MTA and β-alanine. Both these compounds have no functions that can be interrelated with polyamine metabolism.

From an evolutionary point of view, a futile Met-metabolizing pathway, which presumably exists in all cells of the vertebrates, i.e. in the organisms of the highest evolutionary development, seems highly unlikely. It is, therefore, important to find explanations for its functional roles.

Based on some simple assumptions, it seems possible to assign functions to this central scheme of polyamine metabolism, functions which in essence are interrelated with the regulation of polyamine concentrations in individual cells, in organs, and a whole organism.

RESULTS AND DISCUSSION

Regulatory sites of the polyamine metabolic cycle

Three inducible enzymes are involved in the control of polyamine metabolism. The two decarboxylases, L-ornithine decarboxylase (ODC ; E.C. 4.1.1.17) and S-adenosyl-L-methionine decarboxylase (SAM-DC ;

E.C. 4.1.1.50) have been very extensively studied, and their regulation at transcriptional, translational and posttranslational levels is the topic of very numerous papers (for reviews, see Pegg, 1986 ; Seiler and Heby, 1988). As has first been pointed out by Pegg and his collaborators (1981) cSAT induction is a further means of affecting polyamine concentrations. As is the case with ODC and SAMDC, obviously a great variety of factors are capable of inducing cSAT (Seiler, 1987a). Among these the free polyamines are presumably most important (Pegg et al., 1986a). Increase of the concentration of the free polyamines does not necessarily imply an increase of total cellular polyamines. One possibility is the displacement of the polyamines from their binding sites by analogues. However, it seems that there is no unambiguous example for polyamine displacement known at present, although it is likely that polyamine depletion by N^1,N^8-bis-alkyl spermidines (Porter et al., 1985) may be achieved by this mechanism. Another possibility is the decrease in the number of binding sites. The starvation-induced protein and RNA losses in liver, which are accompanied by a proportional loss of spermidine, is presumably the most impressive physiologically relevant example. During starvation the increase of cSAT activity and the excessive formation of N^1-acetylspermidine and N^1-acetylspermine indicate enhanced polyamine degradation (Seiler et al., 1981a).

In a closed metabolic system polyamine concentrations can be enhanced by increasing the activity of the two decarboxylases. The aminopropyltransferases are usually present in excess. In rapidly growing cells, the polyamine pool is divided during cell division among the two daughter cells. Diminution of the polyamine concentration by specific processes seems, therefore, not required and plays a minor role from a quantitative point of view (McCormick, 1978). In non-growing tissues, however, processes alternative to diminution by cell division are needed.

Enhancement of cSAT activity is essential for the active diminution of cellular polyamines. However, assuming that the amount of the synthetic enzymes remains at basal levels, cSAT induction is not a sufficient means for polyamine depletion, because putrescine will be produced from spermidine, which may be reutilized for spermidine synthesis. Enhancement of putrescine formation causes at the same time the activation of SAM-DC, which is present in excess of ODC in most tissues (Raina et al., 1978), but is presumably not fully activated at the low (basal) putrescine concentrations of non-proliferating tissues (Ientile et al., 1985). Since spermidine concentrations are usually high, putrescine and spermidine compete for the available dcSAM and spermidine and spermine are formed at the same time at an enhanced rate. (It should be kept in mind that only the free polyamines are

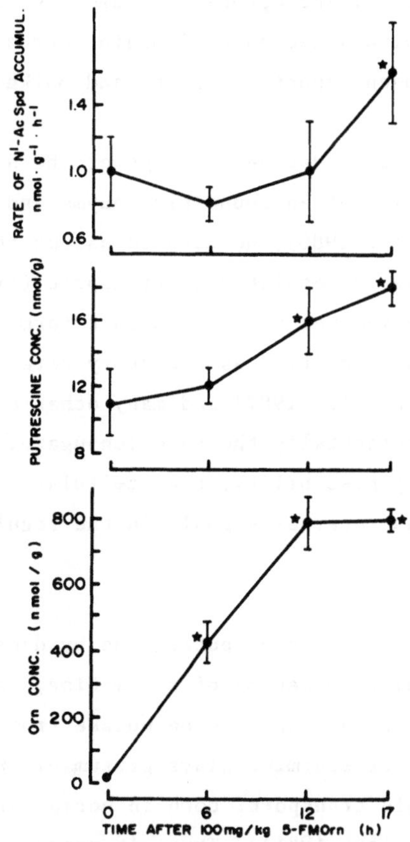

Fig. 2. Enhancement of spermidine turnover in mouse brain due to elevation of putrescine concentration.

Groups of four CD1 albino mice received a single i.p. dose of 5-fluoromethylornithine (5FMOrn), a specific inactivator of L-ornithine:2-oxoglutarate aminotransferase (Daune et al., 1988). At different times after 5FMOrn administration, N,N'-bis-allenylputrescine was administered (50 mg/kg, i.p.) and the rate of N^1-acetylspermidine accumulation in brain was determined during the following 5 h as was described previously (Seiler and Bolkenius, 1985).

An asterisk indicates a statistically significant difference ($p < 0.05$) between treated and control groups.

aminopropylated. Owing to the tighter binding to anionic sites of spermidine, in comparison with putrescine (Bloomfield and Wilson, 1981), putrescine needs not to be present in amounts comparable to those of spermidine, for effective competition for the available dcSAM). As a net effect, the enhancement of polyamine turnover will be observed. This has been shown to occur in mouse brain, when putrescine concentrations were elevated by specific inactivation of ornithine aminotransferase (Fig. 2). The opposite, a decrease of spermidine turnover concomitant with a decrease of brain putrescine concentrations has previously been reported (Seiler and Bolkenius, 1985).

How is the extent of oxidative cleavage of the N^1-acetylpolyamines regulated ? Since PAO is not an inducible enzyme and usually present in excess (Seiler et al., 1980), we have to assume that the control of the extent of splitting is regulated by transport (Fig. 3). This implies that N^1-acetylpolyamines are a transport form. Indeed, N^1-acetylspermidine has been identified as excretory form of baby hamster kidney cells (Wallace et al., 1987) and many other cell types are likely to excrete preferentially the same conjugate. It is presently not known, but a likely possibility, that cellular N^1-acetylpolyamine concentrations serve as internal signals in the regulation of their transport.

The intracellular formation of putrescine by decarboxylation of Orn, or the intracellular formation of spermidine, can in principle be substituted by putrescine and spermidine uptake from the environment. (Elimination or uptake of spermine plays presumably no major role because of its unfavorable transport, both in normal and transformed cells (Bachrach and Seiler, 1981)). Thus, it appears that inward transport is another presumptive regulatory site for the control of intracellular polyamine concentrations. Systematic studies are not yet available, but from the existing preliminary data it is presumed that the N-acetylated polyamines are mainly substrates of polyamine export, whereas putrescine and spermidine may be transported in both ways, as is formulated in Fig. 3.

It is presently not known whether the various polyamine regulatory sites are independently controlled, and if we disregard from the mentioned role of the free polyamines in the regulation of total polyamines, we are ignorant of the factors that enhance or slow down

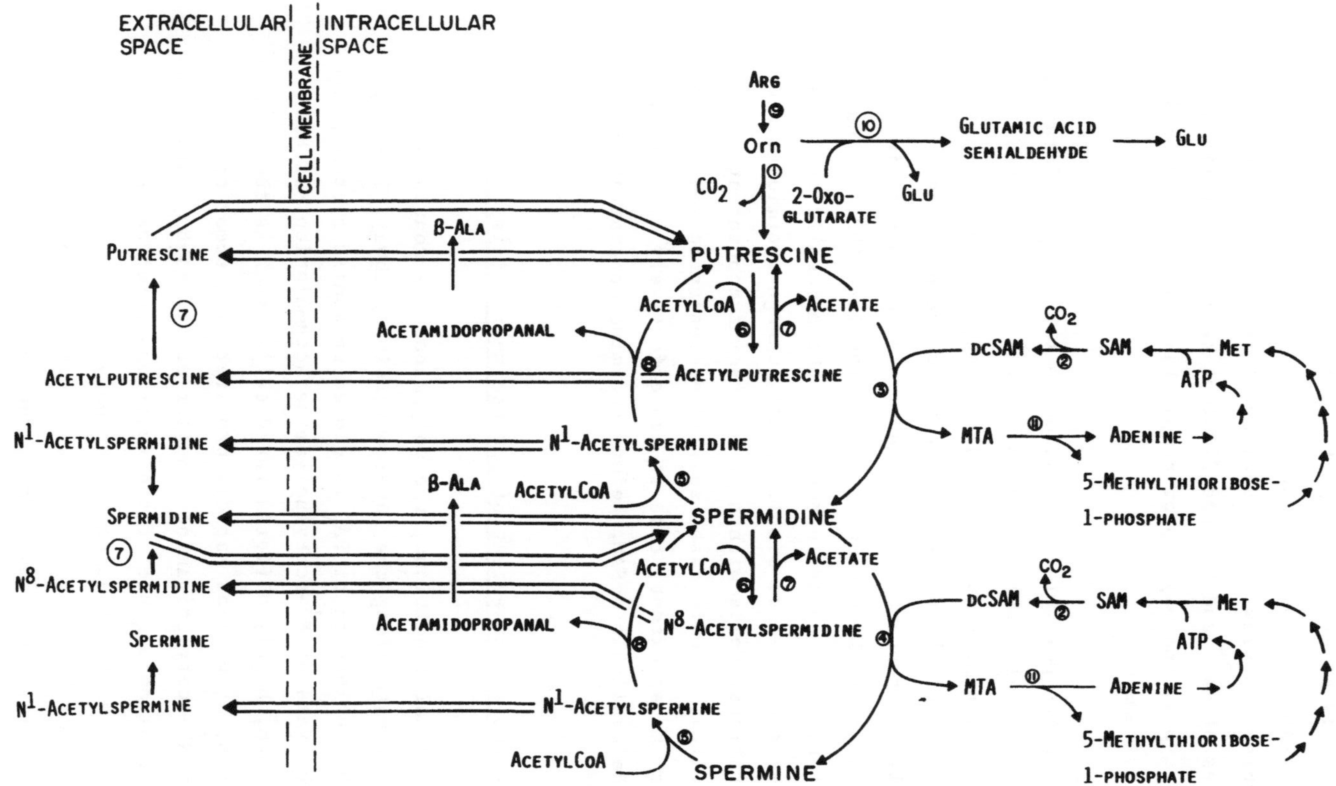

Fig. 3. Interrelations between the polyamine metabolic cycle and cellular polyamine transport. For numbering and abbreviations, see legend to Fig. 1. For the sake of simplicity the reactions involved in terminal polyamine catabolism have been omitted from this scheme. It should, however, be kept in mind that all intermediates of the polyamine metabolic cycle can in principle be oxidatively deaminated and transformed into an amino acid or its lactam and eliminated in this form as well (Seiler, 1987b).

polyamine turnover, polyamine uptake and polyamine elimination under physiological conditions.

Regulation of polyamines by transport is in close analogy to the regulation of inorganic cations. In their case, transport is the major regulatory mechanism. In view of the variety of transport systems and regulatory devices of Ca^{2+} and related divalent cations in different tissues and cell types, one may expect comparably heterogenous polyamine transport systems.

Differences between polyamine regulation in rapidly proliferating cells (tumor cells or embryonal cells) and highly differentiated, non-dividing cells have previously been discussed, and considerations of the potential role of the polyamine metabolic cycle in the rapid adjustment of cellular concentrations to physiological needs, in spite of a slow basal turnover, have been published (Seiler, 1987a). Therefore, it may be sufficient to state here that the metabolic cycle is a means that enables the cells to simultaneously change spermidine and spermine concentrations. Cells with high putrescine concentrations lack this possibility. In this case, spermidine concentration has to be increased first to an extent, that it can effectively compete with putrescine for the available dcSAM, before spermine concentrations are elevated.

<u>Basal polyamine turnover and extent of putrescine reutilisation</u>

Under physiological conditions, the polyamines are continuously degraded at a basal rate along the interconversion pathway. This can be demonstrated by inhibition of PAO in closed metabolic systems, and measurement of the rate of N^1-acetylpolyamine accumulation. It has been shown that both, highly differentiated cells, such as the cells of the mature brain, as well as rapidly proliferating tumor cells produce continuously N^1-acetylpolyamines (Seiler et al., 1985). This demonstrates the general occurrence of the interconversion reaction in mammalian cells. If the elimination of the polyamines from a metabolic system by transport or by irreversible catabolic reactions is small in comparison with the interconversion rate, the rate of N^1-acetylspermidine accumulation is identical with spermidine turnover rate. In Fig. 2, it is shown that the basal spermidine turnover in mouse brain is about 1 $nmol.g^{-1}.h^{-1}$.

Estimates of the extent of putrescine reutilisation are again most easily obtained in closed metabolic systems, as follows from Fig. 1. Inhibition of PAO will produce a decrease of the putrescine pool which is proportional to the quantitative significance of the interconversion reaction sequence in comparison with that of Orn decarboxylation. Quantitative data have been obtained for mouse brain, that can be considered in a first approximation a closed system, as far as polyamine metabolism is concerned. In this case, about 70 % of the putrescine is formed by degradation of N^1-acetylspermidine, and only about 30 % by decarboxylation of Orn (Seiler et al., 1985a). In embryonal brain, with a high rate of de novo polyamine biosynthesis, the proportion of putrescine from the interconversion pathway is small. It gains gradually importance with functional maturation due to the decrease of ODC activity (Bolkenius and Seiler, 1986).

In an intact animal an estimate of the extent of putrescine reutilisation can be obtained from the determination of the urinary polyamines, as follows : The urinary excretion of N^1-acetylspermidine increases, if PAO is inactivated. PAO-inactivation prevents the formation of the putrescine which normally would be reutilised. The amount of additionally excreted N^1-acetylspermidine is identical with the amount of the putrescine that normally re-enters the biosynthetic branch of the polyamine cycle, if we disregard the relatively small amounts of N^1-acetylspermidine and N^1-acetylspermine, which accumulate in the tissues of animals with an inactivated PAO. For rats, the reutilisation of 2 μmol putrescine per animal and day was determined from the excess of N^1-acetylspermidine excretion (Seiler et al., 1985b). This value corresponds to 65 % of the total polyamines excreted by untreated rats, and indicates the quantitative significance of putrescine reutilisation.

Polyamine depletion and polyamine sparing

The formation of a considerable proportion of putrescine along the interconversion pathway in non-proliferating tissues can be expected from the generalization of the observation made with brain. Experimental support for this notion comes from the inefficacy of ODC inhibitors in depleting polyamine pools in non-growing organs, even after prolonged treatment with potent drugs, such as α-difluoromethylornithine (Eflornithine ; DFMO) (Bolkenius and Seiler, 1987 ; Sarhan et al., 1987). Disregarding the presumably incomplete ODC inactivation, the most likely explanation for our observations is that the proportion of putrescine that is formed by decarboxylation of Orn is not only small in brain but in most non-growing tissues, in agreement with the low ODC activities (Raina et al., 1978). From this follows that the actual demand of non-growing tissues for Orn-derived putrescine is usually small. Decarboxylation of Orn has in this case mainly the role of substituting for irreversible polyamine losses. Apparently, this small demand can be supplied from other sources as well, if ODC activity is reduced.

That putrescine reutilisation is an essential putrescine source becomes obvious if DFMO is administered together with an inactivator of PAO, such as N,N'-bis-allenylputrescine. In this case the spermidine concentration can be depleted by about 50 % in nearly all normal tissues (Bolkenius and Seiler, 1987 ; Seiler et al., 1988). Due to the induction of SAM-DC in this situation (Mamont et al., 1982 ; Shirahata and Pegg, 1986) spermine formation from spermidine is enhanced, i.e. the depletion of spermidine is only partly due to its impaired synthesis. The net polyamine depletion in non-proliferating tissues of animals with reduced ODC and PAO activities was only about 30 % (Bolkenius and Seiler, 1987 ; Seiler et al., 1988).

The lowest concentration of spermidine that is attainable with combinations of inhibitors of ODC and PAO depends on the existence of other sources of putrescine, among which the gastrointestinal tract may be important. However, even a complete blockade of all putrescine sources cannot produce a complete polyamine depletion, for the following reason : With the decrease of the polyamine pools the concentration of the free polyamines is gradually diminished to an extent that the acetylation rate comes close to zero, in spite of a reasonably high affinity of spermidine and spermine for cSAT (K_M (µM) 130 and 34 ; Della Ragione and Pegg, 1983), and polyamine turnover comes to a standstill. This is of great physiological importance : Due to the rapid exchange between free and bound polyamines, and the function of the free polyamines and their acetylation products in the control of the cellular polyamine pools, the loss of vital polyamines is prevented, even in a situation of a totally blocked supply with precursors. The slow turnover of spermine in comparison with that of spermidine, as well as its retention within the cells even under extreme pathophysiological conditions, such as starvation (Seiler et al., 1981a) is an essential feature of polyamine sparing. It is directly related with the tight binding of spermine to anionic binding sites. It is presently not known to what extent spermidine can take over physiological functions of spermine and vice versa. Spermine depletion by specific inhibitors of spermine synthase and the concomitant enhancement of the spermidine concentration was without any significant effect on cell proliferation (Pegg et al., 1986b). These observations suggest that spermidine is capable of fulfilling essential functions of spermine in eukaryotes. One may, therefore, ask why spermine was developed during evolution from prokaryotes to eukaryotes. The mentioned advantage of spermine over spermidine in polyamine sparing seems a considerable selective advantage and may be considered as one potential explanation for the existence of spermine.

Urinary polyamine excretion

As is well known, acetylated polyamines are the major urinary excretory products of humans, however, not only the reaction products of cSAT, but also the products of the nuclear acetylCoA:spermidine N^8-acetylstransferase (nSAT), N^8-acetylspermidine and acetylputrescine (Abdel-Monem and Ohno, 1977 ; van den Berg et al., 1986) are excreted. Putrescine and spermidine excretion may serve the same purpose. As a matter of fact, rodents excrete with the urine a large proportion of the polyamines as such, not in a conjugated form (Seiler et al., 1981b). Urinary excretion of the polyamines and their conjugates obviously serves as a means of eliminating irreversibly excessive polyamines from an organism ; it is the alternative method to the oxidative degradation of the polyamines to amino acids, and their excretion, a process that has been designated "terminal polyamine catabolism" (Seiler et al., 1985b).

Owing to the lack of potent cSAT inhibitors, it is not possible to decide whether N^1-acetylation of the polyamines is of vital importance, and it is not clear, whether in the absence of cSAT free putrescine and spermidine could be excreted by humans, as is the case with rodents, or whether terminal catabolic reactions would take over and transform the polyamines into excretable compounds. However, there is evidence that at least in rodents, elimination of spermine is depending on an intact interconversion pathway. If mice were treated with DFMO and N,N'-bis-allenylputrescine over an extended period of time, the blood levels of spermidine remained more or less constant. However, spermine accumulated time-dependently. After 15 days of treatment the blood levels were about 20-times higher than in controls and DFMO-treated animals (Fig. 4) (Claverie et al., 1987). The obvious explanation for this unexpected finding is the following : If due to inactivation of PAO N^1-acetylspermine and N^1-acetylspermidine cannot be degraded, they are released into the circulation. N^1-acetylspermidine is a urinary excretion product. N^1-acetylspermine, however, is deacetylated to spermine. (The hydrolase that is responsible for this reaction has not yet been identified). Since rodents seem not able to excrete either N^1-acetylspermine or spermine (Seiler et al., 1981b), the latter accumulates gradually in the circulation. Owing to the strong binding of polyamines to red blood cells (Chun et al., 1976) one has to assume that the circulating spermine is mainly attached to erythrocytes. Whether it exerts any toxicity is presently not known.

In human urine small amounts of N^1-acetylspermine were detected, and in the urine of patients with Hodgkin's lymphoma even some N^1,N^{12}-diacetylspermine (van den Berg et al., 1986). This indicates a difference between the human and rodent excretory systems. How important this difference is, remains to be established.

Polyamine interconversion and polyamine salvage

It has been mentioned that the proportion of putrescine that originates from the interconversion pathway is small in comparison with Orn decarboxylation in rapidly proliferating cells and tissues. Therefore, little attention has been paid in the past to this putrescine source, for example in the numerous attempts to use polyamine synthesis inhibitors in the therapy of neoplastic diseases (Sunkara et al., 1987 ; Porter and Jänne, 1987 ; Schechter et al., 1987). Moreover, the interconversion pathway may have been considered exclusively as an

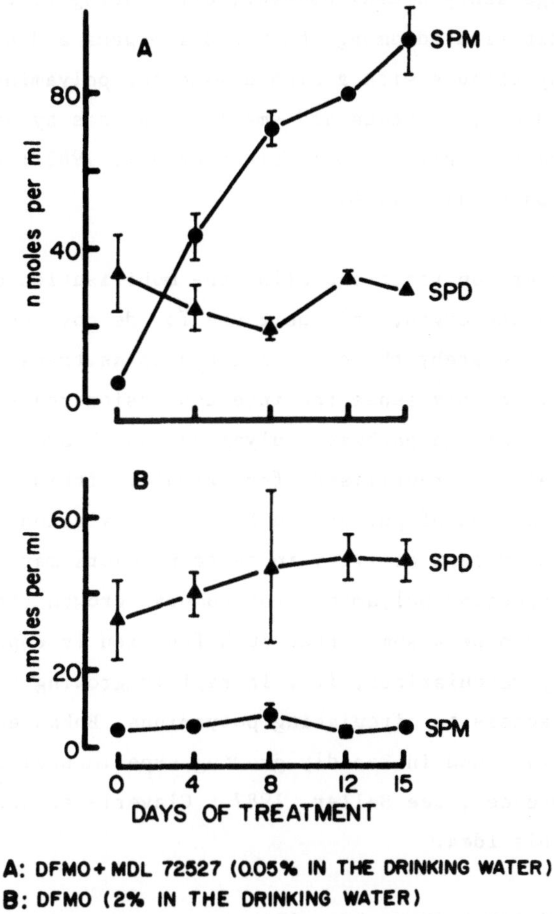

Fig. 4. Polyamines in blood of mice after administration of α-difluoromethylornithine (DFMO) or a combination of DFMO and the polyamine oxidase inhibitor N,N'-bis-allenylputrescine (MDL 72527).

CD1 albino mice received either a 2 % solution of DFMO, or a solution containing 2 % DFMO and 0.05 % MDL 72527 as sole drinking fluid. Blood samples were taken at 10 a.m. and polyamines were determined by HPLC. (Mean values ± S.D. N = 2 ; controls N = 6).

b.d.l. = below the detection limit of the analytical method. (Data from Claverie et al., 1987).

intracellular regulatory cycle. However, circulating putrescine and spermidine are distributed among the various organs and are taken up preferentially by tissues with a high demand for polyamines, as has been shown for tumors, prostate and small intestines by using labelled polyamines (Clark and Fair, 1975 ; Volkow et al., 1983 ; Chaney et al., 1983 ; Heston et al., 1984).

The interconversion reactions allow the mobilisation of spermine and spermidine in one organ, and their use for de novo polyamine synthesis in another, whereby the circulation acts as transport and distribution system. In this sense the interconversion reactions fulfil the function of a salvage pathway. Polyamines which are released from dying cells may also be reutilised, for example, within a tumor, especially if the formation of putrescine by decarboxylation of Orn is impaired. The enhanced cSAT activity in certain tumors may, therefore, be important in securing polyamines for tumoral growth (Bardócz and Weber, 1986). It can be assumed that such function is especially important in poorly vascularized, i.e. in rapidly growing tumors, since these have less access to circulating polyamines. Enhanced cSAT activities were usually found in rapidly growing experimental and human tumors (for references, see Seiler, 1987 ; Claverie et al., 1987), in agreement with this idea.

Inhibition of PAO by N,N'-bis-allenylputrescine significantly improved the antitumoral activities of DFMO and (2R,5R)-6-heptyne-2,5-diamine (MAP) in L1210 leukemia and Lewis lung carcinoma, compared with either drug alone (Claverie et al., 1987). One may conclude from these facts that the interconversion pathway contributes indeed a significant portion of putrescine that is required for tumoral growth in these animal models.

CONCLUSIONS

Long-term treatment of laboratory animals with N,N'-bis-allenylputrescine, i.e. near-complete inactivation of PAO in the entire organism, caused only a small decrease of spermidine concentrations in brain and other tissues (Bolkenius et al., 1985), and did not affect essential physiological functions as appears from the lack of any apparent behavioral changes. The most likely explanation for this observation is that enhanced formation of putrescine by decarboxylation of Orn can usually compensate the polyamine losses due to prevention of putrescine reutilisation. Thus it is obvious that polyamine reutilisation is not of vital importance. However, the role of polyamine interconversion in the regulation of intracellular polyamine metabolism, as well as the global regulation of polyamines of an intact animal by the control of terminal elimination via urinary excretion suggests a considerable physiological significance of the polyamine interconversion pathway, although this is normally not apparent.

REFERENCES

Abdel-Monem, M. M., and Ohno, K., 1977, Polyamine metabolism II : N-(monoaminoalkyl) and N-(polyaminoalkyl) acetamides in human urine, J. Pharm. Sci., 66:1195.

Antrup, H. and, Seiler, N., 1980, On the turnover of polyamines spermidine and spermine in mouse brain and other organs, Neurochem. Res., 5:123.

Bachrach, U., and Seiler, N., 1981, Formation of acetylpolyamines and putrescine from spermidine by normal and transformed chick embryo fibroblasts, Cancer Res., 41:1205.

Bardócz, S., and Weber, G., 1985, Transformation-linked increase in activities of polyamine salvage enzymes in hepatomas. Abstr. N° P2, Intern. Conf. on Polyamines in Life Sciences, Lake Yamanaka, Japan.

Bloomfield, V. A., and Wilson, R. W., 1981, Interactions of polyamines with polynucleotides, in: "Polyamines in Biology and Medicine", D. R. Morris and L. J. Marton, eds., p. 183, Marcel Dekker, New York.

Bolkenius, F. N., and, Seiler, N., 1986, Developmental aspects of polyamine interconversion in rat brain, Int. J. Devl. Neuroscience, 4:217.

Bolkenius, F. N., and Seiler, N., 1987, The role of polyamine reutilisation in depletion of cellular stores of polyamines in non-proliferating tissues, Biochim. Biophys. Acta, 923:125.

Bolkenius, F. N., Bey, P., and Seiler, N., 1985, Specific inhibition of polyamine oxidase in vivo is a method for the elucidation of its physiological role, Biochim. Biophys. Acta, 838:69.

Chaney, J.E., Kobayashi, K., Goto, R., and Digenis, G.A., 1983, Tumor selective enhancement of radioactivity uptake in mice treated with α-difluoromethylornithine prior to administration of ^{14}C-putrescine, Life Sci., 32:1237.

Chun, P. W., Rennert, O. M., Saffen, E. E. and Taylor, J. W., 1976, Effect of polyamines on the electrokinetic properties of red blood cells, Biochem. Biophys. Res. Commun., 69:1095.

Clark, R. B., and Fair, W. R., 1975, The selective in vivo incorporation and metabolism of radioactive putrescine in the adult male rat, J. Nucl. Med. 16:337.

Claverie, N., Wagner, J., Knödgen, B., and Seiler, N., 1987, Inhibition of polyamine oxidase improves the antitumoral effect of ornithine decarboxylase inhibitors, Anticancer Res., 7:765.

Daune, G., Gerhart, F., and Seiler, N., 1988, 5-Fluoromethylornithine, an irreversible and specific inhibitor of L-ornithine:2-oxoacid aminotransferase. Biochem. J., in press.

Della Ragione, R., and Pegg, A.E., 1983, Studies of the specificity and kinetics of rat liver spermidine/spermine N^1-acetyltransferase, Biochem J., 213:701.

Heston, W. D. W., Kadmon, D., Corey, D. F., and Fair, W. R., 1984, Differential effect of α-difluoromethylornithine on the in vivo uptake of ^{14}C-labelled polyamines and methylglyoxal-bis-guanylhydrazone by a rat prostate derived tumor, Cancer Res., 44:1034.

Höltta, E., 1977, Oxidation of spermidine and spermine in rat liver : purification and properties of polyamine oxidase. Biochemistry, 16:91.

Ientile, R., Ruggeri, P., Russo, P., and Macaione, S., 1985, Effect of intraventricular putrescine on adenosylmethionine decarboxylase in rat hypothalamus and caudate nucleus. J. Neurochem., 44:1315.

McCormick, F., 1978, Kinetics of polyamine synthesis and turnover in mouse fibroblasts, Biochem. J., 174:427.

Mamont, P. S., Danzin, C., Wagner, J., Siat, M., Joder-Ohlenbusch, A.-M., and Claverie, N., 1982, Accumulation of decarboxylated S-adenosyl-L-methionine in mammalian cells as a consequence of the inhibition of putrescine biosynthesis, Eur. J. Biochem. 123:499.

Pegg, A. E., 1986, Recent advances in the biochemistry of polyamines in eukaryotes, Biochem. J., 234:249.

Pegg, A. E., Hibasami, H., Matsui, I., and Bethell, D. R., 1981, Formation and interconversion of putrescine and spermidine in mammalian cells. Adv. Enzyme Regul., 19:427.

Pegg, A. E., Sertich, G. J., Kameji, T., Erwin, B. G., and Shirahata, A., 1986a, Regulation of polyamine metabolism by polyamines. Abstr. N° 30, Intern. Conf. on Polyamines in Life Sci., Lake Yamanaka, Japan.

Pegg, A. E., Coward, J. K., Talekar, R. R., and Secrist III, J. A., 1986b, Effects of certain 5'-substituted adenosines on polyamine synthesis : selective inhibitors of spermine synthase, Biochemistry, 25:4091.

Persson, L., Erwin, B. G., and Pegg, A. E., 1985, Spermidine/spermine N^1-acetyltransferase : studies using a specific antiserum, in: "Recent Progress in Polyamine Research", L. Selmeci, M. E. Brosnan, and N. Seiler, eds., p. 287, Akademiai Kiado, Budapest.

Porter, C. W., and Jänne, J., 1987, Modulation of antineoplastic drug action by inhibitors of polyamine biosynthesis, in: "Inhibition of Polyamine Metabolism", P. P. McCann, A. E. Pegg, and A. Sjoerdsma, eds., p. 203, Academic Press, Orlando.

Porter, C. W., Cavanaugh, Jr., P. F., Stolowich, N., Ganis, B., Kelly, E., and Bergeron, R. J., 1985, Biological properties of N^4- and N^1,N^8-spermidine derivatives in cultured L1210 leukemia cells, Cancer Res., 45:2050.

Raina, A., Pajula, R.-L., Eloranta, T., and Tuomi, K., 1978, Synthesis of polyamines and S-adenosylmethionine in rat tissues and tumor cells : Effect of D,L-α-hydrazino- -aminovaleric acid on cell proliferation, Adv. Polyamine Res., 1:75.

Sarhan, S., Knödgen, B., Gerhart, F., and Seiler, N., 1987, Chain-fluorinated polyamines as tumor markers. I. In vivo transformation of 2,2-difluoroputrescine into 6,6-difluorospermidine and 6,6-difluorospermine, Int. J. Biochem., 19:843.

Schechter, P. J., John, L., Barlow, R., and Sjoerdsma, A., 1987, Clinical aspects of inhibition of ornithine decarboxylase with emphasis on therapeutic trials of Eflornithine (DFMO) in cancer and protozoan diseases, in: "Inhibition of Polyamine Metabolism", P. P. McCann, A. E. Pegg, and A. Sjoerdsma, eds., p. 345, Academic Press, Orlando.

Seiler, N., 1981, Amide bond-forming reactions of polyamines, in: "Polyamines in Biology and Medicine" D. R. Morris, and L. J. Marton, eds., p. 127, Marcel Dekker, New York.

Seiler, N., 1987a, Functions of polyamine acetylation, Can. J. Physiol. Pharmacol., 65:2024.

Seiler, N., 1987b, Inhibition of enzymes oxidizing polyamines, in: "Inhibition of Polyamine Metabolism", P. P. McCann, A. E. Pegg and A. Sjoerdsma, eds., p. 49, Academic Pres, Orlando.

Seiler, N., and Bolkenius, F. N., 1985. Polyamine reutilization and turnover in brain, Neurochem. Res., 10:529.

Seiler, N., and Heby, O., 1988, Regulation of cellular polyamines in mammals, Acta Biochim. Biophys. Hungarica.

Seiler, N., Al-Therib, M.-J., Fischer, H. A., and Erdmann, G., 1979, Dynamic and regional aspects of polyamine metabolism in the brain of trout (Salmo irideus Gibb), Int. J. Biochem., 10:961.

Seiler, N., Bolkenius, F. N., Knödgen, B. and Mamont, P., 1980, Polyamine oxidase in rat tissues, Biochim. Biophys. Acta, 615:480.

Seiler, N., Bolkenius, F. N. and Sarhan, S., 1981a, Formation of acetylpolyamines in the liver of fasting animals, Int. J. Biochem., 13:1205.

Seiler, N., Koch-Weser, J., Knödgen, B., Richards, W., Tardif, C., Bolkenius, F. N., Schechter, P., Tell, G., Mamont, P., Fozard, J., Bachrach, U., and Grosshans, E., 1981b, The significance of acetylation in the urinary excretion of polyamines, Adv. Polyamine Res., 3:197.

Seiler, N., Bolkenius, F. N., Bey, P., Mamont, P. S., and Danzin, C., 1985a, Biochemical significance of inhibition of polyamine oxidase, in: "Recent Progress in Polyamine Research", L. Selmeci, M. E. Brosnan, M. E., and N. Seiler, eds., p. 305, Akademiai Kiado, Budapest.

Seiler, N., Bolkenius, F. N., and Knödgen, B., 1985b, The influence of catabolic reactions on polyamine excretion, Biochem. J., 225:219.

Seiler, N., Sarhan, S., Knödgen, B. and Gerhart, F., 1988, Chain-fluorinated polyamines as tumor markers. II. Metabolic aspects in normal tissues, J. Cancer Res. Clin. Oncol., 114:71.

Shirahata, A. and Pegg, A. E., 1986, Increased content of an RNA for a precursor of S-adenosylmethionine decarboxylase in rat prostate after treatment with 2-difluoromethylornithine, J. Biol. Chem., 261:13833.

Sunkara, P. S., Baylin, S. B. and Luk, G. D., 1987, Inhibitors of polyamine biosynthesis : Cellular and in vivo effects on tumor proliferation, in: "Inhibition of Polyamine Metabolism", P.P. McCann, A. E. Pegg, and A. Sjoerdsma, eds., p. 121, Academic Press, Orlando.

Van den Berg, G. A., Muskiet, F. A. J., Kingma, A. W., van der Slik, W. and Halie, M. R., 1986, Simultaneous gas-chromatographic determination of free and acetyl-conjugated polyamines in urine. Clin. Chem., 32:1930.

Volkow, N., Goldman, S. S., Flamm, E. S., Cravioto, H., Wolf, A. P., and Brodie, J. D., 1983, Labelled putrescine as a probe in brain tumors, Science, 221:673.

Wallace, H. M., 1987, Polyamine catabolism in mammalian cells : excretion and acetylation, Med. Sci. Res., 15:1437.

Williams-Ashman, H. G., 1985, Metabolic significance of 5'-deoxy-5'-methylthioadenosine in relation to polyamine turnover in normal and malignant mammalian cells, in: "Recent Progress in Polyamine Research", Selmeci, L., Brosnan, M. E., and Seiler, N., eds., p. 231, Akademiai Kiado, Budapest.

THE BIOLOGICAL FUNCTIONS OF AMINE OXIDASES AND THEIR REACTION PRODUCTS: AN OVERVIEW

B. Mondovì, P. Riccio and E. Agostinelli

Department of Biochemical Sciences and CNR Centre for Molecular Biology, University "La Sapienza" Rome, Italy

INTRODUCTION

Amine oxidases (AO) are enzymes widely distributed among all living organisms.[1] Their widespread occurrence accounts for an undoubtedly relevant biological function in biogenic amine metabolism. AOs represent a class of enzymes heterogenous in structure, catalytic mechanism and mode of substrate oxidation. Mono-, di- and polyamines, as well as several N-acyl amines, are oxidatively deaminated by AOs in a reaction consuming O_2 and H_2O and producing the corresponding aldehyde, the removed amine moiety and H_2O_2 in stoichiometric amounts according to the following equations:

$$R\text{-}CH_2\text{-}NH_2 + O_2 + H_2O \longrightarrow R\text{-}CHO + NH_3 + H_2O_2$$

cleavage at a primary amino group (terminal oxidation reaction).

$$R_1\text{-}CH_2\text{-}NH\text{-}CH_2\text{-}R_2 + O_2 + H_2O \longrightarrow$$

$$\longrightarrow R_1\text{-}CHO + H_2N\text{-}CH_2\text{-}R_2 + H_2O_2$$

cleavage at a secondary amino group (interconversion reaction).

Only di- and polyamines will be concerned herein; for the sake of brevity, the term "polyamine" will be used to include the diamines putrescine and cadaverine, instead of the expression "diamines and polyamines".

Because of the heterogeneity of this class of enzymes, it is useful to provide an unambiguous classification criterion. Referring to the nature of the prosthetic groups, we subdivide AOs into FAD- and Cu-dependent enzymes. To the former subclass belong the mitochondrial monoamine oxidases (MAO) and cytosolic polyamine oxidase (PAO). The serum enzymes (SAO) and diamine oxidases (DAO, histaminase) are instead Cu-AOs. The nomenclature and abbreviations proposed in a recent review on AOs will be employed herein[1] (see table 1).

TABLE 1. Nomenclature, abbreviations, cofactor(s) and tissue distribution of the main amine oxidases.

monoamine oxidase (MAO)	FAD	all tissues (outer mitochondrial membrane)
polyamine oxidase (PAO)	FAD	all tissues (cytoplasm, peroxisomes)
serum amine oxidase (SAO)	Cu, PQQ	serum (some species)
diamine oxidase (DAO)	Cu, PQQ	tissues (cytoplasm, microsomal fraction), biological fluids

Numerous studies have been carried out in order to establish the biological significance of the polyamine oxidative reactions, but no obvious indication of an essential physiological function is available for AOs to date. Actually, prolonged administration of some inhibitors of polyamine oxidation _in vivo_ provokes no striking modification of the overall condition of the animal except an enhanced urinary excretion of both free and acetylated polyamines[2]. In spite of this fact, far from being merely the terminal route for inactivation and elimination, the oxidative reactions represent a means by which the intracellular polyamine concentration undergoes fine regulation, the coarse adjustment being carried out by the biosynthetic ornithine and S-adenosyl-methionine decarboxylases. In addition, polyamine oxidation generates metabolically active reaction products which are in turn capable of influencing, more or less profoundly, other compartments of the cell metabolism. However, oxidation makes possible the recovery of part of the polyamine molecule, providing at the same time a mechanism for their interconversion.

AMINE OXIDASES AND POLYAMINE OXIDATION

AOs have been purified from a variety of bacterial and fungal,[3] plant[4] and animal[5,6] sources. Some of them have been extensively investigated with respect to the structure and mechanistic properties,[5-8] but only in relatively recent times it has become possible to carry out _in vivo_ observations of the AO activity.[2,9-11] The development of sensitive and reliable techniques for biological assays, and the availability of specific AO inhibitors made possible to monitor dynamically the changes of polyamine metabolism in intact animals under several either spontaneously occurring or experimentally induced conditions.

From a physiological point of view, two possible catabolic pathways are distinguishable for polyamines:

terminal oxidation and interconversion reactions[2]. As a general rule, the former reaction is accomplished by the Cu-dependent AOs, and the latter by the FAD-enzymes. By "terminal oxidation", we generally mean that oxidative deamination is carried out at the terminal (primary) amino group, but the same definition also implies, in contrast with the meaning of "interconversion", that this kind of reaction leads to products which cannot be recycled for polyamine resynthesis.

Actually, such a kind of distinction is not absolute; in fact spermine and spermidine, oxidized by Cu-AOs, may undergo spontaneous degradation when not previously further oxidized by aldehyde dehydrogenases, forming spermidine and putrescine, respectively, with production of acrolein.[12, 13]

$H_2N\text{-}(CH_2)_4\text{-}NH\text{-}(CH_2)_3\text{-}NH_2$ (spermidine)

↓ Cu-AO

$H_2N\text{-}(CH_2)_4\text{-}NH\text{-}(CH_2)_2\text{-}CHO \xrightarrow{ADH} H_2N\text{-}(CH_2)_4\text{-}NH\text{-}(CH_2)_2\text{-}COOH$ (putreanine)

↓ spontaneous

$H_2N\text{-}(CH_2)_4\text{-}NH_2 + H_2C{=}CH\text{-}CHO$
(putrescine) (acrolein)

In addition, some bacterial and plant enzymes cleave polyamines at the secondary nitrogen, but form 1,3-diaminepropane as reaction product instead of putrescine which is not liable to being reutilized for resynthesis of higher polyamines.[4, 14,15]

$H_2N\text{-}(CH_2)_4\text{-}NH\text{-}(CH_2)_3\text{-}NH_2$ (spermidine)

↓ PAO

$H_2N\text{-}(CH_2)_3\text{-}CHO + H_2N\text{-}(CH_2)_3\text{-}NH_2$
(4-aminobutyraldehyde) (1,3-diaminepropane)

Disregarding these exceptions of minor importance, this distinction of the polyamine oxidative reactions suggests a different physiological significance for the FAD- and Cu-AOs. The unequal distribution pattern of FAD- and Cu-dependent enzymes in tissues and biological fluids also accounts for a distinct metabolic function.

Cu-AO activity is present at considerably different levels varying the tissues and organs. In contrast with the FAD-enzymes, Cu-AOs show both intra- and extracellular localization, being often present in serum and also in some biological fluids, such as the seminal plasma.[16]

In addition to copper, these enzymes contain a second organic prosthetic group which has been recently identified as pyrroloquinoline quinone (PQQ).[17,18] This characteristic is quite interesting, being Cu-AOs the first eukaryotic

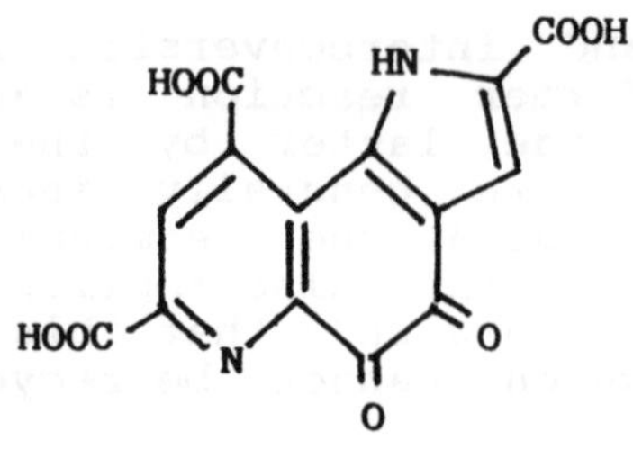

pyrroloquinoline quinone (PQQ)

enzymes known to have PQQ as the cofactor. In a future perspective, this feature may turn of pharmacological interest.

Under physiological conditions, Cu-dependent serum AO is abundant in ruminants, swine, and probably other animal species, but not in horse, rodents and humans. In man, SAO activity is enhanced during pregnancy or under some pathological conditions,[20-23] such as in cancer patients. Heparin administration also is capable of inducing the appearance of DAO activity in the circulation.[24,25]

Serum Cu-AOs oxidatively deaminate diamines and polyamines. Adult bovine serum deaminates only spermine and spermidine, but fetal calf serum contains also a Cu-dependent enzyme active on diamines.[26] Swine SAO instead attacks diamines, and shows also histaminase activity.[27] In contrast, human pregnancy serum appears to contain a single enzyme species capable of oxidizing both di- and polyamines.[28]

Those Cu-AOs showing preferential substrate specificity for diamines, ubiquitously present in the living kingdom, are termed DAOs.[1] The ones from animal sources, generally capable of deaminating histamine as well, are seldom referred to as histaminases. It should the pointed out that a single enzyme species is responsible for both diamine oxidase and histaminase activities.[29] Some DAO activity can be detected nearly in all animal tissues with few exceptions,[5] but is particularly elevated in rapidly proliferating tissues, such as the small intestine mucosa under physiological conditions, fetal tissues, placenta, and organs undergoing hypertrophic, hyperplastic or neoplastic processes.[19-23] A particularly abundant source of DAO is the kidney cortex of some animal species including man. Its localization in the brush border and basal unfoldings of proximal convoluted tubules epithelium[30] would suggest a role as biogenic amine scavenger.

DAO activity level may undergo wide oscillations within the same tissue, depending on the intracellular polyamine concentration. By acting mainly on putrescine, which is the precursor of higher polyamines, DAO appears to play a leading role in the homeostatic control of intracellular polyamine levels.

The FAD-AOs show a more uniform distribution throughout the different tissues and organs. PAO activity is present in all tissues and seems to be localized in the cytosol. In liver only its association with peroxisomes has been reported.[32] N^1-acetylspermine and N^1-acetylspermidine are the best substrates, but the unconjugated polyamines, too, are oxidized, although at a lower rate.[33] The mechanistic properties, being not investigated, are likely to resemble those of MAO.

Mitochondrial MAO also is ubiquitous, and appears to play a major role in the vertebrate brain,[34] where the Cu-AO activity is extremely low. Acetylputrescine, acetylcadaverine and N^8-acetylspermidine are oxidized.[33]

$$\underset{\text{(acetylputrescine)}}{H_3C\text{-}CO\text{-}NH\text{-}(CH_2)_4\text{-}NH_2} \xrightarrow{\text{MAO}} H_3C\text{-}CO\text{-}NH\text{-}(CH_2)_3\text{-}CHO$$

$$\downarrow \text{ADH, deacetylase}$$

$$\underset{\text{(GABA)}}{H_2N\text{-}(CH_2)_3\text{-}COOH} + \underset{\text{(acetic acid)}}{CH_3\text{-}COOH}$$

By catalyzing a terminal oxidation reaction, MAO accomplishes in the brain the function proper to DAO in other organs.

With the exception of putrescine and cadaverine, combined PAO and MAO substrate specificity patterns, being reciprocally complementar, cover practically all the free and acetylated polyamine pool.

TABLE 2. Substrates of the most representative AOs.

	BSAO	SKDAO	RLPAO	MAO
putrescine		*		
acetylputrescine		*		*
cadaverine		*		
acetylcadaverine		*		*
spermidine	*		*	
N^1-acetylspermidine		*	*	
N^8-acetylspermidine	*	*		*
spermine	*		*	
N^1-acetylspermine		*	*	
N^1, N^{12}-diacetylspermine			*	

* : oxidized. BLANK SPACE: not oxidized.
BSAO: bovine serum AO.
SKDAO: swine kidney DAO
RLPAO: rat liver PAO.
MAO: A and B types mitochondrial MAO.
Note that relative substrate affinities are not considered; oxidation rates of AOs differ widely varying the substrate.

It should be mentioned that the <u>in</u> <u>vivo</u> substrate specificity of AOs may be different from that usually found under normal experimental conditions with purified or partially purified enzyme preparations.

Qualitative alteration (transformation) of the catalytic properties of some AOs has been reported after proper chemical treatment.[35]

In vivo transformation of mitochondrial membrane-bound MAO occurs under several circumstances which cause concomitant accumulation of lipid peroxides.[36]

The possible presence of endogenous inhibitors[37] or modifiers, and the effect of different enviromental conditions[38] such as *in vivo* association with membranous structures[39] (smooth endoplasmic reticulum, peroxisomes) should be taken in account.

After endoportal injection of heparin, the guinea pig liver releases DAO in the plasma. Interestingly, a ten-fold increase of the solubilized enzyme's specific activity has been reported with respect to the microsome-bound DAO, suggesting a partial inactivation of the membrane-associated enzyme.[39]

Altered substrate specificity has also been reported for the CH-Sepharose 4B-immobilized bovine SAO.[40] Moreover, the Concanavalin-A-Sepharose-adsorbed swine kidney DAO showed a large affinity change towards substrates with respect to the free enzyme.[40] It appears that enzyme immobilization by chemical or physical binding probably causes a rearrangement of the active site's geometry.

METABOLIC FATE OF THE PRODUCTS OF POLYAMINE OXIDATION

The primary products of polyamine oxidation are represented by aminoaldehydes. These are unstable compounds which may undergo either spontaneous modification, or further transformation by aldehyde-metabolizing enzymes. Aldehyde dehydrogenases and, to a much lesser extent, alcohol dehydrogenase, are responsible for the conversion of the aldehyde group into carboxylic acid or alcoholic function, respectively.

R-CHO (aldehyde) ↗ aldehyde DH ——> R-COOH (carboxylic acid)

R-CHO (aldehyde) ↘ alcohol DH ----> R-CH_2OH (alcohol)

Quantitative estimation of the extent of aldehyde metabolizing enzymes' influence on aldehyde turnover rate may be obtained employing chloral hydrate or acetaldehyde, and phenobarbital or pyrazole as inhibitors of, respectively, aldehyde and alcohol dehydrogenases. Using these inhibitors makes possible to obtain reliable measurements of AO activity on crude tissue homogenates when activity is evaluated by measuring aldehyde formation.

Intermediate and final polyamine oxidation products may

be either excreted or shunted to the citrate cycle. Apart from acetylated polyamines, normal constituents of tissues and urine are also a number of aminoacids derived from polyamine oxidation.[33,34,41]

Putrescine oxidative deamination produces γ-aminobutyric acid (GABA).

$$\underset{\text{(putrescine)}}{H_2N\text{-}(CH_2)_4\text{-}NH_2} \xrightarrow{\text{AO, ADH}} \underset{\text{(GABA)}}{H_2N\text{-}(CH_2)_3\text{-}COOH}$$

Spermidine can be converted into either putreanine or isoputreanine.

$$\underset{\text{(spermidine)}}{H_2N\text{-}(CH_2)_4\text{-}NH\text{-}(CH_2)_3\text{-}NH_2}$$

$$\xrightarrow{\text{AO, ADH}} \underset{\text{(putreanine)}}{H_2N\text{-}(CH_2)_4\text{-}NH\text{-}(CH_2)_2\text{-}COOH}$$

$$\xrightarrow{\text{AO, ADH}} \underset{\text{(N-[3-aminopropyl]-4-amino-butyric acid, isoputreanine)}}{HOOC\text{-}(CH_2)_3\text{-}NH\text{-}(CH_2)_3\text{-}NH_2}$$

Monooxidized spermine forms N^8-(2-carboxyethyl)-spermidine, whereas the dioxidized tetramine leads to spermic acid formation.

$$\underset{\text{(spermine)}}{H_2N\text{-}(CH_2)_3\text{-}NH\text{-}(CH_2)_4\text{-}NH\text{-}(CH_2)_3\text{-}NH_2}$$

$$\xrightarrow{\text{AO, ADH}} \underset{\text{(N8-[2-carboxyethyl]-spermidine)}}{H_2N\text{-}(CH_2)_3\text{-}NH\text{-}(CH_2)_4\text{-}NH\text{-}(CH_2)_2\text{-}COOH}$$

$$\xrightarrow{\text{AO, ADH}} \underset{\text{(spermic acid)}}{HOOC\text{-}(CH_2)_2\text{-}NH\text{-}(CH_2)_4\text{-}NH\text{-}(CH_2)_2\text{-}COOH}$$

The polyamine interconversion reactions generate 3-aminopropionaldehyde which is transformed into β-alanine.[42]

$$\underset{\text{(3-aminopropionaldehyde)}}{H_2N\text{-}(CH_2)_2\text{-}CHO} \xrightarrow{\text{ADH}} \underset{(\beta\text{-alanine})}{H_2N\text{-}(CH_2)_2\text{-}COOH}$$

METABOLIC EFFECTS OF THE PRODUCTS OF POLYAMINE OXIDATION.

It has been observed that oxidized polyamines provoke metabolic effects in some way opposite to those proper of the polyamines themselves.

Impairment of cells growth and division have been reported in the presence of oxidized polyamines,[43–45] both in *vitro* and *in vivo*. An exogenous contribution of DAO inhibits Ehrlich ascites tumors growth in mice.[46] The inhibitory effect has been related to interaction of aminoaldehydes with nucleic acids.[47–49]

Acrolein also is cytotoxic, but its mode of action appears to be different from those of aminoaldehydes.[50–52]

Oxidized polyamines may have a function in modulating the immune response. Incubation of murine lymphocytes with irradiated DAB/2 spleen cells in the presence of polyamines and AO-containing bovine serum reduces DNA synthesis and impairs cytolytic activity.[53] Also, the presence of remarkable AO activity in the maternal part of placenta has been claimed to play a possible role in allowing survival of the fetoplacental unit in spite of its immunological diversity with respect to the mother.[54]

Among the aminoacids occurring as final oxidation products of polyamines, GABA formed from putrescine deserves particular emphasis.[34] Its role in the Central Nervous System as neuromodulator and neurotransmitter has been clearly established, but the extent of AO contribution to its overall production, in addition to the ordinary biosynthesis by glutamate decarboxylation, has not been estimated. It cannot be excluded, however, a possible behavioral effect or contribution to neurological symptoms in patients with brain tumors.

Dioxygen reduction by a two-electron addition is an H_2O_2-producing reaction accomplished by several oxidoreductases. It appears that microsomes account for about 45% of the total H_2O_2 generated in the rat liver.[55] In consideration of the fact that intracellular AO activity is mainly associated with the microsomal fraction, it is very likely that polyamine oxidation in actively proliferating cells is responsible for at least a significant part of the overall H_2O_2 production. Moreover, the uneven compartmentation of H_2O_2-generating reactions may further stress the relative importance of oxidative deamination reactions in the responsibility of H_2O_2-mediated effects.

Actually, an ever-increasing interest has been growing around the metabolic effects of H_2O_2.[56] The oxidative damage is responsible for the toxic effect produced by H_2O_2 on cells. However, it appears that, far from being merely a wasteful byproduct, H_2O_2 possesses functional metabolic importance: evidence is accumulating that would make H_2O_2 an essential metabolite. Among the most recent reports available on this matter, we will recall the following observations.

The addition of H_2O_2 at less than micromolar concentrations is able to enhance human platelet aggregation induced by either soluble or particulate stimulating agents.[57] Moreover, H_2O_2 was found to have an insulin-like effect on several model systems: stimulation of glucose[58] and aminoacid[59] transport and of glycogen synthesis,[60] and inhibition of intracellular protein degradation,[61] have been reported. Stimulation of lipogenesis in fat cells has been observed which is mediated by H_2O_2 formation in the presence of polyamines;[62] the effect is prevented by aminoguanidine or isoniazide, well known AO inhibitors. A spurring effect on lymphocyte blastization has also been ascribed to H_2O_2.[63] In addition, a clear correlation has been evidenced between H_2O_2 and Ca^{2+} flux through the mitochondrial, microsomal vescicle and plasmatic membranes.[64-67]

REGULATION OF AO ACTIVITY

The intracellular polyamine concentration is adjusted either by regulating the level of activity of the biosynthetic and catabolic enzymes,[68] or by controlling the net direction of polyamine acetylation-deacetylation reactions.[69]

The mechanisms by which ODC activity is regulated have been deeply investigated. A very recent report demonstrates that polyamines themselves regulate the translation of the ODC mRNA.[70] On the other hand, little is known on the molecular mechanism by which the homeostatic control on AOs activity levels is exerted. Our present knowledge is limited to some pieces of information on DAO. At any rate, the level of DAO activity is assumed to be indicative of the rate of polyamine catabolism, for it directly controls the concentration of putrescine, which is the focal control point for the biosynthesis of higher polyamines.

Several observations indicate that enhanced ODC activity is usually associated to increased DAO levels.[21, 71-80] In fact, the level of DAO activity is proportional to that of polyamine concentration, and appears to be regulated by *de novo* synthesis of new enzyme,[78,81] rather than by varying DAO turnover. Administration of diamines or spermidine to rats causes an enhancement of DAO activity in the liver, kidney and heart.[82] The involvement of $beta_2$ adrenoceptors in the regulation of DAO activity in the heart and kidney has been suggested.[83-85]

Significant modifications of AO activity occur either in pregnancy or under several pathological conditions. In response to the release of polyamines by fetal tissues, serum DAO increases dramatically during pregnancy reaching, before delivery, the highest values.[19,28] The enzyme is, most probably, of placental origin, for it is abundant in the maternal part of placenta.

Several observations on animal and human neoplasms demonstred the occurrence of high DAO levels both in the tumoral tissue and biological fluids.[22,79,80] In this case, also, the enhancement of AO activity appears to be a cellular response to increased polyamines. In fact, peritumoral tissue also shows, sometimes at a higher level than the tumor itself, enhanced AO activity.[19,80]

PERSPECTIVES

In addition to polyamine oxidation, AOs possess also other functions. They are involved in the inactivation of histamine (DAO),[27] in the cross-linking of collagen and elastin (lysyl oxidase, a Cu-AO)[86] and in the breakdown of biogenic amine neurotransmitters.[87] The development of highly specific inhibitors to be employed in the biochemical and clinical research is therefore of primary importance,[2, 88, 89] to avoid overlapping of inhibitory effect and interference between such different biochemical functions.

The possibility of in vivo enhancing of AO activity by exogenous contribution is also to be investigated. Very recently,[90] it was observed that fusion-mediated microinjection of active AOs into cultured cells, obtained by entrapping the enzymes into reconstituted Sendai virus envelopes, caused inhibition of nucleic acid and protein syntheses. Transformed cells were more sensitive than normal controls, presumably due to higher polyamine content.

Immobilized and liposome-entrapped AOs may be handled, in a near future, as powerful scientific tools for medical and biological applications.

ACKNOWLEDGEMENT

This work was in part supported by the "Ministero della Pubblica Istruzione", by C.N.R., Special Project on "Oncology", Contract n. 87.01363.44, and by AIRC.

REFERENCES

1. B. Mondovi, "Structure and Functions of Amine Oxidases", CRC Press, Boca Raton, FL (1985).
2. N. Seiler, Inhibition of enzymes oxidizing polyamines, in: "Inhibition of Polyamine Metabolism", P. McCann, A. E. Pegg and A. Sjoerdsma eds., Academic Press, NY (1987).
3. H. Kumagai and H. Yamada, Bacterial and fungal amine oxidases, in: "Structure and Functions of Amine Oxidases", B. Mondovi ed., CRC Press, Boca Raton, FL (1985).
4. A. Rinaldi, G. Floris and A. Giartosio, Plant amine oxidases, in: "Structure and Functions of Amine Oxidases", B. Mondovi ed., CRC Press, Boca Raton FL (1985).
5. B. Mondovi and P. Riccio, Animal intracellular amine oxidases, in: "Structure and Functions of Amine Oxidases", B. Mondovi, ed., CRC Press, Boca Raton FL (1985).
6. G. Pettersson, Plasma amine oxidase, in: "Structure and Functions of Amine Oxidases", B. Mondovi ed., CRC Press, Boca Raton FL (1985).
7. A. Finazzi-Agro, Optical and spectroscopic properties of copper-containing amine oxidases, in: "Structure and Functions of Amine Oxidases", B. Mondovi ed., CRC Press, Boca Raton FL (1985).
8. G. Rotilio, Spectroscopic and chemical properties of the amine oxidase copper, in: "Structure and Functions of Amine Oxidases", B. Mondovi ed., CRC Press, Boca Raton FL (1985).
9. H. Dolezalova, M. Stepita-Klauco, J. Kucera, H. Uchimura and M. Hirano, Blood concentrations of monoacyl-cadaverines in schizophrenia, in: "Mass Spectrometry in Drug Metabolism, A. Frigerio and E. L. Ghisalberti

eds., Plenum Press NY (1977).
10. F. N. Bolkenius, P. Bey and N. Seiler, Specific inhibition of polyamine oxidase in vivo is a method for the elucidation of its physiological role, Biochim. Biophys. Acta 838:69 (1985).
11. N. Seiler, F. N. Bolkenius and B. Knodgen, The influence of catabolic reaction on polyamine excretion, Biochem. J. 225:219 (1985).
12. R. A. Alacorn, Acrolein IV. Evidence for the formation of the cytotoxic aldehyde acrolein from enzymatically oxidized spermine or spermidine, Arch. Biochem. Biophys. 137:362 (1970).
13. B. W. Kimes and D. R. Morris, Preparation and stability of oxidized polyamines, Biochim. Biophys Acta 228: 223 (1971).
14. T. A. Smith, Purification and properties of the polyamine oxidase from barley plants, Phytochemistry 11:899 (1972).
15. Y. Suzuki and H. Yanagisawa, Purification and properties of maize polyamine oxidase: a flavoprotein, Plant Cell Physiol. 21:1085 (1980).
16. E. Holtta, P. Pulkkinen, K. Elfving and J. Janne, Oxidation of polyamines by diamine oxidase from human seminal plasma, Biochem. J. 145:373 (1975).
17. C. L. Lobenstein-Verbeek, J. A. Jongejan, J. Frank, and J. A. Duine, Bovine serum amine oxidase: a mammalian enzyme having covalently bound PQQ as prosthetic group, FEBS Lett. 170:305 (1984).
18. P. F. Knowles, K. B. Pandeya, F. X. Rius, C. M. Spencer, R. S. Moog, M. A. McGuirl and D. M. Dooley, The organic cofactor in plasma amine oxidase: evidence for pyrroloquinoline quinone and against pyridoxal phosphate, Biochem. J. 241:603 (1987).
19. C. Maslinski, T. Bieganski, W. A. Fogel and M. A. Kitler, Diamine oxidase in developing tissues, in: "Structure and Functions of Amine Oxidases", B. Mondovi ed., CRC Press, Boca Raton FL (1985).
20. J. Sattler, R. Hesterberg, W. Lorenz, M. Ennis, C. D. Stahlknecht and J. Kusche, Pathophysiological functions of diamine oxidase: an evaluation in animal studies using a probabilistic model with several types of causal relationships, in: "Structure and Functions of Amine Oxidases", B. Mondovi ed., CRC Press, Boca Raton FL (1985).
21. A. Perin, A. Sessa and M. A. Desiderio, Diamine oxidase in regenerating and hypertrophic tissues, in: "Structure and Functions of Amine Oxidases", B. Mondovi ed., CRC Press, Boca Raton FL (1985).
22. S. B. Baylin and G. D. Luk, Diamine oxidase activity in human tumors: clinical and biological significance, in: "Structure and Functions of Amine Oxidases" B. Mondovi ed., CRC Press, Boca Raton FL (1985).
23. G. Bombardieri, A. Milani and L. Rossi, Copper-dependent amine oxidases: clinical aspects, in: "Structure and Functions of Amine Oxidases", B. Mondovi ed., CRC Press, Boca Raton FL (1985).
24. S. B. Baylin, M. A. Beaven, R. M. Krauss and H. R. Keiser, Response of plasma histaminase to small doses of heparin in normal subjects and patients with hyperlipoproteinemia, J. Clin. Invest. 52:1985 (1973).
25. L. D'Agostino, C. Ciacci, B. Daniele, M. V. Barone, R.

Sollazzo and G. Mazzacca, Plasma diamine oxidase (DAO) and heparin, Dig. Dis. Sci. 29:1070 (1984).
26. W. A. Gahl and H. C. Pitot, Polyamine degradation in foetal and adult bovine serum, Biochem. J. 202:603 (1982).
27. F. Buffoni, Histaminase and related amine oxidases, Pharmacol. Rev. 18:1663 (1966).
28. W. A. Gahl, A. M. Vale and H. C. Pitot, Spermidine oxidase in human pregnancy serum, Biochem. J. 201:161 (1982).
29. B. Mondovi, G. Rotilio, A. Finazzi-Agro and A. Scioscia-Santoro, Purification of pig kidney diamine oxidase and its identity with histaminase, Biochem. J. 91:408 (1964).
30. M. P. Argento-Ceru and F. Autuori, Localization of diamine oxidase in animal tissues, in: "Structure and Functions of Amine Oxidases", B. Mondovi ed., CRC Press, Boca Raton FL (1985).
31. N. Seiler, F. N. Bolkenius, B. Knodgen and P. Mamont, Polyamine oxidase in rat tissues, Biochim. Biophys. Acta 615:480 (1980).
32. E. Holtta, Oxidation of spermidine and spermine in rat liver: purification and properties of polyamine oxidase, Biochemistry 16:91 (1977).
33. N. Seiler, Acetylpolyamines as substrates of amine oxidases, in: "Structure and Functions of Amine Oxidases", B. Mondovi ed., CRC Press, Boca Raton FL (1985).
34. N. Seiler and M. J. Al-Therib, Putrescine catabolism in mammalian brain, Biochem. J. 144:29 (1974).
35. V. Z. Gorkin, Qualitative alteration (transformation) in catalytic activity of amine oxidases, in: "Structure and Functions of Amine Oxidases", B. Mondovi ed., CRC Press, Boca Raton FL (1985).
36. V. Z. Gorkin, "Amine Oxidases in Clinical Research", Pergamon, Oxford (1983).
37. F. Buffoni, G. Banchelli, G. Ignesti, R. Pirisino and L. Raimondi, The presence of an inhibitor of benzylamine oxidase in human blood plasma, Biochem. J. 211:767 (1983).
38. P. Guerrieri, A. Finazzi-Agro, M. T. Costa, G. Rotilio and B. Mondovi, Salt effect on diamine oxidase activity, Ital. J. Biochem. 25:160 (1976).
39. W. Schmutzler, O. Goldschmidt, K. P. Bethge and J. Knop, The release of guinea pig liver histaminase and some of its properties, Int. Arch. Allergy 36:45 (1969).
40. R. Stevanato, M. Porchia, O. Befani, B. Mondovi and A. Rigo, Characterization of free and immobilized amine oxidases, submitted for publication.
41. N. Seiler, F. N. Bolkenius and O. M. Rennert, Interconversion, catabolism and elimination of the polyamines, Med. Biol. 59:334 (1981).
42. S. Razin, U. Bachrach and I. Gery, Formation of β-alanine from spermine and spermidine by Pseudomonas Aeruginosa, Nature 181:700 (1958).
43. U. Bachrach, S. Abzug and A. Bekierkunst, Cytotoxic effect of oxidized spermine on Ehrlich ascites cells, Biochim. Biophys. Acta 134:174 (1967).
44. U. Bachrach, Oxidized polyamines, ANN. N. Y. Acad. Sci. 171:939 (1970).
45. J. M. Gaugas and D. L. Dewey, Hog kidney diamine oxidase conversion of biogenic diamines to inhibitors of cell

proliferation, J. Pathol. 134:243 (1981).
46. B. Mondovi, P. Gerosa and R. Cavaliere, Studies on the effect of polyamines and their products on Ehrlich ascites tumours, Agents Actions 12:450 (1982).
47. U. Bachrach and G. Eilon, Interaction of oxidized polyamines with DNA I. Evidence for the formation of cross-links, Biochim. Biophys. Acta 145:418 (1967).
48. G. Eilon and U. Bachrach, Interaction of oxidized polyamines with DNA III. Association with nucleosides, mono- and polynucleotides, Biochim. Biophys. Acta 179:646 (1969).
49. U. Bachrach and S. Persky, Interaction of oxidized polyamines with DNA V. Inhibition of nucleic acid synthesis, Biochim. Biophys. Acta 179:484 (1969).
50. U. Bachrach, S. Don and H. Wiener, Antivirus action of acrolein, glutaraldehyde and oxidized spermine, J. Gen. Virol. 13:415 (1971).
51. K. Nishimura, T. Komano and H. Yamada, Effects of oxidized spermine and acrolein on the transforming activity of T4 DNA, Biochim. Biophys. Acta 262:24 (1972).
52. J. I. Hussain, C. J. Smith and J. C. Allen, Polyamine-mediated inhibition of in vitro cell proliferation is not due to acrolein, Cell Tissue Kinet. 16:583 (1983).
53. W. J. Byrd, D. M. Jacobs and M. S. Amoss, Synthetic polyamines added to cultures containing bovine sera, reversibly inhibit in vitro parameters of immunity, Nature 267:621 (1977).
54. J. M. Gaugas and P. Curzen, Polyamine interaction with pregnancy serum in suppression of lymphocyte transformation, Lancet 1:18 (1978).
55. A. Boveris, N. Oshino and B. Chance, The cellular production of hydrogen peroxide, Biochem. J. 128:617 (1972).
56. T. Ramasarma, Generation of H_2O_2 in biomembranes, Biochim. Biophys. Acta 694:69 (1982).
57. D. Del Principe, A. Menichelli, W. De Matteis, M. L. Di Corpo, S. Di Giulio and A. Finazzi-Agro, Hydrogen peroxide has a role in the aggregation of human platelets, FEBS Lett. 185:142 (1985).
58. M. P. Czech, J. C. Lawrence Jr and W. S. Lynn, Hexose transport in isolated brown fat cells. A model system for investigating insulin action on membrane transport, J. Biol. Chem. 249:5421 (1974).
59. H. Smith-Johannsen, J. F. Perdue, M. Ramjeesingh and A. Kahlenberg, Involvement of membrane sulfhydryls in the activation and maintenance of nutrient transport in chick embryo fibroblasts, J. Supramol. Struct. 7:37 (1977).
60. J. C. Lawrence Jr and J. Larner, Activation of glycogen synthase in rat adipocytes by insulin and glucose involves increased glucose transport and phosphorilation, J. Biol. Chem. 253:2104 (1978).
61. B. A. Helm and M. Gunn, The effect of insulinomimetic agents on protein degradation in H35 hepatoma cells, Mol. Cell. Biochem. 71:159 (1986).
62. J. N. Livingston, P. A. Gurny and D. H. Lockwood, Insulin-like effect of polyamines in fat cells, J. Biol. Chem. 252:560 (1977).
63. A. Finazzi-Agro, P. Conti, M. Hofmann, P. U. Angeletti and G. Rotilio, Activation of lymphocytes by H_2O_2, Agents Actions S 7:220 (1980).
64. J. Satrustegni and C. Richter, The role of hydroperoxides

as calcium release agents in rat brain mitochondria, Arch. Biochem. Biophys. 233:736 (1984).
65. D. P. Jones, H. Thor, M. T. Smith, S. A. Jervell and S. Orrenius, Inhibition of ATP-dependent microsomal Ca^{2+} sequestration during oxidative stress and its prevention by glutathione, J. Biol. Chem. 258:6390 (1983).
66. G. Bellomo, F. Mirabelli, P. Richelmi and S. Orrenius, Critical role of sulfhydryl group(s) in ATP-dependent Ca^{2+} sequestration by the plasma membrane fraction from rat liver, FEBS Lett. 163:136 (1983).
67. S. Baumhutter and C. Richter, The hydroperoxide-induced release of mitochondrial calcium occurs via a distinct pathway and leaves mitochondria intact, FEBS Lett. 148:271 (1982).
68. U. Bachrach and Y. M. Heimer eds., "The Physiology of Polyamines", CRC Press, Boca Raton FL, in press.
69. N. Seiler, Functions of polyamine acetylation, Can. J. Physiol. Pharmacol. 65:2024 (1987).
70. L. Persson, I. Holm and O. Heby, Regulation of ornithine decarboxylase mRNA translation by polyamines, J. Biol. Chem. 263:3528 (1988).
71. D. H. Russell and S. H. Snyder, Amine synthesis in rapidly growing tissues: ornithine decarboxylase activity in regenerating rat liver, chick embryo, and various tumors, Proc. Natl. Acad. Sci. 60:1420 (1968).
72. J. Janne and A. Raina, Stimulation of spermidine synthesis in the regenerating rat liver: relation to increased ornithine decarboxylase activity, Acta Chem. Scand. 22:1349 (1968).
73. H. G. Williams-Ashman, G. L. Coppoc and G. Weber, Imbalance in ornithine metabolism in hepatomas of different growth rates as expressed in formation of putrescine, spermidine and spermine, Cancer Res. 32:1924 (1972).
74. T. G. O'Brien, R. C. Simsiman and R. K. Boutwell, Induction of the polyamine biosynthetic enzymes in mouse epidermis and their specificity for tumor promotion, Cancer Res. 35:2426 (1975).
75. G. Scalabrino, P. Pigatto, M. E. Ferioli, D. Modena, M. Puerari and A. Caru, Levels of activity of the polyamine biosynthetic decarboxylases as indicators of the degree of malignancy of human cutaneous epitheliomas, J. Invest. Dermatol. 74:122 (1980).
76. G. Scalabrino, M. E. Ferioli, D. Modena, M. Puerari and G. Luccarelli, Levels of activity of the polyamine biosynthetic decarboxylases as indicators of the degree of malignancy of human brain tumors, Ital. J. Biochem. 31:60 (1982).
77. G. Quash, T. Keolouangkhot, L. Gazzolo, H. Ripoll and S. Saez, Diamine oxidase and polyamine oxidase activities in normal and transformed cells, Biochem. J. 177:275 (1979).
78. A. Sessa, M. A. Desiderio and A. Perin, Diamine oxidase activity induction in regenerating rat liver, Biochim. Biophys. Acta 698:11 (1982).
79. A. Perin, A. Sessa and M. A. Desiderio, Induction of diamine oxidase activity in some processes of growth, in: "Advances in Polyamine Research" vol. 4, U. Bachrach, A. Kaye and R. Chayen eds., Raven Press NY (1983).
80. B. Mondovi, P. Riccio, A. Riccio and G. S. Marcozzi, Ami-

ne oxidase activity in malignant human brain tumors, *in*: "Advances in Polyamine Research" vol. 4, U. Bachrach, A. Kaye and R. Chayen eds., Raven Press NY (1983).
81. M. A. Desiderio, A. Sessa and A. Perin, Induction of diamine oxidase activity in rat kidney during compensatory hypertrophy, *Biochim. Biophys. Acta* 714:243 (1982).
82. A. Perin, A. Sessa and M. A. Desiderio, Response of tissue diamine oxidase activity to polyamine administration, *Biochem. J.* 234:119 (1986).
83. A. Perin, A. Sessa and M. A. Desiderio, Polyamine levels and diamine oxidase activity in hypertrophic heart of spontaneously hypertensive rats and of rats treated with isoproterenol, *Biochim. Biophys. Acta* 755:344 (1983).
84. M. A. Desiderio, A. Sessa and A. Perin, Regulation of diamine oxidase expression by beta adrenoceptors in normal and hypertrophic rat kidney, *Biochim. Biophys. Acta* 845:463 (1985).
85. M. A. Desiderio, A. Sessa and A. Perin, Involvement of $beta_2$ adrenoceptors in the regulation of diamine oxidase activity in the heart of spontaneously hypertensive rats, *J. Hypertension* 4:S139 (1986).
86. F. Buffoni, Lysyl oxidase (characterization and clinical importance), *in*: "Structure and Functions of Amine Oxidases", B. Mondovi ed., CRC Press, Boca Raton FL (1985).
87. D. M. Kuhn, D. L. Murphy and M. B. H. Youdim, Physiological and clinical aspects of monoamine oxidase, *in*: "Structure and Functions of Amine Oxidases", B. Mondovi ed., CRC Press, Boca Raton FL (1985).
88. W. Bardsley, Inhibitors of copper amine oxidases, *in*: "Structure and Functions of Amine Oxidases", B. Mondovi ed., CRC Press, Boca Raton FL (1985).
89. T. P. Singer, Inhibitors of FAD-containing monoamine oxidases, *in*: "Structure and Functions of Amine Oxidases", B. Mondovi ed., CRC Press, Boca Raton FL (1985).
90. U. Bachrach, I. Ash, L. Abu-Elheiga, M. Hershkovitz and A. Loyter, Fusion-mediated microinjection of active amine and diamine oxidases into cultured cells: effect on protein and DNA synthesis in chick embryo fibroblasts and glioma cells, *J. Cell. Physiol.* 131:92 (1987).

METABOLISM OF 5'-METHYLTHIOADENOSINE IN NORMAL AND TRANSFORMED CELLS

HUMAN 5'-DEOXY-5'-METHYLTHIOADENOSINE PHOSPHORYLASE: KINETIC STUDIES AND CATALYTIC MECHANISM

Vincenzo Zappia, Fulvio Della Ragione, Gabriele Pontoni, Vincenzo Gragnaniello and Maria Cartenì-Farina

Department of Biochemistry of Macromolecules
First Medical School, University of Naples
Via Costantinopoli 16, Naples, Italy

INTRODUCTION

5'-Deoxy-5'-methylthioadenosine (MTA), is a sulfur nucleoside ubiquitously distributed in micromolar amounts in nature (1-4). The molecule originates from S-adenosylmethionine (AdoMet) through several metabolic pathways involving in all instances a nucleophilic attack to the activated methylenic γ-carbon adjacent to the electron withdrawing trivalent sulfur (1-10).

The catabolism of the thioether is due to two distinct enzymatic mechanisms: the first one involving the hydrolytic cleavage of MTA, which leads to adenine and 5-methylthioribose (11-14), and the second one implying the phosphorolytic breakdown of the N-C glycosidic bond to the purine base and 5-methylthioribose-1-phosphate (MTR-1-P) (1,2). Recent distribution studies have demonstrated that the two pathways are not confined to prokaryotes (MTA nucleosidase) and eukaryotes (MTA phosphorylase), as previously hypothesized. Indeed, several data demonstrate that a number of prokaryotes (13,14) have a specific MTA phosphorylase, while plants and some protozoa (15,16) remove the thioether by a nucleosidase.

No other enzymatic mechanism of MTA removal has been so far reported, with the unique exception of deamination by "non specific" enzyme reported solely in some microorganisms (17). It has to be mentioned that in addition to the enzymatic cleavage, MTA is actively excreted into the extracellular compartment as demonstrated in cultures of cancer cells lacking MTA phosphorylase (18) and by its presence in urine (19). The molecule in fact easily permeates biological membranes and a specific facilitated diffusion mechanism not energy-dependent has been reported to be operative (20). In this context it should be also mentioned that significant amounts of MTA sulfoxide are detectable in human urine, thus suggesting that oxidation of the nucleoside could occur _in vivo_ (21).

Recently, growing interest has been paid to MTA phosphorylase, which represents the only enzymatic system responsible for the disposition of the thioether in humans. Indeed, a number of observations make such an enzyme an interesting object of research. A primary role of the phosphorylase is obviously related to the control of the thioether cellular level. This role is of major importance taking into account that the molecule _in vitro_ exerts a significant inhibition on several enzymatic systems (1). Among them, the inhibition of spermine synthase is probably the most important, since the calculated Ki value falls within the range of the cellular levels of MTA (22); the other enzymes being affected to a lesser extent (1). Whether these inhibitions are operative _in vivo_ and in turn could be responsible for the reported cytostatic effect of the molecule is matter of investigation.

Recycling of adenine is of particular importance in conditions of increased polyamine biosynthesis, such as accelerated growth, since MTA cleavage represents in mammals the major non-dietary source of free adenine (5). On the other hand, no data are available from the Literature on the quantitative relevance of methionine formation from MTR-1-P compared to that of the other well-established routes of sulfur recycling, such as methylation of homocysteine (1).

Other two areas of research on MTA phosphorylase may have important clinical and pharmacological implications. The enzyme has been proposed by Savarese and colleagues as a system for cell delivery and activation of pro-drugs (23,24).

Moreover, the absence of MTA phosphorylase activity observed solely in a number of transformed cells is endowed with several important prognostic and therapeutic perspectives (1,25).

In spite of the large number of studies on MTA phosphorylase (26-29), the mammalian enzyme has been purified to homogeneity only recently from human placenta (30). The enzyme is an acidic protein with a M.W. of about 98,000 and is composed by three apparently identical polypeptide chains. The trimeric structure resembles that of the homologous enzyme, purine nucleoside phosphorylase (31), thus leading to the hypothesis of a common ancestor gene for the two phosphorylases. The reaction catalyzed by MTA phosphorylase is fully reversible with an equilibrium constant towards phosphorolysis of 1.39×10^{-2} (39), thus indicating that the synthetic reaction is greatly favoured over the phosphorolysis. Very few studies have been carried out on the regulation and turnover of mammalian MTA phosphorylase with the only exception of the rat liver enzyme, which has been reported to have a half-life longer than 12 hours (3). This finding suggests that the regulation of MTA phosphorylase activity does not primarily rely on rapid changes of the synthesis or degradation of the protein. On the other hand it is entirely possible that the enzyme activity is strictly regulated by substrate concentration.

In order to define the mechanism of the reaction and the sequence of binding of the substrates, a detailed kinetic investigation has been carried out on the purified placental enzyme. The results of this study could be useful in elucidating a possible in vivo regulation of the enzyme activity. Moreover the recognition sites of MTA phosphorylase have been studied by employing both MTA and adenine analogs as substrates or inhibitors of the enzyme. The data obtained will be discussed on the light of the catalytic mechanism.

KINETIC STUDIES

The double reciprocal plot of the initial velocities at variable concentrations of MTA and four fixed concentrations of phosphate yielded a series of lines intersecting to the left of the vertical axis (Fig. 1) . A similar pattern was observed when inorganic orthophosphate was varied at different fixed concentrations of MTA (data not shown).

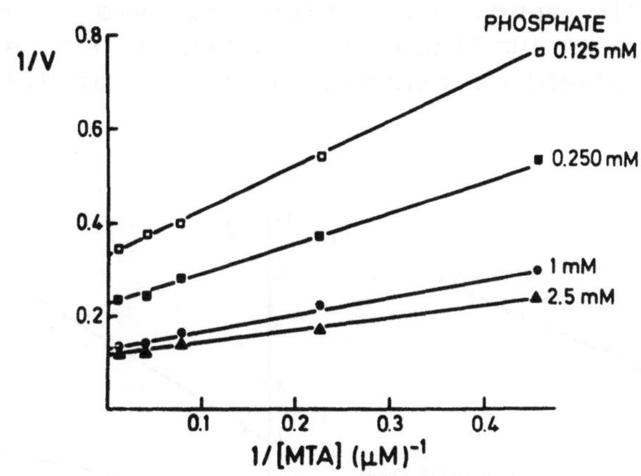

Fig.1. Plot of the reciprocal of the initial velocity (V) versus the reciprocal of MTA concentration at the various indicated fixed levels of phosphate

These results permit us to rule out a ping-pong mechanism and are consistent with a sequential mechanism. In order to distinguish among random, ordered or Theorell-Chance mechanisms, product inhibition studies have been designed. A summary of the results obtained are reported in Table 1.

Table 1. Product inhibition pattern for the phosphorolysis of 5'-methylthioadenosine

Inhibitory Product	Variable Substrate	Type of Inhibition	Inhibition Constant
Adenine	MTA	Competitive	Ki 2.8 μM
Adenine	Phosphate	Non-competitive	Kis 3.3 μM
			Kii 19.2 μM
MTR-1-P	MTA	Non-competitive	Kis 15.2 μM
			Kii 41.8 μM
MTR-1-P	Phosphate	Competitive	Ki 6.2 μM

When MTA was varied at a fixed non-saturating concentration of phosphate, the product adenine competitively inhibited the enzyme, while the phosphorylated sugar exerted a non-competitive inhibition in a similar experimental protocol. When phosphate was varied with MTA as non saturating fixed substrate, MTR-1-P acted as a competitive inhibitor, while the inhibition by adenine was non-competitive. This pattern permits us to exclude an ordered mechanism, as originally proposed by Garbers with the lung enzyme (30), although it does not allow us to distinguish between a rapid equilibrium random mechanism, with two dead-end complexes, and a Theorell-Chance mechanism.

In order to discriminate between these two possibilities, a novel kinetic approach, which employes alternative substrates, has been utilized (32). Arsenate was used instead of phosphate and 5'-chloroadenosine as analog of MTA (Fig. 2).

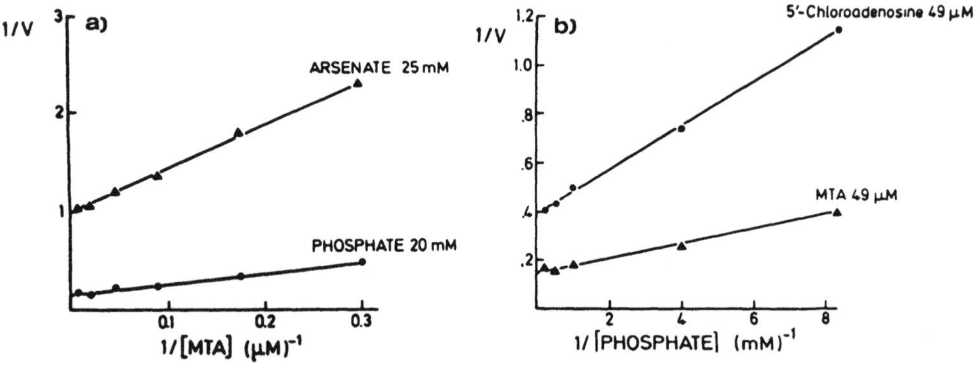

Fig.2. a) Plot of the reciprocal of the initial velocity (V) versus the reciprocal of MTA concentration at saturating levels of phosphate or arsenate; b) plot of the reciprocal of the initial velocity (V) versus phosphate concentration at saturating levels of MTA or 5'-chloroadenosine

As indicated in the figure, in both cases significant changes in slopes and intercepts were observed. These results are clearly diagnostic for a sequential mechanism with a random binding of the two substrates (32).

Fig.3 reports the proposed kinetic scheme; as shown, the free enzyme may bind independently MTA or phosphate, and only after this step the binding with the second substrate is operative. The ternary complex obtained, E-MTA-P, is then converted to E-MTR-1-P-adenine, such interconversion representing the rate-limiting step of the reaction. Finally, the two products are released in a random fashion. The reported occurrence of non-competitive inhibitions by adenine and the phosphorylated sugar indicates the existence of two dead-end complexes.

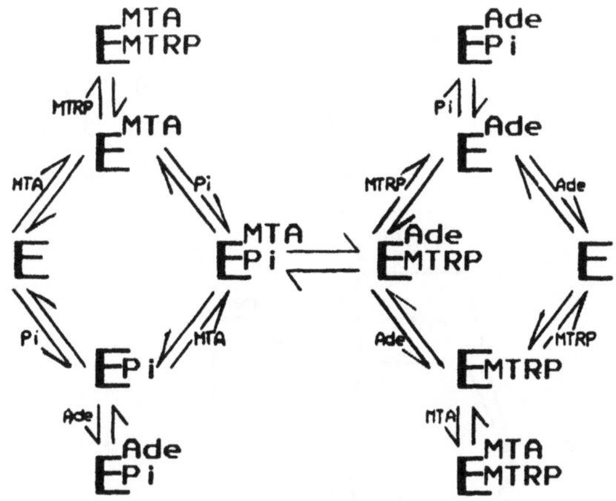

Fig.3. Proposed kinetic mechanism of human MTA phosphorylase

In order to infer information on the <u>in vivo</u> operative conditions of the enzyme, the kinetic constants of the enzyme relative to the four substrates have been compared with their cellular concentrations (Table 2).

Table 2. Kinetic constants and cellular content of MTA phosphorylase substrates

Substrate	Km value (μM)	Cellular content (nmoles/g)
5'- Methylthioadenosine	5	0.5 - 3 *
Phosphate	320	1,000 - 15,000
Adenine	23	30 - 90 **
5-Methylthioribose-1--phosphate	8	< 0.5

* From Refs. 3,4; ** From Ref. 33

The tissue content of MTR-1-P has been estimated by a recently developed isotope dilution procedure. The cellular concentration of the sugar phosphate appears, however, even lower than the limit of sensitivity of the method, which is 0.5 nmoles per g of tissue. Although, on the basis of the equilibrium constant, the synthesis of the thioether appears strongly favoured over the phosphorolytic breakdown, the very low levels of MTR-1-P indicate that the <u>in vivo</u> net MTA formation can be ruled out. Moreover, the reported data permit us to calculate the relative free energy of the already reported enzymatic complexes and therefore to estimate the probability of their formation (Fig. 4).

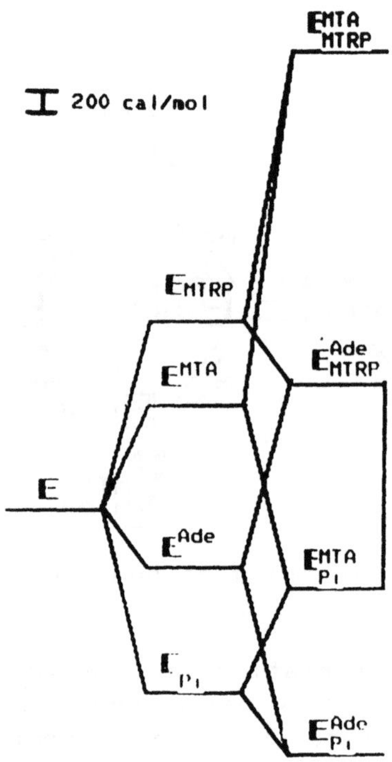

Fig.4. Relative free energies of the postulated enzymatic complexes. The concentration of the substrates employed in the determination of the relative free energies is also reported

Among the various possible routes, the most likely is the formation of an enzyme-phosphate complex which in turn can either generate a productive ternary complex or, more probably, can be turned into an abortive form, which has the lowest energy content.

After the interconversion of the ternary complex, the first product to be released should be MTR-1-P due to the lower free energy of enzyme-adenine complex compared to that of enzyme-MTR-1-P. Finally, the enzyme-adenine complex can either regenerate the free enzyme or interact with phosphate.

Based on these findings, the kinetic scheme previously reported can be modified as shown in Fig. 5 where the thick arrows indicate the favourite routes. Regarding the possible physiological meaning of such a scheme, it is very probable that the adenine levels play a key regulatory role on MTA phosphorylase activity by "freezing" the enzyme in an abortive enzyme-adenine-phosphate complex which accounts for 63% of the enzyme under physiological conditions. A decrease of adenine or a build-up of MTA should result in a rapid release of the enzyme from the inactive form with an increase of the phosphorylase activity. This, in particular, could be of relevance under conditions of increased polyamine biosynthesis or after pharmacological treatment with MTA or its analogs.

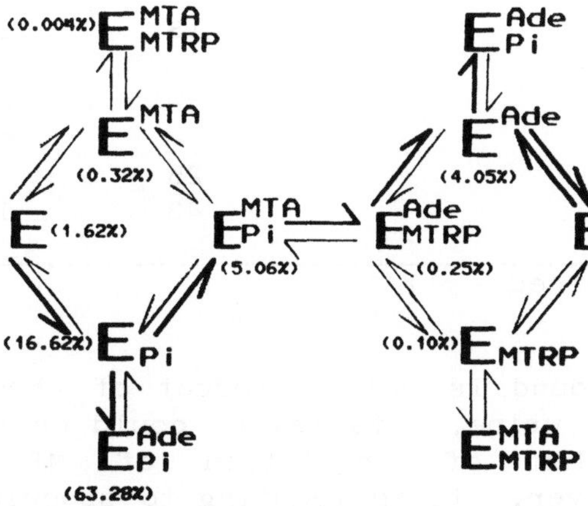

Fig.5. Kinetic mechanism of human MTA phosphorylase at physiological concentrations of substrates. In the brackets are reported the relative amounts of the enzymatic complexes

SUBSTRATE SPECIFICITY AND CATALYTIC MECHANISM

In order to identify the interaction sites and to clarify the catalytic mechanism of human placental MTA phosphorylase, several analogs of MTA and adenine, modified in selected regions, have been tested as substrates (Table 3).

Among the MTA analogs studied, only three are natural compounds, such as MTA sulfoxide, adenosine and the xylosyl derivative of MTA. As mentioned before, MTA sulfoxide has been found in the urine of healthy humans and in increased amounts in that of immunodeficient patients (21). The analog is not a substrate although it acts as a mild competitive inhibitor of the enzyme with respect to MTA (Ki 40 μM).

Table 3. Substrate specificity of human MTA phosphorylase

SUBSTRATE	Km(μM)	Vmax(%)	Vmax/Km
5'-Methylthioadenosine	3.5	100	28
Analogs of 5'-methylthioadenosine			
5'-Chloroadenosine	6	30	5
5'-Cyanoadenosine	N.D.	17	N.D.
5'-Methylthioadenosine sulfoxide	-	<0.05	-
Adenosine	380	6	0.016
2'-Deoxyadenosine	1308	1	0.001
3'-Deoxyadenosine	-	<0.05	-
Arabinoside adenine	-	<0.05	-
5'-Methylthioxylosyladenine	-	<0.05	-
5'-Methylthio-3-deazaadenosine	30	1.1	0.04
5'-Isobutylthio-3-deazaadenosine	95	3.1	0.05
Adenine	18	100	4.76
Analogs of adenine			
1-Methyladenine	-	<0.05	-
N-1-Oxyadenine	-	<0.05	-
2-Chloroadenine	25	9	0.36
2-Methyladenine	43	15	0.35

N.D., not determined

If the compound is not a product of chemical oxidation occurring in the urine, its level could be thought of as a further mechanism of regulation of MTA phosphorylase activity. Moreover, it is tempting to speculate that in MTA phosphorylase-deficient tumors, MTA sulfoxide can be produced intracellularly by microsomal enzymes.

As previously demonstrated by Savarese, adenosine is a substrate of MTA phosphorylase (34). However, its calculated Michaelis constant (Km 380 μM) is more than 30-fold higher than its cellular level (about 10 μM, ref. 35), and the maximal velocity of breakdown is not more than 6% of that of MTA at saturating concentrations. Therefore, it is unlikely that the phosphorylase plays a role in the removal of adenosine. Furthermore, based on the high concentration of ribose-1-phosphate, around 0.5 mM (36), it is entirely possible to hypothesize an enzyme-catalyzed synthesis of adenosine. In order to verify such a possibility, the determination of Km for ribose-1-phosphate appears necessary.

5'-Deoxy-5'-methylthioxylosyladenine is also a naturally occurring compound which has been very recently identified in the mollusc nudibranch *Doris verrucosa* (37,38). The compound is neither a substrate (Table 3) nor an inhibitor of human MTA phosphorylase, indicating the relevance of the chirality of C(3').

All the compounds reported in Table 3 which are not substrates of the placental MTAase are also assayed as inhibitor of the enzyme. Except MTA sulfoxide, none of the analogs significantly interferes with the phosphorylase activity.

By means of the obtained data and those reported in the Literature (23,24,26,29), at least five interaction sites between the enzyme and substrate have been identified (Fig. 6).

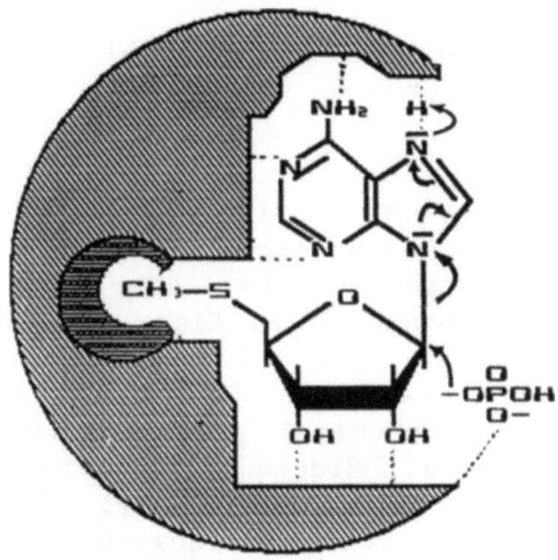

Fig.6. Postulated interaction sites and catalytic mechanism of human placenta MTA phosphorylase

The involvement of the alcoholic functions of ribose moiety and the relevance of the chirality at C(2') and C(3') has been demonstrated by the lack of activity of the xylosylderivative of MTA, 2'- and 3'-deoxyadenosine and arabinoside adenine (Table 3).

The interaction sites of the purine ring include nitrogens 1 and 3, as evidentiated by the inactivity of N-1-methyladenine, N-1-oxyadenine and by the very low substrate efficiency of the 3-deaza derivatives of MTA. In addition, as previously reported (1,2), the 6-amino group plays a relevant role.

The occurrence of a hydrophobic pocket has been proposed on the bases of the data employing 5'-modified analogs. While the lenght of the alkyl chain, at least from 1 to 4 carbon atoms, does not interfere significantly with the binding (1, 31), the increase in the polarity of the 5'-substituent strongly modifies the enzyme-substrate interaction. Indeed a decrease in the substrate efficiency has been evidenced employing 5'-chloroadenosine, 5'-cyanoadenosine and adenosine itself (Table 3).

Finally, it can be proposed a catalytic mechanism triggered by the protonation of nitrogen in position 7, which elicits an electron withdrawal and ultimately stimulates a nucleophilic attack by phosphate on the carbon in 1' of the ribose. This attack takes place through an inversion of configuration at C(1'), which also involves distortion of all bonds within the ribose moiety. In this respect recognition of alcoholic positions in 2' and 3' are of relevance in that they may distort the ring in such a way to assimilate the substrate structure to that of the transition state.

ACKNOWLEDGEMENTS

This work was supported by the National Research Council project "Oncologia", grant n. 86.00618.44.

REFERENCES

1. F. Della Ragione, M. Cartenì-Farina and V. Zappia, 5'-Deoxy-5'-methylthioadenosine: novel metabolic and physiological aspects, in "The Physiology of Polyamines", U. Bachrach and H. Eimer eds., CRC New York, in press
2. H.G. Williams-Ashman, J. Seidenfeld and P. Galletti, Trends in the biochemical pharmacology of 5'-deoxy-5'-methylthioadenosine, Biochem. Pharmacol. 31:277 (1982).
3. J.Seidenfeld, J. Wilson and H.G. Williams-Ashman, Androgenic regulation of 5'-deoxy-5'-methylthioadenosine concentration and methylthioadenosine phosphorylase activity in relation to polyamine metabolism of rat prostate, Biochem. Biophys. Res. Commun. 95:1861 (1980)
4. F. Della Ragione, M. Cartenì-Farina, M. Porcelli, G. Cacciapuoti and V. Zappia, High performance liquid chromatographic analysis of 5'-methylthioadenosine in rat tissues, J. Chromatogr. 226: 243 (1981)
5. N. Kamatani and D.A. Carson, Dependence of adenine production upon polyamine synthesis in cultured hùman lymphoblasts, Biochim. Biophys. Acta 675:344 (1983)
6. S. Nishimura, Y. Taya, Y. Kuchino and Z. Ohashi, Enzymatic synthesis of 3-(3-amino-3-carboxypropyl)uridine in Escherichia coli phenylalanine transfer RNA: transfer of the 3-amino-3-carboxypropyl group from S-adenosylmethionine, Biochem. Biophys. Res. Commun. 57:702 (1974)
7. H.J. Munch and R. Thiebe, Biosynthesis of the nucleoside Y in yeast tRNA-Phe: incorporation of the 3-amino-3-carboxypropyl-group from methionine, FEBS Letters 51:257 (1975)
8. G.L. Stoner and M.A. Eisenberg, Purification and properties of 7,8-diaminopelargonic acid aminotransferase. An enzyme in the biotin biosynthetic pathway, J. Biol. Chem. 250:4029 (1975)
9. Y. Taya, Y. Tanaka and S. Nishimura, Cell-free biosynthesis of discadenine, a spore germination inhibitor of Dictostelium discoideum, FEBS Letters 89:326 (1978)

10. S.K. Shapiro and D.J. Ehninger, The enzymatic decomposition of S-adenosylmethionine, J. Biol. Chem. 233: 631 (1958)
11. J.A. Duerre, A hydrolytic nucleosidase acting on S-adenosylhomocysteine and on 5'-methylthioadenosine, J. Biol. Chem. 237:3737 (1962)
12. F. Della Ragione, M. Porcelli, M. Cartenì-Farina and V. Zappia, Escherichia coli S-adenosylhomocysteine/5'-methylthioadenosine nucleosidase: purification, substrate specificity and mechanism of action, Biochem. J. 232:335 (1985)
13. S. Shimizu, T. Abe, S. Shiozaki and H. Yamada, Catabolism of methylthioadenosine and S-adenosylhomocysteine in microorganisms, in the "International Conference on Polyamines in Life Sciences", Lake Yamanake, Japan, July 14-18 (1986)
14. M. Cartenì-Farina, A. Oliva, G. Romeo, G. Napolitano, M. De Rosa, A. Gambacorta and V. Zappia, 5'-Methylthioadenosine phosphorylase from Caldariella acidophila, Eur. J. Biochem. 101:317 (1979)
15. A.B. Guranowski, P.K. Chiang and G. Cantoni, 5'-Methylthioadenosine nucleosidase: purification and characterization of the enzyme from Lupinus luteus seeds, Eur. J. Biochem. 114:293 (1981)
16. Y. Sugimoto, T. Toraya and S. Fukui, Studies on the metabolic role of 5'-methylthioadenosine in Ochromonas malhamensis, Arch. Microbiol. 108:175 (1976)
17. V. Zappia, P. Galletti, M. Cartenì-Farina and L. Servillo, A coupled spectrophotometric enzyme assay for methyltransferases, Anal. Biochem. 58:130 (1974)
18. N. Kamatani and D.A. Carson, Abnormal regulation of methylthioadenosine and polyamine metabolism in methylthioadenosine phosphorylase-deficient human leukemic cell lines, Cancer Res. 40:4178 (1980)
19. J.S. Mills, G.C. Mills and D.J. McAdoo, Isolation and identification of 5'-methylthioadenosine sulfoxide from human urine, Nucleosides and Nucleotides 2:45 (1983)
20. M. Cartenì-Farina, F. Della Ragione, G. Cacciapuoti, M. Porcelli and V. Zappia, Transport and metabolism of 5'-methylthioadenosine in human erythrocytes, Biochim. Biophys. Acta 727:221 (1983)
21. G.C. Mills, F.C. Schmalstreg and R.M. Goldblum, Urinary excretion of modified purines and nucleosides in immunodeficient children, Biochem. Med. 34:37 (1985)
22 R.L. Pajula and A. Raina, Methylthioadenosine, a potent inhibitor of spermine synthase from bovine liver, FEBS Letters 99:343 (1979)
23. T.M. Savarese, S-H. Chu, M-Y Chu and R.E. Parks Jr, 5'-deoxy-5'-methylthioadenosine phosphorylase-III: role of the enzyme in the metabolism and action of 5'-halogenated adenosine analogs, Biochem. Pharmacol. 34:361 (1981)

24. T.M. Savarese, A.J. Connistra, R.E. Parks Jr, J.A. Secrist III, A.J. Shorknacy and J.A. Montgomery, 5'-deoxy-5'-methylthioadenosine phosphorylase IV: biological activity of 2-fluoroadenine substituted 5'-deoxy-5'-methylthioadenosine analogs, Biochem. Pharmacol. 36:1881 (1987)
25. N. Kamatani, W.A. Nelson-Rees and D.A. Carson, Selective killing of human malignant cells deficient in methylthioadenosine phosphorylase, a purine metabolic enzyme, Proc. Natl. Acad. Sci USA 78:1219 (1981)
26. V. Zappia, A. Oliva, G. Cacciapuoti, P. Galletti, G. Mignucci and M. Cartenì-Farina, Substrate specificity of 5'-methylthioadenosine phosphorylase from human prostate, Biochem. J. 175:1043 (1978)
27. D.L. Garbers, Demonstration of 5'-methylthioadenosine phosphorylase activity in various rat tissues: some properties of the enzyme from rat lung, Biochim. Biophys. Acta 523:82 (1978)
28. G. Cacciapuoti, A. Oliva and V. Zappia, Studies on phosphate-activated 5'-methylthioadenosine nucleosidase from human placenta, Int. J. Biochem. 9:35 (1978)
29. T.M. Savarese, G.W. Crabtree and R.E. Parks Jr, 5'-methylthioadenosine phosphorylase-I: substrate activity of 5'-deoxyadenosine with the enzyme from Sarcoma 180 cells, Biochem. Pharmacol. 30:189 (1981)
30. F. Della Ragione, M. Cartenì-Farina, V. Gragnaniello, M.I. Schettino and V. Zappia, Purification and characterization of 5'-deoxy-5'-methylthioadenosine phosphorylase from human placenta, J. Biol. Chem. 261:12324 (1986)
31. J.D. Stoeckler, R.P.A. Agarwal, K. Schmid and R.E. Parks Jr. Purine nucleoside phosphorylase from human erythrocyte: physicochemical properties of the crystalline enzyme, Biochemistry 17:278 (1978)
32. K. Radika and D. Northrop, A new diagnostic for enzymatic mechanisms using alternative substrates, Anal. Biochem. 141:413 (1984)
33. B-G Chun, W.K. Paik and S. Kim, Simple separation of adenine and adenosyl-sulfur compounds by high performance liquid chromatography, J. Chromatogr. 264:321 (1983)
34. T.M. Savarese, The role of 5'-deoxy-5'-methylthioadenosine phosphorylase in the metabolism and action of 5'-deoxyadenosine, 5'-deoxy-5'-methylthioadenosine, and their analogs in murine tumor cells, Ph. Thesis (1980)
35. D. Fell, L.E. Benjamin and R.D. Steele, Determination of adenosine and S-adenosyl derivatives of sulfur amino acids in rat liver by high performance liquid chromatography, J. Chromatogr. 345:150 (1985)
36. P.L. Ipata and M. Camici, An enzymatic radioactive assay to determine ribose-1-phosphate in tissues, Anal. Biochem. 112:151 (1981)

37. G. Cimino, A. Crispino, S. De Stefano, M. Gavagnin and G. Sodano, A naturally occurring analog of methylthioadenosine (MTA) from the nudibranch mollusc Doris verrucosa, Experientia 42:1301 (1986)
38. M. Porcelli, G. Cacciapuoti, G. Cimino, M. Gavagnin, G. Sodano and V. Zappia, Characterization and biogenesis of 5'-methylthioxylofuranosyladenine, a new natural analog of 5'-methylthioadenosine, in this book

METHYLTHIOADENOSINE (MeSAdo) PHOSPHORYLASE DEFICIENCY IN MALIGNANCY

Dennis A. Carson, Tsutomu Nobori, E. Olavi Kajander, Carlos J. Carrera, Masaru Kubota, and Hisashi Yamanaka

Department of Basic and Clinical Research
Research Institute of Scripps Clinic
10666 North Torrey Pines Road, La Jolla, California

INTRODUCTION

The abnormal growth properties of cancer cells must be mediated finally by changes in specific metabolic pathways. In general, malignant transformation is associated with an increase in the activities of enzymes involved in DNA, RNA, and protein synthesis, and a concomitant decrease in the activities of enzymes that degrade cellular metabolites. However, no specific metabolic pathway has been found that distinguishes normal from malignant cells, or is common to all tumors.

For several years, this laboratory has been studying hereditary abnormalities of purine nucleoside metabolism that impair lymphocyte development and function. The analysis of inborn errors of metabolism has provided essential information concerning the functional roles of biochemical pathways in human cells. It is equally important to determine how somatically acquired defects in purine and polyamine metabolism might alter the biological properties of leukemia cells in a way that facilitated their growth, malignant potential, or response to chemotherapy.

Genetic Basis of MeSAdo Phosphorylase (MeSAdoase) Deficiency

In 1977, Toohey reported the absence of MeSAdoase in four murine malignant hematologic cell lines.[1] MeSAdo is a stoichiometric product, and a potent inhibitor, of the amino-propyltransferase reactions that synthesize spermidine and spermine from decarboxylated S-adenosylmethionine.[2] MeSAdo has also been reported to be an inactivator of S-adenosylhomocysteine hydrolase, and a direct inhibitor of some S-adenosylmethionine dependent transmethylases [reviewed in [3]]. However, MeSAdo does not normally accumulate intracellularly. Rather, in the presence of inorganic phosphate MeSAdoase reversibly cleaves the thioether nucleoside to adenine and 5'-methylthioribose-1-phosphate. The latter compound is recycled to methionine.

Among 31 established human malignant tumor cell lines that were analyzed in this laboratory, 7 (23%) lacked detectable MeSAdoase (< 0.02 nmols/min/mg protein).[4] The enzyme deficient cell lines included 6 hematologic neoplasms and 1 breast cancer. In contrast, all 16 lines of non-malignant origin, derived from lymphoblasts, fibroblasts, and epithelial cells, contained substantial enzyme activity (> 0.2 nmols/min/mg protein).

Two possible explanations for the absence of MeSAdoase among established human malignant tumor cell lines were (i) that the enzyme disappeared during the prolonged maintenance of the malignant cells under artificial growth conditions in tissue culture, and (ii) that the normal cells from which the tumors arose were enzyme deficient. To determine the validity of these competing hypotheses, we measured MeSAdoase levels in fresh cells from 20 patients with leukemia and 9 with solid tumors.[5] As determined by autoradiography of viable cells and by direct enzymatic analysis, samples from one individual with pre-T cell acute lymphoblastic leukemia and one with a common acute lymphoblastic leukemia were MeSAdoase deficient. Interestingly, both these patients were refractory to multiple chemotherapeutic agents. In contrast, normal peripheral blood lymphocytes, thymic lymphocytes, and at least 98% of normal bone marrow cells, had easily detectable MeSAdoase activity. Finally, extracts of normal spleen, colon, lung, breast, thymus, esophagus, and prostate contained abundant levels of the enzyme. Thus, MeSAdoase deficiency is a specific property of certain leukemia cell clones, both *in vivo* and *in vitro*.

Subsequently, we measured levels of the enzyme in interspecies somatic cell hybrids prepared by fusing MeSAdoase deficient mouse L cell lines with human fibroblasts.[6] In the hybrid clones, MeSAdoase activity segregated concordantly with adenylate kinase-1, a marker for human chromosome 9, but not with enzyme markers for any other human chromosome. Importantly, among hybrid clones derived from human fibroblasts with a reciprocal translocation between chromosomes 9 and 17, MeSAdoase activity was confined to cells containing the 9 pter ---> 9 ql2 region. These results indicated that the structural gene for human MeSAdoase, designated MTAP, can be assigned to the 9 pter ---> 9q12 region of human chromosome 9. They raised the possibility that deletions or translocations affecting this chromosomal region might be associated with the enzyme deficient state.

This suggestion was supported by the findings of Chilcote *et al.*[7] These investigators reported that some MeSAdoase deficient malignant clones had cytogenetic abnormalities at the p21-p22 region on the short arm of chromosome 9, a known hereditary fragile site. Thus, it is conceivable that the absence of MeSAdoase indicates the concomitant loss of one or more closely linked, recessive oncogenes that are encoded in the 9p21-9p22 region. Further molecular analyses will be required to test this hypothesis.

Effects of MeSAdo Phosphorylase Deficiency on Cell Metabolism

The effects of the loss of MeSAdoase on cell growth and metabolism cannot readily be ascertained by using naturally enzyme deficient cell lines that may differ in multiple ways from MeSAdoase positive cells. Biochemical genetics provides a means to determine the exact metabolic consequences of MeSAdoase deficiency. Therefore, we perfected a technique for the selection of MeSAdoase deficient mutants, and characterized two MeSAdoase deficient clones derived from the transplantable murine T lymphoma cell line R1.1.[8]

Briefly, MeSAdoase deficient clones were selected by "tritium suicide". Mutagenized cells were briefly suspended in a medium containing 5'-chloro-[2,8-^{3}H] adenosine, a MeSAdo analog. The treated cells were washed, frozen in liquid nitrogen for two weeks to allow radiation damage to accumulate, and finally were thawed and cloned. The death of cells incubated with 5'-Cl[2,8^{3}H] adenosine requires: (i) the transport of the nucleoside into the cells, (ii) its cleavage by MeSAdoase to yield [2,8-^{3}H] adenine, (iii) the conversion of adenine to AMP, and (iv) the further phosphorylation and incorporation of [2,8-^{3}H] AMP into nucleic acid. Cells defective in any step of this pathway would be expected to survive the "tritium suicide" procedure. Variants deficient in each of the four steps listed above were obtained after the final cloning. Remarkably, MeSAdoase deficiency was the most common abnormality, occuring in approximately one-third of the clones. By comparison adenine phosphoribosyltransferase deficient clones were ten-fold less frequent.

The MeSAdoase deficient variants excreted substantial amounts of MeSAdo into the culture medium (as much as 0.32 mol/hr/mg protein). The amount of MeSAdo in the medium of MeSAdoase deficient lymphoma cells eventually rose to 3-5 μM. This concentration of MeSAdo was sufficient to impede eventually the growth of the lymphoma cells. The synthesis and release of MeSAdo was dose-dependently blocked by two inhibitors of polyamine synthesis, α-difluoromethylornithine (DFMO) and methylglyoxal-bis-guanylhydrazone (MGBG).

A mutant lymphoblastoid cell line was selected that was deficient in adenine phosphoribosyltransferase (APRT). The APRT deficient cells generated adenine at a pace equivalent to the rate of MeSAdo formation by cells lacking MeSAdoase. Furthermore, cells deficient in MeSAdoase and APRT did not produce adenine at all. These results established that mammalian cells generate adenine at a considerable rate, and only via the polyamine-MeSAdoase pathway.

Regulation of MeSAdo and L-Homocysteine Production

The consumption of S-adenosylmethionine during polyamine synthesis and transmethylation reactions yields stoichiometric amounts of MeSAdo and S-adenosylhomocysteine, respectively. The S-adenosylhomocysteine subsequently is cleaved to adenosine plus L-homocysteine. Both pathways of S-adenosylmethionine metabolism are essential for cell growth and function, and could play a role in malignant transformation.

Information concerning the regulation of S-adenosylmethionine and polyamine metabolism in viable cells under changing growth conditions is meager. Intracellular polyamine levels, and the activities of ornithine decarboxylase and S-adenosylmethionine decarboxylase, have been reported to change considerably during the cell growth cycle. However, the intracellular polyamine content reflects a balance between synthesis, degradation, excretion, and uptake. Furthermore, measurements of enzyme activities in crude extracts provide only indirect information concerning metabolic rates in intact cells.

We demonstrated that both anchorage-dependent and anchorage-independent MeSAdoase deficient malignant cells excreted MeSAdo and L-homocysteine into the culture medium.[9] MeSAdo and L-homocysteine are products of polyamine synthesis and transmethylation reactions, respectively. Therefore, the measurement of the time-dependent accumulation of MeSAdo and L-homocysteine in the culture medium enabled us to estimate for the first time the actual rates of both polyamine synthesis and transmethylation reactions under dynamic growth conditions.

The two S-adenosylmethionine consuming pathways proceeded quickly, when compared to the cellular S-adenosylmethionine content. The intracellular S-adenosylmethionine pool sufficed for only 5-10 minutes of polyamine synthesis and transmethylation reactions.

The kinetics of polyamine synthesis and transmethylation reactions during the cell growth cycle were strikingly different. MeSAdo production (polyamine synthesis) was maximal early after the release of malignant lymphoblasts from density dependent growth arrest. By contrast, L-homocysteine formation (transmethylation reaction) was proportional to the specific growth rate, and reached a peak during mid-exponential phase, at which time MeSAdo excretion was falling.

Effects of MeSAdo on Polyamine and S-adenosylmethionine Metabolism in Intact Cells

The factors that control the rate of polyamine synthesis in mammalian cells have been studied intensively, but are still not well understood. Literally hundreds of papers have documented that growth promoting stimuli lead to a rapid increase in the activities of ornithine decarboxylase and S-adenosylmethionine decarboxylase [reviewed in [10]]. However, when general protein synthesis increases, enzymes that have a high rate of synthesis and degradation always respond with a rapid and marked rise in activity. By eontrast, proteins with a slow turnover rate increase less in activity, and at a much later time. Hence, the observed elevations in ornithine decarboxylase and S-adenosylmethionine decarboxylase that are induced by growth stimuli are entirely expected, and reveal little specifically about the actual factors that influence the enzyme activities.

The end products of the polyamine biosynthetic pathway are spermidine and spermine, and MeSAdo. The addition of spermidine to mammalian cells has been shown to decrease the activities of ornithine decarboxylase and S-adenosylmethionine decarboxylase. Initially, we noted that as little as 3 μM MeSAdo increased putrescine levels in MeSAdoase deficient lymphoma cells. However, we could not ascertain if the elevation reflected a block in putrescine consumption, or an increase in synthesis. Moreover, we needed to determine whether the accumulation was secondary to changes in sperimidine and spermine pools.

To address these issues, we measured the time dependent effects of MeSAdo on (i) ornithine decarboxylase and S-adenosylmethionine decarboxylase activities, (ii) putrescine and polyamine levels, and (iii) decarboxylated S-adenosylmethionine and S-adenosylhomocysteine levels in MeSAdoase deficient cells.[11] In viable lymphoblasts, the addition of MeSAdo elicited within one hour an increase in the activities of both decarboxylases, and a subsequent rise in putrescine and decarboxylated S-adenosylmethionine. S-adenosylhomocysteine did not rise appreciably. The increased decarboxylase activities occurred prior to any detectable changes in intracellular polyamine pools, but simultaneously with a reduced velocity of spermine synthesis, as estimated by precursor-product incorporation studies with radioactive ornithine. DFMO, an inhibitor of ornithine decarboxylase, did not prevent the MeSAdo-triggered increase in S-adenosylmethionine decarboxylase. MGBG, an inhibitor of S-adenosylmethionine decarboxylase, did not deter the MeSAdo-induced rise in ornithine decarboxylase. The altered enzyme activities were not a non-specific consequence of changes in cell growth kinetics, because (i) they were completed several hours before cell proliferation declined, and (ii) the same changes were observed in secondary mutants impervious to the anti-proliferative effects of MeSAdo.

These experiments suggested that MeSAdo-induced perturbations in the velocity of spermine synthesis, can regulate ornithine decarboxylase and S-adenosylmethionine decarboxylase activities independently of growth state. In the MeSAdoase deficient leukemias, the accumulation of MeSAdo could inappropriately stimulate an increase in the activities of both enzymes. Both ornithine decarboxylase and S-adenosylmethionine decarboxylase are thought to play a role in the initiation of cell proliferation. Conceivably, the induction of the two enzymes by MeSAdo could provide a natural pressure for the selection of MeSAdoase deficient leukemias.

Mechanism of MeSAdo Toxicity

Information concerning the mechanism of MeSAdo toxicity toward mammalian cells is scanty. As discussed earlier, MeSAdo is a potent inhibitor of spermine synthase (Ki about 1-2 μM).[2] These MeSAdo concentrations are routinely achieved during the growth of MeSAdoase deficient cells. One hundred to 1,000 fold higher concentrations of the thioether nucleoside have been reported to block S-adenosylmethionine-dependent transmethylation reactions, either directly or indirectly via inhibition of S-adenosylhomocysteine hydrolase. The true importance of these effects in viable cells has not been established.

To address these issues, we have selected and characterized a mutant clone (designated AKR4) of R1.1 lymphoblasts that is cross resistant to the anti-proliferative actions of MeSAdo, adenosine, and several other adenine nucleosides. The mutants have remained stable during more than one year of continuous passage in tissue culture. Moreover, the mutant cells are also resistant to cycloleucine, ethionine, and other inhibitors of S-adenosylmethionine synthesis.[12]

Biochemical analyses of the resistant cells have shown that the membrane transport of nucleosides and methionine is entirely normal. However, when compared to MeSAdo sensitive lymphoblasts, the mutant cells have 5-7 fold increased levels of S-adenosylmethionine. This change reflects an augmented activity of methionine adenosyltransferase, the enzyme responsible for the synthesis of S-adenosylmethionine from methionine and ATP.

The fact that an increase in cellular S-adenosylmethionine pools could render cells highly resistant to MeSAdo toxicity strongly suggested that the anti-proliferative effects of the thioether nucleoside were due to competitive inhibition of an S-adenosylmethionine-dependent reaction. S-adenosylmethionine functions as an aminocarboxypropyl donor in (i) the modification of several tRNA species, and (ii) the synthesis of diphthamide, an unusual amino acid that is found only in elongation factor-2 (EF-2). MeSAdo is the presumed nucleoside product, and is therefore a potential product inhibitor of these aminocarboxypropyl transfer reactions.

Diphtheria toxin catalyzes the transfer of the ADP-ribose moiety from NAD^+ to diphthamide in EF-2. In the presence of excess diphtheria toxin, one molecular of NAD^+ binds specifically to one residue of diphthamide. The total amount of incorporated $[^3H]NAD^+$ is proportional to the amount of diphthamide in the sample protein. Therefore, we could readily measure the effects of MeSAdo on diphthamide synthesis in intact cells. As measured by susceptibility to diphtheria toxin-induced ADP-ribosylation, MeSAdo inhibited the formation of diphthamide in EF-2.[13] The inhibition was not due to a nonspecific effect on protein synthesis. Indeed, exogenous MeSAdo substantially protected the lymphoma cells from the lethal effects of diphtheria toxin. These results suggest that MeSAdo can specifically modulate the biosynthesis of diphthamide in EF-2 in murine malignant lymphoma cells, and that this effect may be related to the mechanism of MeSAdo toxicity.

Exploiting MeSAdo Phosphorylase Deficiency in Chemotherapy

Although MeSAdo is a potent inhibitor of spermine and spermidine synthesis, cultivation of cells with MeSAdo does not deplete polyamine pools. Instead, MeSAdo treatment triggers a compensatory rise in ornithine decarboxylase and S-adenosylmethionine decarboxylase, that eventually restores polyamine synthesis.

To prevent cellular adaptation to MeSAdo exposure, we co-incubated lymphoma cells with DFMO, a potent inhibitor of ornithine decarboxylase.[14] Even in MeSAdo resistant cell lines, DFMO and MeSAdo acted synergistically to cause severe polyamine depletion and growth inhibition. The MeSAdo-DFMO drug combination would be impractical for chemotherapy, because of the diverse other biochemical effects of the thioether nucleoside. However, it is conceivable that MeSAdo phosphorylase deficient neoplastic cells might display enhanced sensitivity to inhibitors of polyamine synthesis, because of endogenous MeSAdo accumulation. Since MeSAdo phorphorylase deficient malignant cells leak purines extracellularly and cannot recycle MeSAdo to methionine, their growth may also be especially sensitive to chemotherapeutic agents that interfere with purine nucleotide or methionine biosynthesis.[4,15]

ACKNOWLEDGMENTS

Supported by NIH grant GM23200. C.J. Carrera is the recipient of a Clinical Investigator Award (CA01100) from the National Cancer Institute.

REFERENCES

1. J. I. Toohey, Methylthio group cleavage from methylthioadenosine. Description of an enzyme and its relationship to the methylthio-requirement of certain cells in culture, Biochem.Biophys.Res.Commun. 78:1273 (1977).
2. R-L. Pajula, and A. Raina, Methylthioadenosine, a potent inhibitor of spermine synthase from rat brain, FEBS Lett. 99:343 (1979).
3. H. G. Williams-Ashman, J. Seidenfeld, and P. Galletti, Trends in the biochemical pharmacology of 5'-deoxy-5'-methylthioadenosine, Biochem.Pharmacol. 31:277 (1982).
4. N. Kamatani, W. A. Nelson-Rees, and D. A. Carson, Selective killing of human malignant tumor cells deficient in methylthioadenosine phosphorylase, a purine metabolic enzyme, Proc.Natl.Acad.Sci.USA 78:1219 (1981).
5. N. Kamatani, A. L. Yu, and D. A. Carson, Deficiency of methylthioadenosine phosphorylase in human leukemic cells in vivo, Blood 60:1387 (1982).
6. C. J. Carrera, R. L. Eddy, T. B. Shows, and D. A. Carson, Assignment of the gene for methylthioadenosine phosphorylase to human chromosome 9 by mouse-human somatic cell hybridization, Proc.Natl.Acad.Sci.USA 81:2665 (1984).
7. R. R. Chilcote, E. Brown, and J. D. Rowley, Lymphoblastic leukemia with lymphomatous features associated with abnormalities of the short arm of chromosome 9, N.Engl.J.Med. 313:286 (1985).
8. M. Kubota, N. Kamatani, and D. A. Carson, Biochemical genetic analysis of the role of methylthioadenosine phosphorylase in a murine lymphoid cell line, J.Biol.Chem. 258:7288 (1983).
9. T. Iizasa, and D. A. Carson, Differential regulation of polyamine synthesis and transmethylation reactions in methylthioadenosine phosphorylase deficient mammalian cells, Biochim.Biophys.Acta 844:280 (1985).
10. A. E. Pegg, and P. P. McCann, Polyamine metabolism and function, Am.J.Physiol. 243:C212 (1982).
11. M. Kubota, E. O. Kajander, and D. A. Carson, Independent regulation of ornithine decarboxylase and S-adenosylmethionine decarboxylase in methylthioadenosine phosphorylase-deficient malignant murine lymphoblasts, Cancer Res. 45:3567 (1985).

12. E. O. Kajander, M. Kubota, C. J. Carrera, J. A. Montgomery, and D. A. Carson, Resistance to multiple adenine nucleoside and methionine analogs in mutant lymphoma cells with enlarged S-adenosylmethionine pools, Cancer Res. 46:2866 (1986).
13. H. Yamanaka, E. O. Kajander, and D. A. Carson, Modulation of diphthamide synthesis by 5'-deoxy-5'-methylthioadenosine in murine lymphoma cells, Biochim.Biophys.Acta 888:157 (1986).
14. H. Yamanaka, M. Kubota, and D. A. Carson, Synergistic inhibition of polyamine synthesis and growth by difluoromethylornithine plus methylthioadenosine in methylthioadenosine phosphorylase-deficient murine lymphoma cells, Cancer Res. 47:1771 (1987).
15. D. A. Carson, E. H. Willis, and N. Kamatani, Metabolism to methionine and growth stimulation by 5'-methylthioadenosine and 5'-methylthioinosine in mammalian cells, Biochem.Biophys.Res.Commun. 112:391 (1983).

PHYSICO-CHEMICAL AND IMMUNOLOGICAL PROPERTIES OF BOVINE LIVER 5'-DEOXY-5'-METHYLTHIOADENOSINE PHOSPHORYLASE

Fulvio Della Ragione, Adriana Oliva, Massimo Fioretti, Gian Luigi Russo, Rosanna Palumbo and Vincenzo Zappia

Department of Biochemistry of Macromolecules,
First Medical School, University of Naples,
Via Costantinopoli 16, Naples, Italy

INTRODUCTION

The search for specific enzymatic reactions which can clearly differentiate normal mammalian cells from both malignant cells and infective microorganisms represents a major strategy for the development of selective therapeutical approaches. In particular, the intensive research carried out in the past on purine and pyrimidine metabolism has yielded a very rich harvest of drugs effective in the treatment of malignancy as well as of parasitic diseases.

In this context, growing attention has been paid, in recent years, to the enzymatic cleavage of 5'-deoxy-5'-methylthioadenosine (MTA), which represents the major, if not the sole, non-dietary source of free adenine (1). MTA breakdown, which requires either a hydrolytic cleavage catalyzed by specific nucleosidase or a phosphorolytic attack on N-C glycosidic bond due to MTA phosphorylase, may be considered a promising route to selectively interfere with the purine recycling. MTA nucleosidases, with remarkably different substrate specificity, have been demonstrated in some strains of bacteria and protozoa (2-4) as well as in all the plants examined (5,6). MTA phosphorylase activity has been detected in all the mammalian cells and tissues studied (7-12), in insects (13), and in numerous bacteria and protozoa (14,15).

In addition, the finding that MTA phosphorylase activity is absent in a number of mammalian cell lines, while it is always detectable in normal tissues, is endowed with very promising clinical implications including new selective therapeutical strategies (16). MTA phosphorylase (MTAase) was discovered by Pegg and Williams-Ashman in 1969 (7) as an outcome of a study on the synthesis of spermidine and spermine in rat ventral prostate. The activity was found to be strictly dependent on phosphate but the reaction products, namely adenine and 5-methylthioribose-1-phosphate (MTR-1-P), were characterized in detail only in 1979 by Ferro et al. (12).

Following the first report, the enzyme was described and partially characterized in several human and rat tissues (8-10), in human and murine cultured cell lines (11,17), in *Drosophila melanogaster* (13) and more recently in some bacteria and protozoa (14,15). The enzyme, as noted above, plays a key role in the purine metabolism. Moreover, since the carbon skeleton of MTR-1-P (except C-1) is recycled to methionine, an involvement of MTAase in the homeostasis of the amino acid pool is conceivable (18).

Recently, a substrate specificity quite different from the mammalian counterpart has been reported for the enzyme from the protozoa *Trypanosoma cruzi*; therefore it may represent a potential route for a specific chemotherapeutic attack on this and probably other parasites (15).

Although a large number of reports on MTAase have been published, the enzyme has been only recently purified to homogeneity and partially characterized from a mammalian source, namely human placenta (19). However, a more detailed investigation on the enzyme has been hampered by the scarce amount of the protein recovered after the 30,000-fold purification required to obtain the pure phosphorylase. Thus, mammalian MTAase has been prepared from bovine liver, which represents a more suitable source for obtaining substantial quantity of enzyme. The availability of adequate amounts of pure protein has allowed the elucidation of several structural and molecular properties of the enzyme. Furthermore, to gain information on the occurrence of inactive forms of MTAase in malignant cells lacking the enzyme activity, immunological studies have been performed. The following sections report the results of the physico-chemical study as well as preliminary data on immunological properties of the enzyme.

PURIFICATION AND PROPERTIES OF BOVINE LIVER MTA PHOSPHORYLASE

The procedure for the purification of bovine liver MTAase employes several conventional steps taking advantage of the remarkable stability of the enzyme against acid and thermal denaturation (Table 1). Approximately 2 mg of purified enzyme were obtained per Kg of bovine liver in about 13% yield. The final specific activity ranged between 9 and 12 μmol of MTA cleaved per min per mg of protein. This value is similar to that obtained for the placental enzyme. The homogeneity of the purified protein was evaluated by polyacrylamide gel electrophoresis under native and denaturing conditions, as well as by isoelectrofocusing. By both the approaches a single band was observed. In addition, the presence of only one symmetrical boundary in the sedimentation velocity experiments, as well as the linearity of the plot of sedimentation-equilibrium analysis, confirmed that the protein was essentially homogeneous.

Table 1. Purification of 5'-deoxy-5'-methylthioadenosine phosphorylase from bovine liver

	TOTAL PROTEIN (mg)	SPECIFIC ACTIVITY (units*/mg)	YIELD (%)	PURIFICATION (fold)
Supernatant at 15,000 g	203,000	0.0010	100	1
pH 5.2 treatment	119,000	0.0017	99.5	1.7
Heat treatment	66,000	0.0027	88.6	2.7
Ammonium sulfate (45-70%)	39,600	0.0034	66.0	3.4
Acetone (40-60%)	20,600	0.0060	61.0	6.0
DEAE-Sephacel	696	0.17	58.1	170
Chromatofocusing	21	2.1	21.6	2,100
Sephacryl S-200	2.4	10.3	12.3	10,300

* One unit is the amount of enzyme which cleaves 1 μmol of MTA per min at 37°C

In order to investigate the content and behaviour of the two aromatic amino acids, tryptophan and tyrosine, the analysis of the second derivative of the ultraviolet absorption spectrum of MTAase either in native or in denaturing conditions was carried out (Fig. 1).

Indeed the unfolding of the protein, which results in exposure to the polar solvent of chromophores previously buried in non-polar regions of the molecule, allows the estimation of the relative contribution of the aromatic amino acids to the absorption spectrum. By means of the second derivative spectrum of the unfolded enzyme, the ratio between tyrosine and tryptophan content has been estimated to be 1.4. Moreover, by comparing the second derivatives of the spectrum of native and 6 M guanidine-treated enzyme, the percentage of the tyrosyl residues exposed to the solvent has been calculated to be about 38%.

The amino acid composition of pure bovine liver MTAase has been estimated by a single column analysis on an amino acid analyzer after the hydrolysis of the protein for 24, 48 and 72 hrs in 6 M HCl at 110°C. The concentration of threonine, serine, valine and tyrosine are extrapolated values.

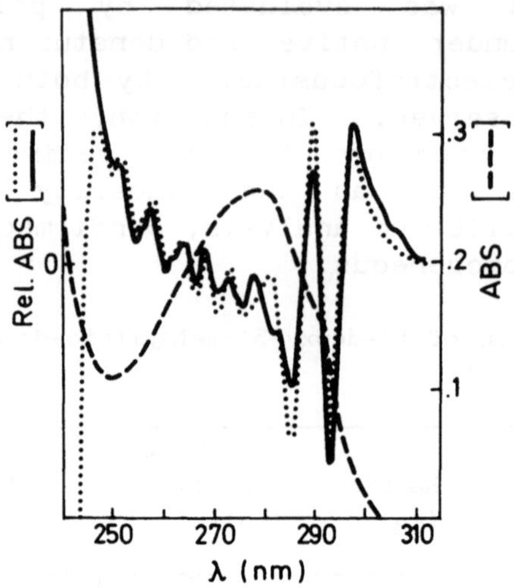

Fig. 1. Absorption spectrum of native bovine liver MTAase (---) and second derivative spectra of native (——) and unfolded (·····) enzyme. The native protein was in 50 mM phosphate, 0.1 M KCl, pH 7.0 and the unfolded protein was in the same buffer plus 6 M guanidine

The content of tryptophan has been determined by employing the calculated ratio between tyrosine and tryptophan (Table 2).

Table 2. Amino acid composition of 5'-deoxy-5'-methylthio-adenosine phosphorylase from bovine liver

AMINO ACID	RESIDUES/MOL	RESIDUE WT %
ASP	75	8.82
THR	51	5.27
SER	84	7.55
GLU	93	12.27
PRO	45	4.47
GLY	75	4.37
ALA	81	5.88
VAL	57	5.78
MET	15	2.01
ILE	57	6.59
LEU	72	8.33
TYR	21	3.50
PHE	27	4.06
HIS	24	3.36
LYS	60	7.86
ARG	36	5.75
TRP	15	2.85
CYS	12	1.27

Fig. 2 reports the equilibrium sedimentation profile at 6,000 and 8,000 rev/min plotted as ln A_{280} versus the square radius.

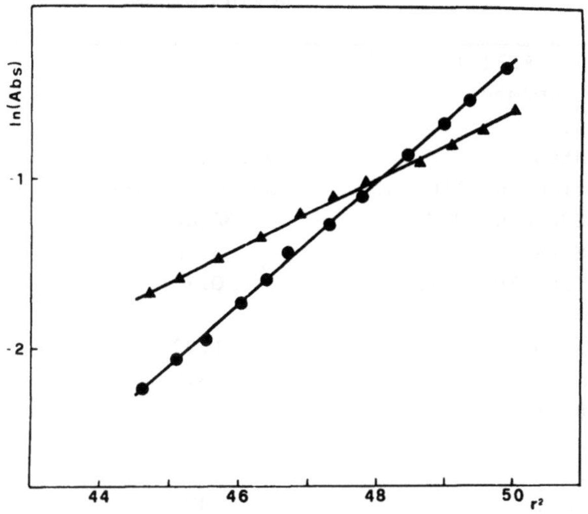

Fig.2. Determination of molecular weight of bovine liver MTAase by sedimentation-equilibrium. The ln of A_{280} in the cell at the distance (r) from the rotation axis was plotted versus the square of r. The data were obtained at 6,000 rev/min (▲—▲) and 8,000 rev/min (●—●)

The least-squares analysis of the data gave a regression coefficient for the straight line higher than 0.99. A molecular weight of 98,000 has been calculated from the slopes. A similar value has been estimated by means of gel filtration on Sephacryl S-200 and Superose 6.

The main physico-chemical properties of bovine MTAase are summarized in Table 3 and compared to those of mammalian purine nucleoside phosphorylase (PNP) reported in the Literature (20).

The sedimentation coefficient was determined at a rotor speed of 40,000 rev/min at a temperature of 20°C. The data have been analyzed by the moving boundary method, plotting the logarithm of distances of the sedimentation-profile inflection point from the rotation axis as a function of time. The $S_{20,w}$ determined by this method was 6.11.

Like the human placenta enzyme, bovine liver MTA phosphorylase is a trimer composed of apparently identical subunits (19). Such a proteic structure, which is quite rare among the enzymes, is also shared by the homologous enzyme PNP (20). The resemblance between MTAase and PNP is not only confined to this finding but can be also observed regarding other hydrodynamic features reported in Table 3. Both proteins have similar frictional ratio, and assuming that the enzymes are non-hydrated prolate ellipsoids, these values indicate an axial ratio of 4.5 for MTAase and 5 for PNP. Moreover, similar apparent specific volume and Stokes radius have been calculated for the two proteins.

Table 3. Physico-chemical properties of mammalian MTA phosphorylase and purine nucleoside phosphorylase

	BOVINE MTAase	HUMAN PNP*
SEDIMENTATION COEFFICIENT $S_{20,w}$ (x 10^{-13} cm/sec)	6.11	5.4
MOLECULAR WEIGHT		
by gel filtration	95,000 + 3,000	
by sedimentation equilibrium	98,000 + 3,000	91,000
SUBUNIT MOLECULAR WEIGHT	32,000	30,000
STOKES RADIUS (Å)	37.2	36.4
APPARENT PARTIAL SPECIFIC VOLUME (ml/g)	0.73	0.73
FRICTIONAL RATIO	1.22	1.25
AXIAL RATIO	4.5	5
SECONDARY STRUCTURE		
% α helix	5	3.2
% β structure	48	32
% random coil	47	65

* From Ref. 20

Analogies are also observable in the content of the various components of the secondary structure as calculated by means of circular dichroism. Indeed a very low amount of α helix and a relative high percentage of random coil have been determined for both the phosphorylases. All together such data can suggest a remarkable similarity in the structure between the two enzymes. However, only the comparison of their amino acid sequences and X-ray analysis of the crystallized proteins could substantiate this view.

IMMUNOLOGICAL STUDIES ON BOVINE MTA PHOSPHORYLASE

Initial attempts to get high titer antisera to native MTAase failed although several animals, namely rabbits, guinea pigs and goats were employed. This observed scarce immunogenicity of the phosphorylase could be tentatively related to structural homologies among the enzymes from several mammalian sources.

Therefore two different strategies, mainly based on coupling of protein with macromolecules have been used to enhance the immunogenicity of MTAase. The first procedure, which employes glutaraldehyde as the coupling agent, allowed us to prepare either soluble polymers of MTAase or hemocyanin-MTAase conjugates. Alternatively, the protein was linked to cyanogen bromide-activated agarose. It is noteworthy, in this respect, that MTAase linked to agarose retains its catalytic activity; therefore it could be employed as immobilized enzyme to prepare MTA or MTR-1-P analogs.

The three different antigens obtained were then injected into rabbits, following standard procedures, and the corresponding sera were tested for their ability to precipitate the bovine liver enzyme (Fig.3).

The titer of the antisera was evaluated by measuring the enzymatic activity that remained in the supernatant after the addition of increasing amounts of the different antibodies. The antiserum against polymeric-MTAase showed a scarce activity indirectly confirming the weak immunogenic power of MTAase by itself. Conversely, the other two antisera were able to precipitate the enzyme and that against hemocyanin-MTAase appeared particularly active.

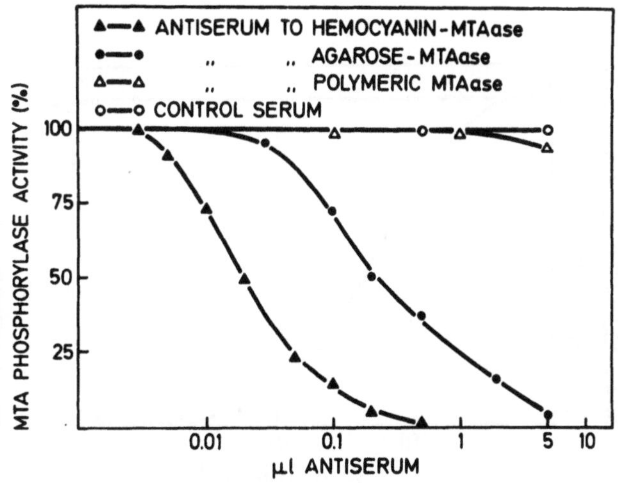

Fig.3. Immunoprecipitating effect of increasing amounts of various antisera on bovine liver MTAase

In order to characterize some of the interactions between the enzyme and the antibodies, the inhibitory effect on the enzyme activity as well as the precipitating ability of the two different antisera produced to hemocyanin-MTAase were investigated.

One of the two antisera was able to precipitate the enzyme and also directly inhibit the enzyme activity (Fig. 4A). This result indicates that the binding with antibodies results in a modification of the enzyme structure with a concomitant loss of MTAase activity. Conversely, a different behaviour has been demonstrated for the other antiserum (Fig. 4B). Indeed, no inhibitory effect was detected while a significantly high precipitating capacity was evidenced. These results confirm the well known high variability of the immune response to an identical immunogen.

When the antibodies were employed to precipitate the phosphorylase activity either from a 15,000g supernatant or from a homogeneous preparation, superimposable results were obtained (data not shown). This finding indicates that no

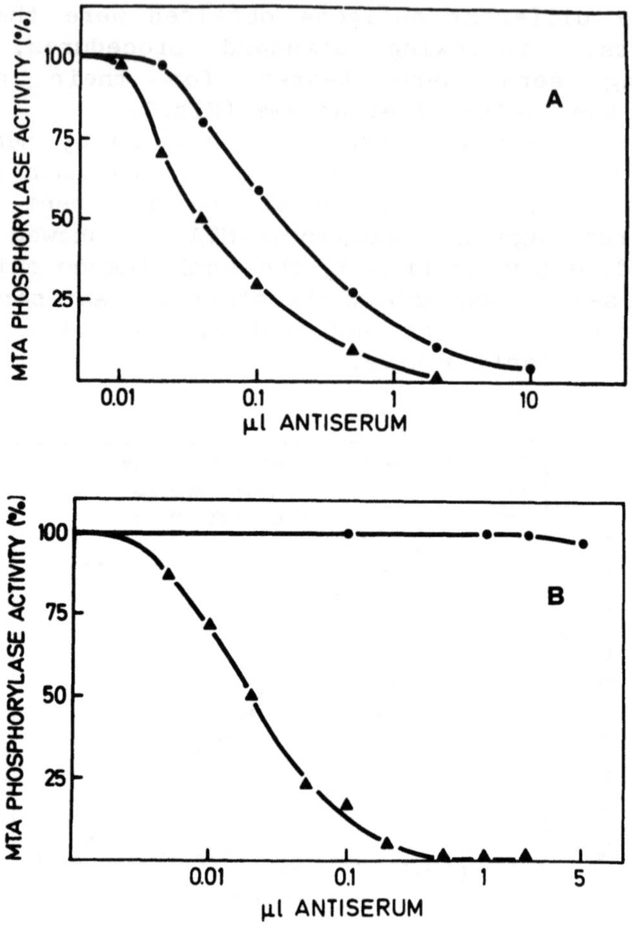

Fig.4. Immunoprecipitating (▲—▲) and inhibitory (●—●) effect of increasing amounts of two polyclonal anti-hemocyanin-MTAase antisera

significant loss of activity occurs during the purification procedure of the bovine enzyme. Thus a quite accurate estimation of the phosphorylase content can be obtained by enzymatic assay taking into account the specific activity of the pure protein.

The antiserum was then tested against MTA-cleaving enzymes from different sources (Fig. 5). The antibodies to bovine liver MTAase were also able to precipitate the human enzyme even if with a lower efficacy. This observation indicates the occurrence of some structural differences between human and bovine enzymes. On the other hand, no precipitation was obtained when Escherichia coli S-adenosylhomocysteine/MTA nucleosidase or Saccharomyces cerevisiae MTA phosphorylase were employed as antigens, thus indicating that these enzymes show a significantly different antigenic structure compared to the mammalian enzyme.

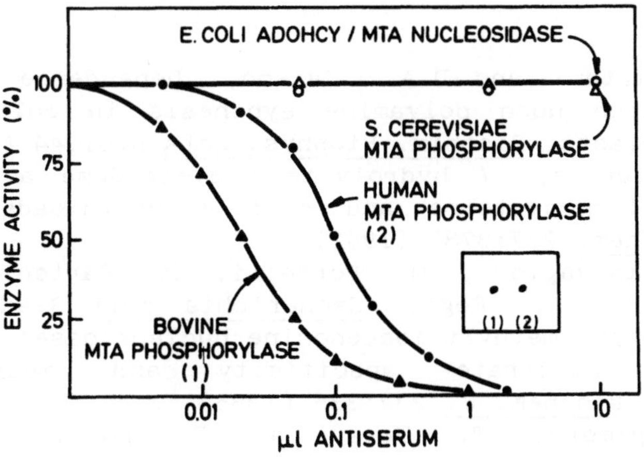

Fig.5. Immunoprecipitation assay on MTA-cleaving enzymes from various sources in the presence of increasing amounts of hemocyanin-MTAase antiserum. In the inset is shown a dot-blot on bovine (1) and human (2) MTAase.

In the inset is reported the result of a dot-blot experiment carried out employing human placenta and bovine liver MTA phosphorylases as antigens. The obtained data confirm the antigenic cross-reactivity between the two mammalian enzymes examined.

Detailed specificity studies are currently under development to give a more accurate immunological picture of the possible evolution of MTA-cleaving enzymes.

In addition, studies are in progress to purify substantial amounts of antibodies by specific immunoaffinity techniques. The purified immunoglobulins could in turn be employed to i) prepare immunoadsorbent matrix for a rapid and quantitative purification of the phosphorylase and ii) develop sensitive immunohistochemical methods for detecting MTAase-deficient cells in solid and liquid tumors.

Finally, the availabilty of high titer polyclonal antibodies and of substantial amounts of pure MTAase should result in a successful cloning of MTAase gene and in the elucidation of its expression in enzyme-positive and -negative cells.

ACKNOWLEDGEMENTS

We thank Drs L. Servillo and A. Giovane for their help in physico-chemical studies and Drs A. Fioretti and L.F. Menna in the preparation of specific antisera. This work was supported by grant n. 86.00618.44 awarded by the National Research Council, project "Oncologia".

REFERENCES

1. N. Kamatani and D.A. Carson, Dependence of adenine production upon polyamine synthesis in cultured human lymphoblasts, Biochim. Biophys. Acta 675:344 (1983)
2. J.A. Duerre, A hydrolytic nucleosidase acting on S-adenosylhomocysteine and on 5'-methylthioadenosine, J. Biol. Chem. 237:3737 (1962)
3. F. Della Ragione, M. Porcelli, M. Cartenì-Farina, V. Zappia, A.E. Pegg, Escherichia coli S-adenosylhomocysteine/5'-methylthioadenosine nucleosidase: purification, substrate specificity and mechanism of action, Biochem. J. 232:335 (1985)
4. Y. Sugimoto, T. Toraya and S. Fukui, Studies on metabolic role of 5'-methylthioadenosine in Ochromonas malhamensis and other microorganisms, Arch. Microbiol. 108:175 (1976)
5. C. Baxter and C.J. Coscia, In vitro synthesis of spermidine in the higher plant, Vinca rosea, Biochem. Biophys. Res. Commun. 59:147 (1973)
6. A.B. Guranowski, P.K. Chiang and G.L. Cantoni, 5'-Methylthioadenosine nucleosidase: purification and characterization of the enzyme from Lupinus luteus seeds, Eur. J. Biochem. 114:243 (1981)
7. A.E. Pegg and H.G. Williams-Ashman, Phosphate-stimulated breakdown of 5'-methylthioadenosine by rat ventral prostate, Biochem. J. 115:241 (1969)
8. D.L. Garbers, Demonstration of 5'-methylthioadenosine phosphorylase activity in various rat tissues: some properties of the enzyme from rat lung, Biochim. Biophys. Acta, 523:82 (1978)
9. V. Zappia, A. Oliva, G. Cacciapuoti, P. Galletti, G. Mignucci and M. Cartenì-Farina, Substrate specificity of 5'-methylthioadenosine phosphorylase from human prostate, Biochem. J. 175:1043 (1978)
10. G. Cacciapuoti, A. Oliva and V. Zappia, Studies on phosphate-activated 5'-methylthioadenosine nucleosidase from human placenta, Int. J. Biochem. 9:35 (1978)
11. T.M. Savarese, G.W. Crabtree and R.E. Parks Jr, 5'-Methylthioadenosine phosphorylase - I: substrate activity of 5'-deoxyadenosine with the enzyme from sarcoma 180 cells, Biochem. Pharmacol. 30:189 (1981)
12. A.J. Ferro, N.C. Wobel and J.A. Nicolette, 5-Methylthioribose -1-phosphate: a product of partially purified rat liver 5'-methylthioadenosine phosphorylase, Biochim. Biophys. Acta 570:65 (1979)
13. L. Shugart, M. Tancer and J. Moore, Methylthioadenosine nucleoside phosphorylase activity in Drosophila melanogaster, Int. J. Biochem. 10:901 (1979)
14. S. Shimizu, T. Abe, S. Shiozaki and H. Yamada, Catabolism of methylthioadenosine and S-adenosylhomocy-

steine in microorganisms, in International Conference on Polyamines in Life Sciences, Lake Yamanaka, Japan, July 14-18 (1986)

15. R.L. Miller, C.L.K. Sabourin and T.A. Krenitsky, Trypanosoma cruzi adenine nucleoside phosphorylase purification and substrate specificity, Biochem. Pharmacol. 36:553 (1987)

16. N. Kamatani, W.A. Nelson-Rees and D.A. Carson, Selective killing of human malignant cells deficient in methylthioadenosine phosphorylase, a purine metabolic enzyme, Proc. Natl. Acad. Sci. USA 78:1219 (1981)

17. T.M. Savarese, S-H. Chu, M-Y Chu and R.E. Parks Jr, 5'-Deoxy-5'-methylthioadenosine phosphorylase -III: role of the enzyme in the metabolism and action of 5'-halogenated adenosine analogs, Biochem. Pharmacol. 34:361 (1985)

18. P.S. Backlund and R.A. Smith, Methionine synthesis from 5'-methylthioadenosine in rat liver, J. Biol. Chem. 256:1533 (1981)

19. F. Della Ragione, M. Cartenì-Farina, V. Gragnaniello, M.I. Schettino and V. Zappia, Purification and characterization of 5'-deoxy-5'-methylthioadenosine phosphorylase from human placenta, J. Biol. Chem. 261:12324 (1986)

20. J.D. Stoeckler, R.P. Agarwal, K.C. Agarwal, K. Schmid and R.E. Parks Jr, Purine nucleoside phosphorylase from human erythrocytes: physicochemical properties of the crystalline enzyme, Biochemistry 17:278 (1978)

EXPLOITATION OF METHYLTHIORIBOSE KINASE IN THE DEVELOPMENT OF ANTIPROTOZOAL DRUGS

John H. Fitchen, Michael K. Riscoe, Adolph J. Ferro

Medical Research Service, Portland V.A. Medical Center
Portland, OR and Epitope, Inc., Beaverton, OR, USA

INTRODUCTION

Parasitic infections such as malaria and giardiasis cause some of the most prevalent diseases of man and produce extensive morbidity and mortality around the world[1]. For example, over 600 million people are chronically infected with Plasmodium falciparum and at least 1.5 million malaria-related deaths occur each year[2,3]. Moreover, it has been estimated that death and illness due to malaria cost third world countries over $2 trillion annually[4]. Despite intensive research efforts, the development of new and effective antimalarial drugs has been difficult because of the close metabolic relationship between protozoa and human cells.

Available drugs for the treatment of malaria are primarily chloroquine, quinine, mefloquine (the quinoline derivatives), pyrimethamine (dihydrofolate reductase inhibitor) and sulfadoxine (dihydropteroate synthase inhibitor)[1]. Although these drugs remain the standard of antimalarial chemotherapy, their usefulness is threatened by the emergence of drug-resistant strains of P. falciparum. Chloroquine resistance of P. falciparum was first noted in 1959 in Thailand and in 1961 in Columbia (for review see 5). By the end of 1984, chloroquine-resistant P. falciparum had spread to over 40 countries in Asia, Africa and South America[5]. In addition, recognition that available antifolate combinations (e.g., pyrimethamine/dapsone) can cause agranulocytosis or Stevens-Johnson syndrome, may seriously limit prophylactic use of these drugs[2]. Current drug therapy, therefore, may soon become ineffective or attended by unacceptable risk.

Giardia lamblia, the etiologic agent of giardiasis, has a worldwide distribution with prevalence estimated as high as 50% in parts of the developing world[7]. It is the leading protozoan cause of diarrhea among travelers[8]. In the United States, giardiasis is the most prevalent parasitic disease and is ranked as the number one cause of diarrheal outbreaks from contaminated water supplies[9]. Giardiasis is also a serious problem in veterinary medicine[10]. Available drugs such as quinacrine, metronidazole, and furazolidone are active against Giardia but have undesirable side effects and none is completely effective[11]. There is also concern that these drugs may be carcinogenic and/or mutagenic[11]. Thus, the need remains for a safe and effective drug for treating this cosmopolitan infection.

A rational approach to the treatment of parasitic diseases has been urged for decades. Numerous researchers have advocated a search for metabolic differences between the host and parasite as potential targets for drug development[12-15]. Effective antiprotozoal drugs would ideally take advantage of physiologic differences between protozoa and human cells.

5'-Deoxy-5'-methylthioadenosine (MTA) is a naturally occurring nucleoside synthesized from S-adenosylmethionine during the production of the polyamines spermidine and spermine[16,17]. Unlike polyamines, MTA does not normally accumulate in cells, but is rapidly degraded through one of two mechanisms (Fig. 1). In mammals[18], MTA is degraded in one step to adenine and 5-methylthioribose-1-phosphate (MTR-1-P) by MTA phosphorylase (Fig. 1, Pathway A). In some other organisms[19], however, MTA is catabolized to MTR-1-P in two steps: first, to adenine and 5-methylthioribose (MTR) via MTA nucleosidase, followed by conversion of MTR to MTR-1-P in an ATP-dependent reaction catalyzed by MTR kinase (Fig. 1, Pathway B). In both cases, MTR-1-P is subsequently recycled to methionine through the intermediate α-ketomethylthiobutyric acid[20-22]. Since MTA nucleosidase and MTR kinase are not present in mammalian cells, we hypothesized that these enzymes were potential targets for the selective killing of organisms which contain them.

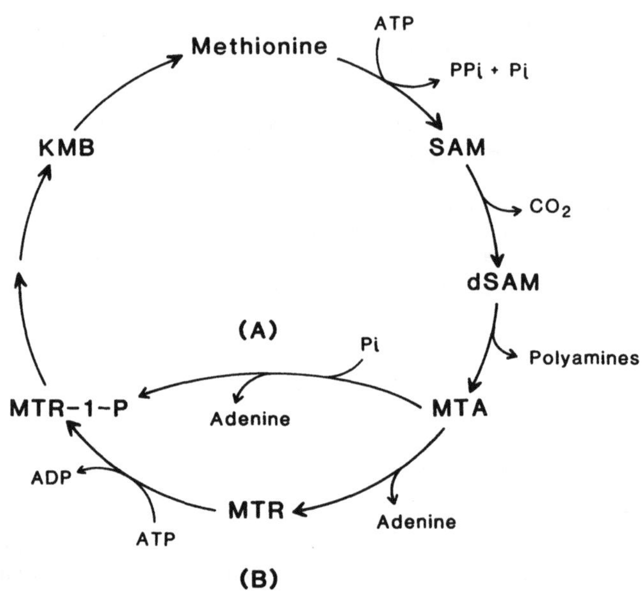

Fig. 1. Methionine recycling pathway. The two alternative pathways for MTA to MTR-1-P are represented by (A) and (B) where (A) is catalyzed by MTA phosphorylase and (B) is catalyzed by MTA nucleosidase and MTR kinase, respectively. Abbreviations: SAM, S-adenosylmethionine; dSAM,, decarboxylated S-adenosylmethionine; MTA, 5'-methylthioadenosine; MTR, 5-methylthioribose; MTR-1-P, 5-methylthioribose-1-phosphate; KMB, α-ketomethylthiobutyric acid.

METHODS AND RESULTS

Enzyme Analysis

Extracts of Plasmodium falciparum, Giardia lamblia, and Ochromonas malhamensis were assayed for MTA phosphorylase[23], MTR kinase[24] and MTA nucleosidase[18] activities. O. malhamensis, a non-pathogenic protozoan, was studied since it has been shown that this organism can proliferate utilizing MTR as a sole source of methionine[25], implying that it contains MTR kinase. Enterobacter aerogenes[26] and Persea americana (avacado fruit[22]) were included as positive controls since each has previously been shown to contain MTR kinase activity (Table 1). Extracts derived from human erythrocytes and liver were also tested. Cell-free extracts from P. falciparum, G. lamblia and O. malhamensis contained significant MTA nucleosidase and MTR kinase activities, but lacked MTA phosphorylase activity. In contrast, human cells contained MTA phosphorylase activity, but were devoid of MTA nucleosidase and MTR kinase activities (Table 1).

Table 1. Enzyme activities in cell-free extracts.

Source of extract	Enzyme activities: MTA nucleosidase[a] (nmol/min/mg)	MTR kinase (pmol/min/mg)	MTA phosphorylase (nmol/min/mg)
E. aerogenes	1.5	3.6	0
P. americana	1.2	2.5	0
Human erythrocytes	0	0	0.3
Human liver	0	0	1.6
P. falciparum[b]	0.3	3.6	0
G. lamblia	2.2	1.0	0
O. malhamensis	2.0	2.0	0

[a]MTA nucleosidase was assayed in the absence of phosphate.
[b]P. falciparum was purified from erythrocytes as described by Sanderson and Walliker[27].

To confirm the presence of MTR kinase, the enzyme reaction mixtures were spotted onto polyethyleneimine-cellulose thin layer chromatography plates and developed in an aqueous solution of 1M LiCl. Radiochromatogram scanning revealed that $[^{14}CH_3]$-MTR incubated with ATP and protozoal extracts (but not with extracts from mammalian cells) was converted to a new radioactive peak which comigrated with authentic MTR-1-P (R_f = 0.50). In similar experiments, the catabolism of MTA to MTR-1-P by mammalian extracts was found to be phosphate-dependent, whereas protozoal extracts exhibited abundant MTA-cleaving activity in the absence of an exogenous phosphate source. Moreover, P. falciparum was tested for growth in methionine-free medium with MTR added as the

Table 2. Effect of MTR Analogs on Growth of Protozoa *in vitro*.

	IC_{50} (μM)[a]		
Compound	O. malhamensis	P. falciparum[b]	G. lamblia
Methylthioribose	> 5,000	> 5,000	> 5,000
Ethylthioribose	75	100	300
Propylthioribose	NT	100	NT
Isopropylthioribose	NT	100	NT
Isobutylthioribose	150	300	2,000
Benzylthioribose	NT	20	NT

[a]Indicated 50% inhibitory concentrations (IC_{50}'s) are the mean of at least two experiments for each organism. Results between experiments did not vary by more than 15%.
[b]ATCC strain #50005. NT = not tested.

sole potential source of methionine. As previously shown with *O. malhamensis*[25], and unlike mammalian cells[28], *P. falciparum* grew readily under these conditions (data not shown). The results demonstrate that these protozoa metabolize MTA in the two-step reaction MTA → MTR → MTR-1-P via MTA nucleosidase and MTR kinase, respectively.

Thus, since MTA nucleosidase and MTR kinase were present in protozoa but not in human cells, these enzymes appeared to be potential targets for the design of new antiprotozoal agents. Since MTA is a substrate for both MTA nucleosidase (protozoa) and MTA phosphorylase (human cells), we focused on MTR kinase and studied analogs of MTR.

Analogs of MTR

We have been examining analogs of MTR in anticipation of finding compounds which selectively inhibit MTR kinase-containing protozoa without affecting mammalian cellular functions. In these efforts, we concentrated first on synthesizing analogs substituted at the 5-position. The general procedure for the synthesis of these analogs involves preparation of the intermediate methyl 2,3-O-isopropylidiene-5-O-p-tolylsulfonyl-ß-D-ribofuranoside from D-ribose as described by Lerner[29]. Displacement of the tosylate group by the appropriate alkylthio group is then effected by stirring in methanol and the protecting groups are removed by dissolving the alkylthio derivative in peroxide-free dioxane. Using this procedure, we have synthesized the ethyl, propyl, isopropyl, butyl, isobutyl, and benzyl derivatives of MTR. The structure and purity of each of these compounds was confirmed by NMR and GC mass spectroscopy and they were then tested for activity against protozoa. The ethyl derivative, ethylthioribose (ETR), was synthesized first and has been most extensively studied.

Effect of MTR Analogs on Protozoa

Initial experiments were performed with the photosynthetic protozoan *Ochromonas malhamensis*. Cultures were initiated at $1x10^5$ organisms/ml and maintained under illumination at room temperature as described by Sugimoto et al[25]. As shown in Figure 2 (next page), ETR was cytocidal to *Ochromonas* producing lysis of organisms after 48 hours of exposure.

The effect of a variety of alkylthio analogs of MTR on *O. malhamensis*, *P. falciparum* and *G. lamblia* is shown in Table 2. For these

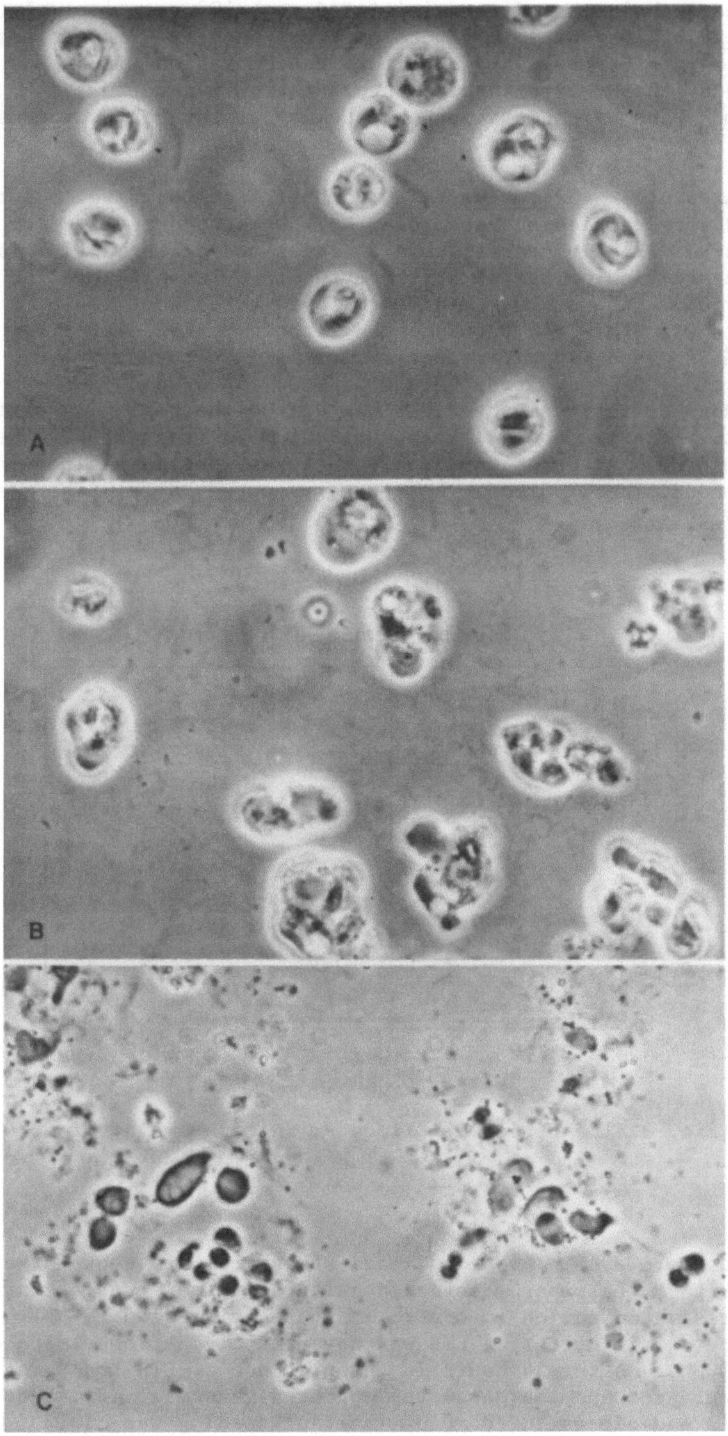

Fig. 2. Effect of ethylthioribose (ETR) on Ochromonas malhamensis. Organisms were grown under standard conditions[25] with or without 1mM ETR. A. Control. B. ETR-treated at 24 hours. C. ETR-treated at 48 hours.

experiments, cultures of the three protozoa were grown according to established techniques and were initiated at $1x10^5$ organisms/ml (O. malhamensis), 1% parasitemia (P. falciparum), and $5x10^4$ organisms/ml (G. lamblia). Fifty percent inhibitory concentration (IC_{50}) was determined by the H^3-hypoxanthine incorporation method for Plasmodium[29], and by enumeration on a hemacytometer for Ochromonas and Giardia.

Cultures of P. falciparum (ATCC #50005; knob-positive, chloroquine-resistant) were carried out according to the methods outlined by Richards and Maples[30]N Briefly, parasitemia of red blood cells was initiated at ~1% by the addition of washed erythrocytes to a heavily infected culture (~10%). The culture medium employed throughout these experiments was RPMI-1640 supplemented with 25mM HEPES (pH 7.2), human serum (10%, A^+) and 7.5 µg/ml gentamycin. All cultures were incubated at 37°C in an atmosphere of 5% CO_2, 5% O_2 and 90% N_2. ETR was added at various concentrations at the the beginning of the experiment. After 24 hours, the spent medium was replaced by fresh medium and drug. After 48 hours, all cultures were maintained in drug-free conditions. Blood smears were prepared daily. The results are given in Figure 3.

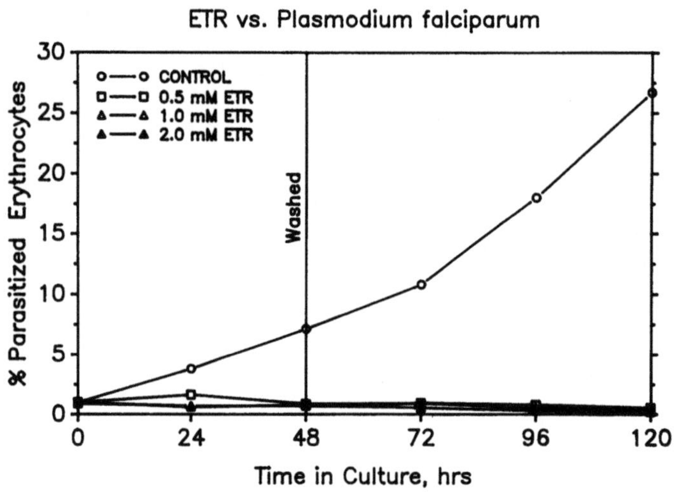

Fig. 3. The Effect of ETR on P. falciparum in vitro.

Microscopic examination of Giesma-stained smears showed that ETR was inhibitory to P. falciparum-infected cultures. There was a continuous and progressive infection in control cultures rising to 20% by 4 days. The anti-parasitic effect of ETR was apparent at the lowest concentration tested (500µM). After 24 hours in the presence of ETR the parasitemia had not increased and the parasites appeared abnormal. By 48 hours the parasitemia had decreased to ~0.3% (initial seeding was ~1.0%). By 72 hours the infection was noted to be less than 1 parasite per 1,000 red cells and the few remaining malarial forms were abnormal (pyknotic). At all concentrations, the parasitemia remained at or below the level of the initial inoculum and only abnormal "crisis" forms could be seen after 24 hours. By 96 hours, all drug-treated cultures appeared to be free of parasites (the infection was judged to be less than 1 parasite per 5,000 cells). Importantly, with all concentrations tested, the parasitemia did not reappear when cultures were continued free of drug for two days after removal of ETR.

In addition to the chloroquine-resistant, knob-positive strain of P. falciparum (ATCC #50005), three additional strains were examined for sensitivity to ETR: FCR-1 (ATCC #30930; chloroquine-resistant), FCR-3 (ATCC #30932; chlorquine-sensitive), and Honduras-1 (ATCC #30950; pyrimethanine-resistant). These experiments were conducted by measuring the reduction in H^3-hypoxanthine incorporation into parasite nucleic acids compared to drug-free controls[31]. All three strains were sensitive to ETR with approximate IC_{50} values in the 100-200μM range (Figure 4).

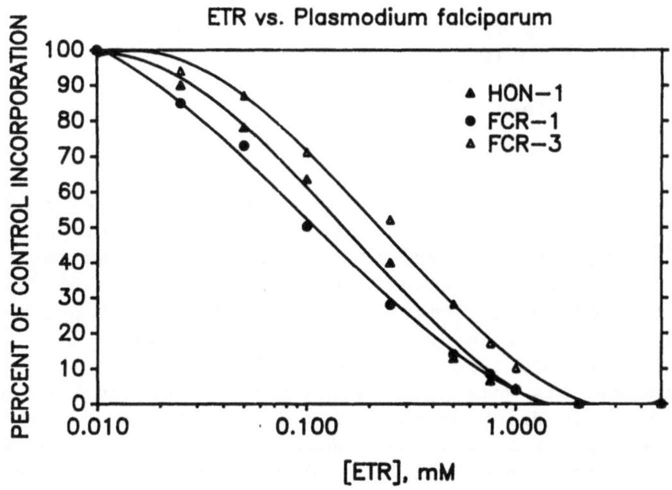

Fig. 4. Effect of ETR on drug-resistant strains of P. falciparum.

Lack of Toxicity of ETR to Human Cells and Intact Mice

The effect of ETR on mammalian systems in vitro and in vivo was also examined. ETR had no effect on the growth of human or murine bone marrow cells or a continuous human cell line in vitro (Table 3A). Oral administration of ETR to intact mice had no demonstrable effect on general appearance, weight, blood counts, serum chemistries or organ histology (Table 3B). In addition, mice given daily intraperitoneal injections of 300μmoles of ETR (2.9g/kg/day; equivalent to 203g/day in an average adult human) for 7 days showed no abnormalities in these parameters (data not shown).

Mechanism of Action of ETR

The most likely mechanism by which ETR produces its growth inhibitory effects are: 1) conversion via MTR kinase and the methionine recycling pathway to ethionine, a toxic analog of methionine[34]; or 2) inhibition of MTR kinase activity thereby preventing recycling of MTR to methionine. In order to explore these possibilities, we assessed the activity of ETR against O. malhamensis over a range of methionine concentrations. As shown in Figure 5, methionine was able to diminish

Table 3. Lack of Effect of ETR on Mammalian Cells *in vitro* and *in vivo*.

A. Effect of ETR on human cells *in vitro*.[a]	ETR Addition	
	None	5mM
Human bone marrow (CFU-GM)	55±6	53±3
Murine bone marrow (CFU-GM)	39±4	39±6
HL-60 cells (growth at 3 days)	295±20%	320±25%
B. Effect of ETR on mice *in vivo*.[b]	**ETR in Drinking Water**	
	None	10mM
General appearance	Normal	Normal
Weight (% change)	+23%	+21%
Hematocrit (%)	43±3	44±3
White blood count (x 10^{-3})	4.6±0.6	4.4±0.4
Serum creatinine (mg/dl)	0.4±0.1	0.3±0.1
Serum alkaline phosphatase	91±14	73±18
Organ histology (liver, kidney, heart, lung, spleen)	Normal	Normal

[a]Human and murine granulocyte-macrophage progenitor cells (CFU-GM) were grown in agar with standard sources of colony stimulating activity[32]. Results are expressed as colonies/10^5 marrow cells plated. The human promyelocytic leukemia cell line (HL-60) was grown in RPMI-1640 medium supplemented with 10% fetal calf serum and antibiotics as described[33]. Numbers indicate mean ± S.D. from 3 experiments for each type of target cell.

[b]*In vivo* experiments were performed with 6-8 week old CF-1 mice given 10mM ETR in drinking water for 3 days (total dose of 240 µmoles or 46 mg per mouse; daily dose of 750 mg/kg per mouse). Numbers indicate mean ± S.D. from 10 mice in the control group and 16 mice in the ETR-treated group seven days after discontinuation of ETR.

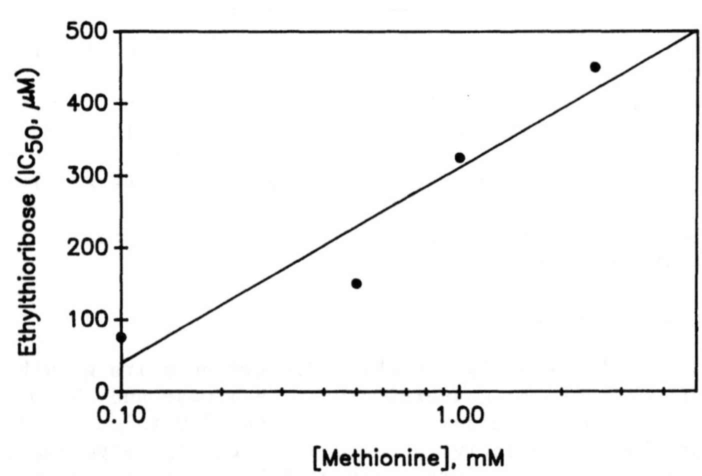

Fig. 5. Effect of methionine concentration on ETR-induced inhibition of *O. malhamensis*.

the inhibitory effect of ETR. We next examined ETR-treated O. malhamensis for the presence of ethionine. For this experiment, organisms were suspended in defined medium containing 0.5mM methionine with or without 0.5mM ETR. After 18 hours, the cells were collected by centrifugation and the amino acids were extracted and derivatized with dabsyl-chloride as described by Chang et al[35]. The derivatized amino acids were then separated by HPLC. Chromatograms showed that a new peak (co-migratory with authentic ethionine) appeared in organisms treated with ETR. Taken together, these results suggest that ETR acts, at least in part, through the generation of ethionine in MTR kinase-containing organisms.

DISCUSSION

In view of the limitation of existing antiprotozoal agents, there is a need for the development of new drugs which capitalize on metabolic differences between protozoa and human cells. We have shown that a metabolic difference exists between certain protozoa (including P. falciparum and G. lamblia) and human cells in the mechanism by which these cells catabolize MTA. In mammalian cells, MTA is degraded in one step to MTR-1-P, whereas in protozoa MTA is metabolized in two steps: first to MTR followed by conversion to MTR-1-P (see Figure 1). Based on this difference, we have synthesized analogs of MTR and have found that they have antiprotozoal activity without apparent adverse effects on human cells or whole animals at similar or higher concentrations. Thus, we have demonstrated that the difference in MTA catabolism can be exploited.

The lack of effective therapy for many protozoal infections and the emergence of drug-resistant parasitic organisms, exemplified by multidrug-resistant strains of P. falciparum[36], underscore the need for mechanistically novel drugs. Since MTR analogs are unrelated in chemical structure to existing antiprotozoal compounds, they could represent a new generation of drugs with a novel mode of action and, therefore, activity against parasites that are resistant to other drugs. Indeed, preliminary experiments with O. malhamensis suggest that ETR, the ethyl analog of MTR, may act as a "pro-drug" which is converted in MTR kinase-containing organisms to ethionine, a toxic analog of the essential amino acid methionine. Whether other analogs or MTR are similarly converted to toxic analogs of methionine remains to be determined.

The most likely mechanisms by which ETR produces its growth inhibitory effects are: 1) conversion via MTR kinase and the methionine recycling pathway to ethioine, a toxic analog of methionine[34]; or 2) inhibition of MTR kinase activity thereby preventing the recycling of MTA to methionine. Further biochemical studies of the levels of methionine and ethionine in ETR-treated cells should lead to a better understanding of the mechanism of action of ETR.

We recognize that the inhibitory concentrations of most MTR analogs against P. falciparum and G. lamblia are greater than those of many available drugs. However, the critical finding of this work is the demonstration of a metabolic difference between certain protozoa and mammalian cells. The analogs reported here are the first of a great number of potential variations designed to exploit this difference. In this regard, it is of note that preliminary experiments suggest that the benzyl derivative of MTR is considerably more potent than ETR against P. falciparum. Additional analogs may demonstrate even greater potency.

Acknowledgements: The authors wish to thank Linda Johnson for outstanding technical assistance; Kevin Olson and Michelle Reynolds for help with preparation of the manuscript; and John Sheley, Paul Conte and Colin Smith for synthesis of analogs. A patent application entitled "Analogs of MTR as medicinals and biocides" has been submitted to the U.S. Patent Office (U.S. Patent Application #823,929).

REFERENCES

1. Wyler, D.J. 1983. Malaria--resurgence, resistance and research. New Eng. J. Med. 308:875.

2. Peters, W. 1985. Chemotherapy of malaria. in Chemotherapy of Parasites. E.A. Denham, ed., Cambridge Univ. Press, Cambridge, p. 705.

3. Warhurst, D.C. 1987. Antimalarial drugs: An update. Drugs 33:50.

4. Cheng, T.C. 1986. Apicomplexa, Microspora, Ascepospara, Myxozoa. in General Parasitiology. Academic Press, Orlando, p. 174.

5. Peters, W. 1970. Chemotherapy and Drug Resistance in Malaria. Academic Press, New York, p. 876.

6. World Health Organization Report. 1985. Seventh Program Report. Tropical Disease Research. Chapter 2, p. 3.

7. Scientific Working Group. 1980. Parasite-related diarrrheas. Bull Wld. Hlth. Org. 58:819.

8. Craun, G.F. 1984. Waterborne outbreaks of giardiasis, Current status. in Giardia and Giardiasis. S.L. Erlandsen and E.A. Meyer, eds., Plenum Press, New York, p. 243.

9. Wolfe, M.S. 1979. Giardia lamblia. in Principles and Practice of Infectious Diseases. G.L. Mandell, R.G. Douglas and J.E. Bennett, eds., John Wiley, New York, p. 2143.

10. Sherding, R.G. 1983. Diseases of the small bowel. in Veterinary Internal Medicine. S.J. Ettinger, ed., W.B. Saunders Co., Philadelphia, p. 1317.

11. Wolfe, M.S. 1984. Viruses, Rickettsiae, Chlamydiae, and Mycoplasmas. in Principles and Practice of Infectious Diseases. G.L. Mandell, R.G. Douglas and J.E. Bennett, eds., Plenum Press, New York, p. 147.

12. Cohen, S.S. 1979. Comparative biochemistry and drug design for infectious disease. Science 205:964.

13. Fairlamb, A.H. 1982. Biochemistry of trypanosomiasis and rational approaches to chemotherapy. Trends Biochem. Sci. 7:249.

14. Opperdowe, F.R. 1983. Toward the development of new drugs for parasitic diseases. in Parasitology: A Global Perspective. K.S. Warren and J.S. Bower, eds., Springer-Verlag, New York, p. 191.

15. Cohen, S.S. 1982. The polyamines as a growth industry. Fed. Proc. Am. Soc. Exp. Biol. 41:3061.

16. Schlenk, F., 1983. Methylthioadenosine. in Advances in Enzymology and Related Area of Moledular Biology. A. Meister, ed., John Wiley and Sons, New York, p. 195.

17. Williams-Ashman, H.G., Seidenfeld, J. and Galletti, P. 1982. Trends in the biochemical pharmacology of 5'-methylthioadenosine. Biochem. Pharm. 3:277.

18. Pegg, A.E. and Williams-Ashman, H.G. 1969. Phosphate-stimulated breakdown of 5-methylthioadenosine by rat ventral prostate. J. Biol Chem. 115:241.

19. Ferro, A.J. and Marchitto, K.S. 1983. 5-Methylthioribose kinase (Enterobacter aerogenes). in Methods in Enzymology. H. Tabor and C.W. Tabor, eds., Academic Press, New York. p. 361.

20. Shapiro, S.K. and Barrett, A. 1981. 5-Methylthioribose as a precursor of the carbon chain of methionine. Biochem. Biophys. Res. Commun. 102:302.

21. Backlund, P.S. and Smith, R.A. 1981. Methionine synthesis from 5'-methylthioadenosine in rat liver. J. Biol. Chem. 256:533.

22. Kushad, M.M., Richardson, D.G., Ferro, A.J. 1983. Intermediates in the recycling of 5-methylthioribose to methionine in fruits. Plant Physiol. 73:257.

23. Ferro, A. J., Barrett, A. and Shapiro, S. K. 1976. Kinetic properties and the effect of substrate analogues on 5'-methylthioadenosine nucleosidase from Escherichia coli. Biochim. Biophys. Acta. 438:487.

24. Ferro, A. J., Barrett, A., and Shapiro, S. K. 1978. 5-Methylthioribose kinase: A new enzyme involved in the formation of methionine from 5-methylthioribose. J. Biol. Chem. 253:6021.

25. Sugimoto, Y., Toraya, T., and Fukui, S. 1976. Studies on metabolic role of 5'-methylthioadenosine in Ochromonas malhamensis and other microorganisms. Arch. Microbiol. 108:175.

26. Kushad, M. M., Richardson, D. G., and Ferro, A. J. 1982. 5-Methylthioribose kinase activity in plants. Biochem. Biophys. Res. Commun. 108:167.

27. Sanderson, A. and Walliker, D. 1981. Enzyme typing of Plasmodium falciparium from African and some other Old World countries. Trans. Roy. Soc. Trop. Med. Hyg. 75:263.

28. Riscoe, M. K. and Ferro, A. J. 1984. 5-Methylthioribose: Its effects and functions in mammalian cells. J. Biol. Chem. 259:5465.

29. Lerner, L. M. 1977. Enantiomeric forms of 9-(5-deoxyerythro-pent-4-enofuranosyl)adenine and a new preparation of 5-deoxy-D-lyxose. Carbohydrate Res. 53:177.

30. Richards, W. H. G., and Maples, B. K. 1979. Studies on Plasmodium falciparum in continuous cultivation. I. The effect of chloroquine and pryimethamine on parasite growth and viability. Ann. Trop. Med. Parasitol. 73:99.

31. Desjardins, R. E., Canfield, C. J., Haynes, J. D., and Chulay, J. D. 1979. Quantitative assessment of antimalarial activity in vitro by a semiautomated microdilution technique. Antimicrob. Agents and Chemother. 16:710.

32. Burgess, A. W., Wilson, E. C., and Metcalf, D. 1986. Stimulation by human placental conditioned medium of hemopoietic colony formation by human marrow cells. Blood 49:573.

33. Collins, S.C. 1987. The HL-60 promyelocytic leukemia cell line: Proliferation, differentiation, and cellular oncogene expression. Blood 70:1233.

34. Alix, J-H. 1982. Molecular aspects of the in vivo and in vitro effects of ethionine, an analog of methionine. Microb. Rev. 46:281.

35. Chang, J-Y., Knecht, R., and Braun, D.G. 1983. Amino acid analysis in the picomole range by precolumn derivatization. in Methods in Enzymology, C.H.W. Hirs and S.N. Timasheff, eds., Academic Press, New York, p. 41.

36. WHO Scientific Group. 1984. Drug resistance in human pathogenic Plasmodia. WHO Tech. Rep. 711:10.

MTA PHOSPHORYLASE IN PROTOZOA:

A POTENTIAL TARGET FOR CHEMOTHERAPEUTIC ATTACK

Richard L. Miller and David P. Toorchen*

Wellcome Research Laboratories
Research Triangle Park, NC 27709

An estimated 1 billion of the world's population are infected by parasitic protozoa. It has only been in the last decade, with the advent of <u>in vitro</u> axenic culture of a few of these organisms, that the hope of the rational development of parasite specific drugs has been expedited.

In general, the reliance of these parasites on the salvage of purines to satisfy their growth requirements makes them excellent candidates for chemotherapeutic attack by purine and purine nucleoside analogs. Many reported studies indicate that each parasite species possesses its own unique set of salvage enzymes which in many cases differ widely from those found in mammalian systems. Three of these parasites, <u>Trypanosoma</u> <u>cruzi</u>, <u>Trypanosoma</u> <u>brucei</u> and <u>Leishmania</u> <u>donovani</u>, the causative agents of Chagas' disease, African sleeping sickness and visceral leishmaniasis respectively, have been extensively studied with regard to their purine salvage pathways[1-3]. The results of these studies have indicated a number of differences in the pathways involved in purine salvage that could potentially be vulnerable to attack. The feasibility of such an approach is borne out in the reported observations that several purine and purine nucleoside analogs possess antiparasitic activity against these organisms (Table I).

A purine salvage pathway that has been largely overlooked in these parasites is that involved in the recycling of 5'-deoxy-5'-methylthioadenosine (MTA), a product of polyamine biosynthesis. The initial step in this pathway is the phosphorolytic cleavage of MTA to adenine and 5-methylthioribose-1-phosphate by MTA phosphorylase. This enzyme has been isolated and studied from a variety of mammalian systems including Sarcoma 180 cells[9-13], HL-60 cells[14], rat lung[15], human placenta[16], human lymphocytes[17], human prostate[18] and human liver[19]. In pioneering studies, Savarese, et al.[9,20] proposed that MTA phosphorylase might serve as a possible route for the delivery of "toxic" purines and pentose compounds into cells. In subsequent studies this hypothesis was supported by their finding that several MTA analogs containing potentially cytotoxic adenine moieties inhibited the growth of human

*Current address: Chemical Abstract Service, P.O. Box 3012, Columbus, OH 43210

Table I. Antiparasitic Activity of Purine and Purine Nucleoside Analogs

Agent	Parasite	Reference
Allopurinol	Trypanosoma cruzi	[4]
	Trypanosoma brucei	[5]
	Leishmania donovani	[6]
Allopurinol riboside	Trypanosoma cruzi	[7]
	Trypanosoma brucei	[5]
	Leishmania donovani	[7]
Thiopurinol	Trypanosoma brucei	[5]
	Leishmania donovani	[8]
Thiopurinol riboside	Trypanosoma brucei	[5]
	Leishmania donovani	[8]
Formycin B	Trypanosoma cruzi	[7]
	Trypanosoma brucei	[5]
	Leishmania donovani	[7]
9-Deazainosine	Trypanosoma cruzi	[7]
	Trypanosoma brucei	[5]
	Leishmania donovani	[7]

carcinoma cell lines possessing MTA phosphorylase but not cell lines lacking the enzyme[11]. With these findings in mind, studies were initiated with the MTA phosphorylase from T. cruzi[21], T. brucei[22] and L. donovani[23]. The substrate specificity of the enzyme from each of these sources was compared with that for the enzyme from human liver[19] and S-180 cells[9-13]. Most of the published studies dealing with substrate specificity have used nucleosides as substrates since the function of the enzyme is thought to be cleavage of MTA. In our studies with the enzyme from T. cruzi[21] and human liver[19], we have also examined the reverse reaction using adenine analogs in the hope of finding differences in the purine binding site of the different enzymes.

Nucleoside Specificity

MTA, adenosine and 2'-deoxyadenosine serve as substrates for MTA phosphorylase from T. cruzi, T. brucei and L. donovani. Initial studies in parasite extracts indicated that MTA, adenosine and 2'-deoxyadenosine cleavage were phosphate dependent[21-23] suggesting that either a single parasite enzyme cleaved the three nucleosides or multiple enzymes were involved. This contrasts to the findings with mammalian cell extracts which exhibit phosphate dependent cleavage of only MTA but not adenosine or 2'-deoxyadenosine. Subsequent studies involving parallel purification of the three activities[21-23], mutual substrate competitive inhibition[21,22] and inhibition by 5'-chloroformycin[21,22] indicated that the three phosphorolytic activities were attributable to a single parasite enzyme species.

The T. cruzi enzyme[21] and the S-180 enzyme[9-13] could accept a variety of groups in place of the 5'-methylthio moiety of MTA and still retain substrate activity. These included ethylthio-, isobutylthio-, iodo-, chloro-, hydrogen and hydroxyl groups. Data obtained with the human liver enzyme indicate that it too could accept these substitutions. Table II contains data comparing the kinetic parameters of those compounds that have been used as substrates for the mammalian and parasite enzymes. These data indicate that the mammalian enzymes have similar substrate specificities. In general, the same can be said for the group of enzymes from parasites. However, comparing the enzymes from the two groups against each other, mammal vs parasite, marked differences are observed. The mammalian enzymes catalyzed the efficient phosphorolysis of MTA, were inefficient at cleaving adenosine, and did not cleave 2'-deoxyadenosine. In contrast, the enzyme from each of the parasites cleaved MTA, adenosine and 2'-deoxyadenosine with high efficiency.

An interesting observation with the mammalian enzymes is that while replacement of the 2'-hydroxyl group of adenosine with a hydrogen completely blocked substrate activity, further replacement of the 5'-hydroxyl group with a hydrogen, ie. 2',5'-dideoxyadenosine, restored substrate activity. This may be due to the compound assuming a different conformation, one of which is amenable to binding to the enzyme. The data in Table II also imply that there is a difference between at least two of the parasite enzymes. The enzyme from T. brucei catalyzed the cleavage of 3'-deoxyadenosine but the enzyme from T. cruzi was incapable of such cleavage.

Table II. Nucleoside Substrates for MTA Phosphorylase[a]

Nucleoside	Human Liver		S-180 Cells		T. cruzi		T. brucei		L. donovani	
	K_m (μM)	Rel. V_m	K_m (μM)	Rel. V_m	K_m (μM)	Rel. V_m	K_m (μM)	Rel. V_m	K_m (μM)	Rel. V_m
MTA	2	100	9	100	3	100	2	100	1	100
Ado	760	62	1400	115	18	109	21	50	47	33
2'-Deoxy Ado		<0.3		-	12	142	77	205	20	250
3'-Deoxy Ado						<0.04	143	16		
5'-Deoxy Ado	3	100	23	180	15	130	22	120	11	135
2',5'-Dideoxy Ado	9	38			5	158		+		
5'-Chloro Ado	1	33	21	46	7	40		+		

[a]Data are from References 10, 19 and 21-23. Abbreviations used are: +, substrate; - non-substrate. Maximal velocities (nmol/min/mg protein) with MTA as substrate for the enzyme from human liver, S-180 cells, T. cruzi, T. brucei and L. donovani were 680, 57, 480, 61 and 640 respectively.

Faced with sorting through these type of data, many investigators have resorted to the comparison of substrate efficiencies (V_m/K_m) of the various compounds. Such a comparison is presented in Figure 1 where the data have been normalized so that all compounds can be compared directly to MTA, the "normal" substrate.

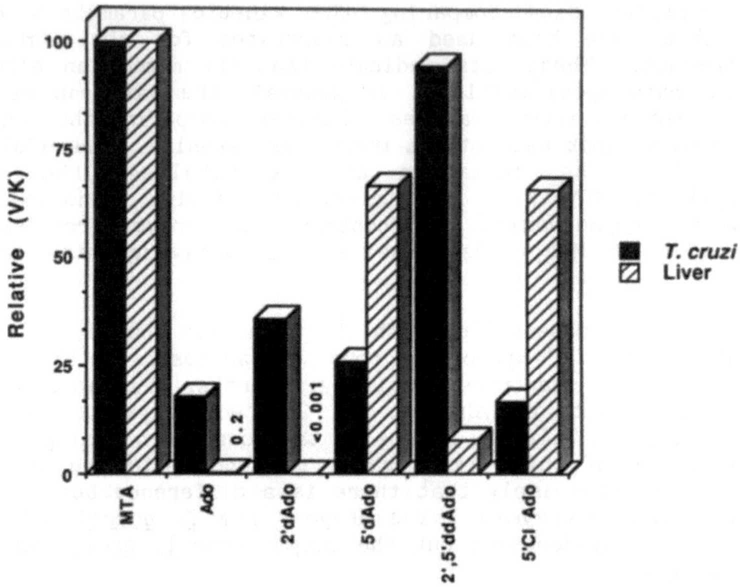

Figure 1. Relative substrate efficiencies (V_m/K_m) for selected nucleosides for the MTA phosphorylases from human liver and T. cruzi.

When examined in this fashion, it is obvious that there exist some striking differences between the two classes of enzymes. Relative to MTA, the human liver enzyme more efficiently cleaves 5'-deoxyadenosine and 5'-chloroadenosine whereas adenosine, 2'-deoxyadenosine and 2',5'-dideoxyadenosine are more efficient substrates for the T. cruzi enzyme.

Purine Base Specificity

All purines that are substrates for the T cruzi MTA phosphorylase are substrates for the human liver MTA phosphorylase. Despite the relatively large amount of data that exists with respect to the nucleoside substrate specificity of MTA phosphorylase from the mammalian and parasitic sources, the only report of the enzyme studied in the reverse direction, ie. nucleoside synthesis, is the T. cruzi enzyme[21]. These studies were facilitated by the fact that the parasite enzyme could use commercially available ribose-1-phosphate as an acceptor of the purine base (Table III). The K_m value obtained for adenine under these conditions was independent of the sugar-phosphate substrate whether it was ribose-1-phosphate, methylthioribose-1-phosphate or 2'-deoxyribose-1-phosphate[21]. However, with the human liver enzyme, adenosine served as a poor substrate (Table II) indicating that ribose-1-phosphate would not be able to be used as the pentosyl

Table III. Purine Substrates for MTA Phosphorylase[a]

Base	Human Liver		T. cruzi	
	K_m (μM)	Rel. V_m	K_m (μM)	Rel. V_m
Adenine	20	100	3	100
2-Aminoadenine	13	65	16	47
2-Fluoroadenine	17	25	1	65
2-Methyladenine	32	122	9	11
2-Hydroxyadenine	350	43	55	5
Purine	140	8	180	34
6-Methylpurine	190	8	220	15
Hypoxanthine		<5		<0.3
1-Deazaadenine	29	220	58	4
3-Deazaadenine		<5		<0.5
7-Deazaadenine		<5		<1.6
8-Azaadenine	11	5	4	3

[a]Data are from References 19 and 21. Maximal velocities (nmol/min/mg) with adenine as substrate for the enzyme from human liver and _T. cruzi_ were 340 and 730 respectively.

donor; 5-methylthioribose-1-phosphate would be required. [^{14}C]5-Methylthioribose-1-phosphate was synthesized and used to obtain the data presented in Table III. Unlike the data with the nucleoside substrates, no purine base was a substrate for one enzyme and not for the other. However, quantitative differences between the enzyme from liver and _T. cruzi_ are apparent. These differences are much more evident when the data is examined in terms of substrate efficiency (Figure 2).

Again as with the nucleoside substrates, marked differences are noted between the liver enzyme and that from _T. cruzi_. Relative to adenine, 2-fluoroadenine is a much better substrate for the _T. cruzi_ enzyme. 2-Aminoadenine, 2-methyladenine and 1-deazaadenine all served as more efficient substrates for the liver enzyme.

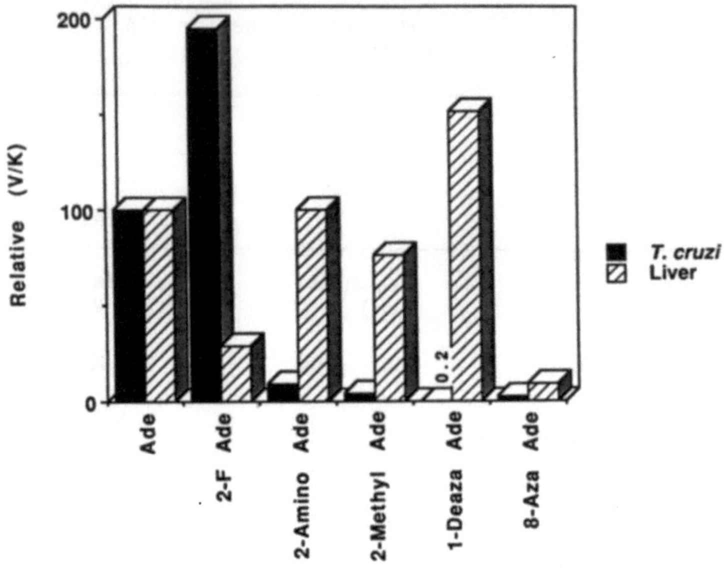

Figure 2. Relative substrate efficiencies (V_m/K_m) for selected purines for the MTA phosphorylases from human liver and T. cruzi.

CONCLUSIONS

Both quantitative and qualitative differences exist between the MTA phosphorylases from parasites and mammals. This implies that it should be possible to develop compounds that will serve as substrates for the parasite enzyme and not the mammalian enzyme. From the data obtained with the MTA phosphorylases from T. cruzi and T. brucei (too little data is presently available for L. donovani) it would appear that the compounds of choice for development would be 2',5'-dideoxy-5'-substituted purine nucleosides. Several advantages can be envisioned for using this class of nucleoside analog for the selective delivery of purines to parasites:

1. 2'-Deoxynucleosides should obviate or cause decreased substrate activity with mammalian MTA phosphorylase (Table II).
2. 6-Amino or 6-aminoalkyl (aryl) substitution on the purine ring should markedly decrease substrate activity with mammalian purine nucleoside phosphorylase[24].
3. 5'-Substitution of the nucleoside would preclude phosphorylation by mammalian nucleoside phosphorylating enzymes, ie. adenosine kinase, deoxycytidine kinase and purine nucleoside phosphotransferase.
4. 2-Substitution on the purine ring and 5'-substitution on the pentose moiety of nucleosides should cause a marked decrease in the ability of these compounds to act as substrates for mammalian adenosine deaminase[25].

Although these studies point to a possible route for the delivery of purine analogs into parasites, this is just the beginning. The purine base released in this selective phosphorolysis must be toxic to the parasite. In the majority of cases, this requires that the base be converted to its nucleoside triphosphate and either interfere with nucleic acid synthesis or be incorporated into the nucleic acids. For this scenario to be operative, many enzymes with different substrate specificities will be required to act upon metabolites of our selected compound. Considering the complexities involved in dealing with not one but many enzyme specificities and our limited knowledge of these enzymes, at present we prefer the course of testing adenine analogs in culture, find those exhibiting growth inhibition of the parasite and then incorporate the toxic purine base into the 2',5'-dideoxy-5'-substituted nucleoside analog.

REFERENCES

1. R. L. Berens, J. J. Marr, S. W. LaFon, and D. J. Nelson, Purine Metabolism in Trypanosoma cruzi, Mol. Biochem. Parasitol. 3:187-196 (1981).
2. W. R. Fish, J. J. Marr, and R. L. Berens, Purine Metabolism in Trypanosoma brucei gambiense, Biochim. Biophys. Acta 714:422-428 (1982).
3. S. W. LaFon, D. J. Nelson, R. L. Berens, and J. J. Marr, Purine and Pyrimidine Salvage Pathways in Leishmania donovani, Biochem. Pharmacol. 31:231-238 (1982).
4. J. J. Marr, R. L. Berens, and D. J. Nelson, Antitrypanosomal Effect of Allopurinol: Conversion in vivo to Aminopyrazolopyrimidine Nucleotides by Trypanosoma cruzi, Science. 201:1018-1020 (1978).
5. W. R. Fish, J. J. Marr, R. L. Berens, D. L. Looker, D. J. Nelson, S. W. LaFon, and A. E. Balber, Inosine Analogs as Chemotherapeutic Agents for African Trypanosomes: Metabolism in Trypanosomes and Efficacy in Tissue Culture, Antimicrob. Agents and Chemother. 27:33-36 (1985).
6. J. J. Marr and R. L. Berens, Antileishmanial Effect of Allopurinol. II. Relationship of Adenine Metabolism in Leishmania Species to the Action of Allopurinol, J. Infect. Dis. 136:724-732 (1977).
7. J. J. Marr, R. L. Berens, N. K. Cohn, D. J. Nelson, and R. S. Klein, Biological Action of Inosine Analogs in Leishmania and Trypanosoma Sp., Antimicrob. Agents and Chemother. 25:292-295 (1984).
8. J. J. Marr, R. L. Berens, D. J. Nelson, T. A. Krenitsky, T. Spector, S. W. LaFon, and G. B. Elion, Antileishmanial Action of 4-Thiopyrazolo(3,4-d)pyrimidine and its Ribonucleoside: Biological Effects and Metabolism, Biochem. Pharmacol. 31:143-148 (1982).
9. T. M. Savarese, G. W. Crabtree, and R. E. Parks Jr., Reaction of 5'-Deoxyadenosine and Related Analogs with the 5'-Methylthioadenosine Cleaving Enzyme of Sarcoma 180 Cells, A Possible Chemotherapeutic Target Enzyme, Biochem. Pharmacol. 28:2227-2230 (1979).
10. T. M. Savarese, G. W. Crabtree, and R. E. Parks Jr., 5'-Methylthioadenosine Phosphorylase-I: Substrate Activity with 5'-Deoxyadenosine with the Enzyme from Sarcoma 180 Cells, Biochem. Pharmacol. 30:189-199 (1981).
11. T. M. Savarese, D. L. Dexter, and R. E. Parks Jr., 5'-Deoxy-5'-methylthioadenosine Phosphorylase-II: Role of the Enzyme in the Metabolism and Antineoplastic Action of Adenine-Substituted Analogs of 5'-Deoxy-5'-methylthioadenosine, Biochem. Pharmacol. 32:1907-1916 (1983).
12. T. M. Savarese, S-H. Chu, M-Y. Chu, and R. E. Parks Jr., 5'-Deoxy-5'-methylthioadenosine Phosphorylase-III: Role of the Enzyme in the Metabolism and Action of 5'-Halogenated Adenosine Analogs, Biochem. Pharmacol. 34:361-367 (1985).

13. T. M. Savarese, A. J. Cannistra, and R. E. Parks Jr., 5'-Deoxy-5'-methylthioadenosine Phosphorylase-IV: Biological Activity of 2-Fluoroadenine-Substituted 5'-Deoxy-5'-methylthioadenosine Analogs, Biochem. Pharmacol. 36:1881-1893 (1987).
14. T. M. Savarese, L. Y. Ghoda, D. L. Dexter, and R. E. Parks Jr., Conversion of 5'-Deoxy-5'-methylthioadenosine and 5'-Deoxy-5'-methylthioinosine to Methionine in Cultured Human Leukemic Cells, Cancer Res. 43:4699-4702 (1983).
15. D. L. Garbers, Demonstration of 5'-Methylthioadenosine Phosphorylase Activity in Various Rat Tissues. Some Properties of the Enzyme from Rat Lung, Biochim. Biophys Acta 523:82-93 (1978).
16. F. D. Ragione, M. Cartenì-Farina, V. Gragnaniello, M. I. Schettino, and V. Zappia, Purification and Characterization of 5'-Deoxy-5'-methylthioadenosine Phosphorylase from Human Placenta, J. Biol. Chem. 261:12324-12329 (1986).
17. M. W. White, A. A. Vandenbark, C. L. Barney, A. J. and Ferro, Structural Analogs of 5'-Methylthioadenosine as Substrates and as Inhibitors of 5'-Methylthioadenosine Phosphorylase and Inhibitors of Human Lymphocyte Transformation, Biochem. Pharmacol. 31:503-507 (1982).
18. V. Zappia, A. Oliva, G. Cacciapuoti, P. Galletti, G. Mignucci, and M. Cartenì-Farina, Substrate Specificity of 5'-Methylthioadenosine Phosphorylase from Human Prostate, Biochem. J. 175:1043-1050 (1978).
19. D. P. Toorchen, and R. L. Miller, (unpublished data)
20. T. M. Savarese, L. Y. Ghoda, and R. E. Parks Jr., Biochemical Considerations in the Design of Analogs of 5'-Deoxy-5'-methylthioadenosine, in: "Development of Target-Oriented Anticancer Drugs," Y.-C. Cheng, B. Goz, M. Minkoff, eds., Raven Press, NY (1983).
21. R. L. Miller, C. L. K. Sabourin, and T. A. Krenitsky, Trypanosoma cruzi Adenine Nucleoside Phosphorylase: Purification and Substrate Specificity, Biochem. Pharmacol. 36:553-560 (1987).
22. L. Y. Ghoda, T. M. Savarese, C. H. Northrup, R. E. Parks Jr., J. Garofalo, L. Katz, B. B. Ellenbogen, and C. J. Bacchi, Substrate Specificities of 5'-Deoxy-5'-methylthioadenosine Phosphorylase from Trypanosoma brucei brucei and Mammalian Cells, Mol. Biochem. Parasitol. 27:109-118 (1988).
23. G. W. Koszalka and T. A. Krenitsky, 5'-Methylthioadenosine (MTA) Phosphorylase from Promastigotes of Leishmania donovani, in: "Purine and Pyrimidine Metabolism in Man. V." W. L. Nyhan, L. F. Thompson, and R. W. E. Watts, eds., Plenum Publishing Company, NY (1986).
24. J. D. Stoeckler, Purine Nucleoside Phosphorylase: A Target for Chemotherapy, in: "Developments in Cancer Chemotherapy," R. I. Glazer, ed., CRC Press, Inc., Boca Raton, FL. (1984).
25. A. Bloch, M. J. Robins, and J. R. McCarthy Jr., The Role of the 5'-Hydroxyl Group of Adenosine in Determining Substrate Specificity for Adenosine Deaminase, J. Med. Chem. 10:908-912 (1967).

CHARACTERIZATION AND BIOGENESIS OF 5'-METHYLTHIOXYLOFURANOSYL ADENINE, A NEW NATURAL ANALOG OF 5'-METHYLTHIOADENOSINE

Marina Porcelli, Giovanna Cacciapuoti, *Guido Cimino, *Margherita Gavagnin, **Guido Sodano and Vincenzo Zappia

Istituto di Biochimica delle Macromolecole, I Facoltà di Medicina e Chirurgia, Università di Napoli
*Istituto per la Chimica di Molecole di Interesse Biologico, CNR, Napoli; **Istituto di Chimica Università della Basilicata, Potenza, Italy

INTRODUCTION

5'-Deoxy-5'-methylthioadenosine (MTA) is a naturally occurring sulfur nucleoside, ubiquitously distributed in nature endowed with antiproliferative activity (1-4). It derives from S-adenosylmethionine (AdoMet) metabolism through several metabolic pathways (1-4). In mammalian tissues the spermidine synthase and spermine synthase reactions represent the quantitatively most important routes for MTA formation (1-4). In spite of the occurrence of multiple biosynthetic pathways, the intracellular concentration of MTA is remarkably low when compared with that of spermidine, spermine and AdoMet (5-8). The thioether indeed does not accumulate intracellular ly because of its rapid cleavage by MTA phosphorylase into adenine and 5-methylthio-ribose-1-phosphate (9).

Recently a naturally occurring analog of MTA, 9-[5'-deoxy--5'(methylthio)- β -D-xylofuranosyl] adenine (xylosyl-MTA), (Fig. 1) has been isolated from Doris verrucosa (10), a Medi-

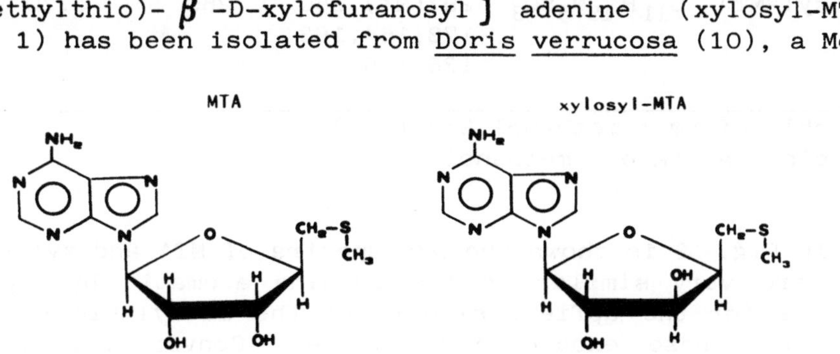

Fig. 1. Structural formulae of MTA and xylosyl-MTA

terranean nudibranch mollusc living in the bay of Naples. In this new sulfur nucleoside, the 5'-deoxy-5'(methylthio)-ribofuranosyl moiety of MTA is replaced by a 5'-deoxy-5'(methylthio)xylofuranosyl residue.

SPECTROSCOPIC AND CHROMATOGRAPHIC CHARACTERIZATION OF XYLOSYL-MTA

Xylosyl-MTA, isolated and purified with the aid of conventional laboratory procedures (10), has been characterized and identified both by high resolution mass spectrometry and proton nuclear magnetic resonance. The mass spectrometry gives the same molecular formula and fragmentation pattern as those obtained with MTA (see Table 1). Furthermore, the xylosyl thioether is characterized by an ultraviolet absorption spectrum identical to that of MTA, with a maximum of 260 nm at pH 7 and of 257 at pH 2, indicating the presence of a 9-substituted adenine nucleus. However, the chromatographic behaviour of xylosyl-MTA on silica gel thin layer chromatography differs significantly from that of MTA in all the systems employed (Table 1).

Table 1. Spectroscopic properties and chromatographic behaviour of MTA and xylosyl-MTA

Product	Molecular formula	MS (m/z)	λmax (nm)	R_f values in systems a	b
MTA	$C_{11}H_{15}N_5SO_3$	297(M^+),250 178,164,163 136,135	260	0.65	0.47
xylosyl-MTA	$C_{11}H_{15}N_5SO_3$	297(M^+),250 178,164,163 136,135	260	0.70	0.60

a = chloroform : methanol 8:2 (v/v)
b = ethyl acetate : methanol 9:1 (v/v)

In Fig. 2 is shown the NMR spectra of MTA and xylosyl-MTA. They are very similar in the adenine aromatic hydrogens as well as for the upfield signals of the methylthio group and for the ribose exocyclic methylene. Conversely, striking differences are observable in the cyclic regions of the ribose, where protons in 2', 3' and 4' as well as the hydroxylic hydrogens in 2' and 3' are significantly shifted. The major distinguishing features are then limited to only three positions of the sugar moiety.

The identification of xylosyl-MTA is confirmed by the total agreement of the NMR spectra of the natural compound with that of a synthetic xylofuranosyl adenine obtained by Montgomery et al. (11). At present, xylosyl-MTA constitutes the first naturally occurring purine carrying a xylose derivative substituent.

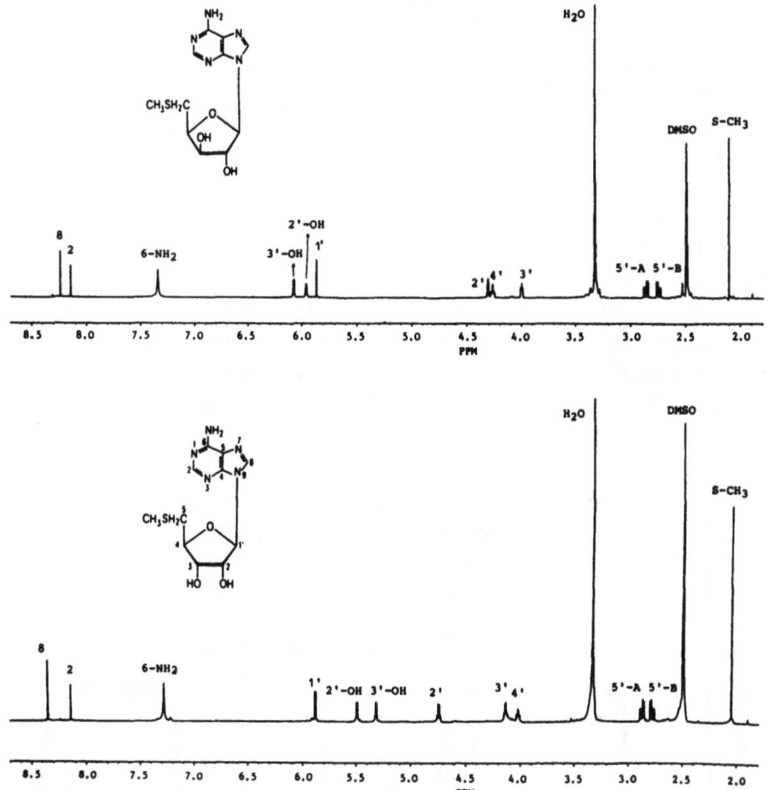

Fig. 2. NMR spectra of xylosyl-MTA and MTA.

HPLC ANALYSIS OF XYLOSYL-MTA

In order to evaluate the cellular concentration of the xylosyl-MTA, an analytical method for the chromatographic separation of MTA and xylosyl-MTA by high performance liquid chromatography has been developed (12). In Fig. 3a is shown a representative chromatogram of a mixture of the two compounds that are completely separated within 15 minutes of analysis; the retention times are 7.8 min and 12.5 min for MTA and xylosyl-MTA respectively. For the evaluation of xylosyl-MTA in Doris verrucosa, two specimens weighing 4-6 gr were homogenized in ice cold 1.5 M perchloric acid (PCA) (1:4, w/v); (Me- ^{14}C)MTA was added as internal standard directly to the samples during homogenization with PCA (2 nmol/gr, 140000 cpm/nmol). After centrifugation, the deproteinized supernatant was neutralized with potassium hydroxide, centrifuged, and an aliquot of this sample equivalent to 20 mg of tissue was analyzed by reversed phase HPLC. A typical chromatogram of a Doris verrucosa extract with a peak identified as xylosyl-MTA according to its retention time is shown in Fig. 3b. The average recovery of (Me- ^{14}C)MTA added to the homogenate was 95%, indicating that the procedure employed does not result in significant loss of xylosyl-MTA.

To confirm the identification of xylosyl-MTA, an aliquot of the sample was incubated for 30 minutes with non-specific adenosine deaminase from Aspergillus oryzae (13) just before the analysis. The chromatogram of the resulting mixture, reported in Fig. 3c, shows the absence of the ultraviolet absorbing material at the retention time corresponding to xylosyl--MTA (12.5 min), and the simultaneous appearance of a new peak with the elution time corresponding to that of xylosylmethylthioinosine (6.5 min). This result allows us to exclude the presence of UV-contaminating compounds in the xylosyl-MTA peak, permitting the quantitative estimation of the molecule. The xylosyl-MTA concentration, averaged from five separate experiments, ranges from 330 to 380 nmoles/gr.

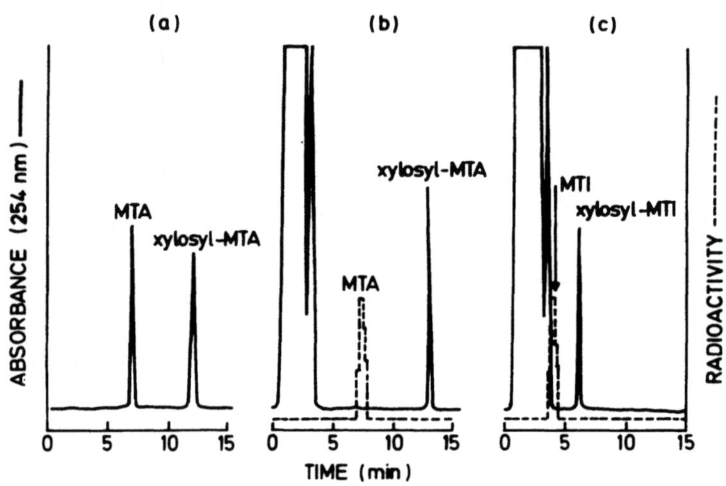

Fig. 3. (a) Separation of MTA and xylosyl-MTA. Column, Ultrasil ODS RP-18; temperature, ambient; eluent 30:70 (v/v) mixture of methanol and 20 mM ammonium formate buffer, pH 7.4; flow rate, 1 ml/min.
(b) Chromatogram of a Doris verrucosa extract.
(c) Chromatogram of the sample shown in Fig. 2b, incubated with adenosine deaminase.

The cellular levels of the xylosyl analog, which are two orders of magnitude higher than that of MTA (<0.5 nmol/g wet tissue), prompted us to investigate whether xylosyl-MTA is a substrate of MTA phosphorylase. Therefore, the molecule was assayed *in vitro* employing MTA phosphorylase from *Doris verrucosa* (specific activity 0.1 nmol/min/ mg of protein). As shown in Fig. 4, while MTA is almost completely converted into adenine, xylosyl-MTA is totally resistant to the enzymatic cleavage. This result justifies the elevated concentration of the molecule found in *Doris verrucosa*. Moreover, the xylosyl-MTA does not exert any inhibitory effect on the enzyme.

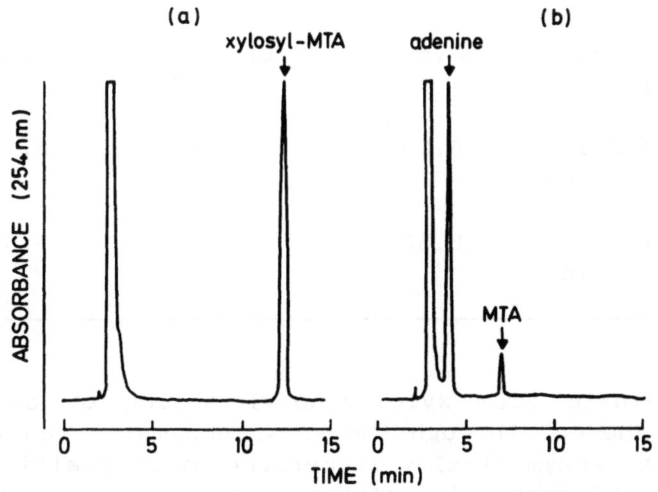

Fig. 4. HPLC analysis of incubation mixtures containing xylosyl-MTA (a) or MTA (b) as substrates of MTA phosphorylase from *Doris verrucosa*.

Xylosyl-MTA and its parent nucleoside have also been assayed as inhibitors for several aminopropyl transferases and the results are reported in Table 2. MTA, as well known, exerts powerful inhibition on the enzyme from *E. coli* and bovine brain as well as on spermidine (spd) and spermine (spm) synthase from *Doris verrucosa*. Conversely, none of these enzymes appears to be inhibited by xylosyl-MTA. In conclusion, polyamine biosynthesis should not be affected by the high cellular amounts of this analog.

BIOGENETIC STUDIES

Preliminary observations suggested that xylosyl-MTA does not derive directly from a dietary source. Three hypothesis concerning the biogenesis can be formulated: i) the molecule

Table 2. Effect of MTA and xylosyl-MTA on propylamino transfer reactions.

Enzymes	nmoles MTA formed/h/mg		
	no addition	+MTA (100 μM)	+xylosyl-MTA (300 μM)
Spd synthase (*D. verrucosa*)	0.25	0.11	0.25
Spm synthase (*D. verrucosa*)	0.54	0.27	0.54
Spd synthase (*E. coli*)	552.60	303.60	552.60
Spd synthase (bovine brain)	6.10	1.08	6.10
Spm synthase (bovine brain)	5.52	0.11	5.52

could originate from xylofuranosyl analog of S-adenosylmethionine (AdoMet) through the known enzymatic reactions; ii) MTA could be enzymatically isomerized at 3'-position; iii) an independent biosynthetic pathway, unrelated to AdoMet and its metabolites, could be operative.

In order to test the validity of the first hypothesis, the occurrence of the xylosyl-analog of AdoMet has been investigated. The sulfonium compound was isolated and purified from *Doris verrucosa* homogenates by the standard chromatographic procedures (12), and the purified material was subjected to the acid hydrolysis at pH 4.5 at 100°C, conditions in which AdoMet and its putative analog are quantitatively cleaved into MTA and xylosyl-MTA respectively (12). Since reversed phase HPLC analysis of the hydrolysis mixture did not reveal the presence of xylosyl-MTA, it is possible to exclude the occurrence of a sulfonium compound modified in the sugar moiety.

To discriminate between the other alternative hypotheses, *in vivo* experiments with labeled precursors have been carried out. In Fig. 5 is summarized the experimental protocol: (*Me*-^{14}C)methionine, (*Me*-^{14}C)MTA and (8-^{14}C)adenine were injected directly into the digestive gland; this treatment apparently does not cause any noxious effect to the mollusc. The treated animals were then kept in a tank containing five liters of oxygenated sea water and were sacrified 1 hour or

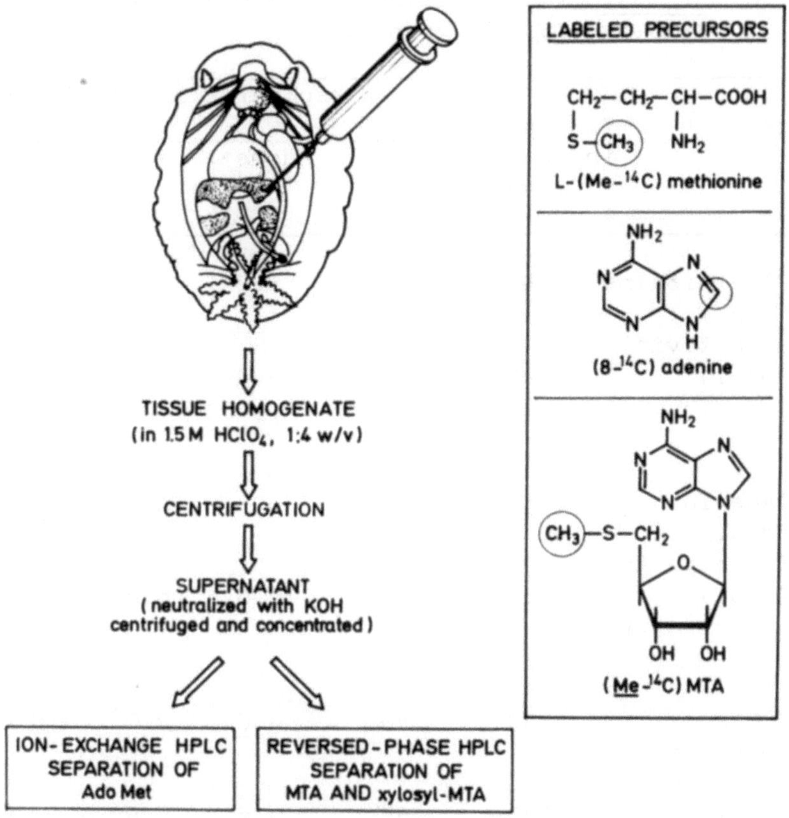

Fig. 5. Procedures employed for biogenetic studies of xylosyl-MTA.

2 hours after the injection. Each experiment was carried out treating three animals and pooling together the homogenates, which were then subjected to the procedure shown in Fig. 5. The results of these experiments are summarized in Table 3. The radioactivity recovered in the animals following the injection as well as the radioactivity recovered in AdoMet, MTA and xylosyl-MTA are expressed as percentage of the total radioactivity administered. Only 65% of the radioactivity administered is still retained 2 hours after the injection in the animals treated with labeled methionine. After 1 hour the majority of radioactivity is associated with AdoMet and only 1.4% is detectable as xylosyl-MTA. After 2 hours, xylosyl-MTA accounts for 8.0% of total injected radioactivity, while MTA is barely detectable. This result is in agreement with the high level of xylosyl-MTA found in Doris verrucosa, suggesting that the xylosyl-thioether can be envisaged as an accumulation product with a very slow turnover rate.

When the animals were injected with adenine no loss of radioactivity was observed, since the majority of the administered radioactivity is still present in the animal even

Table 3. *In vivo* incorporation of labelled precursors into AdoMet, MTA and xylosyl-MTA.

Precursor	Time after injection (min)	recovered radioactivity			
		Total	AdoMet	MTA	xylosyl-MTA
			(%)		
(Me-^{14}C)Methionine ($5x10^6$ d.p.m.; sp. radioactivity 58 mCi/mmol)	60	87	5.0	n.d.	1.4
	120	65	3.4	0.2	8.0
(8-^{14}C)Adenine ($5x10^6$ d.p.m.; sp. radioactivity 54 mCi/mmol)	60	89	0.1	n.d.	1.0
	120	87	n.d.	n.d.	1.4
(Me-^{14}C)MTA ($1.1x10^6$ d.p.m.; sp. radioactivity 55 mCi/mmol)	60	90	n.d.	2.0	16.8
	120	89	n.d.	2.5	34.0

n.d., not detected

2 hours after the treatment. However, because of the multiple biosynthetic pathways involving adenine, no detectable radioactivity appears to be associated with AdoMet and MTA, molecules characterized by a high turnover rate. On the other hand, xylosyl-MTA accounts for 1.4% of the injected radioactivity after 2 hours. From this experimental approach, it is possible to infer that the xylosyl analog derives from methionine and adenine through the usual biosynthetic route in which AdoMet represents the intermediate.

A better understanding of the biogenetic origin of xylosyl-MTA comes from the analysis of the third experiment.

When the animals were injected with (Me-^{14}C)MTA, no radioactivity associated with AdoMet as well as no excretion of labeled compounds is observable. It is worth noting that MTA is rapidly metabolized in that only 2% of the administered radioctivity is associated with this molecule after 1 hour. On the other hand, the high level of radioactivity incorporated into xylosyl-MTA suggests that MTA is a close precursor of the analog. Since after 2 hours xylosyl-MTA increases significantly, despite the low level of radiolabeled

MTA present at 1 hour, it is possible to hypothesize that the conversion of MTA to xylosyl-MTA occurs through intermediate product(s) not yet identified.

A possibility of an enzyme-catalyzed epimerization at 3'-position of the ribose moiety, involving a keto or an epoxy intermediate, is under investigation. In addition, a systematic screening for the distribution of this molecule in different molluscs and other phila, will be beneficial for the elucidation of its physiological role(s).

ACKNOWLEDGEMENTS

We wish to thank A. Crispino for supplying Doris verrucosa and for technical assistance.

REFERENCES

1. V. Zappia, M. Cartenì-Farina, G. Cacciapuoti, A. Oliva and A. Gambacorta, Recent studies on the metabolism of 5'-methylthioadenosine, in: "Natural Sulfur Compounds: Novel Biochemical and Structural Aspects" D. Cavallini G. E. Gaull and V. Zappia eds, Plenum Press, New York (1980).
2. H. G. Williams-Ashman, J. Seidenfeld and P. Galletti, Trends in the biochemical pharmacology of 5'-deoxy-5'-methylthioadenosine, Biochem. Pharmacol. 31:377 (1982).
3. A. E. Pegg and H. G. Williams-Ashman, Pharmacologic interference with enzymes of polyamine biosynthesis and of 5'-methylthioadenosine metabolism, in: "Inhibition of polyamine metabolism: biological significance and basis for new therapies" P. P. McCann, A. E. Pegg and A. Sjoerdsma eds, Academic Press Inc. New York (1986).
4. F. Della Ragione, M. Cartenì-Farina and V. Zappia, 5'-deoxy-5'-methylthioadenosine: novel metabolic and physiological aspects in: "The physiology of polyamines" U. Bacharach and H. Heimer eds, CRC Press Inc. (1988).
5. F. Della Ragione, M. Cartenì-Farina, M. Porcelli, G. Cacciapuoti and V. Zappia, High performance liquid chromatographic analysis of 5'-methylthioadenosine in rat tissues, J. Chromatogr. 526:243 (1981).
6. J. Seidenfeld, J. Wilson and H. G. Williams-Ashman, Androgenis regulation of 5'-deoxy-5'-methylthioadenosine phosphorylase activity in relation to polyamine metabolism of rat liver prostate, Biochem. Biophys. Res. Commun. 95:1861 (1980).
7. E. Usdin, R. T. Borchardt and C. R. Creveling, "Biochemistry of S-Adenosylmethionine and Related Compounds," Macmillan Press LTD, London (1982)

8. U. Bachrach, A. Kaye and R. Chayen, "Advances in Polyamine Research" vol. 4 Raven Press, New York (1982).
9. F. Della Ragione, M. Cartenì-Farina, V. Gragnaniello, M. I. Schettino and V. Zappia, Purification and characterization of 5'-deoxy-5'-methylthioadenosine phosphorylase from human placenta, J. Biol. Chem. 261: 2324 (1986).
10. G. Cimino, A. Crispino, S. De Stefano, M. Gavagnin and G. Sodano, A naturally-occurring analog of methythioadenosine (MTA) from the nudibranch mollusc Doris verrucosa, Experientia 42:1301 (1986).
11. J. A. Montgomery, A. T. Shortnacy and H. J. Thomas, Analogs of 5'-deoxy-5'-(methylthio)adenosine, J. Med. Chem. 17:1197 (1974).
12. M. Porcelli, G. Cacciapuoti, A. Oliva and V. Zappia, High performance liquid chromatographic analysis of a naturally occurring xylosyl-thioether, 9-[5'-deoxy-5'(methylthio)-β-D-xylofuranosyl] adenine, J. Chromatogr. 440:151 (1988).
13. V. Zappia, P. Galletti, M. Cartenì-Farina and L. Servillo A coupled spectrophotometric enzyme assay for methyltransferases, Anal. Biochem. 58:130 (1974).

STUDIES ON HUMAN SERUM 5'-DEOXY-5'-METHYLTHIOADENOSINE PHOSPHORYLASE: MOLECULAR PROPERTIES AND CLINICAL PERSPECTIVES

Gian Luigi Russo, Fulvio Della Ragione, *Riccardo Utili, *Augusto Andreana, *Giuseppe Ruggiero and Vincenzo Zappia

Department of Biochemistry of Macromolecules and *Department of Medical Therapy, First Medical School, University of Naples, Naples, Italy

INTRODUCTION

5'-Deoxy-5'-methylthioadenosine phosphorylase (MTAase) is the only enzyme responsible in eukaryotes for the removal of MTA, a natural sulfur nucleoside produced from S-adenosylmethionine (AdoMet) through several routes[1,2]. The enzyme catalyzes the phosphorolytic breakdown of the N-C glycosidic bond of the thioether leading to adenine and 5-methylthioribose-1-phosphate (MTR-1-P)[3,4]. The carbon skeleton of phosphorylated sugar (except C-1) is then recycled to methionine[5] and the purine base is converted to adenosine 5'-monophosphate by adenine phosphoribosyltransferase[1]: MTAase, therefore, plays a key role in the control of both purine and amino acid pools.

The enzyme has been found in all the mammalian tissues investigated[3,4;6-9], and it is localized almost exclusively in the cytosol compartment[2]. However, low but clearly detectable MTAase activity has been evidenced in some biological fluids including several mammalian sera[10-12] and the secretion of rat ventral and anterior prostate[10]. Although the source of the serum phosphorylase is not known, it is probable that the enzyme is released from the cytoplasm following physiological cell lysis. Thus, it is of interest to determine if acute disease(s) characterized by tissue damage may cause an increase of serum MTAase.

In order to investigate the possible clinical implication(s) of the dosage of MTAase in human serum, the enzyme from this source has been characterized and its properties compared to those of pure human placental phosphorylase[9]. Moreover, the enzyme activity has been determined in sera of both patients with acute hepatitis as well as rats treated with hepatotoxic agents.

The results obtained will be discussed in the light of the potential usefulness of this enzyme as a marker of acute cell damage.

EXPERIMENTAL

Chemicals - AdoMet was prepared from cultures of *Saccharomyces cerevisiae* and isolated by ion-exchange chromatography[13]. S-adenosyl-L-(*Me*-^{14}C)methionine was supplied by the Radiochemical Centre, Amersham, Bucks, U.K. MTA and (*Me*-^{14}C)MTA were prepared from unlabelled and labelled AdoMet and purified by HPLC[14]. (*Me*-^{14}C)MTR-1-P was synthesized from (*Me*-^{14}C)MTA and phosphate in tne presence of pure MTAase[9]. All other chemicals were of the purest commercial grade.

Sample preparation - Serum samples from healthy subjects and patients with hepatitis were prepared using standard procedures and immediately analyzed for the transaminase(s) and MTAase activities. Some samples, stored at -20°C, were assayed after several days to verify the stability of enzyme. No significant decrease of activity was observed up to 10 days of storage.

The rats used were of the Sprague-Dawley strain. Young adult in the 250-350 g weight range were intraperitoneally injected with 30 mmol/kg carbon tetrachloride(CCl_4). Blood plasma was prepared using heparin in amounts that did not inhibit MTAase activity.

MTAase assay - MTAase activity was determined by measuring the formation of (*Me*-^{14}C)MTR-1-P (or its degradation products) from (*Me*-^{14}C)MTA. The standard reaction mixture, unless otherwise stated, contained (in a total volume of 200 µl), 250 mM potassium phosphate buffer, pH 7.4, 5 µM

($\underline{Me}$-^{14}C)MTA (1.5 x 10^7 cpm/μmol), 5 mM dithiothreitol(DTT) and the undialyzed serum. The assay mixture was incubated at 37°C for 1 h and the reaction was stopped by the addition of 50 μl of 3 M trichloroacetic acid. After sample centrifugation at 10,000g per 3 min, a 150 μl aliquot was loaded on a column (0.6 cm x 2 cm) of a Dowex 50-H^+, equilibrated with distilled water. The ($\underline{Me}$-^{14}C)MTR-1-P was eluted into scintillation vials with 2 ml of 10 mM HCl.

__Product identification__ - In these studies, the reaction was stopped by placing the test tube in a boiling water bath. An aliquot of 50 μl of the centrifuged (15,000g, 15 min) assay mixture was then analyzed by both HPLC and TLC methods. In the HPLC analysis, a Perkin-Elmer HPLC system equipped with an anion column Partisil SAX 10 was employed. The elution was carried out with 250 mM ammonium formate, pH 4.4, at a flow rate of 1 ml/min. Fractions of 1 ml were collected into scintillation vials and the radioactivity measured. Under these conditions, MTA was eluted at 6 min and MTR-1-P at 15 min. Analysis by TLC was carried out as described in the reference[9]. An Rf of 0.15 for MTR-1-P and of 0.71 for MTA were determined.

__Molecular Weight Determination__ - The molecular weight of the protein was determined by Sephacryl S-200 gel filtration technique. Sephacryl S-200 (1.5 cm x 95 cm) was calibrated with blue dextran, bovine liver aldolase (M_r 158,000), bovine serum albumin (68,000), equine myoglobin (17,800) and cytochrome c (12,500). The elution was carried out at a flow rate of 2 ml/h with 50 mM potassium phosphate buffer, pH 7.4, 1 mM dithiothreitol, and 200 mM NaCl. Fractions of 1.5 ml were collected and assayed for MTAase activity. The data were plotted in terms of log of molecular weight against Kav.

__Protein determination__ - Protein concentration was estimated according to Bradford[15] using human γ-globulin as standard. The absorbance of column eluates was monitored at 280 nm.

RESULTS AND DISCUSSION

Preliminary experiments were carried out in order to confirm the occurrence of MTAase activity in the human serum and to rule out the possible presence of additional MTA-cleaving enzymes.

Firstly, the products of reaction employing non-hemolyzed human serum as the enzymatic source, were identified by different chromatographic procedures. Analysis of the assay mixture by anion-exchange HPLC procedure allowed the identification of only labelled (Me-^{14}C)MTR-1-P in addition to labelled MTA (Fig.1). A similar result was obtained by TLC analysis (data not shown).

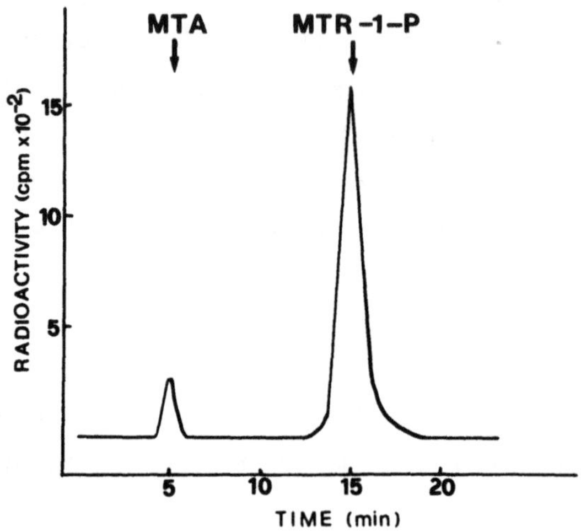

Fig. 1. HPLC analysis of assay mixture containing human serum as MTAase source.
The separation was performed using a Perkin-Elmer HPLC system equipped with an anion column Partisil SAX 10. The elution was carried out with 250 mM ammonium formate, pH 4.4 at a flow rate of 1 ml/min. The assay conditions were as described under the Experimental section.

Moreover, when the serum, freed from phosphate by extensive dialysis against 50 mM Tris/HCl, pH 7.4 and 5 mM dithiothreitol, was employed as the enzymatic source, no cleavage of MTA was observable.

All together these studies clearly indicate the presence of a MTAase activity in the serum.

In order to elucidate some properties of the serum phosphorylase, the effect of dithiothreitol on the enzyme activity was investigated (Table I). Indeed, it is well known that mammalian MTAase requires reducing agents for its maximal activity. Moreover, the effect of DTT was studied at both the physiological serum concentration of phosphate (1 mM) and in the presence of an excess of this ion (250 mM) (Table I). A very low MTAase activity was evidenced, even in the presence of 5 mM dithiothreitol, when the amount of phosphate was left at the physiologi-

Table I. EFFECT OF PHOSPHATE AND DITHIOTHREITOL ON SERUM MTA PHOSPHORYLASE ACTIVITY

ADDITIONS	MTA PHOSPHORYLASE ACTIVITY (units*/ml)
1 mM PHOSPHATE	0.030
1 mM PHOSPHATE + 5 mM DTT	0.033
250 mM PHOSPHATE	0.419
250 mM PHOSPHATE + 5 mM DTT	1.014

* 1 unit is the amount of enzyme which cleaves 1 nmol of MTA per hour at 37°C.
The assay was performed as indicated under Experimental section except that the listed compounds were added at a concentration indicated. The assay mixture contained 100 µl of serum. DTT: dithiothreitol.

cal value. On the other hand, the enzymatic activity increases more than 13-fold at high concentration of phosphate and a further significant build up (30-fold) was noticeable in the presence of 5 mM reducing agent.

These results confirm the requirement of -SH group(s) in the catalitically active MTAase. In addition, it is noteworthy the scarce activity evidenced at physiological concentration of phosphate. Such a result is of particular interest in view of the proposed chemotherapeutic strategies based on the administration of MTA (or its analogs) alone or in association with cytostatic drugs[11]. Indeed, it has been suggested that the occurrence of MTAase in human serum could hamper the efficacy of this therapeutic approach[16]. However, the results obtained indicate that, under physiological conditions, the serum enzyme cannot be responsable for the possible removal of exogenously administered MTA.

The molecular weight of the serum MTAase was then determined employing an enzyme preparation enriched 2-fold by ammonium sulfate precipitation (45-65% saturation). The result obtained indicates that the serum phosphorylase has a M_r of about 95,000 a value in good agreement with that of human placental enzyme[9]. In addition, the serum MTAase was completely precipitated by a specific rabbit antiserum obtained against hemocyanin-MTA phosphorylase[17] (data not shown). This result confirms the similarity of the structure among the enzyme from different sources[17].

In order to select the optimal assay conditions to determine MTAase activity in human serum, the Michaelis constants for both the substrates were calculated employing undialyzed serum as the enzymatic source. The Km for MTA, estimated by Lineweaver-Burk plot, at saturating concentration of phosphate (250 mM), is 0.5 μM (Fig.2), a value that fits very well with those reported for MTAase from other sources[6-10]. The Km value for phosphate, calculated by the same procedure in the presence of 5 μM MTA, is 50 mM (data not shown). This value is significantly higher than that obtained from placental pure MTAase[9] and could explain the low activity detectable at physiological concentration of this ion.

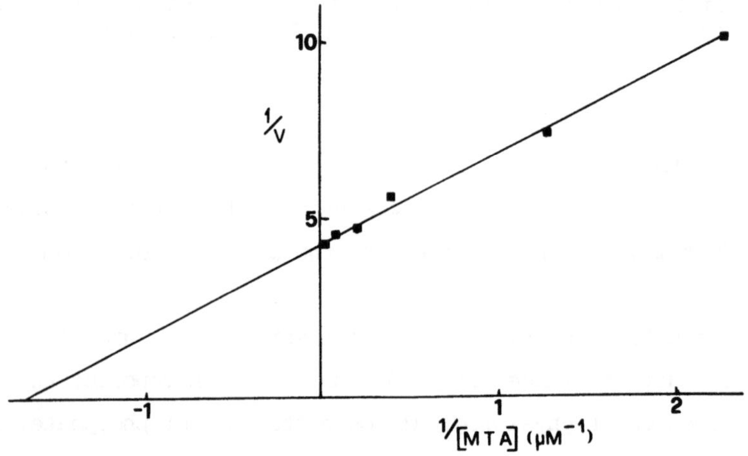

Fig. 2. Effect of MTA concentration on serum MTAase activity. The assay conditions were as described under Experimental section.

The occurrence of clearly detectable MTAase in the human serum and the availability of rapid and reproducible methods of enzymatic assay allowed us to postulate the utilization of the phosphorylase as marker of diseases involving acute cell damage. Therefore, CCl_4, a potent hepatotoxic agent, was injected intraperitoneally to rats and serum MTAase and GOT activities were determined at different times. As shown in Fig. 3, the level of the phosphorylase increases after 12 hrs reaching a peak at about 28 hrs. The GOT activity, used as control, has the same behaviour.

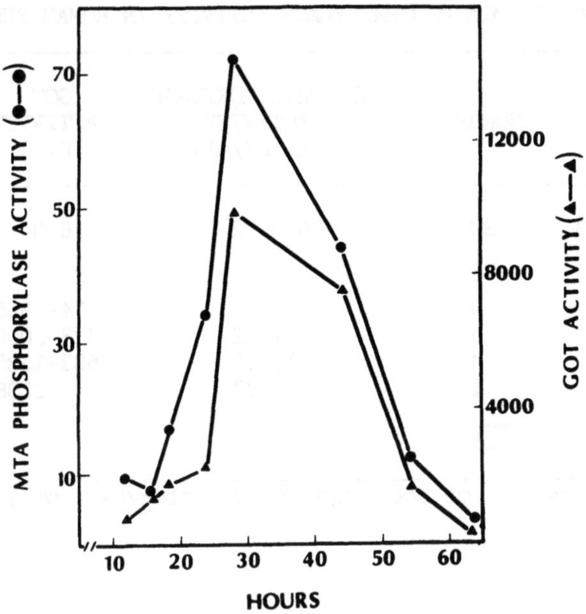

Fig. 3. Time course of serum MTAase and GOT activities in rats intraperitoneally treated with CCl_4 (30 mmol/kg).
(●-●) MTAase activity (nmoles of MTA cleaved/h/ml of serum);
(▲-▲) GOT activity (U/l).

In order to evaluate the clinical implication(s) of this observation, several samples of serum from normal subjects as well as from patients with acute hepatitis have been scrutinized for the enzyme activity (Table II).

A significant increase, ranging from 1.5 to 40-fold, of enzyme activity has been demonstrated in the patients under investigation. Moreover when the MTAase activity was compared to that of GOT and GPT a remarkable correlation was observed.

In preliminary studies, carried out on 4 patients, the activities of serum MTAase and transaminases were monitored for about 20 days (data not shown). The result obtained suggest that the phosphorylase activity brought back to normal value faster than GOT and GPT, thus suggesting that MTAase could be a more sensitive marker of cell recovery after acute damage.

Future studies will be devoted to investigate the clinical value of this marker in acute hepatitis as well as in other pathological conditions characterized by cell lysis namely acute myocardial infarct and progressive muscolar dystrophy.

Table II. MTA PHOSPHORYLASE ACTIVITY IN HUMAN SERUM

	NUMBER	MTA PHOSPHORYLASE ACTIVITY (units*/ml)	GOT ACTIVITY (U/l)	GPT ACTIVITY (U/l)
HEALTHY SUBJECTS	30	0.5- 2	6-36	6-40
PATIENTS WITH ACUTE HEPATITIS	8	3- 6	74- 355	349-1125
	5	7-12	323-1000	950-1500
	4	13-30	683-1585	1309-2108
	1	77	2088	3400
	18			

* 1 unit is the amount of the enzyme which cleaves 1 nmol of MTA per hour at 37°C.

ACKNOWLEDGMENTS

We thank Dr. A. Leonelli for providing us with some of the control sera employed. This work was supported by the National Research Concil project "Oncologia", grant n. 86.00618.44.

REFERENCES

1. F. Della Ragione, M. Cartení-Farina and V. Zappia, 5'-Deoxy-5'-methylthioadenosine: novel metabolic and physiological aspects, in: "The Physiology of Polyamines," U. Bachrach and Y.M. Heimer eds, CRC Press, Boca Raton, in press.
2. H.G. Williams-Ashman, G. Seidenfeld and P. Galletti, Trends in the biochemical pharmacology of 5'-deoxy-5'-methylthioadenosine, Biochem. Pharmacol. 31:277 (1982).
3. A.E. Pegg and H.G. Williams-Ashman, Phosphate-stimulated breakdown of 5'-methylthioadenosine by rat ventral prostate, Biochem. J. 115:241 (1969).
4. V. Zappia, A. Oliva, G. Cacciapuoti, P. Galletti, G. Mignucci and M. Cartení-Farina, Substrate specificity of 5'-methylthioadenosine phosphorylase from human prostate, Biochem. J. 175:1043 (1978).

5. P.S. Backlund and R.A. Smith, Methionine synthesis from 5'-methylthioadenosine in rat liver, J. Biol. Chem. 256:1533 (1981).

6. D.L. Garbers, Demonstration of 5'-methylthioadenosine phosphorylase activity in various rat tissues: some properties of the enzyme from rat lung, Biochem. Biophys. Acta 523:82 (1978).

7. G. Cacciapuoti, A. Oliva and V. Zappia, Studies on phosphate-activated 5'-methylthioadenosine nucleosidase from human placenta, Int. J. Biochem. 9:35 (1978).

8. T.M. Savarese, G.W. Crabtree and R.E. Parks Jr, 5'-Methylthioadenosine phosphorylase-I: substrate activity of 5'-deoxyadenosine with the enzyme from Sarcoma 180 cells, Biochem. Pharmacol. 30:189 (1981).

9. F. Della Ragione, M. Carteni-Farina, V. Gragnaniello, M.I. Schettino and V. Zappia, Purification and characterization of 5'-deoxy-5'-methylthioadenosine phosphorylase from human placenta, J. Biol. Chem. 261:12324 (1986).

10. J. Seidenfeld, J. Wilson and H.G. Williams-Ashman, Androgenic regulation of 5'-deoxy-5'-methylthioadenosine concentration and methylthioadenosine phosphorylase activity in relation to polyamine metabolism of rat prostate, Biochem. Biophys. Res. Commun. 95:1861 (1980).

11. N. Kamatani, W.A. Nelson-Rees and D.A. Carson, Selective killing of human malignant cells deficient in methylthioadenosine phosphorylase, a purine metabolism enzyme, Proc. Natl. Acad. Sci. USA 78:1219 (1981).

12. M.K. Riscoe and A.J. Ferro, 5-Methylthioribose: its effects and function in mammalian cells, J. Biol. Chem. 259:5465 (1984).

13. V. Zappia, C. R. Zydek-Cwick and F. Schlenk, The specificity of S-adenosylmethionine derivatives in methyl transfer reactions, J. Biol. Chem. 244:4499 (1969).

14. F. Della Ragione, M. Carteni-Farina, M. Porcelli, G. Cacciapuoti and V. Zappia, High performance chromatographic analysis of 5'-methylthioadenosine in rat tissues, J. Chromatogr. 226:243 (1981).

15. M.M. Bradford, A rapid and sensitive method for the quantitation of microgram quantities of protein utilizing the principle of protein binding, Anal. Biochem. 72:248 (1976).

16. J.H. Fitchen, M.K. Riscoe, B.W. Dana, H.J. Lawrence and A.J. Ferro, Methylthioadenosine phosphorylase deficiency in human leukemias, Cancer Research 46:5409 (1986).
17. F. Della Ragione, A. Oliva, M. Fioretti, G.L. Russo, R. Palumbo and V. Zappia, Physico-chemical and immunological properties of bovine liver 5'-deoxy-5'-methylthioadenosine phosphorylase, in this book.

POLYAMINES IN CELL GROWTH

AND DIFFERENTIATION

GROWTH-REGULATION OF THE CELLULAR LEVELS AND EXPRESSION OF THE mRNA MOLECULES CODING FOR ORNITHINE DECARBOXYLASE AND S-ADENOSYLMETHIONINE DECARBOXYLASE

David R. Morris and Michael W. White

Department of Biochemistry
University of Washington
Seattle, WA

INTRODUCTION: GROWTH-REGULATION OF POLYAMINE BIOSYNTHESIS

Elevated cellular polyamine levels are required for stimulation of proliferation of normal populations of resting cells (reviewed in Pegg and McCann, 1982; Marton and Morris, 1987). The association between polyamine levels and cell growth has been noted by many workers since the pioneering observations in regenerating liver of Dykstra and Herbst (1965) and Raina and coworkers (1966). Polyamine levels in regenerating liver were shown early to be regulated by the levels of two key enzymes, ornithine decarboxylase (ODC) and S-adenosylmethionine decarboxylase (SDC) (Russell and Snyder, 1968; Janne and Raina, 1968; Russell and Lombardini, 1971; Hannonen et al., 1972); these original observations have since been confirmed in a variety of tissue and cell types (reviewed in Pegg and McCann, 1982; Tabor and Tabor, 1984; Pegg, 1986; White and Morris, 1988). In cases where it has been examined, elevation of ODC and SDC levels is primarily a consequence of increased rates of synthesis of the two proteins (Seyfried et al., 1982; Pegg, 1986). In this paper, we discuss the relative contributions of mRNA level and efficiency of mRNA translation to the regulation of the rates of ODC and SDC synthesis. As well, we consider the mechanisms involved in these two modes of regulation and their relationship to pathways of transmembrane mitogenic signal transduction.

A DISTINCTION: GROWTH CONTROL VERSUS CELL CYCLE CONTROL

A resting population of normal animal cells is arrested with a 2 N content of DNA in a state referred to as "G0". Upon mitogenic activation, resting cells traverse the G1 phase and continue to move through the cell cycle. When one finds that expression of a gene product is stimulated by treatment of cells with a growth factor, the observed regulation could be the result of the transition from G0 to

G1 ("growth control") or it could be due to the fact that cells are moving through the G1 phase of the cycle ("cell cycle control"). Examples of cell cycle regulated genes are those coding for histones. The levels of histone mRNAs are low in resting cells and are elevated as the cells progress through the first cycle after mitogenic activation. In addition to the response to mitogen-activation, histone production is enhanced as each cell in a continuously cycling population traverses the end of G1 phase and initiates DNA replication (for a review, see Schumperli, 1986). Thus, histone synthesis is under "cell cycle control", in that it is stimulated every time a cell passes through a particular point in the cycle and not just in the first cycle as resting cells pass from G0 into G1.

Stimulation of growth-arrested mouse 3T3 cells leads to elevation of the cellular levels of the mRNAs coding for both ODC (Kahana and Nathans, 1984; Liu et al., 1985; Hovis et al., 1986) and SDC (Stimac and Morris, 1987). In order to investigate whether these mRNAs were under cell cycle control, randomly growing cultures of 3T3 cells were fractionated by centrifugal elutriation and the levels of ODC and SDC mRNAs were analyzed (Stimac and Morris, 1987). We found that the cellular contents of these mRNAs in a pure population of early G1 cells were not significantly lower than in other phases of the cell cycle and were elevated compared to those found in growth-arrested "G0" cells. Therefore, the cellular levels of the mRNAs coding for ODC and SDC are growth-controlled, but not regulated during the traverse of the normal cell cycle. These results with the genes coding for these two enzymes of polyamine biosynthesis are similar to those obtained by others with actin (Riddle and Pardee, 1980) and c-myc (Hann et al., 1985; Thompson et al., 1985) and suggest that generally the members of the "competence gene family" (Stiles, 1985) may not be cell cycle regulated.

The relative constancy of ODC and SDC mRNA levels, as mouse 3T3 cells progress through the cycle, is in contrast to earlier measurements of polyamine levels and rates of polyamine synthesis during the cell cycle (reviewed in Pegg and McCann, 1982). Those measurements showed that polyamine levels were high in late G1 and early S phase and were elevated again just prior to division. The discrepancies between cellular levels of polyamines and decarboxylase mRNAs could reflect cycle-dependent modulation of either biosynthetic enzyme activity, rates of polyamine degradation or efficiency of translation of these mRNA molecules.

MITOGENIC SIGNAL TRANSDUCTION AND GROWTH CONTROL OF mRNA LEVELS

The cellular levels of the mRNA molecules coding for ODC and SDC have been shown to be modulated in a variety of growth-regulated systems. In different cell types and with different stimuli, the regulation of ODC message levels has been suggested to be either at transcriptional (Olson and Spizz, 1986; Katz and Kahana, 1987) or post-transcriptional (White et al., 1987b; Rose-John et al., 1987) steps of mRNA metabolism. Regardless of the detailed mechanisms of regulation, the control of expression of ODC and SDC must be coupled to the transmembrane signaling pathways activated by the interaction of mitogen with its receptor on the cell surface. In this section, we discuss these aspects of the regulation of ODC mRNA level specifically in bovine T-lymphocytes. These results are contrasted with conclusions derived from studies of other cell types.

In order to examine the rate of ODC gene transcription, nuclei were isolated from bovine T-cells at various times after mitogenic activation. The level of ODC transcription was monitored by incubation of the isolated nuclei with radioactive substrate and the labeled products were analyzed by hybridization to fragments of an ODC cDNA clone (White et al., 1987b). We found that ODC, as well as several other genes, showed no significant increase in transcription during the period of time that the cellular level of the corresponding mRNA was undergoing large increases. This was in contrast to genes such as c-myc which showed parallel increases in transcription and mRNA levels (White et al., 1987b). Thus, it seems that in normal T-cells there is a group of genes that is regulated in response to mitogenic stimuli by stabilization of a product of transcription. The stabilized product could be either an intranuclear precursor or mature cytosolic message. The half-life of mature ODC mRNA is ca. 2 hr in both resting and mitogen-activated T-cells (M. White and M. Abrahamsen, unpublished results). This result argues for stabilization at the level of intranuclear transcripts, perhaps via a "process versus discard" model (Darnell, 1982).

Rose-John et al. (1987) presented evidence in Swiss 3T3 cells that was consistent with regulation of ODC mRNA level through post-transcriptional stabilization. In contrast, Katz and Kahana (1987) found elevated nuclear transcription of the ODC gene after mitogenic activation of resting Balb/c 3T3 cells. This striking discrepancy between the results obtained by these two groups of workers could be due to different modes of regulation of ODC in the two cell lines employed. Other differences between Swiss and Balb/c 3T3 cells have been noted. In the Balb/c 3T3 cells, elevation of ODC mRNA level in response to mitogenic activation is blocked by cycloheximide (Katz and Kahana, 1987); in contrast, two other groups have found enhancement of ODC mRNA expression by the protein synthesis inhibitor in Swiss 3T3 cells (Stimac and Morris, 1987; Rose-John et al., 1987). Further experiments are required in order to define the origin of these apparent cell-specific differences in the mechanism of ODC regulation.

The activation of several signal transduction pathways has been observed within minutes after binding of a ligand to the T-cell antigen receptor. These putative signals include elevation of cytosolic Ca^{2+} (Tsien et al., 1982), activation of protein kinase C (PKC) (Ledbetter et al., 1987; Grove and Mastro, 1987) and phosphorylation of protein tyrosine residues (Klausner et al., 1987). Activation of phospholipase C is a key signaling process, which catalyzes the hydrolysis of phosphatidyl inositol bisphosphate to two intracellular second messengers, diacylglycerol (DAG) and inositol trisphosphate (IP_3) (Imboden and Stobo, 1985). IP_3 is thought to regulate Ca^{2+} channels in the plasma membrane (Kuno and Gardner, 1987) and DAG synergizes with elevated Ca^{2+} to activate PKC (Wolf et al., 1985).

The involvement of PKC activation in ODC regulation in T-lymphocytes was suspected from the early studies of Kensler et al.(1978) demonstrating the elevation of ODC activity in response to treatment of the cells with phorbol esters. These tumor promoters have subsequently been shown to be activators of PKC (Nishizuka, 1984). As well as elevating enzyme activity, treatment of bovine T-cells with biologically active phorbol esters rapidly increased the cellular level of ODC mRNA (White et al., 1987b). Thus, activation of PKC was sufficient to stimulate ODC gene expression. Recent results from our laboratory have demonstrated that this kinase is essential for coupling expression of this gene to transmembrane signals generated from the activated T-cell receptor. Down-regulation of the cellular level of PKC by prolonged treatment of lymphocytes with high concentrations of phorbol ester (Rodriguez-Pena and Rozengurt, 1985; Blackshear et al., 1985), totally blocked elevation of ODC mRNA in response to phorbol esters or to Con A (Morris et al., 1988). Similarly, treatment of cells with trifluoperazine, an inhibitor of processes mediated by PKC and calmodulin (CaM) (Schatzman et al., 1981), also blocked induction of ODC mRNA by Con A. On the other hand, mitogen-induced elevation of ODC mRNA was unaffected by the CaM antagonists Cd^{2+} and W-7 (White et al., 1987b; Morris et al., 1988). Thus, we conclude that PKC is a necessary component of the signal transduction pathway from the T-cell antigen receptor to the ODC gene; likewise, CaM seems not to be involved in this pathway. Regulation of the ODC gene is in clear contrast to the c-myc gene, which requires both PKC and CaM signals for activation in T-lymphocytes (Morris et al., 1988).

In contrast to T-lymphocytes, where there seems to be an absolute requirement for PKC for ODC induction by mitogenic stimuli, results with non-lymphoid cells suggest that more than one signaling pathway may be coupled to ODC expression. In NIH-3T3 cells, down-regulation of PKC only partially inhibited the expression of ODC in resting cells treated with serum, FGF or PDGF (Hovis et al., 1986). Similar results, suggesting one or more additional pathways of ODC induction, were obtained with Swiss 3T3 cells (McCaffrey et al., 1987). This situation is not unique to expression of the ODC gene, since others have reported the existence of PKC-dependent and -independent pathways of c-myc induction in human fibroblasts (Coughlin et al., 1985) and in Swiss mouse 3T3 cells (Kaibuchi et al., 1986). In considering mechanisms for the involvement of PKC in the regulation of ODC gene expression it is important to recall the experiments of Sussman et al. (1985) with a Balb/c 3T3 cell line defective in $Na^{+}K^{+}Cl^{-}$ cotransport. This mutant line, in contrast to wild type, did not show an elevation of ODC activity in response to phorbol esters, yet it responded normally to serum (presumably via a PKC-independent pathway). Assuming that PKC is the primary site of phorbol ester action in these cells, these results suggest that signaling via the PKC-dependent pathway could involve changes in ion flux across the plasma membrane.

GROWTH REGULATION OF ODC AND SDC TRANSLATION

An elevated rate of protein synthesis is an obligate part of the entry of a growth-arrested cell into a state of proliferation. When resting T-lymphocytes are activated with a mitogenic lectin, there is an early increase in protein synthesis which correlates with the recruitment of translationally inactive ribosomes into polysomes in the absence of new ribosome synthesis (Ahern and Kay, 1975; Cooper et al., 1976; Wettenhall and London, 1974). A similar recruitment of ribosomal subunits has been shown to occur in serum-activated fibroblasts (Johnson et al., 1974). A number of mechanisms have been suggested to be involved in the initial burst of protein synthesis following the addition of a growth signal; these include the utilization of translationally inactive sequestered mRNAs, an increase in the activity of limiting translation factors, and the inactivation of translational inhibitors.

In the studies of Degen et al. (1983) on the expression of cytoskeletal actin in mitogen-activated lymphocytes, the increased rate of actin synthesis was found to be regulated by parallel changes in the level of actin mRNA. Thus, it was concluded that there is no translational component in the regulation of actin expression. Consistent with a lack of translational control, essentially all of the actin mRNA, in both resting and activated-lymphocytes, was found associated with polysomes (Degen et al., 1983). In contrast to actin, ODC and SDC mRNAs are two clear examples of mRNAs which are translationally regulated by a growth signal.

Our studies of ODC expression have revealed major elements of translational control (White et al., 1987a). In resting T-lymphocytes only 55 % of the total ODC mRNA was located on polysomes, with the remainder found on untranslated particles which sediment at 20-30s. During the first 6 hr after a mitogenic stimulus the proportion of ODC mRNA in polysomes decreased to 30 %. This was followed by large shift in the cytoplasmic distribution of ODC mRNA, with the majority of the mRNA (80 %) becoming located in polysomes by 12 hr; this occurs in the absence of any change in the cellular level of ODC mRNA. While there was a large shift in the proportion of ODC mRNA translated, the average number of ribosomes occupying a particular ODC mRNA was unchanged during activation by Con A. The increase in the amount of ODC mRNA associated with polysomes fully accounts for the increase in the rate of synthesis of ODC protein, measured before and after the shift in the distribution of ODC mRNA. Thus, the translation of ODC mRNA is regulated during mitogenesis, through a mechanism which involves the redistribution of the message from untranslated particles into polysomes.

The cellular level of SDC mRNA increases 4-fold by 8 hr after the stimulation of lymphocytes with Con A (Mach et al., 1986). Over this same time period the rate of SDC synthesis was found to increase 8-10 fold revealing a discrepancy between the change in synthetic rate and the change in the level of SDC mRNA. In examining the cytoplasmic distribution of SDC mRNA in resting and mitogen-activated lymphocytes, we have made several interesting observations concerning the translation of this mRNA. Based on the size of polysomes translating the mRNAs encoding hemoglobin and actin, the SDC precursor, which has a molecular weight of 37,000 (Kameji and Pegg, 1987a) would be expected to reside on polysomes larger than 10 ribosomes (Mach et al., 1986). In contrast to this prediction, SDC mRNA in resting lymphocytes is found primarily on monosomes and disomes indicating very inefficient translational initiation (Mach et al., 1986). After the addition of mitogen to the cells, the level of SDC mRNA on polysomes containing an average of 8-10 ribosomes increases 10-fold. As a consequence of this increase, the average polysome containing SDC mRNA, in 8 hr-activated cells, was calculated to contain 2.7 ribosomes, a value two-fold larger than the average polysome size in resting cells. This increase in the number of ribosomes translating SDC mRNA fully accounts for the discrepancy between the elevation in the rate of SDC synthesis and the increase in SDC mRNA level following Con A-stimulation.

In contrast to ODC, we detect little or no SDC mRNA in the mRNP region from either resting or activated lymphocytes. Thus, the behavior of ODC and SDC mRNAs suggest that there exist at least two mechanisms of translational control. In considering established models of translational regulation, which might explain the behavior of ODC and SDC mRNAs, an important model to consider is that of Lodish (1976). Lodish proposed that a general increase in translational initiation should have a large effect on messages which are initiation-limited due to their low affinity for subsaturating initiation factors. An elevation of initiation factor activity should result in an increase in the number of ribosomes translating those messages. An experimental prediction from this model is that messages which are poor competitors should be loaded by ribosomes by selectively inhibiting elongation, creating a situation where elongation is rate limiting (Lodish and Desalu, 1973). The Lodish model has been used to explain the translational differences between alpha and beta-globin mRNAs (Lodish and Desalu, 1973). Treatment of resting lymphocytes with a low concentration of cycloheximide chosen to selectively inhibit elongation, (Lodish and Desalu, 1973; White et al., 1987a) for 15 min, drove all of the SDC mRNA, which is normally found on monosomes, into polysomes of about 10 ribosomes (unpublished results) while this treatment had little effect on polysomes containing actin mRNA (White et al.,1987a). The increase in the polysome size of SDC mRNA in response to both cycloheximide and mitogen demonstrates that SDC mRNA is controlled at the initiation step of translation. Thus, the Lodish model adequately explains the different translational behaviors of actin and SDC mRNAs. The Lodish model, however does not account for the behavior of ODC mRNA. Since there is no effect of mitogen-activation on the size of polysomes containing ODC mRNA, this message must not be regulated at the level of translational initiation. A modification of the Lodish model has been proposed by Thach and coworkers (Walden et al., 1981). The Thach model proposes that mRNAs must first be recruited from an untranslatable pool into a translatable condition by a discrimination step before they may be translationally initiated. In this model, regulation of translation of a particular mRNA may occur at the level of discrimination and/or the level of initiation. The translational behavior of ODC and SDC mRNAs in response to a growth signal appear to fit the Thach model. ODC mRNA, which is redistributed from the untranslated pool into polysomes during mitogenesis, is an example of a mRNA which is regulated at the discrimination step of translation. SDC mRNA is not found in the untranslated pool but instead shows, following the addition of a growth signal, an increase in the number of ribosomes per mRNA and is an example of regulation at the initiation step of the Thach model.

Recently, Kameji and Pegg (1987b) demonstrated that the translation of ODC and SDC mRNAs were inhibited, in a dose dependent manner, by spermidine and spermine in a reticulocyte cell-free translation system. The inhibition by polyamines, appeared to be specific to ODC and SDC, since the translation of a control mRNA, albumin, was unchanged. These results showed, for the first time, that polyamines directly influence the translation of ODC and SDC mRNAs and are in agreement with a number of reports (Kahana and Nathans, 1985; Dircks et al., 1986; Holtta and Pohjanpelto, 1986; Persson et al., 1986; Shirahata and Pegg, 1986), which indicated that the elevation of polyamines in intact cells decreased the rate of synthesis of these proteins without a change in mRNA levels.

It seems unlikely that the translational regulation of ODC and SDC by polyamines is related to the translational changes observed during mitogen-activation. Since the levels of polyamines in resting lymphocytes are low and do not substantially increase until 12 hr after activation (Fillingame and Morris, 1973), one would expect, if a polyamine-mediated mechanism were operative, that ODC and SDC would be maximally translated in resting cells and would begin to be repressed in 12 hr-activated cells; however the opposite is observed. Further, the addition of either putrescine or the polyamine biosynthetic inhibitor, difluoromethylornithine, at the time of mitogenic stimulation, have no effect on the increased translation of either mRNA (unpublished results). Thus, we would conclude that the synthesis of ODC and SDC is under two modes of translational regulation: 1.) feedback inhibition by intracellular polyamine levels and 2.) stimulation during mitogen-activation of resting cells.

SUMMARY AND PERSPECTIVES

After activation of resting cells by mitogens, the rate of synthesis of two key enzymes of polyamine biosynthesis, ODC and SDC, is regulated both by the levels of the two mRNAs and the efficiency with which they are translated.

The cellular content of the two mRNAs is elevated as resting cells exit the G0 state; these changes are unique to the G0/G1 transition, since the levels of these mRNAs remain constant during the progression of growing cells through the cell cycle. Depending on the particular cell type, the content of ODC mRNA seems to be regulated either by transcriptional or by post-transcriptional mechanisms. The mechanism of post-transcriptional regulation of ODC mRNA level is not understood, but it probably does not involve stabilization of mature message. In both T-lymphocytes and in fibroblasts, the regulation of ODC mRNA level is linked to the activation of PKC. Although the mechanism of PKC activation through the mitogenic stimulation of metabolism of the phosphatidyl inositols is reasonably well understood, the number and nature of the steps between PKC and regulation of ODC mRNA expression is totally unknown. The nature of the regulatory mechanisms determining SDC mRNA level and their coupling to mitogenic signals are at this point also undefined.

The regulation of translation of ODC and SDC mRNAs in response to mitogenic stimulation is by two distinct mechanisms. The structural aspects of these two mRNAs that determine interactions with the translational apparatus yielding their unique regulatory behavior is clearly of interest. As well, it will be important to determine whether these regulatory elements are involved in the regulation by polyamines of the translation of these messages. The cellular elements with which these sequences interact will be of obvious importance, as will a definition of how these interactions are modified by growth signals or cellular polyamine levels.

REFERENCES

Ahern,, T., and Kay, J. E., 1975. Protein synthesis and ribosome activation during the early stages of phytohemagglutinin lymphocyte stimulation. Exptl. Cell Res. 92:513-515.

Blackshear, P. J., Witters, L. A., Girard, P. R., Kuo, J. F., Quamo, N., 1985. Growth factor-stimulated protein phosphorylation in 3T3-L1 cells. Evidence for protein kinase C-dependent and independent pathways. J. Biol. Chem. 260:13304-13315.

Boles, T. C. and Hogan, M. E., 1987. DNA structure equilibria in the human c-myc gene. Biochemistry 26:367-376.

Cooper, H. L., Berger, S. L., and Braverman, R., 1976. Free ribosomes in physiologically nondividing cells. Human peripheral lymphocytes. J. Biol. Chem. 251:4891-4900.

Coughlin, S. R., Lee, W. M. F., Williams, P. W., Giels, G. M., and Williams, L. T., 1985. c-Myc gene expression is stimulated by agents that activate protein kinase C and does not account for the mitogenic effect of PDGF. Cell 43:243-251.

Darnell, J. J. E., 1982. Variety in the level of gene control in eukaryotic cells. Nature 297:365-371.

Degen, J. L., Neubauer, M. G., Friezner-Degen, S. J., Seyfried, C. E., and Morris, D. R., 1983. Regulation of protein synthesis in mitogen-activated bovine lymphocytes. Analysis of actin-specific and total mRNA accumulation and utilization. J. Biol. Chem. 258:12153-12162.

Dircks, L., Grens, A., Slezynger, T. C., and Scheffler, I. E., 1986. Posttranscriptional regulation of ornithine decarboxylase activity. J. Cell. Physiol. 26:371-378.

Dykstra, W. G., and Herbst, E. J., 1965. Spermidine in regenerating liver: relation to rapid synthesis of ribonucleic acid. Science 149:428-429.

Fillingame, R. H. and Morris, D. R., 1973. Polyamine accumulation during lymphocyte transformation and its relation to the synthesis, processing and accumulation of ribonucleic acid. Biochemistry 12:4479-4487.

Grove, D. S. and Mastro, A. M., 1987. Changes in protein kinase C and cAMP-dependent kinase in lymphocytes after treatment with 12-O-tetradecanoylphorbol-13-acetate or concanavalin A: quantitation of activities with an in situ gel assay. J. Cell. Physiol. 132:415-427.

Hann, S. R., Thompson, C. B., and Eisenman, R. N., 1985. c-Myc oncogene protein synthesis is independent of the cell cycle in human and avian cells. Nature 314:366-369.

Hannonen, P., Raina, A., and Janne, J., 1972. Polyamine synthesis in the regenerating rat liver: stimulation of S-adenosylmethionine decarboxylase, and spermidine and spermine synthases after partial hepatectomy. Biochim. Biophys. Acta 273:84-90.

Holtta, E. and Pohjanpelto, P., 1986. Control of ornithine decarboxylase in Chinese hamster ovary cells by polyamines. Translational inhibition of synthesis and acceleration of degradation of the enzyme by putrescine, spermidine, and spermine. J. Biol. Chem. 261:9502-9508.

Hovis, J. G., Stumpo, D. J., Halsey, D. L., and Blackshear, P. J., 1986. Effects of mitogens on ornithine decarboxylase activity and messenger RNA levels in normal and protein kinase C-deficient NIH-3T3 fibroblasts. J. Biol. Chem. 261:10380-10386.

Imboden, J. B. and Stobo, J. D., 1985. Transmembrane signalling by the T cell antigen receptor. Peturbation of the T3-antigen receptor complex generates inositol phosphates and releases calcium ions from intracellular stores. J. Exp. Med. 161:446-456.

Janne, J., and Raina, A., 1968. Stimulation of spermidine synthesis in the regenerating rat liver: relation to increase ornithine decarboxylase activity. Acta Chem. Scand. 22:1349-1351.

Johnson, L. F., Abelson, H. T., Green, H., and Penman, S., 1974. Changes in RNA in relation to growth of the fibroblast. I. Amounts of mRNA, rRNA, and tRNA in resting and growing cells. Cell 1:95-100.

Johnson, L. F., Williams, J. G., Abelson, H. T., Green, H., and Penman, S., 1975. Changes in RNA in relation to growth of the fibroblast. III. Posttranscriptional regulation of mRNA formation in resting and growing cells. Cell 4:69-75.

Kahana, C. and Nathans. D., 1984. Isolation of cloned cDNA encoding mammalian ornithine decarboxylase. Proc. Natl. Acad. Sci. USA 81:3645-3649.

Kahana, C., and Nathans, D., 1985. Translational regulation of mammalian ornithine decarboxylase by polyamines. J. Biol. Chem. 260:15390-15393.

Kaibuchi, K., Tsuda, T., Kikuchi, A., Tanimoto, T., Yamashita, T., and Takai, Y., 1986. Possible involvement of protein kinase C and calcium ion in growth factor-induced expression of c-myc oncogene in Swiss 3T3 fibroblasts. J. Biol. Chem. 261:1187-1192.

Kameji, T. and Pegg, A. E., 1987. Inhibition of translation of mRNAs for ornithine decarboxylase and S-adenosylmethionine decarboxylase by polyamines. J. Biol. Chem. 262:2427-2430.

Katz, A. and Kahana, C., 1987. Transcriptional activation of mammalian ornithine decarboxylase during stimulated growth. Mol. Cell. Biol. 7:2641-2643.

Kensler, T. W., Verma, A. K., Boutwell, R. K., and Mueller, G. C., 1978. Effects of retinoic acid and juvenile hormone on the induction of ornithine decarboxylase activity by 12-O-tetradecanoylphorbol-13-acetate. Cancer Res. 38:2896-2899.

Klausner, R. F., Oshea, J. J., Luong, H., Ross, P., Bluestone, J., A., and Samelson, L. E., 1987. T cell receptor tyrosine phosphorylation. J. Biol. Chem. 262:12654-12659.

Kuno, M. and Gardner, P., 1987. Ion channels activated by inositol 1,4,5-trisphosphate in plasma membrane of human T-lymphocytes. Nature 326:301-304.
Ledbetter, J. A., Gentry, L. E., June, C. H., Rabinovitch, P. S., and Purchio, A. F., 1987. Stimulation of T cells through the CD3/T-cell receptor complex: role of cytoplasmic calcium, protein kinase C translocation, and phosphorylation of pp60c-src in the activation pathway. Mol. Cell. Biol. 7:650-656.
Liu, H. T., Baserga, R., and Mercer, W. E., 1985. Adenovirus type 2 activates cell cycle-dependent genes that are a subset of those activated by serum. Mol. Cell. Biol. 5:2936-2942.
Lodish, H. F., 1976. Translational control of protein synthesis. Ann. Rev. Biochem. 45:39-72.
Lodish, H. F., and Desalu, O., 1973. Regulation of synthesis of non-globin proteins in cell-free extracts of rabbit reticulocytes. J. Biol. Chem. 248:3520-3527.
Mach, M., White, M. W., Neubauer, M., Degen, J. L., and Morris, D. R., 1986. Isolation of a cDNA clone encoding S-adenosylmethionine decarboxylase. Expression of the gene in mitogen-activated lymphocytes. J. Biol. Chem. 261:11697-11703.
Marton, L. J., and Morris, D. R., 1987. Molecular and cellular functions of the polyamines. In: "Inhibition of Polyamine Metabolism", P. P. McCann, A. E. Pegg, and A. Sjoerdsma, eds., Academic Press, New York, pp. 79-105.
McCaffrey, P., Ran, W., Campisi, J., and Rosner, M. R., 1987. Two independent growth factor-generated signals regulate c-fos and c-myc mRNA levels in Swiss 3T3 cells. J. Biol. Chem. 262:1442-1445.
Morris, D. R., Allen, M. L., Rabinovitch, P. S., Kuepfer, C. A., and White, M. W., 1988. Mitogenic signaling pathways regulating expression of c-myc and ornithine decarboxylase genes in bovine T-lymphocytes. Biochemistry, in press.
Nishizuka, Y., 1984. The role of protein kinase C in cell surface signal transduction and tumour promotion. Nature 308:693-697.
Olson, E. N., and Spizz, G., 1986. Mitogens and protein synthesis inhibitors induce ornithine decarboxylase gene transcription through separate mechanisms in the BC3H1 muscle cell line. Mol. Cell. Biol. 6:2792-2799.
Pegg, A. E., 1986. Recent advances in the biochemistry of polyamines in eukaryotes. Biochem. J. 234:249-262.
Pegg, A. E. and McCann, P., 1982. Polyamine metabolism and function. Am. J. Physiol. 243 (Cell Physiol. 12):C212-C221.
Persson, L., Holm, I., and Heby, O., 1986. Translational control of ornithine decarboxylase by polyamines. FEBS Lett. 205:175-178.
Raina, A., Janne, J., and Siimes, M., 1966. Stimulation of polyamine synthesis in relation to nucleic acids in regenerating rat liver. Biochim. Biophys. Acta 123:197-201.
Riddle, V. G. H., and Pardee, A. B., 1980. Quiescent cells but not cycling cells exhibit enhanced actin synthesis before they synthesize DNA. J. Cell. Physiol. 103:11-15.

Rodriguez-Pena, A., and Rozengurt, E., 1985. Serum, like phorbol esters, rapidly activates protein kinase C in intact quiescent fibroblasts. EMBO J. 4:71-76.
Rose-John, S., Rincke, G., and Marks, F., 1987. The induction of ornithine decarboxylase by the tumor promoter TPA is controlled at the post-transcriptional level in murine Swiss 3T3 fibroblasts. Biochem. Biophys. Res. Commun. 147:219-225.
Russell, D. H., and Snyder, S. H., 1968. Amine synthesis in rapidly growing tissues: ornithine decarboxylase activity in regenerating rat liver, chick embryo, and various tumors. Proc. Natl. Acad. Sci. USA 60:1420-1427.
Schatzman, R. C., Wise, B. C., and Kuo, J. F., 1981. Phospholipid-sensitive calcium-dependent protein kinase: inhibition by anti-psychotic drugs. Biochem. Biophys. Res. Commun. 98:669-676.
Schumperli, D., 1986. Cell-cycle regulation of histone gene expression. Cell 45:471-472.
Seyfried, C. E., Oleinik, O. E., Degen, J. L., Resing, K., and Morris, D. R., 1982. Purification, properties and regulation of the level of bovine S-adenosylmethionine decarboxylase during lymphocyte mitogenesis. Biochim. Biophys. Acta 716:169-177.
Shirahata, A., and Pegg, A. E., 1986. Increased content of mRNA for a precursor of S-adenosylmethionine decarboxylase in rat prostate after treatment with 2-fluoromethylornithine. J. Biol. Chem. 261:13833-13837.
Stiles, C. D., 1985. The biological role of oncogenes-insights from platelet-derived growth factor: Rhoads Memorial Award lecture. Cancer Res. 45:5215-5218.
Stimac, E. and Morris, D. R., 1987. Messenger RNAs coding for enzymes of polyamine biosynthesis are induced during the G0-G1 transition but not during traverse of the normal G1 phase. J. Cell. Physiol. 133:590-594.
Sussman, I., Prettyman, R., and O'Brien, T. G., 1985. Phorbol esters and gene expression: the role of rapid changes in K^+ transport in the induction of ornithine decarboxylase by 12-O-tetradecanoylphorbol-13-acetate in BALB/c 3T3 cells and a mutant cell line defective in $Na^+K^+Cl^-$ cotransport. J. Cell Biol. 101:2316-2323.
Tabor, C. W., and Tabor, H., 1984. Polyamines. Ann. Rev. Biochem. 53:749-790.
Thompson, C. B., Challoner, P. B., Neiman, P. E., and Groudine, M., 1985. Levels of c-myc oncogene mRNA are invariant throughout the cell cycle. Nature 314:363-365.
Tsien, R. Y., Pozzan, T., and Rink, T. J., 1982. T-cell mitogens cause early changes in cytoplasmic free Ca^{2+} and membrane potential in lymphocytes. Nature 295:68-71.
Walden, W. E., Godefroy-Colburn, T., and Thach, R. E., 1981. The role of mRNA competition in regulating translation. I. Demonstration of competition in vivo. J. Biol. Chem. 256:11739-11746.
Wettenhall, R. E. H., and London, D. R., 1974. Evidence for translational control of "early" protein synthesis in lymphocytes stimulated with concanavaline A. Biochim. Biophys. Acta 349:214-225.

White, M. W., and Morris, D. R., 1988. S-adenosylmethionine decarboxylase: genes and expression. In: "The Physiology of the Polyamines", U. Bachrach and Y. M. Heimer, eds., CRC Press, Boca Rotan, in press.
White, M. W., Kameji, T., Pegg, A. E., and Morris, D. R., 1987. Increased efficiency of translation of ornithine decarboxylase mRNA in mitogen-activated lymphocytes. Eur. J. Biochem. 170:87-92.
White, M. W., Oberhauser, A. K., Kuepfer, C., and Morris, D. R., 1987. Different early-signaling pathways coupled to transcriptional and post-transcriptional regulation of gene expression during mitogenic activation of T lymphocytes. Mol. Cell. Biol. 7:3004-3007.
Wolf, M., LeVine III, H., Stratford May, J. W., Cuatrecasas, P., and Sahyoun, N., 1985. A model for intracellular translocation of protein kinase C involving synergism between Ca^{2+} and phorbol esters. Nature 317:546-549.

STRUCTURE, AMPLIFICATION AND METHYLATION OF ORNITHINE DECARBOXYLASE GENES IN HUMAN MALIGNANT CELLS

Juhani Jänne*, Leena Alhonen*, Ari Hirvonen*, Jarmo Wahlfors*, Riitta Sinervirta*, Terho Eloranta*, Erkki Hölttä** and Arja Kallio***

*Department of Biochemistry, University of Kuopio, P.O.Box 6, SF-70211 Kuopio; Finland,**Department of Pathology, University of Helsinki and ***Orion Genetic Engineering Laboratory, Helsinki, Finland

INTRODUCTION

Ornithine decarboxylase (ODC; EC 4.1.1.17) belongs to those enzymes which in response to metabolic stress are overproduced by exposed tumor cells. Mouse tumor cells easily acquire resistance to α-difluoromethylornithine (DFMO), a mechanism based irreversible inhibitor of ODC by overproducing the enzyme through amplification of transcriptionally active genes (1, 2, 3) or as a results of more efficient transcription or translation at normal gene copy number (4, 5).

DEVELOPMENT OF RESISTANCE TO DFMO IN HUMAN MYELOMA CELLS

We recently selected a human myeloma cell line (Sultan) resistant to 1 mM DFMO (6). This variant cell line, created under the pressure of DFMO, overproduced ODC through gene amplification. As a result of further exposure to increasing concentration of DFMO, these cells (designated as Sultan/20D) now are fully resistant to 20 mM DFMO. The resistance is based on a strikingly enhanced ODC production sufficient to maintain normal polyamine pools in the presence of high concentrations of DFMO.

At about the same time, we started to adapt another human myeloma cell line (Fravel) to grow in the presence of gradually increasing concentrations of DFMO. We selected a variant cell line (Fravel/4D) able to grow in the presence of 4 mM DFMO, a concentration of the drug that swiftly halted the growth of the parental cells. To our surprise, the DFMO-resistant cells did not display abnormally high ODC activity, nor were there any signs of gene amplification or enhanced accumulation of ODC-specific mRNA sequences (7). The variant cells, however, exhibited arginase activity that was roughly 10 times higher than that in the parental cells (7). These cells contained polyamine pools which were reduced by 50 % in comparison with the parental tumor cells. The variant cells were further adapted to grow in the presence of 15 mM DFMO but still not showing any signs of ODC overproduction.

Our recent analyses of the DFMO-resistant mouse and human cell lines indicated that also the Sultan variant cells resistant to 20 mM DFMO showed about 15-fold increase in

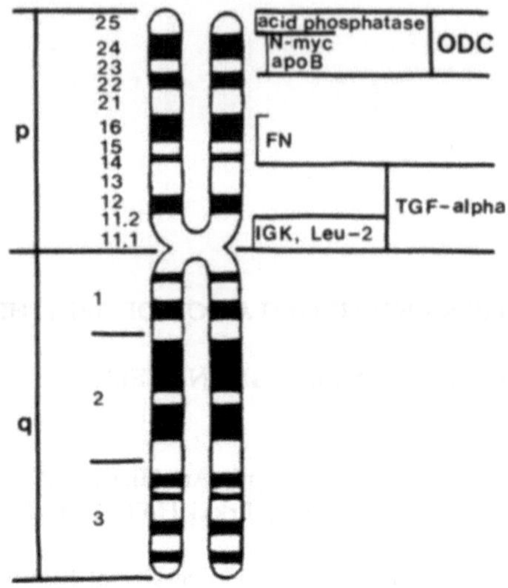

Fig. 1. Some genes residing in the short arm of human chromosome 2.

arginase activity in comparison with the parental cells. Interestingly, out of 7 mouse parental and DFMO-resistant tumor cell line only one (a DFMO-resistant myeloma) contained detectable arginase activity. Thus overproduction of arginase may be a human (or myeloma ?) -specific response to polyamine limitation.

CHROMOSOMAL LOCALIZATION OF AMPLIFIABLE ODC SEQUENCES IN HUMAN MYELOMA CELLS

By using mouse x human somatic cell hybrids, we recently localized the amplifiable ODC genes to human chromosomes 2 and 7 (8). As shown in Fig. 1, ODC genes in chromosome two resides in the short arm of the chromosome together with some other interesting genes such as the sequences for the proto-oncogene N-myc and for the transforming growth factor α (TGF-α).

Even though ODC genes are located in the proximity of the sequences for N-myc, there appears to be no co-amplification of these growth-related sequences as indicated by the unchanged gene copy number for ODC in human neuroblastoma cells with strikingly amplified N-myc sequences (8) and by the lack of amplification of N-myc genes in the DFMO-resistant Sultan cells with amplified ODC sequences (9).

STRUCTURE OF HUMAN ODC GENE

Using partially digested genomic DNA from Sultan/20D cells, we established a genomic library in a λ replacement vector (EMBL3). Out of several positive clones, one (λgODC/H2) was partially characterized. Fig. 2 depicts a partial restriction map of λgODC/H2. This map was constructed by using differential cDNA probes (shown at the bottom of Fig. 2). Some attempts were likewise made to locate the exon/intron structure of the gene. Understandably, this is a rough procedure and only a few exons could be tentatively localized (filled boxes in Fig. 2).

A fragment extending to the Sal I site at 3' end and to the Tha I site in the 5' direction (shown in Fig. 2 as Tha I fragment) was isolated and used for sequencing. At the moment of writing, 11 exons and 10 introns have been sequenced (Wahlfors, Alhonen, Jänne, J., Hickok and Jänne, O.A., unpublished results).

A second fragment extending to the Sal I site at the 3' end and to the Bam HI site in the 5' direction (shown in Fig. 2 as Bam HI fragment) was also isolated by agarose gel electrophoresis and used for subsequent transfection studies.

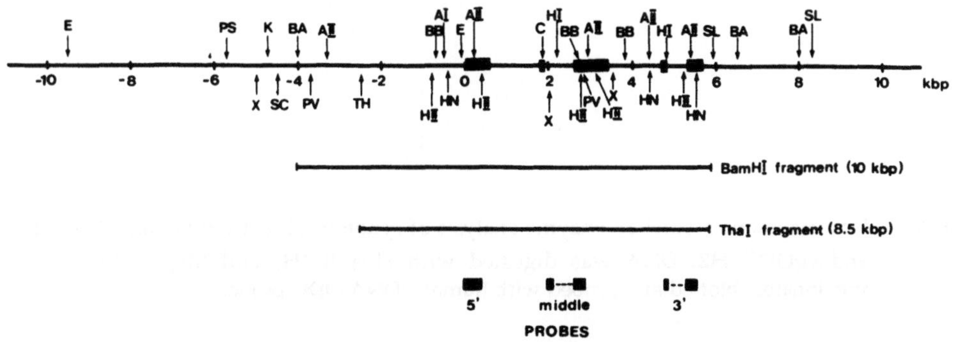

Fig. 2. Partial restriction map of λgODC/H2. The following cleavage sites are shown: E, EcoRI; PS, PstI; X, Xba I; K, Kpn I; SC, Sac I; BA, Bam HI; PV, Pvu I; AI, Ava I; A II, Ava II; HI, Hpa I; H II, Hpa II; TH, Tha I; BB, Bbi II; HN, Hind III; C, Cla 3; SL, Sal I

TRANSFER OF HUMAN ODC GENE INTO HAMSTER CELLS

The Bam HI fragmentof λgODC/H2 (Fig. 2) was used to transfect Chinese hamster ovary (CHO) cells unable to express active ODC as a result of mutagenesis. Digestion of genomic DNA from Sultan/20D cells with the methylation-sensitive Hpa II (H) yielded signals residing in large restriction fragments whereas treatment with Msp I (not sensitive to cytosine methylation) resulted in the appearance of a double signal (2.3 and 2.4 kbp) (9) and a further signal in a fragment of 0.9 kbp (barely visible in the genomic sample (Fig. 3). It is evident from Fig. 3. that ODC gene in the λgODC/H2 clone was totally demethylated as indicated by the identical digestion patterns after Hpa II and Msp I digestions.

The Bam HI fragment was transferred into CHO/ODC$^-$ cells by the standard calcium phosphate precipitation method. The ODC$^+$ cells were selected by virtue of their ability to grow in the absence of exogenous putrescine and cloned. About 40 different ODC$^+$ clones were obtained out of some were characterized in more detail. The presence of human ODC gene in the CHO cells was verified by restriction enzyme analyses by using Hind III (8) and Msp I (9) both giving human-specific signals.

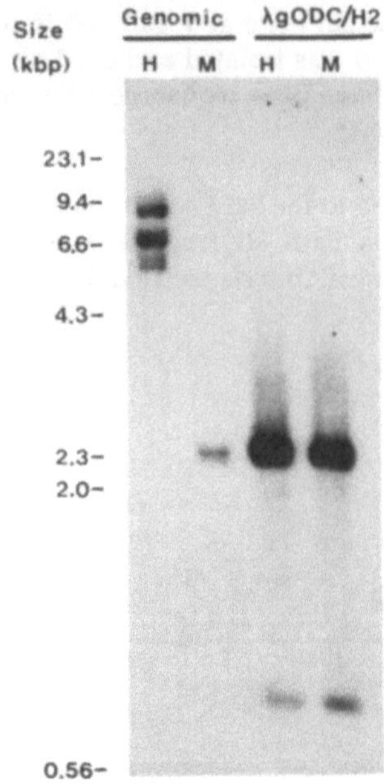

Fig. 3. Isoschizomeric restriction enzyme analysis of genomic DNA from Sultan/20D cells and λgODC/H2. DNA was digested with Hpa II (H) and Msp I (M), size-fractionated, blotted and probed with human cDNA ODC probe.

As shown in Fig. 4, Sultan myeloma cells gave a 5.0-kbp signal when digested with Hind III and a 2.3-2.4-kbp double signal after Msp I digestion. These signals were not present in the hamster cells but were easily detectable in the ODC$^+$ clones.

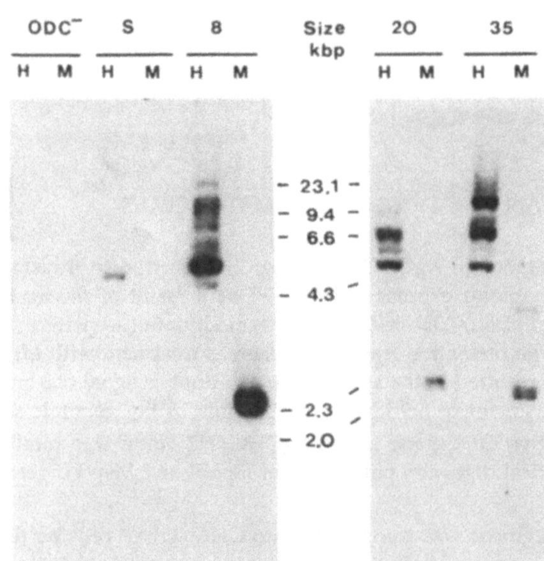

Fig. 4. Restriction enzyme analysis of CHO cells transfected with human ODC gene. H, Hind III; M, Msp I; ODC$^-$, CHO recipient cells; S, Sultan cells; 8,20 and 35, different ODC$^+$ clones.

As indicated in Fig. 4, the apparent gene dosage varied between the ODC-positive clones. Clone 8 contained much more human specific ODC sequences than clones 20 and 35. However, determination of ODC activity in different cell clones revealed that clones 20 and 35 contained enzyme activity that was more than 50-fold higher than that displayed by clone 8. As also shown in Fig. 5, clone 20 expressed ODC-specific mRNA much more effectively than clone 8 despite of the very high gene copy number in the latter clone.

It thus appears that the site of integration rather than the gene dosage most profoundly affects the expression of the ODC sequences.

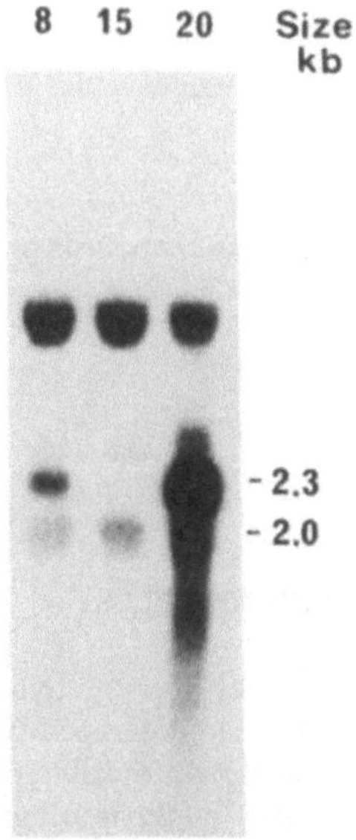

Fig. 5. Accumulation of ODC mRNA in three different ODC+ clones. Total RNA was used for this Northern blot analysis. ODC-specific signals appear close to 2.3 kb size marker. The much larger RNA species probably represent ribosomal RNA.

A restriction enzyme analysis with the methylation-sensitive Hpa II and its methylation-insensitive isoschizomer Msp I revealed interesting difference between the transfected CHO cells containing human ODC gene and the Sultan cells. As shown in Fig. 6, the ODC gene in the transfectant cells (clones 8, 20 and 35) was to a large extent demethylated as Hpa II and Msp I digestion resulted in an almost identical signal pattern in contrast to the Sultan cells. This may indicate that the initially demethylated ODC gene (Fig. 3) had remained demethylated after integration into the genome of the recipient cells. This may suggest that de novo methylation system of DNA is either absent or works very ineffectively in these cells.

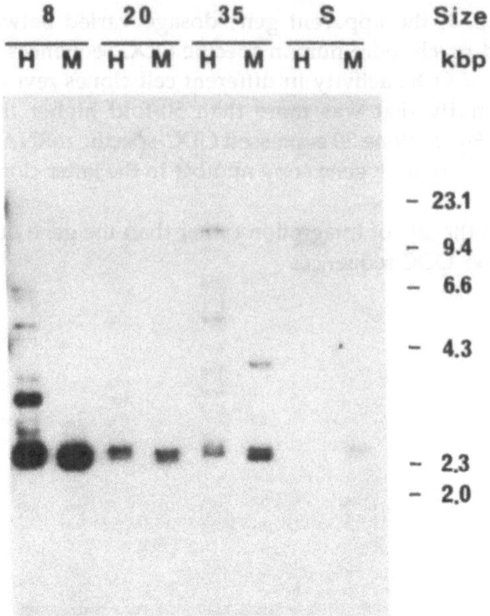

Fig. 6. Restriction enzyme (Hpa II and Msp I) analysis of some ODC$^+$ CHO cells and Sultan (S) cells. H, Hpa II; M, Msp I.

METHYLATION OF ODC GENES IN HUMAN CELLS

Many recent observations suggest that hypomethylation in or around a given gene enhances its transcriptional activity (10-11). We recently selected Sultan myeloma cell variants that were resistant to the antiproliferative effects of dexamethasone (12). Restriction enzyme (methylation sensitive) analysis of genomic DNA isolated from the glucocorticoid-resistant cells indicated that ODC sequences were rendered hypomethylated upon the development of the resistance (14). The hypomethylated status was associated with enhanced accumulation of ODC-specific mRNA (12). We have likewise observed that there exists methylation polymorphism among different human cell lines (9) and that cells derived from lymphocytic leukemia cells are less methylated than human myeloma cells, for instance. As an extension of these studies we have found that in peripheral mononuclear cells obtained from lymphocytic leukemia patients the ODC sequences are less methylated in comparison with cells obtained from healthy volunteers or from patients with some other blood cell malignancies. Interestingly, along the hypomethylation of ODC sequences, which seems to confine to chronic lymphocytic leukemia, also the sequences of Erb-A1 proto-oncogene were distinctly hypomethylated (Lipsanen, Leinonen, Alhonen and Jänne, manuscript submitted for publication).

Although the number of patients still is low, 7 consequtive patients with chronic lymphocytic leukemia all showed hypomethylation of genomic ODC sequences unlike any of the healthy volunteers (Table 1).

Table 1. Gene hypomethylation of ODC and Erb-A1 locus in peripheral mononuclear cells obtained from healthy volunteers and from patients with blood malignancies.

Patient group	Gene hypomethylation ODC	Erb-A1
Healthy volunteers	0/8	0/4
Chronic lymphocytic leukemia	7/7	4/4
Chronic myeloid leukemia	0/2	-
Acute myeloid leukemia	0/1	-
Myeloma	0/1	
Polycythemia vera	0/2	-

Although the significance of ODC gene hypomethylation in certain white cell malignancies remains totally open, this phenomenon may have some diagnostic or prognostic implications in clinical practice.

CONCLUDING REMARKS

Many of the features of the molecular biology of ODC recently uncovered link this enzyme more and more tightly to growth-related proteins. Not only is the amino acid sequence of this enzyme highly concerved but also the genomic organization of ODC gene shows striking similarities in human and mouse cells. The genes for ODC easily undergo amplification in mouse and in man. However, overproduction of ODC based on "conventional" gene amplification is not the only means of tumor cells to acquire resistance to inhibitors of the enzyme. It appears that cells under selection pressure are able to enhance the transcription or translation of the enzyme with unchanged gene copy number. Even more fascinating is the mechanism whereby human myeloma cells develop resistance to DFMO. In the case of Sultan cells the resistance is based on an overproduction of ODC in the presence of an apparent induction of arginase activity, the latter of which seems to be sufficient for Fravel cells to grow in the presence of high concentrations of DFMO.

Even though the significance of gene methylation to its expression remains somewhat controversial, ODC apparently belongs to those proteins whose expression is inhibited by extensive gene methylation.

ACKNOWLEDGEMENTS

The skilful secretarial assistance of Ms. Taru Koponen is gratefully acknowledged. The work in the authors' laboratory has received financial support from N.I.H. (grant number CA 37695), from the University of Kuopio and from Academy of Finland.

REFERENCES

1. McConlogue, L., Gupta, M., Wu, L., and Coffino, P., 1984. Molecular cloning and expression of the mouse ornithine decarboxylase gene. Proc. Natl. Acad. USA 81: 540.
2. Kahana, C., and Nathans, D., 1984. Isolation of cloned cDNA encoding mammalian ornithine decarboxylase. Proc. Natl. Acad. Sci. USA 81: 3645.
3. Alhonen-Hongisto, L., Kallio, A., Sinervirta, R., Seppänen, P., Kontula, K.K., Jänne, O.A., and Jänne, J., 1985. Difluoromethylornithine-induced amplification of ornithine decarboxylase genes in Ehrlich ascites carcinoma cells. Biochem. Biophys. Res. Commun. 126: 734.
4. Alhonen-Hongisto, L., Sinervirta, R., Jänne, O.A., and Jänne, J., 1985. Gene expression of ornithine decarboxylase in L1210 leukaemia cells exposed to DL-2-difluoromethylornithine in the presence of cadaverine. Biochem. J. 232: 605.
5. McConlogue, L., Dana, S.L. and Coffino, P., 1986. Multiple mechanisms are responsible for altered expression of ornithine decarboxylase in overproducing variant cells. Mol. Cell. Biol. 6: 2865.
6. Leinonen, P., Alhonen-Hongisto, L., Laine, R., Jänne, O.A., and Jänne, J., 1987. Human myeloma cells acquire resistance to difluoromethylornithine by amplification of ornithine decarboxylase gene. Biochem. J. 242: 199.
7. Alhonen-Hongisto, L., Leinonen, P., Laine, R., and Jänne, J., 1987. Human myeloma cells acquire resistance to difluoromethylornithine without overproducing ornithine decarboxylase. Biochem. Biophys. Res. Commun. 144: 132.
8. Winqvist, R., Mäkelä, T.P., Seppänen, P., Jänne, O.A., Alhonen-Hongisto, L., Jänne, J., Grzeschik, K.-H., and Alitalo, K., 1986. Human ornithine decarboxylase sequences map to chromosome regions 2pter p23 and 7cen qter but are not amplified with the NMYC oncogene. Cytogenet. Cell Genet. 42: 133.
9. Alhonen-Hongisto, L., Leinonen, P., Sinervirta, R., Laine, R., Winqvist, R., Alitalo, K., Jänne, O.A., and Jänne, J., 1987. Mouse and human ornithine decarboxylase genes. Biochem. J. 242: 205.
10. Fukunaga, R., Matsuyama, M., Okamura, H., Nagata, K., Nagata, S., and Sokawa, Y., 1986. Undermethylation of interferon-gamma gene in human T cell lines and normal T lymphocytes. Nucleic Acids Res. 14: 4421.
11. Doerfler, W., 1983. DNA methylation and gene activity. Ann. Rev. Biochem. 52: 93.
12 Leinonen, P., Alhonen-Hongisto, L., Laine, R., Jänne, O.A., and Jänne, J., 1987. Chronic exposure to dexamethasone induces hypomethylation of ornithine decarboxylase genes in a human myeloma cell line. FEBS Lett. 215: 68.

REGULATION OF POLYAMINE SYNTHESIS IN MAMMALIAN CELLS

Lo Persson[1], Ingvar Holm[2], Louise Stjernborg[2] and Olle Heby[2]

Departments of [1]Physiology and [2]Zoophysiology
University of Lund, S-223 62 Lund, Sweden

INTRODUCTION

Since the polyamines putrescine, spermidine and spermine are essential cellular components for growth and differentiation it is of no surprise that their synthesis is highly regulated. Disregarding arginase, which catalyzes the formation of ornithine, the rate-controlling steps in the biosynthesis of the polyamines are catalyzed by ornithine decarboxylase (ODC) and S-adenosylmethionine decarboxylase (AdoMetDC). Spermidine synthase and spermine synthase are present in relatively large amounts in cells and appear to be regulated by the supply of their substrates. Compared to the synthases the activities of ODC and AdoMetDC are very low, even when fully induced. Both ODC and AdoMetDC have remarkably fast turnover rates. In fact, they are among the shortest known for mammalian enzymes (1,2).

In addition to the possibility of rapidly changing their rate of polyamine synthesis, cells are equipped with an effective pathway for degradation of spermidine and spermine. The first and rate-limiting step in this degradation is an acetylation of the polyamines, which is catalyzed by the inducible enzyme, spermidine/spermine N^1-acetyltransferase (3-5). Also this enzyme has an extremely short half-life (6). Spermidine/spermine N^1-acetyltransferase is rapidly induced by various polyamines (7) and appears to play a role in the cellular protection against the deleterious effects of too high intracellular polyamine concentrations.

Until a few years ago not much was known about the cellular mechanisms by which ODC and AdoMetDC are regulated. Transcriptional, translational as well as post-translational mechanisms have been suggested in a number of reports. However, conclusive data have not been obtained until recently. The preparation of specific antibodies (8-12) to the enzymes followed by the cloning of cDNAs encoding the enzymes (13-19) made it possible to closely study the regulation. Figure 1 represents the current view of the regulation of polyamine synthesis in mammalian cells.

ODC REGULATION

Changes in ODC activity in most cases appear to be mediated by changes in the amount of enzyme protein (12,20). Data on post-translational modifications are sparse and sometimes puzzling. The best known and most interesting example is the presence of an inhibitory protein, named antizyme (21) (see below), which has a remarkably high affinity to ODC (22). Recently, O'Brien et al. (23) reported on the activation of mouse epidermal tumor ODC by GTP. A 50-fold decrease in the K_m for ornithine was observed in the presence of GTP. Whether this activation plays any regulatory role in vivo remains to be established. The sequencing of ODC cDNA revealed a possible site for phosphorylation by casein kinase 2 (15). Experimental data show that ODC from rat heart can be phosphorylated in vitro at this site by casein kinase 2 (24,25). However, the phosphorylation was without effect on the enzyme activity. Whether ODC is phosphorylated in vivo is hitherto unknown. Charge heterogeneity, which could be due to the addition of a phosphate group, has been reported for ODC from mouse kidney (26).

Regulation of ODC Synthesis

Induction of ODC activity by various stimuli has been shown to be, at least partly, correlated with a rise in the cellular level of ODC mRNA (13-16,20,27,28). Whether this is a result of increased transcription of the gene or a stabilization of the message is not fully clarified. Using an in vitro transcription technique Katz and Kahana (27) demonstrated that the induction of ODC in BALB/c 3T3 mouse fibroblasts by serum stimulation was due to an increased transcription. However, using a similar technique, Rose-John et al. (28) showed that the induction of ODC in Swiss 3T3 mouse fibroblasts by a tumor promoter was mainly a post-transcriptional phenomenon (prolongation of the ODC mRNA half-life).

ODC is under strict control by the polyamines. The addition of exogenous polyamines rapidly reduces ODC activity. Part of this decrease can occasionally be explained by an increase in the turnover of the enzyme (see below). However, the major effect of the polyamines appears to be on the synthesis of the enzyme (29). If, on the other hand, polyamine synthesis is restrained by treatment with reversible inhibitors of ODC, e.g. α-hydrazino-ornithine or α-methylornithine, an increase in ODC activity is observed when measured in diluted or dialyzed extracts (30,31). This increase is partly due to a stabilization of the enzyme. The binding of an inhibitor to the active site is believed to render the enzyme less susceptible to proteolytic degradation (32). However, treatment of cells with the irreversible ODC inhibitor α-difluoromethylornithine (DFMO) gives rise to an increase in the cellular content of ODC protein, without any major effect on the stability of the enzyme (20). Thus, the increase in cellular ODC content after treatment with different ODC inhibitors is, at least partly, a result of elevated synthesis. That this is the case has been demonstrated for DFMO which gives rise to a marked increase in ODC synthesis in Ehrlich ascites tumor cells (29).

The rise in ODC synthesis during treatment with DFMO is most probably a compensatory reaction to the decrease in polyamine content. This assumption is confirmed by the finding that provision of polyamines counteracts the effect of DFMO on ODC synthesis (33). Thus, it is apparent that the "normal" cellular content of polyamines exerts an inhibitory effect on ODC synthesis. However, this repression is not maximal since addition of exogenous polyamines inhibits ODC synthesis even further.

Northern analyses of ODC mRNA have revealed that the changes in ODC synthesis, caused by either treatment with inhibitors of polyamine synthesis or provision of exogenous polyamines, are not associated with any changes in the cellular ODC mRNA content (20,29,34-36). These findings imply that the polyamine-mediated control of ODC synthesis is carried out at the translational rather than at the transcriptional level.

ODC mRNA has been reported to be poorly translated in rabbit

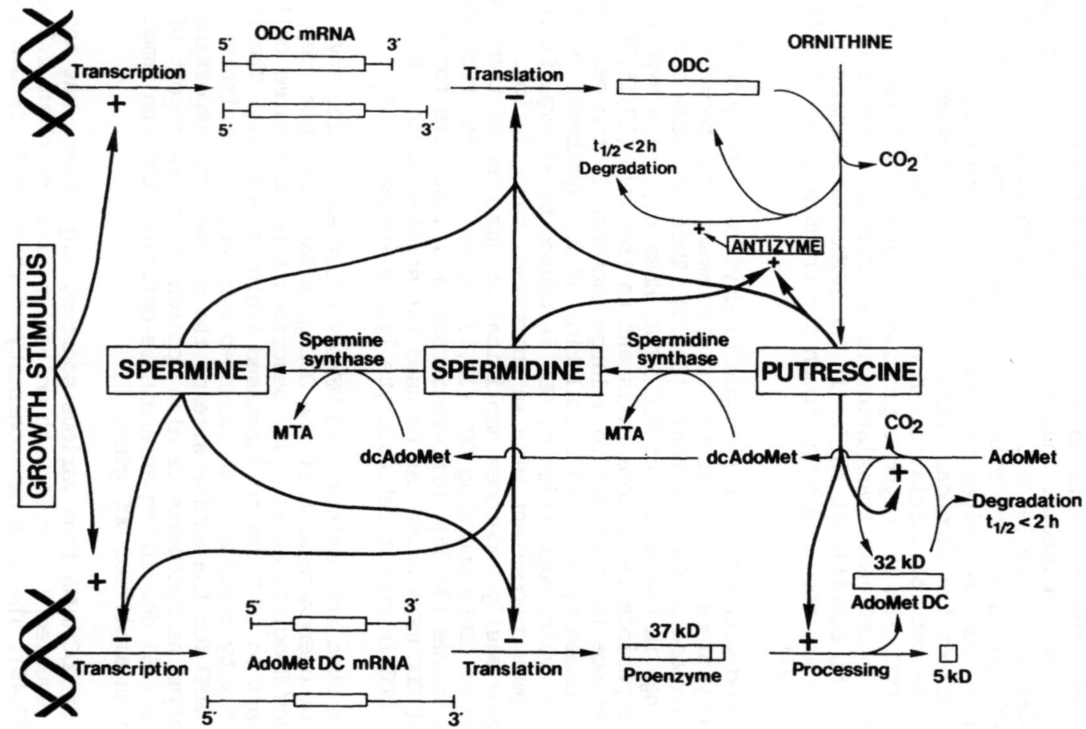

Fig. 1. Regulation of polyamine synthesis in mammalian cells. AdoMet, S-adenosylmethionine; dcAdoMet, decarboxylated S-adenosylmethionine; MTA, 5-methylthioadenosine.

reticulocyte lysate systems (15,37). This finding may be a result of the high levels of spermidine in those systems. By removing polyamines from the lysates, using gel filtration, it was demonstrated by Kameji and Pegg (38) and by Persson et al. (36) that translation of ODC mRNA is more sensitive to the inhibitory effect of polyamines than is translation of total protein. Whether this finding reflects the situation in the cell remains to be established.

The exact mechanism behind the translational control of ODC by polyamines is still unknown. Protein synthesis is usually limited by the rate of initiation, and translational control mechanisms are likely to operate at this level. Molecular analyses of ODC cDNAs have revealed a long non-coding 5'-leader (18,19,39,40). It has been suggested that this leader, which is GC-rich, plays a major role in the translational control of ODC. It is conceivable that the polyamines, which have a strong affinity for nucleic acids, affect the secondary structure of the 5'-leader in a way that interfers with the rate of initiation. However, results from studies of the distribution of ODC mRNA in polysome profiles indicate that the polyamine-mediated changes in ODC mRNA translation is not merely a result of altered initiation (33). No significant shifts in ODC mRNA distribution in the polysome profiles are observed even though the translational efficiency is changed 10-fold (33).

Cells treated with DFMO for a couple of days cease to grow due to the depletion of polyamines. The cells try to circumvent this depletion by a compensatory increase in ODC synthesis (20,29). This increase occurs without any change in the amount of ODC mRNA. However, cells exposed to DFMO for longer time periods can become resistant to the drug (13,15,36,37,41, 42). This resistance is often due to a large increase in the cellular content of ODC caused by a marked stimulation of its synthesis (which can be 100-fold). The elevated expression of ODC is caused by an amplification of the structural gene concomitant with an increased transcription. The fact that the ODC gene easily undergoes amplification has been of tremendous use for the cloning of cDNAs encoding ODC (13,15) as well as for regulatory studies of the enzyme (34,36). DFMO-resistant cell lines are for example rich sources of ODC mRNA which can be used for studies of the in vitro translation of ODC in rabbit reticulocyte lysate systems (36).

Cell lines that are resistant to DFMO without exhibiting any amplification or overexpression of the ODC gene have also been reported (43-45). Alhonen-Hongisto et al. (45) described a human myeloma cell line that is resistant to DFMO due to increased arginase activity. The increase in arginase activity results in an enhanced synthesis of ornithine, which competes with DFMO for the active site of the enzyme. The importance of arginase in polyamine synthesis is also apparent from the report of Hölttä and Pohjanpelto (35) describing an arginase-deficient Chinese hamster ovary cell line with an amplified ODC gene.

Analyses of ODC mRNA from various sources, e.g. mouse, rat and hamster, have revealed 2 species with molecular sizes of about 2.2 and 2.7 kb (14,16,19,35,46). The size heterogeneity of these mRNAs is due to the presence of two polyadenylation sites (47). Interestingly, human cells have been reported to express only a single ODC mRNA species with a molecular size of 2.3 kb (18,42).

Southern analyses of genomic DNA reveal that ODC belongs to a multi-gene family (13,15,42,46). Whether all these genes are structural genes or pseudogenes is not known. However, ODC gene amplification in various ODC-overproducing mouse cell lines appears to involve only one of the genes (giving an EcoRI fragment of 6-7 kb) (13,15,36,48,49). This gene must contain two polyadenylation signals to account for the two different ODC mRNAs (which both are expressed at elevated levels in the ODC-overproducing cells).

Interestingly, one of the genes for ODC has been shown to be co-amplified with the gene for the M2 subunit of ribonucleotide reductase in hamster, mouse and human cells that are resistant to hydroxyurea, indicating a close location of these genes (46,50). The amplified segments of both ODC and the M2 subunit of ribonucleotide reductase are located on hamster chromosome 7, mouse chromosome 12 and human chromosome 2 (46,50). That the amplified ODC gene is functional is indicated by the finding of over-expression of ODC in the hydroxyurea-resistant hamster cell line (46) and the location of the amplified ODC gene to chromosome 2 in an ODC-overproducing human cell line (51,52).

Regulation of ODC Degradation

ODC is reported to have the fastest turnover among mammalian enzymes (1). Its half-life can often be as short as 10 min. However, the mechanism behind this rapid degradation is not well understood. The degradation appears to be non-lysosomal and ubiquitin-independent (53). Neither does ODC follow the "amino-terminal rule" for a rapidly degraded protein (54).

The induction of ODC activity by various stimuli is often accompanied by a stabilization of the enzyme. The degradation of ODC is stimulated by polyamines through a mechanism which appears to be dependent on ongoing protein synthesis (55). Polyamines have been shown to induce the synthesis of a protein, named antizyme, which has an inhibitory effect on ODC (21). Antizyme has been purified to homogeneity by Kitani and Fujisawa (22). The molecular weight determined by gel filtration was 22 kD. The binding (and inhibitory effect) of antizyme to ODC is extremely strong with an equilibrium constant as high as 1.4×10^{11} M^{-1}. It has been suggested and partly experimentally confirmed by Hayashi and coworkers (56) that the antizyme in some way acts as a targeting system for ODC degradation. However, more work is needed to clarify this mechanism.

Like other rapidly degraded proteins, mammalian ODC contains regions that are rich in the amino acids proline, glutamic acid, aspartic acid, serine and threonine ("PEST-regions") (57). Interestingly, it was recently demonstrated that ODC from *Trypanosoma brucei brucei* does not contain these regions and, in fact, is extremely stable (58). Besides confirming the PEST-theory, the latter finding provides an answer to the puzzling question why DFMO is so effective against this parasite. Due to the rapid turnover in mammalian cells DFMO-inactivated ODC is continuously replaced with new, active, enzyme molecules. Hence, large amounts of DFMO are needed to effectively maintain the inhibition of ODC activity. In the trypanosomes, on the other hand, ODC turns over extremely slowly and remains irreversibly inactivated by DFMO for much longer time periods.

Two cell lines with extremely stable ODC (half-lives longer than 5 h) have been isolated (43,59,60). These cell lines were obtained due to their resistance to α-methylornithine and DFMO, respectively. The molecular mechanism(s) behind this stability is hitherto unknown. However, isolation and sequencing of ODC cDNAs from these cells might give valuable information about the underlying cause of this stability.

AdoMetDC REGULATION

Changes in AdoMetDC activity are mainly brought about by changes in the amount of enzyme protein (12). In addition, putrescine exerts a direct stimulatory effect on AdoMetDC by reducing the K_m for the substrate (62).

This mechanism may be of significance for the regulation of AdoMetDC activity in the cell.

Regulation of AdoMetDC Synthesis

In contrast to ODC, AdoMetDC is not dependent on pyridoxal phosphate as a cofactor. Instead, AdoMetDC from E. coli, yeast and mammals relies on a covalently bound pyruvate as a prosthetic group (62-64). Only a few enzymes are known to contain bound pyruvate and AdoMetDC is the only mammalian one. The formation of the pyruvate is best known for histidine decarboxylase from Lactobacillus 30a (65). This enzyme is synthetized as a proenzyme which then is cleaved at a serine-serine linkage yielding a pyruvate-containing α chain and a pyruvate-free β chain of the active enzyme. Recently, Tabor and Tabor (66) demonstrated that the pyruvate moiety in AdoMetDC from E. coli was produced by the cleavage of a lysine-serine peptide bond. Mammalian AdoMetDC is also produced as a proenzyme (37 kD) which then is cleaved into two polypeptides (32 and 5 kD) (67). However, the site for this cleavage is not yet known. Neither is it known whether both these products are subunits of the active enzyme Purified preparations of AdoMetDC have been shown to contain only a 32 kD subunit. However, a subunit with a molecular weight of 5 kD can easily be overlooked.

The processing of the mammalian AdoMetDC proenzyme into the 32 kD subunit has been studied by Pegg and coworkers (67-69). The conversion of the precursor into the enzyme subunit was shown to be stimulated by the presence of putrescine in a reticulocyte lysate system (68). That putrescine stimulates the conversion of the precursor was also suggested by the finding that the proenzyme accumulates in prostates of rats treated with DFMO (69). However, pulse-labeling studies of AdoMetDC synthesis in Ehrlich ascites tumor cells did only demonstrate incorporation of radioactivity into a protein corresponding to the 32 kD subunit (70). This was true even in cells depleted of their putrescine content by DFMO treatment, indicating an extremely rapid turnover of the proenzyme.

Treatment with various inhibitors of ODC is often accompanied by a marked increase in the activity of AdoMetDC (12,71,72). Part of this increase can be explained by a stabilization of the enzyme. However, a large part appears to be caused by elevated synthesis of the enzyme. Pegg and coworkers (67,73) demonstrated that the amount of translatable AdoMetDC mRNA, as measured in a reticulocyte lysate system, is increased after treatment with various inhibitors of the enzymes involved in polyamine synthesis.

Treatment of Ehrlich ascites tumor cells with DFMO induces a marked increase in SAMDC synthesis, as measured by pulse-labeling and immuno-precipitation (70). When the cells are provided with spermidine, in addition to DFMO, AdoMetDC synthesis decreases to undetectable levels. Northern analyses reveal that the changes in AdoMetDC mRNA content are too small to account for the differences observed in synthesis rate; AdoMetDC synthesis increased more than 5-fold after treatment with DFMO, whereas the level of AdoMetDC mRNA increased less than 2-fold (70). Hence, AdoMetDC synthesis is regulated at two different levels by the polyamines: a) the amount of AdoMetDC mRNA and b) the efficiency of AdoMetDC mRNA translation. The latter mechanism is also supported by results from in vitro studies of AdoMetDC mRNA translation (38).

Northern blot analyses show that growth stimulation of various cells is accompanied by an increase in the content of AdoMetDC mRNA (17,74,75). Like ODC mRNA, AdoMetDC mRNA is expressed as 2 species with different molecular sizes (1.7-2.4 kb and 3.0-3.6 kb). Whether these mRNAs also are the products of a single gene and represent multiple polyadenylation signals is not known. Sequencing of AdoMetDC cDNA clones may provide useful information on this subject and for determining the origin of the pyruvate prosthetic group.

Regulation of AdoMetDC Degradation

The turnover of AdoMetDC is also very rapid, with half-lives of less than 1-2 h. Treatment with the reversible inhibitor of AdoMetDC, methylglyoxal bis(guanylhydrazone) (MGBG), gives rise to a dramatic increase in the activity of AdoMetDC (76). All of this increase, however, can be explained by a stabilization of the enzyme (67).

The degradation of AdoMetDC appears to be affected by the cellular content of polyamines. A stabilization of the enzyme is frequently seen in cells depleted of their putrescine and spermidine content by treatment with DFMO (71,72). This change in half-life is counteracted by the provision of exogenous spermidine. Pegg et al. (73) demonstrated that treatment of SV-3T3 cells with the spermine synthase inhibitor, S-methyl-5'-methylthioadenosine, besides reducing the spermine content, induces a rise in AdoMetDC activity which is due to decreased turnover as well as increased AdoMetDC mRNA content. Provision of spermine abolishes the increase in mRNA and counteracts the stabilization of the enzyme.

CONCLUSIONS

The biosynthesis of polyamines is essential for the cell and is thus highly regulated. The polyamines exert a feedback control of their synthesis at many levels. By affecting the synthesis and degradation of the two rate-controlling enzymes, ODC and AdoMetDC, the cellular level of polyamines is quickly adapted to the actual need. The effectiveness and rapidity of this control mechanism is based on the extremely fast turnover of ODC and AdoMetDC.

ACKNOWLEDGEMENTS

Studies carried out by the authors were supported by grants from the Swedish Medical and Natural Science Research Councils.

REFERENCES

1. D. H. Russell and S.H. Snyder, Amine synthesis in regenerating rat liver. Extremely rapid turnover of ornithine decarboxylase, Molec. Pharmacol. 5:253 (1969).
2. P. Hannonen, A. Raina, and J. Jänne, Polyamine synthesis in the regerating rat liver: stimulation of S-adenosylmethionine decarboxylase, and spermidine and spermine synthases after partial hepatectomy, Biochim. Biophys. Acta 273: 84 (1972).
3. I. Matsui and A. E. Pegg, Increase in acetylation of spermidine in rat liver extracts brought about by treatment with carbon tetrachloride, Biochem. Biophys. Res. Commun. 92: 1009 (1980).
4. N. Seiler, Amide-bond-forming reactions of polyamines, in: Polyamines in Biology and Medicine, D. R. Morris and L. J. Marton, eds., p.p. 127-149, Marcel Dekker, New york (1981)
5. I. Matsui, L. Wiegand and A. E. Pegg, Properties of spermidine N-acetyltransferase from livers of rats treated with carbon tetrachloride and its role in the conversion of spermidine into putrescine, J. Biol. Chem. 256: 2454 (1981).

6. L. Persson and A. E. Pegg, Studies of the induction of spermidine/spermine N^1-acetyltransferase using a specific antiserum, J. Biol. Chem. 259: 12364 (1984).
7. A. E. Pegg and B. G. Erwin, Induction of spermidine/spermine N^1-acetyltransferase in rat tissues by polyamines, Biochem. J. 231: 285 (1985).
8. L. Persson, Antibodies to ornithine decarboxylase: immunochemical cross-reactivity, Acta Chem. Scand B36:685 (1982).
9. T. Kameji, Y. Murakami, K. Fujita and S. Hayashi, Purification and some properties of ornithine decarboxylase from rat liver, Biochim. Biophys. Acta 717:111 (1982).
10. J. E. Seely and A. E. Pegg, Changes in mouse kidney ornithine decarboxylase activity are brought about by changes in the amount of enzyme protein as measured by radioimmunoassay, J. Biol. Chem. 258: 2496 (1983).
11. C. E. Seyfried, O. A. Oleinik, J. L. Degen, K. Resing and D. R. Morris, Purification, properties and regulation of the level of bovine S-adenosylmethionine decarboxylase during lymphocyte mitogenesis, Biochim. Biophys. Acta 716: 169 (1982).
12. A. Shirahata and A. E. Pegg, Regulation of S-adenosylmethionine decarboxylase activity in rat liver and prostate, J. Biol. Chem. 260: 9583 (1985).
13. L. McConlogue, M. Gupta, L. Wu and P. Coffino, Molecular cloning and expression of the mouse ornithine decarboxylase gene, Proc. Natl. Acad. Sci. USA 81: 540 (1984).
14. K. K. Kontula, T. K. Torkkeli, C. W. Bardin and O. A. Jänne, Androgen induction of ornithine decarboxylase mRNA in mouse kidney as studied by complemetary DNA, Proc. Natl. Acad. Sci. USA 81: 731 (1984).
15. C. Kahana and D. Nathans, Isolation of cloned cDNA encoding mammalian ornithine decarboxylase Proc. Natl. Acad. Sci. USA 81: 3645 (1984).
16. F. Berger, P. Szymanski, E. Read and G. Watson, Androgen-regulated ornithine decarboxylase mRNAs of mouse kidney, J. Biol. Chem. 259: 7941 (1984).
17. M. Mach, M. W. White, M. Neubauer, J. L. Degen and D. R. Morris, J. Isolation of a cDNA encoding S-adenosylmethionine decarboxylase. Expression of the gene in mitogen-activated lymphocytes, Biol. Chem. 261: 11697 (1986).
18. N. J. Hickok, P. J. Seppänen, G. L. Gunsalus and O. A. Jänne, Complete amino acid sequence of human ornithine decarboxylase deduced from complementary DNA, DNA 6: 179 (1987).
19. H. J. Kranen, L. van de Zande, C. F. van kreijl, A. Bisschop and B. Wieringa, Cloning and nucleotide sequence of rat ornithine decarboxylase cDNA, Gene 60: 145 (1987).
20. L. Persson, S. M. Oredsson, S. Anehus and O. Heby, Ornithine decarboxylase inhibitors increase the cellular content of the enzyme: implications for translational regulation, Biochem. Biophys. Res Commun. 131: 239 (1985).
21. J. S. Heller, W. F. Fong and E. S. Canellakis, Induction of a protein inhibitor to ornithine decarboxylase by the end products of its reaction, Proc. Natl. Acad. Sci. USA 73: 1858 (1976).
22. T. Kitani and H. Fujisawa, Purification and some properties of a protein inhibitor (antizyme) of ornithine decarboxylase from rat liver, J. Biol Chem. 259: 10036 (1984).
23. T. G. O'Brien, O. Hietala, K. O'Donnell and M. Holmes, Activation of mouse epidermal tumor ornithine decarboxylase by GTP: evidence for different catalytic forms of the enzyme, Proc. Natl. Acad. Sci. USA 84: 8927 (1987).
24. F. Meggio, F. Flamigni, C. M. Caldarera, C. Guarnieri and L. A. Pirma, Phosphorylation of rat heart ornithine decarboxylase by type-2 casein kinase, Biochem. Biophys. Res. Commun. 122: 997 (1984).

25. F. Meggio, F. Flamigni, C. Guarnieri and L. A. Pinna, Location of the phosphorylation site for casein kinase-2 within the amino acid sequence of ornithine decarboxylase, Biochim. Biophys. Acta 979: 114 (1987).
26. L. Persson, J. E. Seely and A. E. Pegg, Investigation of the structure and rate of synthesis of ornithine decarboxylase protein in mouse kidney, Biochemistry 23: 3777 (1984).
27. A. Katz and C. Kahana, Transcriptional activation of mammalian ornithine decarboxylase during stimulated growth, Mol. Cell. Biol. 7: 2641 (1987).
28. S. Rose-John, G. Rincke and F. Marks, The induction of ornithine decarboxylase by the tumor promoter TPA is controlled at the post-transcriptional level in murine Swiss 3T3 fibroblasts, Biochem. Biophys. Res. Commun. 147: 219 (1987).
29. L. Persson, I. Holm and O. Heby, Translational regulation of ornithine decarboxylase by polyamines, FEBS Lett. 295: 175 (1986).
30. H. Inoue, Y. Kato, M. Takigawa, K. Adachi and Y. Takeda, Effects of DL-α-hydrazino-δ-aminovaleric acid, an inhibitor of ornithine decarboxylase, on polyamine metabolism in isoproterenol-stimulated mouse parotid glands, J. Biochem. 77: 879 (1975).
31. P. P. McCann, C. Tardif, M.-C. Duchesne and P.S. Mamont, Effect of α-methylornithine on ornithine decarboxylase activity of rat hepatoma cells in culture, Biochem. Biophys. Res. Commun. 76: 893 (1977).
32. R. T. Schimke, Advances in Enzymology 37: Control of enzyme levels in mammalian tissues, 135 (1973).
33. I. Holm, L. Persson, L. Stjernborg, L. Thorsson and O. Heby, unpublished observations.
34. C. Kahana and D. Nathans, Translational regulation of mammalian ornithine decarboxylase by polyamines, J. Biol. Chem. 260: 15390 (1985).
35. E. Hölttä and P. Pohjanpelto, Control of ornithine decarboxylase in Chinese hamster ovary cells by polyamines. Translational inhibition of synthesis and acceleration of degradation of the enzyme by putrescine, spermidine and spermine, J. Biol. Chem. 261: 9502 (1986).
36. L. Persson, I. Holm and O. Heby, Regulation of ornithine decarboxylase mRNA translation by polyamines. Studies using a cell-free system and a cell line with an amplified ornithine decarboxylase gene, J. Biol. Chem. 263: 3528 (1988).
37. L. McConlogue and P. Coffino, A mouse lymphoma cell mutant whose major protein product is ornithine decarboxylase, J. Biol. Chem. 258: 12083 (1983).
38. T. Kameji and A. E. Pegg, Inhibition of translation of mRNAs for ornithine decarboxylase and S-adenosylmethionine decarboxylase by polyamines, J. Biol. Chem. 262: 2427 (1987).
39. C. Kahana and D. Nathans, Nucleotide sequence of murine ornithine decarboxylase, Proc. Natl. Acad. Sci. USA 82: 1673 (1985).
40. M. Gupta and P. Coffino, Mouse ornithine decarboxylase. Complete amino acid sequence deduced from cDNA, J. Biol. Chem. 260: 2941 (1985).
41. J. H. Choi and I. E. Scheffler, A mutant of Chinese hamster ovary cells resistant to α-methyl- and α-difluoromethylornithine, Somatic Cell. Genet. 7: 219 (1981).
42. P. Leinonen, L. Alhonen.Hongisto, R. Laine, O. A. Jänne and J. Jänne, Human myeloma cells acquire resistance to difluoromethylornithine by amplification of ornithine decarboxylase, Biochem. J. 242: 199 (1987).
43. E. Karvonen, L. C. Andersson and H. Pösö, A human neuroblastoma cell line with a stable ornithine decarboxylase in vivo and in vitro Biochem. Biophys. Res. Commun. 126: 96 (1985).

44. L. McConlogue, S. L. Dana and P. Coffino, Multiple mechanisms are responsible for altered expression of ornithine decarboxylase in overproducing variant cells, Mol. Cell. Biol. 6: 2865 (1986).
45. L. Alhonen-Hongisto, P. Leinonen, R. Laine and J. Jänne, Human myeloma cells acquire resistance to difluoromethylornithine without overproducing ornithine decarboxylase, Biochem. Biophys. Res. Commun. 144: 132 (1987).
46. P. R. Srinivasan, P. N. Tonin, E. J. Wensing and W. H. Lewis, The gene for ornithine decarboxylase is co-amplified in hydroxyurea-resistant hamster cells, J. Biol. Chem. 262: 12871 (1987).
47. N. J. Hickok, P. J. Seppänen, K. K. Kontula, P. A. Jänne, C. W. Bardin and O. A. Jänne, Two ornithine decarboxylase mRNA species in mouse kidney arise from size heterogeneity at their 3' termini, Proc. Natl. Acad. Sci. USA 83: 594 (1986).
48. L. Alhonen-Hongisto, A. Kallio, R. Sinervirta, P. Seppänen, K. K. Kontula, O. A. Jänne and J. Jänne, Difluoromethylornithine-induced amplification of ornithine decarboxylase genes in Ehrlich ascites carcinoma cells, Biochem. Biophys. Res. Commun. 126: 734 (1985).
49. L. Alhonen-Hongisto, R. Sinervirta, O. A. Jänne and J. Jänne, Gene expression of ornithine decarboxylase in L1210 leukaemia cells exposed to DL-2-difluoromethylornithine, Biochem. J. 232: 605 (1985).
50. T. L. Yang-Feng, D. E. Barton, L. Thelander, W. H. Lewis, P. R. Srinivasan and U. Francke, Ribonucleotide reductase M2 subunit sequences mapped to four different chromosomal sites in humans amd mice: functional locus identified by its amplification in hydroxyurea-resistant cell line, Genomics 1:72 (1987).
51. R. Winqvist, T. P. Mäkelä, P. Seppänen, O. A. Jänne, L. Alhonen-Hongisto, J. Jänne, K.-H. Grzeschik and K. Alitalo, Human ornithine decarboxylase sequences map to chromosome regions 2pter-p23 and 7cen-qter but are not coamplified with the NMYC oncogene, Cytogenet. Cell. Genet. 42: 133 (1986).
52. L. Alhonen-Hongisto, P. Leinonen, R. Sinervirta, R. Laine, R. Winqvist, K. Alitalo, O. A. Jänne and J. Jänne, Mouse and human ornithine decarboxylase genes. Methylation polymorphism and amplification, Biochem. J. 242: 205 (1987).
53. J. R. Glass and E. W. Gerner, Spermidine mediates degradation of ornithine decarboxylase by a non-lysosomal, ubiquitin-independent mechanism, J. Cell. Physiol. 130: 133 (1987).
54. A. Bachmair, D. Finley and A. Varshavsky, In vivo half-life of a protein is a function of its amino-terminal residue, Science 234: 179 (1986).
55. Y. Murakami, K. Fujita, T. Kameji and S. Hayashi, Accumulation of ornithine decarboxylase-antizyme complex in HMO_A cells, Biochem. J. 225: 689 (1985).
56. S. Hayashi, T. Kameji, K. Fujita, Y. Murakami, R. Kanamoto, K. Utsunomiya, S. Matsufuji, M. Takiguchi, M. Mori and M. Tatibana, Molecular mechanism for the regulation of hepatic ornithine decarboxylase, Adv. Enzyme. Regul. 23: 311 (1985).
57. S. Rogers, R. Wells and M. Rechsteiner, Amino acid sequences common to rapidly degraded proteins: the PEST hypothesis, Science 234: 364 (1986).
58. M. A. Phillips, P. Coffino and C. C. Wang, Cloning and sequencing of the ornithine decarboxylase gene from Trypanosoma brucei. Implications for enzyme turnover and selective difluoromethylornithine inhibition, J. Biol. Chem. 262: 8721 (1987).
59. P. S. Mamont, M. C. Duchesne, J. Grove and C. Tardif, Initial characterization of a HTC cell variant partially resistant to the anti-proliferative effect of ornithine decarboxylase inhibitors, Exp. Cell. Res. 115: 387 (1978).

60. P. P. McCann, C. Tardif, J.-M. Hornsperger and P. Böhlen, Two distinct mechansims for ornithine decarboxylase regulation by polyamines in rat hepatoma cells, J. Cell. Physiol. 99: 183 (1979).
61. A. E. Pegg and H. G. Williams-Ashman, Stimulation of decarboxylation of S-adenosylmethionine by putrescine in mammalian tissues, Biochem. Biophys. Res. Commun. 30: 76 (1968).
62. A. E. Pegg, Evidence for the presence of pyruvate in rat liver S-adenosylmethionine decarboxylase, FEBS Lett. 84: 33 (1977).
63. R. W. Wickner, C. W. Tabor and H. Tabor, Purification of adenosylmethionine decarboxylase from Escherichia coli: evidence for covalently bound pyruvate, J. Biol. Chem. 245: 2132 (1970).
64. M. S. Cohn, C. W. Tabor and H. Tabor, Identification of a pyrovoyl residue in S-adenosylmethionine decarboxylase from Saccharomyces cerevisiae, J. Biol. Chem. 252: 8212 (1977).
65. P. A. Recsei, Q. K. Huynh and E. E. Snell, Conversion of prohistidine decarboxylase to histidine decarboxylase: peptide chain cleavage by nonhydrolytic serinolysis, Proc. Natl. Acad. Sci. USA 80: 973 (1983).
66. C. W. Tabor and H. Tabor, The speEspeD operon of Escherichia coli. Formation and processing of a proenzyme form of S-adenosylmethionine decarboxylase, J. Biol. Chem. 262: 16037 (1987).
67. A. Shirahata and A. E. Pegg, Increased content of mRNA for a precursor of S-adenosylmethionine decarboxylase in rat prostate after treatment with 2-difluoromethylornithine, J. Biol. Chem. 261: 13833 (1986).
68. T. Kameji and A. E. Pegg, Effect of putrescine on the synthesis of S-adenosylmethionine decarboxylase, Biochem. J. 243: 285 (1987).
69. A. E. Pegg, L. Wiest and A. Pajunen, Detection of proenzyme form of S-adenosylmethionine decarboxylase in extracts from rat prostate, Biochem. Biophys. Res. Commun. 150: 788 (1988).
70. L. Persson, L. Stjernborg, I. Holm and O. Heby, unpublished observations.
71. L. Alhonen-Hongisto, Regulation of S-adenosylmethionine decarboxylase by polyamines in Ehrlich ascites-carcinoma cells grown in culture, Biochem. J. 190: 747 (1980).
72. P. S. Mamont, A.-M. Joder-Olenbusch, M. Nussli and J. Grove, Indirect evidence for a strict negative control of S-adenosyl-L-methionine decarboxylase by spermidine in rat hepatoma cells, Biochem. J. 196: 411 (1981).
73. A. E. Pegg, R. Wechter and A. Pajunen, Increase in S-adenosylmethionine decarboxylase in SV-3T3 cells treated with S-methyl-5'methylthioadenosine, Biochem. J. 244: 49 (1987).
74. E. Stimac and D. R. Morris, Messenger RNAs coding for enzymes of polyamine biosynthesis are induced during the G_0-G_1 transition but not during traverse of the normal G_1 phase, J. Cell. Physiol. 133: 590 (1987).
75. J. J. Schultz and A. Shain, Effect of aging on AXC/SSh rat ventral and dorsolateral prostate S-adenosyl-L-methionine decarboxylase and L-ornithine decarboxylase messenger ribonucleic acid content, Endocrinology 122: 120 (1988).
76. A. E. Pegg, A. Corti and H. G. Williams-Ashman, Paradoxical enhancement of S-adenosylmethionine decarboxylase in rat tissues following administration of the specific inhibitor methyl glyoxal bis(guanylhydrazone), Biochem. Biophys. Res. Commun. 52: 696 (1973).

MECHANISMS INVOLVED IN ORNITHINE DECARBOXYLASE INDUCTION BY 12-O-TETRADECANOYLPHORBOL-13-ACETATE, A POTENT MOUSE SKIN TUMOR PROMOTER AND AN ACTIVATOR OF PROTEIN KINASE C

Ajit K. Verma, J.T. Hsieh and R.C. Pong

Department of Human Oncology
University of Wisconsin Clinical Cancer Center
Madison, WI. 53792, U.S.A.

INTRODUCTION

Ornithine decarboxylase (ODC, EC 4.1.1.17) which decarboxylates ornithine to form putrescine, is the key enzyme in mammalian polyamine biosynthesis. ODC is characterized by its inducibility and rapid turnover rate (half life ~ 15 min). ODC activity in quiescent cells is extremely low and is increased within a few hours in response to many different stimuli, including hormones, growth factors, and tumor promoters. In general the induction of ODC activity and subsequent accumulation of putrescine constitute a part of a cascade of biochemical events that accompany cell proliferation and differentiation (1,2).

Among the numerous biochemical changes elicited following a single application of 12-O-tetradecanoylphorbol-13-acetate (TPA) to mouse skin, the induction of epidermal ODC activity is a prominent early event (3). Available data indicate that TPA-induced ODC activity and the resultant accumulation of putrescine play essential roles in mouse skin tumor promotion (4). We analyzed the mechanisms by which TPA treatment may lead to an increase in ODC activity (5-9). Data indicating that TPA-induced ODC activity is the result of an increased synthesis of ODC mRNA regulated at the transcription level, possibly mediated by PKC, are summarized in this chapter.

EXPERIMENTS AND RESULTS

1. TPA-Induced ODC Activity Correlates with the Amount of ODC Protein.

TPA-induced mouse epidermal ODC activity was observed to be proportional to the amount of ODC protein quantified by the [^{3}H]difluoromethylornithine (DFMO)-binding assay (5). In this experiment, soluble epidermal extract was prepared at various times (0,2,4,6,8 and 12 hr) after a single application of acetone or 10 nmol of TPA to mouse skin; ODC activity and the capacity to bind [^{3}H]DFMO were determined. At all time points, ODC activity correlated with [^{3}H]DFMO binding (Fig. 1). DFMO is an irreversible inhibitor of mammalian ODC; it is enzymatically decarboxylated generating an intermediate carbanionic species which, with the loss of fluorine, alkylates a nucleophilic residue at or near the active site, thereby covalently binding the inhibitor to the enzyme. The specificity of [^{3}H]DFMO binding to ODC is well established. Furthermore, Gilmour et al. (6) also found while using polyclonal antibody to ODC that TPA-induced epidermal ODC activity enzyme activity correlates with the levels of ODC protein.

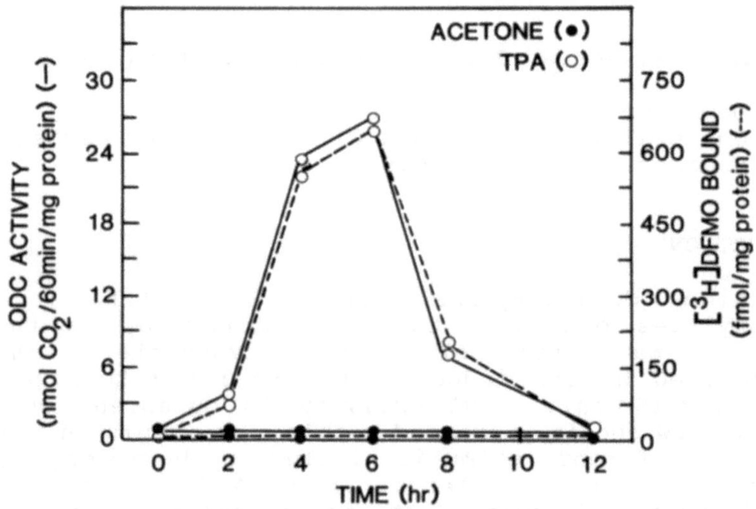

Fig.1. [^{3}H]DFMO binding to soluble epidermal ODC prepared at various times after a single application of acetone or TPA in acetone to mouse skin. Acetone or 10 nmol TPA in acetone was applied to mouse skin. Epidermis from eight mice was scraped (without heat treatment), homogenized, centrifuged, and soluble epidermal ODC was partially purified following ammonium sulfate precipitation and dialysis. [^{3}H]DFMO (26.5 Ci/mmol) binding and ODC activity were assayed at the indicated times after treatment. [^{3}H]DFMO-bound enzyme was immunoprecipitated. The immunoprecipitates were analyzed by SDS-7.5% polyacrylamide tube-gel electrophoresis. The tube-gels were sliced and the radioactivity associated with each gel slice was determined. Each point in the graph ([^{3}H]DFMO) is the total [^{3}H]DFMO counts associated with the [^{3}H]DFMO peak, which was a single peak corresponding to M_r 55000. Equal amounts of protein (0.38 mg) from epidermal extracts were used for [^{3}H]DFMO binding assays.

2. TPA-Induced ODC Activity Correlates with the Level of ODC mRNA.

The effect of TPA on the amount of mouse skin ODC mRNA was determined by hybridization of total cellular RNA from mouse skin against ^{32}P-labelled nick-translated pOD48 (a cDNA clone from ODC mRNA). Dot-blot analysis indicated that a single application of 10 nmol of TPA to mouse skin led to about a 17-fold increase in hybridizable ODC mRNA at 3.5 hr after treatment; the increase in ODC mRNA preceded the TPA-induced increase in ODC activity. We measured the size of ODC mRNA by the Northern blotting procedure. In this experiment, total RNA, prepared from TPA-treated mouse skin, was fractionated by denaturing formaldehyde:agarose-gel electrophoresis. The gel was blotted to GeneScreen and hybridized with nick-translated ^{32}P-labelled pOD48. Application of TPA to mouse skin resulted in an increase in hybridizable-RNA-species containing regions of ODC mRNA, and RNA displayed a single major band of 2.1 kilobases in size. No change was observed in the size of ODC mRNA after TPA treatment of mouse skin (5,7).

Also, we observed, using primary cultures of newborn epidermal cells (MEC) from BALB/c, that a time course and the dose-response curves of ODC induction paralleled that of ODC mRNA induction by TPA in MEC(8). TPA treatment did not elicit any change in the size of ODC mRNA (Fig. 2). The magnitude of ODC induction was proportional to the amount of ODC mRNA increased by TPA (Fig. 3).

3. TPA-induced Accumulation of ODC mRNA is the Result of an Increased Transcription-Initiation.

The possibility was examined that TPA-induced steady-state levels of ODC mRNA may be the result of stimulated transcription of ODC gene. Primary cultures of newborn epidermal cells were used for this investigation. In this experiment, epidermal cells were treated with 2×10^{-7} M TPA in DMSO or vehicle DMSO and were labelled with [^{3}H]uridine during the entire period of TPA treatment. The amount of ODC mRNA was quantified 6 hr after treatment by the DNA-excess filter hybridization technique. Treatment of epidermal cells with DMSO or TPA did not affect the incorporation of [^{3}H]uridine into total cellular RNA. However, treatment of epidermal cells with TPA resulted in about a 3-fold increase in the accumulation of newly synthesized ODC mRNA (Table 1). It is also clear from the results in Table 1 that the specific hybridization of ODC mRNA increased proportionally with a proportional increase in the amount of total cellular RNA used for the hybridization.

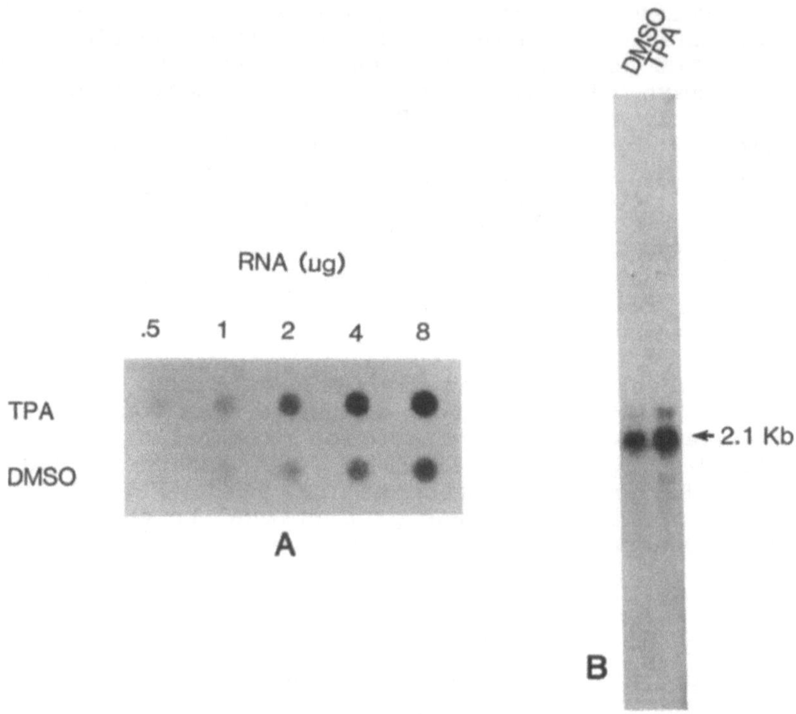

Fig. 2. Effect of TPA on the induction of ODC mRNA in MEC. MEC were treated for 30 min with DMSO or 2×10^{-7} M TPA in DMSO (final DMSO concentration, 0.1%) at 15 hr after plating. Total cellular RNA was isolated at 3 hr, and soluble epidermal ODC activity was determined at 6 hr after DMSO or TPA treatment. For dot-blot analysis, indicated amounts of total RNA were spotted on GeneScreen. For Northern blot analysis, RNA (10 μg) was fractionated by electrophoresis in a denaturing formaldehyde:agarose gel and transferred to GeneScreen. RNA bands containing ODC mRNA were detected by exposing Kodak X-Omat AR film to the washed GeneScreen at -70°C with intensifying screens. The resulting autoradiogram of dot-blot was quantified densitometrically. ODC mRNA (relative induction): DMSO, 1.0; TPA, 5.9. ODC activity (nmol CO_2 per 60 min per mg protein): DMSO, 1.4; TPA, 15.4. Autoradiogram of dot blot (A) and Northern blot (B). Kb, kilobases.

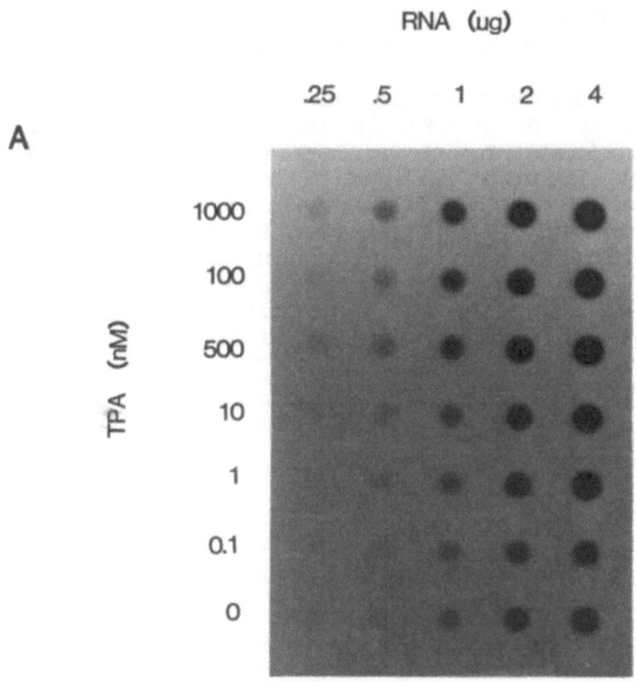

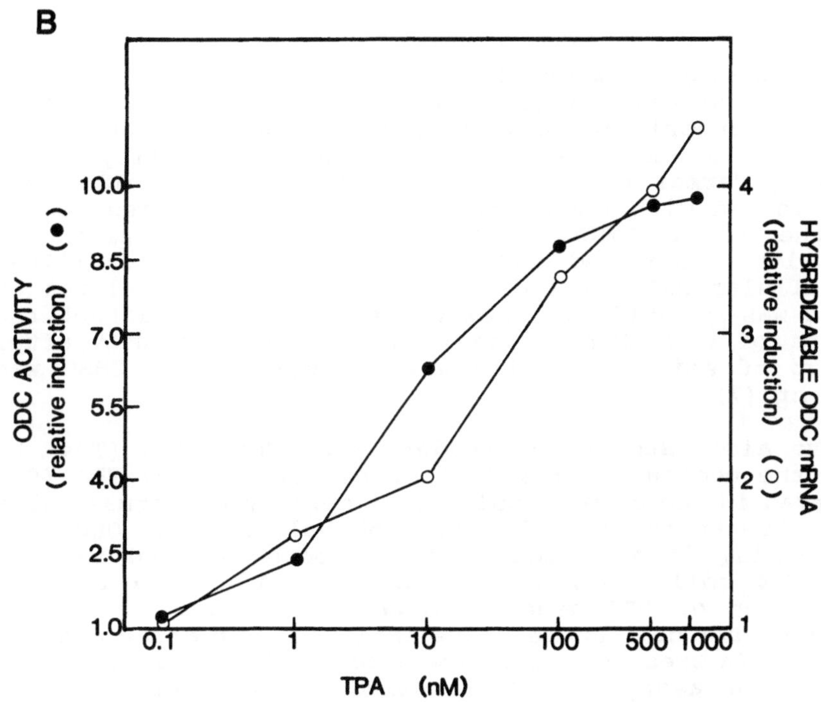

Fig. 3. Effect of TPA concentration on TPA-induced ODC activity and ODC mRNA in MEC. MEC were treated with the indicated concentrations of TPA in DMSO. Induction of ODC mRNA and ODC activity was determined at 6 hr after TPA treatment. Autoradiogram of dot blot (A) and kinetics of TPA-induced ODC and ODC mRNA (B).

Table 1. Effect of TPA on the synthesis of ODC mRNA in newborn mouse epidermal cells.

Primary culture of newborn epidermal cells was treated with DMSO (0.1%), or 2 x 10^{-7} M TPA in DMSO, and labelled with [^{3}H]uridine (250μCi/ml) for 6 hr. Total cellular RNA was isolated and the amount of radiolabelled ODC mRNA was quantified by the DNA-excess filter hybridization technique. An indicated amount of [^{3}H]RNA was used for DNA-excess filter hybridization. The amount of radioactivity bound to the filters with pBR322 was considered nonspecific and that varied from 50 to 61 dpm and was subtracted from the filters with plasmid pOD48 to obtain specific radioactive ODC mRNA - hybridized.

Treatment	Total [^{3}H] RNA used for hybridization (dpm x 10^{-6})	Total RNA (ug)	Specific [^{3}H]ODC mRNA hybridized (dpm)
DMSO	1.69	5	71
	3.38	10	149
TPA	1.23	5	202
	2.45	10	437

To distinguish between the possibility that TPA-induced accumulation of newly synthesized ODC mRNA is the result of increased transcription and/or decreased degradation of ODC mRNA, we determined the effect of TPA on the stability of ODC mRNA in epidermal cells. In this experiment, cells were pulse-labelled with [^{3}H]uridine for 3 hr in the presence of TPA or the vehicle DMSO. At various times after, cells were chased with unlabelled 5mM uridine and 2.5 mM cytidine. Total cellular RNA was isolated and the amount of labelled ODC mRNA was quantified. TPA treatment did not affect the half-life of ODC mRNA. The half-life of ODC mRNA after treatment of epidermal cells with either TPA or DMSO was about 7 hr (7).

We have also used human bladder carcinoma cells (T24) to explore the mechanism of induction of ODC mRNA by TPA (9). As observed in the primary culture of newborn epidermal cells (8), TPA treatment did not alter the half-life of ODC mRNA in T24 cells; it was about 6 hr in both TPA and vehicle treated T24 cells (9). The effect of TPA on the rate of transcription of ODC gene is shown in Table 2. In this experiment, T24 cell nuclei were isolated 4 hr after ethanol vehicle or TPA treatment and were used for the nuclear runoff transcription assays. ^{32}P-labelled RNA transcripts from nuclei were quantified by DNA-excess filter hybridization. TPA treatment resulted in about a 10- to 15-fold increase in transcription rate. Inclusion of α-amanitin, a specific inhibitor of RNA polymerase II, in the nuclear runoff assays inhibited the synthesis of ODC mRNA. Similar results were observed in a repeat experiment. Similarly, Katz and Kahana

(10) have shown that serum-induced ODC mRNA in BALB/c 3T3 cells may be the result of an increase in ODC transcription. However, in other systems using different inducers of ODC, post-transcription regulation has also been shown (11).

Table 2. Effect of TPA on the synthesis of ODC mRNA in T24 cells.

T24 cells were treated with either 0.05% vehicle ethanol or 50 nM TPA for 4 hr, then nuclei were isolated. The rate of ODC gene transcription was determined (9). Different concentrations of the RNA polymerase II inhibitor α-amanitin were added 5 min before addition of [α-^{32}P]UTP. Efficiency of hybridization was determined by adding 10^3 cpm of [^{3}H]cRNA to each assay. The hybridization efficiency varied from 0.15 to 0.45. Total labelled RNA transcript used for hybridization in each assay = 3 x 10^7 cpm.

Treatment	pODC10/2H (cpm)	pBR322 (cpm)	Hybridizable ODC mRNA (cpm)	Transcription rate (ppm)[a]
Ethanol	1212[b]	1097	115	24
Ethanol + α-amanitin (1 ug/ml)	745	621	124	32
TPA	3336	1138	2198	369
TPA + α-amanitin (0.1 ug/ml)	2369	1133	1236	106
TPA + α-amanitin (1 ug/ml)	1354	1129	225	38

[a]Transcription rate = cpm (pODC10/2H - pBR322)/cpm in total RNA x 2.1 kb in ODC mRNA/1.8 kb in cDNA (10^6/efficiency). The length of mature ODC mRNA is 2.1 kb and 1.8 kb is the length of pODC10/2H. Example: In ethanol treatment, the transcription rate = (1212 - 1097)/3 x 10^7 x 2.1/1.8 x 10^6/0.19 = 24.

[b]Each value is the mean of duplicate assays.

4. PKC is Involved in TPA-Induced ODC-gene Expression

Available evidence indicates that PKC is a receptor for phorbol esters (12-14). Phorbol ester receptors from the brain have been purified to homogeneity (13,14). Data from Scatchard analysis indicate that one molecule of tumor promoter may bind to each molecule of PKC in the presence of

Ca^{2+} and phospholipid (12,13). PKC is activated by TPA in vivo and in vitro in the presence of Ca^{2+} and phospholipid (15). PKC forms part of the signal transducing system involving the turnover of inositol phospholipid and is thought to be activated by diacylglycerol produced as a consequence of this turnover. Tumor-promoting phorbol esters appear to interact at the same site as diacylglycerols (16). In its inactive state, PKC is cytoplasmic. PKC is activated when diacylglycerol is generated in the membrane as the consequence of hydrolysis of phosphatidylinositol, which is in turn regulated by the activity of phospholipase C (17). Exposure of various cell types to TPA results in rapid translocation of PKC to the plasma membrane (17,18). It is suggested that the translocation of PKC to the plasma membrane may be an initial step in the mechanism of action of TPA (17,18).

An obligatory role of PKC in mouse skin tumor promotion has not been proven (17). Inhibitors of PKC [such as palmitoylcarnitine (8,19) and quercetin (20)] have been shown to inhibit skin tumor promotion by TPA. However, none of these PKC inhibitors is a specific inhibitor of PKC; these inhibitors appear to interact with the regulatory domain and/or with the phospholipid. Diacylglycerol (DG) is an endogenous activator of PKC and mimics many effects of TPA (8,21,22). Recently, we have shown that DG is a Stage II mouse skin tumor promoter (23). We summarize in this section that activation of PKC may be an early event in ODC gene transcription.

<u>a. DG-induced ODC Activity Correlates with the ODC mRNA Levels in MEC.</u> A time course of the effect of DG on the induction of ODC activity and on the amount of ODC mRNA in MEC is shown in Fig. 4. The effect of DG in MEC exhibited a good correlation between ODC activity and the induction of ODC mRNA (Fig.4). DG at 250 uM concentration in F-12 medium containing 0.3mM Ca^{2+} resulted in the induction of ODC which, at all time points, was proportional to the induction of ODC mRNA. Northern blot analysis of total RNA did not elicit any change in the size of ODC mRNA after treatment of MEC with DG (Fig. 4).

<u>b. Effect of Phospholipase C on the Induction of ODC and ODC mRNA.</u> Treatment of MEC with phospholipase C resulted in a dramatic increase in ODC activity. A peak ODC activity was observed between 4 and 6 hr after phospholipase C treatment. A maximum increase in ODC activity occurred with 0.02 units of phospholipase C per ml; 0.05 units of phospholipase C per ml were less effective than 0.02 units per ml in the induction of ODC activity in MEC. At all the concentrations(0.01 to 0.1 unit/ml), phospholipase A_2 did not affect ODC activity in MEC (Fig. 5). These results are in accord with the findings of Jetten et. al. in tracheal epithelial cells (22). A correlation between phospholipase C-induced ODC activity and ODC mRNA levels was determined in the experiment shown in Fig. 6. Phospholipase C-induced ODC activity accompanied increases in the level of hybridizable ODC mRNA measured by dot-blot analysis. We measured the size of ODC mRNA in phospholipase C-treated MEC. All RNAs displayed a single major band of 2.1 kilobases in size (Fig. 6).

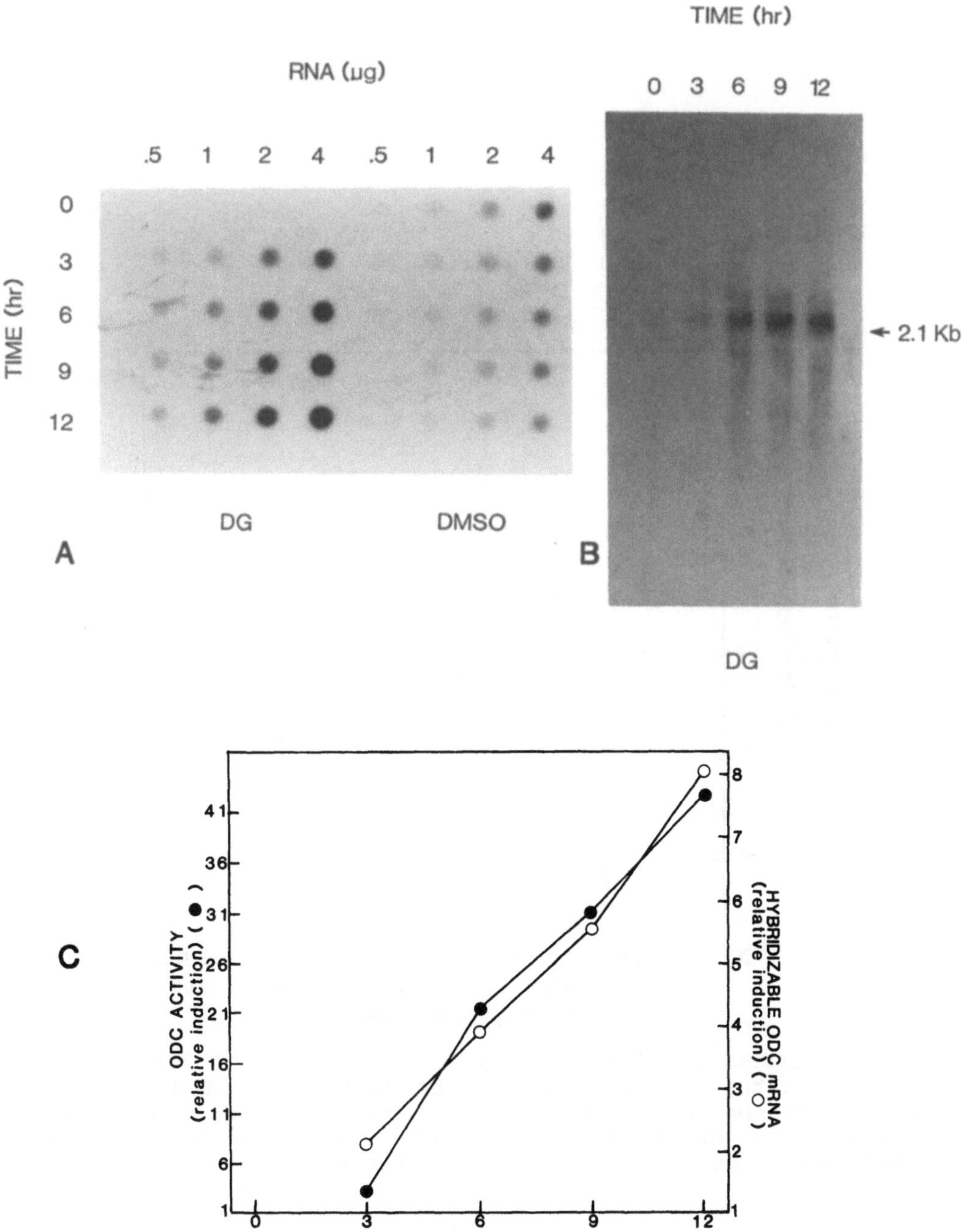

Fig. 4. Effect of DG (1-oleoyl-2-acetyl-rac-glycerol) on ODC and ODC mRNA induction. At 15 hr after plating, cells were washed with phosphate-buffered saline and re-fed with F-12 medium containing the usual concentration of Ca^{2+} (0.3 mM) plus 0.3% bovine serum albumin. DG (250 uM) in DMSO was added to the medium. ODC activity and the amount of ODC mRNA were determined at the indicated times after DG treatment. Autoradiogram of: dot blot (A); Northern blot (B); kinetics of DG-induced ODC and ODC mRNA (C).

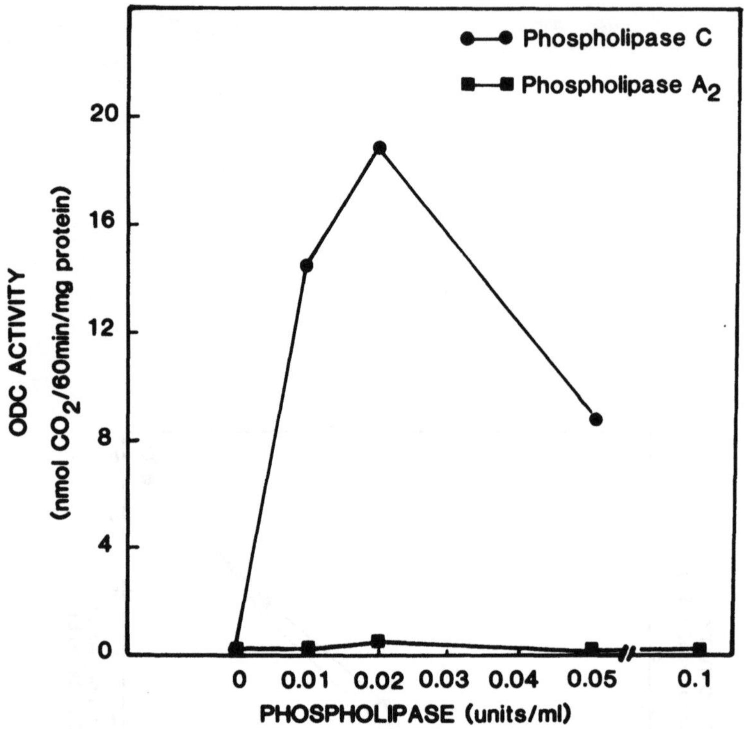

Fig. 5. Effect of phospholipase treatment on ODC induction. At 15 hr after plating, cells were switched to F-12 medium containing the usual concentration of Ca^{2+} (0.3 mM). ODC activity was determined 6 hr after the addition of the indicated concentrations of phospholipase.

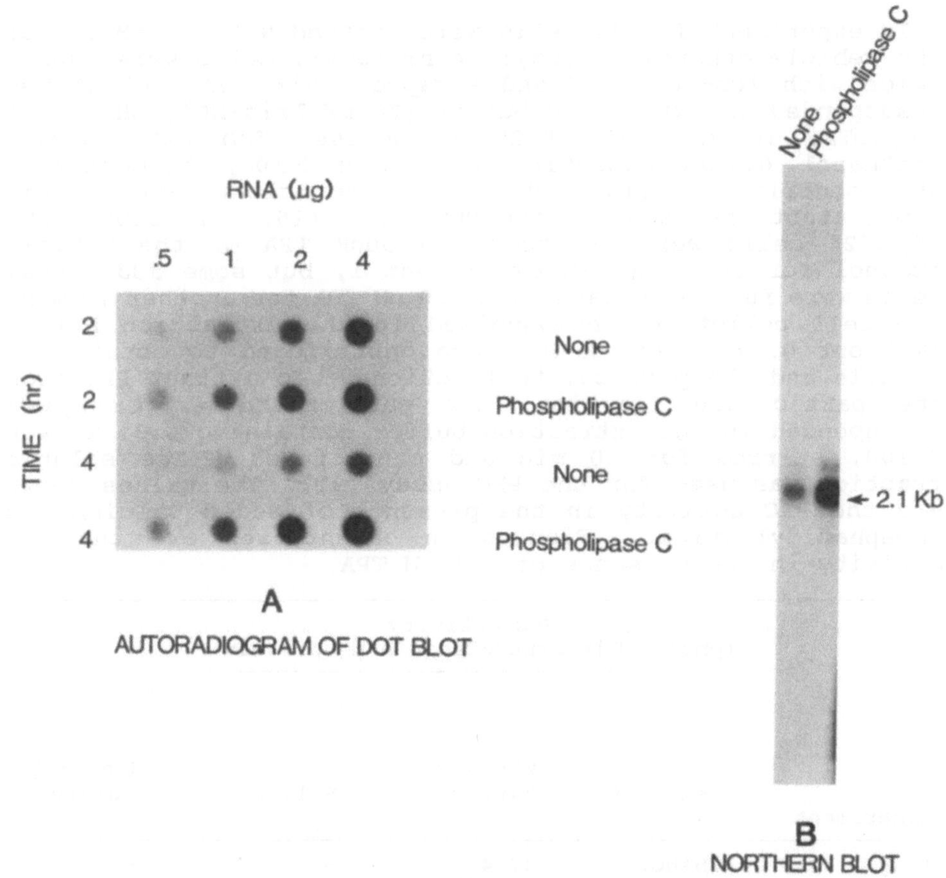

Fig. 6. Effect of phospholipase C on ODC mRNA level in MEC. At 15 hr after plating, cells were switched to F-12 medium containing 0.02 units of phospholipase C per ml. Total cellular RNA was isolated at 2 and 4 hr after treating MEC with phospholipase C. ODC mRNA was quantified by dot-blot analysis, and the size of ODC mRNA was determined by Northern blot analysis. Autoradiogram of: dot blot (A); Northern blot (B).

Table 3. PKC activity in PKC-deficient cells.

In experiment I, T24 cells were treated with 50 nM TPA or the vehicle ethanol (0.05%); 24 hr later, cells were washed twice with ice-cold PBS and scraped. The cell pellet was resuspended in extraction buffer (20 mM Tris-HCl, pH 7.4, 2 mM EDTA, 10 mM EGTA, 0.25 M sucrose, 5mM beta-mercaptoethanol, 0.4 mM leupeptin, 0.3% Triton X-100), homogenized, and centrifuged (100,000 g). The clear whole cell supernatant was used for the PKC assay (16). In experiment II, T24 cells were treated with 50nM TPA or the vehicle ethanol for 24 hr as in experiment I, but some additional cells were further treated with 50 nM TPA for another 10 min. The cell pellet was homogenized in the extraction buffer (without 0.3% Triton X-100) and centrifuged to obtain the soluble and the particulate fractions. To extract PKC from the particulate fraction, the particulate fraction was resuspended in the extraction buffer containing 0.3% Triton X-100, stirred for 30 min and centrifuged. The soluble fraction was used for the PKC assay (9). The values shown are the PKC activity in the presence of added calcium and phosphatidylserine. Figures in parentheses are the PKC activity in the presence of 0.1 uM TPA.

		PKC activity (pmol ^{32}P incorporated/min/mg protein)		
Experiment	Treatment	Whole Cell Extract	Soluble	Particulate
I	ETHANOL	37.4 (73.9)	-	-
	TPA	0.7 (0.0)	-	-
II	ETHANOL	-	45.6	18.8
	TPA	-	0.0	0.0
	ETHANOL + TPA	-	0.0	55.8
	TPA + TPA	-	0.0	0.0

Table 4. Lack of synthesis of ODC mRNA in PKC-deficient T24 cells.

T24 cells were incubated with 50 nM TPA for 24 hr and then replaced with the fresh medium containing the same concentration of TPA or 250 uM DG for another 6 hr. Cells were labelled with [^{3}H]uridine (250 uCi) for 1 hr before harvesting. The same culture plates were used for RNA extraction and ODC activity. An equal amount (16 ug) of [^{3}H]RNA was used for DNA-excess filter hybridization.

Treatment	ODC-activity (nmol CO_2/30 min/mg protein)	Total [^{3}H]RNA added (dpm x 10^{-6})	Specific hybridization (dpm)[a]
ETHANOL	1.03 ± 0.29[b]	3.88	25[c]
TPA	8.13 ± 1.25	2.69	174
DMSO	1.35 ± 0.06	4.32	21
DG	2.95 ± 0.13	4.04	48
TPA-TPA	0.14 ± 0.02	1.96	0
TPA-DG	0.24 ± 0.06	2.43	0
ETHANOL-TPA	9.72 ± 0.02	2.86	169

[a]Specific hybridization represents the dpm bound to the pOD48-containing filter minus the dpm bound to the pBR322-containing filter.
[b]Values are means ± standard error (SE).
[c]Values are means of duplicate assays.

<u>c. ODC mRNA Synthesis in PKC-deficient T24 Cells.</u> We further explored the role of PKC in the transcription-activation of the ODC gene. T24 cells were made PKC deficient by pretreatment with 50nM TPA for 24 hr. PKC activity was not detectable in T24 cells after further treatment with 50 nM TPA (Table 3). As shown in Table 4, the addition of 50 nM TPA or 250 uM DG to PKC-deficient T24 cells induced neither the synthesis of ODC mRNA nor ODC activity. In contrast, when ethanol-pretreated T24 cells were further treated with 50nM TPA, the induction of ODC activity and the synthesis of ODC mRNA were observed to the same degree as were found after treatment of untreated T24 cells with TPA.

<u>5. PKC-Mediated Phosphorylated Polypeptides and ODC Induction.</u>

We have shown in the preceding sections that TPA-induced ODC activity either in primary cultures of newborn epidermal cells or in the T24 cell line correlated with the synthesis of ODC mRNA. TPA treatment did not alter the stability of ODC mRNA but rather increased the rate of transcription of ODC gene in T24 cells. The results (8,9) presented before lend support to the suggestion that PKC activation may be an

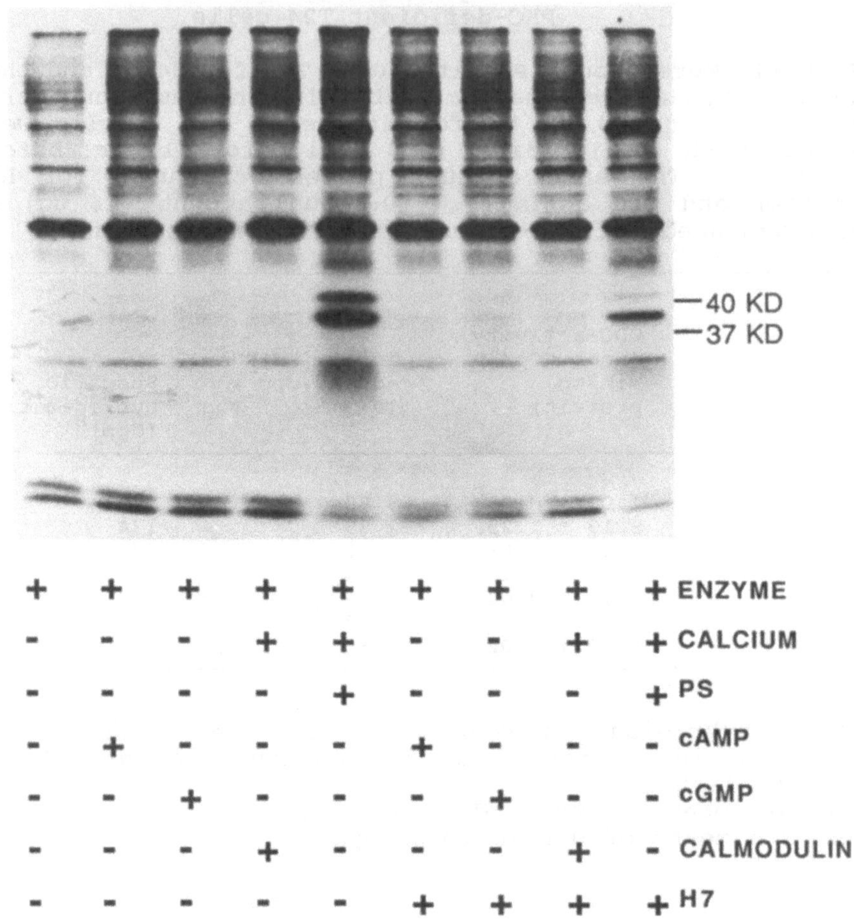

Fig. 7. An autoradiogram showing the phosphorylation of endogenous proteins by the endogenous PKC from the soluble fraction of T24 cells. Unfractionated cytosol (100,000xg) from T24 cells was used for the endogenous protein phosphorylation reaction. The assay mixture contained 20mM Tris/HCl pH 7.4, 10mM magnesium acetate, 0.5 mM EGTA, 0.5 mM DTT, 12.5% BSA, 2.5 mM β-mercaptoethanol and 30 μg cytosol protein in a final volume of 100 μl. After incubation at 37°C for 5 min, the proteins were separated by SDS-PAGE. The separating gel was 10% polyacrylamide with 0.1% SDS. The phosphorylated polypeptides were made visible by exposing Kodak X-Omat AR film to the dried gel at -70°C with intensifying screens.

initial event in ODC mRNA synthesis. We hypothesize that TPA-induced ODC-gene transcription is mediated through interaction of PKC-mediated phosphorylated proteins (trans-acting factors) with cis-acting elements in the 5'-flanking region of ODC gene. As a step towards the search for specific trans-acting factors involved in the regulation of ODC gene transcription, we found that PKC activation specifically phosphorylated 37 and 40 kD polypeptides in T24 cells, and could not be detected by activation of either cAMP-, cGMP- or calmodulin-dependent protein kinases (Fig. 7). The phosphorylation of 37 and 40 kD polypeptides was not observed in PKC deficient cell extract. The phosphorylation of 37 and 40 kD polypeptides in PKC deficient cells could be rescued by purified mouse brain PKC. The phosphorylation of 37 and 40 kD polypeptide was inhibited by the inhibitors of PKC (H7 and staurosporine). It is also noteworthy that the experimental manipulations which abolished PKC activity (e.g., down-regulation of PKC, inhibition of PKC by the PKC inhibitors, H7, palmitoylcarnitine and staurosporine) resulted in inhibition of the phosphorylation of 37 and 40 kD polypeptides and TPA-induced ODC activity in T24 (24). These results indicate that PKC-mediated 37 and 40 kD polypeptides may be involved in TPA-induced ODC gene expression.

SUMMARY AND CONCLUSIONS

ODC, the first enzyme in mammalian polyamine biosynthesis, is rapidly induced in response to a wide variety of growth stimuli. However, there is no single mechanism which may explain the rapid turnover of ODC activity. ODC activity has been shown to be regulated at the level of synthesis (25) and degradation (26), and also by post-translational modifications (27) and an interaction with macromolecules (28). Our results (9) indicate that TPA-induced ODC activity is regulated at the transcriptional level. An initial signal in ODC induction by TPA is not clear. We have suggested that TPA-increased accumulation of epidermal prostaglandins is required, but not sufficient, for ODC induction by TPA (29). Others have suggested the role of lipoxygenase product(s) in ODC induction (20). The role of the microtubule-containing system in regulation of ODC induction has been shown. The involvement of cyclic nucleotides in ODC induction by TPA is controversial (for references, see Ref. 30). Also, generation of free radicals appears to be involved in ODC induction by TPA (31). Data summarized in this chapter indicate that activation of PKC may be an initial step in ODC induction by TPA.

REFERENCES

1. A.E. Pegg, Recent advances in the biochemistry of polyamines in eukaryotes, Biochem. J. 234:249 (1986).

2. A.E. Pegg, Polyamine Metabolism and Its Importance in Neoplastic Growth and as a Target for Chemotherapy, Cancer Res. 48:759 (1988).

3. T.G. O'Brien, R.C. Simsiman, and R.K. Boutwell, Induction of polyamine-biosynthetic enzymes in mouse epidermis by tumor promoting agents, Cancer Res. 35:1662 (1975).

4. A.K. Verma and R.K. Boutwell, Inhibition of carcinogenesis by inhibitors of putrescine biosynthesis, in: Inhibition of Polyamine Metabolism: "Biological Significance and Basis of New Therapies", McCann, P.P., Pegg, A.E., Sjoerdsma, A. eds.,pp 249, Academic Press, Inc., Orlando, Florida (1987).

5. A.K. Verma, D.Erickson, and B.J. Dolnick, Increased mouse epidermal ornithine decarboxylase activity by the tumour promoter 12-0-tetradecanoylphorbol-13-acetate involves increased amounts of both enzyme protein and messenger RNA, Biochem. J. 237:297 (1986).

6. S.K. Gilmour, A.K. Verma, T. Madara, and T.G. O'Brien, Regulation of ornithine decarboxylase gene expression in mouse epidermis and epidermal tumors during two-stage tumorigenesis, Cancer Res. 47:1221 (1987).

7. A.K. Verma, Inhibition of tumor promoter 12-0-tetradecanoylphorbol-13-acetate-induced synthesis of mouse epidermal ornithine decarboxylase messenger RNA and diacylglycerol-promoted mouse skin tumor formation by retinoic acid, Cancer Res. 48:2168 (1988).

8. A.K. Verma, R.C. Pong, and D. Erickson, Involvement of protein kinase C in ornithine decarboxylase gene expression in primary culture of newborn mouse epidermal cells and in skin tumor promotion by 12-0-tetradecanoylphorbol-13-acetate, Cancer Res. 46:6149 (1986).

9. J.T. Hsieh and A.K. Verma, Involvement of protein kinase C in the transcriptional regulation of ornithine decarboxylase gene expression by 12-0-tetradecanoylphorbol-13-acetate in T24 human bladder carcinoma cells, Arch. Biochem. Biophys. 262:326 (1988).

10. A. Katz and C. Kahana, Transcriptional activation of mammalian ornithine decarboxylase during stimulated growth, Mol. Cell. Biol. 7:2641 (1987).

11. F.G. Berger, D. Loose, D.H. Meisner, and G. Watson, Androgen induction of messenger RNA in mouse kidney is posttranslational, Biochemistry 25:1170 (1986).

12. U. Kikkawa, Y. Takai, Y. Tanaka, R. Miyake, and Y. Nishizuka, Protein kinase C as a possible receptor protein

of tumor-promoting phorbol esters, J. Biol. Chem. 258:11442 (1983).

13. J.E. Neidel, L.J. Kuhn, and G.R. Vandenbark, Phorbol diester receptor copurifies with protein kinase C, Proc. Natl. Acad. Sci. USA 80:36 (1983).

14. C.L. Ashendel, J.M. Staller, and R.K. Boutwell, Solubilization, purification, and reconstitution of a phorbol ester receptor from the particulate protein fraction of mouse brain, Cancer Res. 43:4327 (1983).

15. M. Castagna, Y. Takai, K. Kaibuchi, K. Sano, U. Kikkawa, and Y. Nishizuki, Direct activation of calcium-activated, phospholipid-dependent protein kinase by tumor-promoting phorbol esters, J. Biol. Chem. 257:7847 (1982).

16. K.L. Leach and P.M. Blumberg, Modulation of protein kinase C activity and [^{3}H]phorbol-12,13-dibutyrate binding by various tumor promoters in mouse brain cytosol, Cancer Res. 45:1958 (1985).

17. Y. Nishizuka, The role of protein kinase C in cell surface signal transduction and tumor promotion, Nature (Lond.) 308:693 (1984).

18. Y. Nishizuka, Prospectives on the role of protein kinase C in stimulus-response coupling, J. Natl. Cancer Inst. 76:363 (1986).

19. T. Nakadate, S. Yamamoto, E. Aizu, and R. Kato, Inhibition of 12-0-tetradecanoylphorbol-13-acetate-induced tumor promotion and epidermal ornithine decarboxylase activity in mouse skin by palmitoylcarnitine, Cancer Res. 46:1589 (1986).

20. T. Nakadate, S. Yamamoto, M. Ishii, and R. Kato, Inhibition of 12-0-tetradecanoylphorbol-13-acetate-induced epidermal ornithine decarboxylase activity by phospholipase A_2 inhibitors and lipoxygenase inhibition, Cancer Res. 42:2841 (1982).

21. R.C. Smart, M.T. Huang, and A.H. Conney, sn-1,2-Diacylglycerols mimic the effects of 12-0-tetradecanoylphorbol-13-acetate in vivo by inducing biochemical changes associated with tumor promotion in mouse epidermis, Carcinogenesis (Lond.) 7:1865 (1986).

22. A.M. Jetten, B.R. Ganong, G.R. Vandenbark, J.E. Shirley, and R.M. Bell, Role of protein kinase C in diacylglycerol-mediated induction of ornithine decarboxylase and reduction of epidermal growth factor binding, Proc. Natl. Acad. Sci. USA 82:1941 (1985).

23. A.K. Verma, Protein kinase C activator L-α-dioctanoylglycerol: a potent stage II mouse skin tumor promoter, Cancer Res. 48:1736 (1988).

24. R.C. Pong, J.T. Hsieh, M.F. Denning, and A.K. Verma, Protein kinase C-mediated phosphorylation of 37 and 40 kD

polypeptides and increased transcription of the ornithine decarboxylase gene in the T24 human bladder carcinoma cells, Proc. Am. Assoc. Cancer Res. 29:77 (1988).

25. S.K. Gilmour, N. Avdalovic, T. Madara, and T.G. O'Brien, Induction of ornithine decarboxylase by 12-0-tetradecanoylphorbol-13-acetate in hamster fibroplasts, J.Biol.Chem. 260:16439 (1985).

26. J.E. Seely, H. Pösö, and A.E. Pegg, Effect of androgens on turnover of ornithine decarboxylase in mouse kidney, J.Biol.Chem. 257:7549 (1982).

27. D.H. Russell, Posttranslational modification of ornithine decarboxylase by its product putrescine, Biochem. Biophys. Res. Commun. 99:1167 (1981).

28. J.S. Heller, W.F. Fong, and E.S. Canellakis, Induction of a protein inhibitor to ornithine decarboxylase by the end product of its reaction, Proc. Natl. Acad. Sci. USA 73:1858 (1976).

29. A.K. Verma, C.L. Ashendel, and R.K. Boutwell, Inhibition by prostaglandin synthesis inhibitors of the induction of epidermal ornithine decarboxylase activity, the accumulation of prostaglandins, and tumor promotion caused by 12-0-tetradecanoylphorbol-13-acetate, Cancer Res. 40:308 (1980).

30. A.K. Verma and D. Erickson, Induction of mouse epidermal ornithine decarboxylase by the tumor promoter 12-0-tetradecanoylphorbol-13-acetate: dependency on calcium availability, Arch. Biochem. Biophys. 247:272 (1986).

31. W.J. Kozumbo, J.L. Seed, and T.W. Kensler, Inhibition by 2(3)-tert-butyl-4-hydroxyanisole and other antioxidants of epidermal ornithine decarboxylase activity induced by 12-0-tetradecanoylphorbol-13-acetate, Cancer Res. 43:2555 (1983).

POLYAMINES, DNA METHYLATION AND CELL DIFFERENTIATION

Olle Heby[1], Lo Persson[2] and Steven S. Smith[3]

Departments of [1]Zoophysiology and [2]Physiology, University of Lund, S-223 62 Lund, Sweden; and [3]Molecular Surgery, City of Hope National Medical Center, Duarte, CA 91010, USA

INTRODUCTION

Cells that have become depleted of their putrescine and spermidine content, e.g. by treatment with 2-difluoromethyl-ornithine (DFMO), an enzyme-activated irreversible inhibitor of ornithine decarboxylase (Metcalf et al., 1978), grow slowly or not at all (Heby, 1981; Pegg, 1988). In addition to causing growth arrest, polyamine depletion has been found to induce or suppress cell differentiation (Heby et al., 1987). Teratocarcinoma stem cells belong to the category of cells that are induced to differentiate in response to putrescine and spermidine depletion (Heby et al., 1983; 1987; Schindler et al., 1983; Oredsson et al., 1985).

Putrescine and spermidine serve as aminopropyl group acceptors in the consecutive action of spermidine synthase and spermine synthase (Pegg, 1988). When cells are depleted of these polyamines the concentration of decarboxylated S-adenosyl-L-methionine (dcAdoMet), the aminopropyl group donor, increases dramatically (Danzin et al., 1982; Mamont et al., 1982; Pegg et al., 1982; Oredsson et al., 1986). Since the chemical structure of dcAdoMet differs from that of AdoMet (Fig. 1) only in that it lacks the carboxyl group, it is conceivable that dcAdoMet can act as a methyl group donor or as a substrate inhibitor in many of the reactions that are catalyzed by methyltransferases and that use AdoMet as a substrate. Under normal conditions this possibility seems highly unlikely, because the concentration of AdoMet exceeds that of dcAdoMet by 15- to 550-fold (Danzin et al., 1982; Mamont et al., 1982; Oredsson et al., 1986). In putrescine- and spermidine-depleted cells, on the other hand, the concentration of dcAdoMet may be 2- to 25-fold higher than that of AdoMet (Danzin et al., 1982; Mamont et al., 1982; Oredsson et al., 1986).

Methylation of cytosine residues at critical CpG sites in DNA (Fig. 1) can prevent the transcription of mammalian genes (Busslinger et al., 1983; Doerfler, 1983; Razin and Cedar, 1984; Bird, 1986). Conversely, the induction of gene expression in differentiating teratocarcinoma stem cells is accompanied by a genome-wide loss of DNA methylation (Young and Tilghman, 1984). If dcAdoMet can interfere with DNA methylation, the accumulation of large amounts of dcAdoMet may partly explain why DFMO treatment induces differentiation of teratocarcinoma stem cells.

In the present study we address the question whether dcAdoMet may serve as a methyl group donor in reactions catalyzed by human DNA(cytosine-5)methyltransferase (EC 2.1.1.37) and whether dcAdoMet may inhibit DNA methylation at the concentrations reached in polyamine-depleted cells.

DNA METHYLATION

Most cell types seem to be capable of *de novo* methylation of CpGs in their DNA (Fig. 2), early embryonic cells and teratocarcinoma stem cells being particularly active in this respect (Doerfler, 1983; Razin and Cedar, 1984; Bird, 1986). However, little is known about the mechanisms involved. Cells also have mechanisms for copying DNA methylation patterns. Thus, CpG sites are methylated shortly after DNA replication by a maintenance DNA methyltransferase that recognizes the hemimethylated product of

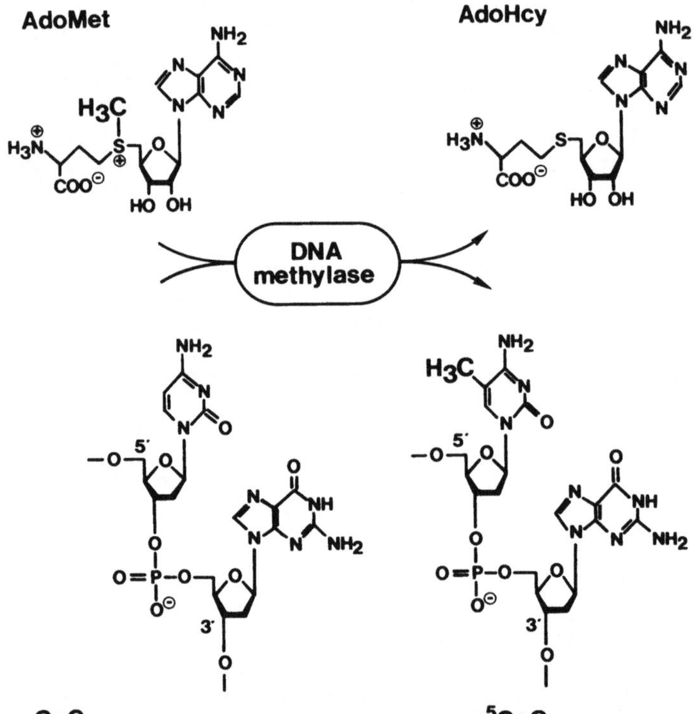

Fig. 1. Methylation of the cytosine residue in a CpG site in DNA, catalyzed by DNA methyltransferase. AdoMet acts as the methyl group donor, yielding S-adenosyl-L-homocysteine (AdoHcy) and m^5CpG as the products.

semi-conservative DNA replication, and catalyzes the transfer of a methyl group from AdoMet to carbon 5 of the appropriate cytosine residue in the newly replicated strand (Fig. 2) (Razin and Riggs, 1980; Doerfler, 1983; Razin and Cedar, 1984; Bird, 1986).

To determine whether dcAdoMet may serve as a methyl group donor in the methylation of cytosine residues in DNA, we synthesized dcAdoMet that was radiolabeled in its methyl group. The transfer of methyl groups to various substrates was studied using the DNA methyltransferase assay described below.

Enzymatic Synthesis and Purification of (Methyl-^{3}H)dcAdoMet

An enzymatic procedure was used to synthesize the biologically active dcAdoMet, because chemical synthesis yields a

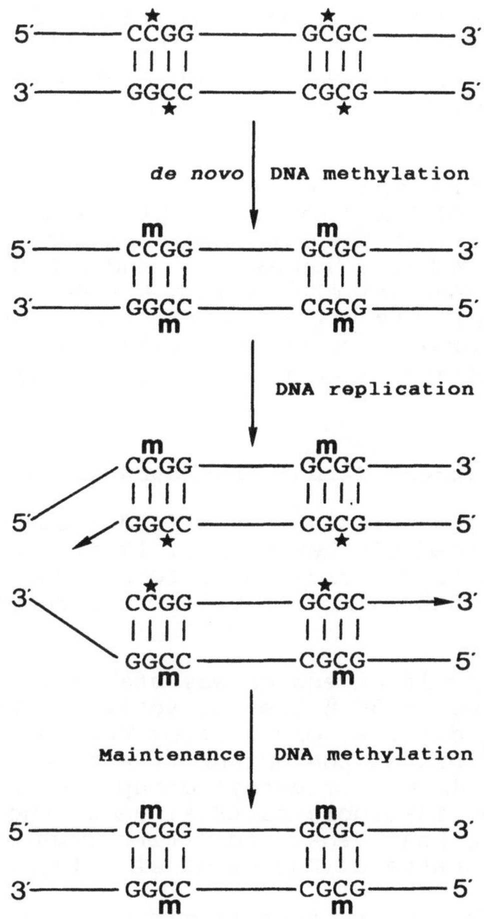

Fig. 2. Replication and methylation of DNA. Sites recognized by *de novo* and maintenance DNA methyltransferases are indicated with stars. The pattern of methylation of the parental strand is conserved in the daughter DNA by the action of a maintenance methyltransferase. It should be noted, however, that there is no direct evidence for distinct *de novo* and maintenance DNA methyltransferases.

racemic mixture. A purified mammalian S-adenosyl-L-methionine decarboxylase (AdoMetDC) was used to decarboxylate (methyl-^{3}H)-AdoMet (specific activity, 15 Ci/mmol, Amersham).

AdoMetDC was purified from rat prostate essentially as described by Pösö and Pegg (1982). One unit of activity was defined as the amount of enzyme which releases 1 nmol of CO_2 per minute under the standard assay conditions.

The reaction mixture contained 0.2 M Tris-HCl (pH 7.5), 5 mM dithiothreitol, 2.5 mM putrescine, 0.1 mM EDTA, 100 µg/ml of bovine serum albumin (fraction V, Sigma), 250 µCi (16.7 nmol) of (methyl-^{3}H)AdoMet and 10 units of AdoMetDC in a final volume of 5 ml. After a 90-min incubation at 37°C, the reaction was terminated by the addition of 1 ml of 2 M $HClO_4$, and the (methyl-^{3}H)-dcAdoMet produced was purified according to one of the methods described by Zappia et al. (1983). The precipitated proteins were removed by centrifugation, and the acidity of the supernatant was reduced to pH 4.5 with 3 M $KHCO_3$. After another centrifugation, to remove the $KClO_4$ precipitate, the supernatant was placed in a boiling water bath for 1 h in order to hydrolyze any residual AdoMet.

The hydrolysate was applied to a Dowex 50, Na^+ column (100-200 mesh, resin bed 1 x 3 cm) previously equilibrated with 0.1 M NaCl. After removal of the hydrolytic products of AdoMet by elution with 0.1 M NaCl (homoserine) and 2 M HCl (5'-methylthioadenosine), dcAdoMet was eluted with 6 M HCl. The fractions containing the majority of the radioactivity were evaporated to dryness. The residues were reconstituted in 0.01 M H_2SO_4 to yield a (methyl-^{3}H)dcAdoMet concentration of 1 mCi/ml.

Purification and Assay of DNA Methyltransferase

DNA methyltransferase was purified from human placenta as previously described (Zucker et al., 1985), except that nucleic acid was removed from hydroxylapatite-concentrated material by passage through DEAE-cellulose prior to Dodecyl-Sepharose column chromatography.

The highly purified enzyme was stable for at least 3 months at -70°C. There was a 50 % loss of activity after 1 year under these storage conditions. One unit of DNA methyltransferase activity was defined as the amount of enzyme which catalyzes the incorporation of 1 pmol of methyl groups per hour into trichloroacetic acid (TCA) insoluble material under the standard assay conditions, using heat-denatured *Micrococcus lysodeikticus* DNA (Sigma) as the substrate (Zucker et al., 1985).

The DNA methylation reaction mixture contained 50 mM Hepes buffer (pH 7.1), 2 mM dithiothreitol, 10 % (v/v) glycerol, 50 mM NaCl, 60 µM spermine·4HCl, 2-12 µM (methyl-^{3}H)AdoMet (specific activity, 15 Ci/mmol), 4 µg/ml of unmethylated or hemimethylated duplex oligodeoxynucleotide, or 4 µg/ml poly(dI-dC)-poly(dI-dC) (Pharmacia), and 25-100 units of DNA methyltransferase (fraction VI) in a final volume of 100 µl. The reaction was initiated after a 15-min preincubation at 37°C by adding the enzyme. After 20 min at the same temperature, the reaction was terminated by adding

5 ml of ice-cold 5% TCA containing 5 mM sodium pyrophosphate (15 min on ice), the precipitate was collected on a Whatman GF/C filter and washed with 2% TCA containing 5 mM sodium pyrophosphate and finally with absolute ethanol. The filter was dried, suspended in Omnifluor/toluene (4 g/l) and counted in a liquid scintillation spectrometer with a 34 % efficiency.

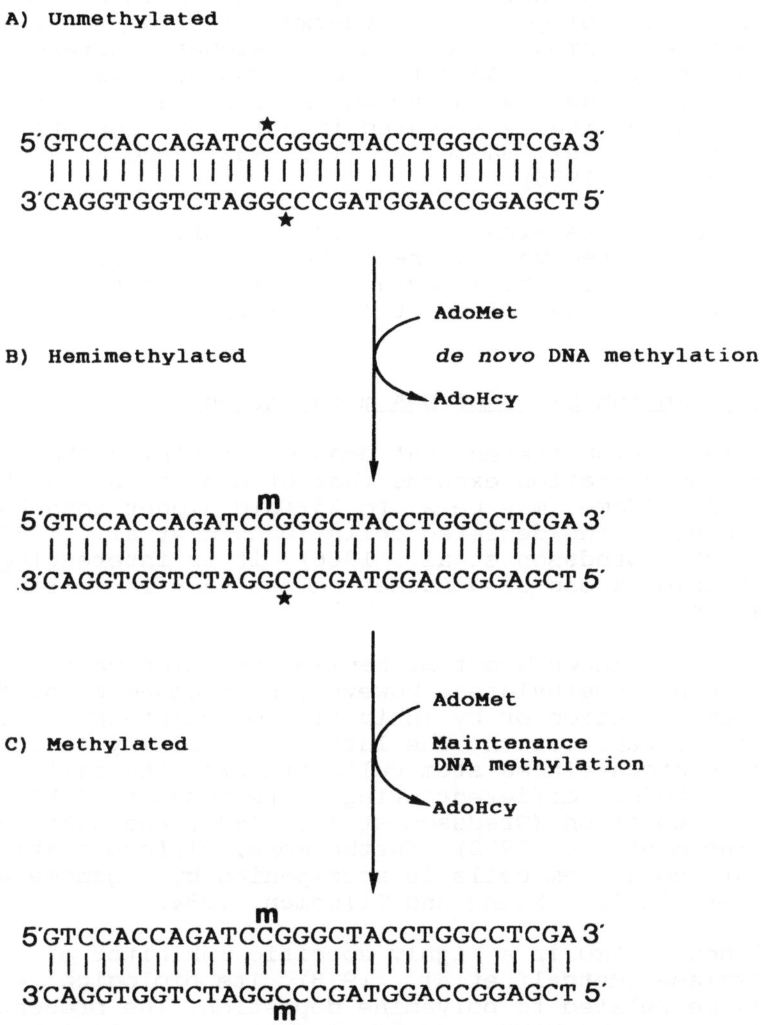

Figure 3. Sequences of the complementary duplex oligonucleotides used as methyl group acceptors. The defined 30-mers were prepared and purified as previously described (Smith et al., 1987). The sequence of each strand was verified. The unmethylated (A) and hemimethylated (B) duplex oligodeoxynucleotides were formed by annealing equimolar amounts of single strands. Restriction analysis was used to confirm the duplex structure. Sites recognized by DNA methyltransferase are marked with stars.

DcAdoMet - a methyl group donor in DNA methylation?

The highly purified DNA methyltransferase from human placenta exhibited a strong preference for the hemimethylated duplex oligodeoxynucleotide substrate (Fig. 3B) over the corresponding unmethylated one (Fig. 3A), a property consistent with that of a maintenance methyltransferase (Smith et al., 1987). As with other DNA methyltransferases (Pedrali-Noy and Weissbach, 1986) poly(dI-dC)-poly(dI-dC), an unmethylated substrate, was a more active methyl group acceptor.

The rate of transfer of methyl groups from dcAdoMet to the synthetic duplex oligo- and polydeoxynucleotide substrates was not significant compared to that from AdoMet. Therefore, it seems highly unlikely that dcAdoMet affects DNA methylation by providing methyl groups, even though the cellular dcAdoMet concentration is dramatically elevated in polyamine-depleted cells (Danzin et al., 1982; Mamont et al., 1982; Pegg et al., 1982; Oredsson et al., 1986).

In experiments with other methyltransferases it has also been shown that removal of the carboxyl group from AdoMet may result in a loss of the methylating ability of the molecule (Zappia et al., 1969; Oliva et al., 1980).

DcAdoMet - an inhibitor of DNA methylation?

Table 1 demonstrates that dcAdoMet inhibits DNA methylation when its concentration exceeds that of AdoMet. Since the concentration of dcAdoMet may be 2- to 25-fold higher than that of AdoMet in polyamine-depleted cells (Danzin et al., 1982; Mamont et al., 1982; Oredsson et al., 1986), it is interesting to note that a 5-fold excess of dcAdoMet causes 96 % inhibition of DNA methylation.

Little is known about mechanisms by which particular regions of DNA become unmethylated. However, such changes could arise by direct demethylation or by inhibition of maintenance activity for one round of replication. The latter possibility is interesting because teratocarcinoma stem cells traverse the cell cycle at least once before differentiating in response to DFMO-mediated polyamine depletion (Oredsson et al., 1985) and dcAdoMet accumulation (Holm et al., 1985). Furthermore, differentiation of teratocarcinoma stem cells is accompanied by a genome-wide loss of DNA methylation (Young and Tilghman, 1984).

Although DFMO is a highly specific inhibitor of ornithine decarboxylase (Metcalf et al., 1978), its biological effects may not only be related to polyamine depletion. The present study indicates that the DFMO-mediated increase in cellular dcAdoMet concentration may inhibit DNA methylation. It has also been shown that dcAdoMet may significantly inhibit other methyltransferases (Zappia et al., 1969), and that it is an active inhibitor of histone acetylation (Pegg et al., 1986).

Table 1. Effects of dcAdoMet on DNA methylation *in vitro*

DcAdoMet/AdoMet ratio	DNA methyltransferase activity*	
	(fmol/min)	(%)
0	780	100
2	631	81
4	196	25
5	29.5	3.8

*The enzyme activity was monitored by measuring the transfer of ^{3}H-methyl groups from AdoMet to acid-precipitable DNA substrates (Zucker et al., 1985; Hardy et al., 1987; Smith et al., 1987). The concentration of AdoMet was kept constant at 2 µM. The synthetic deoxyribonucleotide polymer poly(dI-dC)-poly(dI-dC) was used as the methyl goup acceptor. A similar degree of inhibition by dcAdoMet was seen when a hemimethylated oligodeoxynucleotide (Fig. 3B) was used.

SUMMARY

The cellular concentration of AdoMet is normally very much higher than that of dcAdoMet, the aminopropyl group donor in polyamine synthesis. However, when cells are depleted of their putrescine and spermidine, i.e. the aminopropyl group acceptors, the dcAdoMet concentration increases dramatically, to a level that may greatly exceed that of AdoMet.

Using a highly purified DNA methyltransferase and its preferred substrates, a defined hemimethylated duplex oligodeoxynucleotide or poly(dI-dC)-poly(dI-dC), we demonstrate that dcAdoMet is a poor methyl group donor, and that it starts to inhibit DNA methylation when its concentration exceeds that of AdoMet. At a dcAdoMet/AdoMet ratio of 5:1 there is very little methyl transfer.

This study suggests that the antiproliferative and differentiative effects brought about by inhibitors of polyamine synthesis may be partly attributable to dcAdoMet-mediated inhibition of DNA methylation.

ACKNOWLEDGEMENTS

This work was supported by grants from the Swedish Natural Science Research Council (B-BU 4086/115), the Swedish Medical Research Council (04X-02212), the Gunnar, Arvid and Elisabeth Nilsson Foundation, the Council for Tobacco Research (1571A), the National Institutes of Health (GM-32863) and the City of Hope Cancer Center (CA-335172).

REFERENCES

Bird, A. P., 1986, CpG-rich islands and the function of DNA methylation, Nature, 321:209-213.

Busslinger, M., Hurst, J., and Flavell, R. A., 1983, DNA methylation and the regulation of globin gene expression, Cell, 34:197-206.

Danzin, C., Claverie, N., Wagner, J., Grove, J., and Koch-Weser, J., 1982, Effect on prostatic growth of 2-difluoromethylornithine, an effective inhibitor of ornithine decarboxylase, Biochem. J. 202:175-181.

Doerfler, W., 1983, DNA methylation and gene activity, Ann. Rev. Biochem., 52:93-124.

Hardy, T. A., Baker, D. J., Newman, E. M., Sowers, L. C., Goodman, M. F., and Smith, S. S., 1987, Size of the directing moiety at carbon 5 of cytosine and the activity of human DNA(cytosine-5)methyltransferase, Biochem. Biophys. Res. Commun., 145:146-152.

Heby, O., 1981, Role of polyamines in the control of cell proliferation and differentiation, Differentiation 19:1-20.

Heby, O., Oredsson, S. M., Olsson, I., and Marton, L. J., 1983, A role for the polyamines in mouse embryonal carcinoma (F9 and PCC3) cell differentiation but not in human promyelocytic leukemia (HL-60) cell differentiation, Adv. Polyamine Res., 4:727-742.

Heby, O., Luk, G. D., and Schindler, J., 1987, Polyamine synthesis inhibitors act as both inducers and suppressors of cell differentiation, in: "Inhibition of Polyamine Metabolism", P. P. McCann, A. E. Pegg, and A. Sjoerdsma, eds., Academic Press, New York, pp. 165-186.

Holm, I., Wallon, M., Thorsson, L., and Heby, O., 1985, Methylglyoxal-bis(guanylhydrazone) (MGBG), an inhibitor of S-adenosylmethionine decarboxylase, interferes with α-difluoromethylornithine (DFMO)-induced growth arrest and differentiation of embryonal carcinoma cells, J. Cell Biol., 101:246a.

Mamont, P. S., Danzin, C., Wagner, J., Siat, M., Joder-Ohlenbusch, A.-M., and Claverie, N., 1982, Accumulation of decarboxylated S-adenosyl-L-methionine in mammalian cells as a consequence of the inhibition of putrescine biosynthesis, Eur. J. Biochem., 123:499-504.

Metcalf, B. W., Bey, P., Danzin, C., Jung, M. J., Casara, P., Vevert, J. P., 1978, Catalytic irreversible inhibition of mammalian ornithine decarboxylase (EC 4.1.1.17) by substrate and product analogues, J. Am. Chem. Soc., 100:2551-2553.

Oliva, A., Galletti, P., Zappia, V., Paik, W. K., and Kim, S., 1980, Studies on substrate specificity of S-adenosylmethionine:protein-carboxyl methyltransferase from calf brain, Eur. J. Biochem., 104:595-602.

Oredsson, S. M., Billgren, M., and Heby, O., 1985, Induction of F9 embryonal carcinoma cell differentiation by inhibition of polyamine synthesis, Eur. J. Cell Biol., 38:335-343.

Oredsson, S. M., Kanje, M., Mamont, P. S., Wagner, J., and Heby, O., 1986, Polyamine depletion increases cellular ribonucleotide levels, Mol. Cell. Biochem., 70:89-96.

Pedrali-Noy, G., and Weissbach, A., 1986, Mammalian DNA methyltransferases prefer poly(dI-dC) as substrate, J. Biol. Chem., 261:7600-7602.

Pegg, A. E., 1988, Polyamine metabolism and its importance in neoplastic growth and as a target for chemotherapy, Cancer Res., 48:759-774.
Pegg, A. E., Pösö, H., Shuttleworth, K., and Bennett, R. A., 1982, Effect of inhibition of polyamine synthesis on the content of decarboxylated S-adenosylmethionine, Biochem. J., 202:519-526.
Pegg, A. E., Wechter, R. S., Clark, R. S., Wiest, L., and Erwin, B. G., 1986, Acetylation of decarboxylated S-adenosylmethionine by mammalian cells, Biochemistry, 25:379-384.
Pösö, H., and Pegg, A. E., 1982, Comparison of S-adenosylmethionine decarboxylases from rat liver and muscle, Biochemistry, 21:3116-3122.
Razin, A., and Cedar, H., 1984, DNA methylation in eukaryotic cells, Int. Rev. Cytol., 92:159-185.
Razin, A., and Riggs, A. D., 1980, DNA methylation and gene function, Science, 210:604-610.
Schindler, J., Kelly, M., and McCann, P. P., 1983, Inhibition of ornithine decarboxylase induces embryonal carcinoma cell differentiation, Biochem. Biophys. Res. Commun., 114:410-417.
Smith, S. S., Hardy, T. A., and Baker, D. J., 1987, Human DNA (cytosine-5)methyltransferase selectively methylates duplex DNA containing mispairs, Nucleic Acids Res., 15:6899-6916.
Young, P. R., and Tilghman, S. M., 1984, Induction of α-fetoprotein synthesis in differentiating F9 teratocarcinoma cells is accompanied by a genome-wide loss of DNA methylation, Mol. Cell. Biol., 4:898-907.
Zappia, V., Zydek-Cwick, C. R., and Schlenk, F., 1969, The specificity of S-adenosylmethionine derivatives in methyl transfer reactions, J. Biol. Chem., 244:4499-4509.
Zappia, V., Galletti, P., Oliva, A., and Porcelli, M., 1983, Methods for the preparation and assay of S-adenosyl-(5')-3-methylthiopropylamine (decarboxylated adenosylmethionine, Meth. Enzymol., 94:73-80.
Zucker, K. E., Riggs, A. D., and Smith, S. S., 1985, Purification of human DNA(cytosine-5-)-methyltransferase, J. Cell. Biochem., 29:337-349.

TWO PHASES OF ORNITHINE DECARBOXYLASE ACTIVATION DURING LYMPHOCYTE MITOGENESIS

Tomas Mustelin[i)], Tiina Pessa[i)], Seppo Lapinjoki[ii)], Jukka Gynther[ii)], Tomi Järvinen[ii)], Terho Eloranta[iii)] and Leif C. Andersson[i)]

[i)]Department of Pathology, University of Helsinki, Finland
[ii)]Department of Pharmaceutical Chemistry and [iii)]Department of Biochemistry, University of Kuopio, Finland

Present address of corresponding author: Tomas Mustelin, Scripps Clinic and Research Institute, 10666 North Torrey Pines Road, La Jolla, CA 92037, USA

INTRODUCTION

Ornithine decarboxylase (ODC) has been implicated in the regulation of cell proliferation (Jänne et al., 1978). ODC is the rate-limiting enzyme in the cellular synthesis of the polyamines (Pegg 1986), which are organic cations required for many growth-related cellular functions (Hölttä et al., 1979, Fillingame et al., 1975). The activity of ODC correlates with the growth rate of cells. Highest activities are usually seen in late G_1 phase of the cell cycle and there is a second smaller peak in G_2. Malignant transformation is often accompanied by impaired regulation of ODC (Haddox et al., 1980, Sistonen et al., 1987).

We have studied the induction of ODC activity during stimulation of human lymphocytes. This paper summarizes the current situation of our work.

MATERIALS AND METHODS

Materials

A rabbit antiserum against mouse kidney ODC was kindly provided for by Dr. A. Pajunen (Oulu, Finland). A sample of purified ODC was a kind gift from Dr. L. Persson (Sweden).

Cells

Lymphocytes were isolated from buffy coats of blood units kindly provided for by the Finish Red Cross Blood Transfusion Service, by centrifugation on Ficoll-Isopaque (Pharmacia, Sweden).

The human leukaemia T cell line Jurkat was kept at logarithmic growth in RPMI 1640 medium supplemented with 10% fetal calf serum. The cells were also grown in the presence of rising concentrations of difluoromethylornithine (DFMO).

ODC Assay

Lymphocytes were incubated in RPMI 1640 medium supplemented with 5% serum at 37°C at about 2 x 10^6 cells/ml. The activity of ODC, measured as the liberation of $^{14}CO_2$ from 1-(^{14}C)-ornithine, was determined from supernatants of samples prepared by sonication (Branson Sonifier tip sonicator, 3 x 5sec) and centrifugation at 15 000 x g, 4°C for 30min. The assay mixture contained 50mM Tris/HCl, pH 7.5, 0.1mM EDTA, 40μM pyridoxalphosphate, 78μM L-(^{14}C)-ornithine (Amersham, specific activity 58.6mCi/mmol) and 205μl of supernatant (about 0.2mg protein) in a total volume of 0.25ml. Blancs were prepared by incubating supernatant at 37°C in the presence of 10mM DFMO for 30min. Protein was measured by the Bio Rad protein assay.

Measurement of Polyamines

The amount of sulphosalicylic acid-soluble polyamines were measured by cation exchange chromatography as described earler (Hyvönen et al., 1986).

Measurement of DNA content

The amount of DNA was measured by staining the cells with propidium iodide and analysing the fluorescence in a FACS IV cell sorter.

Gas Liquid Chromatography (GLC) and Mass Spectrometry (MS)

Inositol was detected following acetylation by GLC-MS analysis as described (Mustelin et al., 1987). Deuterated acetylinositol was prepared by the same acetylation procedure using deuterated acetic anhydride.

Immunoprecipitation and Slab Gel Electrophoresis

ODC was immunoprecipitated from ^{3}H-inositol labeled cells and the precipitate analyzed by sodium dodecylsulphate-polyacrylamide gel electrophoresis as described (Mustelin et al., 1987).

RESULTS AND DISCUSSION

Activation of ODC in Lymphocytes

The activity of ODC in resting human lymphocytes was very low, 1.58 $\pm$ 0.60 pmoles $^{14}CO_2$/hour/mg protein (n=15). When these cells were treated with the mitogenic lectin concanavalin A (ConA, 10μg/ml), their ODC activity increased to 5.37 $\pm$ 1.40 pmoles $^{14}CO_2$/hour/mg protein (n=11) (Fig.1). This increase is significant, $p<0.001$ (Students t-test).

These results are in agreement with those of Scott and coworkers (1984, 1985), who showed that mitogenic (but not non-mitogenic) ligands to T cells activate ODC within minutes. They also demonstrated that the rapid activation is insensitive to cycloheximide, but is blocked by inhibitors of cellular energy metabolism and drugs (like cytochalasin D and vinblastine sulphate) which disrupt the cytoskeleton.

The initial rapid increase in ODC activity was followed by a much larger increase starting about 3-4 hours after the addition of ConA, reaching 500-600 pmoles $^{14}CO_2$/hour/mg protein at about 12hours (Fig.2). The magnitude of this rise varied between lymphocytes from different blood donors.

The large increase in ODC activity starting after a lag of several hours is similar to that seen by several other investigators in mitogen-stimulated lymphocytes (Kay and Lindsay, 1973, Korpela et al., 1981, Fidelus et al., 1984). This increase is dependent on de novo synthesis of the enzyme.

The cellular concentration of the polyamines putrescine, spermidine, spermine and N^1-acetylspermidine did not change during the first 30min of lymphocyte activation (Fig. 3). Since the basal ODC activity in lymphocytes is very low, a fourfold increase in ODC activity is not expected to cause measurable changes in total cellular polyamine content. It is possible, however, that the rapid activation of ODC in mitogen-stimulated lymphocytes causes local fluctuations in putrescine concentration, which could be of importance.

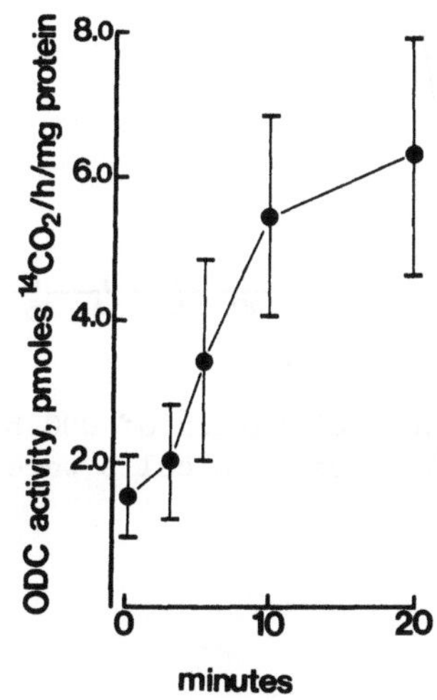

Fig.1. Rapid activation of ODC by ConA. The results are means ± SD from at least three experiments.

Role of G-proteins in ODC Induction

G-proteins are small proteins on the cytoplasmic face of the plasma membrane, involved in signal transduction. They functionally couple surface receptors to their intracellular effector systems. In many cells $AlCl_3$ and NaF act synergistically to activate G-proteins (Sternweis and Gilman, 1982, Blackmore et al., 1985, Cockcroft and Taylor, 1987). The synergism between these salts is thought to arise from the formation of an aluminium tetrafluoride complex, which binds to the nucleotide binding domain of the G-protein. Together with the bound GDP, it forms a GDP-AlF_4 complex, which apparently resembles GTP and activates the G-protein.

Treatment of lymphocytes with $AlCl_3$ (20μM) together with NaF (optimally 10mM) caused a transient about threefold activation of ODC (Fig. 4). Maximum stimulation by AlF_4 was seen at 10-15min and then the activity declined. No stimulation was obtained when chloride, bromide or iodide salts were used instead of fluoride. These results indicate that a G-protein might be involved in the regulation of the rapid activation of ODC by mitogens in human lymphocytes.

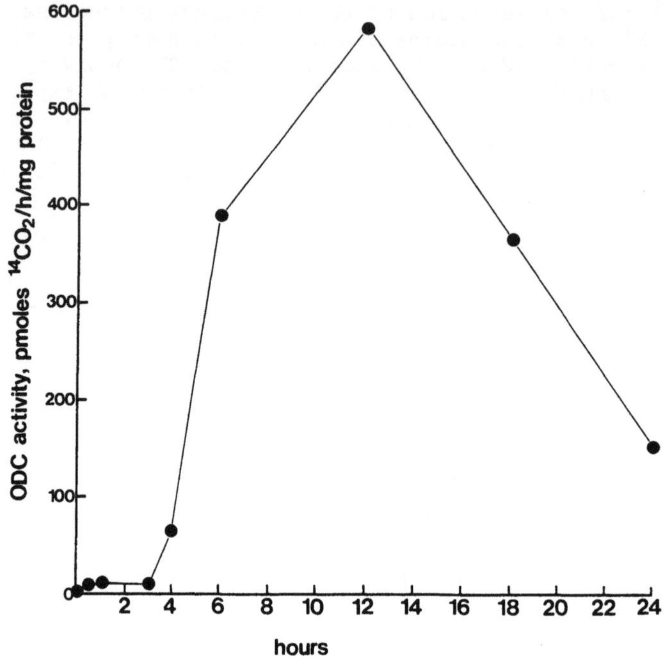

Fig. 2. Time course for the activation of ODC by ConA in human lymphocytes during the first 24 hours. The cells were cultured at 2 x 10^6/ml.

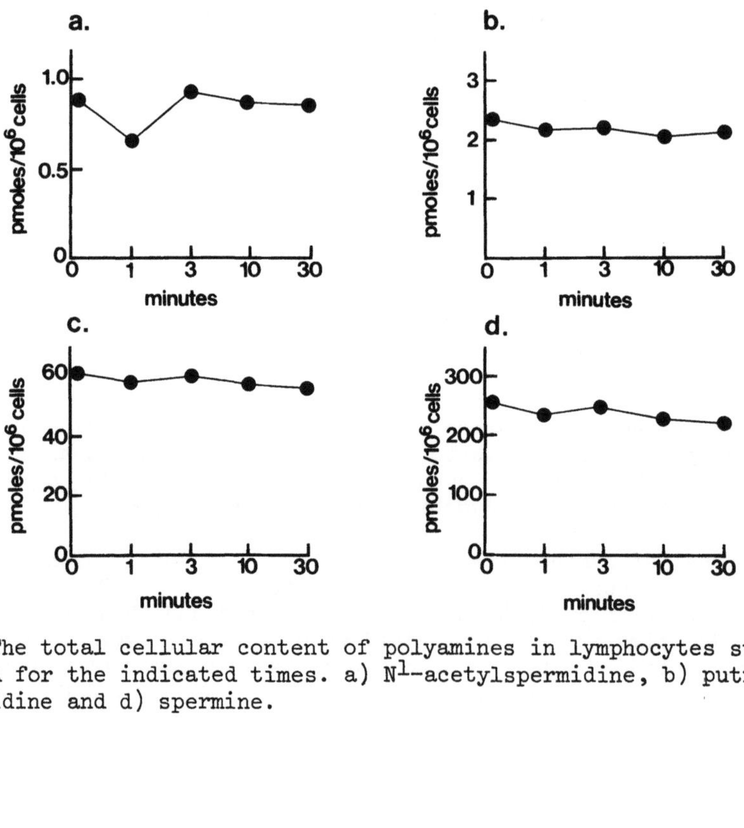

Fig. 3. The total cellular content of polyamines in lymphocytes stimulated with ConA for the indicated times. a) N^1-acetylspermidine, b) putrescine, c) spermidine and d) spermine.

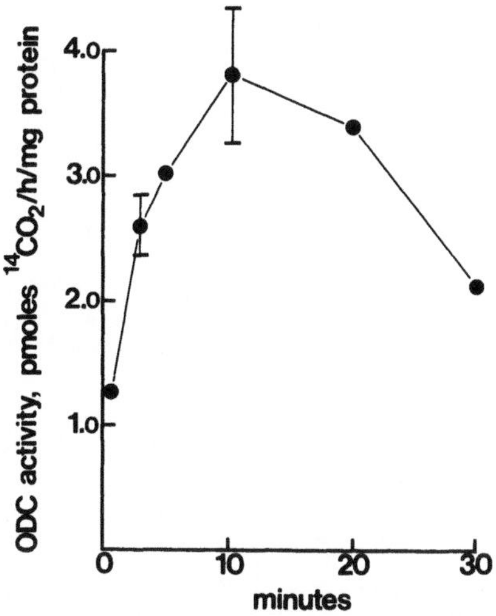

Fig. 4. The effect of $AlCl_3$ (20μM) and NaF (10mM) on ODC activity. Bars indicate SD from three separate experiments.

The stimulatory effect of AlF_4 is in agreement with our earlier finding that non-hydrolyzable analogs of GTP activate ODC within 10min in T lymphocytes (Mustelin et al., 1986a). We have also demonstrated that treatment of lymphocytes with mycophenolic acid, which depletes the cellular stores of guanine nucleotides, blocks the rapid induction of ODC activity by mitogens. The inhibition is reverted by addition of guanine to the medium 10min prior to addition of mitogen (Mustelin 1987).

Treatment of T lymphocyte membranes with cholera toxin and ^{32}P-NAD causes incorporation of label into a 45kD protein (Pessa et al., in press). This protein is apparently identical to G_s, the G-protein which confers the functional (stimulatory) coupling between receptors and adenylate cyclase (Gilman, 1984). When whole lymphocytes were treated for 4h with 5μg/ml of cholera toxin, the subsequent mitogen-induced ODC activation was intact (Fig. 5A). Under these conditions the susceptible G-protein was fully ADP-ribosylated, as indicated by the inability of cholera toxin to incorporate radioactivity from ^{32}P-NAD into membranes isolated from toxin pretreated cells (Fig. 5B). Cholera toxin alone caused a slight increase in basal ODC activity at 6 and 18h.

Under similar conditions pertussis toxin causes ADP-ribosylation of a 41kD G-protein (Pessa et al., in press). In lymphocytes treated for 4h with 0.5μg/ml of pertussis toxin, an intact activation of ODC was obtained following addition of ConA (Fig. 6A). In the isolated membranes from these toxin pretreated cells no further ADP-ribosylation by pertussis toxin could be detected (Fig. 6B), indicating, again, that the substrate was already fully ADP-ribosylated.

We have detected the presence of several GTP-binding proteins in T lymphocyte plasma membranes, using a photoaffinity labeling technique (Pessa et al., in press). One of these could be the G-protein involved in

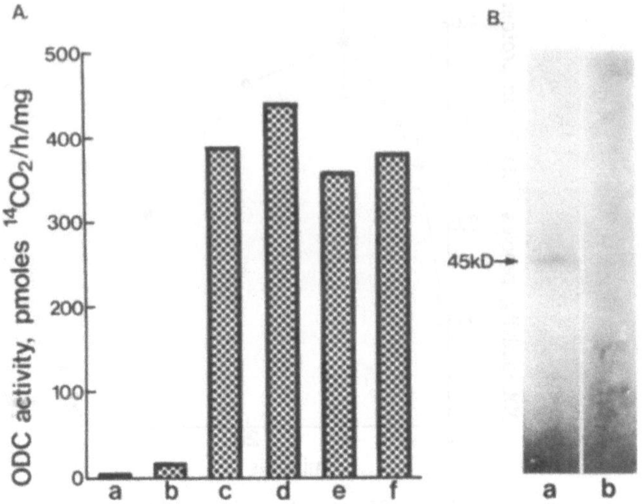

Fig. 5A. Effects of cholera toxin on ODC activation in lymphocytes. a) untreated cells, b) cells treated with cholera toxin alone (18h), c) cells stimulated with ConA for 6h, d) cholera toxin pretreated cells stimulated with ConA, e) cells stimulated with ConA for 18h, f) cholera toxin pretreated cells stimulated with ConA for 18h. B. SDS PAGE of lymphocyte membranes from a) untreated b) cholera toxin pretreated cells, incubated with cholera toxin and ^{32}P-NAD for 15min.

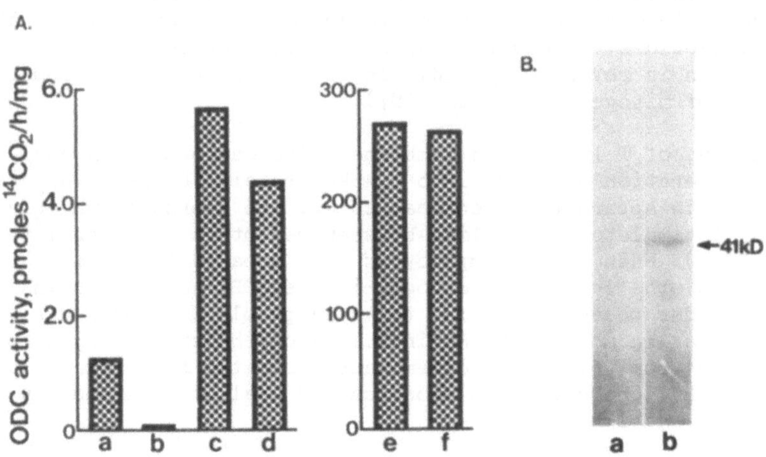

Fig. 6A. Effects of pertussis toxin on ODC activation in lymphocytes. a) untreated cells, b) cells treated with pertussis toxin alone, c) cells treated with ConA for 10min, d) pertussis toxin pretreated cells stimulated with ConA for 10min, e) cells stimulated with ConA for 6h, f) pertussis toxin pretreated cells stimulated with ConA for 6h.
B. SDS page of lymphocyte membranes from a) pertussis toxin pretreated and b) untreated cells, incubated with pertussis toxin and ^{32}P-NAD for 15min.

the rapid activation of ODC by mitogens in human T lymphocytes. The candidates include proteins of relative molecular weights 22kD, 26kD and 31kD. The 22kD protein might belong to the ras proto-oncogene family.

Protein Kinase C is Involved Only In the Second Phase of ODC Activation

Addition of tetradecanoyl phorbol acetate (TPA, 0.1μM) to human lymphocytes did not change their ODC activity within the first 2-3 hours. At 6h there was a severalfold increase in ODC activity (Fig. 7). Since phorbol esters activate a phospholipid and calcium dependent protein kinase (protein kinase C), the lack of effects of TPA on ODC activity during the first hours, indicates that this protein kinase is not involved in the regulation of the rapid protein synthesis-independent activation of ODC in lymphocytes. Protein kinase C might, however, be involved in the second phase of ODC activation, as also suggested by Sasakawa and coworkers (1985).

Effects of Li^+ and inositol on the inducibility of ODC

Treatment of human lymphocytes with 1mM Li^+ for 30min in an inositol-free medium inhibited the rapid increase in ODC activity seen within 10min after the addition of ConA. Addition of 1mM inositol to the medium restored the inducibility of ODC within 10min. Addition of inositol alone did not cause any significant change in ODC activity, but slightly potentiated the effects of ConA (Table I). These results are in agreement with our earlier findings (Mustelin et al., 1986b).

Table I. Effects of Li^+ and inositol on ODC activation

	ODC activity, pmoles $^{14}CO_2$/h/mg
untreated cells	1.58 ± 0.60 (n=15)
cells treated with:	
ConA (10μg/ml, 10min)	5.37 ± 1.40 (n=11)
Li^+ (1mM, 30min)	1.31 ± 0.32 (n=5)
Li^+ (1mM, 30min) + ConA	1.68 ± 0.50 (n=7)
Li^+ (1mM, 30min) + Ino (1mM, 10min) + ConA	5.04 ± 0.70 (n=7)
Ino (1mM, 10min)	1.62 ± 0.45 (n=2)
Ino (1mM, 10min) + ConA	5.56 ± 0.68 (n=2)

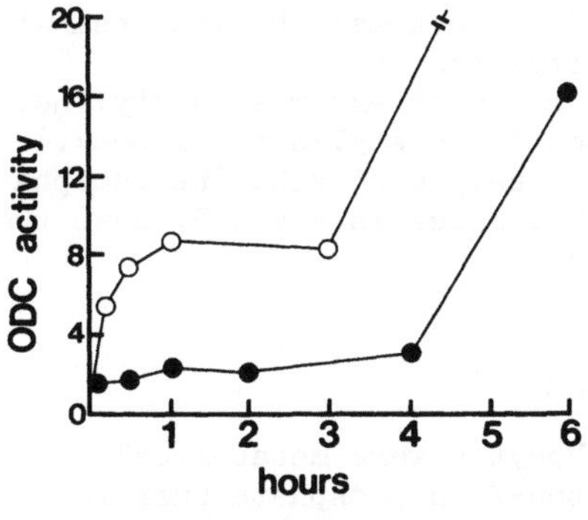

Fig. 7. Effects of TPA (filled symbols) and ConA (open symbols) on ODC activity in human lymphocytes. The activity of ODC is given in pmoles $^{14}CO_2$/h/mg protein.

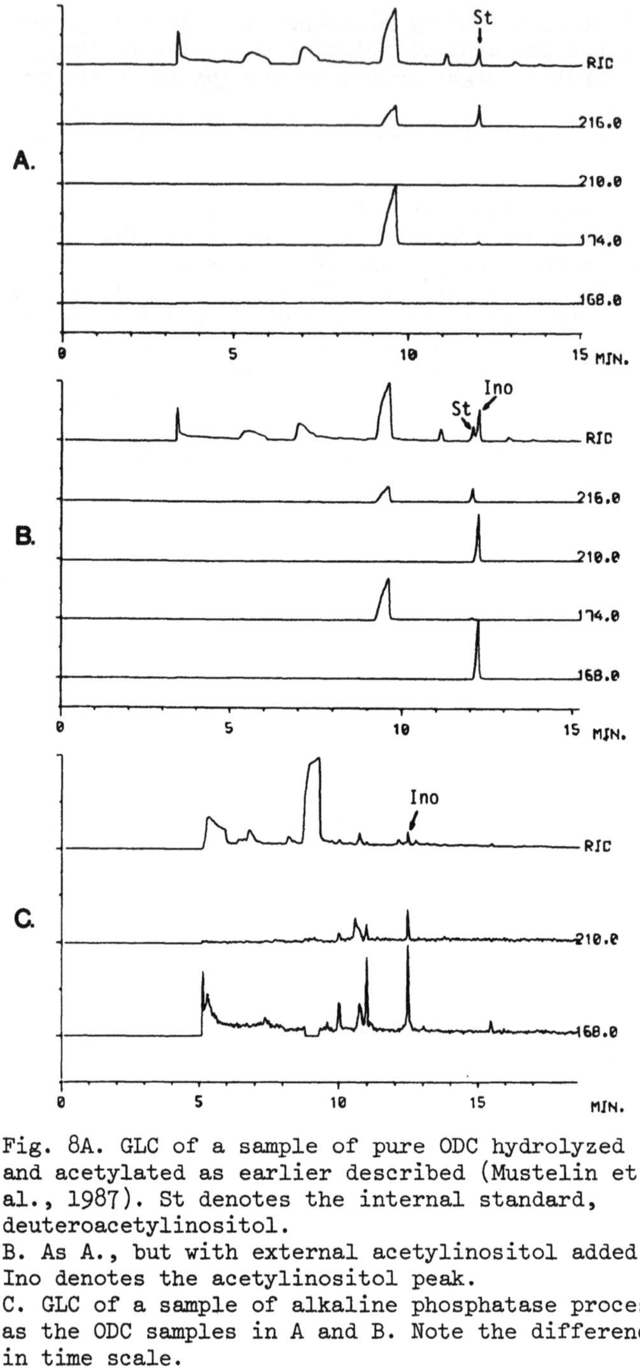

Fig. 8A. GLC of a sample of pure ODC hydrolyzed and acetylated as earlier described (Mustelin et al., 1987). St denotes the internal standard, deuteroacetylinositol.
B. As A., but with external acetylinositol added. Ino denotes the acetylinositol peak.
C. GLC of a sample of alkaline phosphatase processed as the ODC samples in A and B. Note the differences in time scale.

ODC does not contain inositol

Mouse spleen lymphocytes were metabolically labeled with ^{3}H-inositol, lysates made in 1% Triton-X in phosphate buffered saline, ODC immunoprecipitated using a rabbit ODC antiserum and the precipitate analyzed on SDS PAGE as described earlier (Mustelin et al., 1987). No labeled band became apparent on autoradiography of the gels even after extended exposure times (data not shown).

When 50μg of ODC purified to homogeneity (kindly provided for by Dr L. Persson) was hydrolyzed, acetylated and analyzed by gas liquid chromatography-mass spectrometry (GLC-MS) for the presence of covalently bound inositol as earlier described (Mustelin et al., 1987), no indication of its presence was seen either in the total ion graph or in the ions selected for monitoring (m/e 168 and m/e 210). When deuterated acetylinositol was included as an internal standard, it was readily detected at the expected point (Fig. 8A). The area surrounding this peak was analyzed by mass spectrometry using the highest sensitivity of the instrument, but no sign of acetylinositol was found (data not shown). To further confirm the result, some external acetylinositol was added. This gave a peak at the expected area, immediately following that of deuteroacetylinositol (Fig. 8B). In conclusion, no sign of inositol was found in mouse kidney ODC purified to homogeneity.

When alkaline phosphatase, known to contain inositol (Low and Zilversmit, 1980), was processed under identical conditions inositol (as acetyl derivative) was readily detected (Fig. 8C).

Effects of DFMO on the Lymphocytic Leukaemia Line Jurkat

When the human leukaemia T cell line Jurkat was adapted to grow in the presence of difluoromethylornithine (DFMO) by gradually rising the concentration up to 10mM, changes in their morphology occured. There was an accumulation of giant cells many of which were multinucleated (Fig. 9). FACS analysis of their DNA content demonstrated an abundance of polyploidic cells after treatment with 5mM and 10mM DFMO (Fig. 10).

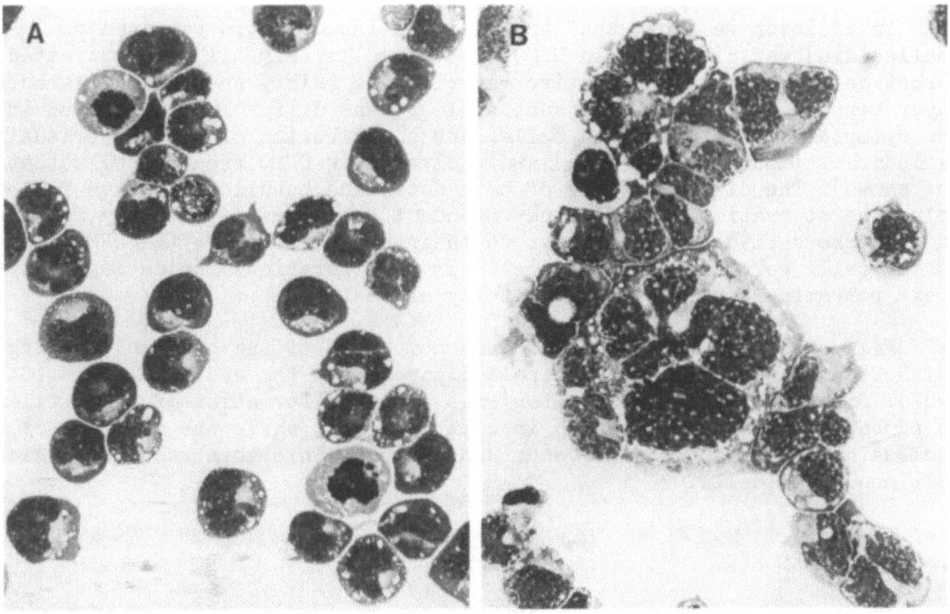

Fig. 9. A. Jurkat cells grown in normal RPMI 1640 medium supplemented with 10% fetal calf serum. B. Jurkat cells adapted to proliferate in the same medium supplemeted with 10mM of DFMO. The cells were stained with May-Grünwald-Giemsa staining. A and B are photographed with the same magnification.

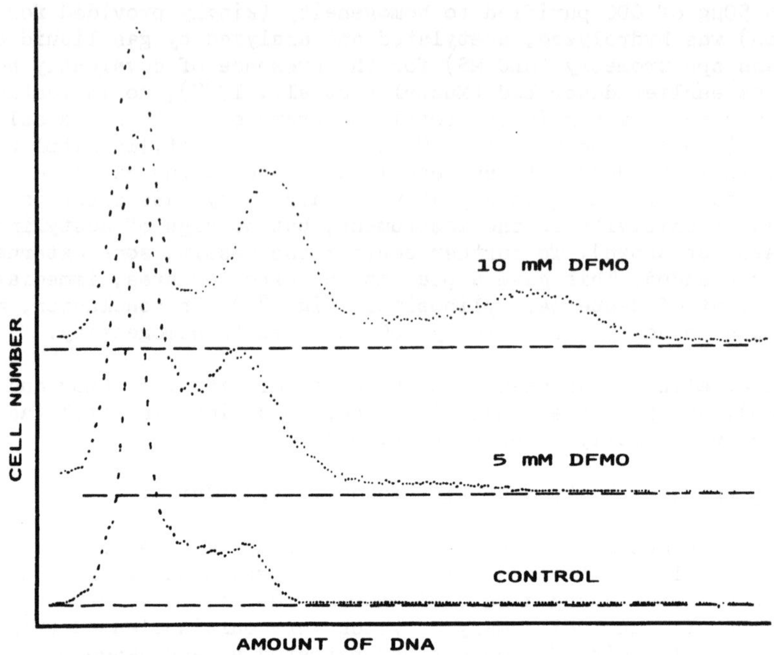

Fig. 10. The cellular DNA content of Jurkat cells adapted to grow in the presence of DFMO (5mM and 10mM). The amount of DNA is given in arbitrary fluorescence units. Control cells were grown in absence of DFMO.

In addition we found that the distribution of actin (stained by phallacidin) was disturbed in DFMO treated cells (Fig. 11). In untreated Jurkat cells phallacidin reactive material was mainly seen as a filamentous layer beneath the plasma membrane, while it was diffusely distributed in the cytoplasm of DFMO treated cells. The distribution of tubulin (visualized by indirect immunofluorescence) was unaltered in DFMO treated cells (data not shown). The dissappearance of both actin and tubulin filaments following polyamine starvation of polyamine-dependent cells was reported by Pohjanpelto and coworkers (1981). In contrast to their cells, however, the DFMO adapted Jurkat cells were able to continue their proliferation, indicating that their polyamine stores were not totally exhausted.

Filamentous actin is an important component of the contractile ring, which forms during cytokinesis (cell cleavage) at the end of mitosis (Geiger, 1983). Since DFMO apparently interferes with the formation of actin filaments it probably causes disturbances in cell division, while the division of the nucleus during telophasis proceeds undisturbed. This then results in giant, multinucleated cells.

CONCLUDING REMARKS

In conclusion, there is a significant, rapid activation of ODC during the first minutes of lymphocyte mitogenesis. This activation is apparently independent of active protein synthesis, indicating regulation of ODC activity at a posttranslational level. It is not yet known if this also operates at later stages of lymphocyte mitogenesis. The significance of the rapid ODC activation is not clear; it does not result in measurable changes in total cellular polyamine content.

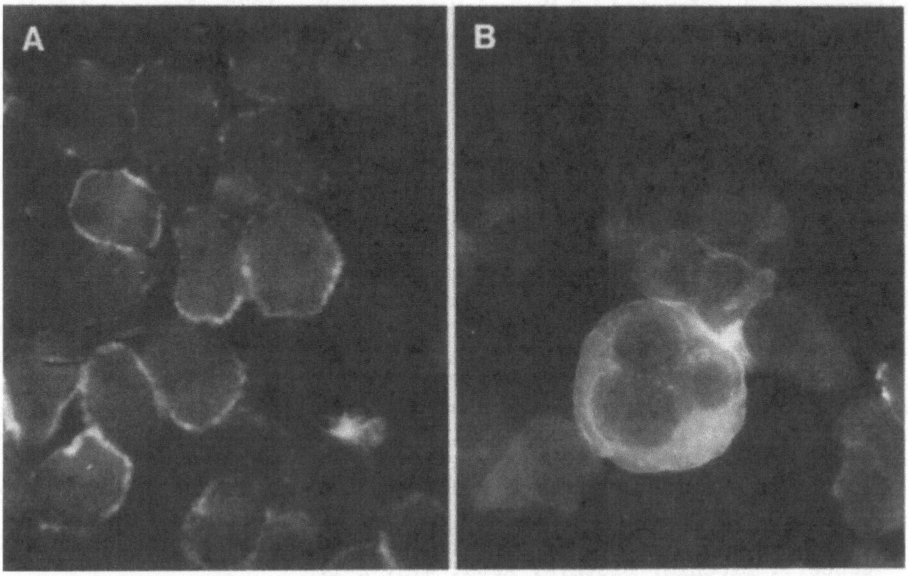

Fig. 11. The organization of actin in A. normal Jurkat cells, B. DFMO adapted Jurkat cells. Actin was visualized by phallacidin and vieved under a fluorescence microscope.

The regulation of the rapid activation of ODC in mitogen-stimulated human lymphocytes seems to involve a G-protein(s). This G-protein is neither a cholera toxin nor a pertussis toxin substrate. There are several candidates for the G-protein involved in mitogen-induced ODC activation, including ras proto-oncogene encoded p21. The metabolism of inositol plays a role for the early activation of ODC, but ODC itself does not contain inositol. Treatment of lymphocytes with DFMO disrupts the normal organization of actin and, apparently as a result, leads to a dissociation between nuclear and cell body division.

ACKNOWLEDGEMENTS

We wish to thank Dr Lo Persson for the kind gift of pure ODC and Dr Antti Pajunen for the ODC antiserum. Ms Hannele Laaksonen and Ms Helena Nousiainen are thanked for technical assistance. Mr Björn Lindroos is thanked for photographical help.

REFERENCES

Blackmore, P.F., Bocckino, S.B., Waynick, L.E. and Exton, J.H., 1985, Role of a guanine nucleotide-binding regulatory protein in the hydrolysis of hepatocyte phosphatidylinositol-4,5-bisphosphate by calcium-mobilizing hormones and the control of cell calcium., J. Biol. Chem., 260: 14477

Cockcroft, S. and Taylor, J.A., 1987, Fluoroaluminates mimic guanosine 5'-(γ-thio)triphosphate in activating the polyphosphoinositide phosphodiesterase of hepatocyte membranes., Biochem. J., 241: 409.

Fidelus, R.K., Laughter, A.H. and Twomey, J.J., 1984, The role of mitogens and lymphokines in the induction of ornithine decarboxylase (ODC) in T lymphocytes. J. Immunol., 132: 1462.

Fillingame, R.H., Jorstad, C.M. and Morris, D.R., 1975, Increased cellular levels of spermidine or spermine are rquired for optimal DNA synthesis in lymphocytes activated by concanavalin A., Proc. Natl. Acad. Sci. USA, 72:4042.

Geiger, B., 1983, Membrane-cytoskeleton interaction., Biochim. Biophys. Acta, 737:305.

Gilman, A.G., 1984, G-proteins and dual control of adenylate cyclase., Cell, 36:577

Haddox, M.K., Magun, B.E. and Russell, D.H., 1980, Ornithine decarboxylase induction during G1 progression of normal and Rous Sarcoma Virus-transformed cells., Cancer Res., 40: 604.

Hölttä, E., Jänne, J. and Hovi, T., 1979, Suppression of the formation of polysomes and macromolecules by DL-α-difluoromethylornithine and methylglyoxal bis (guanylhydrazone) in phytohaemagglutinin-activated human lymphocytes., Biochem. J., 178: 109.

Hyvönen, T., Alakuijala, L., Karvonen, E., Pösö, H., Andersson, L.C. and Eloranta, T.O., 1-aminooxy-3-aminopropane reversibely prevents the proliferation of cultured baby hamster cells and the accumulation and excretion of putrescine, spermidine and acetylated polyamines. Biochem. J., in press.

Jänne, J., Pösö, H. and Raina, A., 1978, Polyamines in rapid growth and cancer., Biochim. Biophys. Acta, 473: 241.

Kay, J.E. and Lindsay, V.J., 1973, Polyamine synthesis during lymphocyte activation. Exp. Cell Res., 77: 428.

Korpela, H., Hölttä, E., Hovi, T. and Jänne, J., 1981, Response of enzymes involved in the metabolism of polyamines to phytohaemagglutinin-induced activation of human lymphocytes. Biochem. J., 196: 733.

Low, M.G. and Zilversmit, D.B., 1980, Role of phosphatidylinositol in attachment of alkaline phosphatase to membranes. Biochemistry, 19: 3913.

Mustelin, T., Pösö, H. and Andersson, L.C., 1986a, Role of G-proteins in T cell activation: non-hydrolysable GTP analogues induce early ornithine decarboxylase activity in human T lymphocytes. EMBO J., 5: 3287.

Mustelin, T., Pösö, H., Iivanainen, A. and Andersson, L.C., 1986b, myo-inositol reverses Li^+-induced inhibition of phosphoinositide turnover and ornithine decarboxylase induction during early lymphocyte activation., Eur. J. Immunol., 16: 859.

Mustelin, T., 1987, GTP-dependence of the transduction of mitogenic signals through the T3 complex in T lymphocytes, indicates the involvement of a G-protein. FEBS Letters, 213: 199.

Mustelin, T., Pösö, H., Lapinjoki, S.P., Gynther, J. and Andersson, L.C., 1987, Growth signal transduciton: Rapid activation of covalently bound ornithine decarboxylase during phosphatidylinositol breakdown., Cell, 49: 171.

Pegg, A.E., 1986, Recent advances in the biochemistry of polyamines in eukaryotes. Biochem. J., 234: 249.

Pessa, T., Mustelin, T. and Andersson, L,C., Identification of GTP-binding proteins in T lymphocyte membranes by photoaffinity labeling with ^{32}P-GTP. Cellular Basis of Immune Modulation, in press.

Pohjanpelto, P., Virtanen, I. and Hölttä, E., 1981, Polyamine starvation causes dissappearance of actin filaments and microtubules in polyamine-auxotropic CHO cells. Nature, 293: 475

Sasakawa, N., Ishii, K., Yamamoto, S. and Kato, R., 1985, Induction of ornithine decarboxylase activity by 1-oleoyl-2-acetyl-glycerol in isolated mouse epidermal cells., Biochem. Biophys. Res. Commun., 128: 913.

Scott, I.G., Pösö, H., Åkerman, K.E.O. and Andersson, L.C., 1984, Mitogens cause a rapid induction of ornithine decarboxylase activity in human T-lymphocytes., Biochem. Soc. Trans., 13: 934.

Scott, I.G., Pösö, H., Åkerman, K.E.O. and Andersson, L.C., 1985, Rapid activation of ornithine decarboxylase by mitogenic (but not non-mitogenic) ligands to human T lymphocytes. Eur. J. Immunol., 15: 783.

Sistonen, L., Keski-Oja, J., Ulmanen, I,. Hölttä, E., Wikgren, B-J.P. and Alitalo, K., 1987, Dose effects of transfected c-Ha-*ras*$^{Val\ 12}$ oncogene in transformed cell clones. Exp. Cell Res., 168: 518.

Sternweis, P.C. and Gilman, A.G., 1982, Aluminium: A requirement for activation of the regulatory component of adenylate cyclase by fluoride., Proc. Natl. Acad. Sci. USA, 77: 6516.

REGULATION OF PROTEIN SYNTHESIS BY POLYAMINES

Kazuei Igarashi, Kiyoshi Ito, Yuko Sakai,
Toshihiro Ogasawara, and Keiko Kashiwagi

Faculty of Pharmaceutical Sciences, Inohana Campus
Chiba University, 1-8-1, Inohana, Chiba 280, Japan

INTRODUCTION

Polyamines have been implicated in numerous growth processes (1,2). Since polyamines are cationic, they probably participate in many cellular processes through their binding to DNA, RNA, and phospholipids (3). The stimulatory influence of polyamines on some kinds of protein synthesis has been established from *in vitro* experiments with both *Escherichia coli* and eukaryotic cell-free systems (4-7) and from *in vivo* experiments with *E. coli* polyamine-requiring mutants and bovine polyamine-deficient lymphocytes (8-11). Findings from these latter studies indicate that an important *in vivo* function of polyamines is to stimulate the synthesis of specific proteins, such as the polyamine-induced protein of *E. coli* (9) and thymidine kinase of bovine lymphocytes (11). These proteins are formed in limited amounts in polyamine-deficient cells.

To clarify the above situation that polyamines play an important role in protein synthesis, we have extended our studies to the followings. Since among macromolecular synthesis protein synthesis is influenced first by polyamines (11-13), we have first attempted to determine what kinds of proteins involved in protein synthesis in *E. coli* are preferentially synthesized in the presence of polyamines. For this objective, three polyamine-requiring mutants were used and the proteins synthesized were analyzed by fluorography of [^{3}H]lysine-labeled proteins precipitated by specific immunoglobulins with the aids of *Staphylococcus aureus* Cowan I cells. Then, we have tried to elucidate the molecular mechanism of polyamine stimulation of the synthesis of thymidine kinase in bovine lymphocytes. We have measured the level of thymidine kinase mRNA by Northern blot hybridization, the rate of transcription of thymidine kinase mRNA in isolated nuclei, the turnover rate of this kinase mRNA, and the mRNA distribution in polysomes in normal and polyamine-deficient cells. Finally, the polyamine effect on globin synthesis was reexamined in a polyamine-free protein synthetic system. The system consisted of globin mRNA, salt-washed ribosomes, partially purified initiation factors, and pH 5 enzymes. The results show that polyamines play an important role for cell growth through their stimulation of the synthesis of specific proteins.

POLYAMINE STIMULATION OF THE SYNTHESIS OF SPECIFIC PROTEINS IN ESCHERICHIA COLI

Since polyamines are thought to be able to enhance cell growth, at least partially, by stimulating the synthesis of some proteins, we have attempted to determine the kinds of proteins preferentially synthesized in the presence of polyamines. The investigation was limited to some of the proteins involved in protein synthesis since among macromolecular synthesis protein synthesis is influenced first by polyamines (11-13). Three polyamine-requiring mutants of *E. coli* (MA261, KK101, and HT283), differing in their generation time, were used for the experiments to generalize the polyamine effect. The increase in growth rate by polyamines in *E. coli* MA261, KK101, and HT283 was 2.0-, 3.7-, and 2.3-fold, respectively under our experimental conditions (Table 1).

The method used for the identification was fluorography of proteins labeled with [^{3}H]lysine. The proteins in the cells were precipitated by immunoglobulin specific for a purified protein with the aid of *S. aureus* Cowan I cells. By two dimensional gel electrophoresis, we have shown previously that polyamine induced (PI) protein is preferentially synthesized in MA261 (9). Thus, we initially investigated whether the synthesis of this protein is stimulated by polyamines in the three mutants. Through the technique of fluorography, polyamines were shown to stimulate PI protein synthesis 2.2- to 4.4-fold in the three mutants (Table 2). The degree of polyamine stimulation in MA261 was 3.8-fold. This value is very similar to that obtained by fluorography of total proteins labeled with [^{14}C]lysine (10).

The synthesis of ribosomal protein S1 was also stimulated significantly by polyamines (Table 2). The degree of its stimulation by polyamines in MA261, KK101, and HT283 was 3.0-, 4.4-, and 2.5-fold, respectively. Analyses by fluorography were performed using the same amounts of radioactivity. At A_{540}=0.2, the stimulation of total protein synthesis by the addition of putrescine was 1.3- to 1.8-fold. Therefore, the actual stimulation of the *in vivo* synthesis of PI protein and ribosomal protein S1 was even greater than the value shown in Table 2.

The synthesis of other ribosomal proteins (S4, L20 and X1) was stimulated slightly (1.2- to 1.6-fold) by putrescine. The synthesis of RNA polymerase was also stimulated slightly by putrescine (Table 2). The degree of stimulation of ßß' subunit synthesis was 1.3- to 1.4-fold, a lower value than the putrescine stimulation of ßß' subunit synthesis obtained by fluorography of total proteins labeled with [^{14}C] lysine (10). Although polyamines can stimulate protein synthesis at the level of initiation (14), the

Table 1. Growth rates and polyamine contents of polyamine-requiring mutants

Strain	Amount of putrescine in medium (μg/ml)	Doubling[a] time (min)	Amount of polyamine (nmol/mg protein)	
			Putrescine	Spermidine
MA261	0	260	0	1.6
	100	128	97	11
KK101	0	496	0	2.4
	100	135	85	10
HT283	0	275	0	0
	100	121	95	0

[a] Doubling time was calculated from the growth curve measured at A_{540}. E. coli MA261 (speA speB speC serA thr leu thi) and HT283 (speA speB speC speD thr pro thi) were kindly provided by Dr. W. K. Maas and H. Tabor, respectively. E. coli KK101 was obtained by nitrosoguanidine treatment of MA261. The detailed procedure will be published elsewhere. For the growth of MA261 and KK101, medium A, containing 4 g of glucose, 7 g of K_2HPO_4, 500 mg of sodium citrate, 1 g of $(NH_4)_2SO_4$, 100 mg of $MgSO_4 \cdot 7H_2O$, 2 mg of thiamin, 10 mg of biotin, and 100 mg each of leucine, threonine, methionine, serine, glycine, and ornithine per liter of water, was used. For the growth of HT283, medium B, containing 4 g of glucose, 6 g of Na_2HPO_4, 3 g of KH_2PO_4, 500 mg of NaCl, 1 g of NH_4Cl, 0.25 g of $MgSO_4 \cdot 7H_2O$, 13 mg of $CaCl_2$, 2 mg of thiamin, and 100 mg each of proline and threonine per liter of water, was used. For polyamine starvation of cells, a culture growing exponentially in medium containing putrescine (100 μg/ml) was diluted in fresh medium lacking putrescine and allowed to grow for 24 h at 37 °C (6 to 7 generations). The resulting polyamine-depleted bacteria were then cultivated in the absence or the presence of 100 μg/ml putrescine. When growth was sufficient to give an A_{540} of 0.2, the cells were harvested by centrifugation at 15,000 x g for 15 min. The cells were washed once with Buffer A (10 mM Tris-HCl, pH 7.5, 10 mM magnesium acetate, 60 mM NH_4Cl, and 6 mM 2-mercaptoethanol), centrifugated as described above, and stored at -80 °C until used. Polyamine levels in E. coli were determined by high performance liquid chromatography as described previously (31). Protein was determined by the method of Lowry et al. (32).

synthesis of initiation factors 2 and 3 is not influenced by polyamines (Table 2). The effect of polyamines on the synthesis of elongation factors was not examined because the polyamine effect on elongation level is small and the synthesis of elongation factor Tu is not influenced by polyamines in a cell-free system (15).

To obtain the more distinct conclusion on polyamine effect on *in vivo* protein synthesis, we have isolated a

Table 2. Effect of putrescine on the amount of proteins synthesized in *E. coli*

Protein	Relative intensity of band (+PUT/-PUT)		
	MA261	KK101	HT283
PI protein[a]	3.79	4.25	2.15
Ribosomal protein			
S1	3.03	4.35	2.48
S4	1.25	1.42	1.24
L20	1.25	1.57	1.21
X1	1.31	1.48	1.31
RNA polymerase			
ββ'	1.33	1.37	1.35
σ	1.15	1.29	1.12
α	1.23	1.32	1.23
Initiation factor			
IF2 α	0.94	0.90	1.03
β	1.01	1.07	1.04
IF3	1.03	1.01	0.95

[a] Polyamine induced protein. Ribosomal proteins (S1, S4, L20, and X1) were purified as described previously (17,33). The X1 protein, a new 50S ribosomal protein (33), according to amino acid sequencing appears to be one of the four new ribosomal proteins found by Wada (34,35). Purification of RNA polymerase and initiation factors (IF2 and IF3) was performed according to the procedure of Burgess and Jendrisak (36) and Hershey et al. (37), respectively. Each purified proteins showed more than 95% homogeneity, as judged by sodium dodecylsulfate(SDS)-polyacrylamide gel electrophoresis. Immunoglobulins for each purified protein were prepared as described previously (38) and partially purified from the antiserum by precipitation with 40% saturated $(NH_4)_2SO_4$. *E. coli* (5 ml aliquot) was grown in the presence or absence of 100 μg/ml putrescine. After the labeling of protein was performed with 50 μCi of [^{3}H]lysine (25 Ci/mmol) for 20 min at A_{540}=0.2, the cells were harvested and washed with Buffer A. The amounts of specified protein synthesized were determined according to the method of Phillipson et al (39). To 400,000 cpm of [^{3}H]lysine-labeled proteins in 0.5 ml of Buffer A, 250 μg of immunoglobulins for each purified protein were added. Then, 0.08 ml of 10% formalin-fixed *S. aureus* Cowan I cells was added. The precipitate was suspended in 0.04 ml of the gel sample buffer (62.5 mM Tris-HCl, pH 6.8, 2% SDS, 10% glycerol, 5% 2-mercaptoethanol, and 0.001% bromphenol blue) and boiled for 4 min. Gel electrophoresis (40) and fluorography (41) were performed with a 30 μl aliquot of the supernatant of the mixture. The relative intensity of the band on fluorography was traced by a Shimazu dual-wave length TLC scanner CS910. The relative intensity of bands was the average value of triplicate experiments.

polyamine - dependent mutant of E. coli (KK101). This attempt was successful since the polyamine effect was most clearly obtained in E. coli KK101 among three polyamine-requiring mutants. In this mutant, the properties of 30S ribosomal subunits have been modified like in another polyamine-dependent mutant. Tabor and his coworkers (16) has reported that streptomycin resistance (rpsL) produces an absolute polyamine-requirement for growth of an E. coli strain unable to synthesize polyamines.

Present experimental data show that the synthesis of ribosomal protein S1 as well as PI protein is stimulated strongly by polyamines in three polyamine-requiring mutants of E. coli. The stimulation of the synthesis of ribosomal protein S1 occurred as early as 20 min after the addition of putrescine, and this was followed by the stimulation of overall protein synthesis by polyamines. The amount of S1 protein in 30S ribosomal subunits from polyamine-sufficient cells has been reported to be much greater than that in 30S ribosomal subunits from polyamine-deficient cells (17). Probably this can be explained by the finding of preferential stimulation of S1 protein synthesis by polyamines and polyamine stimulation of the assembly of S1 protein in S1-deficient 30S particles (17,18). Since S1 protein can stimulate natural mRNA and poly(U) dependent polypeptide synthesis (17,19,20), the increase of S1 protein contents in polyamine-sufficient cells may be one of the reason that polyamine stimulate cell growth.

Since the synthesis of PI protein is also stimulated by polyamines at the early stage after addition of putrescine (9), PI protein is thought to be an important factor for polyamine stimulation of cell growth. We have reported previously that PI protein stimulated MS2 RNA directed synthesis of RNA replicase 2.2-fold in the presence of 1 mM spermidine (9). However, when distribution of PI pro-

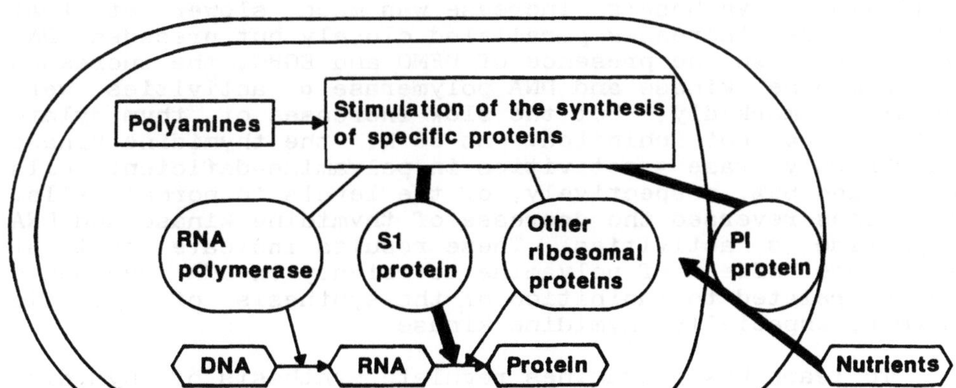

Fig. 1. Relationship between polyamine stimulation of the synthesis of specific proteins and cell proliferation in E. coli.

tein was examined according to the method of Oliver and Beckwith (21), more than 90% of PI protein was located in periplasm fraction. Thus, we infer that PI protein may play an important role for transport of nutrients. Experiments are now in progress to elucidate the physiological function of PI protein.

The stimulation of the synthsis of ribosomal proteins S4, L20, and RNA polymerase appeared at 30 min after the addition of putrescine. The increase in the amount of ribosomal proteins and RNA polymerase probably contributes to the stimulation of overall protein synthesis by polyamines together with the increase in the amount of ribosomal protein S1. A tentative scheme on the relationship between polyamine stimulation of the synthesis of specific proteins and cell proliferation in E. coli is shown in Fig. 1.

POLYAMINE REGULATION OF THE SYNTHESIS OF THYMIDINE KINASE IN BOVINE LYMPHOCYTES

The change in macromolecular synthesis and enzymatic activities concerned with DNA replication during a 40 h incubation of bovine lymphocytes stimulated to proliferate by concanavalin A (Con A) was examined. The cells were made polyamine-deficient by α-difluoromethylornithine (DFMO) and ethylglyoxal bis(guanylhydrazone)(EGBG). Protein synthesis began an accelerated increase at 10 h, and the onset of rapid increase of RNA and DNA sythesis was at 15 h after the addition of Con A. In the presence of DFMO and EGBG, protein synthesis was perturbed before RNA and DNA synthesis were altered. Inhibition of DNA synthesis in polyamine-deficient cells was always greater than that of protein synthesis. It is suggested that the early inhibition of protein synthesis probably influences inhibition of DNA synthesis. By 16 h after Con A addition, the activities of DNA polymerase α, thymidine kinase and thymidylate synthase had started to increase. Thymidine kinase and DNA polymerase α underwent an accelerated increase during the 20 to 36 h incubation, while thymidylate synthase's increase was much slower at that time. These increases paralleled closely but preceded DNA synthesis. In the presence of DFMO and EGBG, the increases of thymidine kinase and DNA polymerase α activities were inhibited markedly, but the slow increases of thymidylate synthase was not inhibited. At 36 h, the thymidine kinase and DNA polymerase α activities in polyamine-deficient cells were 7 and 55%, respectively, of the levels in normal cells. Polyamines reversed the decrease of thymidine kinase and DNA polymerase α activities. These results indicate that at least one effect of polyamine deficiency on DNA synthesis may be related to inhibition of the synthesis of specific enzymes, especially thymidine kinase.

To learn how polyamines regulate synthesis of thymidine kinase, mRNA levels of thymidine kinase were measured through the use of cloned human thymidine kinase cDNA as a probe. Thymidine kinase mRNA levels started to increase at approximately 12 h and reached a maximum at 36 h after the addition of Con A (Fig. 2). The amount of the mRNA at 36 h was 10-fold higher than in resting cells. In the presence of

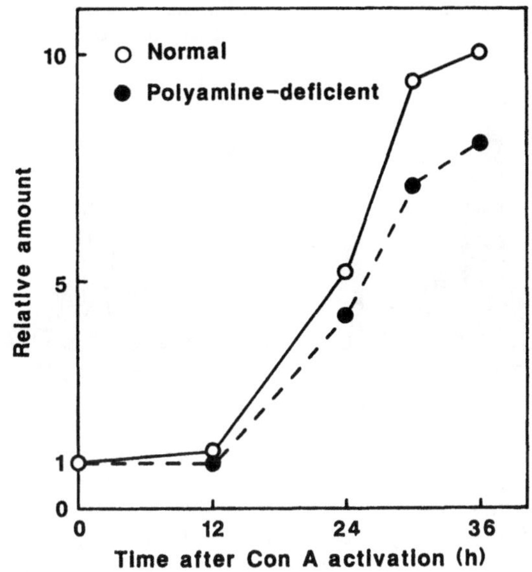

Fig. 2. Northern blot hybridization of thymidine kinase mRNA in bovine lymphocytes treated only with concanavalin A and also with DFMO and EGBG. Bovine suprapharyngeal lymphocytes were prepared, purified, and cultured at 37 °c in a humid incubator in RPMI 1640 medium containing 10% newborn calf serum according to the method of Seyfried and Morris (42). The lymphocytes (3 x 10^6 cells/ml) were activated by addition of concanavalin A at a final concentration of 5 µg/ml. The pH of the medium was adjusted to 7.2 with 24 mM $NaHCO_3$ at an atmosphere of 5% CO_2. The concentration of DFMO and EGBG used were 4 mM and 100 µM, respectively. Bovine lymphocytes (1.5 x 10^8 cells) were homogenized with 2.5 ml of guanidine thiocyanate stock solution (4 M guanidium thiocyanate, 0.5% sodium N-lauroylsarcosine, 25 mM sodium citrate, pH 7.0, and 0.1 M 2-mercaptoethanol). Total RNA of cells was isolated from the guanidine thiocyanate homogenate by ultracentrifugation through a dense cushion of cesium chloride (43). Human thymidine kinase cDNA, kindly supplied by Dr. Deininger, was digested with restriction enzyme BamHI, and the resulting 1.5 Kb fragment was purified on agarose gel electrophoresis (44). The DNA was labeled with [α-^{32}P]dCTP (3000 Ci/mmol) to a specific activity of 2 x 10^9 cpm/µg DNA through the use of a multiprime DNA labeling system. After electrophoresis of the RNA, it was transfered to a Gene Screen membrane. The hybridization of RNA to cDNA probe and its autoradiography was performed according to the method of Ayusawa et al. (45). The amounts of thymidine kinase mRNA were measured by scanning the bands of the autoradiographs with a Shimazu dual-wavelength TLC scanner CS 910. Total RNA from cells activated normally (O——O) and cells treated also with DFMO and EGBG (●– –●) was used for Northern blot analysis.

DFMO and EGBG, the amount of thymidine kinase mRNA at 36 h was approximately 75% of that in cells stimulated normally. These results indicate that synthesis of thymidine kinase mRNA is stimulated significantly by polyamines. The

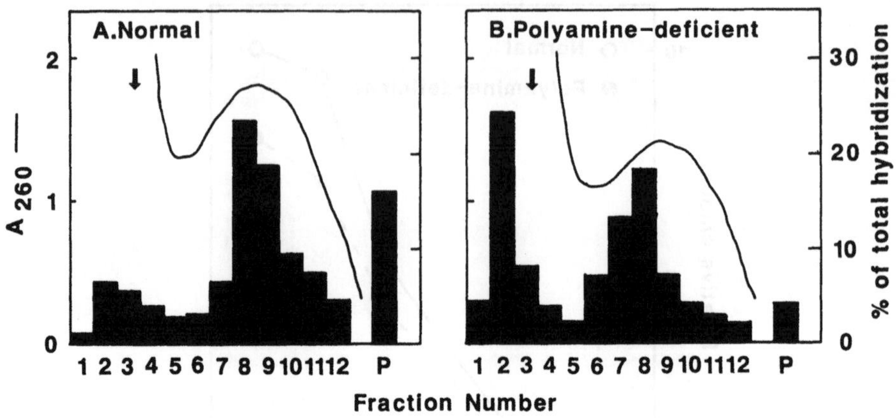

Fig. 3. Distribution of thymidine kinase mRNA in polysomes isolated from normal lymphocytes (A) and DFMO and EGBG treated lymphocytes (B). After a 36 h incubation of cells with concanavalin A , cycloheximide (100 µg/ml) was added to the medium to inhibit protein synthesis. Lymphocytes (10^9 cells) were collected at 5 min after the addition of cycloheximide. Cell lysates were prepared and fractionated on sucrose gradient centrifugation (12 1-ml fractions) according to the method of Mach et al. (46). The RNA in each fraction was extracted by phenol/chloroform/isoamylalcohol (49:50:1) and the amount of thymidine kinase mRNA was determined by Northern blot analysis as described in the legend of Fig. 2. Polysome size was estimated from A_{260}. An arrow indicates the position of 80S ribosomes and P indicates the fraction of precipitate during sucrose gradient centrifugation. Fraction 8 is the position of polysomes containing greater than 4 ribosomes. The amount of hybridization was quantitated from the autoradiographs by densitometoric scanning and expressed as the percent of the total hybridizable material across the gradient.

transcription rate of the mRNA in isolated nuclei was then measured. Incubation of isolated nuclei with [α-^{32}P]UTP and other NTPs can measure elongation of nascent RNAs. Labeled run-off transcripts were isolated and hybridized to membrane filters on which thymidine kinase cDNA was immobilized. The labeled transcripts hybridized to thymidine kinase cDNA were 35% less in nuclei from polyamine-deficient cells than from normal cells. These results show that the synthetic rate for thymidine kinase mRNA in polyamine-deficient cells is 65% of that in normal cells. The turnover rate of thymidine kinase mRNA in normal and polyamine-deficient cells was determined through culturing the cells in medium containing 2 µg/ml actinomycin D. The turnover rate ($t_{1/2}$) of thymidine kinase mRNA in normal and polyamine-deficient cells was 1.7 and 1.9 h, respectively, indicating that the turnover rate of the mRNA in both cells is nearly equal.

Polysomes from lysates of normal and polyamine-deficient cells were fractionated on sucrose gradient centrifugation (Fig. 3). The amount of thymidine kinase mRNA in each fraction was then determined by Northern blot analysis. Approximately 90% of thymidine kinase mRNA was polysome associated in normal cells, while 65% of the mRNA was polysome associated in polyamine-deficient cells. In addition, the size of polysomes associated with thymidine kinase mRNA in polyamine-deficient cells was smaller than that in normal cells (Fig. 3). When normal cells were treated for 30 min with 1 µM pactamycin, an inhibitor of the initiation of protein synthesis, approximately 40% of thymidine kinase mRNA was released from polysomes and size of polysomes associated with the mRNA became smaller. The mRNA was associated with polysomes containing 2 to 4 ribosomes (Fractions 4 to 7). These results suggest that polyamines regulate the synthesis of thymidine kinase at the level of initiation of protein synthesis.

Thymidine kinase activity is related closely to the growth state of cells, being present in rapidly growing, but not in resting cells (22). It has been proven that thymidine kinase activity is strictly dependent on the amount of mRNA existing in cells (23-26). In the case of polyamine-deficient cells, however, the amount of thymidine kinase mRNA present in cells was approximately 75% of that in normal cells, although the enzymatic activity was only 7% of that in normal cells. Therefore, regulation of the synthesis of thymidine kinase by polyamine is very unique. The main regulation of the synthesis of thymidine kinase by polyamines is very likely to be at the level of translation judging from the results presented above. A strong evidence will be obtained if polyamines could stimulate the synthesis of thymidine kinase in mammalian cell-free protein synthetic systems.

NECESSITY OF POLYAMINES FOR MAXIMAL GLOBIN SYNTHESIS IN A PURIFIED SYSTEM

Polyamine effect on globin synthesis was examined with nuclease-treated lysate and a purified system consisting of ribsomes, initiation factors, and pH 5 enzymes. When 25 µl of nuclease-treated lysate was used, the endogenous spermidine and spermine concentration in the reaction mixture (0.05 ml) was 0.2 and 0.02 mM, respectively. The addition of spermidine (0.1 to 0.5 mM) did not stimulate globin synthesis significantly although the optimal magnesium concentration was lowered by spermidine. When a purified system was used, no significant amount of endogenous polyamines was in the reaction mixture. The activity of globin synthesis was stimulated at least 8-fold by 0.5 mM spermidine or 0.08 mM spermine (Fig. 4). The optimal polyamine concentration was then determined in a purified system. Putrescine also stimulated globin synthesis to the same degree as spermidine or spermine. The optimal concentration of putrescine, spermidine, and spermine were 8 to 10 mM, 0.4 to 0.6 mM, 0.08 to 0.1 mM, respectively. The optimal concentration of spermidine was relatively close to the concentration in reticulocyte lysate, since spermidine concentration in the reaction mixture (0.05 ml) containing 25 µl of nuclease-treated lysate was 0.2 mM. The system was completely dependent on globin mRNA, ribosomes, and pH 5 enzymes, and strongly dependent on initiation factors. This suggests that the pH 5 enzymes may contain small amounts of initiation factors.

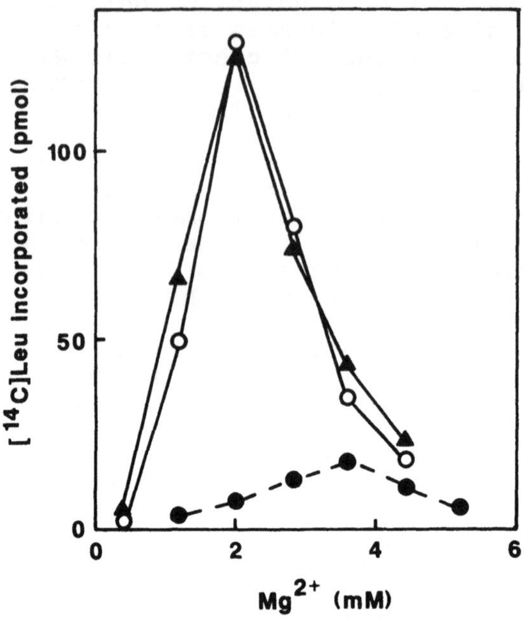

Fig. 4. Effect of polyamines on globin synthesis. Rabbit reticulocyte lysates and nuclease-treated lysates were prepared by the methods of Pelham and Jackson (47). Polysomes and ribosomes were obtained by centrifugation at 150,000 X g for 2.5 h. The sedimented ribosomes were washed with a buffer containing 20 mM Tris-HCl (pH 7.5), 3 mM magnesium acetate, 500 mM KCl, 6 mM 2-mercaptoethanol, and 0.2 M sucrose, centrifuged as above, and dissolved in a buffer containing 20 mM Tris-HCl (pH7.5), 3 mM magnesium acetate, 100 mM KCl, 6 mM 2-mercaptoethanol, and 0.2 M sucrose. The precipitate formed between 25 and 70% ammonium sulfate treatment of the 0.5 M KCl ribosomal wash was a source of the crude initiation factors. Initiation factor (IF) fractions A and B_1 were prepared from crude initiation factors by the method of Schreier and Staehelin (30). Globin mRNA was obtained from polysomes through the use of oligo(dT)-cellulose (48), and further purified by sucrose gradient centrifugation as described previously (49). Peak fractions of 9S were collected. The pH 5 enzymes containing aminoacyl-tRNA synthetase, elongation factors, and tRNA were prepared by the method of Falvey and Staehelin (50). The reaction mixture (0.05 ml) for globin synthesis contained the following : 20 mM Hepes-KOH buffer (pH 7.6), 15 µM hemin, 100 mM potassium acetate, 2 mM dithiothreitol, 1.5 mM ATP, 0.3 mM GTP, 8 mM creatine phosphate, 7.5 µg of creatine kinase, 0.2 µCi of [^{14}C]leucine (specific activity, 300 mCi/mmol), 30 µM each of the 19 other common amino acids, 0.1 A_{260} unit of washed ribosomes, 25 µg of IF A fraction, 12 µg of IF B_1 fraction, 250 µg of pH 5 enzymes, and magnesium acetate and spermidine at the specified concentration. After incubation at 37 °C for 30 min, a 0.04 ml aliquot of each reaction mixture was placed on a paper disc and hot TCA insoluble radioactivity was counted. Globin synthesis was linear until 45 min under these experimental conditions. ○, with 0.5 mM spermidine; ▲, with 0.08 mM spermine; ●, without polyamine.

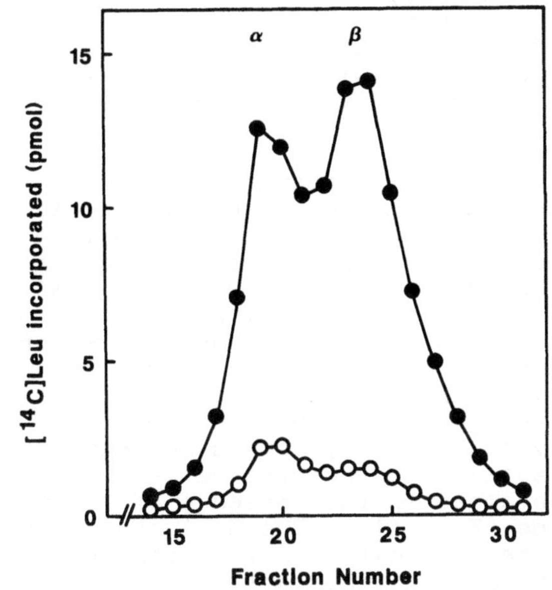

Fig. 5. Separation of α and β chains synthesized in the presence or absence of spermidine. For the separation of α and β chains, a 0.05 ml aliquot of the reaction mixture was used. The α and β chains were separated by CM-cellulose chromatography according to the method of Dintzis (51). The globin was prepared from the 150,000 x g supernatant by acid-acetone treatment (52). Globin (5 mg) was added as a carrier of the ^{14}C-labeled globin synthesized as described above. Formic acid and pyridine were then added to the globin solution to make final concertrations of 0.2 M and 0.02 M, respectively. The globin mixture was applied to a Whatman CM-52 column (0.7 x 6.5 cm), which previously had been equilibrated with 0.2 M formic acid and 0.02 M pyridine. The column was eluted with a linear gradient consisting of 30 ml each of 0.2 M formic acid and 0.02 M pyridine solution and 2 M formic acid and 0.2 M pyridine solution. The radioactivity of 1.5 ml fractions was measured. The incorporation ratio of α to β chains was obtained by calculating the total counts in their corresponding peaks. The ratio was corrected for the content of leucine in each globin chain.
●, with 0.5 mM spermidine; ○, without spermidine.

The ratio of α to β globin chains synthesized in the presence or the absence of spermidine was calculated from the data of [^{14}C]leucine incorporated into α and β globin chains showen in Fig. 5 and the value that leucines in the α and β chains were 17 and 18, respectively (27). The ratio of α to β globin chains synthesized in the presence of spermidine and Mg^{2+} was approximately 1, while the presence of only Mg^{2+} was approximately 1.5. The results suggest that the existence of polyamines in reticulocytes is physiologically important and that the degree of polyamine stimulation differs with the individual proteins synthesized. As shown in the preceding section, synthesis of thymidine kinase in bovine lymphocytes is strongly inhibited with a decrease in polyamine content. In contrast, the synthesis of ornithine decarboxylase was inhibited by polyamines at the translational level (28). The system we have developed here probably is a good system to learn how translation of an individual mRNA is regulated by polyamines.

The efficiency of globin synthesis in two protein sythetic systems was compared at optimal ionic conditions. The amount of [^{14}C] leucine incorporated into globin in a nuclease-treated lysate system was 4.4 pmol/μg RNA, while that in a purified system was 8.0 pmol/μg RNA. The latter corresponded to 132 pmol per 0.1 A_{260} unit of ribosomes incubated at 37 °C for 30 min. Under similar conditions without polyamines, the maximal mRNA dependent globin synthetic activity reported from other laboratories was approximately 40 pmol (29,30). Therefore, the globin synthetic activity in our system is approximately 3.3-fold greater than activities reported ealier. Maximal globin synthesis was obtained in the presence of 1.8 mM Mg^{2+} and 0.5 mM spermidine (Fig. 4). The Mg^{2+} content being 1.55 mM in the rabbit reticulocyte lysate strongly suggests that polyamines play a very important role in _in vivo_ protein synthesis.

SUMMARY

Present experimental data show that the synthesis of ribosomal protein S1 and PI protein was stimulated greatly by polyamines at the early stage after addition of putrescine in polyamine-requiring mutants of _E. coli_. No macromolecular synthesis was stimulated at this stage. Polyamine stimulation of the synthesis of these proteins probably plays an important role for cell growth.

In polyamine-deficient bovine lymphocytes, protein synthesis became perturbed before RNA and DNA synthesis. Among enzymes concerned with DNA replication, thymidine kinase activity was most strongly influenced by polyamines. The activity in polyamine-deficient cells was only 7% of the level in normal cells. Judging from the amount of thymidine kinase mRNA and its distribution in polysomes, it was concluded that polyamines mainly regulate the synthesis of thymidine kinase at the level of initiation of protein synthesis.

A polyamine-free protein synthetic system, established from components of rabbit reticulocytes, consisted of globin mRNA, salt-washed ribosomes, partially purified initiation factors, and pH 5 enzymes. Spermidine added to this system not only lowered the optimal magnesium concentration required for globin synthesis, but it also stimulated the globin synthesis 8- to 10-fold. The optimal spermidine concentration was 0.4 to 0.6 mM, a concentration similar to that in intact rabbit reticulocytes. The ratio of α to β globin chains synthesized in the presence of spermidine and Mg^{2+} was approximately 1.0, while the ratio in the presence of only Mg^{2+} was approximately 1.5. The results strongly suggest that polyamines play an important role in rabbit reticulocyte protein synthesis.

REFERENCES

1. S. S. Cohen, "Introduction to the polyamines," Prentice Hall, Englewood Cliffts, NJ (1971).
2. C. W. Tabor, and H. Tabor, Polyamines, Annu. Rev. Biochem. 53:749 (1984).
3. K. Igarashi, I. Sakamoto, N. Goto, K. Kashiwagi, R. Honma, and S. Hirose, Interaction between polyamines and nucleic acids or phospholipids, Arch. Biochem. Biophys. 219:438 (1982).
4. K. Igarashi, K. Sugawara, I. Izumi, C. Nagayama, and S. Hirose, Effect of polyamines on polyphenylalanine synthesis by Escherichia coli and rat liver ribosomes, Eur. J. Biochem. 48:495 (1974).
5. J. F. Atkins, J. B. Lewis, C. W. Anderson, and R. F. Gesteland, Enhanced differential synthesis of proteins in a mammalian cell-free system by addition of polyamines, J. Biol. Chem. 48:5688 (1975).
6. K. Igarashi, M. Kojima, Y. Watanabe, K. Maeda, and S. Hirose, Stimulation of polypeptide synthesis by spermidine at the level of initiation in rabbit reticulocyte and wheat germ cell-free systems, Biochem. Biophys. Res. Commun. 97:480 (1980).
7. Y. Watanabe, K. Igarashi, and S. Hirose, Differential stimulation by polyamines of phage RNA directed synthesis of proteins, Biochim. Biophys. Acta 656:134 (1981).
8. H. Tabor, and C. W. Tabor, Polyamine requirement for efficient translation of amber codons in vivo, Proc. Natl. Acad. Sci. U.S.A. 79:7087 (1982).
9. K. Mitsui, K. Igarashi, T. Kakegawa, and S. Hirose, Preferential stimulation of in vivo synthesis of a protein by polyamines in Escherichia coli: purification and properties of the specific protein, Biochemistry 23:2679 (1984).
10. K. Mitsui, R. Ohnishi, S. Hirose, and K. Igarashi, Necessity of polyamines for maximum in vivo synthesis of ßß' subunits of RNA polymerase, Biochem. Biophys. Res. Commun. 123:528 (1984).
11. K. Igarashi, and D. R. Morris, Physiological effects in bovine lymphocytes of inhibiting polyamine synthesis with ethylglyoxal bis (guanylhydrazone), Cancer Res. 44:5332 (1984).
12. D. V. Young, and P. R. Srinivasan, Regulation of macromolecular synthesis by putrescine in a conditional Escherichia coli putrescine auxotroph, J. Bacteriol. 112:30 (1972).
13. B. B. Rudkin, P. S. Mamont, and N. Seiler, Decreased protein synthetic activity is an early consequence of spermidine depletion in rat hepatoma tissue culture cells, Biochem. J. 217:731 (1984).
14. K. Igarashi, Y. Watanabe, K. Nakamura, M. Kojima, Y. Fujiki, and S. Hirose, Effect of spermidine on N-formylmethionyl-tRNA binding to 30S ribosomal subunits and on N-formylmethionyl-tRNA dependent polypeptide synthesis, Biochem. Biophys. Res. Commun. 83:806 (1978).
15. Y. Watanabe, K. Igarashi, K. Mitsui, and S. Hirose, Differential stimulation by polyamines of phage DNA-directed in vitro synthesis of proteins, Biochim. Biophys. Acta 740:362 (1983).

16. H. Tabor, C. W. Tabor, M. S. Cohn, and E. W. Hafner, Streptomycin resistance (rps L) produces an absolute requirement for polyamines for growth of an Escherichia coli strain unable to synthesize putrescine and spermidine [Δ(speA-speB)Δ speC], J. Bacteriol. 147:702 (1981).
17. K. Igarashi, K. Kashiwagi, K. Kishida, T. Kakegawa, and S. Hirose, Decrease in the S1 protein of 30S ribosomal subunits in polyamine-requiring mutants of Escherichia coli grown in the absence of polyamines, Eur. J. Biochem. 114:127 (1981).
18. T. Kakegawa, S. Hirose, K. Kashiwagi, and K. Igarashi, Effect of polyamines on in vitro reconstitution of ribosomal subunits. Eur. J. Biochem. 158:265 (1986).
19. G. Van Dieijen, P. H. Van Knippenberg, and J. Van Duin, The specific role of ribosomal protein S1 in the recongnition of native phage RNA, Eur. J. Biochem. 64:511 (1976).
20. T. Yokota, K. Arai, and Y. Kaziro, Studies on 30S ribosomal protein S1 from E. coli. II. Functional properties, J. Biochem. (Tokyo) 86:1739 (1976).
21. D. Oliver, and J. Beckwith, Regulation of a membrane component required for protein secretion in Escherichia coli, Cell 30:311 (1982).
22. L. L. Johnson, G. Rao, and A. Muench, Regulation of thymidine kinase enzyme level in serum stimulated mouse 3T6 fibroblasts, Exp. Cell. Res. 138 (1982).
23. M. Groudine, and C. Casimir, Post-transcriptional regulation of the chicken thymidine kinase gene, Nucl. Acids Res. 12:1427 (1984).
24. P. Stuart, M. Ito, C. Stewart, and S. E. Conrad, Induction of cellular thymidine kinase occurs at the mRNA level, Mol. Cell. Biol. 5:1490 (1985).
25. C. J. Stewart, M. Ito, and S. E. Conrad, Evidence for transcription and post-transcriptional control of the cellular thymidine kinase gene, Mol. Cell. Biol. 7:1156 (1987).
26. A. Kasid, N. E. Davidson, E. P. Gelman, and M. E. Lippman, Transcriptional control of thymidine kinase gene expression by estrogen and antiestrogens in HCF-7 human breast cancer cells, J. Biol. Chem. 261:5562 (1986).
27. A. Efstratiadis, F. C. Kafatos, and T. Maniatis, The primary structure of rabbit ß-globin mRNA as determined from cloned DNA, Cell 10:571 (1977).
28. T. Kameji, and A. E. Pegg, Inhibition of mRNAs for ornithine decarboxylase and S-adenosylmethionine decarboxylase by polyamines, J. Biol. Chem. 262:2427 (1977).
29. R. G. Crystal, A. W. Nienhuis, N. A. Elson, and W. F. Anderson, Initiation of globin synthesis. Preparation and use of reticulocyte ribosomes retaining initiation region messenger ribonucleic acid fragments, J. Biol. Chem. 247:5357 (1972).
30. M. H. Schreier, and T. Staehelin, Initiation of mammalian protein synthesis: the importance of ribosome and initiation factor quality for the efficiency of in vitro systems, J. Mol. Biol. 73:329 (1973).
31. K. Igarashi, K. Kashiwagi, H. Hamasaki, A. Miura, T. Kakegawa, S. Hirose, and S. Matsuzaki, Formation of a compensatory polyamine by Escherichia coli polyamine-

requiring mutants during growth in the absence of polyamines. J. Bacteriol. 166:128 (1986).
32. O. H. Lowry, N. J. Rosebrough, A. L. Farr, and R. J. Randall, Protein measurement with the Folin phenol reagent, J. Biol. Chem. 193:265 (1951).
33. K. Kashiwagi, and K. Igarashi, Nonspecific inhibition of Escherichia coli ornithine decarboxylase by various ribosomal proteins. Detection of a new ribosomal protein possessing strong antizyme activity, Biochim. Biophys. Acta 911:180 (1987).
34. A. Wada, Analysis of Escherichia coli ribosomal proteins by an improved two dimensional gel electrophoresis. I. Detection of four new proteins, J. Biochem. (Tokyo) 100:1583 (1986).
35. A. Wada, and T. Sako, Primary structures of and genes for new ribosomal proteins A and B in Escherichia coli, J. Biochem. (Tokyo) 101:817 (1987).
36. R. R. Burgess, and J. J. Jendrisak, A procedure for the rapid, large-scale purification of Escherichia coli DNA-dependent RNA polymerase involving polymin P precipitation and DNA-cellulose chromatography, Biochemistry 14:4634 (1975).
37. J. W. B. Hershey, J. Yanov, K. Johnston, and J. L. Fekunding, Purification and characterization of protein synthesis initiation factors IF-1, IF-2, IF-3 from Escherichia coli, Arch. Biochem. Biophys. 182:626 (1977).
38. K. Igarashi, K. Terada, Y. Tango, K. Katakura, and S. Hirose, Demonstration by affinity chromatography of the cell-free sythesis of ribonuclease-specific immunoglobulin, J. Biochem. (Tokyo) 77:383 (1975).
39. L. Philipson, P. Anderson, U. Olshevsky, R. Weinberg, and D. Baltimore, Translation of MuLv and MSV RNAs in nuclease-treated reticulocyte extracts; enhancement of the gag-pol polypeptide with yeast suppressor tRNA, Cell 13:189 (1987).
40. U. K. Laemmli, Cleavage of structural proteins during the assembly of the head of bacteriophage T4, Nature 227:680 (1970).
41. R. A. Laskey, and A. D. Mills, Quantitative film detection of ^{3}H and ^{14}C in polyacrylamide gels by fluorography, Eur. J. Biochem. 56:335 (1975).
42. C. E. Seyfried, and D. R. Morris, Methods for the study of the physiological effects of inhibitors of polyamine biosynthesis in mitogen-activated lymphocytes, Methods Enzymol. 94:373 (1983).
43. J. M. Chirgwin, A. E. Przybyla, R. J. Mac Donald, and W. J. Rutter, Isolation of biologically active ribonucleic acid from sources enriched in ribonuclease, Biochemistry 18:5294 (1979).
44. H. D. Bradshow, Jr. and P. L. Deininger, Human thymidine kinase gene: molecular cloning and nucleotide sequence of a cDNA expressible in mammalian cells, Mol. Cell. Biol. 4:2316 (1984).
45. D. Ayusawa, K. Shimizu, H. Koyama, S. Kaneda, K. Takeishi, and T. Seno, Cell-cycle-directed regulation of thymidylate synthetase messenger RNA in human dipioid fibroblasts stimulated to proliferate, J. Mol. Biol. 190:559 (1986).
46. M. Mach, M. W. White, M. Neubauer, J. L. Degen, and D. R. Morris, Isolation of a cDNA clone encoding S-

adenosylmethionine decarboxylase. Expression of the gene in mitogen-activated lymphocytes, J. Biol. Chem. 261:11697 (1986).
47. H. R. B. Pelham, and R. J. Jackson, An efficient mRNA-dependent translation system from reticulocyte lysates, Eur. J. Biochem. 67:247 (1976).
48. A. Krystosek, M. L. Cawthon, and D. Kabat, Improved methods for purification and assay of eukaryotic messenger ribonucleic acids and ribosomes. Quantitative analysis of their interaction in a fractionated reticulocyte cell-free system, J. Biol. Chem. 250:6077 (1975).
49. K. Ito, and K. Igarashi, The increase by spermidine of fidelity of protamine synthesis in a wheat-germ cell-free system, Eur. J. Biochem. 156:505 (1986).
50. A. K. Falvey, and T. Staehelin, Structure and function of mammalian ribosomes. I. Isolation and characterization of active liver ribosomal subunits, J. Mol. Biol. 53:1 (1970).
51. H. M. Dintzis, Assembly of the peptide chains of hemoglobin Proc. Natl. Acad. Sci. U.S.A. 47:247 (1961).
52. A. Rossi-Fanelli, E. Antonini, and A. Caputo, Studies on the structure of hemoglobin. I. Physiological properties of human globin, Biochim. Biophys. Acta 30:608 (1958).

POLYAMINE RECYCLING ENZYMES IN HUMAN CANCER CELLS

Heather M. Wallace, Mark E. Nuttall and Catherine S. Coleman

Clinical Pharmacology Unit, Department of Medicine and Therapeutics and Pharmacology, University of Aberdeen, Polwarth Building, Foresterhill, Aberdeen AB9 2ZD, U.K.

The route of polyamine biosynthesis is a complex, but now well-characterised, pathway controlled at the level of the two decarboxylase enzymes, particularly ornithine decarboxylase (for review see Pegg, 1986). The catabolic pathways are however less well understood. The catabolic pathway should perhaps be more correctly referred to as the retroconversion pathway since catabolism implies breakdown and loss as opposed to the recycling which occurs. Retroconversion and biosynthesis occur in conjunction. For example, in regenerating rat liver Matsui et al. (1981) showed that the early increase in intracellular putrescine was the result of a combination of increased biosynthesis from ornithine and increased breakdown of spermine via acetylation and oxidation. Therefore, both pathways function in rapidly growing cells and they do not occur in isolation at different stages of cell growth.

Excretion of polyamines also occurs in mammalian cells in conjunction with both acetylation (Wallace et al., 1985) and oxidation (Wallace, 1987). Efflux of polyamines is strictly regulated and only occurs when the cell growth rate is decreased. For example, polyamines are excreted from normal cells at confluence (Wallace & Keir, 1981) and from cells in low serum (Melvin & Keir, 1978) but not from serum-stimulated or rapidly growing cells (Wallace & Keir, 1981; 1982). The relationship between acetylation and excretion is not clear but it is possible that acetylation is a prerequisite to the efflux of the polyamines (Wallace et al., 1985).

Acetylation is the rate-limiting reaction in the retroconversion of the polyamines. There are, at least, two acetyltransferase enzymes

Table 1. Effectors of Cytosolic Spermidine/ Spermine Acetyltransferase

Compound	Effect	Reference
Carbon tetrachloride	+	Matsui & Pegg (1980)
Folic acid	+	Matsui & Pegg (1982)
Thioacetamide	+	Matsui et al. (1981)
Methylglyoxal bis (guanylhydrazone)	+	Pegg et al. (1985)
Serum	+	Wallace et al. (1985)
Polyamines	+	Pegg & Erwin (1986)
Dimethylnitrosamine	+	Matsui & Pegg (1982)
Secretin	+	Danzin et al. (1982)
Insulin	+	Grillo (1983)
Sodium butyrate	–	Wallace & Keir (1985)
	+	Matsui et al. (1983)

which will acetylate polyamines. These are physically separated in that there is a nuclear enzyme whose substrate specificity is not fully defined but which includes histones and polyamines (Libby, 1978) and a cytosolic enzyme which has been the centre of much interest recently. In rat liver, the cytosolic acetyltransferase is readily inducible with a number of compounds producing significant increases in the enzyme activity (Table 1). The majority of these compounds are toxic when given in sufficiently high doses and it may be that there is a link between cellular toxicity and increased polyamine acetylation.

Acetylpolyamines are present in cells at low, and often undetectable concentrations, although they have been reported in high concentrations in malignant cells such as breast cancer (Kingsnorth & Wallace, 1985) and colorectal cancer (Takenoshita et al., 1984). They are also major constituents of urine where they are present in high quantities particularly, in patients with malignant disease (Russell et al., 1971). This latter observation provides support for the idea that acetylation and excretion are coordinately linked and perhaps even coordinately regulated. The function of the acetylpolyamine derivatives in cells remains unclear.

Studies using human cancer tissue are limited by both the availability of the tissue and number and type of studies which can be carried out. Tissue culture is a useful alternative system.

Table 2. Human cancer cell lines used in the study

ZR-75-1	breast carcinoma, epithelial, oestrogen responsive, obtained from ascitic effusion (Engel et al., 1978)
T47-D	breast carcinoma, epithelial, oestrogen responsive, obtained from pleural effusion (Keydar et al., 1979)
HT29/219	colon carcinoma, epithelial (Kimball et al., 1981)
HT115	colon carcinoma, epithelial (Ashwell et al., 1982)

Human cancer cells in culture provide a model system for the study of cellular metabolism. In addition they can be used for the preclinical screening of new anticancer drugs and for the study of novel drug combinations. We have used four of these human cancer cell lines (Table 2) to study the retroconversion of the higher polyamines via acetylation and oxidation.

All cell lines were grown routinely in monolayer culture in Dulbecco's modification of Eagle's medium supplemented with 10% (v/v) donor horse serum and 1μM-insulin where indicated. Initially, it was necessary to adapt the cell lines to grow in medium supplemented with horse serum as opposed to foetal calf serum because of the high polyamine oxidase activity in the latter (MacGowan et al., 1987). This was achieved by serial passage via medium supplemented with 5% (v/v) each of foetal calf and horse serum. Cells were grown for a minimum of three passages in horse serum before being used in experiments. Typical growth curves for two of the colorectal cancer cell lines (HT29/219 and HT115) are shown in Fig. 1. These curves do not differ significantly from those obtained in medium supplemented with foetal calf serum. Similarly, the polyamine profiles were characteristic of the cell line and were not affected by the serum type present in the medium. A polyamine profile for HT29/219 cells is shown in Fig.2. In all human cancer cell lines examined spermine was the predominant polyamine. The content of acetyl polyamines was always low and these derivatives were only found in measurable quantities at later times in culture.

Analysis of the catabolic enzyme activities showed that all cell lines tested had polyamine acetyltransferase activity when spermidine was used as the substrate for the reaction (Table 3). The activities in

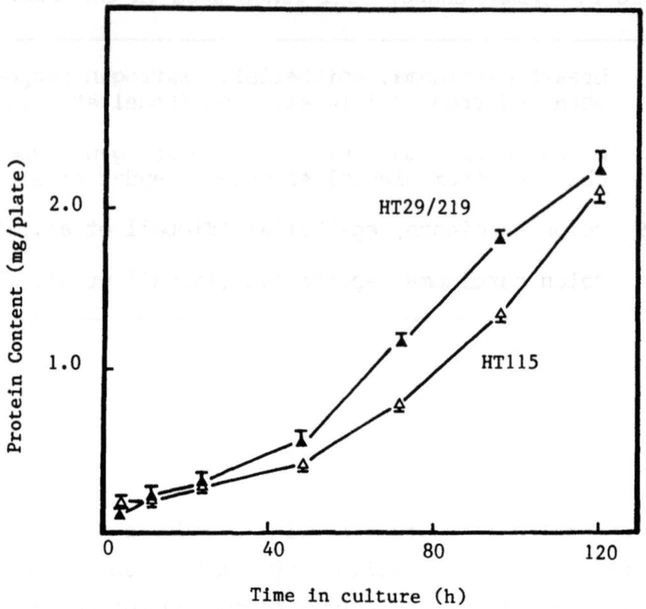

Fig. 1. Growth of human colerectal cancer cells in medium supplemented with horse serum. Cells were grown as described previously (Coleman and Wallace, 1988). Protein content was determined by the method of Lowry et al. (1951). Values are mean ± S.D. (n=3).

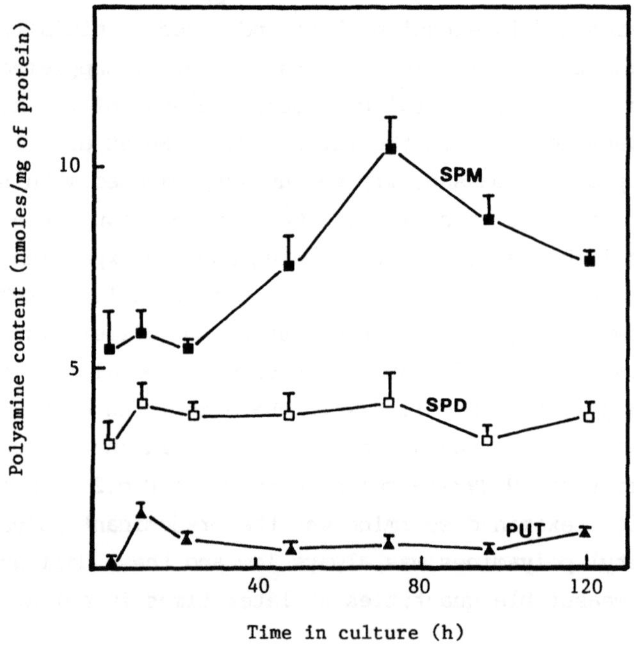

Fig. 2. Polyamine profile of HT29/219 cells. Polyamines were analyzed by h.p.l.c. (Wallace et al., 1988). Values are mean ± S.D. (n=3).

Table 3. Spermidine Acetyltransferase Activity in Human Cancer Cells in Culture

Cell Line	Spermidine Acetyltransferase Activity (pmol acetylspermidine formed/min/mg protein)
Colorectal Cancer Cells	
HT29/219	1.76 ± 0.12
HT115	1.47 ± 0.10
Breast Cancer Cells	
ZR-75-1	0.94 ± 0.14
T47-D	0.96 ± 0.12

Cells were grown in medium supplemented with 10% (v/v) horse serum (and 1μM insulin for T47-D cells) for 72h. Acetyltransferase activity was measured in cell homogenates. Values are mean ± S.D. (n = 3).

the colorectal cancer cell lines were comparable and were higher than those found in the breast cancer cells. Separation of the products of the assay by h.p.l.c. showed that in all cases both N^1- and N^8-acetylspermidine were formed (Table 4). The N^8-acetyl derivative predominated in these untreated cells except in the case of HT115 cells

Table 4. Products of Spermidine Acetylation in Human Cancer Cells in Culture

Cell Line	Acetyl Products		N^1/N^8-Acetyl-spermidine
	N^1-isomer (c.p.m. x 10^{-4}/mg protein)	N^8-isomer	
Colorectal Cancer Cells			
HT29/219	1.81	6.24	0.29
HT115	4.84	4.50	1.08
Breast Cancer Cells			
ZR-75-1	1.10	2.02	0.54
T47-D	3.93	5.35	0.73

Products from the assays described in Table 3 were separated by h.p.l.c. Fractions were collected, diluted and the radioactivity in each fraction determined. Values are the average of two analyses.

where the ratio of N^1/N^8- acetylspermidine was unity. It is not yet clear whether the difference in the ratio of acetyl derivatives between cell lines is meaningful.

Spermidine acetyltransferase (SAT) activity decreased with time in culture in all cell lines (Fig.3). SAT activity is demonstrating an inverse relationship with cell growth since activity is lowest when cell growth is maximal (Fig 3).

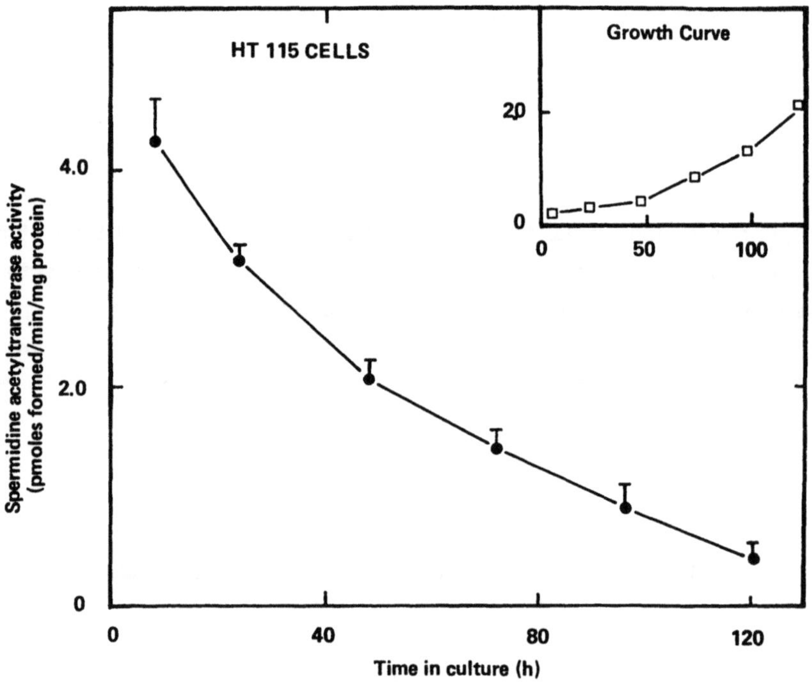

Fig. 3. Change in spermidine acetyltransferase activity in human colon cancer cells with time in culture. Values are mean ± S.D. (n=3).

SAT activity at least in rat liver and kidney can however be stimulated as indicated in Table 1. Methylglyoxal bis(guanylhydrazone) (MGBG), an inhibitor of S-adenosylmethionine decarboxylase, stimulated SAT activity measured in homogenates from all cell lines tested (Fig.4).

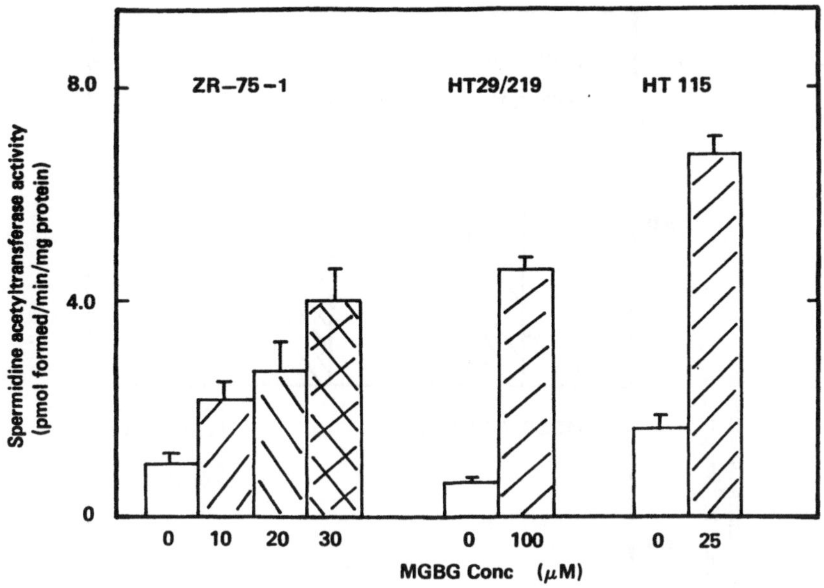

Fig. 4. Effect of MGBG on spermidine acetyltransferase activity in mammalian cells.
Values are mean ± S.D. (n=3).

The concentrations of MGBG used depended on the cell line but in each case those concentrations produced a cytostatic, but not cytotoxic, effect. Both colorectal cancer cell lines were more resistant to MGBG especially the HT29/219 cells where 100μM-MGBG was the concentration required to stop cell growth (Coleman & Wallace, 1988).

Analysis of the reaction products showed that MGBG specifically stimulated the production of N^1-acetylspermidine in the assays from whole cell homogenates (Fig.5). This suggests that MGBG is stimulating or inducing only the acetyltransferase activity which acetylates spermidine in the N^1-position. This agrees with the previous work in rat liver (Pegg et al., 1985) and in baby hamster kidney cells (Wallace et al., 1988) where the drug specifically induced the spermidine N^1-acetyltransferase. In both of these reported cases however it is only the cytosolic enzyme which is stimulated by MGBG.

In order to determine whether this was also the case in these human cancer cells, nuclei and cytoplasm were isolated from HT115 cells (Wallace et al., 1981). Nuclei were examined microscopically in the presence

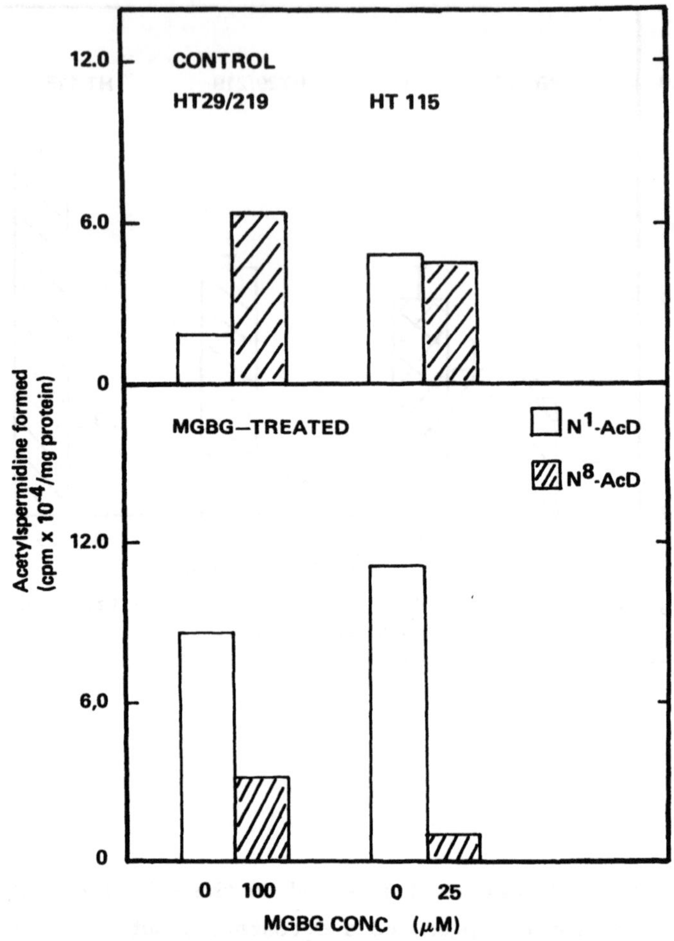

Fig. 5. Effect of MGBG on the production of N-acetylspermidine.
Products from triplicate assays were pooled and separated by h.p.l.c.

of Trypan blue and were found to be translucent and not to incorporate the dye. MGBG stimulated the SAT activity in both the nucleus and cytoplasm by about 1.5 and 2.5-fold respectively (Table 5). Analysis of the products of the reaction by h.p.l.c. showed that in both compartments stimulation was of the production of the N^1-isomer of spermidine although the production of N^8-acetylspermidine was also increased in the cytoplasm (Table 5). It may be then that human cancer cells contain more than one spermidine N^1-acetyltransferase that is inducible by MGBG; one nuclear and one cytoplasmic enzyme. This

Table 5. Effect of MGBG on Spermidine Acetyltransferase Activity in the Nucleus and Cytoplasm from Human Cancer Cells

Treatment	SAT Activity (pmol product formed/ min/mg protein)	Acetyl Products (c.p.m. x 10^{-4}/mg protein)		N^1/N^8-Acetyl Spermidine
	HT115 Cells			
		N^1-isomer	N^8-isomer	
a) Nucleus				
None	20.8 ± 1.4	11.4	38.2	0.3
MGBG	32.7 ± 5.4	29.4	7.6	3.8
b) Cytoplasm				
None	2.0 ± 1.1	5.4	3.8	1.4
MGBG	4.9 ± 1.2	19.9	10.5	1.9

Nuclei and cytoplasm were isolated and SAT activity in each compartment measured. Values are mean ± S.D. (n = 3). Products from triplicate assays were pooled and analysed by h.p.l.c.

suggestion would be consistent with the significantly higher concentrations of acetylpolaymines found in cancer tissue (Kingsnorth & Wallace, 1985; Takenoshita et al., 1984) and in the urine of cancer patients (Russell et al., 1971).

In addition to being able to acetylate spermidine human cancer cells can, to a greater or lesser extent, acetylate spermine and putrescine (Table 6). All cell lines acetylated spermidine but only colorectal cancer cells had detectable activity against spermine. Neither breast cancer cell line acetylated spermine. We have shown in ZR-75-1 cells that spermine is the important polyamine when cell growth is stimulated by oestrogen (Cross & Wallace, 1987). Spermine concentrations increased in response to oestrogen treatment as does cell growth, therefore, perhaps in these cells retroconversion of spermine only occurs in conditions when little or no cell growth is taking place. Only HT115 cells did not acetylate putrescine. All cells did, however, have the capacity to acetylate all three polyamines after MGBG-treatment. Acetyltransferase activity against all substrates was increased in treated cells (Fig.6). Therefore, there is a selectivity in the expression of the enzyme activity. It may be that an inducer is required to activate the expression of the acetyltransferase gene and hence produce enzyme protein.

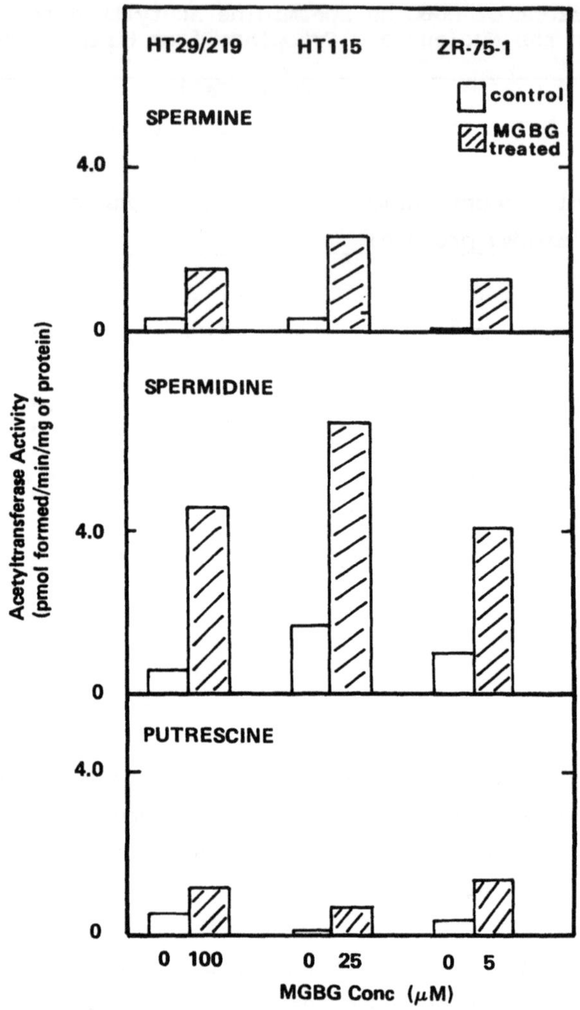

Fig. 6. Effect of MGBG on the acetylation of polyamines.

As regards oxidation of polyamines there was little oxidation of spermine detectable at early times in culture (Table 7). Oxidase activity was however found from 72h onwards with the enzyme activity increasing with time in culture. Treatment of cells with MGBG completely inhibited the oxidation of spermine (Table 7). Since MGBG stimulates acetylation in these cell lines (Table 3) and inhibits

Table 6. Acetyltransferase Activity in Human Cancer Cells

Substrate	Acetyltransferase Activity			
	HT29/219	HT115	ZR-75-1	T47-D
Spermine	0.12 ± 0.03	0.27 ± 0.07	n.d.	n.d.
Putrescine	0.52 ± 0.14	n.d.	0.35 ± 0.06	1.50 ± 0.03

Cells were grown for 72h and acetyltransferase activity measured in cell homogenates. Values are mean ± S.D. (n = 3).

Table 7. Polyamine Oxidase Activity in Human Cancer Cells

Time in culture (h)	HT/29/219 Cells Polyamine Oxidase Activity (Product, % of Total) Control	MGBG-Treated
4	1.1 ± 0.1	0.3 ± 0.1
24	0.3 ± 0.1	n.d.
48	0.1 ± 0.1	1.3 ± 0.3
72	0.9 ± 0.3	n.d.
96	3.1 ± 0.1	n.d.
120	7.4 ± 0.2	n.d.

Polyamine oxidase activity was measured in cell homogenates as described previously (MacGowan et al., 1987). Products were dansylated and separated by T.L.C. and the amount of product (spermidine and putrescine) quantified. Values are mean ± S.D. (n = 3). n.d. not detected.

oxidation (Table 7) one would expect there to be an accumulation of acetylpolyamines within the cell. This does not happen. Therefore the acetyl derivatives must be being lost from the cell. This would provide further evidence for our previous suggestion that acetylation is a prerequisite to excretion (Wallace & Keir, 1981; Wallace et al., 1985; Wallace, 1987) and that excretion is a major mechanism by which the cell can decrease its polyamine content.

ACKNOWLEDGEMENTS

We thank the Medical Research Council, the University of Aberdeen and Grampian Health Board for financial assistance.

REFERENCES

Ashwell, F., Bramwell, M.E. and Harris, H., 1982, A new marker for human cancer cells. The Ca antigen and the Ca1 antibody, Lancet, 2 : 1.

Coleman, C.S. and Wallace, H.M., 1988, Resistance of human colorectal cancer cells to the antimetabolite, methylglyoxal bis(guanylhydrazone), Brit. J. Pharmacol. in press.

Cross, S.J. and Wallace, H.M., 1987, Changes in polyamine concentrations in human breast cancer cells in culture : the effect of serum type and 17β-oestradiol, Biochem. Soc. Trans., 15 : 458.

Danzin, C., Bolkenius, F.N., Claverie, N., Wagner, J. and Grove, J., 1982, Secretion-induced accumulation of N^1-acetylspermidine and putrescine in the rat pancreas, Biochem. Biophys. Res. Commun. 109 : 1234.

Engel, L.W., Young, N.A., Tralka, T.S., Lippman, M.E., O'Brien, S.J. and Joyce, M.J., 1978, Establishment and characterization of three new continuous cell lines derived from human breast carcinoma, Cancer Res. 38 : 3352.

Grillo, M.A., 1983, Acetylation of spermidine in chicken tissues, Advances in Polyamine Res., 4 : 321.

Keydar, I., Chen, L., Karby, S., Weiss, F.R., Delarea, J., Radu, M., Chaitick, S. and Brenner, H.J., 1979, Establishment and characterization of a cell line of human breast carcinoma origin, Europ. J. Cancer, 15 : 659.

Kimball, P.M., Hammonds, L., McKibbin, J.M., Brattain, M.G., Glover, G. and Webb, M., 1981, Proc. Soc. Exptl. Biol. Med. 166 : 107.

Kingsnorth, A.N. and Wallace, H.M., 1985, Elevation of monoacetylated polyamines in human breast cancers, Eur. J. Cancer Clin. Oncol., 21 : 1057.

Libby, P.R. 1978, Calf liver nuclear N-acetyltransferases, J. Biol. Chem. 253 : 233.

Lówry, O.H., Rosebrough, N.J., Farr, A.L. and Randall, R.J., 1951, Protein measurement with the folin phenol reagent, J. Biol. Chem. 193 : 265.

MacGowan, S.H., Keir, H.M. and Wallace, H.M., 1987, Presence of a tissue-type polyamine oxidase activity in mammalian serum, Med. Sci. Res. 15 : 687.

Matsui, I. and Pegg, A.E., 1980, Increase in acetylation of spermidine in rat liver extracts brough about by treatment with carbon tetrachloride, Biochem. Biophys. Res. Commun. 92 : 1009.

Matsui, I. and Pegg, A.E., 1982, Induction of spermidine N^1-acetyltransferase in rat kidney by treatment with folic acid FEBS Lett, 139 : 205.

Matsui, I., Otani, S., Kuramoto, A., Morisawa, S. and Pegg, A.E., 1983, Induction of spermidine N^1-acetyltransferase by sodium n-butyrate and phytohaemagglutinin in bovine lymphocytes, J. Biochem. (Tokyo), 93 : 961.

Matsui, I., Wiegand, L. and Pegg, A.E., 1981, Properties of spermidine N-acetyltransferase from livers of rats treated with carbon tetrachloride and its role in the conversion of spermidine into putrescine, J. Biol. Chem. 256 : 2454.

Melvin, M.A.L. and Keir, H.M., 1978, Polyamine metabolism in BHK-21/C13 cells : Loss of spermidine from cells following transfer to serum-depleted medium, Exptl. Cell Res., 111 : 231.

Pegg, A.E., 1986, Recent advances in the biochemistry of polyamines in eukaryotes Biochem. J. 234 : 249.

Pegg, A.E. and Erwin, B.G., 1988, Regulation of spermidine/spermine N^1-acetyltransferase in L6 cells by polyamines and related compounds, Biochem. J., 238 : 581.

Pegg, A.E., Erwin, B.G. and Person, L., 1985, Induction of spermidine/spermine N^1-acetyltransferase by methylglyoxal bis(guanylhydrazone), Biochim. Biophys. Acta. 842 : 111.

Russell, D.H., Levy, C.C., Schimpff, S.C. and Hawk, I.A., 1971, Urinary polyamines in cancer patients, Cancer Res., 31 : 1555.

Takenoshita, S., Matsuzaki, S., Nakano, G., Kimura, H., Hoshi, H., Shoda, H. and Nakamura, T., 1984, Selective elevation of the N^1-acetylspermidine level in human colorectal adenocarcinomas, Cancer Res. 44 : 845.

Wallace, H.M., 1987, Polyamine catabolism in mammalian cells : excretion and acetylation, Med. Sci. Res. 15 : 1437.

Wallace, H.M., Duff, P.M., Pearson, C.K. and Keir, H.M., 1981, The effect of polyamines on DNA synthesis in vitro, Biochim. Biophys. Acta, 652 : 354.

Wallace, H.M. and Keir, H.M., 1981, Excretion of polyamines from baby hamster kidney cells (BHK-21/C13) : Effect of serum on confluent cell cultures, Biochim. Biophys. Acta, 676 : 25.

Wallace, H.M. and Keir, H.M., 1982, A comparison of polyamine metabolism in normal and transformed hamster cells, Biochem. J., 202 : 785.

Wallace, H.M. and Keir, H.M., 1985, Polyamine acetylation and excretion by mammalian cells in culture, in Recent Progress in Polyamine Research, Selmeci, L., Brosnan, M.E. and Seiler, N., eds., pp 441, VNU, Netherlands.

Wallace, H.M. MacGowan, S.H. and Keir, H.M., 1985, Polyamine metabolism in mammalian cells in culture, Biochem. Soc. Trans., 13 : 329.

Wallace, H.M., Nuttall, M.E. and Robinson, F.C., 1988, Acetylation of spermidine and methylglyoxal bis(guanylhydrazone) in babay hamster kidney cells (BHK-21/C13) Biochem. J. in press.

ESTROGENIC CONTROL OF SPERMIDINE/SPERMINE N^1-ACETYLTRANSFERASE ACTIVITY IN RAT UTERUS

Antonio Perin, Angela Sessa, and M. Alfonsina Desiderio

Institute of General Pathology and Center for Research on Cell Pathology C.N.R., Via Mangialli 31, 20133 Milan, Italy

It is widely accepted that specific intracellular receptor proteins are implicated in the estrogen regulation of gene expression, growth and differentiation in target cells. The unoccupied forms of these receptors, as suggested by Jensen and Jacobson[1] and Gorski et al.,[2] reside primarily in the cytoplasm and translocate to the nucleus after hormone interaction. A revision of this classical "two-step" model has recently indicated that both native and ligand-bound forms of the receptors are localized in the nucleus, as supported by biochemical[3] and immunocytochemical[4] data. The binding of estrogen with receptor leads to a change in the conformational state of the protein, which increases its affinity for nuclear components and DNA.[3,4] A number of nuclear acceptor sites have been proposed, including specific DNA sequences,[5] ribonucleoprotein,[6] basic nonhistone proteins,[7] nuclear matrix,[8] and acidic nonhistone protein DNA complexes.[9] The binding of activated receptor with DNA or other nuclear components triggers the synthesis of specific proteins, growth, and differentiation in responsive tissues.

Experimental data indicate that the native, unactivated receptor is acidic and loses its negative charge after estrogen activation.[10,11] For receptors that are phosphoproteins, such as those for glucocorticoids[12] and progesterone,[13,14] activation seems to be due to a dephosphorylation of the phosphoproteins since the receptors are activated by alkaline phosphatase.[15] Recent studies have demonstrated that polyamines, which at physiological conditions are polycations, interact with receptors and provoke a marked stabilization of the estrogen-receptor complex and an increase in its binding to DNA.[16] These observations suggest that changes in polyamine levels in estrogen-responsive tissues may modulate the estrogenic action of gene expression.

ESTROGENS AND POLYAMINE METABOLISM

The effect of estrogens on polyamine metabolism has been studied mostly on chick oviduct and mammalian uterus, which are principal target organs of these hormones. Other studies regard the involvement of polyamines in estradiol-stimulated proliferation of breast cancer cells.

Estrogens, at concentrations that stimulate uterine growth, have been found to enhance the activity of ornithine decarboxylase (ODC) in chick oviduct[17] as well as ODC and S-adenosylmethionine decarboxylase (SAMD) activities in uterus from immature[18,19] or ovariectomized[17,20] rats, with an increase in polyamine content.[20] These enzyme enhancements are due to enzyme induction since they are prevented by inhibitors of macromolecular syntheses.[17,20]

In human plasma, polyamine levels fluctuate with the menstrual cycle, i.e., they are elevated when estradiol and progesterone levels are high.[21] In rat uterus, ODC[19,22,23] and SAMD[23] activities have been reported to be highest at proestrus and lowest at estrus, and such changes are correlated with the uterine growth observed during the estrous cycle.[23] The uterine content of ODC and polyamines[22] is a function of plasma estradiol levels, which are elevated in proestrus and then decline in the afternoon or in the evening of proestrus.[24-26]

It is noteworthy that ovarian ODC activity in rodents is regulated during estrus by luteinizing hormone.[23,27,28] An increase in size of the ovary during pregnancy is associated with an enhancement in biosynthetic decarboxylases, which are probably under the control of gonadotrophic hormones of placental origin.[29] In pregnant rat uterus, ODC activity increases soon after implantation, whereas diamine oxidase (DAO) activity, which is low before day 10, becomes elevated at this time, reaching high levels comparable with those found in maternal placenta until delivery.[30] It is likely that DAO activity also contributes to control polyamine concentration in pregnant uterus as in other tissues rich in this enzyme activity. Moreover, estradiol causes an increase in DAO activity in uterus of guinea pig.[31]

There is evidence that physiological concentrations of estradiol stimulate growth of cultured cells of some human breast cancers.[32-35] In estradiol-responsive tumor cells, proliferation is associated with the activation of ODC, and α-difluoromethylornithine (DFMO, an inhibitor of ODC) abolishes estradiol-induced growth. These data are in favor of the important role of polyamines in estrogen stimulation of breast cancer cells.[36,37] It is interesting to note that in various human and rat breast cancer cell lines, estrogens stimulate the secretion of growth factor(s),[38,39] which is prevented by DFMO and is restored by the administration of spermidine.[40] This indicates a role of polyamines as mediators of hormonal effects on the growth of endocrine-dependent tumors.

POLYAMINE INTERCONVERSION PATHWAY

The polyamine interconversion pathway contributes to control tissue polyamine content by recycling the four-carbon chain of spermine and spermidine to form putrescine. This pathway results from the combined action of two enzymes, spermidine/spermine N^1-acetyltransferase (SAT) and polyamine oxidase (PAO). SAT, which is the rate-limiting enzyme, catalyzes the acetylation of spermidine and spermine, a metabolic step necessary for the following oxidation via PAO to putrescine.[41,42] SAT is induced or activated by a variety of stimuli,[41,42] such as toxic agents (carbon tetrachloride, thioacetamide, dialkylnitrosamines), drugs (isoprenaline, methylxanthine), vitamins (folic acid, 1,25-dihydroxyvitamin D_3), hormones (dopamine, growth hormone, insulin, glucagon, secretin, parathyroid hormone), growth stimuli (partial hepatectomy, phorbol esters, phytohemagglutinin) as well as physiological and pathological events (fasting, hepatitis, some tumors).[41,42] Treatment of rats with spermidine or spermine also leads to a substantial induction of SAT activity in liver, kidney, lung and pancreas.[43]

To the best of our knowledge, the effect of estrogens on SAT activity in target tissues has not been studied. Here we report our observations on the behavior of SAT activity in rat uterus after 17β-estradiol administration and during the estrous cycle. Cytosolic enzyme activity was assayed with the method of Libby[44] as described by Matsui and Pegg.[45]

STIMULATION OF SAT ACTIVITY BY 17β-ESTRADIOL

In a first group of experiments we studied the effect of 17β-estradiol on SAT activity in uterus from ovariectomized rats. Adult female Wistar rats were ovariectomized and treated with estrogen 20 days after the operation to ensure a complete clearance from the body of hormones produced by the ovaries. Animals were s.c. injected with 50 μg/kg body weight of 17β-estradiol dissolved in sesame seed oil or with an equal volume of the vehicle alone (controls).

The results reported in Fig. 1 show that estradiol caused in a few hours an increase in enzyme activity which was associated with an increase in uterus weight. To study the mechanism of this enzyme stimulation, hormone and inhibitors of macromolecular syntheses were administered contemporaneously. Cycloheximide (2 mg/kg i.p.) or actinomycin D (0.4 mg/kg i.p.) largely prevented the estradiol-dependent enhancement in SAT activity (Fig. 1). This indicates that under our experimental conditions estradiol induces the new synthesis of enzyme protein and suggests that new mRNA synthesis may be necessary for SAT stimulation.

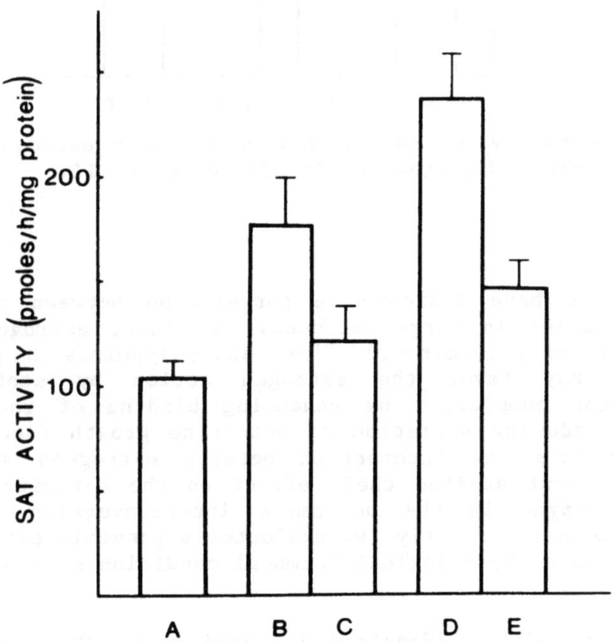

Fig. 1. SAT activity in uterus of ovariectomized rats after administration of 17β-estradiol alone or associated with cycloheximide or actinomycin D. A, Control rats; B, rats injected with 17β-estradiol or C, with 17β-estradiol plus cycloheximide and killed at 8 h; D, rats injected with 17β-estradiol or E, with 17β-estradiol plus actinomycin D and killed at 24 h.

SAT ACTIVITY IN RAT UTERUS DURING THE ESTROUS CYCLE

To verify whether uterine SAT activity is related to physiological changes in the levels of circulating estrogens, we assayed this enzyme activity during the estrous cycle. The results reported in Fig. 2 show that SAT activity was lowest in the estrous phase and highest in the proestrous phase, when there are the highest levels of estrogens in rat plasma.[24-26] It is well known that at proestrous phase uterine weight, polyamine biosynthetic decarboxylases and polyamine content reach their maximal values.[19,22,23]

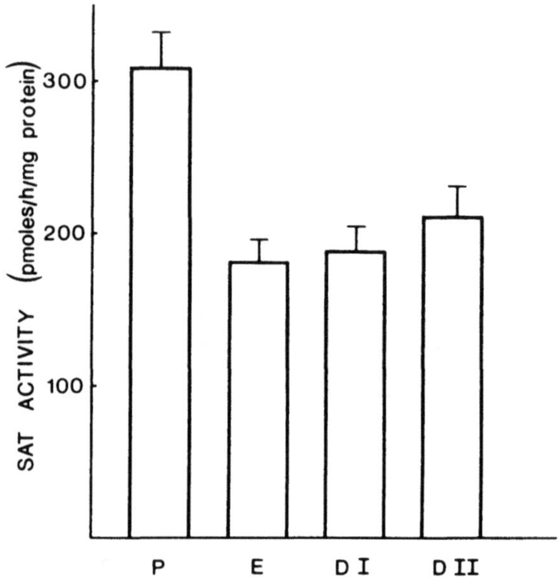

Fig. 2. SAT activity in rat uterus during the estrous cycle. P, Proestrus; E, estrus; DI, diestrus I; DII, diestrus II.

COMMENT

Many studies have indicated a correlation between estrogens and polyamine metabolism in target tissues. In fact, estrogens stimulate the biosynthesis of polyamines,[17-20,22] and polyamines at physiological concentrations may favor the estrogen action by stabilizing the estrogen-receptor complex,[16] by enhancing binding of the complex to DNA,[16] and by inducing secretion of autocrine growth factor(s).[40] To further investigate the interaction between estrogens and polyamine metabolism, we first studied their effect on the activity of SAT, the rate-limiting enzyme in the polyamine interconversion pathway.[41,42] Secondly, and more importantly, we evaluated a possible estrogen control of this enzyme in a physiological hormonal condition such as the estrous cycle.

Experiments with 17β-estradiol indicated that SAT activity increased in uterus (a typical responsive tissue) early after hormone administration by induction of new enzyme protein. Experiments performed during the estrous cycle demonstrated that uterine SAT activity was physiologically regulated by estrogens, since it fluctuated in relation to the changes in plasma estrogen levels. Since the behavior of ODC and SAMD activities is similar to that of SAT in rat uterus during the estrous cycle and after administration of 17β-estradiol,[17-20] it can be concluded that estrogens regulate various key steps of polyamine metabolism by activating biosynthetic and interconversion pathways. SAT activity may thus play a physiological role in the regulation of polyamine content in target cells in relation to hormonal status. Its increase leads to degradation of the higher polyamines to putrescine and may protect the cell against an excess of polyamines, which frequently has an inhibitory action.[46-48]

ACKNOWLEDGMENTS

The study was supported in part by grant no. 212336.44.8706781 from the Italian National Research Council (C.N.R., Rome), Special Project "Oncology", and by the Ministry of Public Education (M.P.I., Rome).

REFERENCES

1. E. V. Jensen and H. I. Jacobson, Basic guides to the mechanism of estrogen action, Recent Prog. Horm. Res. 18:387 (1962).
2. J. Gorski, D. Toft, G. Shyamala, D. Smith, and A. Notides, Hormone receptors: studies on the interaction of estrogen with the uterus, Recent Prog. Horm. Res. 24:45 (1968).
3. W. V. Welshons, M. E. Lieberman, and J. Gorski, Nuclear localization of unoccupied oestrogen receptors, Nature 307:747 (1984).
4. W. J. King and G. L. Greene, Monoclonal antibodies localize oestrogen receptor in the nuclei of target cells, Nature 307:745 (1984).
5. J. P. Jost, M. Geiser, and M. Seldran, Specific modulation of the transcription of cloned avian vitellogenin II gene by estradiol-receptor complex in vitro, Proc. Natl. Acad. Sci. USA 82:988 (1985).
6. T. Ling and S. Liao, Association of the uterine 17β-estradiol receptor complex with ribonucleoprotein in vitro and in vivo, J. Biol. Chem. 249:4671 (1974).
7. G. A. Puca, V. Sica, and E. Nola, Identification of a high affinity nuclear acceptor site for estrogen receptor of calf uterus, Proc. Natl. Acad. Sci. USA 71:979 (1974).
8. E. R. Barrack and D. S. Coffey, The specific binding of estrogens and androgens to the nuclear matrix of sex hormone responsive tissues, J. Biol. Chem. 255:7265 (1980).
9. T. C. Spelsberg, B. A. Littlefield, R. Seelke, G. Martin-Dani, H. Toyoda, P. Boyd-Leinen, C. Thrall, and O. L. Kon, Role of specific chromosomal proteins and DNA sequences in the nuclear binding sites for steroid receptors, Recent Prog. Horm. Res. 39:463 (1983).
10. W. V. Vedeckis, Steroid hormone receptor structure in normal and neoplastic cells, in: "Hormonally Responsive Tumors", V. P. Hollander, ed., Academic Press, New York (1985).
11. W. W. Grody, W. T. Schrader, and B. W. O'Malley, Activation, transformation and subunit structure of steroid hormone receptors, Endocr. Rev. 3:141 (1982).
12. P. R. Housley and W. B. Pratt, Direct demonstration of glucocorticoid-receptor phosphorylation by intact L cells, J. Biol. Chem. 258:4630 (1983).
13. N. L. Weigel, J. S. Tash, A. R. Means, W. T. Schrader, and B. W. O'Malley, Phosphorylation of hen progesterone receptor by cAMP dependent protein kinase, Biochem. Biophys. Res. Commun. 102:513 (1981).

14. J. J. Dougherty, R. K. Puri, and D. O. Toft, Phosphorylation in vivo of chicken oviduct progesterone receptor, J. Biol. Chem. 257:14226 (1982).
15. M. K. Dahmer, P. R. Housley, and W. B. Pratt, Effects of molybdate and endogenous inhibitors on steroid-receptor inactivation, transformation, and translocation, Annu. Rev. Physiol. 46:67 (1984).
16. T. Thomas and D. T. Kiang, Structural alterations and stabilization of rabbit uterine estrogen receptors by natural polyamines, Cancer Res. 47:1799 (1987).
17. S. Cohen, B. W. O'Malley, and M. Stastny, Estrogenic induction of ornithine decarboxylase in vivo and in vitro, Science 170:336 (1970).
18. K. Käpyaho, H. Pösö, and J. Jänne, Role of propylamine transferases in hormone-induced stimulation of polyamine biosynthesis, Biochem. J. 192:59 (1980).
19. A. M. Kaye, I. Icekson, and H. R. Lindner, Stimulation by estrogens of ornithine and S-adenosylmethionine decarboxylases in the immature rat uterus. Biochim. Biophys. Acta 252:150 (1971).
20. D. H. Russell and R. L. Taylor, Polyamine synthesis and accumulation in the castrated rat uterus after estradiol-17β stimulation, Endocrinology 88:1397 (1971).
21. D. W. Lundgren, P. M. Farrell, L. F. Cohen, and J. Hankins, Fluctuations of unbound whole blood polyamine levels during the menstrual cycle, Proc. Soc. Exp. Biol. Med. 152:81 (1976).
22. O. Hernandes, L. M. Ballesteros, D. Mendez, and A. Rosado, Polyamine synthesis and polysome patterns in the rat uterus during the estrous cycle, Endocrinology 92:1107 (1973).
23. H. Nawata, R. S. Yamamoto, and L. A. Poirier, An inverse linear correlation between uterine and ovarian levels of ornithine decarboxylase and S-adenosylmethionine decarboxylase in the rat, Proc. Soc. Exp. Biol. Med. 167:563 (1981).
24. M. S. Smith, M. E. Freeman, and J. D. Neill, The control of progesterone secretion during the estrous cycle and early pseudopregnancy in the rat: prolactin, gonadotropin and steroid levels associated with rescue of the corpus luteum of pseudopregnancy, Endocrinology 96:219 (1975).
25. K. Yoshinaga, R. A. Hawkins, and J. F. Stocker, Estrogen secretion by the rat ovary in vivo during the estrous cycle and pregnancy, Endocrinology 85:103 (1969).
26. R. L. Butcher, W. E. Collins, and N. W. Fugo, Plasma concentration of LH, FSH, prolactin, progesterone and estradiol-17β throughout the 4-day estrous cycle of the rat, Endocrinology 94:1704 (1974).
27. Y. Kobayashi, J. Kupelian, and D. V. Maudsley, Ornithine decarboxylase stimulation in rat ovary by luteinizing hormone, Science 172:379 (1971).
28. L. Persson, K. Isaksson, E. Rosengren, and F. Sundler, Distribution of ornithine decarboxylase in ovaries of rat and hamster during pro-oestrus, Acta Endocrinol. 113:403 (1986).
29. S. K. Guha and J. Jänne, Decarboxylation of ornithine and adenosylmethionine in rat ovary during pregnancy, Acta Endocrinol. 81:793 (1976).
30. S. K. Guha and J. Jänne, The synthesis and accumulation of polyamines in reproductive organs of the rat during pregnancy, Biochim. Biophys. Acta 437:244 (1976).
31. W. A. Fogel, Diamine oxidase (DAO) and female sex hormones, Agents Actions 18:4 (1986).
32. M. E. Lippman and G. Bolan, Oestrogen-responsive human breast cancer in long term tissue culture, Nature 256:592 (1975).

33. M. Lippman, G. Bolan, and K. Huff, The effects of estrogens and antiestrogens on hormone-responsive human breast cancer in long-term tissue culture, Cancer Res. 36:4595 (1976).
34. R. R. Weichselbaum, S. Hellman, A. J. Piro, J. J. Nove, and J. B. Little, Proliferation kinetics of a human breast cancer line in vitro following treatment with 17β-estradiol and 1-β-D-arabinofuranosylcytosine, Cancer Res. 38:2339 (1978).
35. J. C. Allegra and M. E. Lippman, Growth of a human breast cancer cell line in serum-free hormone-supplemented medium, Cancer Res. 38:3823 (1978).
36. A. Manni and C. Wright, Polyamines as mediators of estrogen action on the growth of experimental breast cancer in rats, J. Natl. Cancer Inst. 73:511 (1984).
37. G. Lima and R. P. C. Shiu, Role of polyamines in estradiol-induced growth of human breast cancer cells, Cancer Res. 45:2466 (1985).
38. D. Chalbos, F. Vignon, I. Keydar, and H. Rochefort, Estrogens stimulate cell proliferation and induce secretory proteins in a human breast cancer cell line ($T_{47}D$). J. Clin. Endocrinol. Metab. 55:276 (1982).
39. F. Vignon, M. E. Lippman, H. Nawata, D. Derocq, and H. Rochefort, Induction of two estrogen-responsive proteins by antiestrogens in R_{27}, a tamoxifen-resistant clone of MCF_7 cells, Cancer Res. 44:2084 (1984).
40. A. Manni, C. Wright, P. Feil, L. Baranao, L. Demers, M. Garcia, and H. Rochefort, Autocrine stimulation by estradiol-regulated growth factors of rat hormone-responsive mammary cancer: interaction with the polyamine pathway, Cancer Res. 46:1594 (1986).
41. A. E. Pegg, Recent advances in the biochemistry of polyamines in eukaryotes, Biochem. J. 234:249 (1986).
42. N. Seiler, Functions of polyamine acetylation, Can. J. Physiol. Pharmacol. 65:2024 (1987).
43. A. E. Pegg and G. Erwin, Induction of spermidine/spermine N^1-acetyltransferase in rat tissues by polyamines, Biochem. J. 231:285 (1985).
44. P. R. Libby, Calf liver nuclear N-acetyltransferases: purification and properties of two enzymes with both spermidine acetyltransferase and histone acetyltransferase activities, J. Biol. Chem. 253:23 (1978).
45. I. Matsui and A. E. Pegg, Effect of thioacetamide, growth hormone or partial hepatectomy on spermidine acetylase activity of rat liver cytosol, Biochim. Biophys. Acta, 633:87 (1980).
46. C. W. Tabor and H. Tabor, 1,4-Diaminobutane (putrescine), spermidine, and spermine, Annu. Rev. Biochem. 45:285 (1976).
47. J. Jänne, H. Pösö, and A. Raina, Polyamines in rapid growth and cancer, Biochim. Biophys. Acta 473:241 (1978).
48. C. W. Tabor and H. Tabor, Polyamines, Annu. Rev. Biochem. 53:749 (1984).

ERYTHROCYTE POLYAMINES : THEIR DIAGNOSTIC AND THERAPEUTIC IMPORTANCE IN HUMAN NEURO-ONCOLOGY

J-Ph. Moulinoux*, V. Quemener*, J. Théron**, R. Havouis*, F. Darcel*, Y. Guegan*, P. Courtheoux** and M. Chatel*

Groupe de Recherche en Thérapeutique Anticancéreuse
* CHU of Rennes, ** CHU of Caen, France

INTRODUCTION

In the last few years, the clinical interest in polyamine metabolism in oncology has been amplified and numerous works have been carried out in two main fields: one related to the diagnostic interest in polyamine determination in pathological tissues and in biological fluids of the organism (1), the other concerning the synthesis and clinical use of new drugs which could modify the metabolism of polyamines with consequence for tumor growth(2).

From the first works of Russell (3), clinical research has spread out, but the first symptom of a possible clinical application of polyamine determination in biological fluids is due to Marton (4) who shows that the monitoring of polyamine in CSF allowed a better management in children suffering from medulloblastoma.

Circulating free polyamines being principally transported in blood by erythrocytes (5,6), we were particularly interested in this cell compartment. After observing that in experimental (7,8) and in clinical conditions (7,9,10) red blood cells(RBC) spermidine (Spd) and spermine (Spm) levels appeared as a reliable index of cell proliferation, we were led to investigate the clinical relevance of erythrocyte polyamines (Spd,Spm) levels determination in neuro-oncology (11-15). The mitotic index being extremelylow in normal brain tissue, our markers of cell proliferation could be theoretically used as relevant blood brain tumor markers.

Furthermore, in a patient suffering from a suspected expanding intracranial mass, by comparison with a lumbar puncture necessary for a CSF polyamine levels determination, the innocuous character of a blood puncture (easily performed) could favor the long term follow-up of malignant brain tumor patients. This blood parameter of cell proliferation could eventually serve as an interesting guide for therapeutic decisions in neuro-oncology.

From 1983 to 1988 we tried to give response to the three

following questions: a) before surgery, do RBC polyamine levels help in the diagnosis of a malignant brain tumor? Furthermore, is it possible to precise the histological grade of malignancy by the single consideration of erythrocyte polyamine concentrations? b) after surgery, do RBC polyamine levels improve patient follow-ups, and have they a predictive relapse value preceding the first clinical and CT scan signs of tumor growth? c) taking into account erythrocyte Spd and Spm concentrations, is it possible to increase post-surgical post-radiotherapy chemotherapy efficacy with antiproliferative drugs?

DIAGNOSIS VALUE OF RBC POLYAMINE LEVELS

In a first study involving 73 brain tumor patients undergoing neuro-surgery (17 grade IV, 16 grade III, 5 grade II gliomas, 8 medulloblastomas, 8 metastatic tumors), statistically only grade IV patients (glioblastoma) had abnormally high RBC Spd levels (11). In the long term follow-up of 120 new supratentorial malignant glioma patients (see below), we noticed that pre-surgical grade III and IV tumors were associated with erythocyte Spd levels 2.5 times superior to those present in controls and in grade II glioma patients (Table 1). As normal erythrocyte Spd levels correspond to 8.0 ± 3.0 nmoles/8.10^9 RBC (mean ± 1 SD), we considered as abnormal values, RBC Spd levels superior to 14 nmoles (mean + 2 SD), which means a 5 % inferior statistical risk. Considering the RBC Spd levels frequency distribution in grade II, III and IV glioma patients before surgery, we respectively noticed that 33%, 20% and 28% of erythrocyte Spd values were inferior to 14 nmoles. Whatever the histological grade of malignancy might have been, we never observed statistically abnormal RBC Spm levels, which were always similar to controls (6.2 ± 2.1 nmoles/8.10^9 RBC).

Though we didn't notice statistical difference in erythrocyte polyamine levels in non-tumoral neurological diseases (13), isolated RBC polyamine levels concentrations did not appear by themselves sufficient to establish a definitive diagnosis of the presence of a malignant brain tumor. Nevertheless, in the presence of an intracranial malignant expanding process, high pre-surgical RBC Spd levels are in favor of high grade supratentorial glioma.

ERYTHROCYTE POLYAMINE LEVELS AND TUMOR EVOLUTION

In a second step we tried to evaluate the clinical relevance of estimating the RBC polyamine levels during the postoperative period in patients suffering from supra-tentorial malignant glioma (12-13). This study was based upon 120 patients with histologically confirmed malignant gliomas (WHO classification) divided into 46 grade IV, 45 grade III and 29 grade II. In the above mentioned population, 16 grade IV, 11 grade III and 4 grade II glioma patients were surveyed from the neurosurgical procedure to the clinical recurrence period. 750 polyamine determinations were carried out by an automated reversed phase HPLC analysis (11,12).

Statistically speaking, long term follow-up was first divided into four periods: pre-surgical diagnosis period, post-operative period (corresponding to the first term after surgery), clinical remission period, clinical relapse phase. During diagnosis period each patient was subjected to at least three RBC polyamine determinations.

As shown in Table 1, compared with normal values, only Spd levels were significantly modified during the long term follow-up periods (mean 11 months). It can be observed that abnormally high erythrocyte Spd concentrations were preferentially found during the after surgery first term and during clinical relapse periods, especially in grade III and IV glioma. Post-operative periods (including radiotherapy sessions) were in some cases associated with early increased polyamine levels. Except in the case of early post-surgical relapse, RBC Spd levels normalization statistically appeared three months after the tumor cutting away (Fig. 1).

During remission periods, abnormal Spd levels were only observed in grade IV glioma patients. The clinical remission phase which corresponds to the period preceding the first clinical and/or tomodensitometric (TDM, CT scan) signs of tumor recurrence is certainly one of the most important phases during the

Table 1. RBC polyamine levels during the long term follow-up of supra-tentorial malignant glioma patients.

Clinical status	Histological grade (Patients number)	Spermidine levels mean ± S.D. (polyamine determination number) (range)	Spermine levels
Controls	(45)	8.0 ± 3.0 (45) (2.5 - 14)	6.2 ± 2.1 (45) (2 - 12)
Diagnosis phase	II (14)	11.6 ± 4.5 (42) (4.5 - 26)	7.3 ± 3.1 (42) (3 - 16)
	III (21)	21.9 ± 8.9 (62)* (1 - 42)	7.0 ± 3.5 (62) (1 -19)
	IV (26)	19.5 ± 9.4 (75)* (6 - 49.5)	8.0 ± 4.9 (75) (1 - 20)
First term after surgery	II (4)	17.5 ± 7.7 (9)* (8.5 - 29)	6.9 ± 2.6 (9) (3 - 11)
	III (9)	21.8 ± 9.7 (23)* (5 - 38)	8.0 ± 4.2 (23) (3 - 16)
	IV (19)	18.4 ± 9.7 (55)* (3 - 48)	8.1 ± 4.6 (55) (2 - 22)
Clinical remission phase	II (12)	10.6 ± 4.5 (74) (4 - 26)	6.6 ± 4.0 (74) (0.5 - 27)
	III (18)	11.9 ± 6.9 (199) (2 - 57)	5.8 ± 5.8 (199) (1 - 34)
	IV (18)	15.3 ± 12.7(195)** (2 - 97)	6.7 ± 5.7 (195) (0.5 - 45)
Clinical relapse phase	II (5)	13.5 ± 4.4 (19)*** (6 - 27)	7.5 ± 2.6 (19) (4.5 - 16)
	III (15)	19.5 ± 10.7(73)* (4 - 60)	9.6 ± 6.8 (73) (1.5 - 35)
	IV (28)	20.5 ± 15.7(136)* (3 - 101)	7.9 ± 0.6 (136) (0.5 - 45)

Difference from controls using the non parametric Wilcoxon-signed Rank test is significant with, * $p<10^{-4}$, ** $p<10^{-2}$, *** $p< 5.10^{-2}$.

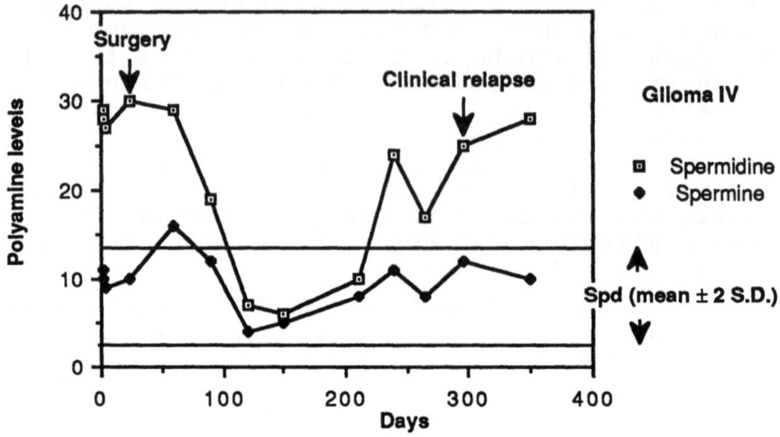

Figure 1. Long term follow-up of Grade IV glioma patient.

long term monitoring of glioma patients. Biologically speaking, this remission period corresponds to a clinically asymptomatic intracranial malignant cell proliferation which leads to the reappearance of a new tumor mass subsequently revealed by usual clinical and/or TDM examination.

In an attempt to dissociate biologically remitting patients from biologically recurring ones, we decided to deem abnormal values, erythrocyte Spd levels above 14 nmoles (normal mean + 2 SD) and RBC Spm concentrations under 2 nmoles (normal mean - 2 SD)(13,14). Abnormally elevated erythrocyte Spm levels being always associated to unusually high RBC Spd levels (higher than 14 nmoles), we didn't take into consideration Spm concentrations superior to 10 nmoles (normal mean - 2 SD).

In 44% grade IV (8/18) and in 50% grade III (7/14) clinically remitting patients (whose polyamine levels related to their clinical remission and relapse were at our disposal) abnormal erythrocyte polyamine values preceded the first clinical and CT scan arguments of tumor progression from one to six months.

Table 2. Erythrocyte polyamine concentrations in biologically remitting and recurring patients

Clinically remitting patients	Histological grade (Patients number)	Spermidine levels mean ± S.D. (polyamine determination number) (range)	Spermine levels
Biological remission	III (14)	9.0 ± 4.0 (73) (2 - 14)	4.6 ± 2.5 (73) (2 - 10)
	IV (18)	8.2 ± 3.1 (83) (2 - 14)	4.7 ± 2.3 (83) (2 - 10)
Biological recurrence	III (14)	18.4 ± 8.0 (102)* (6 - 57)	8.6 ± 5.6 (5.6) (1.5 - 34)
	IV (18)	20.2 ± 14.4 (112)* (4 - 97)	8.5 ± 7.9 (112) (0.5 - 45)

* Difference from biologically remitting patients using the non parametric Wilcoxon-signed Rank test is statistically significant, $p<10^{-4}$.

Furthermore, abnormal erythrocyte polyamine concentrations supported usual arguments of clinical tumor recurrence with the exception of one glioma patient showing a one month delay.

That led us to consider as being in biological relapse, clinically remitting patients who possessed abnormal RBC levels. Hence the biological phase of relapse goes from the discovery of the first abnormal polyamine level to the first clinical argument of tumor progression, whereas the biological remission corresponds to the period starting from the end of the first post-surgical term to the first abnormal RBC polyamine level observed during clinical remission.

The erythrocyte Spd levels thus observed in clinical remission periods of grade III and IV glioma patients having clinically recurred, showed a significant difference between biological remission and relapse periods (Table 2).

Nevertheless, it must be noticed that if in grade IV glioma patients, abnormal RBC polyamine levels preceded in each case clinical and TDM signs of recurrence, in 22% of clinically remitting grade III patients, abnormal erythrocyte polyamine concentrations were not followed by tumor recurrence.

In order to follow tumor proliferation more closely from the measurement of circulating polyamines, we checked in previous studies (12,14) that in clinically recurring patients erythrocyte levels of Spd and Spm were significantly correlated.

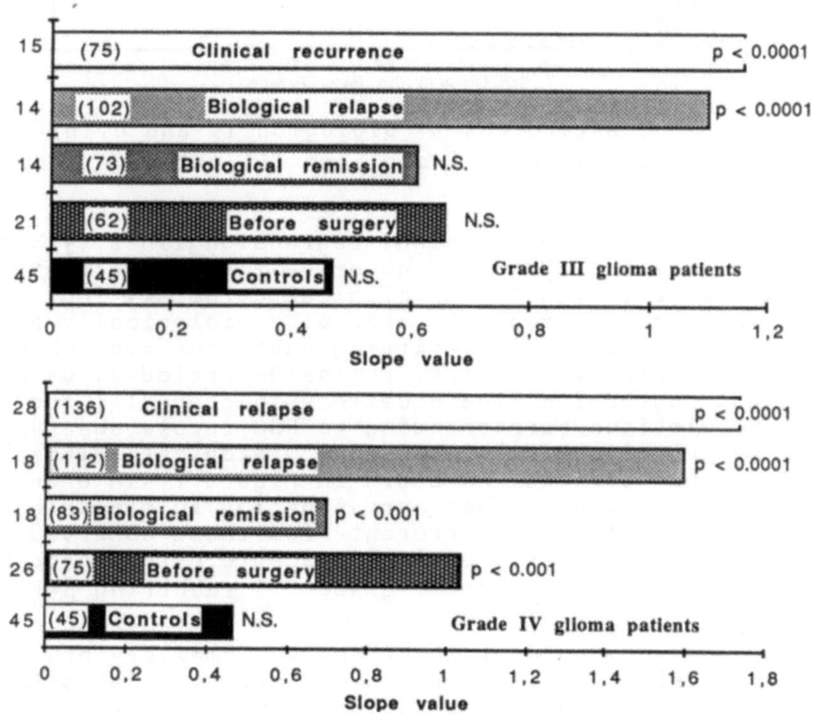

Figure 2. Straight regression line slope values during the long term follow-up of glioma III and IV patients.

Ordinate axis : patient number, () polyamine determination number; abcissa axis : SRL slope value (nmoles Spd = a nmoles Spm + b); p : correlation coefficient significativity.

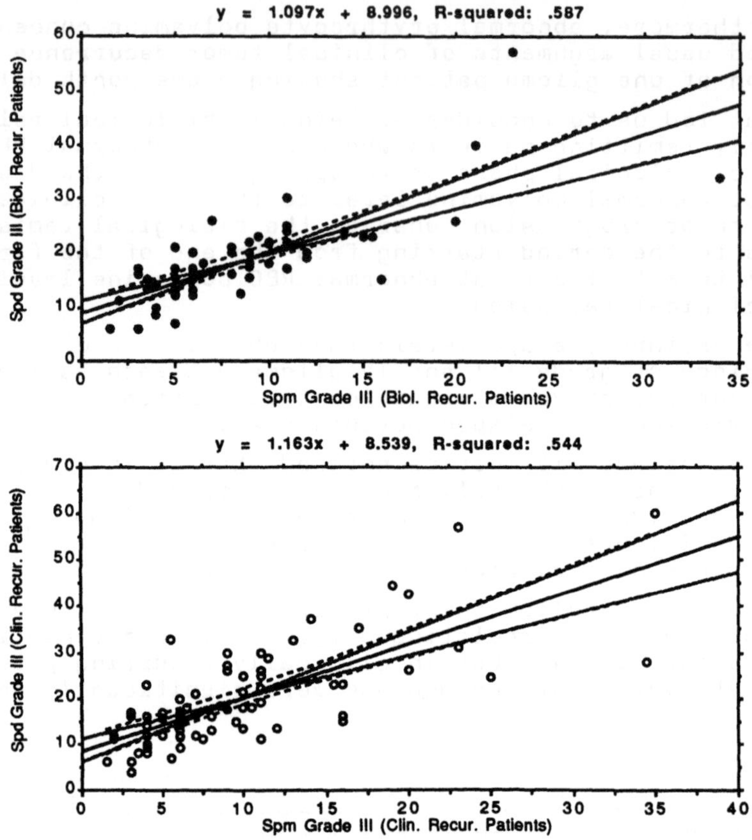

Figure 3. Scattergrams of biologically and clinically recurring glioma III patients.

We calculated individual straight regression lines (y = nmoles of Spd, x = nmoles of Spm) and observed that the slope value was varying in relation to the clinical Karnofsky index (12). In an attempt to differentiate otherwise biologically recurring patients from biologically remitting ones (the two sub-populations corresponding to clinical remission periods), we calculated straight regression line equations in grade III and IV patients populations corresponding to RBC levels observed before surgery, during biological remission, biological relapse and clinical relapse periods. As shown in Figure 2, in case of grade III or IV glioma, slope values corresponding to biological relapse are not statistically different from those observed during clinical relapse. Figure 3 shows the scattergrams corresponding to biologically then clinically grade III recurring patients.

In these conditions, during clinical remission, RBC polyamine levels succeeding to at least one abnormal erythrocyte Spd and/or Spm level (biological relapse), are not statistically different from those observed during clinical relapse periods.

Considering that abnormal RBC polyamine levels a) preceded the first usual argument of tumor progression in more than 40% of clinically remitting patients, b) have accompanied clinical relapse in more than 90% of glioma patients, c) but were present in grade III remitting patients having not clinically recurred (22%), we decided to treat only grade IV clinically remitting patients according to their erythrocyte polyamine levels evolution.

ERYTHROCYTE POLYAMINE LEVELS AS GUIDES FOR POST-SURGICAL INTRA-ARTERIAL CHEMOTHERAPY

A patient should be treated before CT scan signs of malignant glioma recurrence, in other words before the reappearance of a new constituted tumor mass, and the treatment strategy of patients operated for glioblastoma would be more biologically adapted to the intra-cranial hyperplastic level at the time of therapy.

As previously described (15), after cutting away and radiotherapy (40-60 grays), patients received intra-arterial chemo-infusion(s) of BCNU and/or cis-platinum associated or not with a podophyllotoxin derivative (Vehem-Sandoz);furthermore, seven patients received intra-arterial ^{131}Iodine (15 grays).

Table 3. Patients treated by intra-arterial chemotherapy according to their RBC polyamine levels

Name	Age (years)	Post-surgical survival time (months)	Surviving patient (Y/N)	BCNU (1)	Cis-plat. (2)	PC (3)	P sessions number (4)	^{131}Iodine administration (5)
Cha..	44	5	N	B			3	
Fes..	64	5	N			A	2	
Duv..	62	7	N			B	2	
Des..	39	8	Y			C	1	
Oha..	56	8	N			C	1	
Pil..	48	8	N	C	B		4	
Ann..	50	9	Y			C	1	
Cas..	66	12	N			C	3	
Thy..	42	12	N			B	2	
Len..	72	13	N			B	13	1
Rua..	57	13	Y			B	1	
Lar..	50	14	N			A	5	
Xav..	63	15	N	A	A	C	3	
Bor..	54	16	N	A	A	B	5	
Cre..	60	16	N	C	A	B	7	
Lau..	25	16	N	C		A	8	
Tur..	47	16	Y	A		A	6	2
Bar..	49	20	Y			A	3	3
Lec..	60	22	N			C	2	1
Let..	33	24	Y	A		B	2	
Leb..	23	25	Y	A			1	
Vie..	55	25	Y	B	B	B	5	
Duc..	40	26	Y	B	A	B	5	3
Hen..	42	28	N	C		B	6	1
Tre..	31	31	Y	B			2	
Dup..	20	40	Y	C		A	5	2

(1) BCNU (total quantity / patient): A (<200 mg), B (200-400 mg), C (>400 mg).
(2) Cis-platinum (total quantity / patient): A (20 mg), B (200 mg).
(3) Polychemotherapy : A (200 mg BCNU;120 mg cis-platinum;300 mg VM26), B (100 mg BCNU;60 mg cis-platinum;150 mg VM26), C (20 mg BCNU; 12 mg cis-platinum;30 mg VM26).
(4) Polychemotherapy sessions number.
(5) ^{131}Iodine sessions number (15 grays).

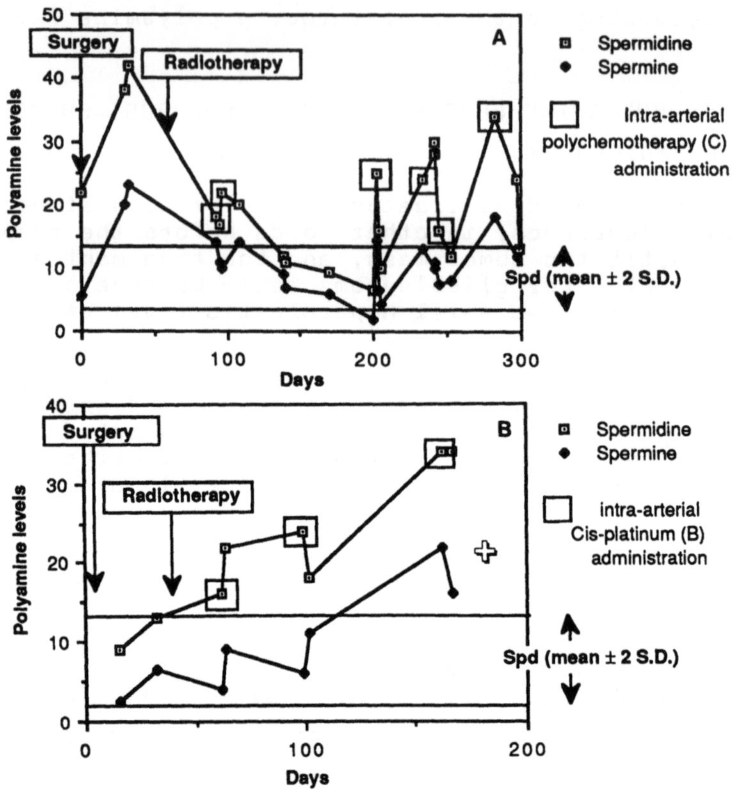

Figure 4. Erythrocyte polyamine evolution in two glioblastoma patients treated by intra-arterial chemotherapy.

Intra-arterial chemotherapy technique consisted in a catheterization of intra-cranial internal carotid with positioning of the silastic catheter above the ophthalmic artery associated or not, depending upon the tumor localization, to the controlateral intra-cranial internal carotid and to the catheterization of the basilar artery (15). Positioning of the catheter above the ophthalmic artery ruled out orbital complications and was preferred to a more distal catheterization of cortical branches, which does not permit the chemo-infusion of the whole vascular territory. In every case intra-arterial chemo-infusion lasted three hours.

In these conditions 26 non hospitalized clinically remitting glioblastoma patients were treated only according to their erythrocyte polyamine levels determined before intra-arterial chemotherapy. In the absence of abnormal RBC polyamine levels, patients were not treated by intra-arterial chemotherapy. Patients were monthly evaluated, including clinical and TDM examinations, and RBC polyamine levels determinations.

Considering Table 3 which summarizes our results, it can be observed that 9 glioblastoma patients (7 being presently alive) survived more than 18 months after neurosurgical procedure, which corresponds to a survival level of 41% (9/22, instead of 9/26, 4 patients being presently alive with a survival time inferior to 18 months, see Table 3).

Having decided to treat all our patients, we have no control group and are unable to produce statistically confirmed results. But, considering that 1) in the absence of radiotherapy, 90% of patients die 6 months after neurosurgery procedure, 2) post-surgical radiotherapy is responsible for a two months' life span increase, 90% of patients dying 8 months after surgical cutting away (10% of them reaching 18 months), 3) best results otained after surgery, radiotherapy and i.v. administration of nitroso-urea derivatives, correspond to 30% of surviving patients 18 months after surgical procedure, our present results (41% of surviving patients 18 months after surgery) lead us to consider our chemotherapy conditions as a new and promising way in the treatment of glioblastoma patients. Nevertheless, this study only based on 26 patients has to be confirmed on a greater number, treated according to RBC polyamine levels.

As always observed in clinical studies concerning survival time of high grade intra-cranial malignant glioma treated patients, the age of the subjects appears as an important parameter (Fig. 5), but was not decisive in our study (Table 3).

In this study no correlation was to be found between the survival time, the quality and the sessions number of intra-arterial chemotherapy (Table 3).

We statistically observed that polychemotherapy sessions were generally linked with a lowering of erythrocyte polyamine levels (Fig. 4A). This post-chemotherapy diminution in RBC polyamine concentrations presently appears rather in treatment responsive tumors, which is not the case in Figure 4B.

In the near future, this blood criteria could favor a more biologically dependent therapy, especially as concerns the quality of anticancer drug to use, according to RBC polyamine levels variations related to the treatment tumor response.

Erythrocyte polyamine levels, associated with other biological parameters, would contribute to the individualization of antiproliferative treatment in glioblastoma patients.

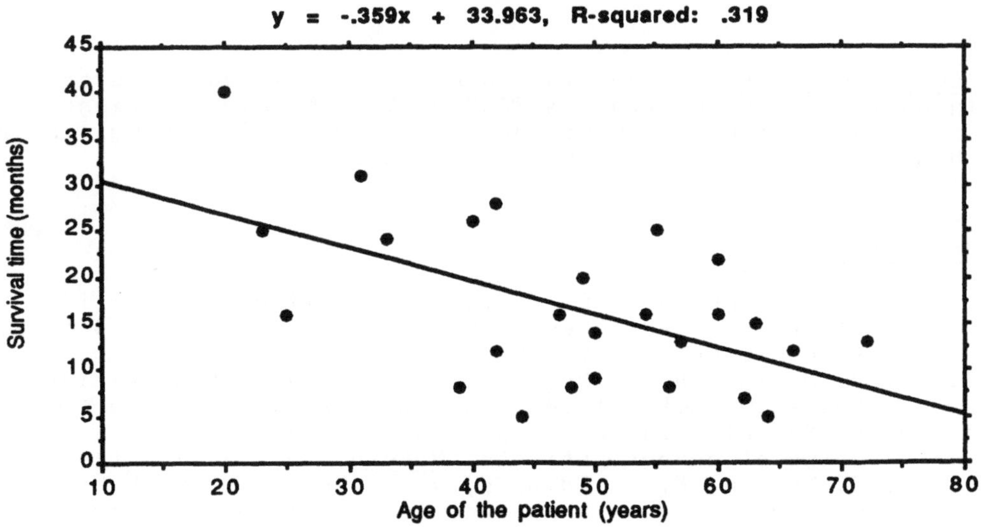

Figure 5. Influence of age on patient survival time.

DISCUSSION

In this study we tried to evaluate the clinical efficacy of erythrocyte polyamine levels determination in neuro-oncology.

As observed in patients undergoing neurosurgery, isolated abnormal RBC polyamine concentrations are not by themselves a diagnosis of intracranial malignant brain tumor. 30% of erythrocyte polyamine levels observed in malignant glioma patients before surgery correspond to normal ones. A possible explanation concerning the presence of normal polyamine levels in patients harboring a CT scan confirmed tumor mass, could be due to the fact that RBC polyamine concentrations depend on intracranial cell proliferation intensity and not on tumor volume.

By contrast, during clinical remission of grade IV glioma patients, the presence of isolated abnormal RBC polyamine levels have to be considered as early biological arguments of tumor growth.

The main problem concerning the erythrocyte polyamine long term follow-up lies in the blood puncture frequency. Our patients were monthly subjected to RBC polyamine level determination, but a better biological monitoring would be certainly obtained if we had at our disposal a simplified analysis method permitting a weekly survey of post-surgical non hospitalized patients. RBC polyamine levels corresponding to an index of cell proliferation, which hasn't been proved as a continuous process up to now, a lot of biological information is regretably lost.

In spite of this technical limitation which presently confines this type of biological analysis to specialized laboratories, RBC polyamine levels determination would favor the individualization of antiproliferative treatment in glioblastoma patients.

Though limited in the number of patients treated, our study may be considered as a new biological approach in neuro-oncology, the aim of which is to adapt therapy to individual intracranial hyperplastic profile.

Acknowledgements : Financial support for this work was provided in part by CNRS (Contract N°8660N0315), the Ligue Nationale Française contre le Cancer and the Foundation Jean Langlois. We thank A. Bernard for manuscript preparation.

References

1. G. Scalabrino and M.E. Ferioli, Polyamine in mammalian tumors, in "Advances in Cancer Research", Vol 35 & 36, Academic Press Inc., New York (1982).
2. P. P. McCann, A. E. Pegg and A. Sjoerdsma, "Inhibition of Polyamine Metabolism", Academic Press Inc., New York (1988)
3. D. H. Russell, Increased polyamine concentrations in the urine of cancer patients, Nature, 233:144-145 (1971).
4. L. J. Marton, M. S. Edwards, V. A. Levin, W. P. Lubich, CSF polyamines : a new and important means of monitoring patients with medulloblastoma, Cancer, 47:757-760 (1981).

5. O. M. Rennert and J. B. Shukla, Polyamine in health and disease, in: "Advances in Polyamine Research", R.A. Campbell, D. R. Morris, D. Bartos, G.D. Davex and F. Bartos, eds., Raven Press, New York (1978).
6. J-Ph. Moulinoux, V. Quemener and G-A. Quash, In vitro studies on the entry of polyamines into normal red blood cells, Biochimie, 66:385-393(1984).
7. J-Ph. Moulinoux, V. Quemener, A-M. Roch, L. Toujas and G-A. Quash, Red blood cell free polymaines in mice bearing the Lewis lung carcinoma (3LL) and in patients with bronchopulmonary cancer, Int. J. Cancer, 34:277-281(1984).
8. J-Ph. Moulinoux, V. Quemener and Y. Chambon, Evolution of red blood cell polyamine levels in partially hepatectomized rat, Eur. J. Cancer, 23:237-244(1987).
9. J-Ph. Moulinoux, D. Delamaire, B. Beau, V. Quemener, P. Brissot, Y. Deugnier, Y. Chambon and M. Bourel, Diagnosis value of erythrocyte free polyamines and histaminemia in malignant hepatic tumors, Clin. Chim. Acta, 145:77-87(1985).
10. V. Quemener, E. Le Gall, C. Edan and J-Ph. Moulinoux, Red blood cell polyamine levels in children with acute leukemia:their clinical interest in disease staging and monitoring therapy efficiency, Cancer J., 4:174-179 (1986).
11. J-Ph. Moulinoux, V. Quemener, M. Chatel and F. Darcel, Polyamines in human brain tumors:a correlative study between tumor, CSF and red blood cell polyamine levels, J. Neuro-Oncology (JNO), 2:153-158(1984).
12. J-Ph. Moulinoux, V. Quemener, H. Hercouet, F. Darcel and M. Chatel, Red blood cell polyamines in malignant glioma patients:spermidine and spermine blood levels, and tumor evolution, in: "Biology of Brain Tumor", Martinus Nijhoff Publishers, Boston(1986).
13. M. Chatel, F. Darcel, V. Quemener and J-Ph. Moulinoux, Red blood cell polyamines as biochemical markers of supratentorial malignant gliomas, Anticancer Res., 7:33-38(1987).
14. J-Ph. Moulinoux, V. Quemener, F. Darcel and M. Chatel, Red blood cell polyamine levels in the long term follow-up of malignant glioma patients, in: "Brain Oncology:Biology, Diagnosis and Therapy", Martinus Nijhoff Publishers, Boston(1987).
15. J. Theron, J-Ph. Moulinoux, A. Casasco, V. Quemener, P. Courtheoux and J-M. Derlon, Intra-arterial chemotherapy with BCNU. Correlations between tumor growth and RBC polyamine levels, in: "Brain Oncology:Biology, Diagnosis and Therapy", Martinus Nijhoff Publishers, Boston(1987).

ABNORMALITIES OF POLYAMINE BIOSYNTHESIS IN SPINAL CORD OF TOTALLY GASTRECTOMIZED RATS

G. Scalabrino, M.E. Ferioli, E. Lorenzini and R. Candiani

Institute of General Pathology and C.N.R. Centre for Research in Cell Pathology, University of Milan, 20133-MILANO, Italy

Wir fühlen dass selbst wenn alle möglichen wissenschaftlichen Fragen beantwortet sind, unsere Lebensprobleme noch gar nicht berührt sind. Freilich bleibt dann eben keine Frage mehr; und eben dies ist die Antwort.

We feel that even when all possible scientific questions have been answered, the problems of life remains completely untouched. Of course there are then no questions left, and this itself is the answer.

L. Wittgenstein, Tractatus Logico-Philosophicus, 6.52.

INTRODUCTION

Hematological and neurological sequelae are the two major manifestations in man of Addisonian pernicious anemia (PA)[1]. The degree of central nervous system (CNS) damage is not always well correlated with the severity of the anemia and therefore, the neurological disease in PA is not secondary to the anemia[2]. The neurological illness in PA chiefly affects the white matter of the dorsal and lateral columns of the spinal cord (SC) in man and can be reproduced experimentally in other mammals[3,4]. It is generally considered that the major neuropathological feature in PA is an uneven and severe "demyelination" (spongiform degeneration)[3,4] of SC.

The first full description of this demyelinating neuropathy was given by Russell, Batten and Collier[5] at the beginning of this century, who named it "subacute combined degeneration" (SCD) of the SC. This term, however, fails to encompass all aspects of the disease, because it does not clearly suggest involvement of peripheral nerves and brain. Although since then many other terms for this neurological disorder have been tentatively introduced, the initial term, i.e., SCD, is generally widely used by most authors and will be used here by us.

The pathogenesis of SCD of the mammalian SC is still substantially a puzzle. It has generally been held that impairment of DNA synthesis accounts for the megaloblastic erythropoiesis but not for the neurological damage in PA, since it was believed that affected nerve cells do not divide and therefore do not synthesize DNA[1,2,6]. Many general hypotheses have been proposed to connect known enzymological functions of some coenzymes with postulated mechanisms of neurological involvement. The most widely accepted envisions abnormalities of the metabolism of myelin lipids as a consequence of impairment of the methyl-malonyl-CoA mutase enzyme, one of the vitamin B_{12}-dependent reactions[7,8]. Understanding the biochemical basis of SCD has also been hampered by the lack of a suitable experimental animal model. The main methods for inducing SCD of SC in experimental animals are reported in Table 1. Until now, the most widely used ex-

TABLE 1

MAIN METHODS TO REPRODUCE SUBACUTE COMBINED DEGENERATION (SCD) OF SPINAL CORD IN EXPERIMENTAL ANIMALS

1) Exposure to nitrous oxide[7,9,10] (monkeys, fruit-bats).
2) Vitamin B_{12}-deficiency through dietary deprivation[8,11-13] (monkeys, fruit-bats).
3) Administration of cycloleucine (an analogue of methionine), which inhibits the biosynthesis of S-adenosyl-L--methionine[14,15] (mice).

perimental model is to expose experimental animals to nitrous oxide (N_2O) and keep them in an N_2O environment[7,9,10,16]. This model is based on the clinical observation that exposure to the anestethic gas N_2O produces neuropathy and megaloblastic changes in the bone marrow in man [17,18], because N_2O oxidizes some forms of vitamin B_{12}[19-21]. Unfortunately, however, rats and mice exposed to N_2O develop neither neurological nor hematological abnormalities[9,20]. Furthermore, even more important, the reduced presence or even the absence of the gastric intrinsic factor (IF) (which are another specific aspect of the pathogenesis of PA) is not mimicked in any of the experimental models.

We have made rats deficient in both vitamin B_{12} and IF through total gastrectomy (TG). Furthermore, we have focused our experimental observations on the neurological damage that can be obtained in the rat CNS in this way and, even more narrowly, only in the SC. The hematological consequences brought in the experimental model were of little or no interest, because it is well known that megaloblastic hemopoiesis in vitamin B_{12} deficiency is specific for humans[20]. We measured the levels of both the polyamine biosynthetic decarboxylases (PBD), L-ornithine decarboxylase (EC 4.1.1.17) (ODC) and S-adenosyl-L-methionine decarboxylase (EC 4.1.1.50) (SAMD), in the SCs of rats totally gastrectomized for one or two months. There was more than one reason to believe that polyamine biosynthesis should be deranged in these SCs: (a) both the PBD are well known and widely used markers of differentiation of mammalian cells, including nerve cell[22,23]; (b) polyamines in mammalian CNS are substantially bound to myelin[24,25]; (c) S-adenosyl-L--methionine (SAM) reverses or, at least, ameliorates many of the effects of vitamin B_{12} deficiency and/or vitamin B_{12} inactivation in rats and humans[19,26] and SAM metabolism is also strictly connected to the polyamine biosynthetic pathway[20]; (d) MTA, another product of the polyamine biosynthetic pathway[22], is by far the most effective compound for restoring folate synthesis in livers of rats in which vitamin B_{12} was inactivated by N_2O[20,26]. We also determined the PBD levels in SCs of rats made deficient in vitamin B_{12} through dietary deprivation and we compared these results with those for SCs of totally gastrectomized rats, to assess the role of vitamin B_{12} in the pathogenesis of SCD. Finally, we did histological studies of SCs of totally gastrectomized rats, to ascertain if and when degenerative changes might begin to occur and if these were concomitant with the changes in polyamine biosynthesis.

MATERIALS AND METHODS

Chemicals. L-[1-^{14}C]-Ornithine monohydrochloride (S.A. 58 mCi/mmol) and S-adenosyl-L-[carboxyl-^{14}C]methionine (S.A. 60 mCi/mmol) were purchased from the Radiochemical Centre (Amersham, Bucks.,U.K.).

Animals. Adult male non-inbred rats (Sprague-Dawley strain), weighing 250-280 g, were used in this study and were obtained from Charles River Italia (Calco, Italy). All rats were housed as previously described[27] in individual cages with screen bottoms. Coprophagy was carefully prevented.

Treatment of animals. Surgery: rats were totally gastrectomized under ketamine chloride anesthesia. After the removal of the stomach, the esophagus and the duodenum were joined by a termino-lateral anastomosis. Control rats were only laparotomized. The totally gastrectomized rats were fed parenterally for the first four-five days after the operation. The laparotomized control rats had the same parenteral nutrition for the same time period after laparotomy. After the period of parenteral nutrition, both the totally gastrectomized rats and the laparotomized rats ate standard pellets. The different times when the rats were killed are reported in Tables 2 and 3. Vitamin B_{12}-deficient diet: the composition of the vitamin B_{12}-deficient diet was essentially that reported by Brink et al.[28], except that a sulfonamide, succinylsulfathiazole(1g%), was added to decrease the number of intestinal bacteria.

Assays of enzyme activities. The ODC and SAMD activities in the 20,000 x g supernatant were determined by the radioisotopic methods routinely used in our laboratory[27]. Each whole SC was processed separately and homogenized 1:3 (w/v) in a medium whose composition has been previously reported[27]. Other organs in which PBD activities were determined were homogenized 1:2 (w/v) in the same medium.

Studies of blood components. Blood from totally gastrectomized rats, laparotomized rats and rats fed the vitamin B_{12}-deficient diet was collected by cardiac puncture, with hemolysis carefully prevented. The levels of vitamin B_{12} and folate in sera of totally gastrectomized rats, of laparotomized rats and of rats fed the vitamin B_{12}-deficient diet were measured by radioimmunoassay. Peripheral blood smears were stained by routine methods and examined for changes in erythrocyte and leukocyte morphology. The hematological course for totally gastrectomized rats was characterized by a marked anisocytosis and a progressive hypochromic anemia, which was first observed one month after TG. Red and white cell counts, hematocrits and hemoglobin concentrations were obtained with the Coulter Model ZBI. The differential counts of leukocytes in rats fed the vitamin B_{12}-deficient diet did not differ quantitatively nor qualitatively from normal (results not reported here). Glycemia, blood ammonia levels, the concentrations of the main electrolytes and of the main proteins in sera of the totally gastrectomized rats or the laparotomized rats during parenteral nutrition were monitored by routine laboratory methods. The values observed for all these hematological parameters in these rats did not differ significantly from normal (results not reported here).

Histological examination. At each experimental killing time after TG or laparotomy, the rats were eviscerated immediately after killing and SCs were rapidly removed. After fixation in 10% formalin for one week, blocks were taken from various levels of the cervical, thoracic and lumbar segments of SC. Each block was embedded in paraffin and 8 μm transverse sections were stained with hematoxylin and eosin, the thionine stain (Nissl's method[29]) and luxol fats blue method according to Klüver and Barrera[30] for myelinated fibers. Microscopic examinations were done without knowledge of which experimental group the SC sections belonged to. Since there were no enzyme changes in SCs of rats fed the vitamin B_{12}-deficient diet (see "results" section), histological examination of these SCs was not done.

TABLE 2

TIME-COURSES OF THE ACTIVITIES OF ORNITHINE DECARBOXYLASE (ODC) AND S-ADENOSYL-L-METHIONINE DECARBOXYLASE (SAMD) IN SPINAL CORD OF RATS KILLED AT DIFFERENT TIMES AFTER LAPAROTOMY (LP), AFTER TOTAL GASTRECTOMY (TG) OR AFTER FEEDING VITAMIN B_{12}-DEFICIENT DIET (Cb1D-D)

Type of treatment	ODC activity (pmol CO_2/mg prot/30 min)	SAMD activity (pmol CO_2/mg prot/30 min)
A NONE	40.1 ± 2.5 (14)	143.8 ± 7.3 (14)
1st MONTH AFTER LP	44.7 ± 3.3 (6)	135.5 ± 10.6 (6)
1st MONTH AFTER TG	37.5 ± 4.3 (8)	119.4 ± 8.3* (8)
2nd MONTH AFTER TG	72.3 ± 6.5**(14)	155.3 ± 9.4 (14)
B NONE	50.6 ± 4.1 (6)	165.1 ± 19.7 (6)
1st MONTH ON Cb1D-D	52.6 ± 5.3 (6)	175.9 ± 13.9 (6)
3rd MONTH ON Cb1D-D	55.7 ± 4.6 (6)	178.3 ± 11.3 (6)

Number of animals in parentheses. Results are mean values ± S.E.M.

* $P < 0.05$, ** $P < 0.01$ vs. both unoperated rats and laparotomized rats (Dunnett's test).

RESULTS

From the data in Table 2A, it appears that there was a moderate, but significant decrease in SAMD activity in the SCs of rats killed one month after TG. One month later, i.e., two months after TG, there was a sharp and significant increase in ODC activity in the SC. Table 2B contains the PBD activity levels in the SCs of rats maintained on the vitamin B_{12}-deficient diet for a period of three months, longer than needed for changes in PBD activities to be seen in SCs of totally gastrectomized rats. Quite differently from what was observed in totally gastrectomized rats, there were no significant changes in either ODC activity or SAMD activity level in the SCs of these vitamin B_{12}-deficient rats after either one month or even after three months of vitamin B_{12} deprivation. Parenthetically, it is worthwhile to mention

TABLE 3

TIME-COURSES OF THE ACTIVITIES OF ORNITHINE DECARBOXYLASE (ODC) AND S-ADENOSYL-L-METHIONINE DECARBOXYLASE (SAMD) IN LIVER OF RATS KILLED AT DIFFERENT TIMES AFTER TOTAL GASTRECTOMY (TG) OR AFTER FEEDING VITAMIN B_{12}-DEFICIENT DIET (Cb1D-D)

Type of treatment	ODC activity (pmol CO_2/mg prot/30 min)	SAMD activity (pmol CO_2/mg prot/30 min)
A		
NONE	38.2 ± 2.7 (6)	192.1 ± 10.1 (6)
1st MONTH AFTER TG	42.2 ± 3.8 (6)	201.9 ± 13.0 (6)
2nd MONTH AFTER TG	36.9 ± 2.1 (6)	189.2 ± 11.8 (6)
B		
3rd MONTH ON Cb1D-D	26.2 ± 2.0*(6)	62.5 ± 5.7**(6)

Number of animals in parentheses. Results are mean values ± S.E.M.
* $P < 0.05$, ** $P < 0.01$ vs. controls (Student t test)

here that both TG and the 3-month deprivation of dietary vitamin B_{12} greatly reduced the levels of vitamin B_{12} but not those of folate in sera of both the groups of rats (results not shown).

The changes in the PBD levels in SCs of totally gastrectomized rats over the experimental period seem to be specific for this organ, since they were not observed in other organs taken from the same rats. We measured PBD activities in livers, thymuses , kidneys and spleens of totally gastrectomized rats and we reported in Table 3A only the data for the livers as a typical example of a PBD pattern with no variations. Table 3B shows the PBD activities in livers of rats fed the vitamin B_{12}-deficient diet for three months. Significant decrease in hepatic ODC activity and an even greater decrease in hepatic SAMD activity occurred.

Histologically, we observed degenerative changes in the white matter of all the different segments, cervical, thoracic and lumbar, of the SCs of 2-month-totally-gastrectomized rats. The demyelinated areas have, characteristically, a spongy appearance in the sections stained with hematoxylin-eosin and with the method of Klüver-Barrera (Figures 1 and 2). The central grey matter both in 2-month-totally-gastrectomized rats and in normal laparotomized rats was intact (see Figures 1 and 2).

DISCUSSION

In the present study we demonstrated that TG causes a sharp increase in ODC activity level in SCs of rats killed two months after operation. Because high levels of ODC activity in mammalian CNS are characteristically observed during fetal development[22], it is conceivable that TG induces fetal-type regressive modifications (which we call "biochemical fetalism") of polyamine biosynthesis in the rat SC. This interpretation is supported by our morphological studies, which revealed a pronounced spongiform demyelination of

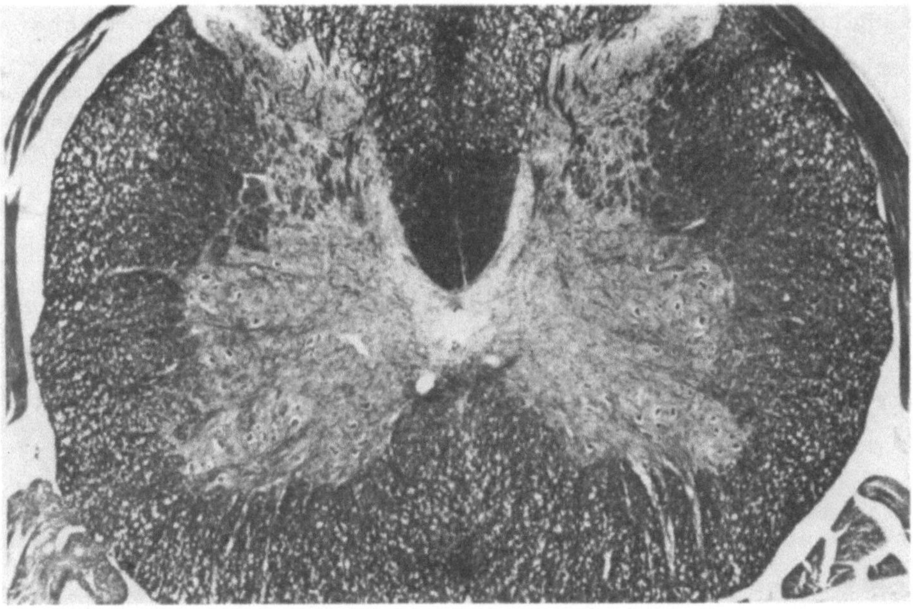

Fig. 1. Transverse section of lumbar segment of the spinal cord of a 2-month-totally-gastrectomized rat. The characteristic spongy degeneration in the white matter is illustrated (Klüver-Barrera; x about 30).

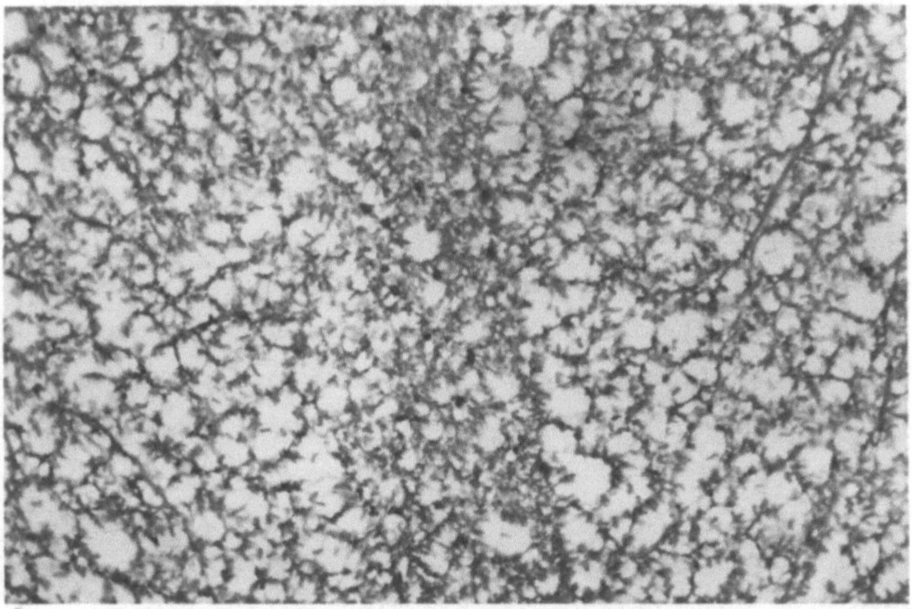

Fig. 2. Higher magnification of the posterior columns seen in Fig. 1. The spongiform appearance is evident (H.E.; x about 180).

the white matter of the whole SC in rats killed two months after TG. It is worthwhile to recall here that myelination is a morphological and biochemical event that characteristically begins in SC during the middle fetal months[31].

Furthermore, our results show that vitamin B_{12}-deficiency by itself is not responsible for the changes in PBD activities in SCs of totally gastrectomized rats, since none of these changes were observed in SCs of rats maintained on the vitamin B_{12}-deficient diet even for a longer time period than the totally gastrectomized rats were kept alive. In keeping with this is the report by Fehling et al.[32], who found no neuropathological changes reminiscent of SCD in SCs of rats after even seven months of feeding a vitamin B_{12}-deficient diet. Again, exposure of rats to N_2O inactivates the methionine synthetase enzyme, but not the methylmalonyl-CoA mutase enzyme[33], which last is one of the steps thought to be involved in the pathogenesis of SCD[7,8]. By contrast, however, another recent report shows that SAM levels in brains of rats exposed to N_2O do not change[34]. Admittedly, the pathogenesis of SCD in both humans and in experimental animals is still an unsolved conundrum, but there is real reason to doubt that vitamin B_{12} has a direct role in the pathogenesis of SCD.

Lastly, previous studies[35,36] have demonstrated that methionine deficiency increases SAMD activity in both cultured cells and in rat liver. It is also known that one of the effects of vitamin B_{12}-deficiency is a drop in hepatic methionine concentration, due to impaired transformation of homocysteine to methionine[19]. It is reasonable to suppose that both our experimental groups of rats (i.e., totally gastrectomized and those on the vitamin B_{12}-deficient diet) would lack available methionine because of malabsorption of vitamin B_{12}. Nevertheless, hepatic SAMD was markedly decreased in rats fed the vitamin B_{12}-deficient diet and remained unchanged in rats totally gastrectomized for two months. Therefore, the connections between the availability of methionine and the regulation of the SAMD activity level in mammalian cells also require further elucidation.

REFERENCES

1. V. Herbert, Biology of disease. Megaloblastic anemias, Lab. Invest. 52:3 (1985).
2. M.M. Wintrobe, "Clinical hematology", Lea & Febiger, Philadelphia (1981), pp. 571-572.
3. L.W. Duchen and J.M. Jacobs, Nutritional deficiencies and metabolic disorders, in: "Greenfield's Neuropathology", J. Hume Adams, J.A.N. Conellis, and L.W. Duchen, eds., Edward Arnold, London (1981).
4. S.S. Pant, A.K. Asbury, and E.P. Richardson Jr., The myelopathy of pernicious anemia, Acta Neurol. Scand. 44 Suppl. 35:8 (1968).
5. J.S.R. Russell, F.E. Batten, and J. Collier, Subacute combined degeneration of the spinal cord, Brain 23:39 (1900).
6. W.S. Beck, The megaloblastic anemias, in: "Hematology", W.J. Williams, E. Beutler, A.J. Erslev, and M.A. Lichtman, McGraw-Hill Book Company, New York (1983).
7. J.M. Scott, J.J. Dinn, P. Wilson, and D.G. Weir, Pathogenesis of subacute combined degeneration: a result of methyl group deficiency, Lancet ii:334 (1981).
8. D.P. Agamanolis, M. Victor, J.W. Harris, J.D. Hines, E.M. Chester, and J.A. Kark, An ultrastructural study of subacute combined degeneration of the spinal cord in vitamin B_{12}-deficient Rhesus monkeys, J. Neuropathol. Exp. Neurol. 37:273 (1978).
9. J.J. Dinn, S. McCann, P. Wilson, B. Reed, D. Weir, and J. Scott, Animal model for subacute combined degeneration, Lancet ii:1154 (1978).

10. J. Van Der Westhuyzen, and J. Metz, Tissue S-adenosylmethionine levels in fruits bats (Rousettus aegyptiacus) with nitrous oxide-induced neuropathy, Br. J. Nutr. 50: 325 (1983).

11. R. Green, A.V. van Tonder, G.J. Oettle, G. Cole, and J. Metz, Neurological changes in fruit bats deficient in vitamin B_{12}, Nature, 254:148 (1975).

12. D.P. Agamanolis, E.M. Chester, M. Victor, J.A. Kark, J.D. Hines, and J.W. Harris, Neuropathology of experimental vitamin B_{12} deficiency in monkeys, Neurology, 26:905 (1976).

13. A.M. Goodman, J.W. Harris, and D. Kiraly, Studies in B_{12}-deficient monkeys with combined system disease. I. B_{12}-deficient patterns in bone marrow deoxyuridine suppression tests without morphologic or functional abnormalities, J. Lab. Clin. Med. 96:772 (1980).

14. W. Jacobson, G. Gandy, and R.L. Sidman, Experimental subacute combined degeneration of the cord in mice. J. Path. 109:Pxiii (1973).

15. G. Gandy, W. Jacobson, and R. Sidman, Inhibition of a transmethylation reaction in the central nervous system-an experimental model for subacute combined degeneration of the cord, J. Physiol. 233:1P (1973).

16. I. Chanarin, Cobalamins and nitrous oxide: a review, J. Clin. Pathol. 33:909 (1980).

17. R.B. Layzer, Myeloneuropathy after prolonged exposure to nitrous oxide, Lancet ii:1227 (1978).

18. N.A. Sourial, J.A.L. Amess, and R.J. Amos, Role of S--adenosylmethionine in DNA synthesis and haemopoiesis, Scand. J. Haematol. 34:303 (1985).

19. I. Chanarin, Megaloblastic anaemia, cobalamin, and folate, J. Clin. Pathol. 40:978 (1987).

20. I. Chanarin, R. Deacon, M. Lumb, M. Muir, and J. Perry, Cobalamin-folate interrelations: a critical review, Blood 66:479 (1985).

21. M. Muir, and I. Chanarin, Conversion of endogenous cobalamins into microbiologically-inactive cobalamin analogues in rats by exposure to nitrous oxide, Br. J. Haematol. 58:517 (1984).

22. G. Scalabrino, M.E. Ferioli, and G. Luccarelli, Polyamine biosynthesis in primary tumors of human central nervous system: review of current knowledge, Progr. Neurobiol. 25:289 (1985).

23. O. Heby, Role of polyamines in the control of cell proliferation and differentiation, Differentiation 19:1 (1981).

24. N. Seiler, and K. Deckardt, Association of putrescine, spermidine, spermine, and GABA with structural elements of brain cells, Neurochem. Res. 1:469 (1976).

25. G.G. Shaw, The polyamines in the central nervous system, Biochem. Pharmacol. 28:1 (1979).

26. J. Perry, I. Chanarin, R. Deacon, and M. Lumb, Chronic cobalamin inactivation impairs folate polyglutamate synthesis in the rat, J. Clin. Invest., 71:1183 (1983).

27. G. Scalabrino, M.E. Ferioli, R. Nebuloni, and F. Fraschini, Effects of pinealectomy on the circadian rhythms of polyamine biosynthetic decarboxylases and tyrosine aminotransferase in different organs of the rats, Endocrinology 104:377 (1979).

28. J.J. Brink, R.A. Beck, J.S. Miller, and S.W. Thenen, Relationship of urinary methylmalonic acid to vitamin B_{12} concentrations and hematologic changes in rats fed vitamin B_{12}-deficient diets, J. Nutr. 110:1338 (1980).

29. "Manual of histologic staining methods of the Armed Forces Institute of Pathology", L.G. Luria, ed., McGraw-Hill Book Company, New York (1968), p. 212.

30. "Manual of histologic staining methods of the Armed Forces Institute of Pathology", L.G. Luria, ed., McGraw-Hill Book Company, New York (1968), p. 203.

31. E.L. House, and B. Pansky, "A functional approach to neuroanatomy", McGraw-Hill Book Company, New York (1967), p. 118.
32. C. Fehling, M. Jägerstad, B. Åkesson, J. Axelsson, and A. Brun, Effects of vitamin B_{12} deficiency on lipid metabolism of the rat liver and nervous system, Br. J. Nutr. 39:501 (1978).
33. R. Deacon, M. Lumb, J. Perry, I. Chanarin, B. Minty, M.J. Halsey, and J.F. Nunn, Selective inactivation of vitamin B_{12} in rats by nitrous oxide, Lancet ii:1023 (1978).
34. M. Lumb, N. Sharer, R. Deacon, P. Jennings, P. Purkiss, J. Perry, and I. Chanarin, Effects of nitrous oxide-induced inactivation of cobalamin on methionine and S-adenosylmethionine metabolism in the rat, Biochim. Biophys. Acta 756:354 (1983).
35. T.O. Eloranta, and A.M. Raina, S-Adenosylmethionine metabolism and its relation to polyamine synthesis in rat liver, Biochem. J. 168:179 (1977).
36. M.J. Tisdale, Effect of methionine deprivation on S-adenosylmethionine decarboxylase of tumour cells, Biochim. Biophys. Acta 675:366 (1981).

ORNITHINE DECARBOXYLASE AND POLYAMINE BIOSYNTHESIS IN PANCREATIC ADAPTATION

Christian Löser, Ulrike Cleffmann, Frauke Alves,
Ulrich R. Fölsch and Werner Creutzfeldt

Div. of Gastroenterology and Endocrinology
Dept. of Medicine, Georg-August-University
D-3400 Göttingen, FRG

SUMMARY

This study was designed to investigate the role of ornithine decarboxylase (ODC) and polyamines in pancreatic adaptation. Cholecystokinin (CCK) is well-known to be a potent trophic stimulus on the pancreas. On the other hand, the oral application of the synthetic trypsin inhibitor camostate results in an extensive release of endogenous CCK in rats. α-difluoromethylornithine (DFMO), an irreversible and specific inhibitor of ODC, was applied simultaneously to elucidate the essential role of polyamines in pancreatic growth. Camostate feeding (200 mg/kg b.wt. orally twice a day) resulted in a rapid elevation of ODC activity already after 2 hours, reaching a maximum after 6 hours (about 200fold above controls) followed by a significant increase in putrescine after 4 hours and spermidine after 24 hours while spermine remained unchanged. The trophic parameters increased as expected in following time-course: thymidine kinase (12 hours), DNA polymerase (12 hours), protein (24 hours), pancreatic weight (24 hours) and DNA (5 days). DFMO (2% in drinking water + 3x300 mg/kg b.wt. i.p. during daytime) was not able to prevent but significantly delayed and reduced the camostate-induced increase in ODC and polyamines as well as the trophic parameters. These data indicate an essential role for ODC and polyamines in camostate-induced pancreatic growth and hormonal mediated pancreatic adaptation.

INTRODUCTION

The naturally occurring polyamines putrescine, spermidine and spermine are ubiquitous normal constituents in all prokaryotic and eukaryotic cells and play an important role in cell proliferation and differentiation (1-3). During intestinal adaptation as observed after jejunectomy (4), pancreatico-biliary diversion (PBD)(5), during recovery after mucosal injury (6) or during lactation (7) the role of polyamines is well-established (for review see 8, 9). However, though increased polyamine concentrations after hormonal stimulation of pancreatic growth were described (10-12), their role in pancreatic adaptation is hitherto less defined.

Cholecystokinin (CCK) is known to be a potent stimulus for pancreatic secretion and growth (13). Chronic oral administration of trypsin inhibitor induce pancreatic hypertrophy and -plasia (12, 14, 15). A large body of evidence indicates that trypsin inhibitors act by inactivating proteolytic enzymes from the small intestine. By an as yet unknown mechanism this causes the release of CCK which stimulates both pancreatic enzyme secretion and growth, indicating a negative feedback control of pancreatic enzyme secretion (16, 17). The synthetic trypsin inhibitor camostate (FOY-305) is a low molecular guanidino acid ester that extensively stimulates pancreatic growth in rats via endogenous CCK release (14, 15, 18).

Therefore, feeding of the non-toxic trypsin inhibitor camostate and the consecutive endogenous release of the growth mediator CCK provides a suitable model to study the role of ODC and polyamine biosynthesis in pancreatic adaptation under approximately physiological conditions.

EXPERIMENTAL DESIGN

Male Wistar rats weighing 180-220 g were treated either with camostate (200 mg/kg b.wt. by oral gavage every 12 hours), camostate plus DFMO (2 % in drinking water + 3x300 mg/kg b.wt. intraperitoneally during daytime), DFMO alone or 0.9 % saline as controls. Each group consisted of 5 animals at each given time-point. DFMO treatment was started 2 days prior to the experiments which were all started at 8:00 a.m.. In order to guarantee comparable metabolic conditions all animals were pair-fed. Non-fasting animals were killed by brief ether anaesthesia and exsanguination after 2, 4, 6, 8, 12 hours and 1, 2, 5, 10, 20 and 30 days of treatment. Blood for CCK determination was drawn from the aorta; the pancreas was immediately excised, weighed and homogenized as previously described (12).

Polyamines were measured by ion-pairing reversed-phase high-performance liquid chromatography (HPLC) and post-column derivatization with o-phthalaldehyde as recently published in detail (19). The enzyme activities of ornithine decarboxylase, thymidine kinase and DNA polymerase were measured according to the methods described by Haarstad et al. (10). Protein was determined according to Bradford (20) using the Bio-Rad reagent (Bio-Rad, Richmond, CA, USA), while DNA was measured using the fluorescent dye H-33258 (Hoechst AG, Frankfurt, FRG) as described by Labarca et al. (21). CCK plasma levels were measured by radioimmunoassay (17).

Camostate FOY-305 (N,N-dimethylcarbamoylmethyl-p-(p-guanidinobenzoyloxy) phenylacetate methanesulfonate) was provided by Sanol Schwarz (Monheim, FRG). α-difluoromethylornithine was kindly supplied by Centre de Recherche Merrell International (Strasbourg, France).

RESULTS AND DISCUSSION

After feeding of the trypsin inhibitor camostate a significant and long-lasting increase in CCK plasma concentrations could be observed prior to and during the increase of trophic parameters like pancreatic weight, DNA, thymidine kinase, DNA polymerase and protein, thus confirming previous studies of other groups (14, 15, 18). CCK is the most powerful hormonal stimulator of pancreatic growth (13, 22). CCK plasma levels in camostate-fed animals were maximal already after 2 hours (27.4 pmol/l), decreased up to 12 hours and remained then at a constant rate between 7-10 pmol/l throughout the 30-day experiment above controls (2.5-4.5 pmol/l).

Control values for ODC, polyamines and all measured trophic parameters (pancreatic weight, protein, DNA, thymidine kinase, DNA polymerase) are given in tables 1 and 2. Camostate feeding resulted in a rapid and highly significant increase in ODC activity with a maximum after 6 hours (about 200fold above controls) and ODC activity remained significantly elevated up to day 30 (fig. 1). The increase in ODC activity of camostate-fed rats was accompanied by significant elevations of putrescine starting after

Table 1. Pancreatic tissue concentrations of ODC, putrescine, spermidine and spermine in rats receiving 0.9 % saline every 12 hours (control group; $\bar{x} \pm SEM$; n = 5)

		Hours after a single administration					Days of treatment					
		2	4	6	8	12	1	2	5	10	20	30
ODC (pmol $^{14}CO_2$ formed/h/mg DNA)	$\bar{x}$	8,53	8,40	8,90	10,30	8,79	6,50	5,53	4,64	7,13	5,45	6,67
	SEM	3,34	2,08	3,00	1,4	1,6	1,47	2,45	1,60	1,90	2,46	2,10
Putrescine (nmol/mg DNA)	$\bar{x}$	10,25	11,70	10,00	8,62	9,78	9,60	6,80	6,57	17,10	14,10	12,2
	SEM	0,3	1,0	0,95	0,99	1,10	1,2	0,3	0,3	1,3	1,9	0,8
Spermidine (nmol/mg DNA)	$\bar{x}$	1874,5	1876,4	1880,1	1757,7	1890,7	1712,4	1804,8	1661,7	2130,2	1903,5	1970,4
	SEM	64,2	58,4	23,3	60,5	66,4	28,4	63,8	72,1	77,9	61,5	55,6
Spermine (nmol/mg DNA)	$\bar{x}$	275,6	296,0	247,0	287,5	301,9	324,7	285,1	283,4	409,8	289,6	347,8
	SEM	28,2	10,1	15,55	8,4	8,4	21,0	24,6	24,6	38,5	14,7	23,9

Table 2. Trophic parameters of the pancreas of the same rats as described in table 1 (control group)

		Hours after a single administration					Days of treatment					
		2	4	6	8	12	1	2	5	10	20	30
Pancreas weight	$\bar{x}$	3,82	3,84	3,73	3,84	3,76	4,10	4,00	3,91	4,37	4,28	4,08
(g/kg b.wt.)	SEM	0,17	0,22	0,10	0,12	0,12	0,20	0,20	0,10	0,30	0,23	0,08
DNA	$\bar{x}$	2,33	2,48	2,61	2,38	2,67	2,80	2,71	2,70	2,74	2,81	2,80
(mg/pancreas)	SEM	0,12	0,16	0,23	0,18	0,17	0,18	0,09	0,08	0,21	0,15	0,10
Protein	$\bar{x}$	172,3	171,6	168,0	178,4	171,7	139,7	146,8	169,4	211,1	253,3	246,1
(mg/pancreas)	SEM	12,8	11,7	5,4	10,0	4,2	4,2	4,3	3,2	12,6	10,2	6,5
DNA polymerase	$\bar{x}$			0,60	0,54	0,48	0,47	0,45	0,45	0,89	0,80	0,56
(pmol ^{3}H-dTTP incorporated/30min/mg DNA)	SEM			0,16	0,08	0,08	0,17	0,05	0,11	0,16	0,08	0,17
Thymidine kinase	$\bar{x}$			64,2	95,2	98,6	101,7	116,3	100,5	82,5	96,3	84,1
(pmol ^{3}H-TMP formed/h/mg DNA)	SEM			10,7	15,9	21,7	10,9	10,5	8,8	16,1	8,7	16,1

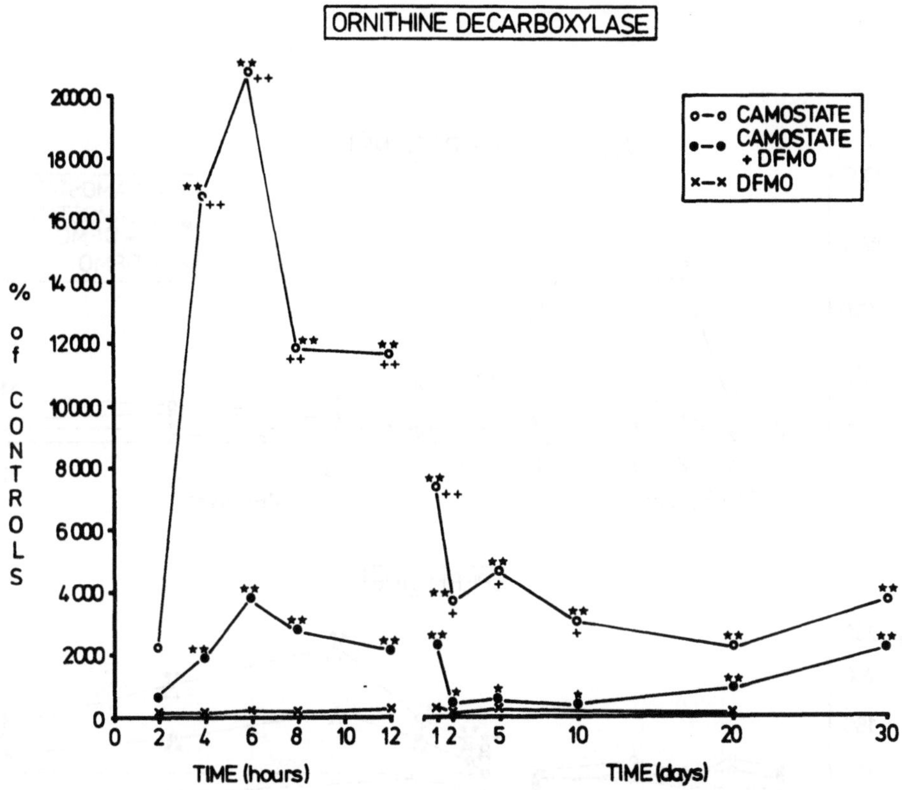

Fig. 1. Ornithine decarboxylase activities in rat pancreas (n=5) after treatment with camostate (200 mg/kg b.wt. orally every 12 hours), camostate plus DFMO (2 % drinking water plus 3x300 mg/kg b.wt. i.p. during daytime) or DFMO alone in percentage of control values (means). Asterisks indicate statistical differences from controls (*p < 0.01; **p < 0.005), Crosses indicate statistical differences of the camostate group versus camostate plus DFMO group (+p < 0.01; ++p < 0.005).

4 hours (maximum after 8 hours about 20fold above controls) and spermidine starting after 24 hours, while spermine concentrations remained unchanged (fig. 2). DFMO treatment was able to delay and significantly reduce the above mentioned increases in ODC, putrescine and spermidine, though a complete inhibition was not achieved (figs. 1 and 2). As far as the trophic parameters are concerned a similar phenomenon was observed. Camostate feeding resulted in an increase in trophic parameters in the following time-sequence: thymidine kinase (12 hours), DNA polymerase (12 hours), pancreatic weight (24 hours), protein (24 hours) and DNA (5 days)(fig. 3). Simultaneous administration of DFMO could not prevent the increase in these parameters but delayed and reduced them (fig. 3).

Since rats mainly drink during nighttime, DFMO administration in drinking water only proved to be less effective and therefore additional doses of 3x300 mg DFMO/kg b.wt. i.p. were given during daytime as previously pointed out (23, 24). The average daily DFMO consumption per animal amounted to about 3.4 g/kg b.wt./day. In accordance with previous findings (24, 25) this

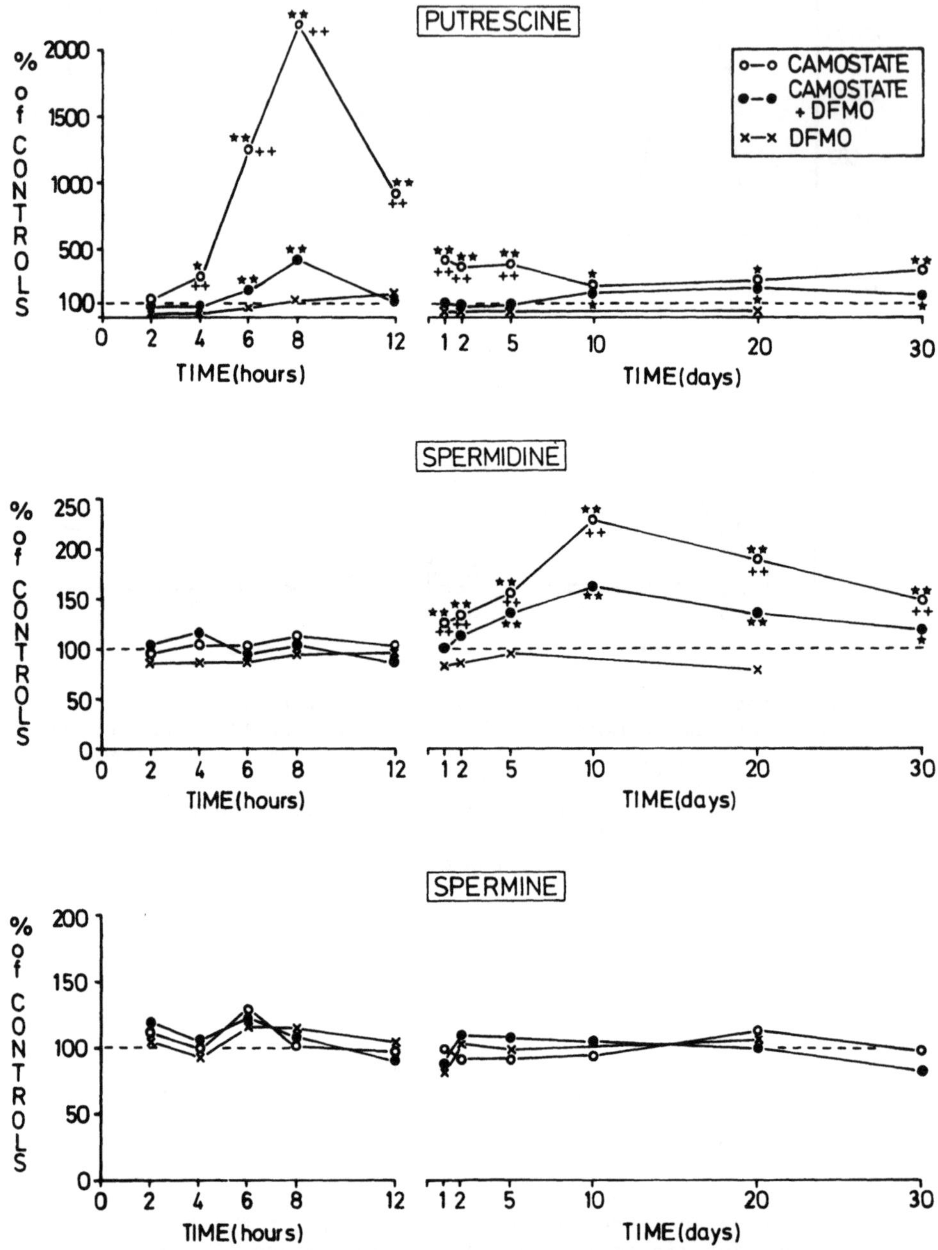

Fig. 2. Putrescine, spermidine and spermine concentrations in rat pancreas (n = 5) after treatment with camostate (200 mg/kg b.wt. orally every 12 hours), camostate plus DFMO (2 % in drinking water plus 3x300 mg/kg b.wt. i.p. during daytime) or DFMO alone in percentage of control values (means). Symbols as in fig. 1.

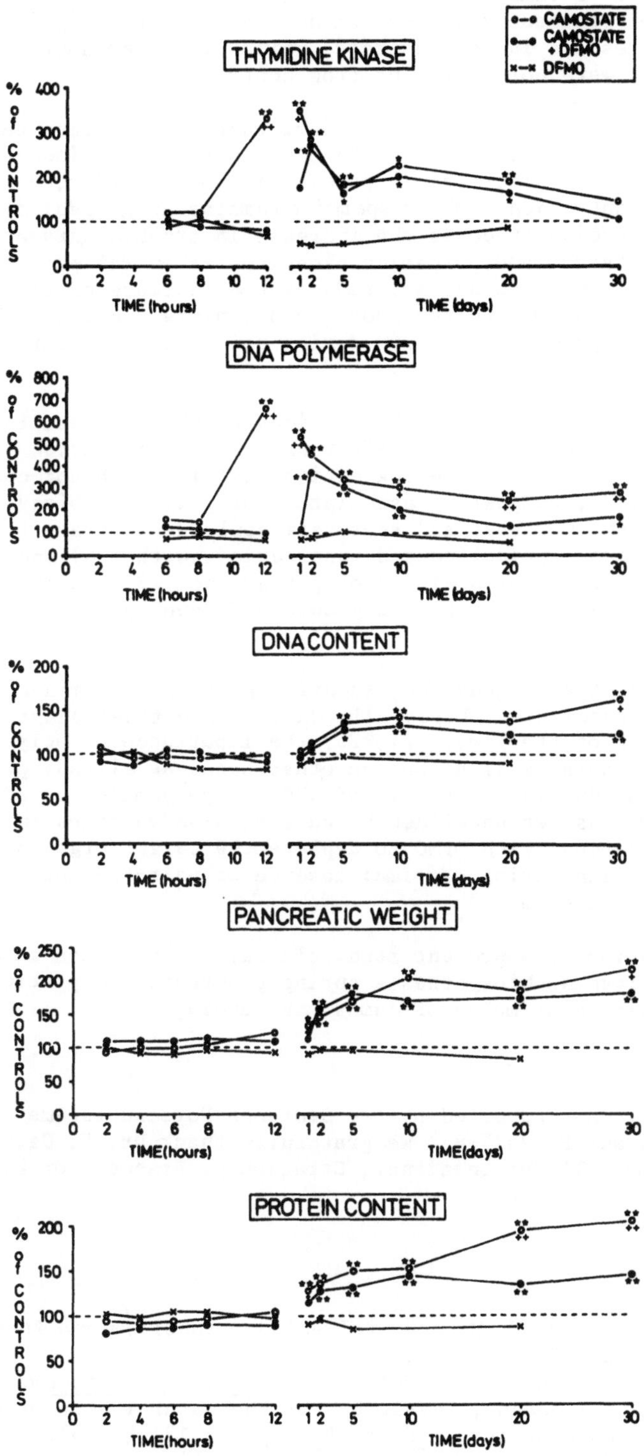

Fig. 3. Thymidine kinase, DNA polymerase, DNA content, pancreatic weight and protein content in rat pancreas (n = 5) after treatment with camostate, camostate plus DFMO or DFMO alone in percentage of control values (means). Symbols as in fig. 1.

dosage proved to be remarkably non-toxic, particularly diarrhoea was not observed. DFMO treatment alone caused an obvious but insignificant decrease in ODC and putrescine in rat pancreas, while all other above mentioned trophic parameters were not different from controls.

The results of our experiments show, that 1) camostate feeding causes a powerful endogenous CCK release, 2) camostate feeding induces via CCK release a significant increase in pancreatic ODC activity, putrescine and spermidine concentrations, while spermine remains unchanged, 3) the increase in ODC and putrescine precedes the increase in trophic parameters, 4) DFMO not only inhibits ODC and polyamine biosynthesis in rat pancreas in vivo, but also significantly delays and reduces the increase of trophic parameters and 5) a detailed view into the short- and long-term kinetics of polyamine biosynthesis and pancreatic growth during pancreatic adaptation in response to camostate feeding is provided.

Our data confirm previous observations which described increasing pancreatic polyamine concentrations after exogenous stimulation with CCK-like peptides (10, 11) or after feeding active trypsin inhibitor-containing raw soybean flour (12), suggesting association of polyamine biosynthesis with pancreatic growth. Morisset and Benrezzak could demonstrate that DFMO reduced the CCK-like peptide-induced pancreatic growth (26) and that this effect is reversible by the addition of putrescine (27). Furthermore, it was shown that DFMO reduces pancreatic growth and development in neonatal rats (28).

In various other organs the essential role of polyamines in trophic events is well-documented. Especially during intestinal adaptation (4-9) and regeneration of the liver (24, 25) the importance of polyamine biosynthesis is clearly characterized. On the other hand, as already pointed out by Luk and Yang in 1987 (9), the role of ODC and polyamines in adaptation of the pancreas was as yet undefined because of disadvantages in study design (i.e. no pretreatment with DFMO to deplete the target organ or insufficient DFMO administration during daytime; absence of pair-feeding; lack of short- and long-term studies etc.).

In conclusion, the present study clearly demonstrates an essential role for ODC and polyamine biosynthesis during pancreatic adaptation in response to the CCK-mediated stimulus of camostate feeding.

ACKNOWLEDGEMENT

This study was supported by the Deutsche Forschungsgemeinschaft, 5300 Bonn 2 (grant Fo 73/7-8). We gratefully thank Dr. F. Gerhart (Centre de Recherche Merrell International, Strasbourg, France) for kindly supplying DFMO.

REFERENCES

1. A. E. Pegg, and P. P. McCann, Polyamine metabolism and function, Am. J. Physiol. 243:C212 (1982).
2. C. W. Tabor, and H. Tabor, Polyamines, Ann. Rev. Biochem. 53:749 (1984).
3. A. Raina, and I. Jänne, Physiology of the natural polyamines putrescine, spermidine and spermine, Medical Biology 53:121 (1975).
4. G. Luk, and S. Baylin, Polyamines and intestinal growth - increased polyamine biosynthesis after jejunectomy, Am. J. Physiol. 245:G656 (1983).
5. M. Hosomi, F. Lirussi, N. H. Stace, S. Vaja, G. M. Murphy, and R. H. Dowling, Mucosal polyamine profile in normal and adapting (hypo and hyperplastic) intestine: effects of DFMO treatment, Gut 28, s. 1:103 (1987).

6. G. Luk, L. Marton, and S. Baylin, Ornithine decarboxylase is important in intestinal mucosal maturation and recovery from injury in rats, Science 210:195 (1980).
7. P. Yang, S. B. Baylin, and G. D. Luk, Polyamines and intestinal growth: absolute requirement for ODC activity in adaptation during lactation, Am. J. Physiol. 247:G553 (1984).
8. R. H. Dowling, M. Hosomi, N. H. Stace, F. Lirussi, B. Miazza, H. Levan, and G. M. Murphy, Hormones and polyamines in intestinal and pancreatic adaptation, Scand. J. Gastroenterol.20, suppl. 112:84 (1985).
9. G. D. Luk, and P. Yang, Polyamines in intestinal and pancreatic adaptation, Gut 28, suppl. 1:95 (1987).
10. H. Haarstad, A. Winnberg, and H. Petersen, Effects of a cholecystokinin-like peptide on DNA and polyamine synthesis in the rat pancreas, Scand. J. Gastroenterol. 20:530 (1985).
11. J. Morisset, and O. Benrezzak, Polyamines and pancreatic growth induced by caerulein, Life Sci. 35:2471 (1984).
12. Chr. Löser, U. R. Fölsch, D. Mustroph, P. Cantor, U. Wunderlich, and W. Creutzfeldt, Pancreatic polyamine concentrations and cholecystokinin plasma levels in rats after feeding raw or heat-inactivated soybean flour, Pancreas 3:285 (1988).
13. U. R. Fölsch, K. Winckler, and K. G. Wormsley, Influence of repeated administration of cholecystokinin and secretin on the pancreas of the rat, Scand. J. Gastroenterol. 13:663 (1978).
14. B. Göke, H. Printz, I. Koop, U. Rausch, G. Richter, R. Arnold, and G. Adler, Endogenous CCK release and pancreatic growth in rats after feeding a proteinase inhibitor (camostate), Pancreas 1:509 (1986).
15. M. Otsuki, A. Ohki, Y. Okabayashi, I. Suehiro, and S. Baba, Effect of synthetic protease inhibitor camostate on pancreatic exocrine function in rats, Pancreas 2:164 (1987).
16. D. S. Louie, D. May, P. Miller, and C. Owyang, Cholecystokinin mediates feedback regulation of pancreatic enzyme secretion in rats, Am. J. Physiol. 250:G252 (1986).
17. U. R. Fölsch, P. Cantor, H. M. Wilms, A. Schafmayer, H. D. Becker, and W. Creutzfeldt, Role of cholecystokinin in the negative feedback control of pancreatic enzyme secretion in conscious rats, Gastroenterology 92:449 (1987).
18. J. R. Wisner, R. E. McLaughlin, K. A. Rick, S. Ozawa, and I. G. Renner, Effects of L-364,718 a new cholecystokinin receptor antagonist, on camostate-induced growth of the rat pancreas, Gastroenterol. 94:109 (1988).
19. Chr. Löser, U. Wunderlich, and U. R. Fölsch, Reversed-phase liquid chromatographic separation and simultaneous fluorimetric detection of polyamines and their monoacetyl derivatives in human and animal urine, serum and tissue samples - An improved, rapid and sensitive method for routine application, J. Chromatogr, in press (1988).
20. M. M. Bradford, A rapid and sensitive method for the quantitation of microgram quantities of protein utilizing the principle of protein-dye binding, Anal. Biochem. 72:248 (1976).
21. C. Labarca, and K. Paigen, A simple, rapid, and sensitive DNA assay procedure, Anal. Biochem. 102:344 (1980).
22. U. R. Fölsch, Regulation of pancreatic growth, Clin. Gastroenterol. 13: 679 (1984).
23. J. C. Romijn, C. F. Verkoelen, and T. A. W. Splinter, Problems of pharmacokinetic studies on alpha-difluoromethylornithine in mice, Cancer Chemother. Pharmacol. 19:30 (1987).
24. H. Pösö, and A. E. Pegg, Effect of α-difluoromethylornithine on polyamine and DNA synthesis in regenerating rat liver. Reversal of inhibition of DNA synthesis by putrescine, Biochim. Biophys. Acta 696:179 (1982).
25. G. D. Luk, Essential role of polyamine metabolism in hepatic regeneration. Inhibition of DNA and protein synthesis and tissue regeneration by difluoromethylornithine in the rat, Gastroenterology 90:1261 (1986).

26. O. Benrezzak, and J. Morisset, Effects of α-difluoromethylornithine on pancreatic growth induced by caerulein, Regul. Pept. 9:143 (1984).
27. J. Morisset, and O. Benrezzak, Reversal of α-difluoromethylornithine inhibition of caerulein-induced pancreatic growth by putrescine, Regul. Pept. 11:201 (1985).
28. J. Morisset, and G. Grondin, Implication of ornithine decarboxylase and polyamines in pancreatic growth of neonatal rats, Pancreas 2:303 (1987).

POST-SYNTHETIC MODIFICATIONS

OF PROTEINS INVOLVING POLYAMINES

TRANSGLUTAMINASES AND THEIR REGULATION: IMPLICATIONS FOR POLYAMINE METABOLISM

Peter J. A. Davies, E. Antonio Chiocca, James P. Basilion, Susmita Poddar and Joseph P. Stein

Departments of Pharmacology and Medicine
The University of Texas Medical School at Houston
Houston, Texas 77225, U.S.A.

INTRODUCTION

Transglutaminases are a group of enzymes that catalyze the post-translational modification of protein-bound glutamine residues. It appears that the major physiological role of these enzymes is to cross-link proteins by promoting the formation of isopeptide bonds between protein-bound glutamine and lysine residues. These enzymes also catalyze the covalent conjugation of polyamines to proteins and this latter activity has led to speculation about their involvement in the biology of polyamine action. In this report we would like to present a brief review of the theoretical and experimental evidence for considering transglutaminases as components in polyamine metabolism.

THE TRANSGLUTAMINASE REACTION

To appreciate the potential role of transglutaminases in polyamine metabolism, it is necessary to understand the basic two-step reaction catalyzed by these enzymes (Fig. 1, for a detailed review see Ref. 1). Transglutaminases are related to a group of enzymes (mostly proteases) than contain an unusually reactive cysteine residue within their active site. In the first step of the transglutaminase reaction, this activated thiol reacts with the γ-carboxamide group of a protein-bound glutamine residue, resulting in the elimination of ammonia and the formation of an activated thiol ester enzyme-substrate complex (Step 1 in Fig. 1). This step is entirely analogous to the first step in the reaction of thiol proteases (such as the calpains or papain) with their substrates. The thiol ester intermediate is then subject to nucleophilic attack by a primary or secondary amine (Step 2, Fig. 1). The products of this second step are the substrate glutamine residue with a substituted γ-carboxamide group and the regenerated active site thiol.

The key to understanding the biology of transglutaminases lies in understanding the difference between the specificity of these enzymes for their glutamine and amine substrates. Transglutaminases are extremely selective in terms of the glutamine residues with which they will react. A basic feature of transglutaminases that distinguishes them from other enzymes of glutamine metabolism is that they react only with protein bound glutamine residues. Furthermore, not all protein-bound glutamine residues

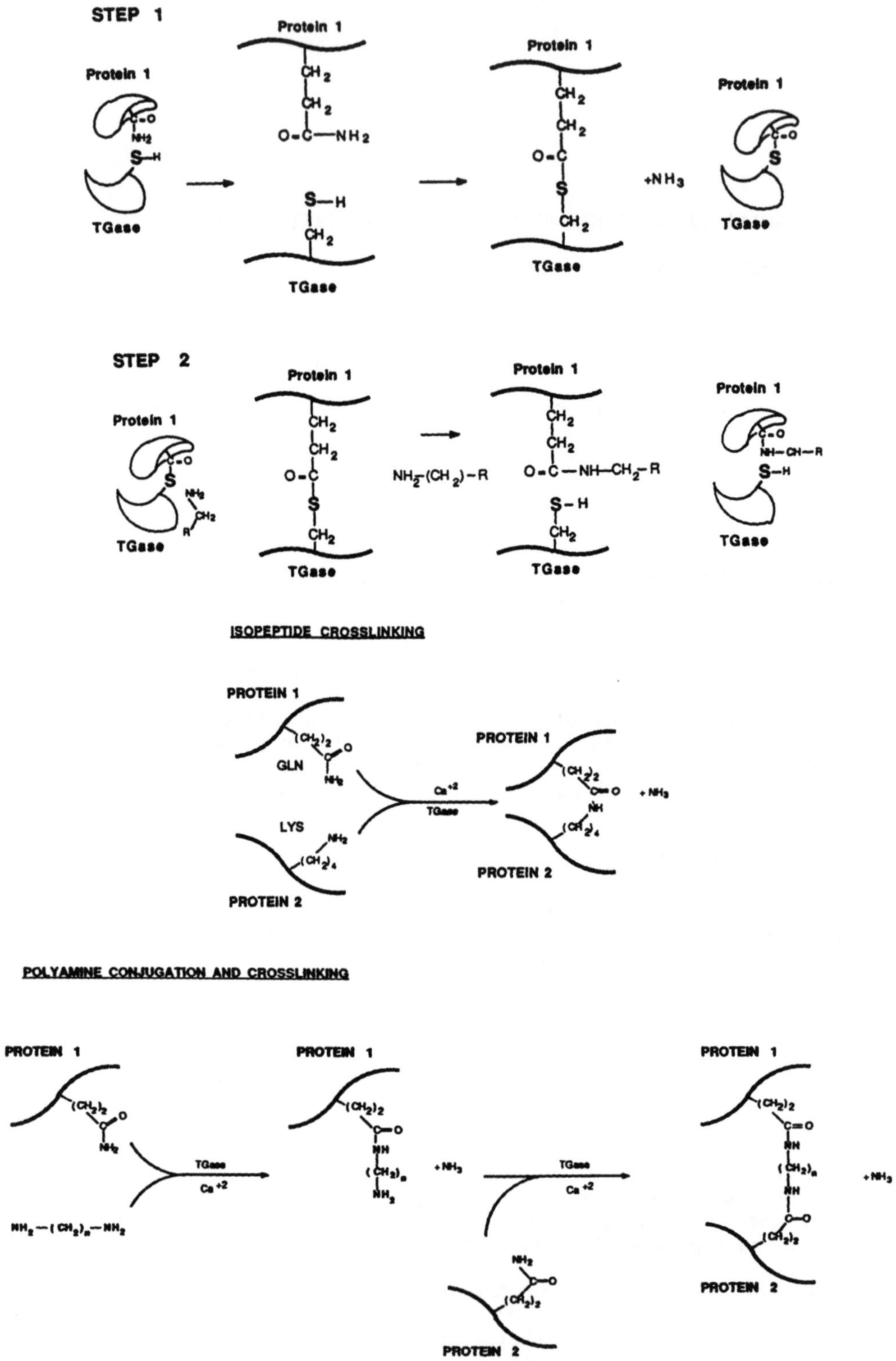

Figure 1. The Transglutaminase Reaction

are substrates for transglutaminase activity; most proteins do not serve as substrates and among those that are substrates only a single glutamine residue is usually subject to modification. Comparison of the amino acid sequence flanking the reactive glutamine in a number of transglutaminase substrates has failed to reveal a consensus recognition sequence. Although the substrate specificity of transglutaminases remains to be defined, it seems likely that it depends on a combination of sequence and conformational parameters. The important implication of this selectivity is that transglutaminases are not indiscriminate in the proteins they modify, and it is reasonable to anticipate the transglutaminase-mediated post-translational modifications are likely to be of functional significance.

In contrast to their high degree of specificity for glutamine substrates, transglutaminases are much less stringent in terms of the types of amines that may serve as substrates (2,3). It is this relaxed specificity that gives rise to the diverse products that may be produced by transglutaminase activity. There is considerable evidence to suggest that the preferred amine substrate for most transglutaminases is a protein-bound lysine residue. The condensation of the epsilon amino group of the lysine with the γ-carboxamide group of the glutamine residue leads to the formation of two proteins cross-linked by an ε-(γ-glutamyl)lysyl isopeptide bond (Fig. 1). Free lysine is a poor substrate for this reaction. The structural determinants that dictate the selection of one lysyl residue over another are not known.

Even though free lysine is not a good substrate for the transglutaminase reaction, free polyamines are excellent substrates (4). Putrescine, spermine and spermidine are readily conjugated to proteins by transglutaminases in vitro. The product of this reaction is γ-glutamyl-putrescine, -spermine or -spermidine (Fig. I). The fact that the Km of the polyamines in the transglutaminase reaction is in the range of concentrations that might be achieved in vivo, has led to the suggestion that this reaction could be of physiological significance (5). The condensation of a polyamine with a specific glutamine residue could lead to the formation of a modified protein with altered structural or catalytic properties. In addition, the conjugated polyamine may serve as a substrate for a second conjugation reaction since it contains a terminal amino group (6). Reaction of this amine with the second glutamine residue leads to the formation of a covalent cross-link between the two polypeptide chains. In this case the proteins are linked not by a glutamyl-lysyl isopeptide bond but by an N,N-bis(γ-glutamyl)polyamine cross-bridge (Fig. 1). Thus, polyamine conjugation has the potential not only for modifying specific glutamine residues but also of participating in protein cross-linking reactions.

FORMATION OF PROTEIN-POLYAMINE CONJUGATES IN VIVO

The preceding discussion has set forward the reasons for believing that transglutaminases could catalyze the conjugation of polyamines to proteins. Table I summarizes the studies that have suggested that this reaction actually occurs in vivo.

There is considerable evidence to suggest that polyamines may be conjugated to extracellular proteins. Several years ago Rennert and his colleagues showed that amniotic fluid contained polyamines that could only be released from proteins by acid hydrolysis (7). The conditions required to liberate the polyamines suggested that they were covalently bound to specific proteins although these studies did not identify the biochemical

TABLE I
POLYAMINE - PROTEIN CONJUGATES IN VIVO

Tissue	Polyamine-Protein Conjugates	Reference
1. EXTRACELLULAR		
Human Amniotic Fluid	γ-Glu-spermine, -spermidine and putrescine	Rennert et al. (7)
Coagulated Rat Seminal Plasma	γ-Glu-spermidine and -spermine Bis-γ-Glu-spermidine and spermine	Williams-Ashman (8); Folk et al. (13)
2. INTRACELLULAR		
Rat Liver Nuclei	γ-Glu-Putrescine, -spermine and -spermidine	Haddox and Russell (10)
Rat Liver, Testis and Kidney	γ-Glu-spermine, -spermidine bis-γ-Glu-spermine	Beninanti et al. (11)
Aplysia Neuron	γ-Glu-putrescine, -spermidine	Ambron and Kremzner (12)
Activated Lymphocytes	γ-Glu-putrescine, -spermidine	Folk et al. (13)
CHO cells	γ-Glu-putrescine, -spermidine	Beninanti et al. (14)

nature of the protein-polyamine conjugate. Furthermore, the source of the conjugated polyamines and the enzymatic process that resulted in their formation was not ascertained.

A much more detailed analysis of extracellular polyamine conjugates has come from the study of the clotting reactions of seminal plasma conducted by Williams-Ashman and his colleagues (8). Rodent seminal plasma contains high concentrations of both spermine and spermidine as well as one or more specific extracellular transglutaminases. When rodent seminal plasma coagulates, as it does in vivo to form the cervical plug, there is extensive protein cross-linking. Hydrolysates of coagulated rodent plasma revealed large amounts of glutamyl-lysine isopeptide bonds (up to 9 moles per 100,000g) as well as γ-glutamyl-spermine, -spermidine and N,N-bis(γ-glutamyl)spermine (13). The specific identification of the glutamyl conjugates, coupled with the demonstration of active transglutaminases in situ provides strong evidence that in this particular reaction polyamines are incorporated into proteins in vivo. The consequences of polyamine conjugation in seminal plasma are not well known although Williams-Ashman has suggested that the monosubstitution reaction may compete with isopeptide formation and act to retard the coagulation reaction (9). It is also possible that the cross-links introduced by polyamine conjugation (N,N-bis(γ-glutamyl)spermine) introduce steric or mechanical constraints on the clots that are different from those produced by glutamyl-lysine cross-links.

Several laboratories have identified protein-polyamine conjugates that may be formed inside cells. Russell and Haddox showed that purified liver nuclei contained small amounts of conjugated putrescine, spermine and spermidine (10). Furthermore, manipulations that altered the intracellular transglutaminase activity of the tissue were associated with corresponding increases in conjugate formation. Recently, Beninanti and her co-workers have measured the amount of conjugated polyamines in several tissues including liver, testis and kidneys (11). Proteolytic digests of each of these tissues has been shown to contain measurable levels of glutamyl-spermidine and -spermine and N,N-bis(γ-glutamyl) spermine. The identity of these conjugates was confirmed by both high pressure liquid chromatography and enzymatic analysis. Glutamyl-putrescine was not detected in the tissue extracts.

The measurement of polyamine conjugates in tissue extracts does not rigorously establish that the conjugates were generated intracellularly. This issue is more easily addressed in cultured cells. Ambron and Kremzner approached the problem by microinjecting radio-labelled putrescine into individual neurons of the sea slug Aplysia (12). These studies showed that the injected putrescine was rapidly incorporated into proteins. Enzymatic digestion of the extracts identified γ-glutamyl putrescine and γ-glutamyl-spermidine as the most abundant products.

An alternative approach to injecting radio-labelled polyamines into cells is to radiolabel the endogenous polyamine pool by incubation of the cells with tritiated putrescine. This procedure is technically difficult since the uptake of exogenous putrescine is carrier mediated and quite limited in most cells and endogenous pools of polyamines may be large. Nonetheless, two laboratories have succeeded in identifying conjugated polyamines in cultured cells using this approach. Folk and his colleagues showed several years ago that the incubation of activated lymphocytes with labelled putrescine led to the formation of low but detectable amounts of γ-glutamyl-putrescine and possibly γ-glutamyl-spermidine (13). The identification of the latter compound was complicated by the incorporation of putrescine into hypusine, a metabolite of lysine, that co-chromatographed with γ-glutamyl-spermidine. A similar set of metabolic labelling studies was recently reported by Beninati and her co-workers using Chinese Hamster ovary (CHO) cells (14). In order to detect polyamine conjugates these workers reduced the endogenous pool of polyamines by treatment of the cells with DFMO and MGBG. Under these conditions they were able to detect radiolabelled glutamyl-spermidine and glutamyl-putrescine as well as lesser amounts of the N,N-bis(γ-glutamyl) polyamine conjugates. The identity of these products was confirmed by hydrolysis and enzymatic digestion. These authors showed that increasing the transglutaminase activity of the cells, by pre-treatment with butyrate, resulted in a large increase in conjugate formation.

These studies have provided firm documentation for the presence of protein-conjugated polyamines in living cells. Although several studies have attempted to identify the intracellular proteins that serve as substrates for the conjugation reaction (12-14), none have, as yet, been identified. The failure of several groups to detect such proteins suggests that many proteins may be subject to modification. Alternatively the polyamines may be incorporated preferentially into proteins that are already extensively cross-linked and hence not readily fractionated by SDS-gel electrophoresis.

TRANSGLUTAMINASES AND POLYAMINE CONJUGATION

The studies on the level of conjugated polyamines in cells and tissues has shown a direct correlation between the formation of these compounds and the level of transglutaminase activity. This suggests that factors that affect transglutaminase activity could have very direct

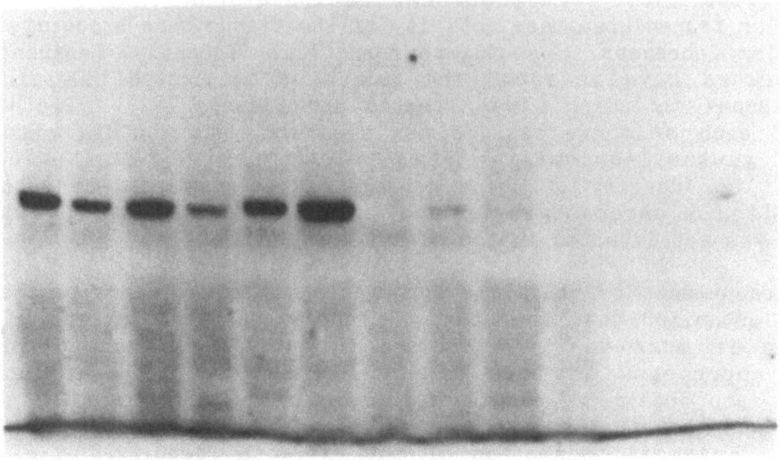

Figure 2. Tissue Distribution of Mouse Tissue Transglutaminase. 50 g of mouse liver (Lv), spleen (S), kidney (K), testis (T), heart (H), lung (L), thyroid (Ty), thymus (Tm), brain (B) and eye (E) and 5 ng of purified guinea pig liver tissue transglutaminase (gpl TGase) were fractionated on a 7.5% polyacrylamide gel and Western blotted with an affinity purified antibody to tissue transglutaminase. Bound immunoglobulin was detected with rabbit anti-goat immunoglobulin followed by 1251-labelled protein A and autoradiography.

effects on the levels of conjugated polyamines in cells and tissues. We have been interested for several years in the factors that regulate the activity of the intracellular transglutaminases of normal and transformed cells. In these studies we have focussed on the regulation of the enzyme tissue transglutaminase. Our reason for selecting this enzyme is that it appears to be the most ubiquitous of the intracellular transglutaminases. Although several distinct intracellular transglutaminases have been identified, most of these enzymes are distributed in a very tissue specific manner. The membrane-bound transglutaminase of keratinocytes, for instance, is found only in the upper layers of the epidermis and other cornified epithelia (15). Hair follicle transglutaminase, as its name implies, appears to be restricted to specific regions of the hair follicle (16). Tissue transglutaminase, on the other hand, is found in many cells and tissues. The enzyme itself is a single polypeptide chain of 80,000 daltons that appears to be distributed entirely in the cytosolic compartment of cells. There is no evidence that the enzyme is secreted under physiological conditions.

The tissue distribution of tissue transglutaminase suggests that it is very likely to be involved in the conjugation of polyamines to proteins. This enzyme is particularly abundant in the liver, kidneys and testis, tissues that have been shown to contain significant levels of conjugated polyamines (Fig. 2). CHO cells also contain significant levels of the enzyme. Beninati has shown that CHO cells contain substantial amounts of protein-conjugated polyamines and induction of the enzyme is associated with increased levels of these conjugates (14).

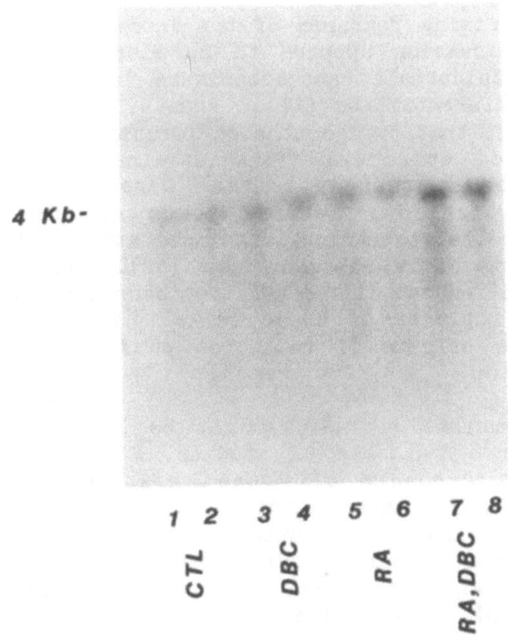

Figure 3. Northern gel analysis of tissue transglutaminase mRNA from Chinese Hamster ovary cells. RNA was prepared from control CHO cells (lanes 1 and 2), and from cells treated with 1 mM dibutyryl cyclic AMP (lanes 3 and 4), 1 μM retinoic acid (lanes 5 and 6) and retinoic acid plus dibutyryl cyclic AMP (lanes 7 and 8) for 17 h. The RNA was probed with a antisense cRNA probe to mouse tissue transglutaminase.

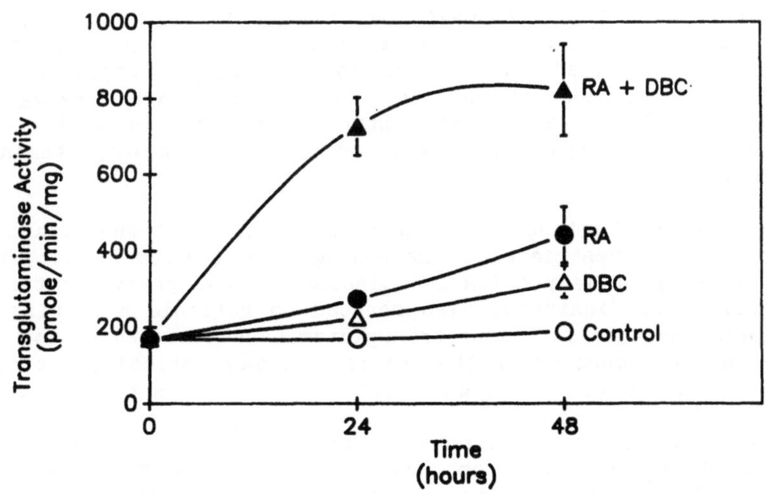

Figure 4. Transglutaminase activity in CHO cells. CHO cells were treated for 0, 24 and 48 hours with 1 mM dibutyryl cyclic AMP, 1 μM retinoic acid or retinoic acid plus cyclic AMP and the transglutaminase activity was determined.

One of the striking features of the intracellular transglutaminases is that their expression appears to be closely linked to growth and differentiation. Epidermal transglutaminase is induced in keratinocytes during terminal differentiation (17). Agents, such as phorbol esters or high calcium medium, that trigger the differentiation of these cells also act as inducers of epidermal transglutaminase expression (17). The expression of tissue transglutaminase also appears to be linked to cellular differentiation. Several years ago Birckbichler and his colleagues reported that the transglutaminase activity of many transformed and rapidly growing cells was very low (18). Agents that suppressed cellular growth and induced differentiation caused marked increases in the transglutaminase activity of these cells. We have been particularly interested in the effects of retinoids on tissue transglutaminase in cultured cells.

The suggestion that retinoids might be linked to transglutaminase activity comes from two sets of studies. Lichti and her co-workers observed that the exposure of mouse keratinocytes to retinoic acid resulted in a marked increase in intracellular transglutaminase activity (17). These workers initially assumed that this was due to induction of the epidermal transglutaminase but later showed that the increase in enzyme activity was in fact due to the induction of tissue transglutaminase (19). Scott et al. also observed that the treatment of CHO cells with very high levels of retinoic acid resulted in transient increases in intracellular transglutaminase activity (20). Our own interest in the effect of retinoids on the expression of this enzyme comes from studies on the induction of transglutaminase in mouse macrophages. The activation of macrophages in vivo is associated with very large increases in intracellular transglutaminase activity (21,22). Macrophages contain two transglutaminases, factor XIII an extracellular transglutaminase that is presumably synthesized in the macrophage and secreted into the serum (23), and tissue transglutaminase (24). We used purified polyclonal antibodies to tissue transglutaminase to show that the increased intracellular transglutaminase activity in activated cells was due to a specific induction of tissue transglutaminase (24,25). The link to retinoid action came with the demonstration that this induction could be reproduced in vitro by exposing macrophages to physiological levels of retinoic acid (26). In a series of studies we subsequently showed that retinoic acid induced the enzyme by producing a rapid and dramatic increase in the transcriptional activity of the tissue transglutaminase gene (27-29). These experiments showed that retinoic acid acted as an acute, direct and specific regulator of tissue transglutaminase gene expression.

The ability of retinoic acid to regulate tissue transglutaminase gene expression is not restricted to macrophages or myeloid cells but appears to be a property of many cells and tissues. Human promyelocytic leukemia (HL-60) cells, for instance, also respond to retinoic acid with an acute and dramatic increase in tissue transglutaminase gene expression (30). In these cells the induction of the enzyme is synergistically potentiated by exposure of the cells to agents that elevate intracellular cyclic AMP (30,31). Chinese hamster ovary cells also respond to retinoic acid with increased expression of tissue transglutaminase. Figure 3 shows a Northern blot of total RNA from control CHO cells and from cells that have either been treated with retinoic acid alone, dibutyryl cyclic AMP alone or a combination of retinoic acid plus cyclic AMP. It is apparent from inspection of the bands in this figure that retinoic acid alone has a small inductive effect, but in the presence of the cyclic AMP analogue, it produces a significant increase in the levels of the tissue transglutaminase mRNA. The induction of transglutaminase mRNA is reflected in a retinoid-dependent increase in transglutaminase activity

(Fig. 4). Recently Lichti et al. have reported that this induction can be observed in vivo. Topical administration of retinoic acid to mice produced a marked increase in the tissue transglutaminase activity of the skin (32).

CONCLUDING REMARKS

What then is the link between retinoid-induced expression of transglutaminase activity in cells and the formation of protein-polyamine conjugates? At this point we reach the limits of experimental information and must move into the realm of speculation. There are many lines of evidence to suggest that one of the primary roles of retinoids is to arrest the proliferation and induce the terminal differentiation of certain types of cells (33). It seems likely that the induction of tissue transglutaminase is an integral part of the terminal differentiation program of these cells. Several studies have suggested that the activation of transglutaminases is a component of the senescence of many cells and may be involved in the process of programmed cell death (apoptosis) (34,35). We speculate that the irreversible conjugation of polyamines to proteins, as well as protein cross-linking, may be an integral part of the process of cellular senescence and death.

ACKNOWLEDGMENTS

The authors would like to express their appreciation to Ms. Nubia Alban and Ms. Mary Sobieski for technical assistance in the completion of these studies. They also are grateful for the assistance of Ms. Joan Jennings in the completion of the manuscript. The studies described herein were supported in part by research grants DK27078 and CA41829 from the National Institutes of Health. PJAD is an Established Investigator of the American Heart Association.

REFERENCES

1. J. E. Folk and J. S. Finlayson, The ε-(γ-glytamyl)lysine crosslink and the catalytic role of transglutaminases, Adv. Protein Chem. 31:1-133 (1977).
2. L. Lorand, K. N. Parameswaran, P. Stenberg, Y. S. Tong, P. T. Velasco, N. A. Jonsson, L. Mikiver, and P. Moses, Specificity of guinea pig liver transglutaminase for amine substrates, Biochem. 18:1756-1765 (1979).
3. J. Shrode and J. E. Folk, Stereochemical aspects of amine substrate attachment to acyl intermediates of transglutaminases, J. Biol. Chem. 254:653-661 (1979).
4. D. D. Clarke, M. J. Mycek, A. Neidle, and H. Waelsch, The incorporation of amines into protein, Arch. Biochem. Biophys. 79:338-354 (1959).
5. H. G. Williams-Ashman and Z. N. Cannelakis, Polyamines in mammalian biology and medicine, Perspect. Biol. Med. 22:421-453 (1979).
6. J. Shrode and J. E. Folk, Transglutaminase-catalyzed cross-linking through diamines and polyamines, J. Biol. Chem. 253:4837-4840 (1978).
7. O. M. Rennert, W. Y. Chan, and G. Griesman, Polyamine-peptide conjugates: proposed function, Physiol. Chem. Physics. 12:441-450 (1980).

8. H. G. Williams-Ashman, Transglutaminases and the clotting of mammalian seminal fluids, Mol. Cell Biochem. 58:51-61 (1984).
9. L. Lorand and H. G. Williams-Ashman, Clotting of proteins by extracellular transglutaminases, Acta Biochim. Biophys. Hungar. 22:337-349 (1987).
10. M. K. Haddox and D. H. Russell, Differential conjugation of polyamines to calf nuclear and nucleolar proteins, J. Cell. Physiol. 109:447-452 (1981).
11. S. Beninanti, M. Piacentini, M. P. Argento-Ceru, S. Russo-Caia, and F. Autuori, Presence of di- and polyamines covalently bound to protein in rat liver, Biochim. Biophys. Acta 841:120-126 (1985).
12. R. T. Ambron and L. T. Kremzner, Post-translational modification of neuronal proteins: Evidence for transglutaminase activity in R2, the giant cholinergic neuron of aplysia, Proc. Natl. Acad. Sci. U.S.A. 79:3442-3446 (1982).
13. J. E. Folk, M. H. Park, S. I. Chung, J. E. Schrode, E. P. Lester, and H. L. Cooper, Polyamines as physiological substrates for transglutaminase, J. Biol. Chem. 255:3695-3700 (1980).
14. S. Beninati, M. Piacentini, E. T. Cocuzzi, F. Autuori, and J. E. Folk, Covalent incorporation of polyamines as γ-glutamyl derivatives into CHO proteins, Biochim. Biophys. Acta. 952:325-333 (1988).
15. N. L. Parenteau, A. Pilato, and R. H. Rice, Induction of keratinocyte type I transglutaminase in epithelial cells of the rat, Differentiation 33:130-138 (1986).
16. N. Martinez, H. C. Kim, J. E. Girard, T. P. Nigra, D. H. Strong, S. I. Chung, and J. E. Folk, Epidermal and hair follicle transglutaminases, J. Biol. Chem. 263:4236-4241 (1988).
17. S. H. Yuspa, T. Ben, H. Hennings, and U. Lichti, Phorbol ester tumor promoters induce epidermal transglutaminase activity, Biochem. Biophys. Res. Commun. 97:700-708 (1980).
18. P. J. Birckbichler, G. R. Orr, E. Conway, and M. K. Patterson, Transglutaminase activity in normal and transformed cells, Cancer Res. 37:1340-1344 (1977).
19. U. Lichti, T. Benn, and S. H. Yuspa, Retinoic acid induced transglutaminase in mouse epidermal cells is distinct from epidermal transglutaminase, J. Biol. Chem. 260:1422-1426 (1985).
20. K. F. Scott, F. L. Meyskens, Jr., and D. H. Russell, Retinoids increase transglutaminase activity and inhibit ornithine decarboxylase activity in Chinese hamster ovary cells and in melanoma cells stimulated to differentiate, Proc. Natl. Acad. Sci. U.S.A. 79:4093-4097 (1982).
21. G. Schroff, C. Neumann, and C. Sorg, Transglutaminase as a marker for subsets of murine macrophages, Eur. J. Immunol. 11:637-642 (1981).
22. R. W. Leu, M. J. Herriott, P. E. Moore, G. R. Orr, and P. J. Birckbichler, Enhanced transglutaminase activity associated with macrophage activation, Exp. Cell Res. 141:191-199 (1982).
23. R. Adany, A. Belkin, T. Vasilevskaya, and L. Muszbek, Identification of blood coagulation factor XIIIa in human peritoneal macrophages, Eur. J. Cell Biol. 38:171-173 (1985).
24. M. P. Murtaugh, K. Mehta, J. Johnson, M. Myers, R. L. Juliano, and P.J.A. Davies, Induction of tissue transglutaminase in mouse peritoneal macrophages, J. Biol. Chem. 258:11074-11081 (1983).
25. P. J. A. Davies, W. T. Moore, Jr., and M. P. Murtaugh, Retinoid-regulated gene expression in normal and leukemic myeloid cells, BioEssays 1:160-165 (1984).
26. W. T. Moore, Jr., M. P. Murtaugh, and P. J. A. Davies, Retinoic acid-induced expression of tissue transglutaminase in mouse peritoneal macrophages, J. Biol. Chem. 259:12794-12802 (1984).
27. M. P. Murtaugh, O. Dennison, J. P. Stein, and P. J. A. Davies, Retinoic acid-induced gene expression in normal and leukemic myeloid cells, J. Exp. Med. 163:1325-1330 (1986).

28. E. A. Chiocca, P. J. A. Davies, and J. P. Stein, The molecular basis of retinoid action: transcriptional regulation of tissue transglutaminase gene expression in macrophages, J. Biol. Chem. In Press.
29. E. A. Chiocca, P. J. A. Davies, and J. P. Stein, The regulation of tissue transglutaminase gene expression as a molecular model for retinoid effects on proliferation and differentiation, J. Cell. Biochem. In Press.
30. P. J. A. Davies, M. P. Murtaugh, W. T. Moore, Jr., G. S. Johnson, and D. Lucas, Retinoic acid-induced expression of tissue transglutaminase in human promyelocytic leukemia (HL-60) cells, J. Biol. Chem. 260:5166-5174 (1985).
31. M. P. Murtaugh, W. T. Moore, Jr., and P. J. A. Davies, Cyclic AMP potentiates the retinoic acid-induced expression of tissue transglutaminase in peritoneal macrophages, J. Biol. Chem. 261:614-621 (1986).
32. U. Lichti and S. H. Yuspa, Modulation of tissue and epidermal transglutaminases in mouse epidermal cells after treatment with 12-O tetradecanoylphorbol-13-acetate and/or retinoic acid in vivo and in culture, Cancer Res. 48:74-81 (1988).
33. R. Lotan, Effects of vitamin A and its analogs (retinoids) on normal and neoplastic cells, Biochim. Biophys. Acta, 605:33-91 (1980).
34. L. Lorand and S. M. Conrad, Transglutaminases, Mol. Cell. Biochem. 58:9-35 (1984).
35. L. Fesus, V. Thomazy, and A. Falus, Induction and activation of tissue transglutaminase during programmed cell death, FEBS Letters 224:104-108 (1987).

PURIFICATION AND STRUCTURAL CHARACTERIZATION OF IN VITRO SYNTHESIZED (γ-GLUTAMYL)SPERMIDINE CONJUGATES OF A MAJOR PROTEIN SECRETED FROM THE RAT SEMINAL VESICLES

R. Porta*, C. Esposito*, S. Metafora#, A. Malorni## , P. Pucci** and G. Marino**

Departments of *Biochemistry and Biophysics, and of **Organic and Biological Chemistry, University of Naples, Naples; #International Institute of Genetics and Biophysics, Naples; ##CNR Institute of Chemistry of Molecules of Biological Interest, Arco Felice, Naples, Italy

INTRODUCTION

Very little is known about the physiological meaning of spermidine (Spd) and spermine (Spm) occurrence in mammal seminal plasma where polyamines are secreted in large amounts by the prostate (1-3). One postulated role for semen polyamines is to serve as acyl acceptor substrates for transglutaminase (TGase; EC 2.3.2.13). In fact, it has been demonstrated that rat seminal clot proteins have covalently bound polyamines and upon proteolytic digestion released (γ -glutamyl)polyamine derivatives (4). Moreover, N-mono- and N,N-bis-(γ -glutamyl)polyamine derivatives were found after proteolysis of rat vesicular secretion proteins that have been incubated with coagulating gland extracts (5). Some of us have recently shown the ability of one of the major proteins secreted from the rat seminal vesicle epithelium to covalently bind radioactive Spd in the presence of either purified guinea pig liver TGase or crude secretory fluid produced by the rat coagulating gland (6). Such a protein, named SV-IV, is the fourth major band observed on the SDS-PAGE pattern of the proteins present in the rat seminal vesicle secretion. The sequence of its 90 amino acids (Mr= 9,758) and the main features of the gene coding for it are known (6-9). We have demonstrated that SV-IV possesses immunosuppressive and anti-inflammatory activities (6, 10, 11). Microgram amounts of SV-IV are able to inhibit both the mitogen-induced lymphocyte blastogenesis and the mixed lymphocyte reaction; the observed anti-inflammatory activity, comparable to that of dexamethasone, seems to be due to the block of arachidonic acid cascade at level of the enzyme phospholipase A_2.

In the present paper we report the results of our studies devoted both to isolate the Spd-SV-IV conjugates synthesized in vitro by TGase and to determine by Fast Atom Bombardment Mass Spectrometry (FABMS) (12) the glutamine residues of SV-IV recognized by the enzyme as acyl donor sites.

SEPARATION OF (γ-GLUTAMYL)SPERMIDINE CONJUGATES OF SV-IV BY FAST PROTEIN LIQUID CHROMATOGRAPHY (FPLC)

Panel A of Fig.1 shows that when native SV-IV was passed through a cation exchange column under the reported experimental conditions only one molecular species was detected in the elution pattern. In contrast, three additional protein peaks (peaks 2, 3 and 4), more retained than native SV-IV (peak 1), were eluted from the column when the protein was previously incubated in the presence of TGase, Ca^{2+} and high

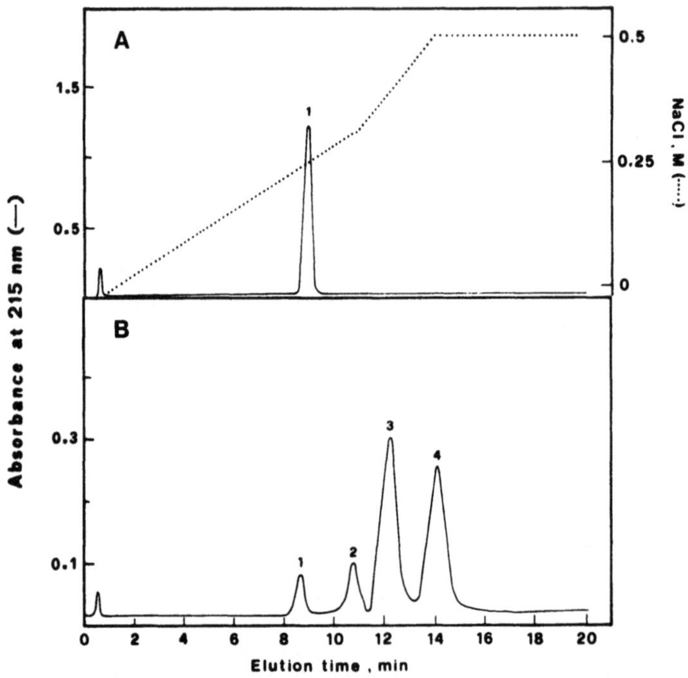

Fig. 1. FPLC pattern of native SV-IV (Panel A) and of SV-IV previously treated with active TGase and Spd (Panel B).
0.15 mg of SV-IV were incubated at 37° C for 4 h in 0.1 ml of 125 mM Tris-HCl buffer, pH 8.0, containing 10 mM dithiothreitol, 2.5 mM $CaCl_2$ and 0.2 M Spd. 10 ug of purified guinea pig liver TGase were added at the beginning of incubation; another addition of the same amount of enzyme was performed after 2 h of incubation. Native SV-IV and TGase-treated SV-IV were injected onto the column (Mono S HR 5/5, Pharmacia Fine Chemicals) previously equilibrated with 50 mM sodium phosphate buffer, pH 7.6 (buffer A). The column was washed with 1 ml of buffer A. Elution was then performed at a flow rate of 1 ml/min with a linear gradient of 0-60% buffer A containing 0.5 M sodium chloride (buffer B) followed by a second linear gradient of 60-100% buffer B.

concentrations of Spd (Fig. 1, Panel B). These results are in line with the assumption that the polyamine binding to SV-IV increases the number of free amino groups of the protein, thus increasing its pI. SDS-PAGE analysis of the collected fractions indicated that peaks 2, 3 and 4 contained protein species having Mr very similar to that of native SV-IV (data not shown). Therefore, these data suggest that (i) no intermolecular cross-bridges are formed in the presence of Spd, (ii) TGase produces different N-mono-(γ-glutamyl)Spd derivatives of SV-IV.

FABMS CHARACTERIZATION OF (γ-GLUTAMYL)SPERMIDINE CONJUGATES OF SV-IV

FABMS technique allowed us to obtain the finger-print of native SV-IV by direct analysis of the peptide mixture generated by enzymatic cleavage of the protein by trypsin and Staphylococcus aureus (V-8 strain) protease. To characterize the TGase-modified forms of SV-IV contained in peaks 3 and 4 obtained by FPLC (see Fig. 1) we separately collected both peaks and subjected the two samples to proteolysis before FABMS analysis. The partial spectrum shown in Panel B of Fig. 2 corresponds to the protein contained in peak 3, whereas the partial spectrum reported in Panel A refers to unmodified SV-IV. The comparative analysis of the spectra shows the disappearance of the signals of the glutamine containing peptides 84-89 and 84-90 and the appearance of two new signals corresponding to the covalent adducts between one Spd molecule and 84-89 or 84-90 fragments, respectively.

The partial spectrum reported in Panel B of Fig. 3 corresponds to the protein species contained in peak 4 eluted from FPLC . It shows the disappearance of all signals of the glutamine containing peptides. In fact, in this case also 7-12 fragment of SV-IV was no more detectable, whereas three new signals appeared: they correspond to the covalent adducts between one Spd molecule and 84-89, 84-90 or 7-12 fragments, respectively. All these data clearly demonstrate that (i) the modified form of SV-IV occurring in peak 3 is characterized by one Spd molecule covalently bound to Gln-86, (ii) the modified form of SV-IV occurring in peak 4 is characterized by two Spd molecules respectively bound to Gln-9 and Gln-86 residues of the protein. Fig. 4 shows the sequences of unmodified and modified glutamine containing fragments of SV-IV detected by FABMS together with the molecular weights of their protonated derivatives (MH^+).

FABMS CHARACTERIZATION OF (γ-GLUTAMYL)SPERMIDINE CONJUGATES OF BrCN FRAGMENTS OF SV-IV

Since SV-IV is endowed with only one methionine residue (Met-70) it was possible to obtain two fragments of the protein (F1, 1-70; F2, 71-90) each containing one acyl donor site (Gln-9 and Gln-86, respectively). F1 and F2 were separated by HPLC and singly treated with active TGase and Spd as described for SV-IV in the legend to Fig. 1. The structural analysis of the modified and unmodified BrCN fragments was then performed by FABMS. The partial spectra of the peptide mixtures generated from the proteolysis of TGase-treated F1 and F2 (Figs. 5 and 6, Panels B) show the disappearance of the signals of the glutamine containing peptides 7-12 and 84-90 and the appearance of new signals corresponding to the covalent adducts between these peptides and one Spd molecule. These results indicate that also the two BrCN fragments of SV-IV are able to act as acyl donor substrates of TGase, thus confirming that both glutamine residues of the protein are effective polyamine binding sites.

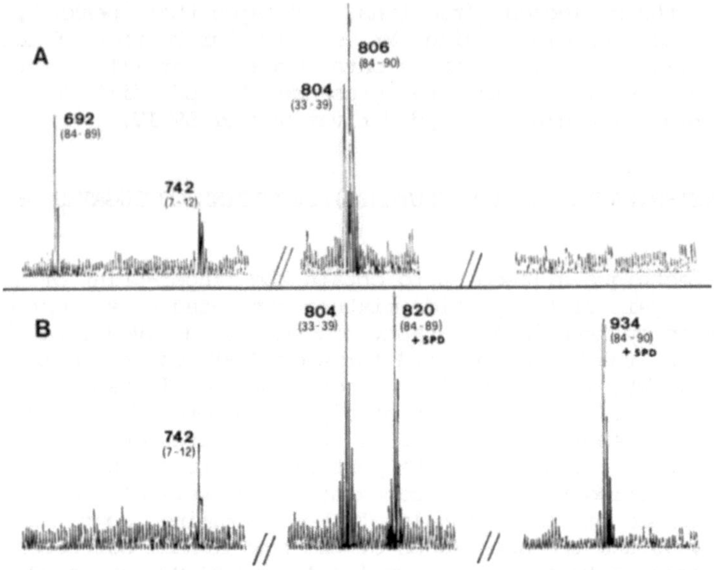

Fig. 2. Partial FAB mass spectra of the proteolytic digests of native SV-IV (Panel A) and of SV-IV previously treated with active TGase and Spd and eluted as peak 3 from FPLC (Panel B).

The treatment of SV-IV with TGase and Spd, as well as FPLC separation of the reaction products, were performed as described in the legend to Fig. 1. Native and TGase treated SV-IV were loaded onto Sep-Pak C-18 cartridges equilibrated in 0.1 % trifluoroacetic acid (TFA). Each cartridge was washed with 5 ml of 0.1 % TFA and the protein fractions, eluted with 0.1 % TFA/Acetonitrile (30/70 v/v), were pooled and evaporated to dryness. Tryptic digestion of the samples was carried out in 0.4 %ammonium bicarbonate buffer, pH 8.5, for 4 h at 37 °C. At the end of incubation the reaction mixtures were heated at 100 °C for 5 min and then evaporated to dryness. The dry samples were redissolved in 0.4 % ammonium bicarbonate buffer, pH 8.0, and treated with V-8 protease at 40 °C for 6 h. At the end of incubation the samples were evaporated to dryness, then redissolved in 20 ul of 5 % acetic acid, and finally their aliquots were directly loaded onto a glycerol-coated probe tip. FAB mass spectra were recorded on a Kratos MS 50 double focusing mass spectrometer equipped with an M-Scan FAB gun, using a xenon as primary ionizing beam (9 KeV, 2.5 uA).

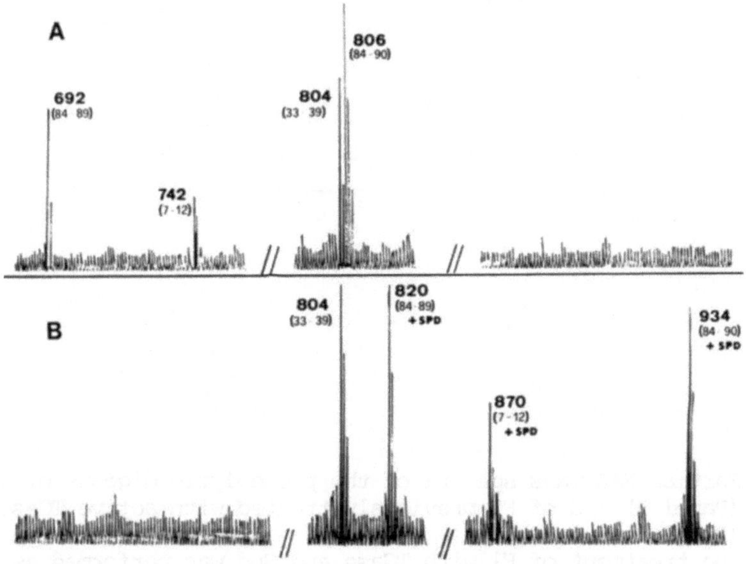

Fig. 3. Partial FAB mass spectra of the proteolytic digests of native SV-IV (Panel A) and of SV-IV previously treated with active TGase and Spd and eluted as peak 4 from FPLC (Panel B). Experimental details are described in the legend to Fig. 2.

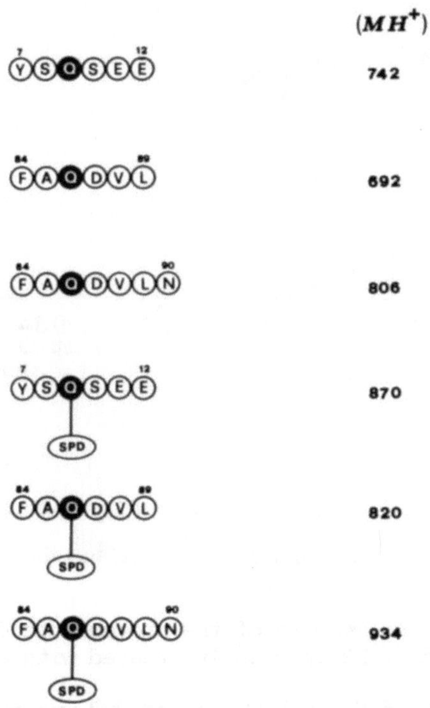

Fig. 4. Amino acid sequences and relative MH^+ of glutamine containing fragments of SV-IV and of their (γ-glutamyl)Spd derivatives detected by FABMS.

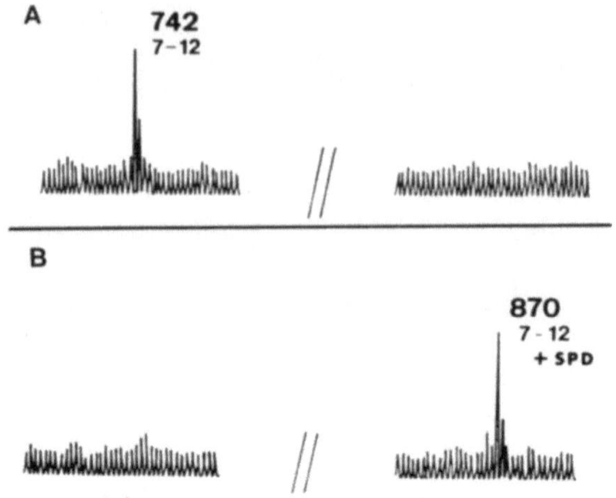

Fig. 5. Partial FAB mass spectra of the proteolytic digests of native F1 (Panel A) and of F1 previously treated with active TGase and Spd (Panel B).
The treatment of F1 with TGase and Spd was performed as described for SV-IV in the legend to Fig. 1. The reaction was stopped by adding 0.2 ml of TFA and the resulting sample was passed through a Sep-pak C-18 cartridge. The latter procedure, as well as tryptic and V-8 protease digestion of both native and modified F1 and FABMS analyses, were carried out as described for SV-IV in the legend to Fig. 2.

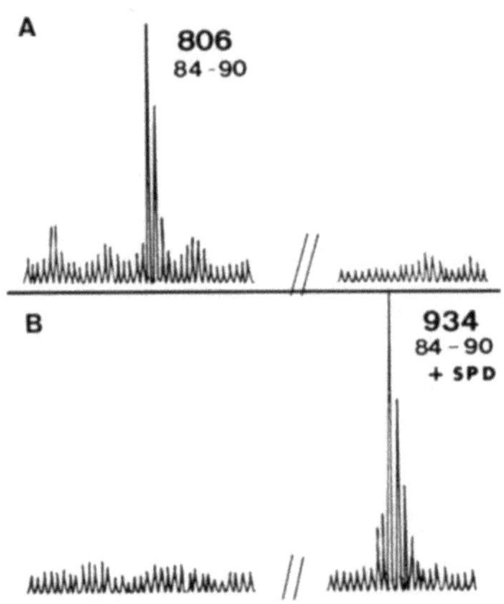

Fig. 6. Partial FAB mass spectra of the proteolytic digests of native F2 (Panel A) and of F2 previously treated with active TGase and Spd (Panel B).
The treatment of F2 with TGase and Spd was performed as described for SV-IV in the legend to Fig. 1. Sep-Pak C-18 purification of the reaction mixture, as well as tryptic digestion of the sample and FABMS analyses, were carried out as described in the legend to Fig. 2.

REFERENCES

1. O. Rosenheim, The isolation of spermine phosphate from semen and testis, Biochem. J. 18:1253 (1924).
2. H. G. Williams-Ashman, A. E. Pegg and D. H. Lockwood, Mechanisms and regulation of polyamine and putrescine biosynthesis in male genital glands and other tissues of mammals, Adv. Enzyme Regul. 7:291 (1969).
3. H. G. Williams-Ashman, A. Corti and A.R. Sheth, Formation and functions of aliphatic polyamines in the prostate gland and its secretions, in "Normal and Abnormal Growth of the Prostate", Goland M., ed., Thomas, Springfield (1975).
4. H. G. Williams-Ashman and Z.N. Canellakis, Transglutaminase-mediated covalent attachment of polyamines to proteins: mechanisms and potential physiological significance, Physiol. Chem. Phys. 12:457 (1980).
5. J. E. Folk, M.H. Park, S.I. Chung, J. Schrode, E.P. Lester and H.L. Cooper, Polyamines as physiological substrates for transglutaminase, J. Biol. Chem. 255:3695 (1980).
6. R. Porta, C. Esposito, P. Persico, G. Peluso and S. Metafora, Transglutaminase-catalyzed crosslinking of an immunosuppressive and anti-inflammatory protein secreted from the rat seminal vesicles, Adv. Exp. Med. & Biol. 231:153 (1988).
7. Y. E. Pan and S. S. Li, Structure of secretory protein IV frcm rat seminal vesicles, Int. J. Pept. Prot. Res. 20:177 (1978).
8. S. J. Higgins and J. M. Burchell, Effects of testosterone on messenger ribonucleic acid and protein synthesis in rat seminal vesicle, Biochem. J. 174:543 (1978).
9. M. C. Ostrowski, M. K. Kistler and W. S. Kistler, Purification and cell-free synthesis of a major protein from rat seminal vesicle secretion. A potential marker for androgen action, J. Biol. Chem. 254:383 (1979).
10. S. Metafora, G. Peluso, G. Ravagnan, V. Gentile, A. Fusco, R. Porta, Factors modulating immunocompatibility of spermatozoa: role of transglutaminase and SV-IV, one of the major proteins secreted from the rat seminal vesicle epithelium, in: "Morphological Basis of Human Reproductive Function", Spera G. and de Kretser D.M., eds., Plenum Press, New York and London (1987).
11. S. Metafora, G. Peluso, G. Ravagnan, P. Persico, C. Esposito and R. Porta, Immunosuppressive and anti-inflammatory properties of the major protein secreted from the rat seminal vesicle epithelium, Biochem. Pharmacol., in press.
12. A. Dell and M. Panico, Fast Atom Bombardment mass spectrometry of biomolecules, in: "Mass Spectrometry in Biomedical Research", S. Gaskell, ed., Wiley, Chichester (1986).

COVALENT POLYAMINE-PROTEIN CONJUGATES: ANALYSIS AND DISTRIBUTION

Simone Beninati and J.E. Folk

National Institute of Dental Research, NIH
Laboratory of Cellular Development and Oncology
Bethesda, Maryland 20892

INTRODUCTION

Most suggestions as to the biological functions of the polyamines, putrescine, spermidine, and spermine, are based on the observed non-covalent binding of these polycations to nucleic acids, proteins, and phospholipids.[1-3] It was recently shown that polyamines are also present in mammalian tissues and body fluids in covalent association with proteins. Protein modifications in which structural elements of polyamines are involved occur by two pathways. In one of these the amino acid hypusine [N^{ε}-(4-amino-2-hydroxybutyl)lysine] is formed through transfer of the butylamine moiety of spermidine to the ε-amino group of a protein lysine residue and through subsequent hydroxylation.[4] In the other polyamines are attached in covalent amide linkage to the γ-carboxyl groups of protein glutamic acid residues. Conjugation of the amines in this manner is catalyzed by transglutaminases, Ca^{2+}-dependent enzymes which promote exchange of primary amines for ammonia at the carboxamide groups of certain glutaminyl residues.[5,6] A number of transglutaminases have been identified and they are found widely distributed in mammalian cells and in biological fluids.[7,8] These enzymes are responsible for production of ε-(γ-glutamyl)lysine crosslinks that connect protein chains and play a central role in such extracellular events as fibrin clot stabilization and seminal plug formation.[7,8]

Recently, a number of reports have appeared in which polyamine-protein conjugates were identified and in which it was demonstrated directly that the polyamines are coupled to protein in γ-glutamylamine

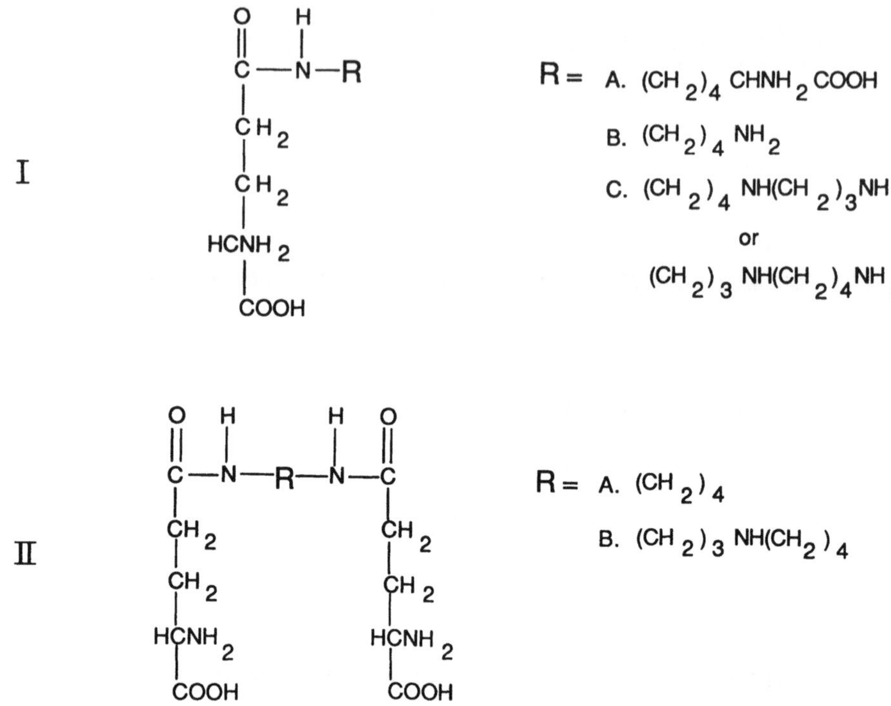

Fig. 1. Structures of γ-glutamylamines

linkage.[5,9-15] In each of these reports proof of the covalent linkage was provided through chromatographic identification of the γ-glutamylpolyamine in exhaustive proteolytic digests of the protein fraction.

ISOLATION, IDENTIFICATION, AND QUANTITATION OF γ-GLUTAMYLPOLYAMINES

In order to determine the type and quantity of conjugation of a polyamine to a protein in a biological system it is essential to isolate a simple compound that contains the bond connecting the protein and the polyamine, define its structure, and measure its amount. Indeed, identification and measurement of ε-(γ-glutamyl)lysine crosslinks between protein chains employs basically this methodology.[16] Because all bonds produced by the action of transglutaminases are in the form of amides of the γ-carboxyl group of protein-bound glutamic acid and because these γ-amides are labile to many procedures employed to cleave peptide bonds, special means must be used to preserve the γ-amides during degradation of proteins in which they occur. The method used is to hydrolyze the peptide bonds by exhaustive proteolytic digestion thus liberating the γ-glutamylamines (Fig. 1) which are resistent to this process, and to isolate, identify, and measure the freed γ-glutamylamines. It has been shown using a model system that the proteolytic digestion procedure provides quantitative release of γ-glutamylpolyamines from protein.[5]

An automated ion exchange chromatographic procedure designed for separation of γ-glutamylpolyamines[5] proved useful for identification and quantitation of the radiolabeled forms of these compounds in digests of polyamine-protein conjugates produced in biological systems to which radioactive polyamines were added. This ion exchange system, however, was found not to be fully satisfactory for isolation of unlabeled forms of γ-glutamylpolyamines from digests, because only the most basic γ-glutamylpolyamines were completely separated from the amino acids in the digests. Even without complete separation of the polyamine derivatives from amino acids, it was possible to identify them from their elution positions and to estimate their amounts by measurement of the polyamines produced either upon acid hydrolysis or by their treatment with γ-glutamylamine cyclotransferase, an enzyme specific for release of amines from γ-glutamylamines.[17,18] We have recently developed a reverse-phase high-performance liquid chromatographic method for direct identification and measurement of γ-glutamylamines. This procedure entails precolumn derivatization with o-phthalaldehyde and linear gradient chromatography on a C_{18} column and provides complete isolation of the derivatives of the γ-glutamylamines from the amino acid derivatives when applied to partially separated materials in fractions from ion exchange.[19] In Fig. 2 is shown a chromatogram of γ-glutamylamine derivatives using this high-performance liquid chromatographic procedure.

γ-GLUTAMYLPOLYAMINES IN MAMMALIAN TISSUES

Surprisingly high levels of polyamines, observed to be associated with the acid-insoluble protein fractions of rat liver, kidney, and testis, provided impetus for a carefully conducted investigation of the nature of the protein-polyamine associations.[11] All of the polyamines covalently bound to liver protein were shown to be linked by γ-glutamyl bonds. However, only 14% of the putrescine, 89% of the spermidine, and 84% of the spermine associated with protein were calculated to be covalently attached. The remainer of each, about 9 nmol putrescine, 9 nmol spermidine, and 10 nmol spermine per g liver, most of which was released as free polyamine by proteolytic digestion, was thought to be tightly linked to the protein fraction by electrostatic and/or hydrophobic forces. Because of this tight noncovalent binding, it was pointed out that proof for the occurrence of covalent polyamine-protein conjugates could only be obtained

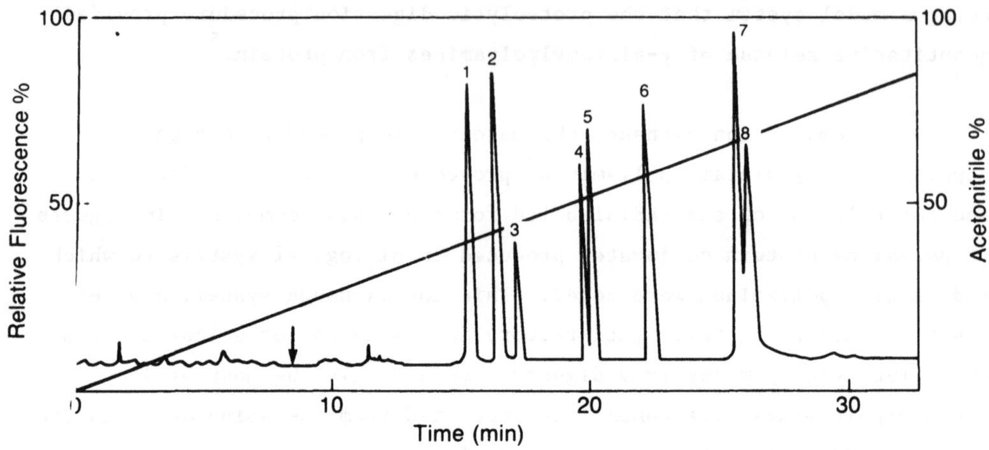

Fig. 2. High-performance liquid chromatography of the o-phthalaldehyde derivatives of some γ-glutamylamines. Chromatography was conducted using a linear gradient from 10 mM sodium acetate, pH 5, to 10 mM sodium acetate, pH 5, -acetonitrile (20/80 v/v) over a period of 33 min at a flow rate of 1 ml per min. The derivatives are those of: (1) N^1,N^8-bis(γ-glutamyl)-spermidine, (2) N^1,N^4-bis(γ-glutamyl)putrescine, (3) ε-(γ-glutamyl)lysine, (4) N^1-(γ-glutamyl)spermidine, (5) N^8-(γ-glutamyl)spermidine, (6) N-(γ-glutamyl)putrescine, (7) lysine, (8) hypusine. The arrow denotes the position that salts present in the derivatization mixtures are eluted. The structures of the γ-glutamylamines are shown in Fig. 1.

in the form of direct and unequivocal identification of the γ-glutamylamine linkage.

Examination of freshly isolated rat hepatocytes for polyamine-protein conjugates represented an extension of the above study.[13] The finding of γ-glutamyl derivatives of all polyamines in proteolytic digests of the protein fraction from hepatocytes served to define this main cell type of liver as at least a part of the source of the protein-bound γ-glutamyl-polyamines seen earlier in whole liver tissue.[11] The observation of significant differences in the levels of the individual γ-glutamylpoly-amines in hepatocytes and in whole liver was taken as indication of a distribution of these derivatives in liver components.

γ-GLUTAMYLPOLYAMINES IN CULTURED CELLS

The first direct evidence for conjugation of polyamines to protein through γ-glutamyl linkage was obtained in human peripheral lymphocytes in culture with the use of radiolabeled putrescine.[5] It was reasoned that lymphocytes provided an ideal model system for this study, because soon after mitogen stimulation these cells display an increased influx of Ca^{2+} [20] and enchanced transglutaminase activity.[21] This is soon followed by an increase in polyamines,[22] thus supplying each of the ingredients for Ca^{2+}-dependent enzyme-mediated incorporation of polyamines. Small amounts of N-(γ-glutamyl)putrescine (Fig. 1 B) and N^1-and N^8-(γ-glutamyl)-spermidines (Fig. 1 C) were identified in proteolytic digests of lymphocyte proteins. However, hypusine was by far the major radiolabeled component found in the protein fraction.[23]

Similar patterns of radiolabeled components were seen in digests of the protein fractions from Chinese hamster ovary cells that had been cultured in the presence of [^{3}H]putrescine.[14] These cells were cultured without other additions to the medium and with additions of various combinations of agents that enhance transglutaminase activity and inhibit enzymes of polyamine metabolism. The results are shown in Table I. Radiolabeled hypusine was found in the protein fraction from cells cultured under each condition. In addition those cells treated with both a specific ornithine decarboxylase inhibitor, difluoromethylornithine (DFMO) and methylglyoxal bis(guanylhydrazone) (MGBG), an inhibitor of both S-adenosylmethionine decarboxylase and diamine oxidase, and those treated with DFMO, MGBG, and sodium butyrate, an agent that increases cellular transglutaminase activity, (Table I, cultures 7 and 8, respectively) also contained measurable amounts of γ-glutamylpolyamines in their protein fractions. This is clearly evident in the polyacrylamide gel electrophoretic patterns shown in Fig. 3. Hypusine was found only in the band corresponding in molecular weight to eukaryotic initiation factor 4D (M.W. 18,000), the hypusine-containing protein[23], and this was the only band seen in untreated cells (Culture 1). In treated cells (cultures 7 and 8) several additional bands are apparent in both the stacking gels and the running gels and these were found to contain putrescine and/or spermidine.

In a recent investigation of polyamine metabolism in primary cultures of mouse epidermal cells,[15] it was observed that with growing cells (those at low Ca^{2+} both with and without retinoic acid) radioactivity from polyamines was covalently incorporated into the protein fraction primarily in the form of hypusine (Table II). In cells induced to undergo terminal differentiation (cornified cell envelope formation) by raising the Ca^{2+} level, however, γ-glutamyl-bound polyamines were found to be the main

Table I. DISTRIBUTION OF RADIOACTIVITY IN COMPONENTS OF THE TRICHLOROACETIC ACID-INSOLUBLE FRACTION FROM CHO CELLS TREATED WITH VARIOUS COMPOUNDS

Culture No.:	1		3		4		7		8	
Addition:	none		DFMO		butyrate + DFMO		DFMO + MGBG		butyrate + DFMO + MGBG	
	cpm	pmol	cpm	pmol	cpm	pmol	cpm	pmol	cpm	pmol
Hypusine	11625	29.5	63400	8.0	54450	8.6	51535	18.5	5590	2.8
N^1,N^4-Bis(γ-glutamyl) putrescine	250	-	300	-	800	-	910	-	560	-
N^1-(γ-Glutamyl) putrescine	-	-	-	-	-	-	455	-	21830	2.4
N^1,N^8-Bis(γ-glutamyl) spermidine	-	-	-	-	-	-	1590	-	4230	2.0
N^1-(γ-Glutamyl) spermidine / N^8-(γ-Glutamyl) spermidine	590	-	300	-	-	-	11140	4.0	4410	2.2
γ-Aminobutyric acid	-		2425		3100		-	-	-	-
Di- and polyamines	420		3825		2800		4020		1375	
Other	-		11200		25000		6670		8000	

The radioactivities and amounts are given in units in the total 5 mg samples employed for analysis. The amounts were calculated using the specific radioactivities of the component polyamines; that used for hypusine calculation was that of its immediate precursor, spermidine. Data are the average of three determinations which differed by less than 25%.

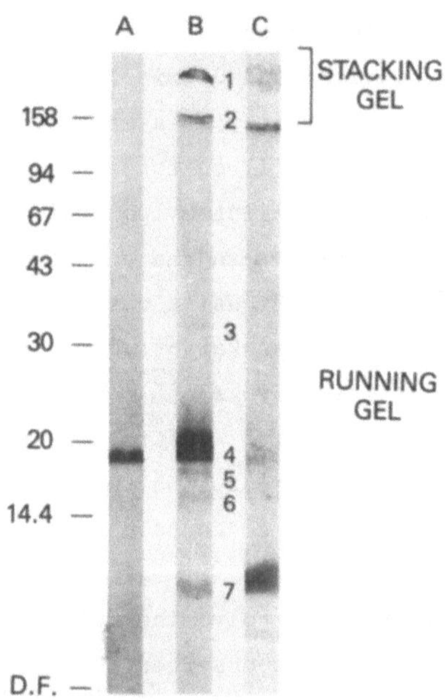

Fig. 3. Fluorogram from a polyacrylamide slab gel showing radioactivity incorporation into the proteins of Chinese hamster ovary cells. Cells from: A) Culture 1, Table I; B) Culture 7, Table I; C) Culture 8, Table I. D.F., dye front.

TABLE II. DISTRIBUTION OF RADIOACTIVITY FROM [^{3}H]PUTRESCINE IN THE PROTEIN FRACTION OF MOUSE EPIDERMAL CELLS IN CULTURE: EFFECT OF CALCIUM AND RETINOIC ACID.

Radiolabeled component	Radioactivity of component from protein of cells cultured with		
	Low Ca^{2+}	High Ca^{2+}	Retinoic acid
	cpm/mg protein		
N^1,N^4-bis(γ-Glutamyl)putrescine	ND	9,310	ND
N^1,N^8-bis(γ-Glutamyl)spermidine	2,230	27,520	ND
N^1-(γ-Glutamyl)putrescine	928	39,750	ND
Hypusine	10,530	4,560	15,100
N^1- and N^8-(γ-Glutamyl)spermidines[a]	1,860	35,300	990

[a] The N^1- and N^8-(γ-glutamyl)spermidines were not separated by the procedure used. They are reported together.

The values given are the mean of three determinations with S.D. less than 10%. ND, not detected.

radioactive components of protein (Table II). A low molecular weight material isolated from the protein fraction of differentiated cells was found to contain γ-glutamylputrescine and γ-glutamylspermidines (Fig. 1 IB and IC, respectively), but no bis(γ-glutamyl)polyamines (Fig. I II). Both mono- and bis(γ-glutamyl)polyamines were observed in digests of the protein fraction containing the cornified cell envelopes. These findings provided a basis for speculation on the role of polyamines in envelope formation and prompted an investigation that led to the discovery of bis(γ-glutamyl)-polyamine crosslinks in normal cell envelopes and the variation of these crosslinks in envelopes from diseased skin.

POLYAMINE-DERIVED CROSSLINKS IN NORMAL AND PSORIATIC HUMAN EPIDERMAL CELL ENVELOPES

The coordinate program of terminal differentiation in epidermis yields a superficial layer of tough dead squamae well suited to the protective function of the integument. These squamae consist largely of insoluble disulfide-bonded keratin tonofilaments, but also contain a so-called cornified envelope immediately beneath the plasma membrane. This cornified envelope, which contributes to the coherence, plasticity, and resistence of the skin, consists of protein that is resistent to keratinolytic agents and organic solvents, but is degraded by proteolytic enzymes.[25] The high degree of chemical stability of the cell envelope has been attributed to its ε-(γ-glutamyl)lysine crosslinks, the production of which is catalyzed by transglutaminase(s).[26]

We have found that cell envelopes contain another crosslink that in all probability is formed through transglutaminase action.[12] This additional linkage, bis(γ-glutamyl)spermidine (Fig. II B), occurs at about one-half the level of ε-(γ-glutamyl)lysine in cornified envelopes from callus of normal individuals (Fig. 4). Although envelopes isolated from scales of psoriatic patients were found to be crosslinked with ε-(γ-glutamyl)-lysine bonds to about the same extent as envelopes from normal persons, their bis-(γ-glutamyl)spermidine crosslinking was observed to be much higher than that in normals. Envelopes from uninvolved areas, i.e., callus, of psoriatic patients, however, were found to display patterns of cross-linking very similar to those of envelopes of normal individuals.

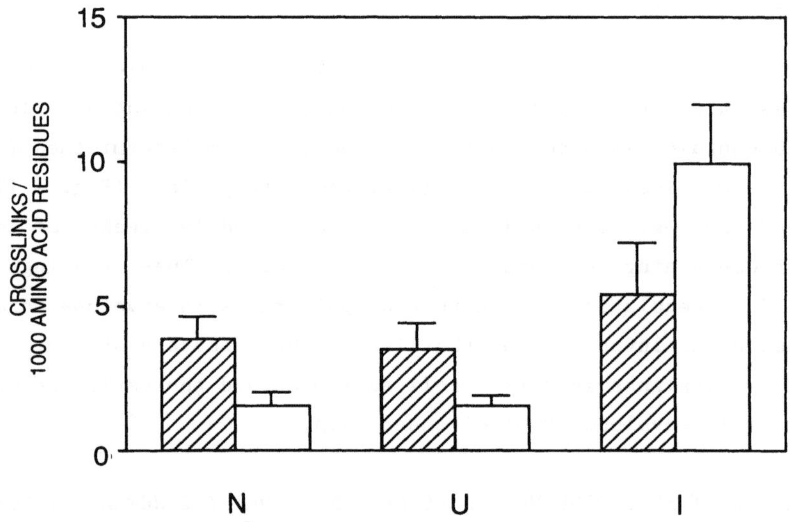

Fig. 4. ε-(γ-Glutamyl)lysine (hatched bars) and bis(γ-glutamyl)spermidine (open bars) crosslink levels in cornified envelopes isolated from specimens of callus and psoriatic scales. N = envelopes from callus of normal individuals, U = envelopes from uninvolved areas (callus) of psoriatic patients, I = envelopes from scales of psoriatic patients. The values are averages from 10 individuals.

CONCLUDING REMARKS

Although earlier studies with mammalian tissues and cells supplied proof for the occurrence of covalent polyamine-protein conjugates and demonstrated that bonding occurs in these conjugates through γ-glutamyl linkage, they provided no information as to the biological significance of these covalent reactions and very little information about the proteins involved. The identification of bis(γ-glutamyl)spermidine in human epidermal cell envelopes reported here is the first evidence of polyamine crosslinking in cell envelopes and, indeed, is the first report of polyamine involvement in crosslinking in any defined biological structure. Because the common genetically determined chronic disease psoriasis is characterized by a disordered epidermal differentiation and elevated cutaneous levels of polyamines,[27] we chose this disease as one in which to study possible abberations in cell envelope γ-glutamyl crosslinking. Indeed, it bacame evident that the crosslinking patterns of envelopes from callus of normal individuals and those from scales of psoriatic patients were very different, whereas there were little, in any, differences between envelope crosslinking in normal and uninvolved psoriatic skin (Fig. 4).

It seems almost certain that the bis(γ-glutamyl)spermidine crosslink, like the ε-(γ-glutamyl)lysine crosslink, is a product of transglutaminase action. Despite their likely common enzymatic origin, the two crosslinks bear several striking differences: (i) the polyamine crosslink has a more extended structure than the ε-(γ-glutamyl)lysine linkage and, thus allows greater distance between bonded protein chains. This may, therefore, permit chain attachments at positions too far apart for crosslinking through the lysine-derived linkage. (ii) The transglutaminases act on primary amines; secondary amines do not function as substrates.[8] Therefore, the secondary amino group of spermidine remains unreacted during crosslinking. It is possible that the positive charge contributed by this secondary amino group has some specific biological function. (iii) For crosslinking between protein chains by ε-(γ-glutamyl)lysine one chain must provide an available lysine residue and the other a reactive glutamine. For bis(γ-glutamyl)spermidine crosslinking spermidine serves as the amine functionality and thus each protein chain need only contribute reactive glutamine residues.

It is well established that both normal human skin[28] and psoriatic lesions[27] contain levels of spermine equal to or higher than those of spermidine. In addition, it is known that spermine effectively serves as a substrate for transglutaminases <u>in vitro</u> to form the bis(γ-glutamyl)-spermine crosslink between model protein molecules.[5,29] The finding of γ-glutamylspermidine crosslinks, but no γ-glutamylspermine crosslinks in epidermal cell envelopes must result from the failure of spermine to enter the catalytic reaction, either because of its absence from the site of catalysis or because of its inability to serve as a cosubstrate with envelope macromolecular precursors. Exclusion of spermine from the reaction site could occur due to specific compartmentalization within the cell, whereas failure of spermine to serve as a substrate with precursor proteins might result from structural features of these proteins that preclude its proper approach to the enzyme catalytic site. In either case the fact that bis(γ-glutamyl)spermidine, but not bis(γ-glutamyl)spermine is found in cell envelopes keynotes a specificity in crosslinking. Whether this specificity resides primarily in the envelope precursors, the enzyme(s) catalyzing the reaction, or the state of the polyamines is not immediately apparent.

Certainly the variations shown in Fig. 4 in epidermal cell envelope crosslinking must be construed as evidence for a direct association of polyamine metabolism with psoriasis. The obvious questions that arise are: is there a causal relationship between psoriasis and abnormal cell envelope crosslinking? and, if so, is a change from normal crosslinking a predisposing cause for production of lesions in psoriatic patients? The answers to these questions must await the results of experiments designed to test the effects on crosslinking of clinical treatments for psoriasis and their effects on cutaneous levels of polyamines and activities of the enzymes that catalyze crosslinking. Even more important in answer to the above questions may be the complete identification of the cellular envelope precursor proteins and an understanding of the relationships between envelope protein compositions and crosslink patterns. In this regard, it was recently suggested that cell envelopes are composed of those proteins available to the crosslinking enzyme(s) at the moment that it becomes activated and, therefore, that the compositions of the envelopes are not strictly determined.[30]

REFERENCES

1. S. S. Cohen, The functions of the polyamines. Adv. Polyamines Res. 1:1 (1978).
2. C. W. Tabor and H. Tabor, 1,4 Diaminobutane (putrescine), spermidine and spermine, Ann. Rev. Biochem. 42:285 (1976).
3. G. Scalabrino and M. E. Ferioli, Polyamines in mammalian tumors. Part I, Adv. Cancer. Res. 35:151 (1981).
4. M. H. Park, A. Abbruzzese and J. E. Folk, Post-translational formation of hypusine: biogenesis of translation initiation factor eIF-4D, in: "Advances in Post-translational Modifications of Proteins and

Ageing", (V. Zappia, P. Galleti, R. Porta, and F. Wold, eds.) p 633, Plenum Press, London (1988).

5. J. E. Folk, M. H. Park, S. I. Chung, J. Schrode, E. P. Lester and H. L. Cooper, Polyamines as physiological substrates for transglutaminases, J. Biol. Chem. 255:3695 (1980).
6. H. G. Williams-Ashman and Z. N. Canellakis, Transglutaminase-mediated covalent attachment of polyamines to proteins: mechanisms and potential physiological significance, Physiol. Chem. & Physics 12:457 (1980).
7. J. E. Folk, Transglutaminases, Ann. Rev. Biochem. 49:517 (1980).
8. J. E. Folk and J. S. Finlayson, The ε-(γ-glutamyl)lysine crosslink and the catalytic role of transglutaminases, Adv. Protein Chem. 31:1 (1977).
9. R. T. Ambron and L. T. Kremzner, Post-translational modification of neural proteins: evidence for transglutaminase activity in R2, the giant cholinergic neuron of Aplysia, Proc. Natl. Acad. Sci. USA 79: 3442 (1982).
10. L. Cariello, J. Wilson and L. Lorand, Activation of transglutaminase during embryonic development, Biochemistry 23:6843 (1984).
11. S. Beninati, M. Piacentini, M. P. Argento-Ceru', S. Russo-Caia and F. Autuori, Presence of di- and polyamines covalently bound to protein in rat liver, Biochim. Biophys. Acta 841:120 (1985).
12. S. Beninati, N. Martinet, T. Nigra, G. Peck and J. E. Folk, Spermidine as a covalent crosslinking component of normal and psoriatic human cell envelopes, J. Cell Biol. 105:82a (1987).
13. M. Piacentini and S. Beninati, γ-Glutamylamine derivatives in isolated rat hepatocyte proteins, Biochem. J. 249:813 (1988).
14. S. Beninati, M. Piacentini, E.R. Cocuzzi, F. Autuori and J. E. Folk, Covalent incorporation of polyamines as γ-glutamyl derivatives into CHO cell protein, Biochim. Biophys. Acta 952:325 (1988).
15. M. Piacentini, N. Martinet, S. Beninati and J. E. Folk, Free and protein-conjugated polyamines in mouse epidermal cells, J. Biol. Chem. 263-3790 (1988).
16. A. G. Loewy, The N^{ε}-(γ-glutamic)lysine crosslink: Method of analysis, occurrence in extracellular and cellular protein, Methods Enzymol. 107:241 (1984).
17. M. L. Fink, S. I. Chung and J. E. Folk, γ-Glutamylamine cyclotransferase: specificity toward ε-(L-γ-glutamyl)-L-lysine and related compounds, Proc. Natl. Acad. Sci. USA 77:4564 (1980).
18. M. L. Fink and J. E. Folk, γ-Glutamylamine Cyclotransferase (Rabbit Kidney), Methods Enzymol. 44:347 (1983).

19. S. Beninati, N. Martinet and J. E. Folk, High-performance liquid chromatographic method for the determination of ε-(γ-glutamyl)-lysine and mono- and bis-γ-glutamyl derivatives of putrescine and spermidine, J. Chrom. 443:329 (1988).

20. G. Allwood, G. L. Asherson, M. J. Davey and P. J. Goodford, The early uptake of radioactive calcium by human lymphocytes treated with phytohaemagglutinin, Immunology 21:509 (1971).

21. A. Novogrodsky, S. Quittner, A. L. Rubin and K. H. Stenzel, Transglutaminase activity in human lymphocytes: early activation by phytomitogens, Proc. Natl. Acad. Sci. USA 75:1157 (1978).

22. R. H. Fillingame and D. R. Morris, Polyamine accumulation during lymphocyte transformation and its relation to the synthesis, processing, and accumulation of ribonucleic acid, Biochemistry 12:4479 (1973).

23. M. H. Park, H. L. Cooper and J. E. Folk, Identification of hypusine, an unusual amino acid, in a protein from human lymphocytes and of spermidine as its biosynthetic precursor, Proc. Natl. Acad. Sci. USA 78:2869 (1981).

24. H. L. Cooper, M. H. Park, J. E. Folk, B. Safer and R. Braverman, Identification of the hypusine-containing protein Hy+ as translation initiation factor eIF-4D, Proc. Natl. Acad. Sci. USA 80:1854 (1983).

25. A. G. Matolsty and C. A. Balsamo, A study of the components of cornified epithelium of human skin, J. Biophys Biochem. Cytol. 1:339 (1955).

26. R. H. Rice and H. Green, The cornified envelope of terminally differentiated human epidermal keratinocytes consists of crosslinked protein. Cell 11:417 (1977).

27. P. Bohlen, J. Grove, M. F. Beya, J. Koch-Weser, M. H. Henry and E. Grosshans, Skin polyamine levels in psoriasis: the effect of dithranol therapy, Eur. J. Clin. Invest. 8:215 (1978).

28. P. El Baze, G. Milano, P. Verrando, N. Renee and J. P. Ortonne, Polyamine levels in normal human skin, Arch. Dermatol. Res. 275:218 (1983).

29. J. Schrode and J. E. Folk, Transglutaminase-catalyzed crosslinking through diamines and polyamines, J. Biol. Chem. 253:4837 (1978).

30. S. Michel, R. Schmidt, S. M. Robinson and U. Reichert, Morphological and biochemical characterization of the cornified envelopes from human keratinocytes of different origin, J. Invest. Dermatol. 87:156 (1986).

SPERMIDINE IN MAMMALIAN LYMPHOCYTES AND SEA URCHIN EMBRYOS: UPTAKE AND LABELING OF MACROMOLECULES

Zoe Nakos Canellakis

West Haven Veterans Administration Medical Center, West Haven CT 06516, and the Department of Pharmacology, Yale University School of Medicine, New Haven, CT 06510

With special thanks and recognition for active participation by Frederick W. Scalise, Leslie L. Marsh, Yukari C. Manabe, and Elinor La through whose efforts the diversity of this research was possible.

INTRODUCTION

We have studied the uptake and attachment of spermidine and some of its metabolites to cellular macromolecules in a variety of cell systems. We have selected systems which demonstrate significant uptake of polyamines when the polyamine is presented in the external environment of the cell. We chose to work with cells which are capable of increasing their metabolic activity in a manner approaching synchrony in response to a biological stimulus. In the presence of exogenous radioactive spermidine and a stimulus provoking focused cell function the cell takes up the spermidine and actively utilizes it in a biosynthetic manner. We have traced the metabolic fate of the radioactive spermidine under these conditions. The cells which we have primarily studied are young sea urchin embryos and mammalian lymphocytes. The mammalian lymphocytes we have studied include murine splenic lymphocytes, normal human peripheral lymphocytes and peripheral lymphocytes from patients with chronic lymphocytic leukemia (CLL). These represent a continuation of earlier studies where we showed that these cells can be activated by a B-cell mitogen and this activation can be abrogated by simultaneous exposure to the diacetyl derivative of putrescine or of its analogue 1,6 diaminohexane (1-3). The sea urchin embryo studies are a continuation of earlier investigations where we documented the covalent binding of spermidine to a unique protein in very young sea urchin embryos (4). Both mitogen activation and the event of fertilization can be thought of as agents inducing "natural" synchrony and consequent increased uptake and utilization of exogenous polyamines.

Certain parallels between the two biological systems as we observe them will be described. No attempt is made in this report to be inclusive in describing studies from other laboratories of polyamines associated with macromolecules. Thus, this is in the form of a progress report of our current experiments.

METHODS

Cell preparation: Murine splenic lymphocytes (female BALB/c, 4 months old) were prepared and cultured according to our published procedures (1). Human peripheral lymphocytes from normal subjects and from patients with CLL were also prepared according to our usual manner (2). Viability of cells used for lymphocyte culture always exceeded 97% at the onset of the experiments. Sea urchin embryos (*Strongylocentrotus purpuratus*) were obtained, fertilized and cultured in artificial sea water at 4°C (4). Whereas CLL-lymphocytes are primarily B-cells, populations of lymphocytes from normal subjects are primarily T-cells. Murine spleen generally contains B-cells and T-cells in approximately equal numbers. No attempt has been made to separate subpopulations of lymphocytes in these studies.

Measurement of proliferative capacity, viability of lymphocyte cultures, and incubations: Cells were evaluated in every instance for both proliferative capacity and viability by methods described in the original research paper as referenced. Cell cultures were maintained as controls or exposed to optimal levels of the relevant mitogen: (10 μg/ml) *E. coli* lipopolysaccharide (LPS), a specific B-cell mitogen for murine cultures; 0.01% *Staphylococcus aureus* Cowan strain I (SAC) for human lymphocyte cultures; (0.5μg/ml) of phytohemagglutinin (PHA), a specific T-cell mitogen for murine cultures; HMBA addition when indicated: 3 mM HMBA. Incubations were carried out for 48 hours for murine lymphocyte cultures and for 72 hours for human lymphocyte cultures.

Uptake and metabolism of ^{3}H polyamine: Cells prepared as described above and fertilized or cultured in the absence or presence of the cell mitogen and/or HMBA as indicated, were exposed to ^{3}H-putrescine or ^{3}H-spermidine (10 μCi/ml, 0.3 μM, 2,3-^{3}H(N)-putrescine or 1,8-methylene-^{3}H-spermidine). After the incubation period appropriate to the biological system under study the cells were harvested and washed exhaustively until free of extracellular radioactivity. Cells were then precipitated with 10% trichloroacetic acid (TCA) in order to separate the acid-soluble and the acid-insoluble fractions. Radioactivity in each fraction was measured by liquid scintillation counting. The protein-free acid-soluble fractions were analyzed by HPLC (5). The acid-insoluble fraction was analyzed by an HPLC molecular sieve method which we have recently developed in our laboratory.

HPLC analyses:

1. Cation exchange: HPLC analysis (5): Ultrasil SCX 10 µm cation exchange column, 25 cm x 4.6 mm, flow rate 0.8 ml/min eluted with a 30 min gradient: 0.01-1.0 M Na+ (acetate, nitrate), pH 4.6 buffer; acid-soluble fraction applied directly; post-column derivatization with orthophthalaldehyde for quantitative determinations and fluorescent identification of standards; scintillation counting of effluent for measurement of radioactivity. When recovery of the radioactive sample was desired for further analysis, elution was performed with an ammonium formate gradient: 0.01-1.0 M ammonium formate, pH 4.4. Evaluation of uptake of HMBA was made as follows: free 1,6 diaminohexane was measured both before and after hot acid hydrolysis of the acid-soluble fraction of cells. The difference between the two determinations represents amounts present as the mono or the diacetyl derivative of the diamine.

2. Molecular sieve analysis of macromolecules: For molecular sieving, samples containing approximately 50,000 cpm in the acid-insoluble fraction were prepared for analysis. The sample was washed exhaustively with 5% TCA until radioactivity of the washes reached background levels. The residual pellet was then extracted with solubilization buffer (0.1 M sodium phosphate, pH 6.8 containing 0.05% sodium dodecyl sulfate [SDS]), and when necessary, the sample was concentrated (Amicon, Centricon-10 microconcentrator). There was no radioactivity present in the salt fraction of the concentrate. The concentrated sample was applied directly to a TSK 2000SW column, 60cm x 7.5mm (MicroPak TSK Exclusion Column, Toyo Soda Manufacturing Co., Ltd, Tokyo, obtained from Varian), flow rate 0.8 ml/min and eluted with 0.1 M sodium phosphate, pH 6.8 buffer containing 0.05% SDS. The TSK 2000SW column is a silica-based, rigid, hydrophilic column with spherical particles of 10-13 µm diameter. We have observed quantitative recovery of radioactivity and high reproducibility in column performance for over 300 cycles of continuous use. The protein content of the sample is regularly monitored at 280 nm. When chemical analysis of the effluent peaks was desired, small fractions of 0.2 min were taken, desalted using the Amicon microconcentrator, acid hydrolyzed and analyzed on the cation HPLC column described above. Radioactive eIF-4D purified from Chinese Hamster ovary cells was kindly supplied by Myung Hee Park.

RESULTS

Significant amounts of radioactive N^1-acetylspermidine are formed in both murine and human lymphocytes and the developing sea urchin embryo cultured in the presence of exogenous ^3H-putrescine or ^3H-spermidine (Table I) (6,7, sea urchin manuscript in preparation). In murine splenic lymphocytes there is a small but significant conversion of both ^3H-putrescine and ^3H-spermidine to N^1-acetylspermidine: 6.9 % of the radioactivity appears as N^1-acetylspermidine with ^3H-putrescine as precursor and 5.4 % with ^3H-spermidine as precursor. The ratio of intracellular radioactive spermidine/ N^1-acetylspermidine is 6.6 with ^3H-putrescine as precursor and 16.8 with ^3H-spermidine as precursor. In the sea urchin embryo radioactive label in N^1-acetylspermidine increases from 2.5 % of the total intracellular acid-soluble radioactivity in the 4 hour embryo to 14.0 % in the 28 hour embryo. In human lymphocytes we again find that N^1-acetylspermidine is a major metabolite. Radioactivity in the N^1-acetylspermidine derived from ^3H-spermidine reaches levels of up to 20 % during mitogen activation in both normal and CLL lymphocytes. Addition of HMBA causes a significantly diminished level of conversion of ^3H-spermidine to N^1-acetylspermidine in CLL lymphocytes.

In these studies we have observed a real difference in the intracellular metabolic fate of ^3H-putrescine and of ^3H-spermidine. Whereas ^3H-putrescine is readily oxidized to ^3H-GABA and also metabolized to ^3H-spermidine intracellularly, ^3H-spermidine seems to be conserved in activated murine lymphocytes (6). For example, after 48 hours of culture only 12 % of radioactive label initially present as ^3H-putrescine remains as labeled putrescine, whereas 72 % of radioactive label initially present as ^3H-spermidine remains as labeled spermidine. In the sea urchin at 28 hours of embryogenesis although 60% of the ^3H-spermidine is metabolized, there is no detectable metabolism to either putrescine or GABA. In human lymphocytes we find similarly that N^1-acetylspermidine is a significant metabolite and spermidine is again conserved; 45-55% of the intracellular ^3H-spermidine remains as radioactive spermidine after 72 hours of culture (7). Radioactivity in N^1-acetylspermidine derived from ^3H-spermidine reaches levels of up to 20% during mitogen activation in both normal and CLL lymphocytes. Little conversion to putrescine is noted (less than 5%), although significant levels of both radioactive N-acetylputrescine and GABA appear (15-20%). Both oxidative degradation and metabolism to spermidine are possible explanations for the persistence of generally low intracellular levels of radioactive putrescine as compared to spermidine in these systems.

Table I: N^1-ACETYLSPERMIDINE
(Radioactivity)

SOURCE	PERCENT*
Murine spleen (BALB/c):	
activated by B cell mitogen: 48 hours	
Precursor: ^{3}H-putrescine	6.9
^{3}H-spermidine	5.4
Sea urchin embryos (<u>S. purpuratus</u>):	
4 hour embryo:	2.5
28 hour embryo:	14.0
Human peripheral lymphocytes: 72 hours	
normal: control	11.4
B-cell mitogen	17.6
chronic lymphocytic leukemia:	
control	9.6
B-cell mitogen	20.6
HMBA	2.6

* Radioactivity in N^1-acetylspermidine conjugate as percent of total acid-soluble, cell-associated radioactivity. Precursor: ^{3}H-spermidine, 10 μCi/ml, 0.3 μM spermidine in extracellular medium; when ^{3}H-putrescine was added (as noted) it was added at the same level as ^{3}H-spermidine (6,7, and sea urchin manurscript in preparation).

HMBA is taken up by the cell and metabolized (Table II). In murine splenic lymphocytes cultured in the presence of 3 mM HMBA, levels of 0.34 - 0.57 n moles of 1,6 diaminohexane/10^6 cells appear intracellularly. The bulk of this exists as 1,6 diaminohexane and the rest represents acetyl derivatives. Similar levels of intracellular 1,6 diaminohexane are seen in Friend erythroleukemia cells induced to differentiate by HMBA. These levels correspond to intracellular concentrations of 0.18 mM and 0.60 mM in murine spleen and in Friend erythroleukemia cells respectively, concentrations which are of the same order of magnitude as putrescine in these cells (6,8). In contrast with murine spleen and Friend erythroleukemia cells, CLL cells take up very little exogenous HMBA (7), and none is detectable as the free amine. Only 0.08 n moles HMBA/10^6 cells, which corresponds to 8×10^{-3} % uptake, become cell associated. This level is in contrast to the n mole intracellular level of the normally occurring polyamines. By contrast, the developing sea urchin embryos regularly take up large amounts of HMBA (sea urchin manuscript in preparation). In the sea urchin embryo there is no intracellular hydrolysis of HMBA.

Table II: 1,6 DIAMINOHEXANE
(Cell-associated)

	n moles/10^6 cells	Percent as free diamine
Sea urchin (S. purpuratus): 3 mM HMBA		
embryos from 10^6 eggs: 4 hours	0.0 (340)*	0
: 28 hours	0.0 (520)	0
Friend erythroleukemia cells:		
4 mM HMBA, 72 hours	0.31	
Murine splenic lymphocytes: 48 hours		
control	0.0 (0.0)	
3 mM HMBA	0.28 (0.36)	79
LPS	0.0 (0.0)	
LPS + 3 mM HMBA	0.25 (0.34)	72
PHA	0.0 (0.0)	
PHA + 3 mM HMBA	0.36 (0.57)	63
Human peripheral blood		
lymphocytes: 72 hours		
Chronic lymphocytic leukemia		
3 mM HMBA	0.0 (0.08)	0

* parentheses indicate values after hot acid hydrolysis; sensitivity of measurement is less than 0.007 n moles/10^6 cells.

We have used HPLC-molecular sieving to fractionate proteins labeled by radioactive polyamines during sea urchin embryogenesis (sea urchin manuscript in preparation) and in lymphocyte cultures (6,7) which have been exposed to exogenous ^{3}H-polyamine. Three major fractions can be distinguished (Table III). The larger macromolecules elute earlier, from 16'-20', with a peak of radioactivity at 16.5'. Smaller molecules elute later, with a peak at 21'. In lymphocytes we regularly note a distinct, smaller peak at 19'. The pattern of elution of protein as measured by ultraviolet absorption at 280 nm coincides with the pattern of elution of radioactivity. We have found it convenient to define the ratio of radioactivity in the major peaks eluting with peaks at 16.5' and 21' as 16.5'/21'. These peaks can be separated and, therefore, further characterized. Typical elution profiles comparing normal lymphocytes with CLL-lymphocytes are the basis on which the 16.5'/21' ratio has been determined (Fig. 1). The radioactive peak eluting at 21' coincides with the elution profile of the eukaryotic initiation factor eIF-4D which contains hypusine and is the only unique protein known to date which is post-translationally modified by polyamines (9).

Table III: SPERMIDINE LABELED MACROMOLECULES
(HPLC Molecular Sieve Peaks)
(% of total protein-associated counts)

CELLS	Time of elution 16.5'	19'	21'	Ratio 16.5'/21'
Sea Urchin embryos				
4 hr whole embryo				
Control	65	-*	27	2.4/1.0
HMBA	60	-	32	1.9/1.0
28 hr whole embryo				
Control	25	-	70	0.4/1.0
HMBA	30	-	59	0.5/1.0
4 hr embryo				
cytoplasm	70	-	20	3.5/1.0
nucleus	31	-	69	0.5/1.0
24 hr embryo				
cytoplasm	32	-	51	0.6/1.0
nucleus	6	-	85	0.1/1.0
CLL-lymphocytes				
Control	46	18	29	1.6/1.0
Mitogen activated	45	17	33	1.4/1.0
HMBA	45	19	30	1.5/1.0
Lymphocytes from normal subjects				
Control	36	17	39	0.9/1.0
Mitogen activated	34	15	39	0.9/1.0
Murine spleen cells				
Mitogen activated	34	24	34	1.0/1.0

15,000 to 20,000 cpm were applied to the HPLC molecular sieve column. * No radioactive 19' peak was observed in sea urchin embryo studies.

It is characteristic of sea urchin embryogenesis that there is a progressive change in the ratio of 16.5'/21' peaks from a high value of 2.4/1.0 in the 4 hour embryo to a low ratio of 0.5/1.0 by 28 hours. Evaluation of this ratio during cell fractionation studies of the developing sea urchin embryo identified the cytoplasm as the site of the highest ratio throughout the first 24 hours of embryonic development and also verified the progressive drop in this ratio as embryogenesis proceeds. In CLL-lymphoctyes the ratio is 1.5/1.0 in contrast to a ratio of 0.9/1.0 which we observe in lymphoctyes from normal subjects. Murine lymphoctyes have a ratio of 1.0/1.0.

We have verified our earlier observations that an unique 30 kD protein containing spermidine appears early in sea urchin embryogenesis by examining the radioactive components of the individual radioactive peaks of fractionated macromolecules (4, sea urchin manuscript in preparation). Macromolecules containing exclusively spermidine are present in both the 16.5' peak and the 21' peak early in embryogenesis.

As embryogenesis proceeds hypusine is seen in both the 16.5' peak and the 21' peak. Our present finding that hypusine is also a component of the larger macromolecule fraction is, to our knowledge, the first time that hypusine has been identified as associated with macromolecules distinct from eIF-4D. Using gel electrophoretic analysis we have identified 3 isoforms of a 18 kD protein which appear during sea urchin embryogenesis with isoelectric points of 6.0, 6.3, and 6.5 (10). Any possible relationship between these isoforms and the hypusine-containing larger macromolecules in the 16.5'peak or the hypusine-containing smaller macromolecules in the 21' peak remains to be evaluated.

Molecular sieve analysis of proteins from several experiments with sea urchin embryos using inorganic ^{32}P in either in vivo growth situations or with sea urchin embryo homogenates incubated with inorganic ^{32}P in vitro showed radioactive peaks in the region of both the larger and the smaller macromolecules. We have observed a constant $^{32}P/^{3}H$ ratio in these peaks which is indicative of a unique protein species: these latter findings describe work in progress at this time.

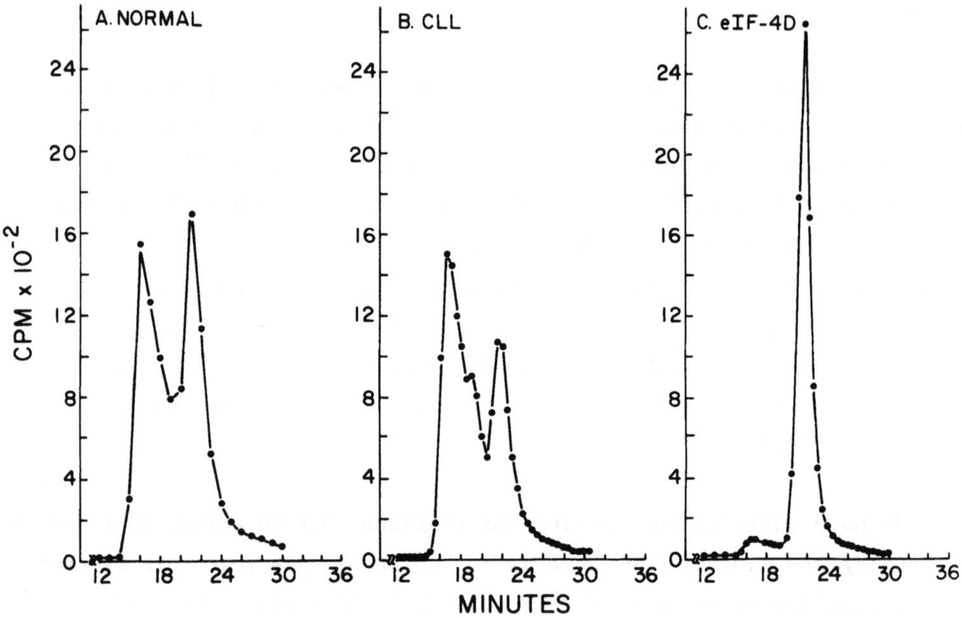

Fig 1. Macromolecular sieve analysis of spermidine labeled macromolecules. A: Activated normal human lymphocytes; B: activated CLL lymphycytes; C: eIF-4D.

DISCUSSION

In our studies to date we have shown that acetylated polyamines are modulators of cell function. The naturally occurring monoacetylputresine, TMBA (tetramethylene*bis*acetamide, diacetylputrescine), HMBA (hexamethylene*bis*acetamide), and the two biologically known isomeric monoacetylspermidines, N^1-acetylspermidine and N^8-acetylspermidine participate in a feedback mechanism which regulates ornithine decarboxylase activity (11,12). TMBA and HMBA, which our laboratory among others has shown are inducers of differentiation of Friend erythroleukemia cells in culture, are taken up by the cell and metabolized (8,13,14). TMBA and HMBA are specific inhibitors of B-lymphocyte activation and function (1,2), and they also inhibit uptake of exogenous polyamines by cultured lymphocytes (6). In addition, we have shown that both TMBA and HMBA inhibit expression of the c-myc oncogene (15).

In general, it is accepted that acetylpolyamines are intermediates in the cycling of polyamines which facilitates maintenance of intracellular polyamine levels; and monoacetylputrescine is an intermediate in the cellular biosynthesis of γ-aminobutyric acid (GABA) (16-18). Acetylputrescine has been found in association with the membrane fraction of cells (8,14). Both of the monoacetylated spermidines have been found in significant amounts in urine, but until recently very little of these derivatives has been detected in animal tissues, although the enzymes responsible for their biosynthesis are now well characterized (16). Some acetylspermidine has been reported in various normal tissues (19-21), tumor tissues (22-23) and in various drug induced situations (24-25). N^1-acetylspermidine has now been identified in high levels in lymphocytes (6,7,9).

Our studies have noted conservation of a high level of intracellular spermidine and a possibly regulated conversion to N^1-acetylspermidine which is consistent with a cell modulatory function for N^1-acetylspermidine. We propose that acetylation of spermidine underlies a focused cell function for the polyamines and their derivatives. It is timely to consider alternatives to traditional ideas that acetylation of polyamines only provides a suitable substrate for oxidases or facilitates export from the cell through obviation of cationic charges.

TMBA and HMBA, which are inducers of differentiation of Friend erythroleukemia cells in culture (8,13,14), are taken up by the cell and metabolized (6,8,14). We have also shown that TMBA and HMBA are specific inhibitors of B-lymphocyte activation and function in murine splenic lymphocytes (1,6), lymphocytes from blood of normal human subjects and patients with CLL (7). HMBA inhibits the uptake of both putrescine and spermidine when it is added to the exogenous medium in cell culture of either human or murine lymphocytes (6,7). In addition, HMBA causes a drop in the conversion of ^{3}H-spermidine to N^{1}-acetylspermidine in CLL cells (7). In these studies viability remains uncompromised. We do not know how this phenomenon of inhibition of uptake correlates with the known effect of HMBA in inhibiting proliferation and function of B-lymphocytes. A site of action at the cell surface is suggested by the observations thus far. Thus, HMBA seems to have diverse roles. The underlying mechanism is unclear, but several observations from our studies may be relevant in seeking explanations for the function of HMBA.

We have partially resolved, by HPLC-molecular sieving, macromolecules which become labeled by radioactive polyamines and their derivatives. The peaks of radioactive macromolecules have been identified by their elution time and a ratio describing the relationship between the two major peaks has been defined (16.5'/21'). The ratio seems to characterize the cellular site (i.e. nucleus) of the macromolecules and perhaps also the cell type (i.e. CLL-lymphocytes, normal lymphocytes, murine splenic lymphocytes, developing sea urchin embryo). The data we have obtained suggest to us that the ratio may be an index of cell proliferation and function.

ACKNOWLEDGEMENT

We wish to acknowledge the Veterans Administration Merit Review Program. We thank Georgi Danley for help in manuscript preparation.

REFERENCES

1. J.L. Ryan, P.K. Bondy, L. Gobran, Z.N. Canellakis, Acetylated Diamines Inhibit Endotoxin-Induced Lymphocyte Activation. J Immunol 132:1888-1891, (1984).
2. J. Lacy, W.C. Summers, Z.N. Canellakis, Effects of Diacetyl Diamines on in-vitro activation and Proliferation of Human B Lymphocytes. J Immunol 235:3772-3776 (1985).
3. Z.N. Canellakis, P.K. Bondy, C.S. Portlock, Acetyl Polyamines Inhibit Activation of Chronic Lymphocytic Leukemia Cells in vitro. Clin Res 35:628A, (1987).

4. Z.N. Canellakis, P.K. Bondy, A.A. Infante, Spermidine is bound to a unique protein in early sea urchin embryos. Proc Natl Acad Sci USA 82:7613-7615, (1985).
5. P.K. Bondy, Z.N. Canellakis, High-Performance Liquid Chromatography in the Separation and Measurment of di-and Polyamines and Their Derivatives and Specific Preparation of Isomers of their Monoacetyl Derivatives. J Chromatogr 244:371-379, (1980).
6. L.L. Marsh, P.K. Bondy, Z.N. Canellakis, Polyamines in Murine Splenic Lymphocytes (submitted).
7. Z.N. Canellakis, C.S. Portlock, P.K. Bondy, Polyamines in Chronic Lymphocytic Leukemia (submitted).
8. Z.N. Canellakis, P.K. Bondy, Diacetylputrescine Induces Differentiation and is Metabolized in Friend Erythroleukemia Cells. In: Advances in Polyamine Research 4:769-778, (1983).
9. J.E. Folk, H.P. Myung, C. Scc II, J. Schrode, E.P. Lester, H.L. Cooper, Polyamines as Physiological Substrates for Transglutaminases. J Biol Chem 255:3695-3700 (1980).
10. F.W. Scalise, A.A. Infante, Z.N. Canellakis, Features of Uptake and Utilization of Spermidine in the Sea Urchin Strongylocentrotus Purpuratus (submitted).
11. Z.N. Canellakis, L.A. Lande, P.K. Bondy, Factors Modulating the Activity of Ornithine Decarboxylase in Rat HTC Cells. Med Biol 59:300-307, (1981).
12. Z.N. Canellakis, Effects of Acetylated Polyamines on Ornithine Decarboxylase in Rat HTC Cells. Biochem Biophys Res Commun 100:929-933 (1981).
13. R.C. Reuben, R.L. Wife, R. Breslow, R.A. Rifkind, P.A. Marks, A New Group of Potent Inducers of Differentiation in Murine Erythroleukemia Cells. Proc Nat Acad Sci 73:862-866, (1976).
14. Z.N. Canellakis, L.L. Marsh, P. Young, P.K. Bondy, Polyamine Metabolism in Differentiating Friend Erythroleukemia Cells. Cancer Res. 44:3841-3845, (1984).
15. G.G. Luk and Z.N. Canellakis, Diacetylputrescine and Its Analog Suppress c-myc Expression and Activation of B-Lymphocytes. (submitted).
16. C.W. Tabor, H. Tabor, Polyamines. Ann Rev Biochem 53:749-790, (1984).
17. A.E. Pegg, P.P. McCann, Polyamine Metabolism and Function. Am J Physiol 243:C212-C221, (1982).
18. N. Seiler, Functions of Polyamine Acetylation. Can J Physiol Pharmacol 65:2024-2035 (1987).

19. S. Matsuzaki, K. Hamana, K. Imai, K. Matsuura, Occurrence in High Concentrations of N^1-Acetylspermidine and sym-Homospermidine in the Hamster Epididymis. Biochem Biophys Res Commun 107:307-313 (1982).
20. N. Seiler, F.N. Bolkenius, B. Knodgen, K. Heaegele, The determination of N^1-acetylspermine in Mouse Liver. Biochem Biophys Acta 676:1-7 (1981).
21. N. Seiler, F.N. Bolkenius, S. Sarhan, Formation of Acetylpolyamines in the Liver of Fasting Animals. Int J Biochem 13:1205-1214 (1981).
22. H. Yamazaki, S. Matsuzaki, T. Tsukahara, H. Kurihara, Elevation of N^1-acetylspermidine in Brain Tumor Tissue. International Conference on Polyamines in Life Sciences. Lake Yamanaka, Japan. Abstr. No. P70 (1986).
23. S. Takenoshita, S. Matsuzaki, G. Nakano, H. Kimura, H. Hoshi, H. Shoda, T. Nakamura, Selective Elevation of the N^1-acetylspermidine Level in Human Colorectal Adenocarcinomas. Cancer Res 44:845-847 (1984).
24. C. Stefanelli, D. Carati, C. Rossoni, F. Flamigni, C.M. Caldarera, Accumulation of N^1-acetylspermidine in Heart and Spleen of Isoprenaline-treated Rats. Biochem J 237:931-934 (1986).
25. K. Hamana, S. Matsuzaki, Elevation of Acetylpolyamines in Mouse Liver, Serum and Urine after Drug-induced Hepatic Injury and in Human Hepatitis. International Conference on Polyamines in Life Sciences. Lake Yamanaka, Japan. Abstr. No. P67 (1986).

BIOSYNTHESIS OF HYPUSINE IN eIF-4D PRECURSORS

M.H. Park[*], E.C. Wolff[*], A. Abbruzzese[+] and J.E. Folk[*]

[*]National Institute of Dental Research, NIH, Bethesda, Maryland 20892
[+]Department of Biochemistry and Biophysics, University of Naples, Italy

INTRODUCTION

An unusual basic amino acid hypusine was discovered in 1971 in extracts of bovine brain by Shiba and coworkers, who determined its chemical structure as N^{ε}-(4-amino-2-hydroxybutyl)lysine(1). Its distribution in various animal tissues as the free amino acid(2) and as a component of protein(3) was subsequently reported. Initial attempts to isolate any specific protein that contained hypusine were unsucessful, however(3). Ten years after the discovery of hypusine, we observed radiolabeling of a single cellular protein (M_r 18,000, pI 5.3) upon incubation of human peripheral lymphocytes with [1,8-^{3}H]spermidine or [2,3-^{3}H]putrescine in the presence of the mitogen, phytohemagglutinin and identified the polyamine-derived radioactive component of this protein as hypusine(4).

In most mammalian cells examined, hypusine occurs predominantly in one highly conserved cellular protein which was identified as eukaryotic translation initiation factor 4D (eIF-4D)(5). In this protein, hypusine is formed by post-translational modification of one specific lysine residue(6). Furthermore, eIF-4D preparations from rabbit reticulocytes(5) or human red blood cells(7) each contain approximately 1 mole per mole of hypusine. Hypusine formation was also detected in lower eukaryotes, yeasts(8,9), Drosophila melanogaster(8), Neurospora crassa(10), and plants(9) but not in bacteria(8,9). Hypusine synthesis increases with the rise in protein synthesis in human peripheral lymphocytes upon mitogen treatment(11) and correlates in general with cellular proliferation in various cultured cells(12-14). These findings led to the suggestion that the conversion of lysine to hypusine may play a role in the regulation of cell growth, perhaps through control of protein synthesis. Although eIF-4D has been considered as an initiation factor principally on the basis of its ability to stimulate methionyl-puromycin synthesis(15), its mode of action and its role in the translation of natural messenger RNA has not as yet been thoroughly studied.

In various rapidly proliferating mammalian cells, the post-translational synthesis of hypusine represents the major covalent structural contribution of polyamines to macromolecules. In view of the potential significance of this unique protein modification in mediating effects of polyamines on cellular proliferation, we commenced a systematic study on the mechanism and regulation of hypusine biosynthesis.

BIOSYNTHESIS OF HYPUSINE

Early studies of hypusine biosynthesis were carried out in rats by intraperitoneal injection of radiolabeled lysine, putrescine, or spermidine. Detection of a radioactive protein component with properties similar to those of hypusine led to the preliminary suggestion that lysine and polyamines are possible precursors(2,3). Definitive evidence that hypusine is derived by modification of a lysine residue was provided by our finding that in cultured mammalian cells, radiolabel from [4,5-^{3}H]lysine is incorporated into the lysine moiety of hypusine (scheme 1)(16). In addition, our early studies showed that either [2,3-^{3}H]putrescine or [1,8-^{3}H] spermidine supplied radiolabel in the 4-amino-2-hydroxybutyl moiety of hypu-

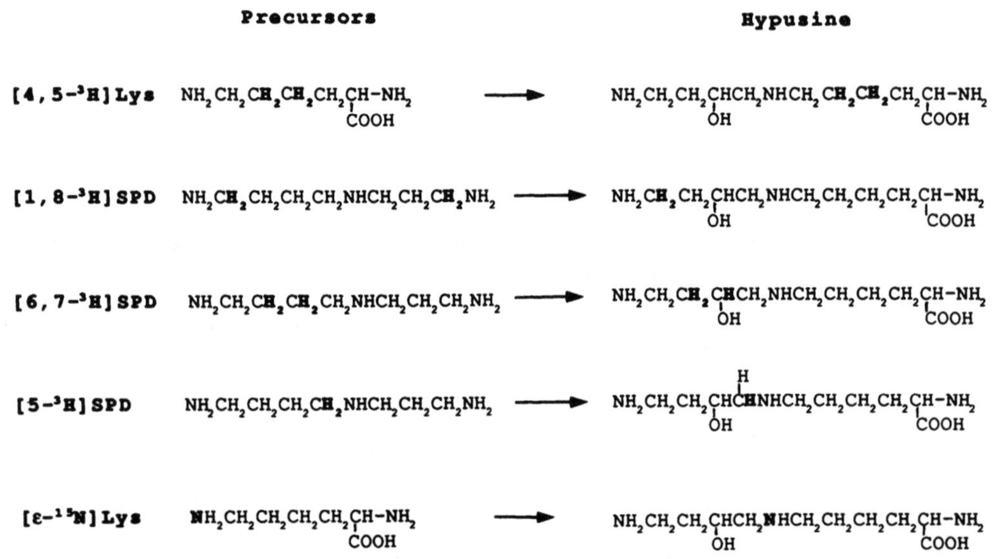

Scheme 1. Labeling of hypusine with specifically isotopically labeled precursors.
The positions of tritium in [^{3}H]spermidine, [^{3}H]lysine, [^{3}H]hypusine and of ^{15}N in [^{15}N]lysine and [^{15}N]hypusine are indicated by bold face type. The position of radiolabel in [^{3}H]hypusine was determined by oxidative cleavage by periodic acid and permanganate and identification of the labeled oxidation products as described earlier(4).

sine(4). First indications of the direct amine precursor role of spermidine were obtained from experiments in which intracellular conversion of putrescine to spermidine was inhibited(4). Further evidence for the precursor role of spermidine and for the specific alignment of carbons in the 4-amino-2-hydroxybutyl moiety was obtained with the use of specifically radiolabeled spermidines (scheme 1) (17). Extension of these studies using a mixture of [5,8-^{14}C]spermidine and various [^{3}H]spermidines revealed that half of the ^{3}H on carbon 5 was abstracted during transfer of butylamine moiety, whereas all ^{3}H on carbon 6,7,8 was retained in the intermediate, deoxyhypusine(18).

Experiments performed to determine the source of the secondary amine nitrogen in hypusine revealed that cells grown with [ε-^{15}N]lysine yielded hypusine enriched in ^{15}N (scheme 1) whereas those in which cellular spermidine was replaced with [4-^{15}N]spermidine did not(17). This information concerning the origins of the atoms of hypusine led us to propose a catalytic mechanism involving enzymatic oxidative cleavage at the secondary amine group of spermidine and transfer of the 4-carbon moiety (scheme 2) (18).

When Chinese hamster ovary (CHO) cells are cultured with [^{3}H]spermidine in the presence of the metal chelator, α,α,-dipyridyl, deoxyhypusine [N^ε-(4-aminobutyl)lysine], the unhydroxylated form of hypusine, accumulates in protein(16). That the normal biosynthesis of hypusine proceeds through hydroxylation of deoxyhypusine (scheme 2) was supported by evidence that the conversion of deoxyhypusine to hypusine in protein was seen both in intact cells and in cell free lysates(16). Deoxyhypusine hydroxylase that catalyses this final step (scheme 2) has been partially purified from rat testis(19). Certain catalytic characteristics of this enzyme indicate that it is a unique hydroxylase distinct from the lysyl and prolyl hydroxylases(20).

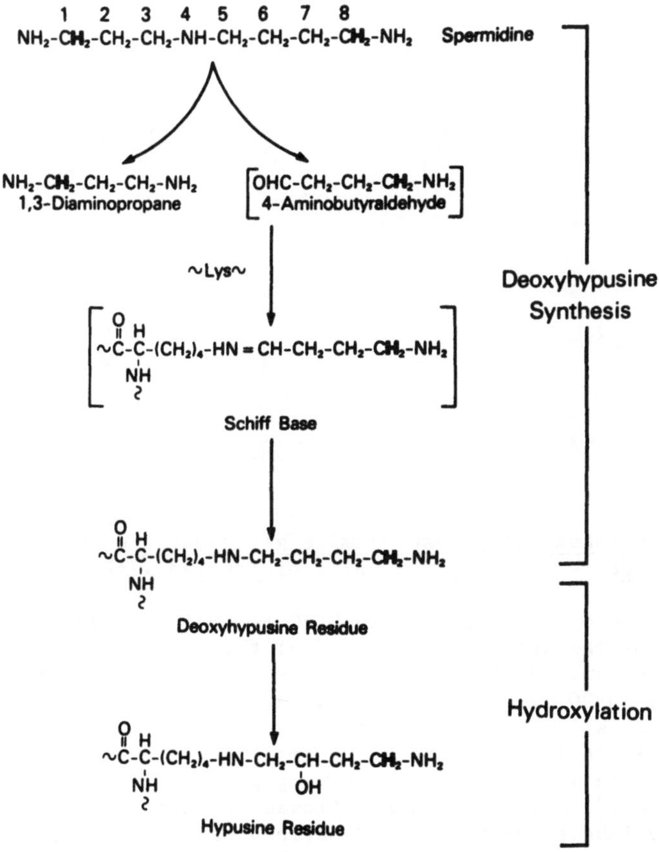

Scheme 2. Proposed pathway of hypusine biosynthesis
The positions of tritium labeling of [1,8-^{3}H]spermidine and its metabolites are indicated by bold face type. ~Lys~ denotes the lysine residue of the protein substrate (eIF-4D precursor). The postulated transient intermediates are in parentheses. Reprinted from Ref. 27.

REGULATION

The close relationship between cellular growth and hypusine synthesis raises the possibility that the post-translational events that lead to hypusine formation may play an important role in the regulation of cellular proliferation. The rate of hypusine formation may depend on one or more of the following: the concentration of the protein substrate, the level of intracellular spermidine, or the activities of the enzymes involved. In order to assess the relative importance of these factors, we examined the effects of translation inhibition and spermidine depletion on the biosynthesis of deoxyhypusine + hypusine(21). In control cells, inhibition of translation by cycloheximide resulted in the complete arrest of new synthesis of hypusine, consistent with the reported inhibition by cycloheximide of labeling of eIF-4D with [^{3}H]spermidine(22). The total dependence of the modification upon new synthesis of protein substrate in these control cells suggests that hypusine is synthesized de novo and that there is no pathway for turnover of the 4-amino-2-hydroxybutyl moiety of hypusine. Furthermore, this finding implied a lack of accumulation of the protein substrate for deoxyhypusine synthesis in untreated cells. In cells that had been previously depleted of spermidine by use of α-difluoromethylornithine (DFMO)(23), an inhibitor of ornithine decarboxy-

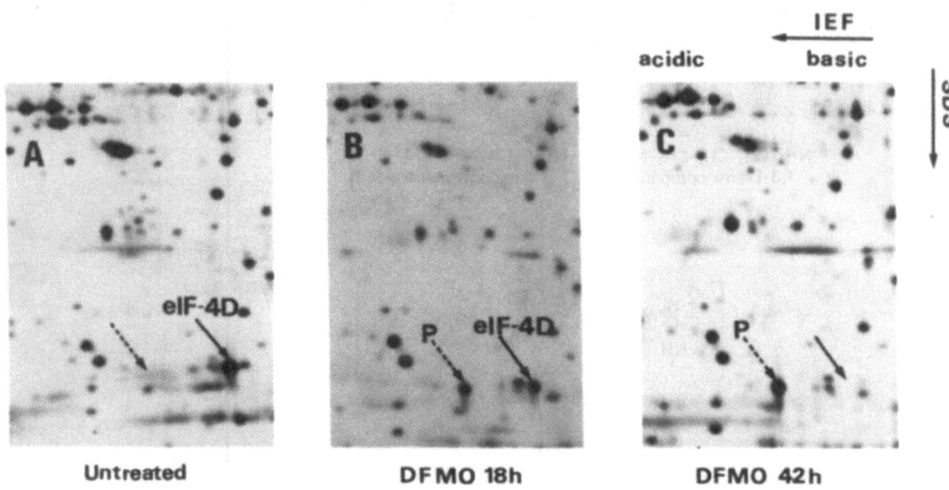

Fig. 1. Two-dimensional polyacrylamide gel separation of proteins from untreated CHO cells (A) and from cells pretreated with DFMO for 18 h (B) or for 42 h (C).
Cells, untreated or pretreated with DFMO (4 mM), were incubated for 2 h in the labeling medium which contained 5% of the normal level of lysine, 3% fetal bovine serum, and 250 μCi/ml [^{3}H]lysine without or with DFMO. Approximately 10-20 μg of labeled cellular proteins were separated by isoelectric focusing on gels containing pH 5 to 7 ampholytes followed by electrophoresis on 12.5% SDS polyacrylamide gels, by the method of Garrels(25). The [^{3}H]hypusine-containing protein isolated from CHO cells incubated with [^{3}H]spermidine was employed to identify eIF-4D on the map of [^{3}H]lysine-labeled proteins as described earlier(5). Portions of the fluorogram of the gels surrounding eIF-4D are shown. The positions of eIF-4D and Protein P identified as the eIF-4D precursor (lysine form), are inducated by the solid and broken arrows, respectively. The levels of spermidine were 14, 2, 0.2 nmol/mg protein in cells untreated, treated with DFMO for 18 h, and for 42 h, respectively.

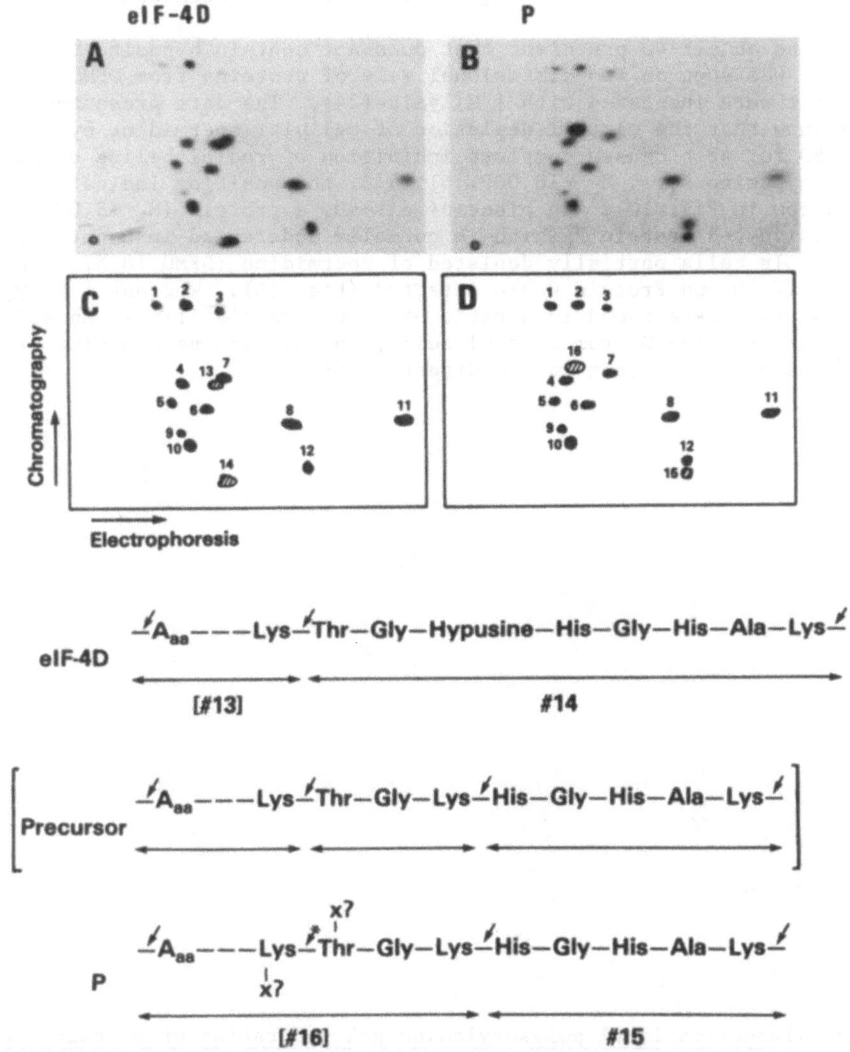

Fig. 2. Patterns of radioactivity on peptide maps prepared from tryptic digests of [^{3}H]lysine-labeled eIF-4D (A,C) and of [^{3}H]lysine-labeled Protein P (B,D).

The radiolabeled areas corresponding to the positions of eIF-4D and Protein P were excised from 6 gels each of proteins from untreated and DFMO-treated CHO cells (Fig. 1), respectively. The protein in the gel pieces was digested in 50 mM ammonium bicarbonate buffer containing 20 µg/ml trypsin for 18 h at 37°C. Peptide maps were prepared on 0.2 mm layers of silica gel 60 on 20 x 20 cm plastic sheets (E. Merck) as described earlier(6). Fluorograms of peptide maps are shown (A and B). In the traced pattern (C and D), the peptides common to both digests (No. 1 through No. 12) are denoted as solid spots; the peptides present in only one of the digests (No. 13 through No. 16) are indicated by shaded spots. The tryptic cleavage sites are indicated by arrows.

lase, the initial regulatory enzyme for polyamine biosynthesis, formation of deoxyhypusine + hypusine did occur in the presence of cycloheximide(21). This finding is a strong indication that there is a pool of the protein substrate that accumulated during the depletion of spermidine.

Indeed an eIF-4D precursor that does not contain hypusine or deoxyhypusine, was seen on two-dimensional gels of proteins from DFMO-treated cells that were incubated with $[^3H]$lysine(24). The data presented in Fig. 1 show that the virtual depletion of cellular spermidine by treatment with DFMO for 42 h causes complete inhibition of radiolabeling of mature eIF-4D (hypusine form, M_r ~18,000, pI ~5.3, the position indicated by the solid arrow in Fig. 1C). In place of eIF-4D, a protein (M_r ~18,000, pI ~ 5.1), designated Protein P which is normally undetected in control cells, is seen. In cells partially depleted of spermidine (DFMO 18 h), labeling of both eIF-4D and Protein P are detected (Fig. 1B). Whereas $[^3H]$hypusine and $[^3H]$lysine were found in a ratio of approximately 1:14 in an acid hydrolysate of eIF-4D from control cells, no radioactive hypusine or deoxyhypusine was present in the digest of Protein P.

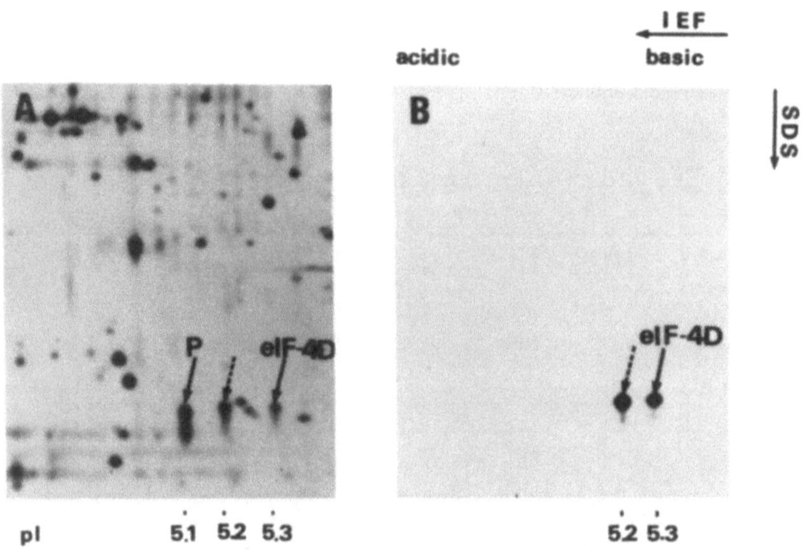

Fig. 3. Two-dimensional polyacrylamide gel separation of proteins from DFMO-treated cells after their replenishment with spermidine.

A: CHO cells were pretreated with DFMO for 42 h, labeled with $[^3H]$-lysine as in Fig. 1, washed free of $[^3H]$lysine and DFMO, and incubated in medium containing 2.5 µM unlabeled spermidine, 5 µCi/ml $[^3H]$spermidine and cycloheximide (100 µg/ml) for an additional 2 h. The positions of eIF-4D and Protein P are indicated by the solid arrows; the additional precursors at pI 5.2 by the broken arrow. Protein P contained 3,300 cpm of $[^3H]$lysine only. The labeled area at pI 5.2 (broken arrow) contained 1350 cpm, 900 cpm, and 190 cpm of $[^3H]$lysine, $[^3H]$deoxyhypusine and $[^3H]$-hypusine, respectively. The position of eIF-4D at pI 5.3 contained 580 cpm, 50 cpm and 750 cpm of $[^3H]$lysine, $[^3H]$deoxyhypusine and $[^3H]$hypusine, respectively.

B: CHO cells were pretreated with DFMO for 42 h, washed free of DFMO, and incubated in medium containing 50 µCi/ml $[^3H]$spermidine, and cycloheximide (100 µg/ml) for an additional 2 h. The labeled area at pI 5.2 (broken arrow) contained 6258cpm of $[^3H]$deoxyhypusine and 663 cpm of $[^3H]$hypusine. The area of eIF-4D at pI 5.3 contained 356 cpm of $[^3H]$-deoxyhypusine and 3580cpm of $[^3H]$hypusine.

The structural relationship of Protein P to eIF-4D was examined by two dimensional peptide mapping of the tryptic fragments of the two [^{3}H] lysine-labeled proteins(24). It is evident from comparison of the two maps (Fig. 2) that these proteins are closely related. Fourteen distinct [^{3}H]lysine-labeled peptides are detected in both digests. Twelve peptides (Fig. 2C and 2D, peptides No. 1 through No. 12, solid spots) occupy identical positions on both maps. Two peptides of eIF-4D (No. 13 and No. 14, Fig. 2C) are not seen in the digest of Protein P. Instead, two different peptides (No. 15 and No. 16, Fig. 2D) are detected. In a previous communication, we identified peptide No. 14 as the hypusine-containing tryptic fragment of eIF-4D(6) and determined its amino acid sequence as Thr-Gly-Hypusine-His-Gly-His-Ala-Lys(7). In an eIF-4D precursor in which lysine has not been converted to deoxyhypusine or hypusine, (Fig. 2, precursor in parenthesis), trypsin would cleave the bond between Lys and His to generate two new fragments Thr-Gly-Lys and His-Gly-His-Ala-Lys. Indeed, a synthetic pentapeptide His-Gly-His-Ala-Lys was found to comigrate with peptide No. 15. The tripeptide Thr-Gly-Lys, however, was not detected in the digest of Protein P. This finding leads to the suggestion that an additional post-translational process occurs. This modification, which may involve either the threonine residue or the lysine residue on the amino side of the threonine, results in prevention of cleavage of the peptide bond by trypsin. This process may be reversed after biosynthesis of deoxyhypusine or hypusine and thus allow tryptic cleavage to form threonine-amino-terminal peptide No. 14.

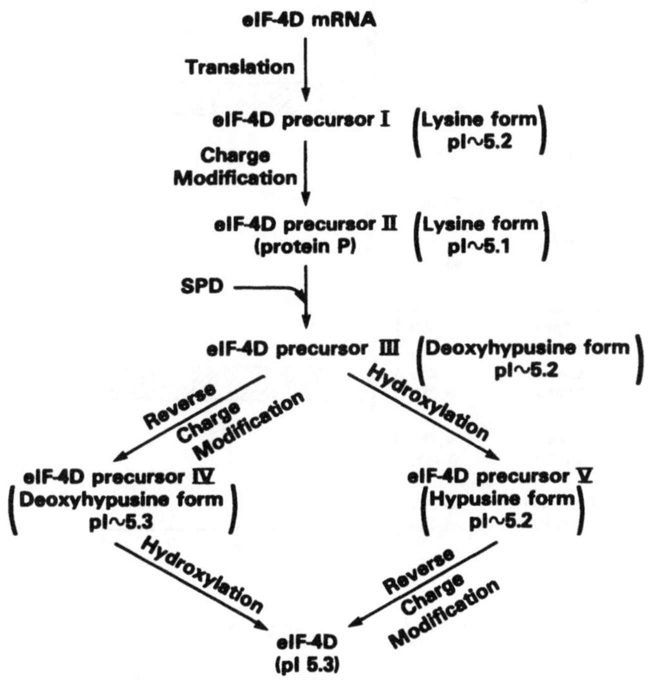

Scheme 3. Proposed pathway of eIF-4D biogenesis

The scheme is based on evidence for an additional modification in Protein P (Fig. 2), on the detection of two deoxyhypusine-containing intermediates (pI 5.2 and 5.3, of Fig. 3) corresponding to precursors III and IV, and of two hypusine-containing proteins (pI 5.2 and 5.3, Fig. 3) corresponding to precursor V and mature eIF-4D.

Further evidence for the additional modification is provided in an experiment shown in Fig. 3. When spermidine is rapidly replenished in cells pretreated with DFMO (42h), there is conversion of Protein P (pI ~5.1) to eIF-4D (pI ~5.3) and to previously undetected intermediate forms (pI ~5.2, broken arrows, Fig. 3 A and B). Both labeled hypusine and deoxyhypusine were found at the positions indicated by the broken arrows (pI ~5.2) and by the solid arrows (pI~5.3). It is clear from this finding that separation of the forms at pI 5.2 and pI 5.3 is not due to the hydroxylation of deoxyhypusine but rather is the result of an unknown charge modification. Although the nature of this reversible charge modification is as yet not known, it is concluded that biogenesis of eIF-4D involves several steps in addition to deoxyhypusine synthesis and its hydroxylation (scheme 3).

CELL-FREE SYNTHESIS OF DEOXYHYPUSINE

Earlier attempts to develop a cell-free system for deoxyhypusine synthesis were unsucessful primarily because of the absence of the protein

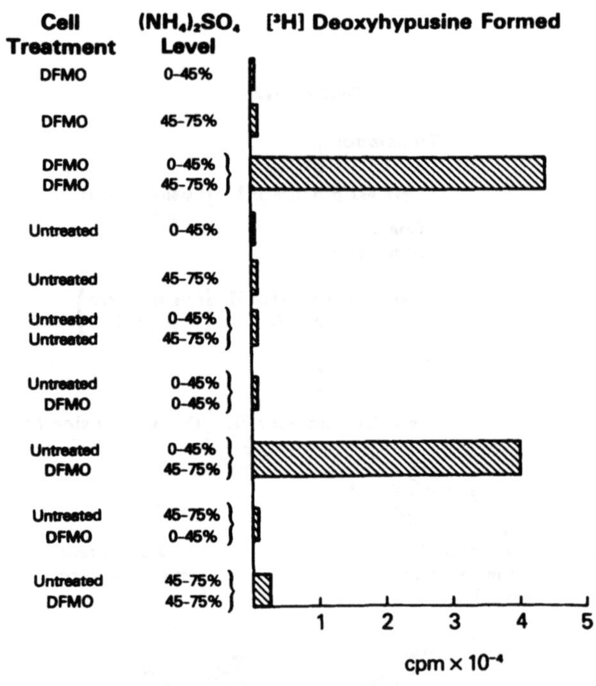

Fig. 4. Deoxyhypusine synthesis in single and combined ammonium sulfate fractions

The ammonium sulfate fractions were prepared from untreated and DFMO-treated CHO cells and deoxyhypusine synthesis was measured as described (27). The reaction mixtures contained single or different combination of 0-45% and 45-75% ammonium sulfate fractions from untreated or DFMO-treated cells, [^{3}H]spermidine (1.5 μM) and 0.5 mM DTT in 0.5 M glycine NaOH buffer, pH 9.5. After incubation at 37°C for 2 h, the amount of [^{3}H]deoxyhypusine in protein was measured as described (27). Reprinted from Ref. 27.

substrate from cells and tissues that contain the normal level of polyamines. The data in Fig. 1 and Fig. 2 suggest that an eIF-4D precursor (M_r~18,000, pI~5.1) which contains no hypusine or deoxyhypusine accumulates only upon depletion of spermidine. Murphey and Gerner recently accomplished cell-free synthesis of deoxyhypusine in lysates of DFMO-treated hepatoma tissue culture (HTC) cells and showed that the reaction is critically dependent upon pH(26). We have confirmed this finding using DFMO-treated CHO cells, and in addition have separated the enzyme(s) and the protein substrate(27). The data in Fig. 4 illustrate a clean simple separation and demonstrate that the failure of the lysate of control cells to synthesize deoxyhypusine is due to the absence of protein substrate. Only two combinations of ammonium sulfate fractions (Fig. 4) gave efficient deoxyhypusine synthesis: the 45-75% fraction from DFMO-treated cells with the 0-45% fraction from either untreated cells or DFMO-treated cells. In view of previous evidence for the presence of an eIF-4D precursor in DFMO-treated cells but not in untreated cells, it is clear that the 45-75% fraction from DFMO-treated cells contains the protein substrate and that the 0-45% fractions of both untreated and DFMO-treated cells contain the enzymatic activity for deoxyhypusine synthesis.

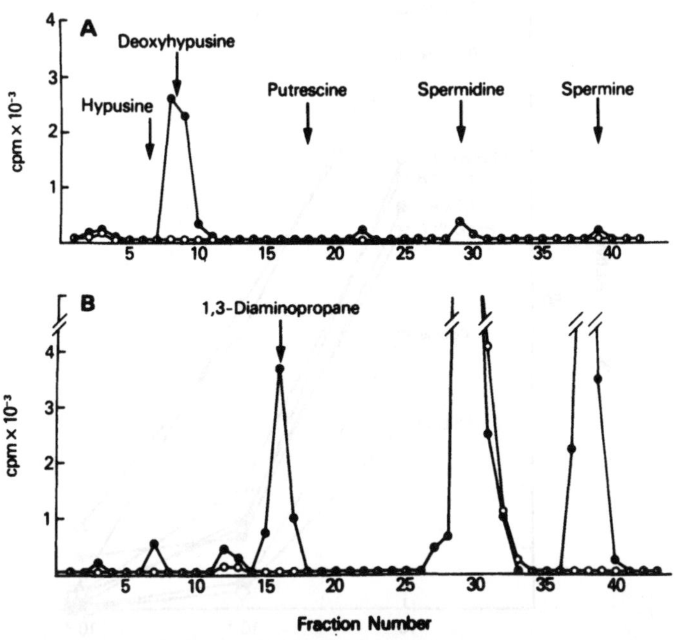

Fig. 5. Ion exchange chromatographic separation of radiolabeled components in protein (A) and TCA supernatant (B)

The reactions were carried out with combinations of 0-45% and 45-75% ammonium sulfate fractions from either untreated (o) or DFMO-treated (•) CHO cells, as described in Fig. 4. After incubation, the reactions were terminated by the addition of TCA. Ion exchange chromatographic separation of acid hydrolysate of TCA precipitates (A) and TCA supernatant (B) were carried out as described (27). The radioactivity in the protein hydrolysate was corrected for losses of [^{3}H]deoxyhypusine-containing protein during precipitation, washing and transfer to permit direct comparison with part B. Radioactivity in fractions 1-20 of TCA supernatant (B) was corrected for impurities present in commercial [^{3}H]spermidine. The radioactivities in spermidine and spermine were 156,800 cpm and 17,430 cpm, respectively, in the reaction mixture containing DFMO-treated cell fractions (•) and that in spermidine was 181,300 cpm in mixture from untreated cell fraction (o).

Only the butylamine portion of spermidine is utilized as a structural component for deoxyhypusine synthesis. The data presented in Fig. 5 provide strong evidence that transfer of the butylamine moiety from spermidine to a specific lysine residue to form deoxyhypusine results in the formation of free 1,3-diaminopropane (scheme 2). Fig. 5A shows $[^3H]$deoxyhypusine as the single major radioactive component in the acid hydrolysate of the protein fraction of the cell-free reaction mixture. The trichloroacetic acid (TCA) supernatant from the same reaction mixture contains, in addition to $[^3H]$spermidine and $[^3H]$spermine, a labeled material that chromatographs at the position of 1,3-diaminopropane (Fig. 5B -●-). The amount of radioactivity in this component was found to be approximately the same as that in the protein-bound deoxyhypusine. The identity of this component as 1,3-diaminopropane was confirmed by comparison of the properties of its dansyl and benzoyl derivatives with those of authentic 1,3-diaminopropane.

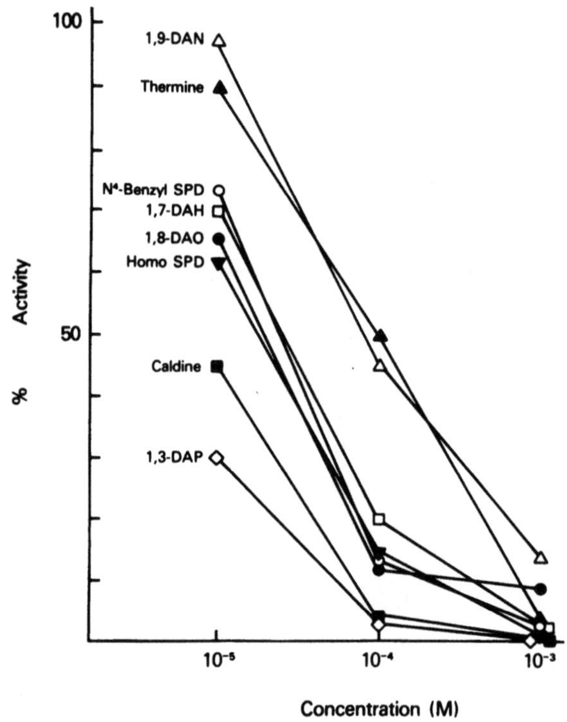

Fig. 6. Inhibition of deoxyhypusine synthesis.

The reactions were carried out with 0-45% and 45-75% ammonium sulfate fractions from DFMO-treated cells as outlined in Fig. 4 in the absence or presence of the indicated concentrations of the following amines:

1,9-DAN, 1,9-diaminononane, $NH_2(CH_2)_9NH_2$;
Thermine, $NH_2(CH_2)_3NH(CH_2)_3NH(CH_2)_3NH_2$
N^4-BenzylSPD, N^4-benzylspermidine; $NH_2(CH_2)_3N(Bz)(CH_2)_4NH_2$;
1,7-DAH, 1,7-diaminoheptane, $NH_2(CH_2)_7NH_2$;
1,8-DAO, 1,8-diaminooctane, $NH_2(CH_2)_8NH_2$;
HomoSPD, homospermidine, $NH_2(CH_2)_4NH(CH_2)_4NH_2$;
Caldine, $NH_2(CH_2)_3NH(CH_2)_3NH_2$;
1,3-DAP, 1,3-diaminopropane, $NH_2(CH_2)_3NH_2$;

Reprinted from Ref. 27

In Fig. 6 are shown the inhibitory effects of various amines on deoxyhupusine synthesis. Compounds closely related in structure to spermidine, e.g., caldine, homospermidine, N^4-benzylspermidine, a spermine analogue thermine, and diamines with the two primary amino groups spaced at a similar distance to that in spermidine, e.g., 1,7-diaminoheptane, 1,8-diaminooctane, and 1,9-diaminononane, strongly inhibited deoxyhypusine formation in a cell-free system. Diamines with 10 or more methylenes (1,10-diaminodecane, and 1,12-diaminododecane) and those with 6 or less methylenes (diaminoethane, 1,5-diaminopentane, and 1,6-diaminohexane) were found to be poor inhibitors (< 40% inhibition at 10^{-3}M). A notable exception was 1,3-diaminopropane which displayed a stronger inhibition than any of the other amines tested. Monoamines, e.g. n-propylamine, n-heptylamine, n-octylamine and n-nonylamine, were ineffective as inhibitors.

CONCLUDING REMARKS

We have shown with the use of specifically labeled precursors that the hypusine intermediate, deoxyhypusine, is produced in mammalian cells by transfer of the butylamine moiety from the secondary amine nitrogen of spermidine to the nitrogen at the ε position of a specific lysine residue in the eIF-4D precursor Protein P (scheme I). Protein P, which accumulates only in cells depleted of spermidine, was isolated by two-dimensional gel electrophoresis (Fig. 1). Comparison of the tryptic peptide map of [^{3}H]lysine-labeled eIF-4D and that of similarly labeled Protein P provided evidence that another posttranslational process is involved in the biogenesis of eIF-4D (Fig. 2). There is indication that this additional process which is separate from, but probably essential for, butylamine transfer is reversed after formation of deoxyhypusine or hypusine (scheme 3). From the location of previously undetected eIF-4D intermediates (pI 5.2, Fig. 3) on two-dimensional gels, it was evident that a modification in protein charge occurs in this process.

We have separated the protein substrate, and the enzyme(s) required for cell-free synthesis of deoxyhypusine (Fig. 4) and have demonstrated that 1,3-diaminopropane is the cleavage product formed in conjunction with butylamine transfer from spermidine to produce deoxyhypusine (Fig. 5). Preliminary inhibition studies showed effective inhibition of deoxyhypusine formation by 1,3-diaminopropane, by several spermidine analogs, and by diamines structurally related to spermidine (Fig. 6).

These recent findings reveal important additional features of the pathway of eIF-4D biogenesis. Clearly this key pathway upon which polyamines may enter into regulation of mammalian cellular growth involves at least one reversible step in addition to butylamine transfer and subsequent hydroxylation. That this protein charge modification step, which appears to be essential for butylamine transfer, may be the prime regulatory point for eIF-4D production is worthy of serious consideration. Present knowledge of the structural requirements for reversible inhibition of butylamine transfer provide the basis for design of mechanism-based enzyme inactivators. Although these inactivators should be useful for investigating the biological role of hypusine, special attention must also be now directed toward the additional charge modification step in eIF-4D production.

REFERENCES

1. T. Shiba, H. Mizote, T. Kaneko, T. Nakajima, Y. Kakimoto, and I. Sano, Hypusine, a new amino acid occurring in bovine brain: Isolation and structural determination, Biochim. Biophys. Acta 244:523 (1971).
2. T. Nakajima, T. Matsubayashi, Y. Kakimoto, and I. Sano, Distribution of hypusine, N^6-(4-amino-2-hydroxybutyl)-2-6-diaminohexanoic acid, in mammalian organs, Biochim. Biophys. Acta 252:92 (1971).
3. N. Imaoka, and T. Nakajima, Hypusine, N^6-(4-amino-2-hydroxybutyl)-2-6-diaminohexanoic acid, in tissue proteins of mammals, Biochim. Biophys. Acta,320:97 (1973).
4. M.H. Park, H.L. Cooper, and J.E. Folk, Identification of hypusine, an unusual amino acid, in a protein from human lymphocytes and of spermidine as its biosynthetic precursor, Proc. Natl. Acad. Sci. USA, 78:2869 (1981).
5. H.L. Cooper, M.H. Park, J.E. Folk, B. Safer, and R. Braverman, Identification of the hypusine-containing protein Hy^+ as translation initiation factor eIF-4D, Proc. Natl. Acad. Sci. USA, 80:1854 (1983).
6. M.H. Park, S.I. Chung, H.L. Cooper, and J.E. Folk, The mammalian hypusine containing protein, eukaryotic initiation factor 4D: Structural homology of this protein from several species, J. Biol. Chem. 259:4562 (1984).
7. M.H. Park, T.Y. Liu, S.H. Neece, and W.J. Swiggard, Eukaryotic initiation factor 4D: Purification from human red blood cells and the sequence of the amino acids around its single hypusine residue, J. Biol. Chem. 261:14515 (1986).
8. E.D. Gordon, R. Mora, S.C. Meredith, C. Lee and S.L. Lindquist, Eukaryotic initiation factor 4D, the hypusine containing protein is conserved among eukaryotes. J. Biol. Chem. 262:16585 (1987).
9. M.H. Park Unpublished results.
10. K.Y. Chen and Y.C. Yang, Spermidine labels a 21,000-dalton protein in Neurospora Crassa both in vivo and in vitro, The FASEB J. 2:A1563 (1988).
11. H.L. Cooper, M.H. Park, and J.E. Folk, Posttranslational formation of hypusine in a single major protein occurs generally in growing cells and is associated with activation of lymphocyte growth, Cell 29:791 (1982).
12. K.Y. Chen, An 18,000 dalton protein metabolically labeled by polyamines in various mammalian cell lines, Biochim. Biophys. Acta 856: 395 (1983).
13. B.M. Torrelio, M.A. Paz, and P.M. Gallop, Cellular proliferation and hypusine synthesis, Exp. Cell Res. 154:454 (1983).
14. E.W. Gerner, P.S. Mamont, A. Bernhardt and M. Siat, Post-translational modification of the protein-synthesis initiation factor eIF-4D by spermidine in rat hepatoma cells, Biochem. J. 239:379 (1986).
15. A.A.M. Thomas, R. Benne. and H.O. Voorma, Initiation of eukaryotic protein synthesis, FEBS Letters 128:177 (1981).
16. M.H. Park, H.L. Cooper, and J.E. Folk, The biosynthesis of protein-bound hypusine [N^ε-(4-amino-2-hydroxybutyl)lysine]: Lysine as the amino acid precursor and the intermediate role of deoxyhypusine [N^ε -(4-aminobutyl)lysine], J. Biol. Chem. 257:7217 (1982).
17. M.H. Park, D.J. Liberato, A.L. Yergey, and J.E. Folk, The biosynthesis of hypusine [N^ε -4-amino-2-hydroxybutyl)lysine]: Alignment of the butylamine segment and source of the secondary amino nitrogen, J. Biol. Chem. 259:12123 (1984).
18. M.H. Park, and J.E. Folk, Biosynthetic labeling of hypusine in mammalian cells: carbon-hydrogen bond fissions revealed by dual labeling. J. Biol. Chem. 261:14108 (1986).
19. A. Abbruzzese, M.H. Park, and J.E. Folk, Deoxyhypusine hydroxylase from rat testis: Partial purification and characterization, J. Biol. Chem. 261:3085 (1986).

20. K.I. Kivirikko, and R. Myllyla, Post-translation enzymes in the biosynthesis of collagen: Intracellular enzymes, Methods Enzymol. 82:245 (1982).
21. M.H. Park, Regulation of biosynthesis of hypusine in Chinese hamster ovary cells: Evidence for eIF-4D precursor polypeptides. J. Biol. Chem. 262:12730 (1987).
22. R.F. Duncan and J.W.B. Hershey, Changes in eIF-4D hypusine modification or abundance are not correlated with translational repression in HeLa cells, J. Biol. Chem. 261:12903 (1986).
23. B.W. Metcalf, P. Bey, C. Danzin, M.J. Jung, P. Casara, and J.P. Vevert, Catalytic irreversible inhibition of mammalian ornithine decarboxylase (E.C. 4.1.1.17) by substrate and product analogues, J. Am. Chem. Soc. 100:2552 (1978).
24. M.H. Park, Identification of an eukaryotic initiation factor 4D precursor in spermidine-depleted Chinese hamster ovary cells. J. Biol. Chem. 263:7447 (1988).
25. J.I. Garrels, Two-dimensional gel electrophoresis and computer analysis of proteins synthesized by clonal cell lines, J. Biol. Chem. 254:7961 (1979).
26. R.J. Murphey and E.W. Gerner, Hypusine formation in protein by a two-step process in cell lysates, J. Biol. Chem. 262:15033 (1987).
27. M.H. Park and E.C. Wolff, Cell-free synthesis of deoxyhypusine: Separation of protein substrate and enzyme and Identification of 1,3-diaminopropane as a product of spermidine cleavage. J. Biol. Chem. (in press).

HYPUSINE BIOSYNTHESIS IN PROTEIN AND ITS BIOLOGICAL CONSEQUENCES

Roberta J. Murphey, Margaret E. Tome and Eugene W. Gerner

Departments of Radiation Oncology and Biochemistry
University of Arizona Health Sciences Center
Arizona Cancer Center
Tucson, Arizona 85724 USA

INTRODUCTION

In 1971, the amino acid known as hypusine was discovered in free form in bovine brain (Shiba et al., 1971). The structure of hypusine was established and confirmed as N^6-(4-amino-2-hydroxybutyl)-2,6-diamino hexanoic acid (Shiba et al., 1982; Tice and Ganem, 1983). Since its discovery, hypusine has been found as a free amino acid (Nakajima et al., 1971) and bound to protein (Sano et al., 1984). Protein-bound hypusine primarily occurred in a protein, $M_r \simeq 18{,}000$, which was originally observed in lectin-stimulated lymphocytes and Chinese hamster ovary (CHO) cells (Park et al., 1981; Cooper et al, 1982). The synthesis of protein-bound hypusine coincides with an increase in protein synthesis (Cooper et al., 1982) and this biochemical process was observed in several mammalian cell lines (Chen, 1983; Torrelio et al., 1984). Because of the ubiquity of the hypusine modification in an $M_r \simeq 18{,}000$ protein and its conservation among eukaryotes (Park et al., 1984a; Gordon et al., 1987), the biophysical characteristics of the protein were compared to those of known eukaryotic protein translation initiation factors. The $M_r \simeq 18{,}000$ protein modified by hypusine was identified as the putative eukaryotic protein synthesis initiation factor eIF-4D (Cooper et al., 1983).

HYPUSINE BIOSYNTHESIS IN PROTEIN

Biosynthesis of hypusine in protein involves a multi-step reaction. First, a butylamine moiety is transferred from the polyamine spermidine to the ϵ-amino group of lysine in protein to form deoxyhypusine (Imoaka et al., 1973; Park et al., 1982). The second step is hydroxylation of the second carbon on this added group (Park, 1982) to form hypusine. To elucidate the enzyme(s) and mechanisms involved in deoxyhypusine synthesis, a cell-free system was designed (Murphey and Gerner, 1987). Hypusine biosynthesis in the lysate involved two-pH sensitive reactions. Deoxyhypusine was formed from a protein-bound lysine at pH $\simeq$ 9.5 and hypusine was formed from protein-bound deoxyhypusine at pH $\simeq$ 7.1 (Fig. 1A and B). In addition, the cell-free assay required endogenous spermidine depletion to support the modification. This condition led to accumulation of the unmodified form of eIF-4D as suggested by earlier studies (Gerner et al., 1986) and subsequently shown directly by Park (1988). The protein modified in this cell-free system was indistinguishable from eIF-4D modified by spermidine in intact cells.

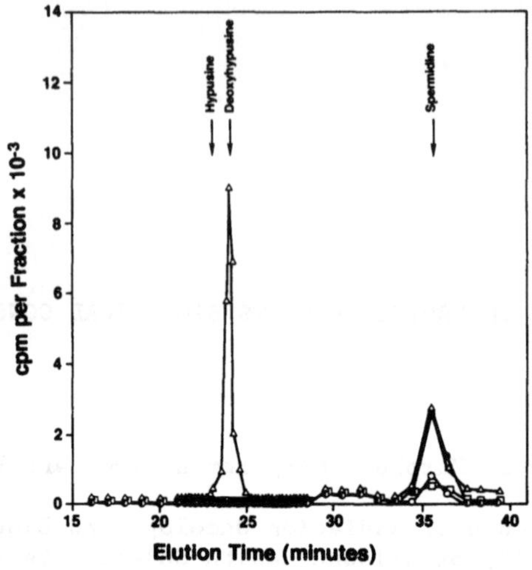

Figure 1A. Effects of Reaction pH on Synthesis of Peptide-bound Deoxyhypusine in Cell Lysates. Lysates from cells treated for 48 hr with α-difluoromethylornithine (5 mM) were initially adjusted to different pH conditions: 7.1 (●), 8.0 (O), 9.5 (Δ) to 11.0 ([]). All samples contained the same amount of protein and were incubated for 6 hr in the presence of [1,8-^{3}H]spermidine. Shown is the elution of radiolabel after separation of the hydrolyzed protein sample by HPLC. The early fractions (elution times less than 15 min) contained no radiolabel. (Reproduced with permission, Murphey and Gerner, 1987)

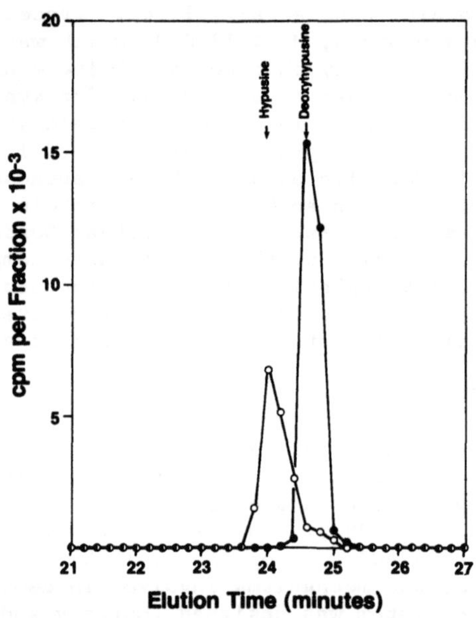

Figure 1B. Effect of pH Shift on the Synthesis of Peptide-bound Hypusine and Deoxyhypusine. Duplicate lysates from α-difluoromethylornithine-treated cells were initially adjusted to pH 9.5 and incubated in the presence of [1,8-^{3}H]spermidine for 6 hr. At that time, one sample was shifted to pH 7.1 (O) while the other remained at pH 9.5 (●) for another 18 hr. This figure shows the elution of radiolabel in hydrolyzed protein after separation on HPLC. All other fractions did not have detectable radiolabel. (Reproduced with permission, Murphey and Gerner, 1987)

Polyamine Specificity

Early investigations into the substrate responsible for the butylamine moiety contribution indicated the polyamine spermidine. Labelled polyamines spermidine and the spermidine precursor, putrescine, were used, in addition to a potent inhibitor of s-adenosylmethione decarboxylase, MGBG (Corti et al., 1974) to study substrate specificity for hypusine biosynthesis. The inhibitor, in the presence of labelled putrescine, blocked hypusine biosynthesis in lymphocytes. On the other hand, if the inhibitor was present, along with labelled spermidine, hypusine biosynthesis was not affected. By blocking conversion of putrescine to spermidine, hypusine was not synthesized if exogenous spermidine was not supplied to the cells. Several amines have been used to compete with this polyamine as a substrate for hypusine. N^1-acetylspermidine, added to hepatoma tissue culture (HTC) cells, was unable to compete with the labelled spermidine in this regard. Inhibition of polyamine oxidase when spermidine was added to polyamine - depleted cells did not affect hypusine synthesis, ruling out putrescine as the natural polyamine substrate. Spermine, the final polyamine in this biosynthetic pathway, was also a poor substrate for hypusine synthesis (Gerner et al., 1986).

Wolff and Park (1988) used the cell-free system to study other spermidine analogs of varying carbon lengths in an attempt to inhibit deoxyhypusine synthesis. Diaminoheptane, diaminooctane and caldine competitively inhibited the synthesis of deoxyhypusine but the diamines; cadavarine, diaminohexane, diaminodecane and diaminododecane were ineffective. One diamine, diaminopropane, markedly inhibited the synthesis of deoxyhypusine. It was possible that this inhibition resulted from a negative feedback mechanism, because diaminopropane is a proposed product of hypusine synthesis. The data, in sum above indicate that the butylamine portion of hypusine is derived from spermidine.

Protein Substrate

Protein-bound hypusine was identified in several eukaryotic cell lines. Studies in human lymphocytes and CHO cells revealed the hypusine-containing protein had a $M_r \simeq 17,000-19,000$ and $pI \simeq 5.1-5.5$ (Cooper et al., 1982). Because the protein was ubiquitous, and correlated with protein synthesis and cell growth (Cooper et al.,1982; Chen,1983; Torrelio et al., 1984), it was compared to protein translation initiation factors. The hypusine-containing protein was identical to eIF-4D (Cooper et al., 1983).

To study the function of eIF-4D, researchers have attempted to isolate unmodified eIF-4D in several cell lines. Park et al. (1984a) studied the ratio of the modified to unmodified protein in CHO cells. Edman degradation of radiolabelled hypusine in eIF-4D showed that the label was released as a single peak. For every mole of eIF-4D, there was a stoichiometric modification of the third amino acid (lysine) in protein (Park et al., 1984a; Park et al., 1986). All of the eIF-4D appeared to be modified in log phase CHO cells. When polyamine depleted HTC cells were exposed to exogenous polyamines, the initial rate of hypusine synthesis was 5-10 times greater than those cells that had normal levels of polyamines (Gerner et al., 1986). In this same study, the rate of hypusine synthesis was dependent on the final endogenous content of spermidine (Fig. 2) Therefore, the protein substrate seemed to accumulate due to the lack of polyamine availability. In addition, only lysates from spermidine-depleted cells supported the synthesis of deoxyhypusine in the cell lysate assay (Fig. 3). The substrate which accumulated was probably unmodified eIF-4D. The recent studies by

Park (1987, 1988) investigated eIF-4D and hypusine levels in CHO cells, and substantiate this hypothesis. She found that depletion of spermidine reduced deoxyhypusine and hypusine synthesis. Cycloheximide reduced the synthesis rate in only those cells that were not polyamine depleted or retained 10% of their spermidine pool. Cells depleted of spermidine using a longer treatment time with α-difluoromethylornithine (DFMO) showed the accumulation of a precursor protein which could be modified by

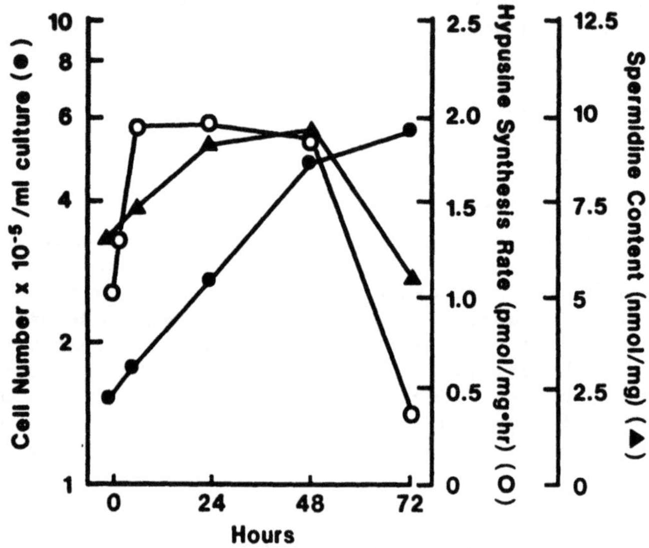

Figure 2. Relationship Between Hypusine Formation, Growth Rate and Spermidine Concentration. High-density cell cultures ($8x10^5$ cells/ml) were diluted into fresh medium at a cell density of 1.5×10^5 cells/ml. At the indicated times, samples were pulse-labelled for 1 hr with 2 μCi of radioactive spermidine (44.3 Ci/mmol)/ml. At the end of the pulse-labelling, cell number (●), spermidine content (▲) and protein-bound hypusine formation (O) were determined as described in the Experimental section. (Reproduced with permission, Gerner et al., 1986)

the addition of spermidine. The modification in these samples was not affected by cycloheximide since the DFMO treatment had built up a pool of unmodified protein prior to cycloheximide treatment. Thus, hypusine biosynthesis is not dependent on new protein synthesis, as long as the precursor protein is available. All of this evidence indicates an unmodified protein accumulates in spermidine depleted cells.

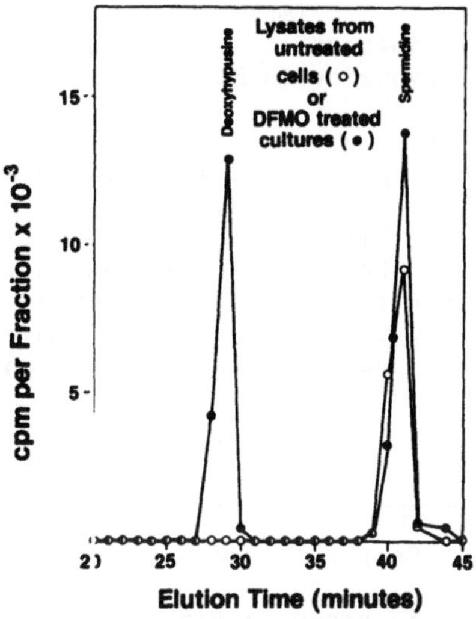

Figure 3. Effect of α-difluoromethylornithine Treatment on Cells to be Used as a Source of Protein in the Cell Lysate System. HTC cells were cultured in the presence (●) or absence (O) of 5 mM α-difluoromethylornithine for 48 hr. Lysates were prepared with glycine-NaOH buffer at pH 9.5. [1,8-^{3}H]spermidine (10 μCi/ml) was added to each sample and incubated for 24 hr. Sample proteins were analyzed by HPLC. The cpm/fraction are normalized to show values for equivalent protein contents (1 mg) of α-difluoromethylornithine treated and untreated cell lysates. (Reproduced with permission, Murphey and Gerner, 1987)

ENZYMES OF HYPUSINE BIOSYNTHESIS

There appeared to be two reactions involved in the first step of hypusine biosynthesis. The enzyme(s) responsible for the synthesis of deoxyhypusine catalyzed the cleavage and transfer of the butylamine moiety of spermidine to the ϵ-amino group of lysine in eIF-4D. The enzyme(s), although unknown, required a pH $\simeq$ 9.5 (Murphey and Gerner, 1987) later found to be closer to pH $\simeq$ 9.1 (unpublished), is(are) greater than $M_r \simeq 25,000$ as evidenced by Amicon $M_r \simeq 25,000$ molecular sieve cones (unpublished data), and may require NAD^+ (Chen, 1987). Several inhibitors were used in an attempt to block deoxyhypusine synthesis in whole cells (Gerner et al, 1986), and in the cell lysate system (Table 1). The most effective inhibitor in blocking deoxyhypusine was the plant amine oxidase inhibitor, guazatine. Guazatine blocked polyamine oxidase of oat seedlings (Smith, 1983) and inhibited spermidine conversion to Δ^1-pyrroline, 1,3-diaminopropane and H_2O_2. Spermidine dehydrogenase of Serratia marcescens (Bachrach, 1962; Campello et al., 1965) also catalyzed the spermidine conversion to Δ^1-pyrroline and other products. The Δ^1-pyrroline product could be an intermediate to deoxyhypusine synthesis. Since guazatine effectively inhibits deoxyhypusine synthesis, this result suggested that Δ^1-pyrroline might be a transient intermediate between spermidine and deoxyhypusine. Recent evidence (manuscript in preparation) indicates that deoxyhypusine synthesis(s) is an oxygen-dependent process, implying that a spermidine oxidase, rather than a dehydrogenase, is involved in this step.

TABLE 1
Effect of Various Inhibitors of Amine Oxidases on Formation of Deoxyhypusine in HTC Cell Lysates

HTC cells were grown in the presence of α-difluoromethylornithine (5 mM) for 48 hr. Lysates were prepared as described and partitioned, such that each reaction tube contained approximately 3 mg of protein (corresponding to about 10^7 cells). [1,8-^{3}H]Spermidine (10 μCi/ml) and the designated amine oxidase inhibitor at the specified concentration were added to the various tubes and incubated for 24 hr. The samples were analyzed by HPLC as described. This table presents the percent inhibition of deoxyhypusine formation in protein relative to a control reaction containing no inhibitor.

Inhibitor	Class	Concentration μM	% Inhibition
Diethyldithiocarbamate	Copper chelator	100	20
		1000	0
Iodoacetamide	SH group reagent	10	16
		100	6
MDL-72.521	Polyamine oxidase inhibitor	100	0
Guazatine	Plant amine oxidase inhibitor	10	15
		100	67
		1000	100

(Reproduced from Murphey and Gerner, 1987, with permission)

Recent data reported by Chen et al. (1987) indicated the cell lysate assay required NAD$^+$ in mouse neuroblastoma cells and that NAD$^+$ eliminated the pH dependence of this reaction. Preliminary studies from our lab (not shown) indicate that deoxyhypusine formation in HTC lysates is increased by NAD$^+$ at concentrations as low as 150 μM. We do not yet have information on the effect of this NAD$^+$ concentration on the pH dependence of deoxyhypusine formation. Since the enzyme(s) involved in deoxyhypusine require(s) oxygen and apparently NAD$^+$, is (are) blocked by guazatine and involve(s) a transfer reaction, we refer to it as a spermidine oxidase-deoxyhypusine synthetase complex.

The final step in hypusine biosynthesis involves the hydroxylation of deoxyhypusine. The enzyme that catalyzes this step, deoxyhypusine hydroxylase, has been partially purified (Abbruzzese et al., 1986). The hydroxylation was inhibited by metal chelators α,α-dipyridyl and Desferal (Park, 1982) and hydralazine (Paz et al, 1984). The inhibition by α,α-dipyridyl was reversed by cobalt but not by iron (Abbruzzese et al., 1986). The hydroxylase was found to be different from hydroxylating enzymes of propyl and lysyl procollagen processing since it failed to stoichiometrically decarboxylate α-ketogluturate. It appears deoxyhypusine hydroxylase uses a distinct catalytic mechanism which is specific for the synthesis of hypusine. Other studies using α,α-dipyridyl or hydralazine to inhibit deoxyhypusine hydroxylase in HTC cells indicate the hydroxylation does not affect the stability of the protein. The turnover of deoxyhypusine-containing eIF-4D was identical to control samples of hypusine-containing eIF-4D (Murphey et al., 1986).

BIOLOGICAL CONSEQUENCES

Since the identification of the hypusine-containing protein as eIF-4D (Cooper et al., 1983), no specific function for the protein has been reported. The only known function for eIF-4D is the stimulation of the simulated first peptide bond formation in eukaryotic protein translation measured by by methionyl puromycin formation. The activity of this reaction is only enhanced 2 to 3 fold and eIF-4D does not stimulate general protein synthesis (Thomas et al., 1981). The putative initiation factor is 10 times more abundant than other initiation factors (Thomas et al., 1979) and has a long half-life, greater than 24 hours (Gerner et al., 1986). Hypusine-containing eIF-4D is approximately 0.1% of HTC cell protein (unpublished data). Hypusine biosynthesis in eIF-4D correlates with protein synthesis and cell growth (Cooper et al., 1982; Chen, 1983; Torrelio et al., 1984; Gerner et al., 1986) but this would be expected of prevalent proteins. In addition to the protein prevalence, it is highly conserved throughout eukaryotic species (Park et al. 1984; Gordon et al., 1987). In prokaryotes, the spermidine label is incorporated into a protein of different molecular weight. When cell fractionation studies were done, hypusine-containing eIF-4D was not associated with lipids or DNA (unpublished data). Therefore, hypusine-containing eIF-4D is a prevalent, highly conserved protein among eukaryotic species. The modification occurs rapidly, taking place on ribosomes or soon after translation by a two-step process. The modified protein, or one containing deoxyhypusine has a long half-life. The function of either the unmodified or modified protein has not yet been conclusively elucidated.

Because little is known about the function of hypusine in eIF-4D, or of eIF-4D itself, we chose to study a mutant cell line with an altered spermidine-dependent process, since spermidine is the polyamine precursor in hypusine synthesis. The mutant cell line, HMO_A (HTC α-methylornithine series A cell) was generated by selection of HTC cells for growth in α-methylornithine (α-MO) (Mamont et al., 1978). The HMO_A cell line is resistant to the anti-proliferative effects of the ODC inhibitor and the half-life of ODC is increased from 20-30 minutes to several hours (Pritchard et al., 1982). ODC degradation is a spermidine-dependent process (Glass and Gerner, 1987). Thus, we decided to compare the parental cell line, HTC, with the variant HMO_A, for hypusine levels in protein. Results to be published elsewhere showed that radiolabelled hypusine from cell-free protein hydrolysates was reduced in HMO_A cells by about 75% compared to HTC cells. Polyacrylamide gel electrophoresis of these same cell populations supported this data, since the radioactive hypusine in eIF-4D of HTC cells was approximately 4 times that of the HMO_A cells. Since there was less hypusine and hypusine-containing eIF-4D in HMO_A cells versus HTC cells, mixing experiments using partially purified preparations of enzymes and substrates from both cell lines that were combined to analyze hypusine synthesis. Preliminary results suggest that reduced levels of hypusine in HMO_A cells are due to lower deoxyhypusine synthesizing enzyme activity rather than an altered protein substrate. Therefore, it appears the HMO_A cell line is altered in two spermidine dependent processes, ODC degradation and hypusine biosynthesis. The latter appears to be due to a change in the enzyme(s) responsible for deoxyhypusine synthesis.

CONCLUSIONS

Hypusine biosynthesis proceeds by a multi-step mechanism (Park et al., 1984b). The first reaction catalyzes the synthesis of deoxyhypusine-containing eIF-4D and the second reaction involves the hydroxylation of deoxyhypusine to synthesize hypusine in this protein. These two reactions apparently have unique pH optima, pH $\simeq$ 9.5 for deoxyhypusine formation from a protein bound lysine residue, and pH $\simeq$ 7.1 for hypusine formation from protein-bound deoxyhypusine. Deoxyhypusine formation is an oxygen- and NAD^+-dependent reaction, which can be inhibited by guazatine. These findings suggest that deoxyhypusine formation may actually involve two distinct activities. One of these appears to be an oxidase acting on spermidine, possible producing Δ^1-pyrroline as a transient intermediate, and a second may be an NAD^+ requiring activity responsible for the attachment of the butylamine moiety of spermidine to the ϵ-amino group of the protein-bound lysine. While the function of elF-4D (either modified or not) is unknown, the recent observation that hypusine formation is decreased in the rat hepatoma cell line HMO_A, compared to the parental HTC cell line may be helpful in determining the physiological role of this unique, polyamine-mediated protein modification.

ACKNOWLEDGEMENTS

We thank Dr. Stephen Carper and Dr. David Fuller for comments concerning preparations of this review. The excellent secretarial assistance of Sally Anderson and Robin Dyer is greatly appreciated. This work was supported by grants from the USPHS National Cancer Institute, CA-17232 and CA-30052.

REFERENCES

Abbruzzese, A., Park, M.H. and Folk, J.E., 1986, Deoxyhypusine hydroxylase from rat testis. Partial purification and characterization. J. Biol. Chem. 261, 3085-3089.

Abbruzzese, A., Park, M.H. and Folk, J.E., 1985, Deoxhypusine Hydroxylase. Distribution and partial purification from rat testis. J. Cell Biology 44, 1487.

Bachrach, U., 1962, Spermidine oxidase from Serratia marcescens. J. Biol. Chem. 237, 3443-3447.

Campello, A.P., Tabor, C.W. and Tabor, H., 1965, Resolution of spermidine dehydrogenases from Serratia marcesens: requirements for flavin adenine dinucleotide and an additional electron carrier. Biochem. Biophys. Res. Commun. 19, 6-9.

Chen, K.Y., 1983, An 18,000 dalton protein metabolically labeled by polyamines in various mammalian cell lines. Biochem. Biophys. Acta 756, 395-402.

Chen, K.Y. and Dou, Q.P., 1988, NAD^+ stimulated the spermidine-dependent hypusine formation on the 18 kDa protein in cytosolic lysates derived from NB-15 mouse neuroblastoma cells. FEBS Lett. 229, 325-328.

Cooper, H.L., Park, M.H. and Folk, J.E., 1982, Post-translational formation of hypusine in a single major protein occurs generally in growing cells and is associated with activation of lymphocyte growth. Cell 29, 791-797.

Cooper, H.L., Park, M.H., Folk, J.E., Safer, B. and Braverman, R., 1983, Identification of the hypusine-containing Hy^+ as translation initiation factor eIF-4D. Proc. Natl. Acad. Sci. USA 80, 1854-1857.

Corti, A., Dave, C., Williams-Ashman, H.G., Mihich, E. and Schenone, A., 1974, Specific inhibition of the enzymatic decarboxylation of S-adenosylmethionine by methylglyoxal bis (guanylhydrazone) and related substances. Biochem. J. 139, 351-357.

Gerner, E.W., Mamont, P.S., Bernhardt, A. and Siat, M., 1986, Post-translational modification of the protein-synthesis initiation factor eIF-4D by spermidine in rat hepatoma cells. Biochem. J. 239, 379-386.

Glass, J.R. and Gerner, E.W., 1987, Spermidine mediates degradation of ornithine decarboxylase by a non-lysosomal, ubiquitin-independent mechanism. J. Cell Phys. 130, 133-141.

Gordon, E.D., Mora, R., Meredith, S.C., Lee, C. and Lindquist, S.L., 1987, Eukaryotic initiation factor 4D, the hypusine-containing protein, is conserved among eukaryotes. J. Biol. Chem. 262, 1-5.

Imaoka, N. and Nakajima, T., 1973, Hypusine, N^6-(4-amino-2-hydroxybutyl) -2,6-diaminohexanoic acid, in tissue proteins of mammals. Biochim. Biophys Acta 320, 97-103.

Mamont, P.S., Duchesne, M.-C., Grove, J. and Bey, P., 1978, Anti-proliferative properties of DL-α-difluoromethylornithine in cultured cells. A consequence of the irreversible inhibition of ornithine decarboxylase. Biochem. Biophys. Res. Commun. 81, 58-66.

Mamont, P.S., Duchesne, M.-C., Grove, J. and Tardi, F.C., 1978, Initial characterization of a HTC cell variant partially resistant to the anti-proliferative effects of ornithine decarboxylase inhibitors. Exp. Cell Res. 115, 387-393.

Murphey, R.J. and Gerner, E.W., 1987, Hypusine formation in protein by a two-step process in cell lysates. J. Biol. Chem. 262, 15033-15036.

Murphey, R.J., Tome, M.E. and Gerner, E.W. (1986) Post-translational modification of eIF-4D by Spermidine. J. Cell Biology 103, 327a.

Nakajima, T., Matsubayashi, T., Kakimoto, Y. and Sano, I., 1971, Distribution of hypusine, N^6(4-amino-2-hydroxybutyl) 2,6-diaminohexanoic acid, in mammalian organs. Biochem. Biophys. Acta 252, 92-97.

Park, M.H., 1987, Regulation of biosynthesis of hypusine in Chinese hampster ovary cells. Evidence for eIF-4D precursor polypeptides. J. Biol. Chem. 262, 12730-12734.

Park, M.H., Chung, S.I., Cooper, H.L. and Folk, J.E., 1984a, The mammalian hypusine-containing protein, eukaryotic initiation factor 4D. Structural homology of the protein from several species. J. Biol. Chem. 259, 4563-4565.

Park, M.H., Liberato, D.J., Yergay, A.L. and Folk, J.E., 1984b, The biosynthesis of hypusine (N^ϵ-(4-amino-2-hydroxybutye)lysine). Alignment of the butylamine segment and source of the secondary amino nitrogen. J. Biol. Chem. 259, 12123-12127.

Park, M.H., Cooper, H.L. and Folk, J.E., 1981, Identification of hypusine, an unusual amino acid, in a protein from human lymphocytes and of spermidine as its biosynthetic precursor. Proc. Natl. Acad. Sci. USA 78, 2869-2873.

Park, M.H., Cooper, H.L. and Folk, J.E., 1982, The biosynthesis of protein-bound hypusine (N^ϵ-(4-amino-2-hydroxybutyl)lysine). Lysine as the amino acid precursor and the intermediate role of deoxyhypusine (N^ϵ-(4-aminobutyl)lysine). J. Biol. Chem. 257,7217-7222.

Park, M.H., Liu, T.-Y., Neece, S.H., Swiggard, W.J., 1986, Eukaryotic initiation factor 4D purification from human red blood cells and the sequence of amino acids around its single hypusine residue. J. Biol. Chem. 261, 14515-14519.

Paz, M.A., Torrelio, B.M. and Gallop, P.M., 1984, Hydralazine inhibition of the post-translational hydroxylation of deoxyhypusine, a polyamine-derived amino acid. Biochem. Pharm. 33, 779-785.

Pritchard, M.L., Pegg, A.E., and Jefferson, L.S., 1982, Ornithine Decarboxylase from hepatoma cells and a variant cell line in which the enzyme is more stable. J. Biol. Chem. 257, 5892-5899.

Sano, A., Miyake, M. and Kakimoto, Y., 1984, A rapid and sensitive method for the determination of hypusine in proteins and its distribution and developmental changes. Biochim. Biophys Acta 800, 135-139.

Shiba, T., Akiyama, H., Umeda, I., Okada, S. and Wakamiya, T., 1982, Synthesis and stereochemistry of hypusine, a new amino acid in bovine brain. Bull. Chem. Soc. Jpn. 55, 899-903.

Shiba, T., Mizole, H., Kaneko, T., Nakajima, T., Kakimoto, Y. and Sano, I., 1971, Hypusine, a new amino acid occurring in bovine brain. Isolation and structural determination. Biochem. Biophys. Acta 244, 523-531.

Smith, T.A., 1983, Polyamine oxidase (oat seedlings). Meth. in Enz. 94, 311-314.

Thomas, A.A.M., Benne, R. and Voorma, H.O., 1981, Initiation of eukaryotic protein sythesis. FEBS Lett. 128, 177-185.
Thomas, A., Goumans, H., Amesz, H., Benne, R. and Voorma, H.O., 1979, A comparison of initiation factors of eukaryotic protein synthesis from ribosomes and from postribosomal supernatant. Eur. J. Biochem. 98, 329-337.
Tice, C.M. and Ganem, B., 1983, Chemistry of naturally occurring polyamines. &. Selective functionalization of hydroxyputrescine. J. Org. Chem. 48, 5043-5048.
Torrelio, B.M., Paz, M.A. and Gallop, P.M., 1984, Cellular proliferation and hypusine synthesis. Exp. Cell Res. 154, 454-463.
Wolff, E.C. and Park, M.H., 1988, In vitro synthesis of deoxyhypusine in eIF-4D precursor. FASEB J. 2, A1563.

DEOXYHYPUSINE HYDROXYLASE

A. Abbruzzese[*], V. Liguori[*] and M.H. Park[+]

[*]Dipartimento di Biochimica e Biofisica, Universita di Napoli, Italy
[+]National Institute of Dental Reserach, NIH, Bethesda Maryland 20892

INTRODUCTION

The unusual basic amino acid hypusine [N^{ε}-(4-amino-2-hydroxybutyl) lysine] is a derivative of lysine which was originally isolated from a trichloroacetic acid-soluble extract of bovine brain (1). Subsequently it was found that in its free form hypusine concentration is highest in the brain (2), whereas, as a protein component, testes are the richest source (3). Furthermore, in developing rats, the hypusine concentration in brain protein is higher in the first 2 weeks of postnatal life and then steadily decreases until adulthood (3). In this organ the amino acid is also present bound to γ-aminobutyric acid (GABA) (4).

Hypusine synthesis involves the transfer of 4-aminobutyl moiety from the polyamine spermidine to a specific lysine residue to form protein-bound deoxyhypusine and subsequent hydroxylation of this intermediate (5, 6). The final step of the post-translational modification is catalyzed by deoxyhypusine hydroxylase (7), a metal requiring oxygenase, distinct from other reported hydroxylases (8). The content of deoxyhypusine in the protein fraction of cells or tissues normally is very low ($<$1% of total hypusine), but a significant accumulation of deoxyhypusine occurs upon inhibition of the hydroxylase by treatment with metal chelators (6).

The post-translational modification of one specific cellular protein, the eIF-4D precursor, through contribution of the polyamine spermidine represents an important function of polyamines. The correlation between hypusine synthesis and proliferation of various mammalian cells in culture (9, 10, 11, 12) prompted us to investigate the regulation of deoxyhypusine hydroxylase in relation to proliferation and development. In this article, partial purification and characterization of deoxyhypusine hydroxylase and its level in developing rat brain and cultured cells are reported.

PREPARATION OF SUBSTRATE

Since deoxyhypusine hydroxylase does not act on the free amino acid, deoxyhypusine, and no synthetic substrate is available, it was necessary to prepare the biosynthetically radiolabeled protein substrate from cultured cells. For this purpose, Chinese hamster ovary (CHO) cells were cultured in the presence of [2,3-^{3}H]putrescine (4 μCi/ml) and α,α-dipyridyl

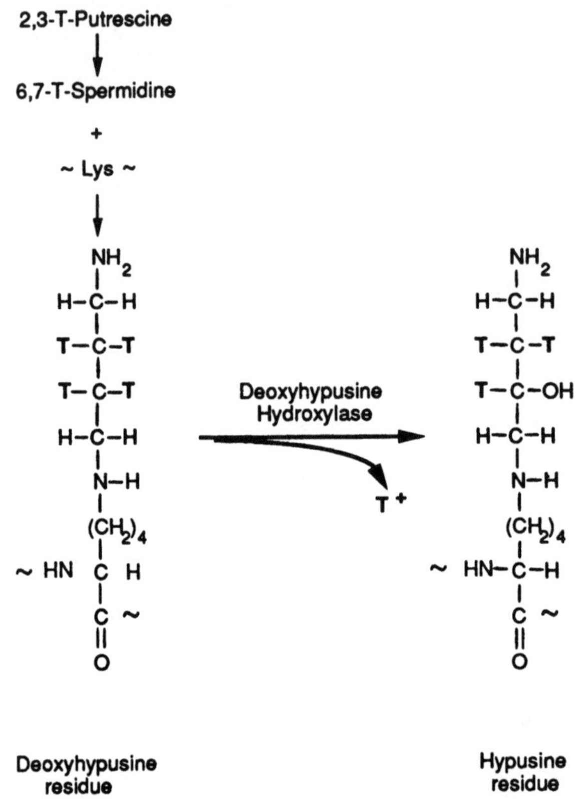

$$T^{+} + H_2O = THO + H^{+}$$

Scheme 1. Biosynthesis of the deoxyhypusine residue and the hypusine residue; tritium distribution from [2,3-^{3}H]putrescine. ~Lys~ and T denote the lysine residue in polypeptide linkage and ^{3}H, respectively.

(100 μM). This level of α,α-dipyridyl was added to prevent completely the conversion of deoxyhypusine to hypusine. After 24 h, the cells were collected and washed. The proteins in a lysate prepared from these cells were fractionated with ammonium sulfate in the presence of nonradioactive polyamines, the 40 to 70% fraction extensively dialysed in the presence of nonradioactive polyamines, and finally dialysed free of polyamines. More than 90% of the total trichloroacetic acid (TCA) precipitable radioactivity of crude cell lysate was recovered in the 40 to 70% ammonium sulfate fraction. Most of the radioactivity (>95%) in the substrate protein fraction was in the form of deoxyhypusine and this fraction was essentially free of enzyme activity. The specific radioactivity of substrate was estimated by fluorometric determination of deoxyhypusine in an acid hydrolysate of substrate (7).

ASSAY

The direct method for hypusine and deoxyhypusine determination by ion exchange chromatographic separation (7, 13) provides accurate measurement of the degree of hydroxylation of the labeled protein (scheme 1, scheme 2 D). However, this assay is time-consuming and inconvenient for analysis of a large number of samples. For this reason, three simple

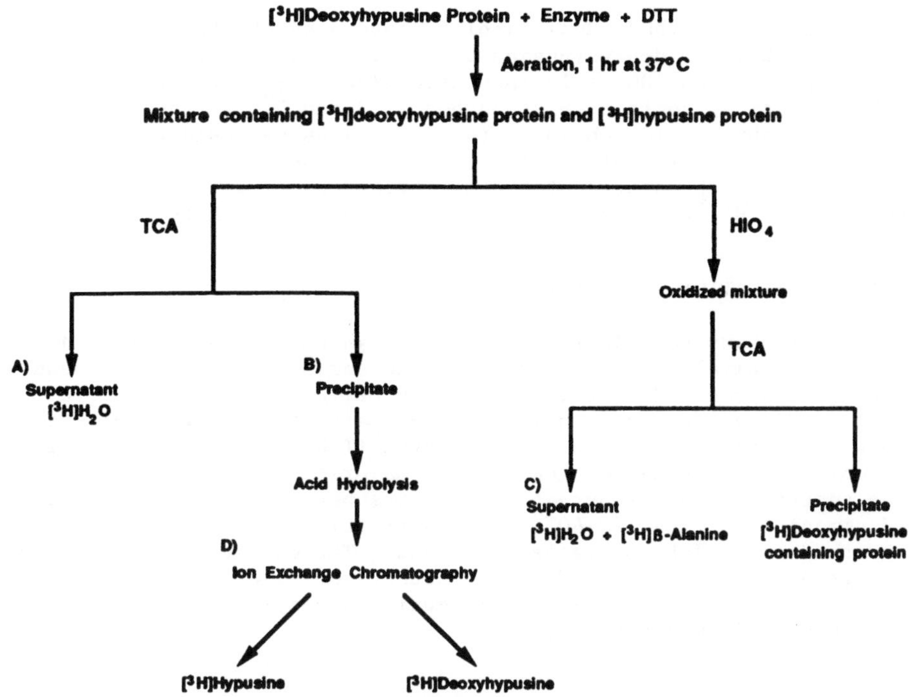

Scheme 2. Different assay methods of deoxyhypusine hydroxylase activity.
A. Measurement of tritium release.
B. Use of dual-labeled protein substrate and determination of $^{3}H:^{14}C$ ratio in protein fraction.
C. Release of radioactivity from the side chain of [^{3}H]hypusine residue by periodate oxidation.
D. Direct quantitation of [^{3}H]labeled hypusine and deoxyhypusine in acid hydrolysate of protein fraction.

indirect methods that do not require acid hydrolysis of proteins were developed. Good correspondence of each of these assays with the direct method was observed (14).

The first method depends upon measurement of the tritiated water formed from the specifically radiolabeled protein substrate (scheme 1). According to the scheme, one fourth of tritium atoms in [^{3}H]deoxyhypusine is released during hydroxylation. The measurement of radioactivity in the TCA-supernatant (scheme 2, A) is rapid, simple and sensitive when relatively pure enzyme is used. With a crude enzyme preparation, nonspecific degradation of the labeled protein substrate by proteases may cause a high background. This problem can be circumvented by separation of tritiated water from peptides or amino acids on small Dowex-50 columns. However, this method allowed only 30 - 50% recovery of the stoichiometric amount of tritiated water.

The second method is based on the specific periodate cleavage of the hypusine residue due to the presence of vicinal amino alcohol groups in the side chain. Upon oxidation with sodium periodate, the radioactivity in the 4-amino-2-hydroxybutyl moiety of hypusine residue is released into the TCA-supernatant, whereas the radioactivity in the unreacted deoxyhypusine residue remains in the protein substrate (scheme 2, C).

In the third method, dual-labeled protein substrate with ^{14}C and ^{3}H on the side chain of deoxyhypusine is used. Double-labeled protein substrate prepared from CHO cells incubated with 1,4-^{14}C putrescine and 2,3-^{3}H putrescine, contain ^{3}H at carbon 2, 3 and ^{14}C at the 1,4 positions of the 4-aminobutyl side chain of the deoxyhypusine residue. During hydroxylation, one fourth of the tritium is released whereas all ^{14}C is retained. Thus, the degree of hydroxylation can be estimated from the change in the ^{3}H:^{14}C ratio in the protein fraction (scheme 2, B).

PARTIAL PURIFICATION OF ENZYME

Table I shows the distribution of deoxyhypusine hydroxylase activity in different animal organs and cells. The enzyme activity was found in all animal tissues and cultured mammalian cells tested, but not in bacteria. Since rat testis homogenate contains the highest activity, this organ was chosen as a source of the enzyme (Table II). Upon subcellular fractionation of tissue homogenate, the majority of activity was recovered in the post-microsomal supernatant. This supernatant was treated with solid ammonium sulfate (20-40% cut) and protein in this fraction was chromatographed on a DEAE ion exchange column. The enzyme, eluted with a linear gradient of 0-0.7 M KCl, was further purified by gel filtration. This four step procedure resulted in enriched enzyme preparations that were free of protease activity.

CATALYTIC CHARACTERISTICS OF THE ENZYME

The enzyme showed a pH optimum between 7.0 and 7.5 for catalysis in 20 mM Tris buffer. Higher levels of ionic strength caused reduction in rates of catalysis. A sulfhydryl compound was required for activity of the partially purified enzyme. Dithiothreitol at 5 mM and above provided

TABLE I. Distribution of deoxyhypusine hydroxylase activity in different animal tissues.

The organs were from healthy male animals. The enzyme activity in the tissue extract was measured by periodate oxidation method.

Animal tissues	Specific activity units/mg protein	Enzyme content units/g tissue
Rat testis	7.9	270
Rat epididymis	7.1	110
Rat heart	3.9	170
Rat kidney	3.9	250
Rat liver	3.7	250
Rat lung	1.8	70
Rat brain	1.3	38
Rat pancreas	1.3	41
Guinea pig testis	7.9	264
Guinea pig kidney	4.1	240
Guinea pig liver	3.8	222
Guinea pig brain	1.1	41
Rabbit testis	7.4	242
Rabbit kidney	3.6	232
Rabbit liver	3.5	219
Chick embryo	7.8	

An enzyme unit is defined as the amount of enzyme that catalyzes the formation of 1 pmole of [^{3}H]hypusine in 150 min under the standard assay (7).

the maximum attainable activity. Several metal chelating compounds, e.g., α,α-dipyridyl, Desferal, picolinic acid and o-phenanthroline caused significant inhibition of deoxyhypusine hydroxylase. The potent inhibition by o-phenanthroline (10 μM) was relieved by addition of 30 μM cobaltous acetate or 5 μM ferrous sulfate. However 5 μM $FeSO_4$ alone exerted significant inhibition of enzyme activity. Addition of mixtures of substances known to be cofactors for α-ketoglutarate-dependent dioxygenases, i.e., α-ketoglutarate, ascorbic acid, and Fe^{2+} failed to bestow activity on partially purified enzyme in the absence of dithiothreitol. Furthermore, no decarboxylation of α-ketoglutarate was seen during hypusine formation. By varying the ^{3}H-labeled substrate concentration between 12 and 160 nM, an apparent Michaelis constant of 52±12 nM and a maximum velocity for partially purified enzyme of 6.25±0.67 pmol min^{-1} (per mg) were obtained.

DEOXYHYPUSINE HYDROXYLASE ACTIVITY IN DEVELOPING RAT BRAIN AND CULTURED CELLS

Since hypusine biosynthesis is closely related to cellular growth processes (9,10,11,12), we examined the levels of deoxyhypusine hydroxylase activity in rat brain during development. This organ is of particular interest because it is an especially rich source of both hypusine (2) and spermidine (15). Furthermore, the concentration of hypusine in rat brain proteins in developing rats is relatively high in the first 2 weeks of postnatal life and then decreases until adulthood (3). The majority (>80%) of deoxyhypusine hydroxylase activity in this tissue was present in the post-microsomal supernatant (16). Fig. 1 shows the specific activity and enzyme content of rat brain as a function of age. It was observed that the specific activity of the enzyme in the supernatant fractions increased after birth and reached maximal levels in 4-6 days. Thereafter, the activity gradually decreased up to 19 days of age. The content of enzyme activity reached maximum on the third day of postnatal life, then decreased to a stable value after 20 days. After this period the specific activity and enzyme content remained constant at approximately 45% of the maximum level throughout adulthood.

TABLE II. Purification of deoxyhypusine hydroxylase

The purification was from 100 g of rat testis. Reprinted from Ref. 7.

Step	Fraction	Protein mg	Specific Activity units/mg	Yield %
	Homogenate	6900	8	
1.	Clarified supernatant	3500	14	89
2.	20-40% $(NH_4)_2SO_4$	1260	31	71
3.	DEAE-cellulose	121	212	47
4.	Sephacryl S-300	36	416	27

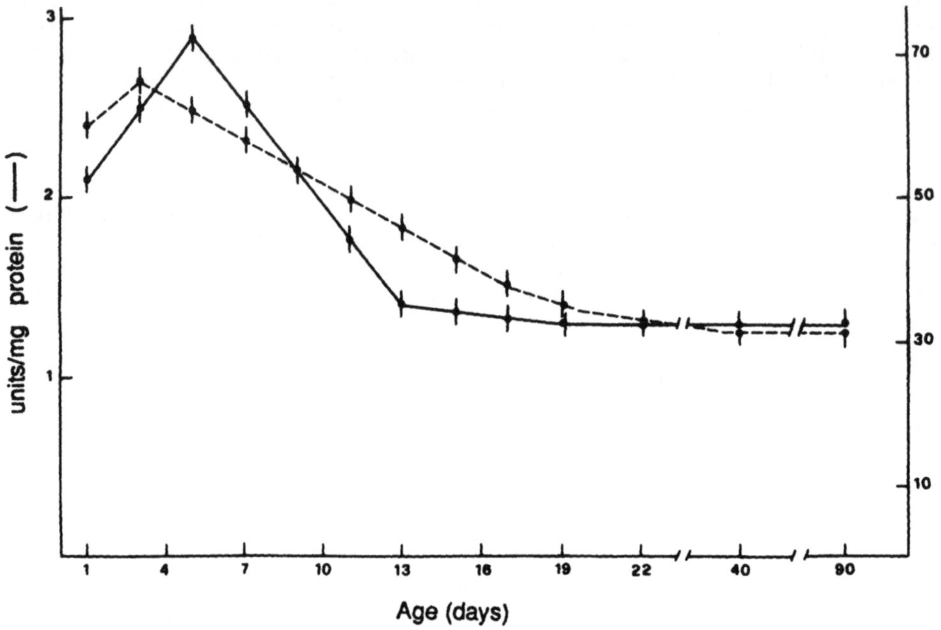

Fig. 1. Specific activity (—) and content (---) of deoxyhypusine hydroxylase in developing rat brain.
Deoxyhypusine hydroxylase activity in the post-microsomal supernatant was measured by periodate oxidation method (scheme 2C) as function of age. Each point represents the mean of triplicate assays performed on six animals. Reprinted from Ref. 16 with permission of the copyright owner.

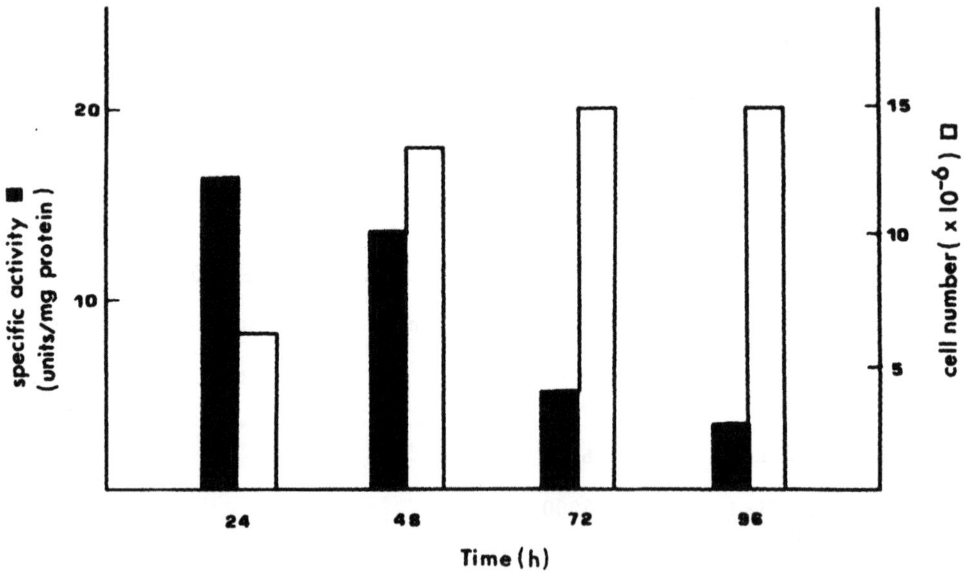

Fig. 2. Deoxyhypusine hydroxylase in cultured CHO cells.
CHO cells were seeded at 2 x 10^{6} cells/100 mm dish in α-modification of Eagle's medium supplemented as described (6). After 24, 48, 72 and 96 h, cells were harvested, the enzyme activities in the lysates were measured by the periodate oxidation method.

Deoxyhypusine hydroxylase was also studied in mammalian cells as a function of growth. In CHO cells, the specific activity of deoxyhypusine hydroxylase was high in exponentially growing cells but gradually decreased as cells reached confluency (Fig. 2). In cultured lymphocytes, deoxyhypusine hydroxylase specific activity increased 4-fold upon treatment of the cell with the mitogen phytohemaggulutinin for 20 h (17).

CONCLUDING REMARKS

The exact role of eIF-4D in protein synthesis and cellular proliferation is as yet unknown and the significance of hypusine formation awaits the elucidation of the physiological function of this protein. The fact that the deoxyhypusine-containing eIF-4D precursor does not normally accumulate in cultured cells or tissues implies that hydroxylation of deoxyhypusine is required for activity of this protein.

The hydroxylation reaction occurs by a unique catalytic mechanism different from either lysyl or prolyl hydroxylases. This is evidenced by the apparent lack of requirement for deoxyhypusine hydroxylase of known cofactors of the α-keto acid dioxygenases, i.e., Fe, α-ketoglutarate and ascorbic acid. The only apparent cofactor required for the partially purified enzyme is sulfhydryl compounds. The enzyme probably contains a tightly associated cofactor, metal ion, since hydroxylation is inhibited by several metal chelating compounds.

Deoxyhypusine hydroxylase activity is present in all animal tissues and eukaryotic cells examined. In the proliferative organ testis and in chick embryo, the highest specific activity is observed, and there appears to be a correlation between hypusine production, deoxyhypusine hydroxylase and cell growth in rat brain during development and in cultured cells. Although the hydroxylation is presumably not the normal rate-limiting step in hypusine biosynthesis, there may exist a general coordinate control of steps involved in hypusine synthesis, i.e., the translation of eIF-4D mRNA, deoxyhypusine synthesis, and its hydroxylation. The close relationship between hypusine synthesis and cell growth suggests that eIF-4D may play a specific role in cellular proliferation possibly through regulation of protein synthesis. In this regard, the post-translational synthesis of hypusine in eIF-4D precursor may represent a key pathway by which polyamines enter the regulation of cellular proliferation.

REFERENCES

1. T. Shiba, H. Mizote, T. Kaneko, T., Nakajima, Y. Kakimoto, and I Sano, Hypusine, a new amino acid occurring in bovine brain: Isolation and structural determination, Biochim. Biophys. Acta 244:523 (1971).
2. T. Nakajima, T. Matsubayashi, Y. Kakimoto, and I. Sano, Distribution of hypusine, N^6-(4-amino-2-hydroxybutyl)-2-6-diaminohexanoic acid, in mammalian organs, Biochim. Biophys. Acta 252:92 (1971).
3. A. Sano, M. Miyake and Y. Kakimoto, A rapid and sensitive method for the determination of hypusine in proteins and its distribution and developmental changes, Biochim. Biophys. Acta 800:135 (1984).
4. A. Sano, K. Kotani and Y. Kakimoto, Isolation and identification of α-(γ-aminobutyryl)hypusine, J. Neurochem. 46:1046 (1986).
5. M.H. Park, H.L. Cooper, and J.E. Folk, Identification of hypusine, an unusual amino acid, in a protein from human lymphocytes and of spermidine as its biosynthetic precursor, Proc. Natl. Acad. Sci. USA, 78:2869, (1981).
6. M.H. Park, H.L. Cooper, and J.E. Folk, The biosynthesis of protein-

bound hypusine [N^{ε}-(4-amino-2-hydroxybutyl)lysine]:Lysine as the amino acid precursor and the intermediate role of deoxyhypusine [N^{ε}-(4-aminobutyl)lysine], J. Biol. Chem. 257:7217 (1982).
7. A. Abbruzzese, M.H. Park, and J.E. Folk, Deoxyhypusine hydroxylase from rat testis: Partial purification and characterization, J. Biol. Chem. 261:3085 (1986).
8. K.I. Kivirikko, and R. Myllyla, Post-translation enzymes in the biosynthesis of collagen: Intracellular enzymes, Methods Enzymol. 82:245 (1982).
9. H.L. Cooper, M.H. Park, and J.E. Folk, Posttranslational formation of hypusine in a single major protein occurs generally in growing cells and is associated with activation of lymphocyte growth, Cell, 29:791 (1982).
10. K.Y. Chen, An 18,000 dalton protein metabolically labeled by polyamines in various mammalian cell lines Biochim. Biophys. Acta. 856:395 (1983).
11. B.M. Torrelio, M.A. Paz, and P.M. Gallop, Cellular proliferation and hypusine synthesis, Exp. Cell Res. 154:454 (1983).
12. E.W. Gerner, P.S. Mamont, A. Bernhardt and M. Siat, Post-translational modification of the protein-synthesis initiation factor eIF-4D by spermidine in rat hepatoma cells, Biochem. J. 239:379 (1986).
13. M.H. Park, H.L. Cooper, and J.E. Folk, Chromatographic Identification of Hypusine [N^{ε}-(4-amino-2-hydroxybutyl)lysine] and deoxyhypusine [N^{ε}-(4-aminobutyl)lysine], Methods in Enzymology 94:458 (1983).
14. A. Abbruzzese, M.H. Park, and J.E. Folk, Indirect assays for deoxyhypusine hydroxylase using dual-label ratio changes and oxidative release of radioactivity, Anal. Biochem. 154:664 (1986).
15. N. Seiler, Polyamine metabolism and function in brain, Neurochem. Int. 3:95 (1981).
16. A. Abbruzzese, Developmental pattern for deoxyhypusine hydroxylase in rat brain, J. Neurochem. 50:695 (1988).
17. A. Abbruzzese, T. Isernia, V. Liguori, and S. Beninati, Polyamine-dependent post-translational modification of protein and cell proliferation, Perspectives in Polyamine Research (A. Perin, G. Scalabrino, M.E. Feriolo, eds) (in press).

INTERACTIONS OF POLYAMINES

WITH MACROMOLECULES

EFFECT OF POLYAMINES ON PROTEIN KINASE C ACTIVATION PROCESS

Maria Stella Moruzzi, Giorgio Piccinini, *Bruna Tadolini, Maria Giuseppina Monti, §Bruno Barbiroli and Gabriele Mezzetti

Istituto di Chimica Biologica, Università di Modena; *Dipartimento di Biochimica, Università di Bologna; §Cattedra di Biologia Molecolare, Università di Bologna

INTRODUCTION

Protein phosphorylation and dephosphorylation are major mechanisms for signal transduction in biological systems (Krebs, 1985) and are fundamental in the regulation of many cellular activities (Rubin and Rosen, 1975). The Ca^{2+} and phospholipid-dependent protein kinase C (PKC) is a key regulatory enzyme believed to be involved in basic cell functions such as proliferation and differentiation (Schwantke et al., 1985). Activation of PKC by diacylglycerol (DAG) has linked the activity of this enzyme to the function of many stimuli acting via phosphatidylinositol turnover (Berridge, 1987). Indeed DAG has been postulated to be the second messenger able to activate PKC without a change in cytosolic Ca^{2+} concentration. Biologically active phorbol esters which usually act as tumor promoters also activate PKC and are presumed to exert their functions in this way; indeed this enzyme appears to be a cellular phorbol ester receptor (Ashendel, 1985).

PKC exists in different compartments of the cell and its intracellular distribution may vary rapidly between soluble and membrane-bound forms (Kraft and Anderson, 1983; Hirota et al., 1985). It is generally believed that the translocation process is closely related to the activation of the enzyme itself, since its physiological regulation requires a specific interaction with membrane phospholipids (Bell, 1986; Nishizuka, 1986). However, so far, the intimate mechanism of PKC binding to membranes and its activation by phospholipids and Ca^{2+} are not well understood.

Since as a prerequisite for the physiological response the enzyme requires association to cell membrane, we have undertaken a study concerning the molecular interactions which are involved in the PKC activation process. The identification of modulators could be very helpful to understand the mechanism by which this enzyme is regulated *in vivo*.

Ubiquitous polyamines (i.e. spermine, spermidine, putrescine) are normal cellular constituents, each having a polybasic character which gives them an affinity for acidic substances (Tabor and Tabor, 1984). As far as protein kinases are concerned, a pletora of effects have been ascribed to these polycations in several experimental systems. Quite different mechanisms of action may be operative in these kinase-directed reactions. It has been suggested that polyamines may affect protein phosphorylation either directly, acting mainly on the enzyme and/or substrates (Mezzetti et al., 1985; Hathaway and Traugh, 1982), or indirectly, interacting with other cellular effectors (Mezzetti et al., 1986; Job et al., 1979). In dealing with protein kinase C Qi et al. (1983) and Thams et al. (1986) have shown that these polycations can inhibit native protein kinase C activity from various sources in vitro; hovewer the site of action, as well as the intimate mechanism by which polyamines exert this effect, is still obscure.

In the present paper we report that polyamines, at physiological concentrations, are effective in preventing PKC association to membranes. At first the studies were conducted examining the association of PKC to inside-out erythrocyte vesicles. Further investigations, employing phospholipid vesicles of defined composition, have been performed in order to understand the basis of molecular mechanism of polyamine inhibition of PKC activation.

EXPERIMENTAL

Materials

2-Mercaptoethanol, dithiothreitol, EDTA, EGTA, ATP, fatty-acid bovine gamma globulins, phosphatidylserine (PS), phosphatidylcoline (PC), 1,2--diolein, polyamines, PDBu and histone (type III-S) were purchased from Sigma Chem. Co, U.S.A. DEAE-cellulose and phosphocellulose P11 were obtained from Whatman, UK. $[^3H]$PDBu (12,5 Ci/mmol), $[\gamma-^{32}P]$ATP(2,900 Ci/mmol) and Protosol were from New England Nuclear, Germany. All other reagents were of commercial origin of the highest grade of purity.

Methods

Protein kinase C isolated from Sprague-Dawley rat brain was routinely purified by the method of Walsh et al. (1984) modified as previously described in detail (Moruzzi et al., 1987). During purification protein kinase C activity was assayed by the method of Kikkawa et al. (1982) using histone III-S as the substrate.

Unilamellar phospholipid vesicles made of different percentage of PS and PC were prepared by a published method (Ohnishi and Ito, 1974) and resuspended in 20 mM Tris-HCl buffer pH 7.5. Phosphorus determination was performed by the method of Marinetti (1962).

The amount of protein kinase C associated with the phospholipid vesicles was estimated by measuring its $[^3H]$PDBu-binding activity (Leach and Blumberg, 1985). The binding assay was carried out in 1.5 ml Eppendorf microcentrifuge tubes containing, in a total volume of 0.25 ml: 20 mM

Tris-HCl pH 7.5, 1 mM dithiothreitol, bovine gamma globulines 10 mg/ml, 0.1 mM EGTA, 0.01-0.05 ml of protein kinase C (9 nmol of ^{32}P incorporated into histone III-S/min. per ml.) being protein concentration approximately 0.3 mg/ml. As indicated, phospholipid vesicles and 40 nM $[^3H]$PDBu were added and incubated in the presence or absence of 500-fold excess of radioinert PDBu. $CaCl_2$ was added to give 1 µM free Ca^{2+} concentration together with other additions as indicated. The suspension was incubated at 30°C for 30 min. The tubes were then chilled and proteins precipitated by addition of 0.187 ml of cold buffered 35% polyethylene glycol (wt/vol). Samples were kept for 15 min in ice and then centrifuged for 15 min at 12,000 rpm in a Beckman Microfuge 12. A 0.1 ml aliquot of the supernatant was removed for determination of the actual free $[^3H]$PDBu concentration. The pellet was dissolved with 0.8 ml of Protosol and thereafter counted for radioactivity in a toluene-based scintillation cocktail (Mezzetti et al., 1980). Determinations were performed in triplicate and specific binding was calculated as the difference between total binding and non-specific binding (i.e. in the presence of excess of radioinert PDBu). Analysis of specific binding data was performed by the method of Scatchard (1949) and regression analysis was used for curve fitting.

The aggregation experiments were conducted under the standard conditions used for the association of protein kinase C to phospholipids. The aggregation of the vesicle suspension in the presence of various amounts of spermine was monitored by the turbidity changes of the solution at 400 nm. The A_{400} change obtained 5 min after spermine addition was plotted as a function of the polyamine concentration (Tadolini et al. 1985).

Inside-out vesicles from human erythrocytes were prepared by the method of Sarkadi et al. (1980) and association studies of protein kinase C to membrane preparation was performed as previously described in detail (Moruzzi et al., 1987).

The concentration of protein was determined as described by Lowry et al. (1951).

RESULTS

During the last years many studies have investigated the requirements for PKC activity. This enzyme is normally present in an inactive form, and a complex constituted by diacylglycerol or phorbol ester, Ca^{2+}, membrane phospholipids and the enzyme itself is required for its activation. However the intimate mechanism which underlies PKC association to membranes is a much more complex process which involves other regulatory cellular components and their physical relationship with certain specific domains of cell membranes.

It is well known that positively charged polyamines are capable of interacting with a variety of cellular polyanions including membrane phospholipids (Tadolini et al., 1985; Chung et al., 1985) and that such polycations can affect the intracellular translocation of several enzymes such as phosphatidate-phosphohydrolase (Martin-Sanz et al., 1985) and hexokinase (Kurokawa et al., 1983).

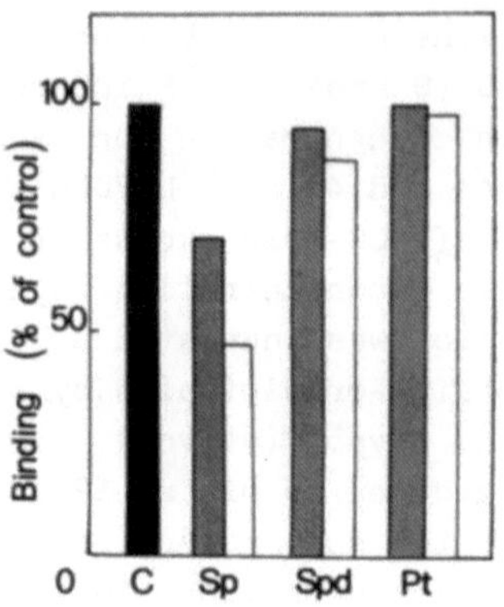

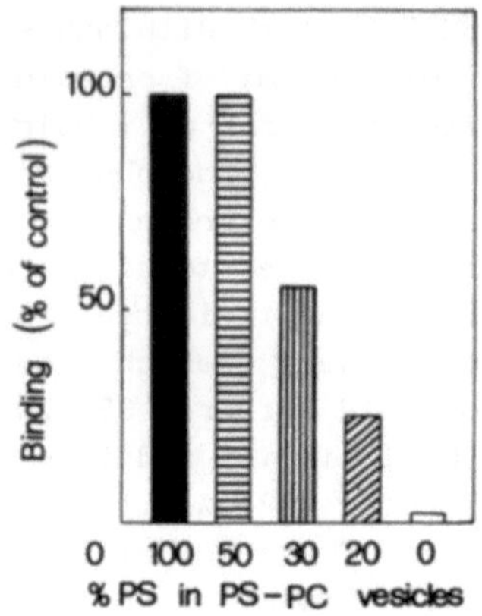

Fig. 1. (left) Effect of polyamines on the association of protein kinase C inside-out erythrocyte membrane vesicles.

Enzyme preparation (100 μl; 0.3 mg/ml) was associated with membranes (100 μg of protein) in the presence of either 50 μM (cross-hatched bars) or 100 μM (open bars) of indicated polyamines. Results are expressed as percentage of the amount of $[^3H]$PDBu bound in the absence of polyamines (1.4 pmol). Results are means for four separate experiments.

Fig. 2. (right) Protein Kinase C binding to phospholipid vesicles as a function of PS content.

Enzyme preparation (20 μl; 0.28 mg/ml) was associated with phospholipid vesicles of various PS content. The phospholipid vesicles contained: 100%, 50%, 30%, 20%, 0% PS, PC being the other phospholipid present. The PS concentration was 50 μM in every case. Results are expressed as percentages of the amount of $[^3H]$ PDBu bound in presence of vesicles made of 100% PS (2.6 pmol). Results are means for three separate experiments.

These considerations prompted us to study whether these naturally occurring polybasic amines could interfere with PKC activation process. To this aim we utilized an opportune model system consisting of purified PKC from rat brain and human inside-out erythrocyte vesicles. This experimental system is believed to be a good model to study physiologically relevant enzyme-lipids interactions (Wolf et al., 1985b). This is particularly true for PKC since human erythrocyte membranes lack endogenous enzyme, thus allowing accurate estimations of PKC associated to membranes under defined experimental conditions. PKC-membrane complex was quantitatively assayed as phorbol ester receptor, i.e. for its $[^3H]$ PDBu-high affinity binding activity, in the presence of physiological concentrations of Ca^{2+} (Wolf et al., 1985a).

As shown in Fig. 1 the addition of 50-100 μM spermine, during the step of PKC association to membranes, dramatically decreased $[^3H]$PDBu-binding activity, suggesting that this polyamine could effectively interfere in PKC activation process. The inhibitory effect shown by polyamine in this reaction appears to be quite specific since it is dose-dependent and a precise chemical structure of the interacting polycation is required for optimal activity. In fact the tetramine (spermine) was a much better effector than a triamine (spermidine) and a diamine (putrescine). In the experiments reported the concentration of enzyme protein as well as the amount of membrane was always equilibrated in the various assays and a saturating concentration of radioactive ligand was used so that changes in $[^3H]$PDBu bound to membranes were representative of changes of phorbol ester binding activity by PKC associated to membrane.

Recently it has been reported that polyamines at physiological concentrations bind to liposomes forming tight complexes with acidic phospholipids, particularly with phospatidic acid (PA) and phosphatidylserine (PS) (Tadolini et al., 1985; Meers et al., 1986). In this light it is conceivable that spermine may exert its effect through an interaction with specific acidic domains present on the membranes perturbing the lipid bilayer and making it unsuitable for PKC binding.

From a molecular point of view the interaction of polyamines with the different phospholipids of cell membrane is largely unknown. It has been reported that membranes of subcellular organelles have different phospholipid compositions, besides having phospholipid asymmetry between the two layers (Nilsson and Dallner, 1977) and that membrane enzymes are very often surrounded by a domain of well defined and constant phospholipid composition necessary for their function (Gazzotti and Peterson, 1977). The pos-

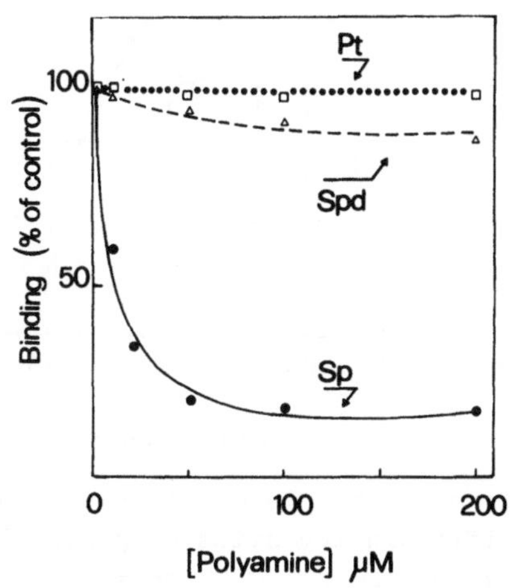

Fig. 3. Effect of polyamines on protein kinase C binding to phospholipid vesicles.
Enzyme preparation (20 μl; 0.3 mg/ml) was associated with 50% PS vesicles (100 μM total phospholipid concentration) in the absence or presence of various concentrations of spermine (Sp; ●) spermidine (Spd; △) and putrescine (Pt; □). Results are expressed as percentage of the amount of $[^3H]$ PDBu bound in the absence of polyamines (2.9 pmol). Results are means for four separate experiments.

sibility that this property is also shared by PKC and in the perspective to contribute to the elucidation of the molecular mechanisms involved in PKC--phospholipid-spermine interactions, we have undertaken a study concerning the use of phospholipid vesicles of defined composition in the place of the human erythrocyte membranes.

Since it is well established that PS is the most effective phospholipid in activating PKC we have studied the PDBu-binding activity in the presence of liposomes containing different percentage of PS. As shown in Fig. 2 the activation of PKC, as measured by radioactive PDBu-binding activity, was clearly affected by the composition of the liposome itself. More specifically, in the presence of 100% and 50% PS liposomes, PDBu binding is directly related to the amount of PS in the assay mixture. On the contrary when 30% and 20% PS liposomes are assayed, PDBu binding as a function of PS content is greatly decreased. PC liposomes do not allow the formation of the PDBu/PKC complex.

However data obtained in other experimental systems indicate that PKC activation in the presence of PS containing liposomes is enhanced by increasing the PS content up to 20%; a higher density of PS does not facilitate further PKC binding to liposomes (Bazzi and Nelsestuen, 1987). Nevertheless, in our assay conditions, the enzyme was catalytically active and

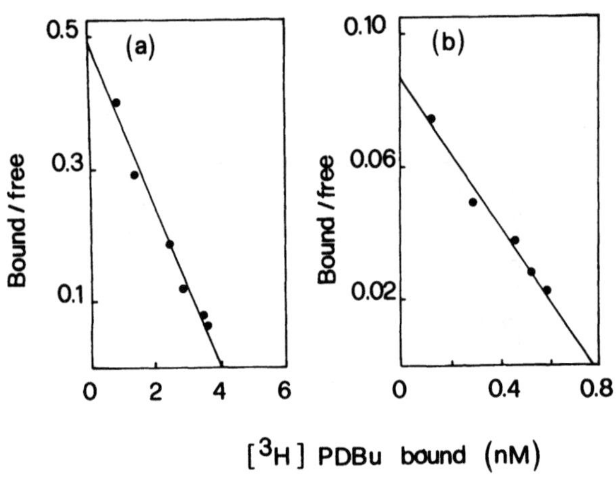

Fig. 4. Effect of polyamines on dissociation constant and number of $[^3H]$PDBu specific binding sites on phospholipid vesicles. Enzyme preparation (20 µl; 0.27 mg/ml) was associated with 50% PS vesicles (100 µM total phospolipid concentration) in the absence (Panel a) or presence (Panel b) of 50 µM spermine with increasing concentrations of $[^3H]$ PDBu. Incubation conditions and procedures used to determine specific binding were as described in the Experimental section. Specific binding data were subjected to Scatchard analysis; the apparent dissociation constant (K_d) and the maximal number of specific binding sites (N_{max}) were calculated. Results are means of four separate experiments.

addition of EGTA caused a partial dissociation of the PKC from liposome. Similar lack of total reversibility was observed in the brain where nearly one third of PKC activity is membrane bound and can be extracted only in the presence of detergent. In addition, the binding activity appeared highly specific, since $[^3H]$PDBu bound to protein kinase C was almost completely displaced in the presence of excess radioinert PDBu. The complex that we measure may thus be of physiological relevance. For all subsequent experiments phospholipid vesicles containing 50% PS were utilized.

The effect of polyamine concentrations on the binding reaction of PKC to liposomes was then studied (Fig. 3). As it appears, the addition of 10--200 μM spermine to the standard assay containing 50% PS liposomes greatly inhibited the binding reaction. This effect of spermine was dose-dependent and selective since no significant inhibition was detected when other polyamines were present at the same concentrations. In addition, the differences between the three polyamines confirm that their effects are not the result of simple charge neutralization. This experiment was performed at a single (40 nM) concentration of radioactive phorbol ester predetermined to saturate PKC whose apparent dissociation equilibrium binding constant was approximately 8 nM. Therefore it was important to assess the possibility that the inhibition of $[^3H]$PDBu binding observed in the presence of spermine was the result of the decrease in the amount of PKC binding sites available or was due to a lowered PKC-ligand affinity.

Fig. 4 shows such experimental data obtained by saturation binding analysis performed at various $[^3H]$PDBu concentrations (0.5-80 nM) either in the presence or absence of 20 μM spermine. Incubations were allowed to

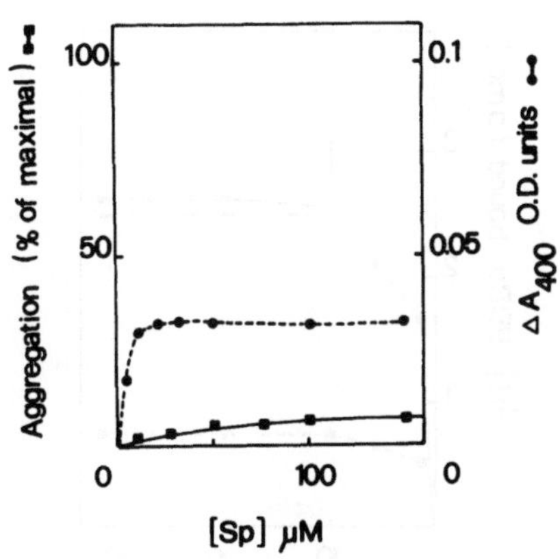

Fig. 5. Aggregation of phospholipid vesicles by spermine. 50% PS vesicles (100 μM total phospholipid concentration) were incubated for 5 min at 25°C with increasing concentrations of spermine. The extent of liposome aggregation was evaluated by measuring turbidity changes as described in the Experimental section. Data are expressed as percentages of maximal aggregation (A_{400}=2.183) and as rate of aggregation (ΔA_{400}/min). Results are means for three separate experiments.

proceed for 30 min at 30°C, conditions that allow equilibrium binding. The results were analyzed by the Scatchard plot (Scatchard, 1949), which gave an estimate of the number of PKC binding sites (N_{max}) and also allowed assessment of any change in the PKC-ligand affinity (K_d). The results of Fig. 4 show that K_d of PKC-ligand was not significantly affected by the presence of spermine varying from 8.1 to 8.9 nM. On the contrary the N_{max} was dramatically diminished from 111 pmol to 22 pmol/mg of protein in the presence of spermine, suggesting that a decrease in PKC molecule number bound to liposomes is a result of spermine action.

The decreased N_{max} observed in the presence of spermine may be due to polyamine induced aggregation of liposomes resulting in a decreased availability of binding sites for PKC. In fact it is well known that the interaction of polyamines with vesicles containing acidic phospholipids could result in an extensive aggregation of liposomes (Schuber et al., 1983). Although the rate of aggregation in the presence of the spermine concentrations active on PKC is high, this phenomenon could not account for the dramatic inhibitory effect observed. In fact, as shown in Fig. 5, under the standard conditions used for binding assay, the extent of liposome aggregation was very low, ranging from 3 to 5% of the maximal aggregation of the vesicle suspension (100%) as obtained by addition of PEG.

In the experiment of Fig. 6 is depicted a time-course study of association of PKC to 50% PS liposomes, either in the absence or presence of 20 µM spermine. In both cases liposome association of the enzyme was a rapid phenomenon being almost complete after 7-8 min, and it remained constant for at least 30 min more. Spermine was added to the assay in two different fashions, i.e. when the PKC liposome-complex formation was in equilibrium

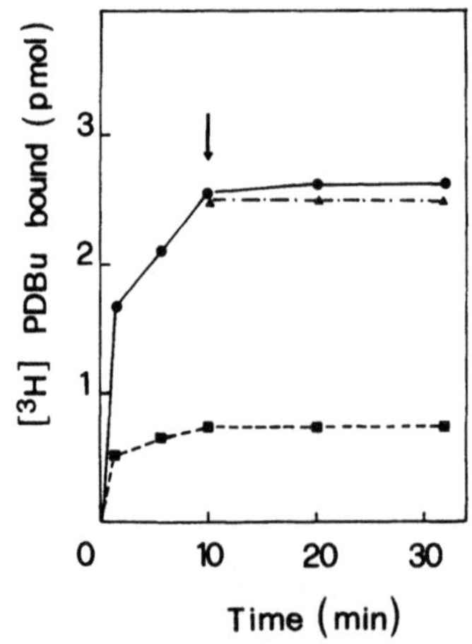

Fig. 6. Time course of association of PKC with phospholipid vesicles. 50% PS vesicles (100 µM total phospholipid concentration) were incubated with 20 µl enzyme preparation (0.27 mg/ml) for various times at 30°C in the absence (●) or presence of 50 µM spermine either added initially (■) or after 10 min of incubation (▲). Results are means for three separate experiments.

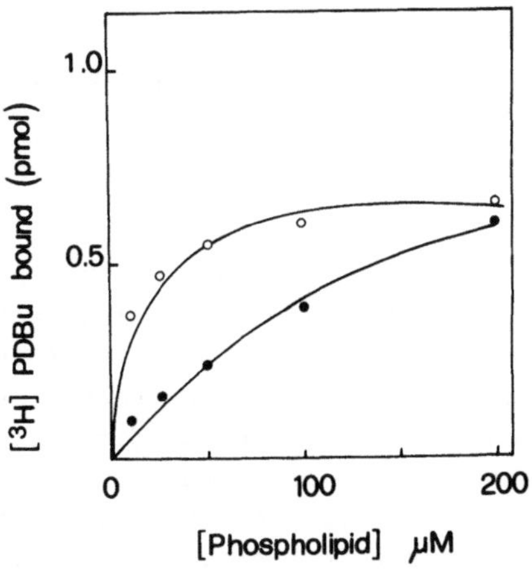

Fig. 7. Reversibility of spermine inhibition by phosphatidylserine. Increasing amounts of 50% PS vesicles were associated with 20 μl of enzyme preparation (0.05 mg/ml) in the absence (O) or presence (●) of 20 μM spermine. Results are means for four separate experiments.

state (10 min) or just before the beginning of the association reaction (time 0). From the data shown in Fig. 6 it is clearly apparent that spermine effect was showable only when the polycation was added at the beginning of the incubation period. In this case the principal action of polyamine was a dramatic inhibition of the maximal extent of the PDBu-binding activity, suggesting that spermine can exert its effect by an initial interaction with the acidic phospholipids of the liposome, forming complexes that prevent the subsequent association of PKC. This hypothesis is further strengthened by the finding that when the same amount was added at the saturation equilibrium of the association reaction no significant variation of the kinetic parameters was observed. This indicates that when PKC-phospholipid complex is already formed, polyamine, at the concentration used, is not able to displace PKC from the complex.

To further investigate the basis of molecular mechanism of polyamine action we examined whether the inhibitory level of spermine was somehow related to the concentration of the acidic phospholipid component of liposome. From the data reported in Fig. 7 it appears that increasing contents of PS markedly reduced the inhibitory effect of this polycation. In fact 20 μM spermine was active at low PS concentrations, being this effect overcome by an excess of phospholipids in the assay. The reversibility of polyamine inhibition by PS indicates that spermine exerts its action by decreasing the availability of PS in the binding site for PKC thereby interfering with the function of its regulatory domain (s).

CONCLUDING REMARKS

This discussion has advanced and to some extent documented that spermine may effectively regulate the number of molecules of PKC that can associate to membrane phospholipids, the step that is required for fully activation of this enzyme. This can be of great importance in the living cell where changes in polyamine content accompany important variations in cellular functions.

An initial assumption that can be made is that polyamine with its positive charges exerts its effect through an interaction with acidic components of the membrane. In fact electrostatic interactions of spermine with acidic phospholipids can drastically decrease the pool of charged PS involved in the formation of specific PKC binding sites. The tetramine spermine lining up to the membrane may bind with its charges to PS molecules, thereby forming a compact complex which limitates the availability of this phospholipid. The aliphatic hydrophobic regions of the polyamine may assure its specific partition into the membrane lipid so that amino-groups can be correctly approached by three to four negatively charged PS head groups. In addition the great conformational flexibility of the polyamine could favour the formation of such multiple complexes resulting in neutral clusters with no longer binding properties for PKC.

The characteristics of spermine action described in this paper may illustrate how polyamine may interact with plasma membrane *in vivo*. In this light it is important to emphasize that beside sphingosine (Hannun et al., 1986) spermine differs from the majority of known PKC modulators *in vitro* since it is an important and natural component of the cells. Polyamine levels may be modulated in response to either intra or extracellular signals thus functioning as a regulated effector of protein kinase activity. Spermine interacting with specific phospholipids components can thus hinder or preclude the critical association of PKC with appropriate elements of the bilayer, therefore a key role of polyamine in modulation of transmembrane signaling may emerge. These considerations suggest a new aspect of molecular control which should now be investigated at the cellular level.

Acknowledgments

We are most grateful to Professor Giovanni Moruzzi for his valuable advice and continuous encouragement in this research field.

REFERENCES

Ashendel, C.L., 1985, The phorbol ester receptor: a phospholipid--regulated protein kinase, *Biochim. Biophys. Acta*, 822:219.

Bazzi, M.D., and Nelsestuen, G.L., 1987, Association of protein kinase C with phospholipid vesicles, *Biochemistry*, 26:115.

Bell, R.M., 1986, Protein kinase C activation by diacylglycerol second messengers, *Cell*, 45:631.

Berridge, M.J., 1987, Inositol trisphosphate and diacylglycerol: two interacting second messengers, *Ann. Rev. Biochem.*, 56:159.

Chung, L., Kaloyanides, G., McDaniel, R., McLaughlin, A., and

McLaughlin, S., 1985, Interaction of gentamicin and spermine with bilayer membranes containing negatively charged phospholipids, Biochemistry 24:442.
Gazzotti, P., and Peterson, S.W., 1977, Lipid requirement of membrane--bound enzymes, J. Bioenerg. Biomemb. 9:373.
Hannun, Y.A., Loomis, C.R., Merrill Jr., A.H., and Bell, R.M., 1986, Sphingosine inhibition of protein kinase C activity and of phorbol dibutyrate binding in vitro and in human platelets, J. Biol. Chem., 261:12604.
Hathaway, G.M., and Traugh, J.A., 1982, Casein kinases-multipotential protein kinases, Curr. Topics Cell. Reg. 21:101.
Hirota, K., Hirota, T., Aguilera, G., and Catt, K.J., 1985, Hormone--induced redistribution of calcium-activated phospholipid-dependent protein kinase in pituitary gonadotrophs, J. Biol. Chem., 260:3243.
Job, D., Pirollet, F., Cochet, C., and Chambaz, E.M., 1979, Interaction of a casein kinase (G-type) with a specific endogenous inhibitor, FEBS Lett., 108:508.
Kikkawa, U., Takai, Y., Minakuchi, R., Inohara, S., and Nashizuka, Y., 1982, Calcium-activated, phospolipid-dependent protein kinase from rat brain, J. Biol. Chem., 257:13341.
Kraft, A.S., and Anderson, W.B., 1983, Phorbol esters increase the amount of Ca^{2+}, phospholipid-dependent protein kinase associated with plasma membrane, Nature, 301:621.
Krebs, E.G., 1985, The phosphorylation of proteins: a major mechanism for biological regulation, Biochem. Soc. Trans. 13:813.
Kurokawa, M., Yokoyama, K., and Ishibashi, S., 1983, Polyamines stimulate the binding of hexokinase type II to mitochondria, Biochim. Biophys. Acta, 759:92.
Leach, K.L., and Blumberg, P.M., 1985, Modulation of protein kinase C activity and $[^3H]$phorbol 12,13-dibutyrate binding by various tumor promoters in mouse brain cytosol, Cancer Res., 45:1958.
Lowry, O.H., Rosebrough, N.J., Farr, A.L., and Randall, R.J., 1951, Protein measurement with the folin phenol reagent, J. Biol. Chem., 193:265.
Marinetti, G.V., 1962, Chromatographic separation, indentification and analysis of phosphatides, J. Lipid. Res., 3:1.
Martin-Sanz, P., Hopewell, R., and Brindley, D.N., 1985, Spermine promotes the translocation of phosphatidate phosphohydrolase from the cytosol to the microsomal fraction of rat liver and it enhances the effects of oleate in this respect, FEBS, 179:262.
Meers, P., Hong, K., Bentz, J., and Papahadjopoulos, D., 1986, Spermine as a modulator of membrane fusion: interactions with acidic phospholipids, Biochemistry, 25:3109.
Mezzetti, G., Moruzzi, M.S., Capone, G., and Barbiroli, B., 1980, Polyamine binding by a cytoplasmic factor in the duodenal mucosa of new-born chick, Biochem. Biophys. Res. Commun., 97:222.
Mezzetti, G., Moruzzi, M., Monti, M.G., Piccinini, G. and Barbiroli, B., 1985, Polyamine-sensitive protein kinase from chick intestinal mucosa, Mol. Cell. Biochem., 66:175.
Mezzetti, G., Moruzzi, M., Piccinini, G., Monti, M.G., and Barbiroli, B., 1986, Modulation of cyclic nucleotide-independent protein

kinase from ckich intestine by naturally occurring polyamines and mucopolysaccharides, Mol. Cell. Biochem., 70:141.
Moruzzi, M., Barbiroli, B., Monti, M.G., Tadolini, B., Hakim, G., and Mezzetti, G., 1987, Inhibitory action of polyamines on protein kinase C association to membranes, Biochem. J., 247:175.
Nilsson, O.S., and Dallner, G., 1977, Transverse asymmetry of phospholipids in subcellular membranes of rat liver, Biochim. Biophys. Acta, 464:453.
Nishizuka, Y., 1986, Studies and perspectives of protein kinase C, Science, 233:305.
Ohnishi, S., and Ito, T., 1974, Calcium-induced phase separation in phophatidylserine-phosphatidylcoline membranes, Biochemistry, 13:881.
Qi, D., Schatzman, R.C., Mazzei, G.J., Turner, R.S., Raynor, R.L., Liao, S., and Kuo, J.F., 1983, Polyamine inhibit phospholipid--sensitive and calmodulin-sensitive Ca^{2+}-dependent protein kinases, Biochem. J., 213:281.
Rubin, C.S., and Rosen, O.M., 1975, Protein phosphorylation, Ann. Rev. Biochem., 44:831.
Sarkadi, B., Szasz, I., and Gardos, G., 1980, Characteristics and regulation of active calcium transport in inside-out red cell membrane vesicles, Biochim. Biophys. Acta, 598:326.
Scatchard, G., 1949, The attractions of proteins for small molecules and ions, Ann. N.Y. Acad. Sci., 51:660.
Schuber, F., Hong, K., Düzgünes, N., and Papahadjopoulos, D., 1983, Polyamines as Modulators of membrane fusion: aggregation and fusion of liposomes, Biochemistry, 22:6134.
Schwantke, N., Le Bouffant, F., Dorèe, M., and Le Peuch, C.J., 1985, Protein Kinase C: properties and possible role in cellular division and differentiation, Biochimie, 67:1103.
Tabor, C.W., and Tabor, H., 1984, Polyamines, Ann. Rev. Biochem., 53:749.
Tadolini, B., Cabrini, L., Varani, E., and Sechi, A.M., 1985, Spermine binding and aggregation of vesicles of different phospholipid composition, Biogenic Amines, 3:87.
Thams, P., Capito, K., and Hedeskov, C.J., 1986, An inhibitory role for polyamines in protein kinase C activation and insulin secretion in mouse pancreatic islets, Biochem. J., 237:131.
Walsh, M.P., Valentine, K.A., Ngai, P.K., Carruthers, C.A., and Hollenberg, M.D., 1984, Ca^{2+}-dependent hydrophobic-interaction chromatography, Biochem. J., 224-117.
Wolf, M., Cuatrecasas, P., and Sahyoun, N., 1985a, Interaction of protein kinase C with membranes is regulated by Ca^{2+}, phorbol esters, and ATP, J. Biol. Chem., 260:15718.
Wolf, M., Le Vine III, H., May Jr., W.S., Cuatrecasas, P. and Sahyoun, N., 1985b, A model for intracellular translocation of protein kinase C involving synergism between Ca^{2+} and phorbol esters, Nature, 317:546.

INTERACTION OF POLYAMINES WITH PHOSPHOLIPIDS: SPERMINE AND Ca^{2+} COMPETITION FOR PHOSPHATIDYLSERINE CONTAINING LIPOSOMES

Bruna Tadolini and Gabriele Hakim

Dipartimento di Biochimica
Università di Bologna
Bologna, Italy

INTRODUCTION

The ability of polyamines to interact with biological membranes was first described by Tabor who showed that polyamines aggregate subcellular organelles and stabilize mitochondria and protoplasts (1). Since then many reports pointed out a possible involvement of polyamines on membrane structure, metabolism and function.

In general the mechanism by which polyamines exert their action is the formation of polyamine-phospholipid complexes. In vitro studies have shown that polyamines can bind phospholipid vesicles with in order of effectiveness spermine> spermidine> putrescine (2,3). The binding strength of the complexes with spermine depends on the type of acidic phospholipid, i.e. phosphatidic acid (PA)> phosphatidylserine (PS)> phosphatidylinositol> cardiolipin (4,5). Spermine strongly interacts with polyphosphoinositides-containing liposomes (5-8). The binding strength of spermine to mixed vesicles depends on the density of acidic phospholipid on the liposomes (4,9). Determination of the stoichiometry of the complex demonstrates that different stoichiometries of acidic phospholipid-spermine complex exist, and depend on the relative ratios of the reactants in solution (4,5,9). The spermine lies flat on the surface in the complex formed at a high ratio of acidic phospholipid to spermine, the spermine molecule would be equivalent to a point tetravalent charge cross-linking the acidic phospholipid components of the liposome (4,9). At low ratios of acidic phospholipid to spermine the spermine molecule seems to bind perpendicularly to the vesicle surface: in this complex spermine can bridge acidic phospholipids on different vesicles grafting them into high molecular weight aggregates (4,5,10-12). Only when bound to vesicles containing a high proportion of PA, does spermine appear to perturb and release the water of hydration with its hydrocarbon segments (13). The binding of polyamine to liposomes appears to be a charge interaction as demonstrated by the fact that zwitterionic vesicles, made of phosphatidylcholine (PC) and phosphatidylethanolamine (PE), do not appreciably adsorb this polycation (4,9-11,14).

From these *in vitro* studies it appears that polyamines, in particular spermine, behave as counterions to acidic phospholipids. However, *in vivo* it is conceivable that the simple neutralization of charges may also be accomplished by other inorganic cations. The high and variable number of charges (3 and 4 for spermidine and spermine respectively), the charge distance as well as the flexibility of the overall structure of the molecule, render the polyamines able to fulfill other functions that would be impossible for inorganic cations. It is reasonable to think that *in vivo* polyamines may be specifically involved in reactions such as bridging of distant anionic charges and clustering of single diffusible anionic charges.

The relationship among polyamines, inorganic cations and membranes has attracted much attention; in particular the relationship between polyamines and Ca^{2+}, due to the importance that this inorganic cation appears to have in the control of cellular functions. It was proposed that signal reception could elicit a cation-exchange reaction whereby the polyamines bind to anionic sites in the plasma membrane and competitively displace bound calcium as free cytosolic Ca^{2+} (15). However polyamines, in particular spermine, appear to participate to the homeostasis of cytosolic Ca^{2+} by decreasing its concentration. In fact spermine increases the Ca^{2+} sequestering ability of mitochondria (16) and increases the Ca^{2+}-induced Ca^{2+} release from sarcoplasmic reticulum (17). Competition between polyamines and Ca^{2+} has also been suggested as a mechanism to modulate protein kinase C (PKC) activation. Polyamines were shown to inhibit PKC *in vitro* (18-20). It was suggested that the positively charged polyamines could remove Ca^{2+} from the PKC binding site on the membrane (18). The binding site was proposed to be a cluster formed by four PS molecules and Ca^{2+}(21). Conversely Ca^{2+}influx could decrease polyamine inhibition of PKC by displacing the polyamine (19). However other authors showed that spermine inhibition of PKC was non-competive with respect to Ca^{2+} (18,20). Also a synergic effect of polyamines and Ca^{2+} on membrane fusion has been reported. Polyamines at physiological concentrations were unable to fuse liposomes of pure PS or PA and mixture of PA with PC. However in their presence a large decrease in the threshold Ca^{2+} concentration required for fusion was observed (11). A drastic increase in the rate of Ca^{2+}-induced fusion occurred also in the presence of liposomes containing phosphatidylinositol 4,5-diphosphate (5). A similar synergism was observed when liposomes composed of mixtures of acidic phospholipids and a high mole fraction of PE were used. These vesicles however, could be induced to fuse by polyamines in the absence of Ca^{2+} (11). Also the nature of the negatively charged phospholipid present in the liposome influences the spermine effect on Ca^{2+}-induced fusion of liposomes. Spermine has a small effect on the fusion of PS liposomes as opposed to the large effect on PA liposomes (5). These results were ascribed to an enhancement of calcium binding to PA and inhibition to PS, during the polyamine-induced aggregation of the liposomes (5).

We have tried to contribute to the knowledge of the relationship between spermine and calcium by studying both their competition in the binding to PS liposomes and formation constant of their complexes with the same liposomes.

EXPERIMENTAL

Materials

Egg yolk phosphatidylcholine, phospatidylserine, 1,2-diolein (DAG), spermine tetrahydrochloride, phorbol-12,13-dibutyrate (PDBu), arsenazo III were purchased from Sigma Chemical Co. (U.S.A.); [^{14}C] spermine and $^{45}CaCl_2$ from Radiochemical Centre Amersham (U.K.), heparin-Sepharose CL-6B from Pharmacia Fine Chemicals (Sweden). All other chemicals used were of the highest purity available.

Methods

Unilamellar phospholipid vesicles were prepared by a published method (22) in 5 mM Mops buffer, pH 7. Phosphorus determination was by the method of Marinetti (23). The external/total ratio of the negative phospholipid in vesicles was determined by a published spectrophotometric method (24).

The binding of [^{14}C] spermine to phospholipid vesicles was determined by a cation-exchange-resin method previously described (4) and similar to the anion-exchange-resin method developed by Walaas (25). The binding of $^{45}Ca^{2+}$ to phospholipid vesicles was determined by a modification of the cation-exchange-resin method used for the study of [^{14}C] spermine binding: spermine was substituted by 7 μM Ca^{2+} in the presence of a double amount of heparin-Sepharose CL-6B. Equations and derivations described by Nakai and Glismann (26) were used to interpret our experimental data.

Free Ca^{2+} was determined by measuring the absorbance at 655 nm of the samples containing 50 μM arsenazo III in 10 mM Mops buffer, pH 7.4 following the method of Kendrick et al. (27). Arsenazo III has been previously purified using Chelex 100 ion-exchange-resin. Starting $CaCl_2$ concentration was 50 μM; aliquots of vesicle suspension were added in a 1 ml volume sample, and the dilutions were taken into account.

RESULTS AND DISCUSSION

To study the interaction of Ca^{2+} with phospholipid vesicles we utilized the dye arsenazo III. It has been demonstrated that this dye binds Ca^{2+} and the stoichiometry is one-to-one (27). The K_d of the complex is 0.74 x 10^{-6} M. The K_d is such that the dye does not significantly displace Ca^{2+} from complexes of possible physiological significance and can measure, quantitatively micromolar ionized Ca^{2+}

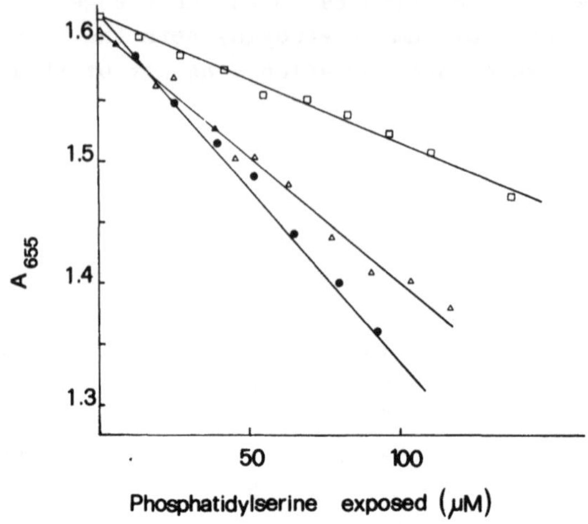

Fig. 1. Calcium binding to liposomes containing different percentages of PS.
Removal of Ca^{2+} from Ca^{2+}-arsenazo III complex by successive addition of 100% PS (●), 50% PS (△) and 25% PS (□) liposomes to the reaction mixture.

concentrations (27). Upon Ca^{2+} addition, the dye, forms a complex whose absorbance spectrum differs from that of the free dye. The new spectrum has a maximum at 655 nm whose intensity is directly related to the amount of Ca^{2+} added in free form to the solution. When a compound with an affinity for Ca^{2+} comparable to that of the dye is added to the solution, it competes with the dye for Ca^{2+} and removes it from the coloured complex. Increasing inhibition of Ca^{2+} binding to the dye can be seen as a function of the concentration of the compound and its affinity for Ca^{2+}.

Fig. 1 shows the effect of liposomes containing different percentages of PS, the reminder of the lipid being PC, on the Ca^{2+} arsenazo III complex. The different slope of the curves shows that the liposomes tested (100, 50, 25% PS) bind Ca^{2+} with different affinities. We have thus studied the effect of different concentrations of spermine on the binding of Ca^{2+} to both 100 and 25% PS liposomes (Fig. 2 a,b). In the presence of spermine (5 μM) the plot of the absorbance at 655 nm versus the amount of 100% PS vesicles added yields two linear portions. The first shows no change in absorbance, while the second is similar to the vesicle alone. Spermine apparently binds to 100% PS vesicles and masks the negative charges so that there is no removal of Ca^{2+} from the dye until an amount of PS vesicles in excess of the polyamine is added. At higher spermine concentrations a decrease in the initial absorbance of the Ca^{2+}-arsenazo III complex is observed as the polyamine binds the dye, although weakly. When the vesicles are added, the polyamine, by binding the phospholipid releases the dye and this results in an increased absorbance. However this phenomenon does not mask the titration of the polyamine by liposomes. A similar pattern is observed when spermine effect on Ca^{2+} binding to 25% PS vesicles is studied : proportionally increasing amounts of negative charges are blocked by the

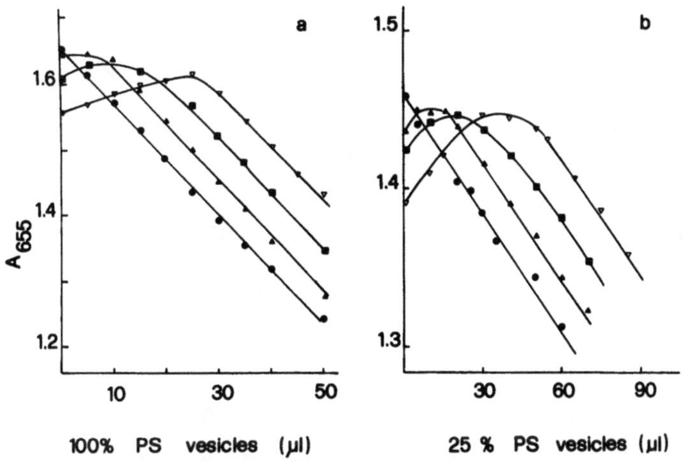

Fig. 2. Effect of spermine on calcium binding to PS containing liposomes.
Removal of Ca^{2+} from Ca^{2+}-arsenazo III complex by successive addition of either 100% PS liposomes (panel a) or 25% PS liposomes (panel b) to reaction mixtures containing 0 (●), 5 µM (▲), 10 µM (■) and 15 µM (▽) spermine.

polyamine resulting in a regular shift of the curves to the right. These results indicate that spermine binds PS liposomes with much higher affinity than Ca^{2+}. We have then tried to determine the formation constants of these complexes.

The ability of different phospholipid vesicles to complex either spermine or Ca^{2+} was inferred from their ability to remove these cations from their complexes with heparin-Sepharose. First we studied spermine binding to phospholipid vesicles containing different percentages of PS, the reminder of the lipid was PC. The liposomes tested (100, 50, 25% PS) showed different affinities for spermine as judged from the molarity values of acidic phospholipid causing the removal of 50% spermine bound to the resin (Fig.3a). The secondary plot of the data produced straight lines with a positive slope (Fig. 3b). These results indicate that spermine complexes with these vesicles successively to give rise to two different species. The interactions of the plots with the ordinate are similar and represent a measure of the formation constant (K_1) of the acidic phospholipid-spermine complex formed at low ratio of these reagents in the reaction mixture (4). The K_1 s amount to 5.9-6.5 x $10^4 M^{-1}$. The slope of the curves, by contrast, differs. As the slope represents a measurement of $K_1 . K_2$, the formation constant, K_2, appears to be decreased by the decreased density of PS in the liposome. K_2 is the formation constant of the complex formed at a high phospholipid/spermine ratio (4). The K_2 s calculated, are 3.3 x 10^5 M^{-1}, 2.2 x 10^5 M^{-1} and 1.6 x 10^5 M^{-1} for 100, 50 and 25% PS vesicles, respectively.

Another series of experiments was carried out to examine the effect of the density of PS in the liposomes on the binding of Ca^{2+}. As shown in Fig. 4a, the binding of Ca^{2+} to heparin-Sepharose was less and less

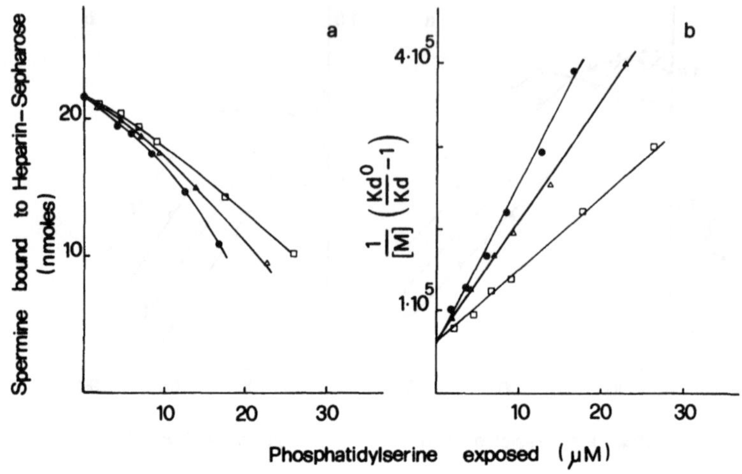

Fig. 3. Spermine binding to liposomes containing different percentages of PS.
a) Inhibition of spermine binding to heparin-Sepharose by increasing concentrations of 100% PS (●), 50% PS (△) and 25% PS (□) liposomes. b) Secondary plots of the data in (a).

affected as the PS density on the liposome decreased. The secondary plot of the data gave straight lines with a positive slope (Fig. 4b). The K_1s are similar and amount to 5.2-6.4 x 10^4 M^{-1} whereas the K_2 s significantly differ. The K_2 s calculated are 6.5 x 10^4 M^{-1}, 2.5 x $10^4 M^{-1}$ and 8.2 x 10^3 M^{-1} for 100, 50 and 25% PS vesicles, respectively.

Experiments were also conducted to assess the possibility that the lipophilic activators of PKC, i.e. diacylglycerol and phorbol ester, may

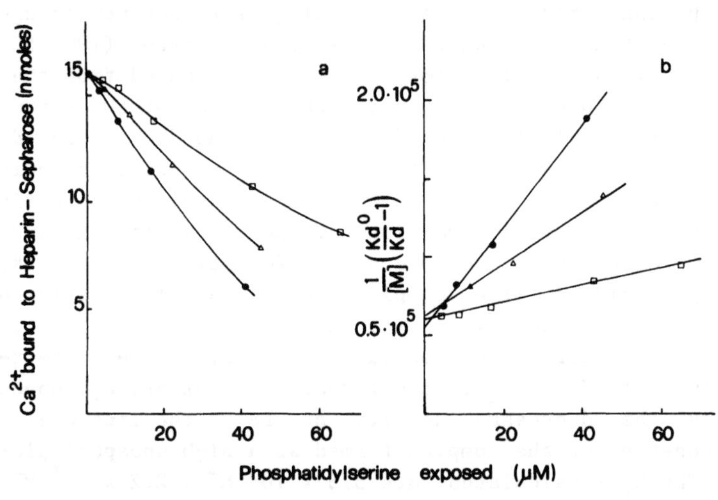

Fig. 4. Calcium binding to liposomes containing different percentages of PS.
a) Inhibition of calcium binding to heparin-Sepharose by increasing concentrations of 100% PS (●), 50% PS (△) and 25% PS (□) liposomes. b) Secondary plots of the data in (a).

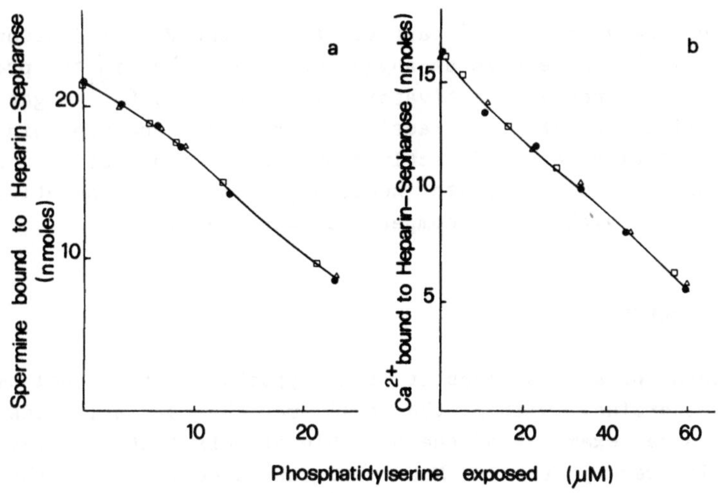

Fig. 5. Spermine and calcium binding to 50% PS liposomes. Inhibition of either spermine (panel a) or calcium (panel b) binding to heparin-Sepharose by increasing concentrations either of 50% PS liposomes in the absence (●) or presence of 100 nM PDBu (△) or of 50% PS liposomes containing 10% DAG (□).

modulate Ca^{2+} and/or spermine binding to the liposomes. In this study we utilized 50% PS liposomes. Fig. 5 a,b shows that the presence of 100 nM PDBu does not interfere in the removal of both spermine and Ca^{2+} from the resin by the 50% PS liposomes. Also 50% PS liposomes containing 10% DAG compete with the resin for spermine and Ca^{2+} in a manner indistinguishable from that of 50% PS liposomes.

The results clearly indicate that spermine interacts with PS-containing liposomes more strongly than Ca^{2+}. In the presence of these vesicles, spermine preferentially forms, with the acidic phospholipids, the complex characterized by the K_2 formation constant. Spermine, aligning parallel to the vesicle surface, causes a clustering of PS molecules. The decrease of PS density marginally affects the formation of this cluster. In contrast, only in the presence of 100% PS vesicles, the acidic phospholipid-Ca^{2+}complex preferentially formed is that characterized by the K_2 formation constant. At lower PS densities the K_2 s are significantly decreased, and thus Ca^{2+} preferentially binds to one head group. This is in agreement with results obtained by McLaughlin et al. (28) who showed that Ca^{2+} does bind to one head group, although the simultaneous existance of some 2:1 complexes, i.e. one Ca^{2+} complexed to two adjacent PS head groups was not excluded. However a clustering of negatively charged phospholipids by Ca^{2+}, resulting in phase separations, was observed only at very high concentrations of this cation (29).

These results suggest that Ca^{2+} binding to membranes carries out functions, i.e. dehydration of membrane surface (30), different from the possible clustering function of spermine. This hypothesis could explain the synergism of spermine and Ca^{2+} on liposome fusion. For this reason it seems unlikely that Ca^{2+} participates in the PKC activation process by clustering PS molecules as suggested (21) either in the presence or absence of the lipophilic activators of the enzyme. Ca^{2+} might directly bind PKC affecting its interactions with the phospholipid blayer. Spermine inhibition of PKC activation would thus still be due to sequestration of PS substrate necessary for the generation of PKC binding sites but without any competition with Ca^{2+}.

CONCLUDING REMARKS

The data presented support the hypothesis that polyamines, in particular spermine, within the cell may be strongly adsorbed to membranes. A few examples of the possible effects that they may exert in this location were presented. Many more can be derived from the relevant literature which is accumulating and it is imaginable that still many more will be found when further research is conducted on the influence of polyamines in the expanding field of the molecular biology of membranes.

REFERENCES

1. C.W.Tabor. The stabilizing effect of spermine and related amines on mitochondria and protoplasts, Biochem. Biophys. Res. Commun. , 2:117 (1960).
2. B. Tadolini, L. Cabrini, L. Landi, E. Varani and P. Pasquali. Polyamine binding to phospholipid vesicles and inhibition of lipid peroxidation. Biochem. Biophys. Res. Commun ., 122:550 (1984).
3. B. Tadolini, L. Cabrini, E. Varani, L. Landi, P. Pasquali and A.M. Sechi. Polyamine interactions with phospholipid vesicles, in: "Recent Progress in Polyamine Research" L. Selmeci, M.E. Brosnan, N. Seiler eds. Akademiai Kiado, Budapest p. 217 (1985).
4. B. Tadolini, L. Cabrini, E. Varani and A.M. Sechi. Spermine binding and aggregation of vesicles of different phospholipid composition, Biogenic Amines , 3: 87 (1985).
5. P. Meers, K. Hong, J. Bentz and D. Papahadjopoulos. Spermine as a modulator of membrane fusion: interaction with acidic phospholipids. Biochemistry , 25:3109 (1986).
6. B. Tadolini and E. Varani. Interaction of spermine with polyphosphoinositide containing liposomes and myo-inositol 1,4,5 triphosphate, Biochem. Biophys. Res. Commun. , 135:58 (1986).
7. L. Tilley, B.Tadolini and C. Oriol-Audit. Calcium- and polyamine induced aggregation of phosphatidylserine- and polyphosphoinositide - containing phospholipid vesicles. Biogenic Amines , 5:69 (1988).
8. M.W. Yung and C. Green. The binding of polyamines to phospholipid bilayers, Biochem. Pharmacol. , 35:4037 (1986).
9. L. Chung, G. Kaloyanides, R. McDaniel, A. McLaughlin and S.

McLaughlin. Interaction of gentamicin and spermine with bilayer membranes containing negatively charged phospholipids, Biochemistry , 24:442 (1985).
10. K. Hong, F. Schuber and D. Papahadjopoulos. Polyamines. Biological modulators of membrane fusion, Biochim. Biophys. Acta , 732: 469 (1983).
11. F. Schuber, K. Hong, N. Düzgüneş and D. Papahadjopoulos. Polyamines as modulators of membrane fusion: aggregation and fusion of liposomes. Biochemistry , 22:6134 (1983).
12. S. Ohki and J. Duax. Effects of cations and polyamines on the aggregation and fusion of phosphatidylserine membranes. Biochem. Biophys . Acta., 861:177 (1986).
13. B. Tadolini, E. Varani and L. Cabrini. The influence of poly(ethylene glycol) 6000 on spermine-induced aggregation of liposomes. Biochem. J. , 236:651 (1986).
14. K. Igarashi, I. Sakamoto, N. Goto, K. Kashiwagi, R. Honma and S. Hirose. Interaction between polyamines and nucleic acids or phospholipids. Arch. Biochem. Biophys. , 219:438 (1982).
15. H. Koenig, A. Goldstone and C.Y. Lu. Polyamines regulate calcium fluxes in a rapid plasma membrane response. Nature, 305:530 (1983).
16. C.V. Nicchitta and J.R. Williamson. Spermine. A regulator of mitochondrial calcium cycling. J. Biol. Chem. , 21:12978 (1984).
17. P. Palade. Drug-induced Ca^{2+} release from isolated sarcoplasmic reticulum. III Block of Ca^{2+}-induced Ca^{2+} release by organic polyamines. J. Biol. Chem ., 262:6149 (1987).
18. D. Qi, R.C. Schatzman, G.J. Mazzei, R.S. Turner, R.L. Raynor, S. Liao and J.F. Kuo. Polyamines inhibit phospholipid-sensitive and calmodulin-sensitive Ca^{2+} -dependent protein kinases. Biochem. J. , 213:281 (1983).
19. P. Thams, K. Capito and C.J. Hedeskov. An inhibitory role for polyamines in protein kinase C activation and insulin secretion in mouse pancreatic islets. Biochem. J. , 237:131 (1986).
20. M.S. Moruzzi, B. Barbiroli, M.G. Monti, B. Tadolini, G. Hakim and G. Mezzetti. Inhibitory action of polyamines on protein kinase C association to membranes. Biochem. J. , 247:175 (1987).
21. Y.A. Hannun, C. R. Loomis and R. M. Bell. Protein kinase C activation in mixed micelles. Mechanistic implications of phospholipid, diacylglycerol, and calcium interdependences. J. Biol. Chem. , 261:7184 (1986).
22. C. Huang. Studies on phosphatidylcholine vesicles. Formation and physical characteristics. Biochemistry , 8:344 (1969).
23. G.V. Marinetti. Chromatographic separation, identification and analysis of phosphatides. J. Lipid Res ., 3:1 (1962).
24. S. Massari, D. Pascolini and G. Gradenigo. Distribution of negative phospholipids in mixed vesicles. Biochemistry , 17: 4465 (1978).
25. E. Walaas. Stability constants of metal complexes with mononucleotides. Acta Chem. Scand ., 12:528 (1958).

26. C. Nakai and W. Glinsmann. Interactions between polyamines and nucleotides. Biochemistry , 16: 5637 (1977).
27. N.C. Kendrick, R.W. Ratzloff and M.P. Blaustern. Arsenazo III as an indicator of ionized calcium in physiological salt solutions: its use for determination of the CaATP dissociation constant, Anal. Biochem. , 83:433 (1977).
28. S. McLaughlin, N. Mulrine, T. Gresalfi, G. Vario and A.C. McLaughlin. Adsorption of divalent cations to bilayer membranes containing phosphatidylserine. J. Gen Physiol. , 77:445 (1981).
29. H.J. Galla and E. Sackmann. Chemically induced lipid phase separation in model membranes containing charged lipids: a spin label study. Biochim. Biophys. Acta. , 401:509 (1975).
30. A. Portis, C. Newton, W. Pangborn and D. Papahadjopoulos. Studies on the mechanism of membrane fusion: evidence for an intermembrane Ca^{2+} phospholipid complex, synergism with Mg^{2+}, and inhibition by spectrin. Biochemistry , 18:780 (1979).

BIDIRECTIONAL TRANSPORT OF SPERMINE ACROSS THE INNER MEMBRANE OF LIVER MITOCHONDRIA

A. Toninello, D. Siliprandi, P. Castagnini, M. C. Novello and N. Siliprandi

Centro Studio Fisiologia Mitocondriale CNR e
Dipartimento di Chimica Biologica 35131 Padova

INTRODUCTION

Polyamines as well as being involved in cellular growth and differentiation and in protein phosphorylation (1), are able to preserve the permeability properties of mitochondrial membranes.

Long ago Tabor (2) recognized that spermine and spermidine prevent the swelling of mitochondria suspended in a hypotonic medium and later Phillips and Chaffees(3) found that spermine prevents the loss of respiratory control in heat aged liver mitochondria and also restores the phosphorylative capacity of deenergized mitochondria.

Subsequently we found that spermine fully prevents the fall of membrane potential ($\Delta\psi$), the release of endogenous Mg^{2+} and adenine nucleotides and the efflux of accumulated Ca^{2+} induced in liver mitochondria exposed to damaging concentration of Ca^{2+} and phosphate (4). Furthermore spermine fully restores membrane potential of deenergized mitochondria provided that ATP is also added. (4). Similarly spermine protects the energy linked processes of mitochondria exposed to hydroperoxides. Since the damaging action of these compounds is supressed either by removing Ca^{2+} from the medium by EGTA or preventing its transport by ruthenium red, the protective and restorative action of spermine may be referred also in the case of hydroperoxides to a control on Ca^{2+} flux across the inner membrane.

Indeed Akerman has observed that spermine inhibits Ca^{2+} uptake by interferring with Ca^{2+} binding to low affinity sites of mitochondrial membrane (5).

Conversely Nicchitta and Williamson (6) found that spermine by acting on both the Ca^{2+} uniporter and the ruthenium red insensitive Ca^{2+} efflux system, accelerates the rate of Ca^{2+} cycling in liver mitochondria. At the same conclusion arrived Lenzen et al. (7) showing that in liver, heart and brain mitochondria spermine decreases the Km for the Ca^{2+} uniporter.Owing to this influence on Ca^{2+} transport spermine may be an important factor in the homeostasis of Ca^{2+} in both extra and intramitochondrial space.

Another interesting function of spermine and spermidine is the stimulation of the import in rat liver mitochondria of extramitochondrial protein, namely the precursor of ornitine carbamoyl-transferase (8).

The action of polyamines on mitochondrial function has generally been attributed to their capability to bind to membrane anionic sites. It appeared infact unlikely that strongly cationic compounds might permeate into the matrix space (9). Recently we have shown that at physiological concentrations spermine may be transported into the inner mitochondrial space provided that transmembrane potential ($\Delta\Psi$) is increased above a critical level.

The conditions allowing bidirectional spermine transport across the inner mitochondrial membrane are described in the present report.

SPERMINE INFLUX

We have previously demonstrated that spermine is transported inside the mitochondria in a process driven by respiration provided that phosphate is present in the medium (10). Uncouplers or respiratory chain inhibitors prevent spermine uptake and induce the release of much of the accumulated oligoamine. The influx of spermine shows a saturation kinetics with an apparent Km of 0.5 mM, a value in the range of cytosolic concentration of the oligoamine. Neither Ca^{2+} addition, nor mitochondrial Ca^{2+} chelation by EGTA affected the rate and the extent of spermine uptake. Moreover ruthenium red, inhibitor of Ca^{2+} uniporter, was also without effect on spermine influx, thus indicating that Ca^{2+} movements do not affect oligoamine transport. On the contrary magnesium ions inhibit spermine uptake and, if added to mitochondria after spermine has been taken up, induce a partial release of accumulated oligoamine (10).

Spermine uptake by respiring mitochondria is dependent on Pi and the uptake of Pi is enhanced by the concurrent transport of the oligoamine (10). Considering that spermine under physiological conditions is largely protonated, the effect of Pi should be attributed to the generation of intramitochondrial negativity which facilitates the electrophoretic influx of the oligoamine. In turn the entrance of spermine in the matrix space decreases $\Delta\Psi$ and shifts the protonmotive force to a higher Δ pH, thus favouring Pi influx.

That spermine uptake is driven by the increased $\Delta\Psi$ is also confirmed by addition of acetate, another permeant anion, capable to increase, even at a lower extent than Pi, transmembrane potential (11). The same result has been obtained with nigericin in the absence of any permeant anion. Indeed acetate, Pi, nigericin share the common effect to increase $\Delta\Psi$ above the maximum values obtained with succinate energization (from 180 mV to 190,200, 210 mV respectively).

The time course of spermine uptake by rat liver mitochondria at different gradients of membrane electrical potential is shown in fig. 1. The results clearly indicate the sharp dependency of spermine uptake on $\Delta\Psi$. As shown in the insert of fig. 1 the rates of spermine uptake plotted against $\Delta\Psi$ exhibit a non linear relationship. This behaviour falls under the rubric of non ohmic conductance, described in mitochondria by Nichols (12). It may be observed that spermine uptake by RLM is governed by a narrow range of $\Delta\Psi$ (from 180 to 210 mV).

It may be concluded that spermine is transported electrophoretically into the mitochondrial matrix provided that a very high transmembrane potential (above 180 mV) is achieved. A similar behaviour has been demonstrated for inorganic monovalent cations (13).

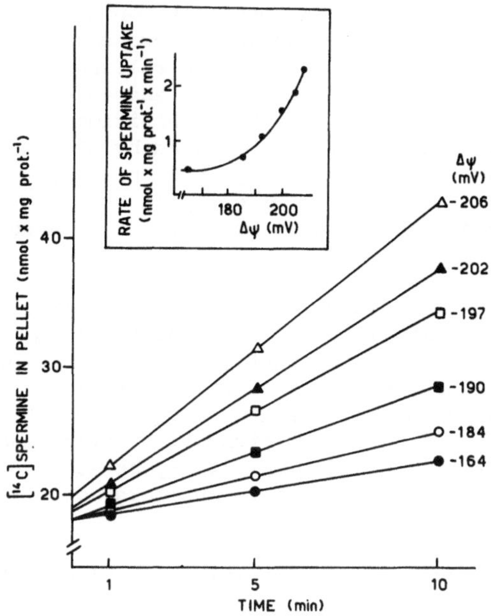

Fig. 1. Dependence of spermine uptake on ΔΨ.
Rat liver mitochondria (1 mg prot./ml) were incubated at 20° C in the standard medium (200 mM sucrose, 10 mM Hepes (pH 7.4), 5 mM sodium succinate and 1.25 μM rotenone) containing 1 mM sodium phosphate and 1 mM [^{14}C] spermine. ΔΨ was manipulated (13) by including limiting amounts of FCCP (0-60 nM) and was monitored with the TPP^+ selective electrode. The highest values of ΔΨ were obtained by adding nigericin (0.33 μg/mg).

SPERMINE EFFLUX

In physiological conditions in order to avoid limitless accumulation of spermine a separate mechanism for its release from the matrix space is required. Infact it is highly improbable that mitochondria could utilize for spermine extrusion the same pathway of its influx: the high negative membrane potential would prevent the outward diffusion.

When mitochondria preloaded with [^{14}C] spermine and Pi are suspended in a new standard medium a rapid, partial release of accumulated spermine is observed (fig. 2). Since this release is enhanced by external spermine

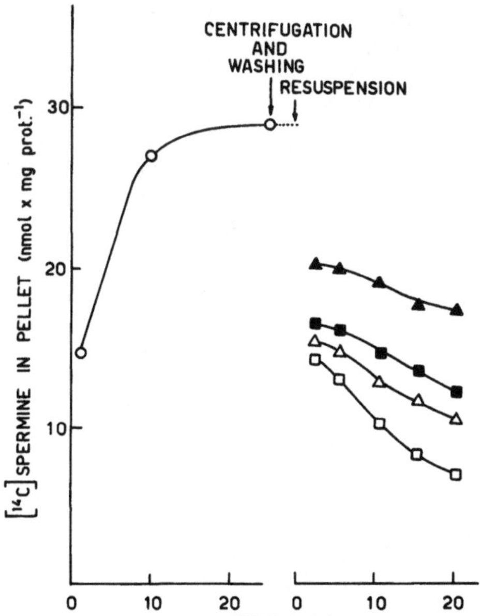

Fig. 2. Exchange between accumulated and external spermine.
Rat liver mitochondria (1 mg prot./ml) were incubated in the standard medium containing 1 mM [^{14}C] spermine and 1 mM Pi. After 25 minutes of incubation, mitochondrial suspensions were centrifuged at 3° C for 2 minutes at 17,300 x g. The supernatants were discarded, and the pellets were washed three times with cold standard medium and resuspended at 25° C in standard medium (■-■), in standard medium plus 1 mM Pi (▲-▲), in standard medium plus 1 mM Pi and 1 mM spermine (△-△) and in standard medium plus 1 mM spermine (□-□)

and it is also observed in uncoupled mitochondria, it probably represents loss of [^{14}C]spermine bound to the surface of mitochondrial membranes. In the energized mitochondria this nearly instantaneous spermine release was followed by a slow lasting release which was inhibited by Pi and enhanced by the presence of spermine in the incubation medium. The inhibition by Pi could be explained by the higher electronegativity produced by the anion inside the mitochondria. The increased electronegativity would prevent the efflux of positively charged spermine. The higher rate of spermine efflux induced by external spermine might reflect an exchange between accumulated and external spermine.

Fig. 3 shows that addition of N-ethilmaleimide (NEM) to mitochondria loaded with spermine and Pi induces a concomitant efflux of the two ions. Furthermore addition of NEM decreased $\Delta\Psi$ and concurrently increased ΔpH (results not reported). Indeed by blocking the influx of Pi on the Pi/H^+ symporter, NEM prevents the attainment of the high $\Delta\Psi$ consequent to the entry of Pi into the mitochondrial membrane. As the protonmotive force ($\Delta\mu H^+$) did not result significantly changed after NEM addition, it may be assumed that spermine is transported outwards the mitochondria in an energy dependent process driven by ΔpH.

The same results were obtained by addition of mersalyl which blocks the Pi transporter and the Pi/dicarboxilate antiporter. In this case it was possible to induce the reuptake of both the released compounds by addition of dithioerythritol (DTE). This reagent, by removing mersalyl, allows a full restoration of the conditions promoting spermine influx (data not shown).

The energy dependent influx and efflux of spermine across the inner

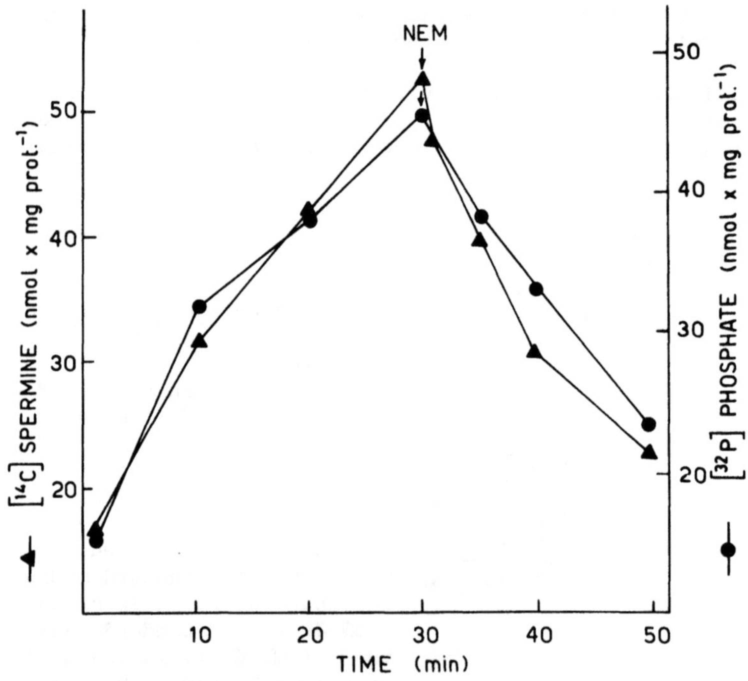

Fig. 3. Efflux of spermine induced by NEM.
Rat liver mitochondria (1 mg prot./ml) were incubated in the standard medium containing 1 mM [^{32}P] Na-phosphate and 1 mM [^{14}C] spermine, 50 µM NEM was added.

mitochondrial membrane are consequently governed by a balance between the two components of the protonmotive force: $\Delta\Psi$ and ΔpH. A high $\Delta\Psi$ drives spermine uptake, a high ΔpH polyamine efflux.

The efflux of spermine might occur electroneutrally concurrently with negatively charged Pi. In this regard it has to be mentioned that Pi might be transported electrophoretically via the inner membrane anion channel described by Garlid (14) which unlike the classical Pi/H^+ symporter is activated and not inhibited by thiols reagents (15,16). Alternatively, an electroneutral release of spermine from mitochondria may be explained by the presence of a cation/H^+ antiporter responsive to ΔpH; in this case spermine can be placed in the context of other cations undergoing uniport/antiport cycling (17).

All together the results reported in the present paper indicate that spermine is transported bidirectionally across the inner mitochondrial membrane. Both the inward and outward transports are energy linked, depending the former on mitochondrial $\Delta\Psi$ and the latter on mitochondrial ΔpH.

REFERENCES

1. C. W. Tabor and H. Tabor, Polyamines, Ann. Rev. Biochem. 53:749 (1984).
2. C. W. Tabor, The stabilizing effect of spermine and related amines on mitochondria and protoplasts, Biochem. Biophys. Res. Comm. 2:117 (1960).
3. J. E. Phillips and R. R. J. Chaffee, Restorative effects of spermine on oxidative phosphorylation and respiration in heat aged mitochondria, Biochem. Biophys. Res. Comm. 108:176 (1982).
4. A. Toninello, F. Di Lisa, D. Siliprandi and N. Siliprandi, Protective and restorative effects of spermine on the membrane potential of rat liver mitochondria, in: "Advances in Polyamines in Biomedical Science", C. M. Caldarera and U. Bachrach, ed., CLUEB, Bologna (1984)
5. E. O. Akerman, Effect of Mg^{2+} and spermine on the kinetics of Ca^{2+} transport in rat liver mitochondria, J. Bioenerg. Biomembr. 9:65 (1977).
6. C. V. Nicchitta and J. R. Williamson, Spermine a regulator of mitochondrial calcium cycling, J. Biol. Chem. 259:12978 (1984).
7. S. Lenzen, R. Hickethier and U. Panten, Interactions between spermine and Mg^{2+} on mitochondrial Ca^{2+} transport, J. Biol. Chem. 261:16478 (1986).
8. C. Gonzalez- Bosch, V. J. Miralles, J. Hernandez-Yago and S. Grisolia, Spermidine and spermine stimulate the transport of the precursor of ornithine carbamoyltransferase into rat liver mitochondria, Biochem. Biophys. Res. Comm. 149:21 (1987).
9. A. Huunan-Seppala, Cation permeability induced by spermine and polybrene in rat liver mitochondria, J. Bioenerg. Biomembr. 2:197 (1971).
10. A. Toninello, F. Di Lisa, D. Siliprandi and N. Siliprandi, Uptake of spermine by rat liver mitochondria and its influence on the transport of phosphate, Biochim. Biophys. Acta 815:399 (1985).
11. A. Toninello, G. Miotto, D. Siliprandi, N. Siliprandi and K. D. Garlid, On the mechanism of spermine transport in liver mitochondria, J. Biol. Chem. submitted (1988).
12. D. G. Nicholls, The influence of respiration and ATP hydrolysis on the proton-electrochemical gradient across the inner membrane of rat liver mitochondria as determined by ion distribution, Eur. J. Biochem. 50:305 (1974).
13. G. C. Brown and M. D. Brand, Changes in permeability to protons and other cations at high proton motive force in rat liver mitochondria, Biochem. J. 234:75 (1986).
14. K. D. Garlid and A. D. Beavis, Evidence for the existence of an inner

membrane anion channel in mitochondria, Biochim. Biophys. Acta 853: 187 (1987).

15. K. M. Scott, V. A. Knight, C. T. Settlemire and G. P. Brierley, Differential effects of mercurial reagents on membrane thiols and on the permeability of the heart mitochondria, Biochem. 9:714 (1970).

16. K. Bogucka and L. Wojtczak, On the mechanism of mercurial-induced permeability of the mitochondrial membrane to K^+, FEBS Lett. 100:301 (1979).

17. K. D. Garlid, in: "Integration of mitochondrial function", J. J. Lemasters, C. R. Hackenbrock, R. G. Thurman and H. V. Westerhoff, ed., Plenum Publ. Corp., New York, in the press.

POLYAMINES STIMULATE THE "IN VITRO" TRANSPORT OF THE PRECURSOR OF ORNITHINE CARBAMOYLTRANSFERASE INTO RAT LIVER MITOCHONDRIA

José Hernández-Yago, Carmen González-Bosch, Vicente J. Miralles, María-Jesús Marcote and Santiago Grisolía

Instituto de Investigaciones Citológicas de la Caja de Ahorros de Valencia
Amadeo de Saboya, 4. 46010-Valencia. Spain

INTRODUCTION

Most mitochondrial proteins are coded by nuclear genes, synthesized outside mitochondria, and imported into these organelles in a post-translational event (1). The import process usually involves the following steps: 1) synthesis of a precursor polypeptides with an amino-terminal extension (signal peptide), on free cytoplasmic polysomes, 2) binding of the precursor to a receptor-like component on the mitochondrial surface, 3) energy-dependent translocation across both mitochondrial membranes, possibly through "contact sites" between outer and inner membranes, 4) removal of the amino-terminal extension by a chelator-sensitive protease in the mitochondrial matrix, and 5) mature proteins location into their functional sites. Exceptions to this sequence of events include apocytochrome c and all proteins of the outer membrane which are synthesized without a leading peptide and transported into mitochondria without need of a membrane potential.

Mitochondrial mature proteins turn over with characteristic half-lives. The life cycle of a mitochondrial protein of cytosolic origin is depicted in Fig. 1. Since during steady state conditions the overall concentration of protein per cell volume remains essentially constant, protein synthesis and degradation must be closely interrelated. We have for some time been interested in clarifying the factors that regulate the turnover of mitochondrial proteins; we have evidence that their concentration in mitochondria may be regulated by the import of their precursors from the cytosol. We are currently studying the "in vitro" transport of ornithine carbamoyltransferase (OCT) and this paper presents some of our recent findings.

SYNTHESIS "IN VITRO" OF THE PRECURSOR OF ORNITHINE CARBAMOYLTRANSFERASE (pOCT)

A main difficulty to work with the cytosolic precursor of ornithine carbamoyltransferase (pOCT), a mitochondrial matrix protein of the urea cycle, is that its half-life in the cytosol is very short ($<$5 min.) and therefore, its concentration in tissues is extremely low (2). This prevents its isolation from the cytosol. To obtain large amounts of pOCT we have used the specific rat liver cDNA (a gift of Dr. G. Shore,

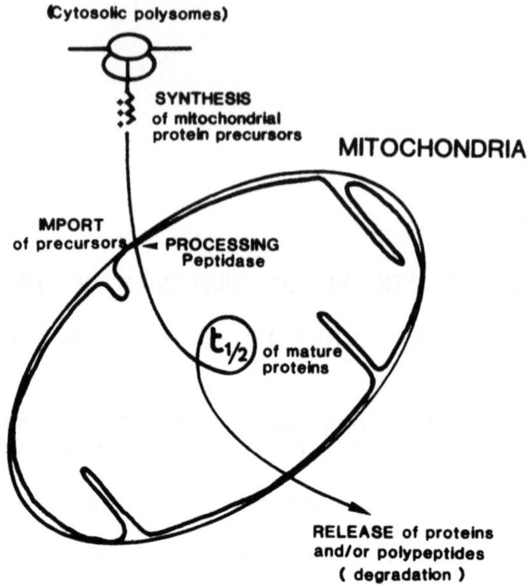

Fig. 1. Cycle of a mitochondrial protein of cytosolic origin.

Mc Gill University/Montreal). This cDNA is cloned in the transcription SP6 system, and the specific mRNA obtained is translated in an "in vitro" rabbit reticulocyte lysate system, consisting of 5 µl of a nuclease treated rabbit reticulocyte lysate plus a translation mixture, in a final volume of 14.25 µl, containing 22 mM Hepes pH 7.6, 76 mM KCl, 0.6 mM magnesium acetate, 0.7 mM ATP, 0.3 mM GTP, 15 mM creatine phosphate, 1.5 mg/ml creatine phosphokinase, 2.2 mM dithiothreitol, 0.5 mM spermidine, 27 µM each of an aminoacid mixture minus methionine and 20 µCi of (^{35}S)methionine and ca. 25 ng pOCT-mRNA. Incubation was carried out at 30 ºC, for 60 min., and at the end of the reaction, the mixture was made to contain 12 µg/ml of cycloheximide.

The major polypeptide synthesized by this system is the precursor of OCT, that can be distinguished from its mature form by its different molecular weight, due to the presence of the signal peptide. This is shown in Fig. 2.

PREVIOUSLY REPORTED FACTORS FOR THE "IN VITRO" TRANSPORT OF MITOCHONDRIAL pOCT

It has been reported that transport of pOCT into mitochondria requires, in addition to energized mitochondria, soluble cytosolic factor(s) present in the reticulocyte lysate, and low molecular weight components from the mixture used for the "in vitro" synthesis of protein precursors (3,4). Miura et al. (3) reported that magnesium and potassium ions and a trypsin-sensitive, heat-unstable and no dyalizable component of the reticulocyte lysate, were required for the transport. Also, Argan et al. (4), showed that a stimulatory factor for the transport was retained by Sephadex G-25. On the other hand, Firgaira et al. (5) indicated the possible implication of an RNase-sensitive component of the reticulocyte lysate. However, there is some controversy regarding the number and nature of the factors present in the transport medium, involved in the transport of pOCT into mitochondria.

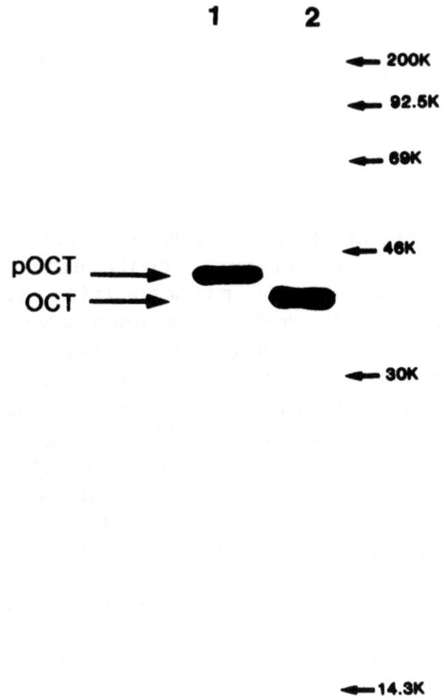

Fig. 2. "In vitro" synthesis of rat liver pOCT. Lane 1: "in vitro" products synthesized in a cell-free system, by translation of the pOCT mRNA transcript; Lane 2: (^{14}C)methylated rat liver OCT.

TRANSPORT EXPERIMENTS

The standard transport medium used in our experiments is described in the legend of Fig. 3. Under our conditions we confirmed that components of both, reticulocyte lysate and translation mixture, are required for pOCT transport. As shown in Fig. 3, the absence of the components of the translation mixture (Lanes 1,2), decreases the binding of pOCT to mitochondria. Their presence results in a higher binding of pOCT (Lanes 5,6), but they are not enough to permit transport, which requires also the presence of reticulocyte lysate (Lanes 3,4). The different mobilities of the bands in lanes 2, 4 and 6 with respect to lanes 1, 3 and 5 are due to the different amount of mitochondrial protein; this does not affect the transport efficiency.

POLYAMINES STIMULATE THE "IN VITRO" TRANSPORT OF pOCT INTO MITOCHONDRIA

Polyamines plus Reticulocyte Lysate

The analysis of the effect of the different components of the translation mixture, on the transport of pOCT into mitochondria, in the presence of reticulocyte lysate, revealed that: a) Spermidine at 0.5 mM has a stimulatory effect on the transport of pOCT into mitochondria (Fig. 4, lane 2); b) Potassium plus magnesium ions stimulate this process (Fig. 4, lane 6), as was described by Miura et al. (3); and

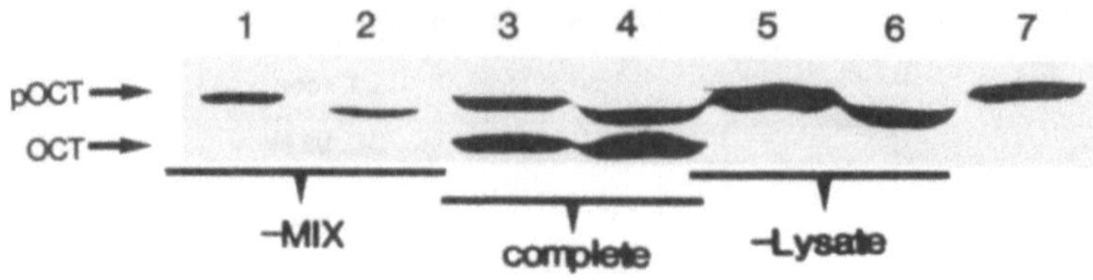

Fig. 3. Requirement of translation medium for pOCT transport.
2 µl of the translated mixture (containing pOCT), were incubated with different amounts of mitochondria, in a final volume of 100 µl. The transport medium we used consisted of: 5 mM Hepes pH 7,6, 0.25 M sucrose an other components as indicated. After incubation for 30 min, at 30 ºC, mitochondria were sedimented. The transport was estimated by analyzing the products associated to mitochondria by SDS-polyacrilamide gel electrophoresis and fluorography. Lane 1: rabbit reticulocyte lysate and 200 µg of mitochondrial protein. Lane 2: as lane 1 but 400 µg of mitochondrial protein. Lane 3: rabbit reticulocyte lysate, the translation mixture (except ^{35}S-methionine, pOCT-mRNA and lysate) and 200 µg of mitochondrial protein. Lane 4: as lane 3 but 400 µg of mitochondrial protein. Lane 5: as in lane 3 minus reticulocyte lysate. Lane 6: as lane 5 but 400 µg of mitochondrial protein. Lane 7: pOCT synthesized "in vitro".

c) the other components - ATP, GTP, creatine phosphate, creatine phosphokinase (Fig. 4, lane 3), dithiothreitol (Fig. 4, lane 4), and aminoacid mixture (Fig. 4, lane 5)- at the same concentrations as in the translation mixture, have no effect on the transport of pOCT.

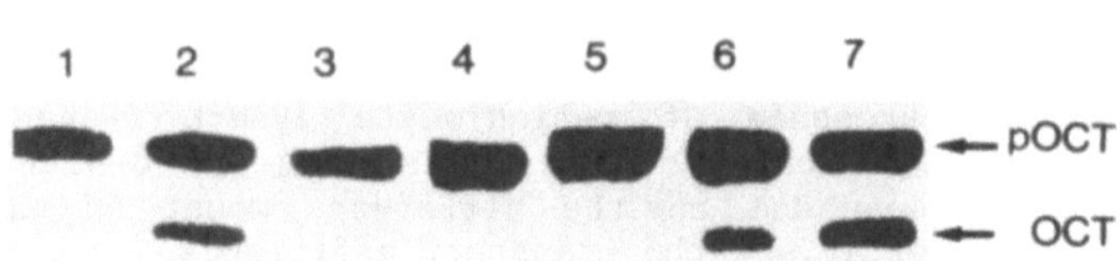

Fig. 4. Spermidine stimulates the transport of pOCT into mitochondria.
2 µl of the cell-free translated mixture (containing pOCT) were incubated in a final volume of 100 µl, with 400 µg of mitochondrial protein, 20 µl of reticulocyte lysate (145 mg/ml) and other additions as indicated. Lane 1: no additions. Lane 2: 0.5 mM spermidine. Lane 3: 0.7 mM ATP, 0.3 mM GTP, 15 mM creatine phosphate, 1.5 mg/ml creatine phosphokinase. Lane 4: 2.2 mM dithiothreitol. Lane 5: 27 µM aminoacid mixture. Lane 6: 0.6 mM magnesium acetate and 76 mM potassium chloride. Lane 7: the translation mixture components. The gel was dried and treated for fluorography.

This is the first time that polyamines have been implicated in the transport of a mitochondrial protein precursor into mitochondria (6). For this reason we were interested in clarifying the stimulatory effect of spermidine on the transport of pOCT. We studied whether this effect was dependent on the concentration of spermidine in the transport medium, and also whether the transport was stimulated by other related polyamines. As it is shown in Fig. 5., both binding and transport of pOCT increased as the concentration of spermidine increased till it reached 4mM, in the presence of reticulocyte lysate. Addition of this polyamine at higher concentration (20 mM), completely inhibited the transport although higher binding of pOCT to mitochondria was observed.

Both binding and transport of pOCT were also stimulated in a concentration-dependent manner, to a maximum at 1 mM spermine. Indeed, at 4 mM, spermine markedly stimulated the binding of pOCT to mitochondria, but transport was inhibited. When spermine was used at 20 mM, binding and transport were completely inhibited.

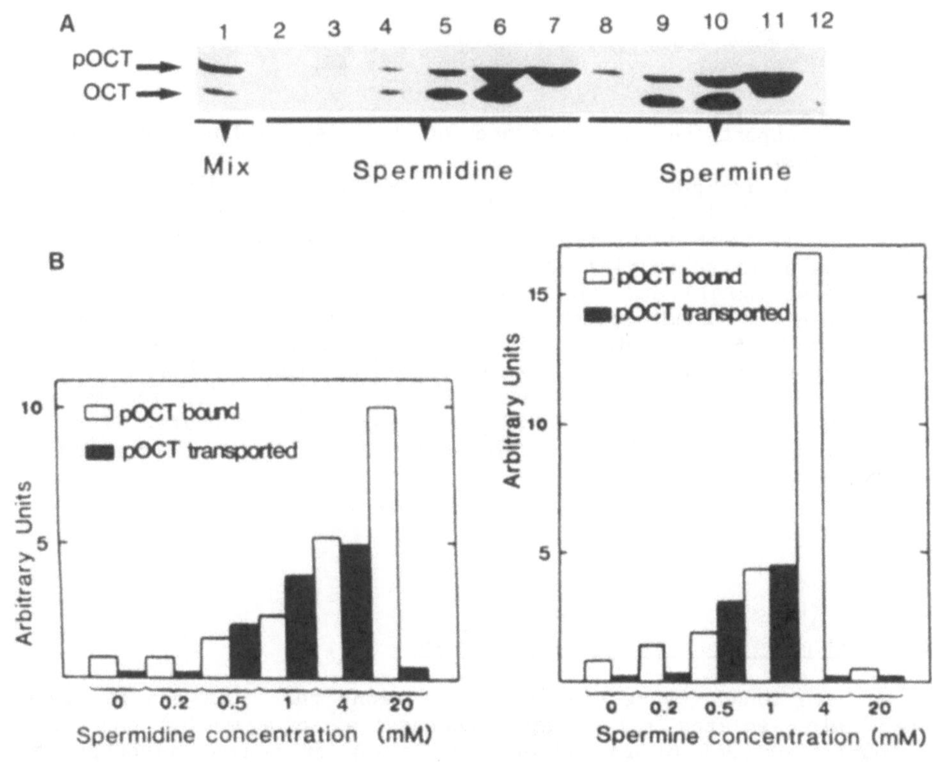

Fig. 5. Effect of different concentrations of polyamines on pOCT transport

Panel A: 2 µl of the translated mixture (containing pOCT) were incubated in a final volume of 100 µl, with 400 µg of mitochondrial protein, 20 µl of rabbit reticulocyte lysate (145 mg/ml) and other additions as indicated. Lane 1: the translation mixture components. Lane 2: no additions. Lanes 3-7: 0.2, 0.5, 1, 4 and 20 mM spermidine. Lanes 8-12: 0.2, 0.5, 1, 4 and 20 mM spermine. The gel was dried and treated for fluorography. Panel B: the binding and transport of pOCT into mitochondria, presented in the Panel A, was quantitated densitometrically.

The inhibitory effect at high concentrations of spermine and spermidine has been described also in other processes regulated by polyamines (7). The mechanism of the inhibition it is not known.

Since magnesium ions stimulated pOCT transport in the presence of lysate, and it has been reported that magnesium ions and polyamines are both required in some cellular functions (8), we studied the effect of different concentrations of these ions on the transport of pOCT in the presence and absence of spermidine. Increasing concentrations of magnesium ions augmented the stimulatory effect on the pOCT transport (Fig. 6, lanes 8-10). However, the stimulatory effect of spermidine at 4 mM was inhibited by the addition of magnesium ions, in the range of concentrations used (Fig. 6, lanes 3,4), suggesting, therefore, an antagonistic effect, as it has been already reported for other metabolic processes, such as cell-free protein synthesis (9-11).

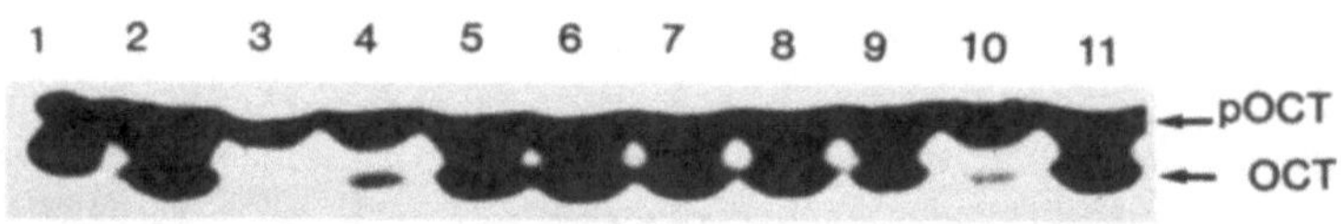

Fig. 6. Transport in the presence of both spermidine and magnesium ions.
Lane 1: pOCT synthesized "in vitro" and (^{14}C)-labeled OCT. Lanes 2-11: 2 µl of the cell-free translated mixture (containing pOCT) were incubated, in a final volume of 100 µl, with 400 µg of mitochondrial protein, 20 µl of rabbit reticulocyte lysate (145 mg/ml) and additions as follow. Lane 2: 1 mM spermine. Lane 3: 2 mM spermidine and 1.5 mM magnesium acetate. Lane 4: 2 mM spermidine and 0.75 mM magnesium acetate. Lane 5: 4 mM spermidine. Lane 6: 2 mM spermidine. Lane 7: 1 mM spermidine. Lane 8: 3 mM magnesium acetate. Lane 9: 1.5 mM magnesium acetate. Lane 10: 0.75 mM magnesium acetate. Lane 11: translation mixture. The gel was dried and treated for fluorography.

<u>Polyamines plus rat liver cytosolic extract</u>

We have found (manuscript in preparation) that pOCT can be transported into isolated rat liver mitochondria in a medium containing a cytosolic extract of rat liver, consisting of a post mitochondrial supernatant (PMS). This is a more physiological medium than the commonly used rabbit reticulocyte lysate; it also requires the presence of low molecular weigth components of the translation mixture (Fig. 7). The stimulatory activity of the cytosol on the import is inhibited by trypsin treatment, slightly decreased by RNase treatment but is not lost by dyalisis or by heating at 100 ºC.

It was found that the addition of spermidine and spermine, at physiological concentrations, to the cytosolic extract, show a stimulatory effect on transport similar to that reported above using a medium containing reticulocyte lysate.

Polyamines in the absence of both reticulocyte lysate and cytosolic extract

The effect of polyamines by themselves on the transport of pOCT was tested. Both, the binding and the transport of pOCT were increased, in a concentration-dependent manner, when spermidine was added to the transport medium in the absence of reticulocyte lysate or rat liver cytosol (Fig. 8). The presence of higher than physiological concentrations, in this case, does not produce an inhibitory effect on the transport of pOCT. Under the same conditions, spermine and putrescine alone, also stimulate transport.

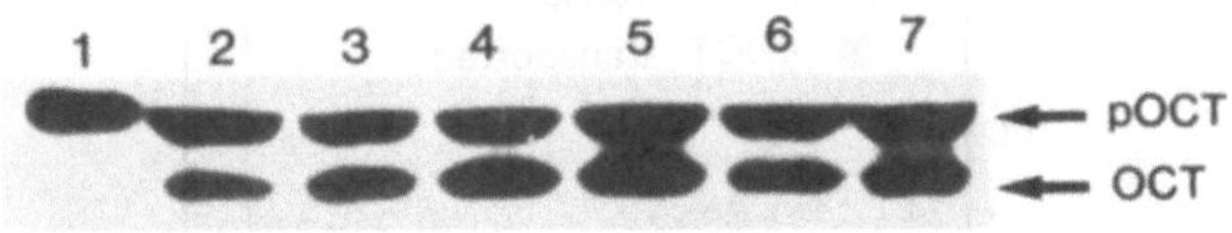

Fig. 7. Import of pOCT can be carried out with the cytosol of rat liver.
2 µl of the cell-free translated mixture (containing pOCT) were incubated with 400 µg of mitochondrial protein, the translation mixture and differents amounts of post-mitochondrial supernatant (40 mg/ml of protein) or reticulocyte lysate (145 mg/ml protein), in a final volume of 100 µl. Lane 1: "in vitro" synthesyzed pOCT. Lane 2: 1.45 mg of rabbit reticulocyte lysate. Lane 3: as in lane 2 but 2.90 mg of rabbit reticulocyte lysate protein. Lane 4: as in lane 2 but 4.35 mg of rabbit reticulocyte lysate protein. Lanes 5-7: as in lanes 2-4 but 0.4, 0.8 and 1.2 mg, respectively, of post--mitochondrial supernatant protein instead of rabbit reticulocyte lysate. The gel was dried and treated for fluorography.

These results indicate that polyamines are enough to permit the "in vitro" transport of pOCT. Obviously, other cytosolic factors may be also involved which, perhaps, modulate their stimulatory effect.

Ultrastructural Changes in Isolated Mitochondria by Polyamines

In order to establish possible mechanisms of action of these polyamines on the transport of pOCT, we have studied the ultrastructure of mitochondria by electron microscopy, incubated in the presence and absence of polyamines. Mitochondria incubated in the presence of polyamines show an orthodox configuration, with the outer and inner membranes closer with respect to mitochondria that are incubated in the absence of polyamines Fig. 9, (upper). In the last conditions, mitochondria show a condensed configuration and an evident separation between both membranes (Fig. 9, lower).

These results suggest that the polyamines reduce the distance between outer and inner membranes, facilitating the translocation of pOCT across both membranes. An interesting possibility is that polyamines can affect the "contact sites" between both membranes (12,13), proposed as possible routes for transport of mitochondrial protein precursors across mitochondrial membranes (14).

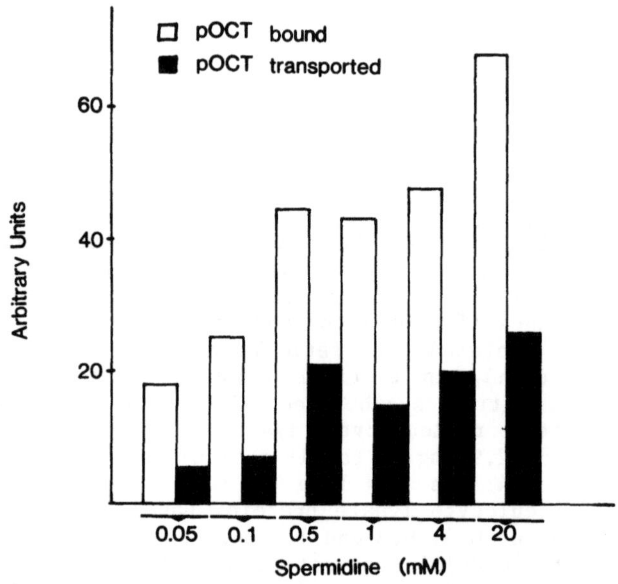

Fig. 8. Polyamines alone stimulate the transport of pOCT. 2 µl of the translated mixture (containing pOCT) were incubated with 400 µg of mitochondrial protein, in a final volume of 100µl. The transport medium contains 2.5 mM Hepes pH 7.4, 0.25 M sucrose and different amounts of spermidine (0.05, 0.1, 0.5, 1, 4 and 20 mM). The mitochondrial pellet was analyzed by SDS-polyacrilamide gel electrophoresis and further fluorography. Finally, the binding and transport of pOCT were quantitated densitometrically.

DISCUSSION

It is not known at present what is the mechanism of action of polyamines on the "in vitro" transport of pOCT. Possibly polyamines

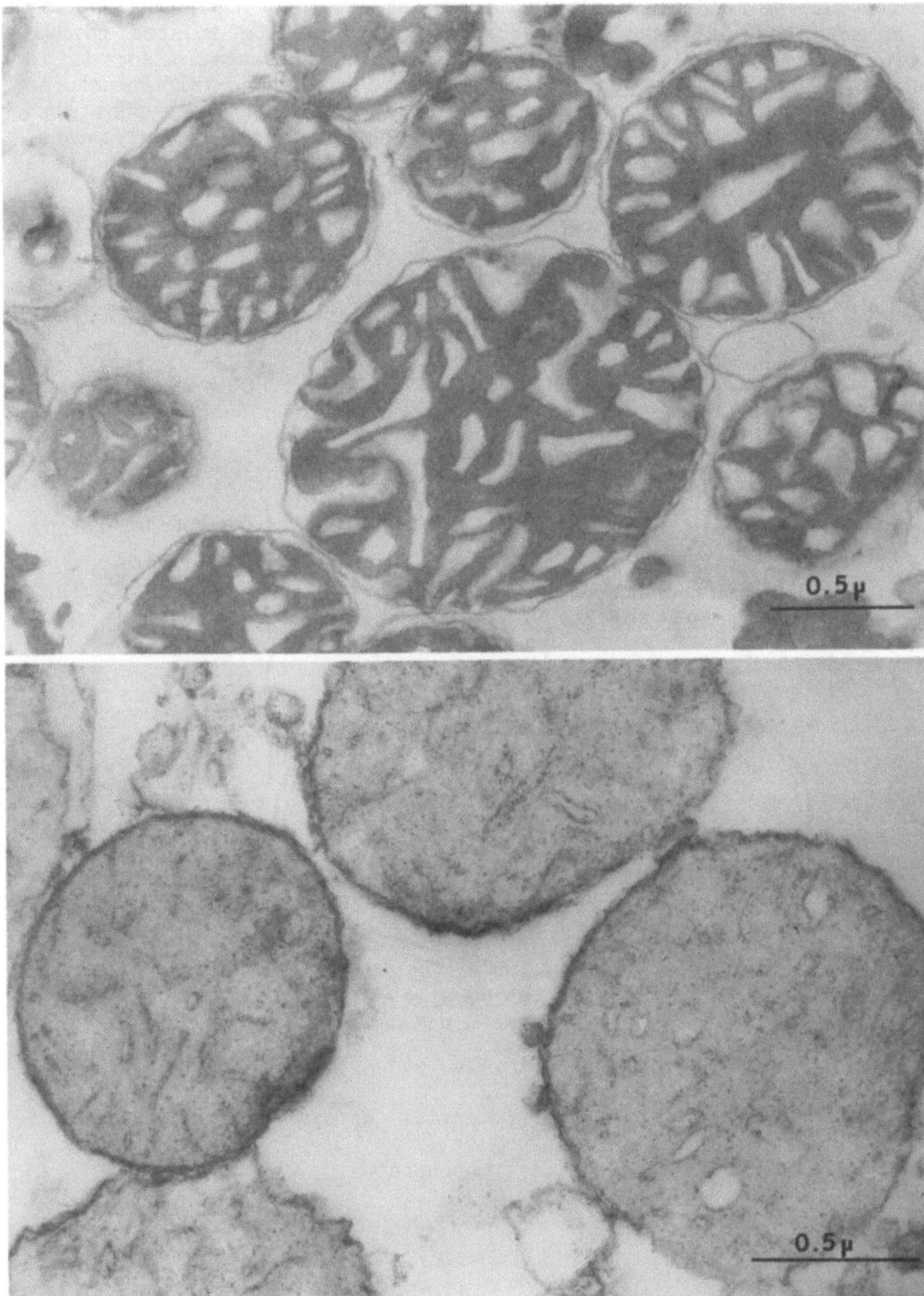

Fig. 9. Electron micrographs of isolated rat liver mitochondria incubated for 30 min at 30 ºC in a transport medium containing 2.5 mM Hepes pH 7.4, o.25 M sucrose in the absence (upper) or presence (lower) of 4 mM spermidine.

act at different levels in the pOCT transport process, affecting the precursor polypeptide, favouring a conformation competent for the transport, and/or acting on the mitochondria. In this sense the amphiphilic nature of these molecules, similar to the structure of the signal peptide of mitochondrial protein precursors (15), may be related to the translocation of the polypeptide across both membranes. Also, it is possible that these molecules affect some mitochondrial structures related to the binding or translocation of pOCT. The fact that both mitochondrial membranes are closer in the presence of polyamines, suggest that the formation or stabilization of "contact sites" by these molecules is responsible for their effect on the transport of pOCT.

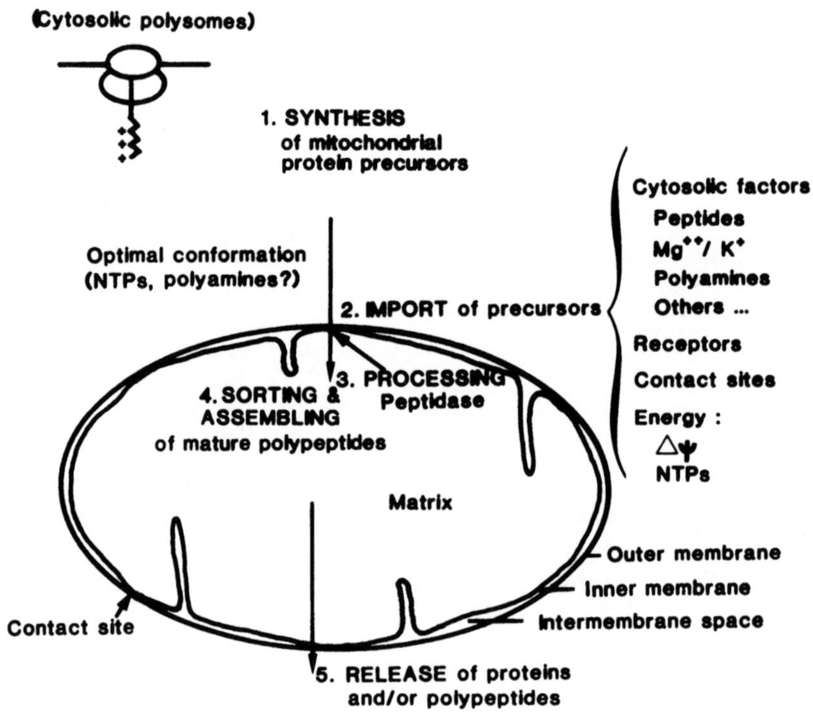

Fig. 10. Scheme of synthesis, import and processing of mitochondrial protein precursors.
$\Delta\Psi$= membrane potential across the inner mitochondrial membrane.
NTPs= nucleoside triphosphates.

The effect of polyamines on the transport of mitochondrial protein precursors other than pOCT, with or without leading peptide, remains to be investigated, particularly their physiological "in vivo" significance, which may be an exciting future field.

A resume of the factors thus far suggested to be involved in the transport of mitochondrial protein precursors is depicted in Fig. 10.

CONCLUSIONS

1. The polyamines spermine, spermidine and putrescine, stimulate at physiological concentrations the "in vitro" transport of pOCT into rat liver mitochondria.

2. The presence of polyamines in the transport medium induces outer and inner membranes to approximate; this may, perhaps, facilitate the formation and/or stabilization of contact sites between both membranes, which have been proposed as transport routes for mitochondrial protein precursors.

ACKNOWLEDGEMENTS

We are particularly grateful to Dr. Nguyen and Dr. G. C. Shore, for supplying the clone pMN152 used to obtain pOCT-mRNA. This work has been supported by the Comision Asesora de Investigación Científica y Técnica of Spain (2386/83 & 0547/84), the U.S.-Spain Joint Committee for Scientific and Technological Cooperation, the Fondo de Investigaciones Sanitarias, the Institució Valenciana d'Estudis i Investigació and the IIC-KUMC International Cytology Program. V.J. Miralles is now a postdoctoral fellow at the University of Medicine & Dentistry of New Jersey (USA). C. González-Bosch is a "Giménez del Río" fellow and M. J. Marcote is a "Reina Doña Sofía fellow of the IIC.

REFERENCES

1. R. Hay, P. Bohni and S. Gasser, How mitochondria import proteins, Biochim. Biophys. Acta 779: 65-87 (1984).
2. M. Mori, T. Morita, F. Ikeda, Y. Amaya, M. Tatibana and P. Cohen, Synthesis, intracellular transport, and processing of the precursor for mitochondrial ornithine transcarbamylase and carbamoyl--phosphate synthetase I in isolated hepatocytes, Proc. Natl. Acad. Sci. USA 78: 6056-6060 (1981).
3. C. Argan, C. J. Lusty and C. G. Shore, Membrane and cytosolic components affecting transport of the precursor for ornithine carbamyltransferase into mitochondria, J. Biol. Chem. 258: 6667-6670 (1983).
4. S. Miura, M. Mori and M. Tatibana, Transport of ornithine carbamoyltransferase precursor into mitochondria, J. Biol. Chem. 258: 6671-6674 (1983).
5. F. Firgaira, J. Hendrick, F. Kalousek, J. Kraus and L. Rosenberg, RNA required for import of precursor proteins into mitochondria, Science 226: 1319-1322 (1984).
6. C. González-Bosch, V. J. Miralles, J. Hernández-Yago and S. Grisolía, Spermidine and spermine stimulate the transport of the precursor of ornithine carbamoyltransferase into rat liver mitochondria, Biochem. Biophys. Res. Commun. 149: 21-26 (1987).
7. C. Tabor and H. Tabor, Polyamines, Annu. Rev. Biochem. 53: 749-790 (1984).
8. E. Canellakis, D. Viceps-Madore, D. A. Kyriakidis and J. S. Heller, The regulation and function of ornithine decarboxylase and of the polyamines, Curr. Top. Cell. Regul. 15: 155-202 (1979).
9. A. Toninello, F. Di Lisa, D. Siliprandi, and N. Siliprandi, Uptake of spermine by rat liver mitochondria and its influence on the transport of phosphate, Biochim. Biophys. Acta 815: 399-404 (1985).
10. S. Lenzen, R. Hickethier and U. Panten, Interactions between spermine and Mg^{2+} on mitochondrial Ca^{2+} transport, J. Biol. Chem. 261: 16478-16483 (1986).

11. J. Phillips, and R. R. J. Chaffee, Restorative effects of spermine on oxidative phosphorylation and respiration in heat-aged mitochondria, Biochem. Biophys. Res. Commun. 108: 174-181 (1982).
12. G. Knoll and D. Bricza, Changes in freeze-fractured mitochondrial membranes correlated to their energy state: dynamic interactions of their boundary membranes, Biochim. Biophys. Acta 733: 102-110 (1983).
13. R. Van Venetie and A. J. Verkleij, Possible role of non-bilayer lipids in the structure of mitochondria: a freeze-fracture electron microscopy study, Biochim. Biophys. Acta 692: 397-405 (1982).
14. M. Schleyer and W. Neupert, Transport of proteins into mitochondria: translocational intermediates spanning contact sites between outer and inner membranes, Cell 43: 339-350 (1985).
15. D. Roise, S. J.,Horvath, J. M. Tomich, J. H. Richards and G. Schatz, A chemically synthesized pre-sequence of an import mitochondrial protein can form an amphipathic helix and perturb natural and artificial phospholipid bilayers, EMBO J. 5: 1327-1334 (1986).

UPTAKE OF POLYAMINES BY HUMAN LYMPHOCYTES AND THEIR EFFECT ON LACTATE FORMATION FROM GLUCOSE

Luisa Fasulo, Barbara Fulgosi, Sebastiano Colombatto, and Maria Angelica Grillo

Dipartimento di Medicina e Oncologia Sperimentale - Sezione di Biochimica. Università di Torino Torino (Italy)

INTRODUCTION

The existence of a polyamine transport system has been demonstrated in both prokaryotes and eukaryotes. However, although its physiological significance has been suggested by several findings[1] the system is poorly known. It has been usually studied in tumor cells, in the aim of modifying the intracellular polyamine concentration, but seldom in normal cells. According to Seppänen et al.[2] exogenous polyamines are taken up by mammalian cells extremely slowly, with only negligible changes in their intracellular pools. However, erythrocytes take up polyamines and transport them[3,4]. The ability of other blood cells, such as lymphocytes, to do so is not known. These cells, in contrast with erythrocytes, are able to synthesize polyamines, but it is not known whether they are also able to metabolize exogenous polyamines. According to Kay and Lindsay[5], stimulated lymphocytes take up putrescine, but only a very small proportion of the labelled compound enters the cells within 3 hours. No data are available for spermidine and spermine. Moreover, it is not known whether polyamines accumulate inside these cells, and whether they have an effect on the cellular metabolism. This would be important in vivo, as polyamine concentration is modified in blood in several conditions.

We have therefore studied the uptake of polyamines by lymphocytes. To start the study of polyamine effect on cellular metabolism, we investigated their effect on the rate of lactate formation from glucose. It is shown that 1) the transport of polyamines is Na^+-dependent and requires the energy of ATP; 2) polyamines inhibit the rate of lactate formation.

RESULTS

When ^{14}C-polyamines are added to lymphocytes in culture, they are slowly taken up. Maximal radioactivity in the cells is reached after 4 hours, being, however, higher with spermidine than with spermine or putrescine (Fig. 1). With spermine, the uptake depends on the concentration up to 0.1 mM; at higher concentrations, the amount taken up by the cell does not increase any more. The amount of radioactivity incorporated is maximal at pH 7.3, much lower at higher pH (8.0) (Fig. 2).

In order to show whether the incorporated polyamines are metabolized, the cells were extracted with $HClO_4$. After centrifugation, the supernatant was dansylated and fractionated by HPLC. The peaks of putrescine, spermidine, spermine and N^1-acetylspermidine were collected and counted. The

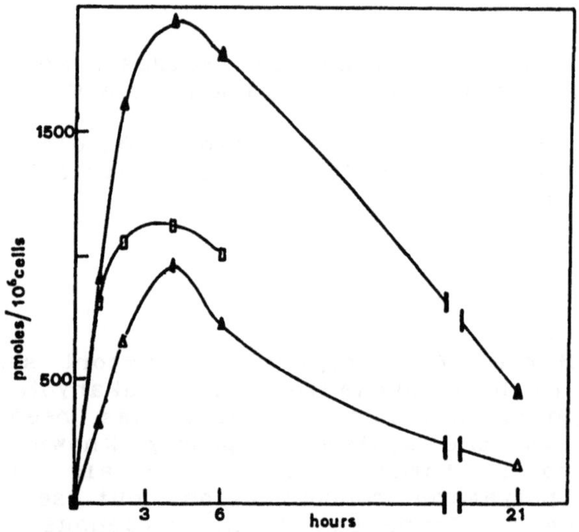

Fig. 1. Putrescine, spermidine and spermine incorporation in normal human lymphocytes.
Cells were incubated with 0.1 mM radioactive putrescine, spermidine or spermine for the time shown and radioactivity incorporated was measured.
△——△ , spermine; ▲——▲ , spermidine; □——□ , putrescine.

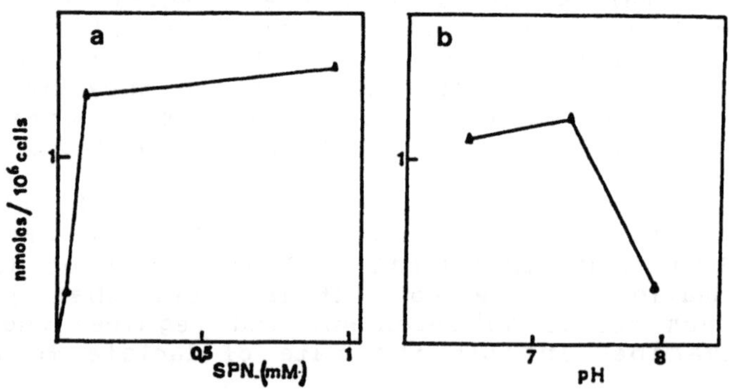

Fig. 2. Effect of concentration and pH on spermine uptake
a) Cells were incubated with different concentrations of ^{14}C-spermine. b) cells were suspended in PBS, containing 8 mM Hepes and 2 % albumin at the pH shown and incubated with 0.1 mM ^{14}C-spermine. After 4 hours (a) or 2 hours (b) radioactivity incorporated was measured.

results obtained with spermidine are shown in Fig.3. After 4 hours incubation most of the radioactivity is still present in the spermidine itself, and a small part is in N^1-acetylspermidine. After 6 hours, putrescine and spermine are also measurable. In those incubated with spermine, spermidine is already detectable after 1 hour, and its amount increases

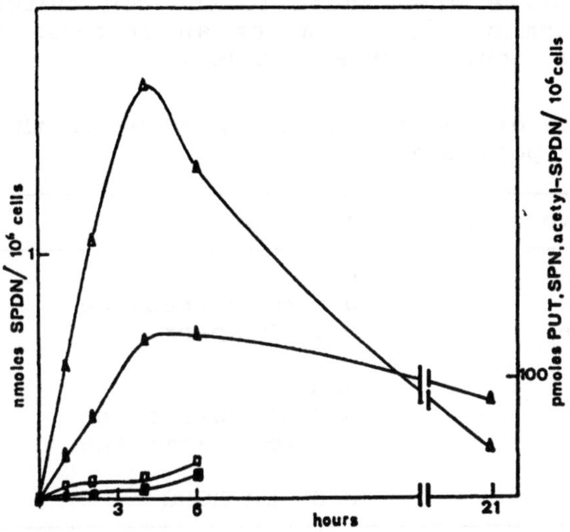

Fig. 3. Radioactive compounds in lymphocytes incubated with spermidine.
Cells were incubated with 0.1 mM ^{14}C-spermidine for the time shown, then extracted with 0.2 N $HClO_4$. After dansylation, polyamines were separated by HPLC[6]. Peaks containing the polyamines and the acetylpolyamines were collected and counted. △——△, spermidine; ▲——▲, N^1-acetylspermidine; □——□, putrescine; ■——■, spermine.

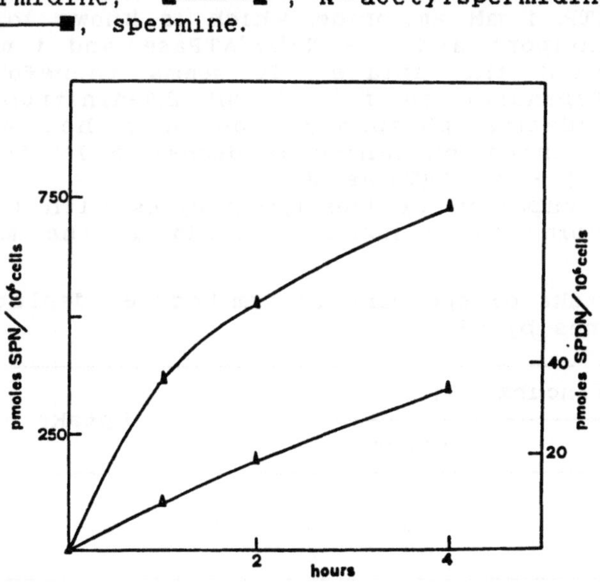

Fig. 4. Radioactive compounds in lymphocytes incubated with spermine.
Experimental conditions as in Fig. 3., except that incubation was performed with ^{14}C-spermine. △——△, spermidine; ▲——▲, spermine.

steadily for 4 hours (Fig. 4).

The uptake of spermidine is strongly counteracted by spermine (0.1 mM). Spermidine at the same concentration partially counteracts uptake of spermine, and also methylglyoxal bis(guanylhydrazone) (MGBG) (0.05 mM), which is supposed to be taken up by the same transport system.Putrescine has almost no effect (Table 1). Difluoromethylornithine (DFMO) has no effect on the uptake when added to the cell culture at the same time as ^{14}C-spermine; but promotes an increase of the uptake, when added 48 hours earlier (Table 2).

Table 1. Uptake of spermidine or spermine in the presence of other polyamines

Radioactive amine	Addition	Uptake
spermidine	none	100
spermidine	0.1 mM putrescine	89
spermidine	0.1 mM spermine	16
spermine	none	100
spermine	0.1 mM putrescine	113
spermine	0.1 mM spermidine	78
spermine	0.05 mM MGBG	78

Uptake was measured by incubating the cells with 0.1 mM ^{14}C-spermidine or ^{14}C-spermine. After 4 hours, the cells were collected and radioactivity incorporated was measured.

0.15 mM ouabain, an inhibitor of Na^+,K^+-ATPase, decreased the uptake of spermine by 56 %, while increasing the concentration of ATP. 1 mM amiloride, which is known to inhibit both the Na^+/H^+ antiport and the Na^+/K^+ATPase, and 1 mM vanadate completely block the uptake. It seems therefore that the uptake is dependent on Na^+. 1 mM 2,4-dinitrophenol, which uncouples oxidative phosphorylation, also has an effect on both ATP concentration (which is decreased by 67 %) and spermine uptake (- 43 %)(Table 3).

After incubation of the lymphocytes with 0.1 mM spermidine or spermine for 4 hours, to build up the maximum polya-

Table 2. Uptake of spermine in lymphocytes depleted of polyamines by DFMO

Preincubation		Uptake of spermine
with	hours	
-	-	100
6 mM DFMO	0	100
6 mM DFMO	48	164

Cells were preincubated with DFMO for 48 hours; then ^{14}C-spermine was added (0.1 mM final concentration) and uptake was measured 4 hours later. Controls were incubated in the same way, but without additions, or with DFMO added at the same time as spermine.

mine concentration inside the cells, these were used for measuring glycolytic activity. Lactate formation was compared with that of cells cultured in the same way, but without the polyamine. As shown in Table 4, both spermidine and spermine markedly inhibit the formation of lactate. If the activity is measured in the presence of rotenone (5 μg/ml), which is known to inhibit NADH reoxidation, a higher amount of lactate is obtained (+ 80%) but polyamines give the same inhibition (data not shown). If the polyamines are not preincubated with the cells, but added at the same time as ^{14}C-glucose, the inhibition is much lower, but still evident.To show whether the different inhibition with cells preincubated or not with spermine is due to the different cell polyamine concentration, the experiment has been repeated with a soluble enzyme preparation. In the presence of 4 mM spermine, the inhibition is

Table 3. Effect of ouabain, 2,4-dinitrophenol, vanadate and amiloride on spermidine and spermine uptake

Addition	Uptake spermidine	Uptake spermine	ATP (nmoles/10^6 cells)
none	100	100	2.1
0.15 mM ouabain	48	44	3.1
1 mM 2,4-dinitrophenol	60	57	0.7
1 mM vanadate		0	
1 mM amiloride		0	

Cells were incubated in the presence or in the absence of the inhibitors at the final concentration shown for 30 min; then ^{14}C-spermidine or ^{14}C-spermine (0.1 mM final concentration) was added and uptake measured after 4 hours. Cells incubated in the same conditions were used for measuring ATP concentration[7].

lower than that observed with the whole cells (47 % instead of 78 %) (Table 5). The effect of spermine is unchanged in the presence of catalase (20 μg/ml), which would destroy the H_2O_2 formed during oxidation of the polyamine (data not shown).

Table 4. Effect of preincubation with spermidine or spermine on the inhibition of lactate formation

Addition	Preincubation	Lactate (nmoles/10^6 cells/hr)	Inhibition %
spermidine	-	31.8	28
spermidine	4 hours	11.2	75
spermine	-	36.6	18
spermine	4 hours	9.8	78

Cells preincubated for 4 hours in the presence or in the absence of 0.1 mM spermidine or spermine were added with ^{14}C-U-glucose and incubated for further 2 hours. Lactate was then measured in the medium[8].

Table 5. Effect of spermine on the glycolytic rate of a soluble enzyme preparation of lymphocytes

Addition	Inhibition %
1 mM spermine	29
4 mM spermine	47

Cells homogenized in an Ultra-Turrax were added with spermine at the final concentration shown. The amount of lactate formed from ^{14}C-U-glucose was measured[8].

DISCUSSION

Our findings show that polyamines are taken up by lymphocytes, and accumulated and partially interconverted inside these cells. Their cellular level can therefore be modulated not only by ornithine decarboxylase and S-adenosylmethionine decarboxylase activities, but also by the uptake from the external medium. Understanding of the transport mechanism is therefore very important for all cells, and also for lymphocytes, as in blood polyamine concentration can be markedly altered in several conditions.

The polyamine uptake by lymphocytes occurs in a concentration, time and pH dependent manner. The two polyamines seem to share the same transport system, as one inhibits the uptake of the other, in contrast with putrescine, which has almost no effect. The affinity of the system seems higher for spermine, as this is a stronger inhibitor of spermidine uptake, than spermidine is of spermine uptake. Several experiments have therefore been performed with spermine only.

The transport is dependent on Na^+, as shown by its inhibition by ouabain, amiloride and vanadate, all inhibitors of Na^+/K^+ ATPase. Na^+-dependent systems have also been shown in other cells: in rat hepatocytes, an Na^+-dependent and an Na^+-independent mechanism of transport are present[9]; in Ehrlich ascites carcinoma cells[10] an Na^+-dependent system able to transport MGBG, a structural analogue of spermidine, supposed to share a common transport system with the natural polyamines. In contrast with the tumor cells[10], however, energy production is also required, as shown by the inhibition by dinitrophenol. In mouse neuroblastoma cells a putrescine transport system has been studied. This also shares some characteristics common to other known sodium-dependent systems[11]. Our system, however, seems specific for spermidine and spermine.

Lymphocytes depleted of polyamines by DFMO take up spermine faster than normal cells, as it occurs in hepatoma cells[12]. However the calculated normal cell concentration (2.5 to 5 mM) is higher than that previously calculated for tumor cells[2]. Therefore, in contrast with what is usually believed, the amount accumulated is by no means negligible.

Inside the lymphocytes, both spermidine and spermine are partially metabolized: from the former N^1-acetylspermidine is accumulated, from the latter, spermidine. The interconversion pathway is therefore fully operative. Nevertheless, most of the polyamine accumulates unchanged.

The high concentration of spermidine inside the cell could have an effect on the cellular metabolism, as polyamines are known to affect several enzyme activities, such as protein kinases[13,14] and protein phosphatases[15-17]. To test whether this is the case, we have measured the glycolytic activity in cells incubated in the presence and in the absence of spermidine or spermine for 4 hours. The rate of lactate formation is markedly reduced by the polyamines. The reduction is much smaller, when the polyamine is added at the same time as ^{14}C-glucose, showing that a higher concentration of the polyamine inside the cell is needed to have marked inhibition. The lower amount of lactate measurable in the presence of spermidine or spermine is not due to a higher rate of pyruvate degradation, as the same inhibition is observed in the presence and in the absence of rotenone. An effect on glucose uptake could also explain the results.

In lymphocytes, the effect of polyamines on glucose metabolism is completely different from that observed in other cells, such as adipocytes, where they stimulate both glucose transport and oxidation. In renal brush border membrane vesicles[18], too, they have been shown to facilitate glucose transport. The insulinomimetic effect of polyamines on fat cells has since been shown to be mediated by H_2O_2[19], as catalase abolished the effect. The effect promoted in lymphocytes, however, is not.

Taken together, these results may be said to show that polyamines are taken up by lymphocytes with an active transport system. Moreover, they can promote modifications which influence the cellular functions. The resulting effect depends on the cell type. In lymphocytes, they inhibit glucose metabolism.

REFERENCES

1. Kano, K. and Oka, T. Polyamine transport and metabolism in mouse mammary gland. General properties and hormonal regulation. J. Biol. Chem. 251: 2795 (1976).
2. Seppänen, P., Alhonen-Hongisto, L. and Jänne, J. Death of tumor cells in response to the use of a system of stimulated polyamine uptake for the transport of methylglyoxal bis(guanylhydrazone) Eur. J. Biochem. 118: 571 (1981).
3. Tadolini, B., Hakim, G., Orlandini, G. and Casti A. Intracellular location of polyamines associated to red blood cells. Biochim. Biophys. Res. Commun. 134: 1365 (1986).
4. Moulinoux, J.-P., Le Calve, M., Quemener, V. and Quash, G. In vitro studies on the entry of polyamines into normal red blood cells. Biochimie 66: 385 (1984).
5. Kay, J. E. and Lindsay, V. J. Control of ornithine decarboxylase activity in stimulated human lymphocytes by putrescine and spermidine. Biochem. J. 132: 791 (1973).
6. Stefanelli, C., Carati, D. and Rossoni, C. Separation of N^1- and N^8-acetylspermidine isomers by reversed-phase column chromatography after derivatization with dansyl chloride. J. Chromat. 375:49 (1986).
7. Lemasters, J. J. and Hackenbrok, C. R. Firefly luciferase assay for ATP production by mitochondria in: " Methods in Enzymology", vol. 57, p. 36 (1978).

8. Probst, I., Schwartz, P. and Jungermann, K. Induction in primary culture of "gluconeogenic" and "glycolytic" hepatocytes resembling periportal and perivenous cells. Eur. J. Biochem. 126: 271 (1982).
9. Auberger, P., Samson, M. and Le Cam A. Inhibition of hormonal induction of tyrosine aminotransferase by polyamines in freshly isolated rat hepatocytes. Biochem. J. 214: 679 (1983).
10. Seppänen, P. Some properties of the polyamine deprivation-inducible uptake system for methylglyoxal bis(guanylhydrazone) in tumor cells. Acta Chem. Scand. 35: 731 (1981).
11. Rinehart, C. A. and Chen, K. Y. Characterization of the polyamine transport system in mouse neuroblastoma cells. Effects of sodium and system A amino acids. J. Biol. Chem. 259: 750 (1984).
12. Mamont, P. S., Duchesne, M. -C., Grove, J. and Bey, P. Anti-proliferative properties of DL-α-difluoromethylornithine in cultured cells. A consequence of the irreversible inhibition of ornithine decarboxylase. Biochim. Biophys. Res. Commun. 81: 58 (1978).
13. Atmar, V. J., Kuehn, G. D. and Casillas, E. R. A polyamine-dependent protein kinase from bovine epididymal spermatozoa. J. Biol. Chem. 256: 8275 (1981).
14. Qi D. F., Schatzman, R. C., Mazzei, G. J., Turner, R. S., Raynor, R. L., Liao S. and Kuo, J. F. Polyamines inhibit phospholipid-sensitive and calmodulin-sensitive Ca^{++}-dependent protein kinases. Biochem. J. 213: 281 (1983).
15. LeCam, A., Auberger, P., Samson, M. and Le Cam, G. Insulin regulation of protein phosphorylation in hepatocytes. Studies using two effectors: amiloride and natural aliphatic polyamines. Biochimie 67: 1125 (1985).
16. Fisher, M. J., Dickson, A. J. and Pogson, C. I. The polyamine-dependent modulation of phenylalanine hydroxylase phosphorylation state and enzyme activity in isolated liver cells. Biochem. J. 237: 277 (1986).
17. Thomas, A. P., Diggle, T. A. and Denton, R. M. Sensitivity of pyruvate dehydrogenase phosphate phosphatase to magnesium ions. Similar effects of spermine and insulin. Biochem. J. 238: 83 (1986).
18. Elgavish, A., Wallace, R. W., Pillion, D. J. and Meezan, E. Polyamines stimulate D-glucose transport in isolated renal brush-border membrane vesicles. Biochim. Biophys. Acta 777: 1 (1984).
19. Livingston, J. N., Gurny, P. A. and Lockwood, D. H. Insulin-like effects of polyamines in fat cells. Mediation by H_2O_2 formation. J. Biol. Chem. 252: 560 (1977).

THEORETICAL AND EXPERIMENTAL CHARACTERIZATION OF POLYAMINE/ DNA INTERACTIONS

Burt G. Feuerstein, Hirak S. Basu, and Laurence J. Marton

Department of Laboratory Medicine and the Brain Tumor Research Center of the Department of Neurological Surgery, School of Medicine, University of California, San Francisco, CA 94143 USA

INTRODUCTION

For several years, work in our laboratory has been directed toward an understanding of the interactions of polyamines and nucleic acids. The initial impetus to this research was the finding that the cytotoxicity of antineoplastic agents such as cis-platinum and the 2-chloroethylnitrosoureas are altered in tumor cells in which intracellular levels of polyamines had been depleted by pretreatment with α-difluoromethylornithine (1,2). It is thought that these agents kill cells by forming DNA interstrand crosslinks that prevent replication and lead to cell death. One possible explanation for the effects of polyamine depletion on cytotoxicity is that the conformation of intracellular DNA is altered, which would affect the mechanism of crosslink formation.

MODELING B-DNA/SPERMINE INTERACTIONS

Polyamines are known to affect the conformation of DNA in at least two ways. One is the condensation of DNA (3,4) and the other is induction of the transition from B-DNA (a right-handed helix) to Z-DNA (a left-handed helix) in suitable base sequences (5). In our first approach to understanding polyamine-DNA interactions, we modeled the effect of spermine binding to B-DNA (6).

The initial modeling studies used molecular mechanics calculations. In this method (7), a collection of atoms is described by quantifying the energies of bonded (covalent) and nonbonded (electrostatic and van der Waals) interactions. An initial best structure is estimated based on either experimental results (X-ray diffraction studies, NMR studies, etc.) or emperical values obtained from theoretical studies. The energies of the interactions based on the position of the atoms in the initial structure are computed to build a force field, after which the atomic positions are iteratively optimized (the atoms are moved slightly many times) in order to find the most stable (lowest energy) local positions of the atoms (the minimum energy state). Our initial modeling studies (8) showed that spermine interacts well with the major groove of alternating purine/pyrimidine sequences of B-DNA and that this interaction can greatly affect the conformation of DNA. Models of spermine (with a primary amine group marked N_2') docked into the major groove of $d(AT)_5dAT_5$ in the B conformation before (left) and after (right) energy minimization are shown in Figure 1. Note that the major groove, marked by squares in each model, decreases in size, which causes the phosphate groups to approach each other, while the minor groove, marked by circles, increases in size.

These changes in groove size force the helical axis of the DNA to bend; alterations in sugar puckering decrease the intrastrand phosphate distance. Thus, the interaction of spermine with alternating purine/pyrimidine sequences appears to cause appreciable changes in DNA structure, including a bend in the helical axis, that might correlate with changes in structure, such as DNA condensation, found experimentally.

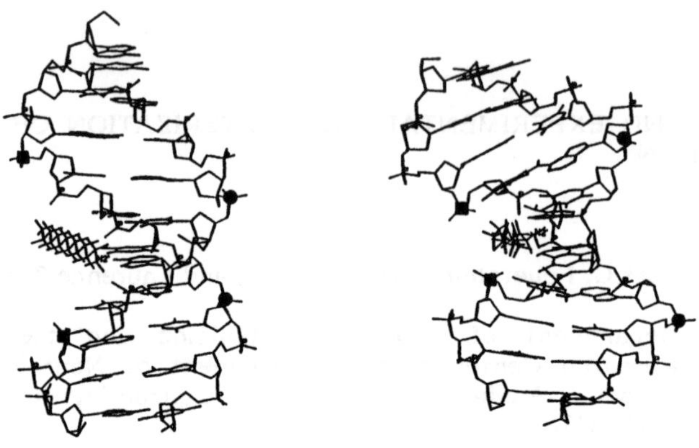

Figure 1. Spermine in the major groove of $d(AT)_5d(AT)_5$ before (left) and after (right) energy minimization. Spermine is marked by a terminal primary amino group. DNA phosphorus atoms are marked by P's.

Based on these initial results, our studies were extended to include positioning of spermine along the phosphate backbone, within the minor groove, and bridging the minor groove (the Liquori/Tsuboi model [9,10]) of B-DNA (Figure 2). We examined several examples of repetitive sequences, both heteropolymers (alternating purine/pyrimidine sequences) and homopolymers (purines in one strand and pyrimidines in the other) in each of these positions. Results of modeling studies with these structures showed that the most favorable position for spermine was within the major groove of alternating purine/pyrimidine sequences. This position produced the most stable complexes between spermine and DNA and the most dramatic changes in DNA tertiary structure (11).

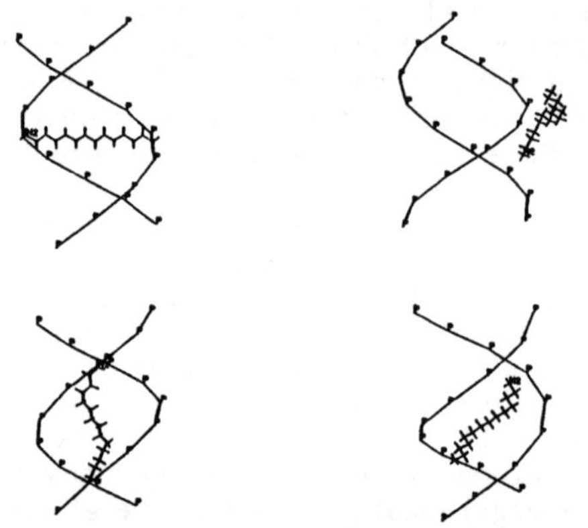

Figure 2. Positioning of spermine and DNA (shown only as the phosphorus atoms in the backbone). Four possible positions for binding were examined. Upper left shows spermine interacting across the major groove. Upper right shows spermine interacting along the phosphate backbone. Lower left shows spermine interacting across the minor groove. Lower right shows spermine interacting within the minor groove.

These results were extended using molecular dynamics (7). In this method, the same description of interactions are used to describe a force field, but kinetic energy is introduced into the simulation and Newtonian mechanics are used to predict the paths of each atom. This approach allows simulation of molecular interactions in time. We examined the interactions of spermine in the major groove of the heteropolymer $d(GC)_5$-$d(GC)_5$ and the homopolymer $d(G)_{10}d(C)_{10}$ and confirmed that the interaction of spermine with the heteropolymer is more favorable (12).

The changes in DNA structure predicted by our studies are interesting for several reasons. First, the decrease in the size of the major groove (with the close approach of phosphate groups across the major groove), the increase in the size at the minor groove, and the alteration of sugar puckering and intrastrand phosphate distance are characteristic of the A-DNA conformation; it has been reported that polyamines do stabilize DNA in the A-form (13). Second, the Z-conformation also contains similar alterations in sugar puckering and intrastrand phosphate distance; polyamines stabilize the Z conformation as well. It is interesting that although the interstrand phosphate distance is narrowed in the Z conformation, the phosphate groups approach one another over the minor groove, which is narrow and deep, instead of over the major groove, which is narrow and deep in the A conformation. The major groove does not exist as a groove in Z-DNA, but appears as a convex surface on the outside of the helix. Thus, polyamines stabilize conformations with narrow and deep grooves and short interphosphate distances.

The prediction of a spermine-induced bend in DNA at specific sequences could have important implications for stabilization of the tertiary structure of the genome. Transcription may be regulated in part by tertiary structure, and involvement of polyamines in the maintenance of it could provide a basis for their involvement in gene expression. Spermine appears to be important for packaging of viral DNA (14,15), and packaging of eukaryotic DNA from supercoils through nucleosomes and chromosomes may be profoundly affected by polyamines.

PROTON EXCHANGE STUDIES

As an experimental probe of the interaction of B-DNA with spermine, we examined the base pairing characteristics of poly d(G-C).poly d(G-C) and poly d(A-T).poly d(A-T) by hydrogen-deuterium exchange kinetics (16), which provides a measure of how quickly both imino and amino protons of base pairs exchange with solvent water (Figure 3). The exchange rate of the amino proton was not affected by spermine; the exchange rate of the imino proton, however, varied inversely with spermine concentration in calf thymus DNA, poly(dG-dC). poly(dG-dC), and poly(dG-^{5}medC).poly(dG-^{5}medC), but that it varied directly with spermine concentration in poly (dAdT).poly (dAdT). Imino exchange rates varied inversely with magnesium concentration in all polymers.

The inverse relation between cation concentration and imino proton exchange rates parallels known enhancement of thermal stabilization of the double helix caused by spermine and magnesium ions. It would be expected that stabilization of the helix would decrease disruption of base pairs, which would limit the access of solvent and decrease the imino proton exchange rate. The direct relation between imino proton exchange rate for poly(dA-dT).poly(dA-dT) and spermine concentration, however, indicates that there is an increased accessibility of these protons to solvent water. Our modeling studies indicated that spermine binding widens the minor groove. For poly(dG-dC).poly(dG-dC), the imino proton is protected from solvent exchange by hydrogen bonds in both the major and minor grooves. In poly(dA-dT).poly(dA-dT), however, widening the minor groove in the absence of a minor-groove hydrogen bond may increase accessibility of imino protons to solvent and the rate of imino proton exchange is increased. The experimental results are consistent with a bend in the helical axis of B-DNA that was predicted by our calculations (8).

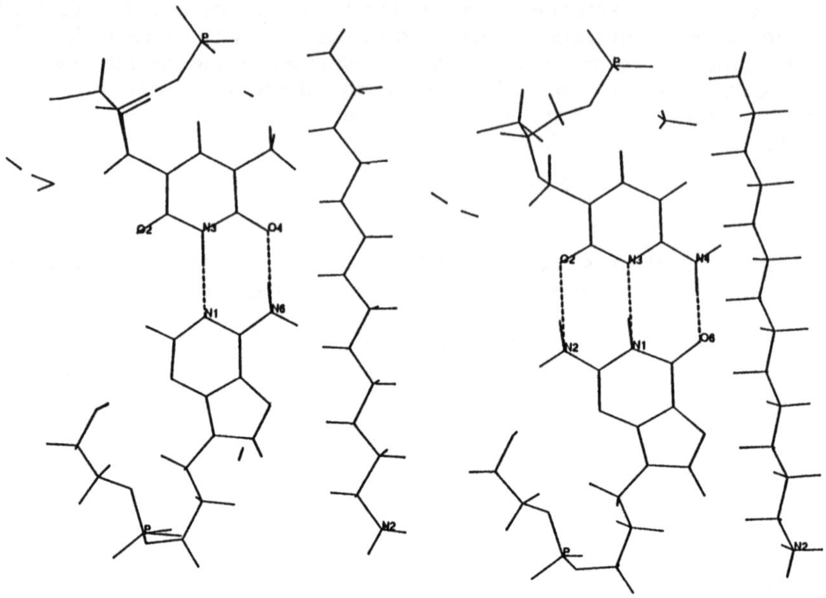

Figure 3. Base pairing in DNA. The guanine/cytosine base pair is on the right and the adenine/thymine base pair (AT) is on the left. Two amino protons are located on both N6 of adenine and N2 of guanine; one imino proton is located on N3 of thymine and N1 of guanine. Spermine is located in the major groove, to the right of each base pair; the minor groove is on the left. The imino proton is protected on both major and minor grooves sides in a GC base pair, but is unprotected on the minor groove side of the AT base pair. Widening the minor groove might allow better access of solvent to the imino proton for unprotected AT than for protected GC.

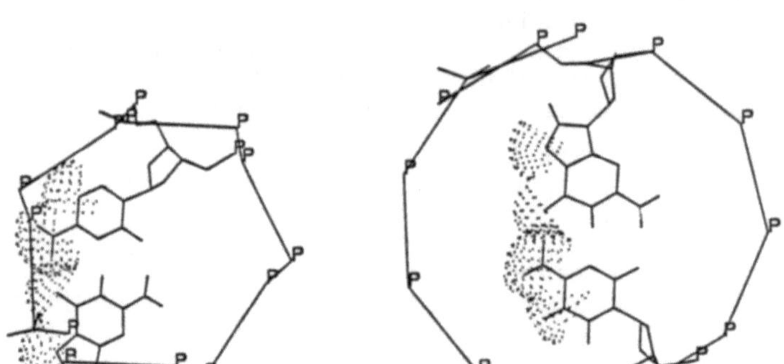

Figure 4. Views of B (right) and Z (left) DNA along the helical axis. In each molecule, the phosphate backbone is represented by phosphorus atoms only, and one base pair is shown. The major groove is marked by the van der Waals surfaces of N7 and O6 of guanine and of N4 and C5 of cytosine. Because the bases in Z-DNA are displaced toward the outside of the helix, the major groove is really a convex surface, while in B-DNA it forms a major groove.

STUDIES WITH Z-DNA

Z-DNA was characterized in 1979 by Rich's group during their pioneering work on X-ray crystallography of oligonucleotides. The Z-form is a left-handed helix with a dinucleotide repeat; the base pairs are displaced to the edge of the helix, producing a narrow, deep minor groove and a wide major convex surface (Figure 4). In contrast, B-DNA is a right-handed helix with a single nucleotide repeat; the base pairs are in the middle of the helix, resulting in a wide major groove and a narrower minor groove (17). Polyamines stabilize Z-DNA better than B-DNA. In an attempt to understand this stabilization, we studied the competitive binding of spermine and antibodies directed to different sites of Z-DNA (18). One antibody was directed against the phosphate backbone, one against the major convex surface, and one against elements of both. In the presence of spermine, the antibody to elements of both the major convex surface and the phosphate backbone bound less well, and the antibody to the phosphate backbone was not affected. These results suggest that spermine binds to some part of the major convex surface of Z-DNA, which confirms X-ray diffraction studies by Quigley _et al._ (19). It is interesting that the antibody with activity against the major convex surface bound better in the presence of spermine. This may be caused by a direct change in Z-DNA structure or by local environmental changes related to an asymmetric distribution of polyamines in the proximity of DNA.

STUDIES WITH ANALOGS OF POLYAMINES

The structures of polyamines clearly affect DNA binding. Vertino _et al._ (20) and Thomas _et al._ (21) found that DNA binding was affected by relatively minor structural changes in analogs of spermidine. In our initial studies (22) we used pentamines with either an _n_-propyl or _n_-butyl group between primary and/or secondary amines. We determined the effects of these compounds on the B-Z transition, aggregration of DNA, and the melting temperature of DNA; apparent binding constants were calculated from melting temperatures. Results confirmed the importance of variation in structure on these physiologic functions. In addition, we found no clear relation between activity in one interaction and activity in another; for instance, a compound that caused a strong elevation in melting temperature did not affect DNA aggregration. We have begun testing a series of analogs of spermine and spermidine in which the structures are altered systematically. We are testing the analogs _in vitro_ to determine the effects of changes in polyamine structure on DNA, and we plan to correlate the _in vitro_ results with cellular functions such as growth.

Acknowledgments. Supported in part by NIH grants CA-41757 (to B.G.F.), Program Project Grant CA-13525, and National Cooperative Drug Discovery Group Grant CA-37606 (to L.J.M), and the Aaron Silvera Cancer Research Fund. We thank Peter Kollman for use of the AMBER program and Robert Langridge for use of the MIDAS program and the facilities of the Computer Graphics Laboratory of the Department of Pharmaceutical Chemistry, University of California, San Francisco (supported in part by NIH Grant RR-1081). We acknowledge the contributions of our colleagues N. Pattabiraman, R.H. Shafer, D. Zarling, H. Schwietert, R.J. Bergeron, K. Samejima, W.P. Lubich, and D.F. Deen.

REFERENCES

1. D.T. Hung, D.F. Deen, J. Seidenfeld, and L.J. Marton, Sensitization of 9L rat brain gliosarcoma cells to 1,3-_bis_ (2-chloroethyl)-1-nitrosourea by α-difluoromethylornithine, an ornithine decarboxylase inhibitor. _Cancer Res._, 41:2783 (1981).

2. S. M. Oredsson, D.F. Deen, and L.J. Marton, Decreased cytotoxicity of _cis_-diamminedichloroplatinum (II) by α-difluoromethylornithine depletion of polyamines in 9L rat brain tumor cells _in vitro_. _Cancer Res._, 42:1296 (1982).

3. L.C. Gosule and J. A. Schellman, DNA condensation with polyamines. J. Mol. Biol., 121: 311, 327 (1978).

4. R. Marquet, C. Houssier, and E. Fredericq, An electro-optical study of the mechanisms of DNA condensation induced by spermine, Biochem. Biophys. ACTA, 825:365 (1985).

5. M. Behe and G. Felsenfeld, Effects of methylation on a synthetic polynucleotide: the B-Z transition in poly(dG-^{5}medC). Proc. Natl. Acad. Sci. U.S.A., 78:1619 (1981).

6. B.G. Feuerstein and L.J. Marton, Specificity and binding in polyamine/nucleic acid interactions, in: "The Physiology of Polyamines," U. Bachrach and Y. Heimer, eds, CRC Press, Boca Raton (in press).

7. U.C. Singh, P.K. Weiner, J. Caldwell, and P.A. Kollman, Program AMBER UCSF, Version 3.0, Univesity of California, 1986.

8. B.G. Feuerstein, N. Pattabiraman, and L.J. Marton, Spermine DNA interactions: A theoretical study. Proc. Natl. Acad. Sci. U.S.A., 83:5948 (1986).

9. A.M. Liquori, L. Constantino, V. Crescenzi, V. Elia, E. Giglio, R. Puliti, M. DeSantis-Savino, and V. Vitigliano, Complexes between DNA and polyamines: A molecular model. J. Mol. Biol., 24:113 (1967).

10. M. Tsuboi, On the melting temperature of nucleic acids in solution, Bull. Chem. Soc. Jpn., 37:1514 (1964).

11. B.G. Feuerstein, N. Pattabiraman, and L.J. Marton, Molecular mechanics of the interactions of spermine with DNA (submitted).

12. B.G. Feuerstein, N. Pattabiraman, and L.J. Marton, Molecular dynamics of spermine binding to DNA (submitted).

13. E.E. Minyat, V.I. Ivasnov, A.M. Kritzyn, L.E. Minchenkova, and A.K. Schyolkina, Spermine and spermidine-induced B to A transition of DNA in solution. J. Mol. Biol., 128:397 (1978).

14. W.C. Earnshaw and S.R. Casjens, DNA packaging by the double stranded DNA bacteriophages, Cell, 21:319 (1980).

15. K.A. Marx and G.C. Ruben, A study of ØX-174 DNA torus and lambda DNA torus tertiary structure and the implications for DNA self assembly, J. Biol. Struct. Dynam., 4:23-39 (1986).

16. H.S. Basu, R.H. Shafer, and L.J. Marton, A stopped-flow H-D exchange kinetics study of spermine-polynucleotide interactions, Nuc. Acid Res., 15:5873 (1987).

17. A. Rich, A. Nordheim, and A.-H. Wang, The chemistry and biology of left-handed Z-DNA, Ann. Rev. Biochem., 53:791 (1984).

18. H.S. Basu, B.G. Feuerstein, D.A. Zarling, R.S. Shafer, and L.J. Marton, Recognition of Z-RNA and Z-DNA determinants by polyamines in solution: experimental and theoretical studies, J. Biomol. Struct. Dyn. (in press).

19. G.J. Quigley, The interactions of ions, water, and drugs with DNA, in: "Molecular Structure and Biological Activity," J.F. Griffith and W.L. Duax, eds., Elsevier Biomedical Press, Amsterdam, 1982, p. 317.

20. P.M. Vertino, R.J. Bergeron, P.F. Cavanaugh, and C.W. Porter, Structural determinants of spermidine/DNA interactions, Biopolymers, 26:691 (1987).

21. T.J. Thomas and R.P. Messner, Structural specificity of polyamines in left-handed Z-DNA formation, J. Mol. Biol., 201:463 (1988).

22. H.S. Basu and L.J. Marton, The interaction of spermine and pentamines with DNA, Biochem. J., 244:243 (1987).

EFFECT OF SPERMINE ON TRANSFER RNA AND TRANSFER RNA-RIBOSOME INTERACTIONS

Željko Kućan,[1] Tatjana Naranđa,[1] Miroslav Plohl,[2] Vesna Nöthig-Laslo,[2] and Ivana Weygand-Đurašević[1]

[1]Department of Chemistry, Faculty of Science, University of Zagreb, Strossmayerov trg 14, and [2]Rugjer Bošković Institute, Bijenička cesta 54, YU-41000 Zagreb, Yugoslavia

INTRODUCTION

Translation of genetic message requires a coordianted interplay of more than a hundred kinds of macromolecules. Besides the ribosomes, which are multicomponent ribonucleoprotein particles by their own nature, molecules of transfer RNA, aminoacyl-tRNA synthetases, soluble protein factors and mRNA are involved in the process. Ribosomes provide an unspecific stage for the codon-anticodon interaction and catalyze the peptide bond formation. Molecules of tRNA play a crucial and highly specific role in the over-all process: they are recognized by specific aminoacyl-tRNA synthetases to be charged with their cognate amino acids; the resulting aminoacyl-tRNAs are then brought, in the form of ternary complexes with GTP and the elongation factor Tu, to the decoding or A site of the ribosome. There they are screened for the proper codon-anticodon matching and only the correct aminoacyl-tRNA is allowed to enter the transpeptidation reaction. The resulting peptidyl-tRNA is then translocated to the peptidyl or P site to serve as the peptidyl donor in the next round of translation. The process in its simplified form is shown in Figure 1.

Spermine and spermidine are known to stabilize the conformation of tRNA and stimulate its aminoacylation.[1-5] They also stimulate the over-all rate of protein synthesis and are included into the reaction mixture for fast translation in vitro.[6] While the first two points had been studied extensively, studies of the role of polyamines in various stages of translation on the ribosomes were nearly neglected. Our own findings that the structural stabilization of yeast $tRNA^{Tyr}$ can be achieved by a fixed small number of spermine molecules,[7] while the aminoacylation of the same tRNA is not strictly dependent on spermine, except at very low Mg^{2+} concentration, prompted our search for other steps in translation which would possibly parallel the structural effects more closely. Indeed, such steps were identified. They are strongly stimulated by small, tRNA-specific number of spermine molecules per each tRNA.

STABILIZATION OF tRNA STRUCTURE BY SPERMINE

Theoretical calculation and experimental findings suggest that monotonous polyanionic structures like DNA do not bind spermine by simple sequence-independent charge neutralization. This is even less so with tRNA molecules, which in their tertiary structure approach the architectural complexity of proteins.[8,9] In addition to purine and pyrimidine bases involved in classical and non-classical pairing, the ribose 2'-hydroxyl, which distinguishes RNA from DNA, provides a donor group for additional hydrogen bonding with phosphate oxygens, playing so a key role in stabilizing the tRNA structure. Spermine and Mg^{2+} provide additional stabilizing effects. The combined data from two independent crystallographic studies,[10,11] of yeast $tRNA^{Phe}$ locate three spermines in the structure of this tRNA. One spermine is found in the deep pocket formed by bending of the D-stem nucleotides towards those of the variable loop. By interposing a string of positive charges which neutralize the negative charges of phosphates P9, P10 P11 of one chain and P46 and P47 of the

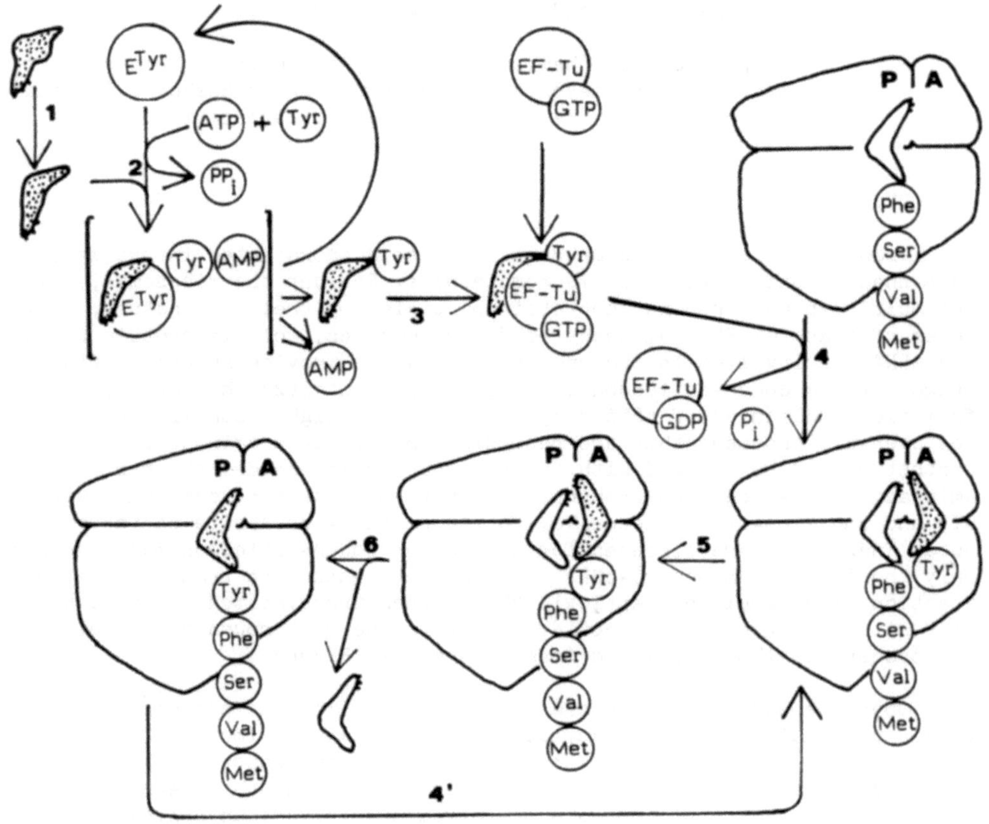

Figure 1. Involvement of tRNA in protein synthesis. The life cycle of $tRNA^{Tyr}$, on which most of the authors' work has been performed, is shown in this schematic presentation. E^{Tyr} and EF-Tu stand for tyrosyl-tRNA synthetase and elongation factor Tu, respectively. The molecule of mRNA is ommitted from the scheme. The following steps are shown: 1, stabilization of tRNA tertiary structure; 2, aminoacylation; 3, formation of the Ef-Tu GTP tRNA ternary complex; 4, screening of ternary complexes at the A site for proper codon-anticodon matching, accompanied by GTP hydrolysis; 5, transpeptidation (intrinsic enzymatic activity of the ribosomes); 6, translocation (mediated by the elongation factor G and accompanied by GTP hydrolysis). For $tRNA^{Tyr}$, the full cycle would be completed after another round of translation (steps 4, 5, and 6).

other, this spermine obviously stabilizes the central part of the tRNA structure.[10] Second spermine appears to form hydrogen bonds to four different phosphate residues on both sides of the deep groove of the double helix in the anticodon stem and also to O6 of a guanine residue. This interaction draws the opposite sides of the major groove about 3Å closer than they would be in the absence of the spermine, causing a significant bend of the helix axis.[10] The third spermine is apparently found in the major groove of the helix formed jointly by the acceptor stem and the T stem;[11] it may also stabilize the central part of the tRNA structure.

Calculations of molecular electrostatic potential of $tRNA^{Phe}$ in its crystalline conformation have predicted the above mentioned three strong binding sites for spermine and two additional weak binding sites in the major grooves of the TΨ stem and the anticodon stem.[12] Data for other tRNAs are not available, but one should note that spermine has been indispensable for obtaining crystals of various tRNAs. Hence, it probably plays a stabilizing role in the crystal structure of these (or possibly all) tRNA, though the exact location and the number of its specific binding sites will almost certainly depend on the nucleotide sequence of individual tRNAs, and the number of nucleotides in their variable regions.

Moving from tRNA crystals to tRNA in solution, one finds in the existing literature many more data and much less precision. Without exception, a stabilizing role of polyamines (together with Mg^{2+}) in maintaining the biologically active tRNA structure is suggested. Two classes of polyamine binding sites were usually determined by direct physical methods like equilibrium dialysis or equilibrium gel filtration. It was suggested that two to five spermine molecules bind very strongly and in a cooperative manner, while many others (up to a total of 16) bind less tightly.[13,14] Some of these early experiments should be re-interpreted, since the binding of polycations to tRNA must be sensitive to ionic strength in a manner described for divalent cations.[15] Only the experiments conducted at sufficiently high ionic strength can reveal specific binding sites. Instead of measuring the number of spermine molecules bound to tRNA, one may also measure a structural feature in the tRNA as a function of added spermine molecules. An excellent example are the high-resolution NMR experiments[16] demonstrating the formation of two secondary base pairs upon the binding of three spermine molecules to $yeast^{Phe}$. Significantly, this number coincides with the number of spermines found in the crystal structure[10,11] and the number of calculated spermine binding sites.[12]

In search for specific information about the binding of spermine to t-RNA in solutions, we have utilized yeast $tRNA^{Tyr}$, spin-labelled in N^6-(Δ^2-isopentenyl)adenosine moiety (i^6A) at the position 37, i.e. next to the anticodon and belonging to the stacked sequence starting at the first base of the anticodon (34) and extending through the anticodon stem to the central part of the molecule. Structure of the nitroxide spin label attached to i^6A and two characteristic electron-spin resonance (ESR) spectra are shown in Fig. 2. When the lebel is attached to a free nucleoside it shows a symmetrical triplet spectrum, while the spectrum of spin-labelled tRNA is asymmetric and typical of a rapidly tumbling nitroxide.[16] The rotational correlation time τ describing the rate of nitroxide tumbling can be calculated from the spectral linewidths. For spin-labelled $tRNA^{Tyr}$ τ is close to 1 ns, which is an order of magnitude higher than τ for the free spin label, but at least one order of magnitude lower than expected for the motion of the tRNA molecule as a whole. Hence, the spectrum reflects the motion of the label around an axis fixed to the anticodon loop of $tRNA^{Tyr}$ and should be sensitive both to the conformational changes of this part of the tRNA molecule and to the contacts with other structures.

For instance, the spectrum is impaired when $tRNA^{Tyr}$ is bound to ribosomes[17], but not when it forms anticodon:anticodon pair with $tRNA^{Val}$.[18]

ESR spectra of spin-labelled tRNA, measured as a function of temperature, reveal a change in the activation energy of spin-label motion at some critical temperature, t_{cr}, caused by a conformational change in the tRNA molecule.[16] Spermine increases t_{cr} by 10°C. Addition of the first four spermine molecules to $tRNA^{Tyr}$ causes no effect; five or more spermines per $tRNA^{Tyr}$ are needed for this conformational stabilization.[19]

Since Mg^{2+} also stabilizes the tRNA structure, we have paid a particular attention to the distinction between genuine effects of spermine and those where spermine could be substituted by divalent cations. As one approach, we have studied the binding of Mn^{2+}, a paramagnetic substitute for Mg^{2+}, to yeast $tRNA^{Tyr}$ in the presence and absence of spermine. Taking into account the neutralization of four negative charges on tRNA by each spermine and using the electrostatic model[15], we concluded[19] that in fact spermine created two to three new binding sites on the molecule of $tRNA^{Tyr}$. On the other hand, the initial binding of Mn^{2+} to $tRNA^{Tyr}$ in the presence of spermine affects the motional properties of the spin-label attached to i^6A, indicating individually specific roles of divalent cations and polyamines in maintaining the tRNA structure.[20] All these effects of spermine are lost upon dissection of the tRNA on smaller fragments or upon extensive C→U modification preventing the formation of native structure.[20] Spermine does not stimulate the binding of Mn^{2+} to single or double stranded polyribonucleotides. Hence, the enhanced binding of divalent cations to tRNA in the presence of spermine must be related to the unique tertiary structure of tRNA.[20] Both the divalent cations and spermine cooperate in maintaining this structure. The number of spermine molecule needed for structural stabilization is characteristically small: three for yeast $tRNA^{Phe}$ and five for $tRNA^{Tyr}$.

EFFECT OF SPERMINE ON THE AMINOACYLATION OF tRNA

Aminoacylation of tRNA is the first reaction in the course of protein synthesis (step 2 in Fig. 1). It is accompanied by ATP hydrolysis to AMP and PP_i. In this reaction, catalyzed by a specific aminoacyl-tRNA synthetase, the amino acid is both activated for protein synthesis and adapted for the reading of one or several of its specific codons on mRNA. It has been known for a long time that Mg^{2+} is required for this reaction,[21]

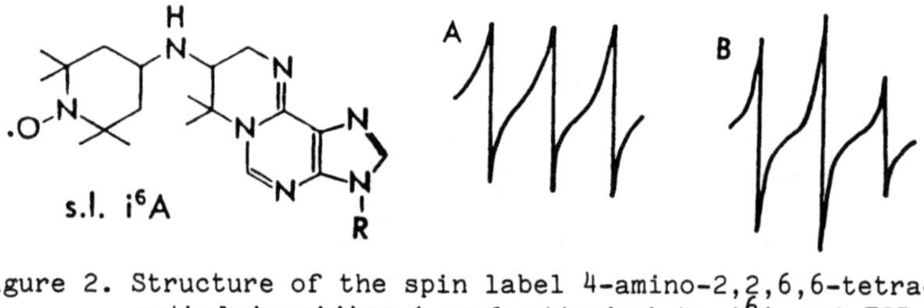

Figure 2. Structure of the spin label 4-amino-2,2,6,6-tetramethylpiperidine-1-oxyl attached to i^6A and ESR spectra of the same spin label attached to the free nucleoside i^6A (a) and the i^6A in the position 37 of yeast $tRNA^{Tyr}$ (b).

while polyamines have a pronounced additional effect.[4,5] In spite of relative simplicity of the aminoacylation reaction, the stimulatory effects of Mg^{2+} and spermine could not be always ascribed to the binding of these cations to tRNA, since they also bind ATP, which is unavoidably present in the reaction mixture. To complicate the matter even further, the ATP Mg^{2+} is active, but the ATP spermine complex is not.[22] Hence, the aminoacylation experiments at low spermine/tRNA ratios are not feasible.

Detailed studies of the effect of spermine or spermidine on aminoacylation of tRNA have revealed variable response of individual tRNAs. Our own experiments[7] on the aminoacylation of yeast $tRNA^{Tyr}$ catalyzed by homogeneous tyrosyl-tRNA synthetase,[23] have shown no influence of spermine on the K_m for the tRNA, while the rate of aminoacylation was stimulated at low Mg^{2+} concentrations; nevertheless, the rate of the reaction never exceeded the rate obtained at the optimal concentration of Mg^{2+} alone. This is in a good agreement with some tRNA-synthetase pairs, but not with the others, where the reaction rate was higher in the presence of spermine than at any concentration of Mg^{2+} alone,[5,24] or those where spermine reverses the inhibition caused by the excess of Mg^{2+}. At the moment this variability of the tRNA response to spermine cannot be ascribed to any known feature of the structures of individual tRNAs.

SPERMINE STIMULATES THE CODON-DIRECTED BINDING OF tRNA TO THE RIBOSOMAL A AND P SITES

Aminoacyl-tRNA forms ternary complexes with GTP and the elongation factor Tu (EF-Tu). we are not aware of any systematic studies on the possible role of polyamines in this interaction (step 3, in Fig. 1). However, we have performed numerous experiments of the role of spermine in the next steps, i.e. in the binding of the ternary complex aminoacyl-tRNA EF-Tu GTP to the ribosomal A site, and also in stabilization of the peptidyl.tRNA interaction with the ribosomal P site. In these experiments, $tRNA^{Phe}$ and $tRNA^{Tyr}$ derivatives were bound to the ribosomal sites. These two tRNAs were used because there are aboundant crystallographic and other structural data about the former, and because of our own solution studies and experience with the latter. These tRNAs can be conveniently matched to codons UUU (phenylalanine) and UAU (tyrosine); therefore, we have prepared a synthetic polynucleotide, poly(U_{11},A), which contained a suitable proportion of the two codons, and used it as a substitute for mRNA. In addition to fully active ribosomes, pure EF-Tu had to be prepared for these experiments and used to form the ternary complex with aminoacyl-tRNA and GTP, which will bind the A site when added to the ribosomes. For the P site binding, acetyl-aminoacyl tRNA is a perfect substitute for peptidyl-tRNA. One version of this experiment is depicted in Fig. 3. In the other, the phenylalanyl-tRNA is acetylated and therefore directed to the P site. While the A site binding of the ternary complex is a genuine step of protein synthesis, the P-site binding of extraneous acetyl-aminoacyl-tRNA (or peptidyl-tRNA) is not: in the course of protein synthesis, the P-site is filled exclusively by translocation of peptidyl-tRNA mRNA complex from the A site (cf. step 6, in Fig. 1).

Both binding sites in fully active ribosomes can be almost completely saturated with $tRNA^{Phe}$ derivatives either with poly(U) or poly(U_{11},A) as the synthetic messenger. On the contrary, the binding of $tRNA^{Tyr}$ derivatives is considerably less efficient even after long incubations (Table 1). Spermine has a miraculous effect on this binding: the ribosomal sites can be almost completely filled in its presence. Detailed kinetic studies have shown that the initial rates of binding were several fold faster when spermine was added to tRNA prior to the binding to either site. This was also true for $tRNA^{Phe}$ derivative either with poly(U_{11},A) or even poly(U) as the messenger (Fig. 1).

Table 1. Effect of Spermine on the Binding of $tRNA^{Tyr}$ and $tRNA^{Phe}$ Derivatives to A and P Sites of Poly(U_{11},A)-Programmed E. coli Ribosomes

	A Site (From Ternary Complex)	P Site* (Acetyl Derivative)
Saturation with $tRNA^{Phe}$	73%	80%
$tRNA^{Phe}$ + SP	97%	96%
$tRNA^{Tyr}$	9%	16%
$tRNA^{Tyr}$ + SP	89%	91%
Optimum SP/$tRNA^{Phe}$	3%	3%
SP/$tRNA^{Tyr}$	5%	5%

*The P site was always filled before the addition of aa-tRNA EF-Tu GTP ternary complex. The imput ratio of ribosome:acetyl-aa-tRNA:ternary complex was always 1:1:1.

Addition of spermine did not change the optimum magnesium concentration for the binding, showing again that spermine and divalent cations do not substitute for each other but act in their specific ways. But most significantly, the number of spermine molecules which had to be added to tRNA for the best binding was rather small and specific for the tRNA species, not for the ribosomal site: the optimal spermine/tRNA ratio was 3 for $tRNA^{Phe}$ and 5 for $tRNA^{Tyr}$. Of course, these are once more the exact ratios causing structural stabilization of the two tRNAs and, for $tRNA^{Phe}$, the ratios found in crystal structure[10,11] and predicted by theoretical

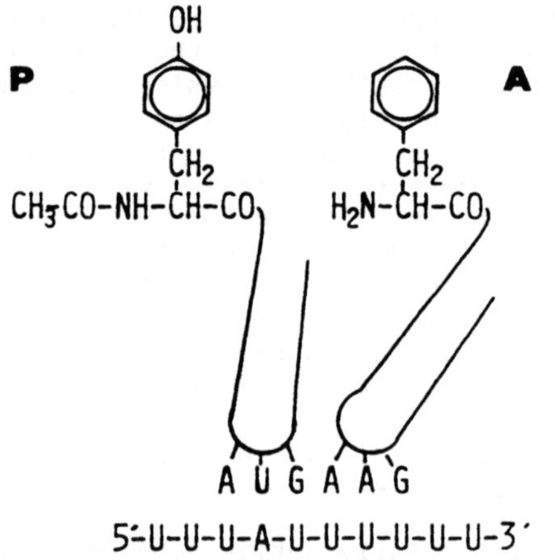

Figure 3. Acetyl-|^{14}C|Tyr-$tRNA^{Tyr}$ can substitute for peptidyl-tRNA in the P site of poly(U_{11},A)-programmed ribosomes. |^{3}H|-Phe-$tRNA^{Phe}$ shown in the decoding or A site, has to be brought there in the form of a ternary complex with the elongation factor Tu and GTP.

calculations.[12] This coincidence suggests that spermine stimulates the codon-directed tRNA-ribosome interaction by acting on tRNA; it also suggests that spermine may do so by stabilizing the biologically active tRNA conformation. Additional support for such a conclusion comes from the experiments in which the potential spermine-binding components were individually preincubated with spermine prior to allowing them to interact. Stimulation of the final binding of acetyl-tyrosyl-tRNA to the poly(U_{11},A)-programmed ribosomes was the best when the tRNA, or tRNA and the polynucleotide, were preincubated with spermine. Preincubation of the polynucleotide alone had little effect, while preincubation with the ribosomes was inhibitory!

SPERMINE PREVENTS SOME ERRONEOUS tRNA-RIBOSOME INTERACTIONS

In addition to staging the specific steps of the protein synthesis, governed by proper codon:anticodon interactions as implicated (though not explicitly shown) in Figure 1, the ribosome may work erroneously both in vivo and in vitro. One such example is misreading of the genetic code. The natural fidelity of translation can be impaired by several ribosome-binding drugs, like streptomycin,[25] or by mutations resulting in altered ribosomal proteins.[26] In the presence of streptomycin UUU codons in poly(U) can be read as leucine (codons UU^{A}_{G}), so the third (or the "wobble") base of the codon is misread. Polyamines are known to reduce such misreading by on order of magnitude.[27] The same UUU codon can be misread as isoleucine (codons AU^{U}_{C}), i.e. the first letter of the code is misread in this case. It can be easily shown that spermine reduces the misincorporation of isoleucine into the polypeptide product in vitro. We have traced this erroneous reading of the code to the ternary complex:ribosomal A site interaction. Only a few molecules of spermine per tRNA are needed to reduce the misreading. We would like to explain this effect by structural stabilization of the tRNA by spermine. Once the tRNA molecule, and in particular its anticodon region, is stabilized in the native conformation, it becomes incapable of the wrong base pairing; hence, the fidelity of translation is increased.

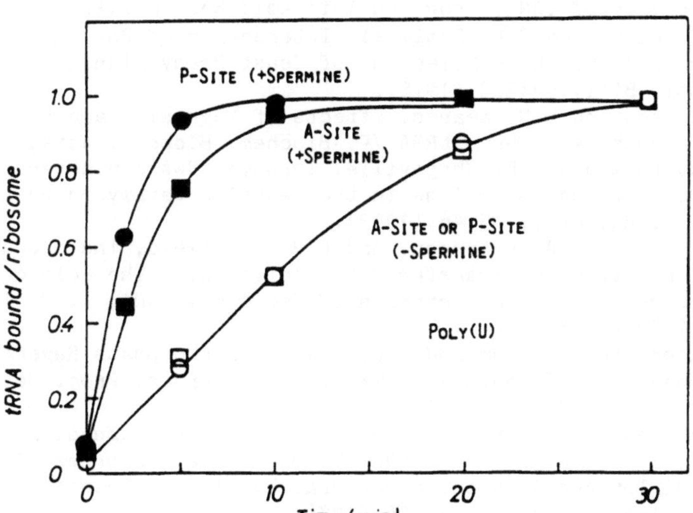

Figure 4. Time course of the binding of Phe-$tRNA^{Phe}$ and Ac-Phe-$tRNA^{Phe}$ to the A and P sites, respectively, of poly(U)-programmed ribosomes. The other site was unoccupied. Ribosomes and tRNA derivatives were present in equimolar ratios.

Another example of erroneous function of the ribosome is the messenger-free polymerization of lysine. Lysine (codons AA^{A}_{G}) can be polymerized into poly(Lys) even without its usual synthetic messenger, poly(A). This strange reaction must rely upon $tRNA^{Lys}$, since the misacylated species, Phe-$tRNA^{Lys}$, is a perfect donor for the messenger-free synthesis of poly(Phe).[28] The messenger-free synthesis of poly(Lys) requires the elongation factor G, GTP, and higher Mg^{2+} concentration than normal. While the poly(A)-directed synthesis of poly(Lys) is stimulated by spermine, the messenger-free synthesis is strongly inhibited. In neither case the addition of spermine changes the optimal Mg^{2+} concentration for the reaction. We could trace down this effect to the interference of spermine with the A-site binding. Again, this effect can be explained by assuming that the spermine-free $tRNA^{Lys}$ has a rather flexible, ill-defined tertiary structure which will be recognized by ribosome as a tRNA, but will be able to escape the screening point for the codon-anticodon matching.[29] The more rigid tRNA structure stabilized by spermine cannot avoid the contact with the codon before proceeding to the next step of protein synthesis.

All data presented in this short review point to the fact that polyamines stabilize the structure of tRNA. They also stimulate the proper, codon-directed tRNA ribosome interactions, while inhibiting some unspecific and/or erroneous interactions. Both structural and the functional effects of spermine occur at similar and rather low spermine/tRNA ratios. None of the effects described can be observed by substituting Mg^{2+} for spermine. Hence, polyamines must play a specific role in maintaining the proper, biologically active tRNA conformation. By doing so, they play a very important role in protein synthesis process by increasing its efficiency and accuracy.

REFERENCES

1. T.T. Sakai and S.S. Cohen, Effects of Polyamines on the Structure and Reactivity of tRNA, Prog. Nucleic Acid Res. Mol.Biol.,17:15 (1976).
2. A.H. Schreier and P.R. Schimmel, Interaction of Polyamines with Fragments and Whole Molecules of Yeast Phenylalanine-Specific tRNA, J. Mol. Biol.,93:323 (1975).
3. P.H. Bolton and D.R. Kearns, Effects of Magnesium and Polyamines on the Structure of Yeast $tRNA^{Tyr}$, Biochem. Biophys. Acta, 477:10 (1977).
4. S. Chousterman and F. Chapeville, Tyrosyl-tRNA Synthetase of E. coli B. Role of Magnesium Ions in the Reaction Catalyzed by the Enzyme, Eur. J. Biochem., 35:46 (1973).
5. T.N.E. Lövgren, A. Peterson, and R.B. Loftfield, The Mechanism of Aminoacylation of Transfer Ribonucelic Acid. The Role of Magnesium and Spermine in the Synthesis of Isoleucyl-tRNA, J. Biol. Chem., 253:6702 (1978).
6. C. Jelenc and C.G. Kurland, Nucleoside Triphosphate Regeneration Decreases the Frequency of Translation Errors, Proc. Natl. Acad. Sci. USA, 76:3174 (1979).
7. M. Plohl and Ž. Kućan, Effects of Spermine and Magnesium Ions on the Aminoacylation of Yeast $tRNA^{Tyr}$, Biochemie, in press (1988).
8. G.J. Quigley and A. Rich, Structural Domains of Transfer RNA Molecules, Science, 194:796 (1976).
9. J.D. Robertus, J.E. Ladner, J.T. Finch, D. Rhodes, R.S. Brown, B.F.C. Clark, and A. Klug, Structure of Yeast Phenylalanine tRNA at 3Å resolution, Nature, 250:546 (1974).
10. G.J. Quigley, M.M. Teeter, and A. Rich, Structural Analysis of Spermine and Magnesium Ion Binding to Yeast Phenylalanine Transfer RNA, Proc. Natl. Acad. Sci. USA, 75:64 (1978).

11. S.R. Holbrook, J.L. Sussman, R.W. Warant, and S.H. Kim, Crystal Structure of Yeast Phenylalanine Transfer RNA II. Structural Features and Functional Implications, J. Mol. Biol., 123:631 (1978).
12. K. Zakrzewska and B. Pullman, Spermine-Nucleic Acid Interactions: A Theoretical Study, Biopolymers, 25:375 (1986).
13. T.T. Sakai, R. Forget, I. Jacqueline, C.E. Freda, and S.S. Cohen, The Binding of Polyamines and of Ethidium Bromide to tRNA, Nucleic Acid Res., 2:1005 (1975).
14. Y. Takeda and T. Ohnishi, Polyamines and Protein Synthesis. IV. Stimulation of Aminoacyl tRNA Formation by Polyamines, Biochem. Biophys. Res. Commun., 37:917 (1969).
15. J.L. Leroy and M. Gueron, Electrostatic Effects in Divalent Ion Binding to tRNA, Biopolymers, 16:2429 (1977).
16. I. Weygand-Đurašević, V. Nöthig-Laslo, J.N. Herak, and Ž. Kućan, Conformational Changes in Yeast $tRNA^{Tyr}$ Revealed by EPR Spectra of Spin-Labelled N^6-(Δ^2-Isopentenyl)-adenosine Residue. Biochim. Biophys. Acta, 479:332 (1977).
17. I. Weygand-Đurašević, V. Nöthig-Laslo, and Ž. Kućan, Involvement of the 3' Side of the Anticodon Loop of Yeast $tRNA^{Tyr}$ in Messenger-Free Binding to Ribosomes. An Electron-Spin Resonance Study, Eur. J. Biochem., 139:541 (1984).
18. I. Weygand-Đurašević, T.A. Cruse, and B.F.C. Clark, The Influence of Elongation Factor-Tu GTP and Anticodon-Anticodon Interaction on the Anticodon Loop Conformation of Yeast $tRNA^{Tyr}$, Eur. J. Biochem., 116:59 (1981).
19. V. Nöthig-Laslo, I. Weygand-Đurašević, T. Živković, and Ž. Kućan, Binding of Spermine to $tRNA^{Tyr}$ Stabilizes the Conformation of the Anticodon Loop and Creates Binding Sites for Divalent Cations, Eur. J. Biochem., 117:263 (1981).
20. V. Nöthig-Laslo, I. Weygand-Đurašević, and Ž. Kućan, Structural Changes of Yeast $tRNA^{Tyr}$ Caused by the Binding of Divalent Ions in the Presence of Spermine, J. Biomol. Struct. Dynamics, 2:941 (1985).
21. C.C. Allende, J.E. Allende, M. Gatica, J. Celis, G. Mora, and M. Matamala, The Aminoacyl Ribonucleic Acid Synthetase I. Properties of the Threonyladenylate-Enzyme Complex, J. Biol. Chem., 241:2245 (1966).
22. E. Holler, Isoleucyl Acid Synthetase of E. coli B. Effects of Magnesium and Spermine on the Amino Acid Activation Reaction, Biochemistry, 12:1142 (1973).
23. Ž. Kućan and R.W. Chambers, Purification of Tyrosine-tRNA Ligase from Saccharomyces cerevisiae S288C, J. Biochem. (Tokyo), 73:811 (1973).
24. A.C.Carr, G.L. Igois, G.R. Penzer, and J.A. Plumbridge, The Effect of Spermine and Mg^{2+} on the Catalytic Mechanism of Isoleucin:tRNA Ligase, Eur. J. Biochem., 54:169 (1975).
25. S. Pestka, Inhibitors of Protein Synthesis, in "Molecular Mechanism of Protein Synthesis", H. Weissbac and S. Pestka, eds., Academic Press, New York (1977).
26. R. Rosset and L. Gorini, A Ribosomal Ambiguity Mutation, J. Mol. Biol., 39:95 (1969).
27. K.A. Abraham, S. Olsnes, and A. Pihl, Fidelity of Protein Synthesis in vitro is Increased in the Presence of Spermidine, FEBS Lett., 101:93 (1979).
28. G.Z. Yusupova, N.V. Belitsina, and A.S. Spirin, Template-Free Ribosomal Synthesis of Polypeptide Chains from Aminoacyl-tRNA, FEBS Lett., 206:42 (1986).
29. Ž. Kućan, On the Role of Spermine in Protein Synthesis, in "The Roots of Modern Biochemistry", H. Kleinkauf, H. von Döhren, and L. Jaenicke, eds. Walter de Gruyter, Berlin (1988).

SPERMINE INHIBITION OF BASAL AND STIMULATED ADENYLATE CYCLASE IS MEDIATED BY THE INHIBITORY GTP-BINDING PROTEIN (G_i)

Carlo Clo, Benedetta Tantini*, Pietro Sacchi and Claudio M. Caldarera

Dipartimento di Biochimica, Centro Studi e Ricerche sul Metabolismo Cardiaco, Università di Bologna, e *Istituto di Citomorfologia, C.N.R., Bologna, Italy

INTRODUCTION

In the past few years, substantial evidence has accumulated suggesting that a family of heterotrimeric (α, β, γ-subunits) guanine nucleotide binding proteins (G-proteins) may mediate a number of transmembrane signaling events. These include regulation of membrane-bound adenylate cyclase[1,2,3,4] the breakdown of phosphatidylinositol biphosphate by phospholipase C,[5] the activation of cGMP-phosphodiesterase by rhodopsin[6] and the conductance of some ionic (K^+, Ca^{2+}) channels[7]. Recent evidence also suggests that G-proteins may be regulated by phosphorylation by protein kinase C[8] and the insulin receptor tyrosyl kinase.[9]

Important insights into the pivotal role of G-proteins in coupling receptors to appropriate signal generator systems are mainly derived from studies on adenylate cyclase (AC).[2,3,4,10] The (AC) system is composed of receptor binding site (inhibitory and stimulatory) coupled to the catalytic (C) moiety by either a stimulatory (G_s) or an inhibitory (G_i) G-protein, which is activated once GTP is bound to the α-subunit and βγ-complex is released. This process is enhanced by a stimulatory or inhibitory hormone-receptor interaction. Activation is terminated by the hydrolysis of GTP to GDP by a GTPase inherent to both G-proteins, followed by recombination of α with βγ. Non hydrolyzable analogs of GTP, as guanosine 5'-(γ-thio) triphosphate (GTPγS) or guanosine 5'-(β-γ-imido) triphosphate (Gpp(NH)p), also activate G_s or G_i but, owing to their resistance to GTPase, cause persistent activation or inhibition of AC. Besides, the rate of activation of G_s and G_i is closely dependent on the amount of cations,[3,10,11,12] as Mg^{2+} and Mn^{2+}. A crucial role in defining the organization and function of AC is played by membrane phospholipids which are considered a further component of the system.[13]

Evidence has been provided by us that polyamines (PA) as spermine (SPM), spermidine (SPD) and putrescine (PTC), at μmolar doses, early and rapidly depress AC activity.[14] This occurs both in intact cells and in membranes, by lowering the affinity of the enzyme for ATP and its catalytic property.

The order of potency of the amines: SPM > SPD > PTC, probably reflects a different capacity to interact with the negatively charged groups of phospholipids[15] involved in the AC system. The goal of this work was to gain information on a possible action of PA (spermine) on the G_i component of AC, by using quiescent chick embryo heart cells (HC) in culture and the hearts from chick embryos (CEH) as source of the enzyme.

EXPERIMENTAL PROCEDURE

Primary HC ,obtained from 14 day old chick embryos, were grown as monolayer as described.[16] At confluency, the cells were fed for 20 hrs with a serum-free medium containing or not pertussis toxin (0.1 μg/ml) and added with SPM (1μM) for the last 30 min of incubation. Four plates were pooled for each experimental condition. In other experiments the hearts from 14 day old chick embryos were used as source of AC.

Cultured cells or excised hearts were rinsed twice with ice-cold 0.85% NaCl and twice with homogenization buffer (10 mM Tris-HCl, 0.25 M sucrose, 1 mM $MgCl_2$,1 mM EDTA, 1 mM DTT), pH 7.4, and homogenized in a Polytron. After centrifugation for 10 min at 300 g, the supernatants were recentrifuged at 12,000 g for 10 min and the pellet (enzyme) was resuspended in 50 mM Tris-HCl, pH 7.4, containing 0.15 M sucrose,1 mM EDTA, 1 mM DTT and 10 mM theophylline. Proteins were assayed by the method of Bradford.[17] For the assay of AC activity,the standard reaction mixture consisted of 50 mM Tris-HCl (pH 7.5), 5 mM $MgCl_2$, 5 mM theophylline, 15 mM phosphocratine, 3 U creatine-phosphokinase, 1.5 mM ATP and 20-40 μg of enzyme, in a final volume of 200 μl. When indicated, $MgCl_2$ was 2.5 or 10 mM. Forskolin, dissolved in ethanol (0.1% in the reaction), SPM, $MnCl_2$, Gpp(NH)p and GTP γ S were added alone or in certain combinations as indicated in the legends. Reactions, carried out at 37°C, were initiated by the addition of ATP and terminated by adding 300 μl of 7.5 mM EDTA, boiling and centrifuging at 2,000 g for 10 min. Aliquots of the supernatants were assayed for cAMP formed by using a specific radioimmunoassay kit from Amersham. Specific activity of AC is expressed as pmoles cAMP/mg prot/min and it is corrected for the effect of ethanol if present in the reaction. Data are the mean of 6 determinations from 3 different samples, differing by no more than 10%.

RESULTS AND DISCUSSION

The relevance of Mg^{2+} in the control of basal and stimulated AC has been well established[3,10,12] althoug the exact mechanism and sites of its action are not completely clarified. The conversion of G_s or G_i to the active state requires not only GTP but also Mg which probabily promotes the dissociation of α from $\beta\gamma$ subunits. The affinity of G_s for Mg^{2+} is lower than that of G_i and increasing concentrations of the ion lead to an increased proportion of total active G_s. Fig. 1 shows that the activity of AC from CEH is linear up to 10 min and it is closely dependent on the amount of Mg^{2+} in the assay. A progressive time-dependent inhibition of AC activity occurs in the presence of 1 μM SPM in a manner inversely related to Mg^{2+} concentration (2.5, 5 and 10 mM). Similarly, increasing amounts (1 to 10 mM) of Mg^{2+}, while progressively fa-

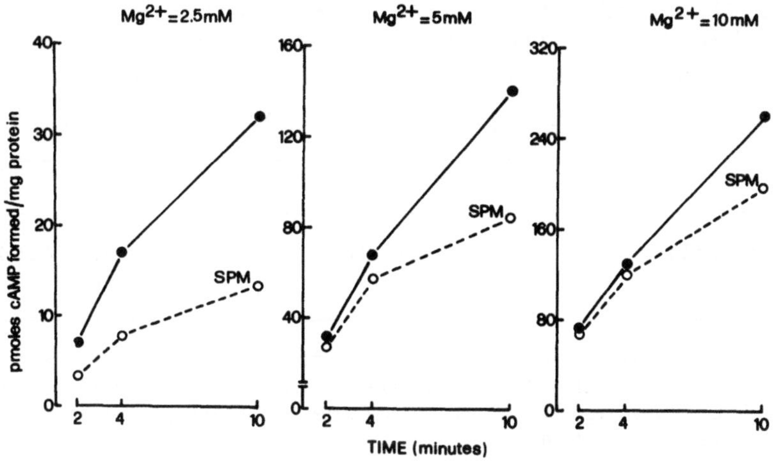

Fig.1. Influence of Mg^{2+} on SPM-induced inhibition of basal AC from CEH. The enzyme was incubated in the presence of 2.5, 5 or 10 mM $MgCl_2$ with (O) or without (●) 1 μM SPM.

vor G_s-mediated stimulation of basal AC activity by GTP γS or Gpp(NH)p, progressively overcome or completely prevent (10 mM Mg^{2+}) the ability of SPM to counteract within the first 4 min the two GTP analogs (not shown). Thus, these findings suggest the existance of a close interplay between Mg^{2+} and SPM in the promotion of active state of G_i and G_s.

The possibilty that SPM-induced inhibition of AC is mediated by G_i is supported by the fact that (Table 1) in the presence of 2 mM $MnCl_2$, a concentration able to prevent the formation of G_i-C complex,[18] SPM fails to reduce basal activity.

Table 1. Influence of Mn^{2+} on Spermine-Induced Inhibition of Chick Embryo Heart Adenylate Cyclase Activity.

Addition (10 min)	Adenylate Cyclase Activity (% of Control) -$MnCl_2$	+$MnCl_2$ (2 mM)
None (Control)	100(27.2)[a]	100(31.2)[a]
SPM (1 μM)	41	95

[a] pmoles cAMP formed/mg prot/min

Similarly, SPM is uneffective on the AC of HC cultures treated for 20 hrs with pertussis toxin (Table 2), which is known to prevents activation of G_i by catalyzing the ADP ribosilation of the α-subunit.[3,19]

Further clues as to a direct action on G_i by SPM come from the data of in vitro experiments on cardiac AC stimulated by GTP γS, Gpp(NH)p or forskolin (FORSK).This diterpene is a very powerful activator of all AC systems studied,[3,20,21] except that of sperm cells.[22] While it may be that FORSK acts by directly

Table 2. Effect of SPM on Basal Adenylate Cyclase Activity of Heart Cells Treated With or Without Pertussis Toxin

Addition (30 min)	Adenylate Cyclase Activity (% of Control)	
	- Pertussis Toxin	+ Pertussis Toxin
None (Control)	100(19.0)[a]	100(24.0)[a]
SPM (1 μM)	45	98

Confluent cultures were incubated for 20 hrs in a serum-free medium with or without pertussis toxin (0.1 μg/ml). Spermine was added to the cells 30 min before stopping the experiments. [a]pmoles cAMP/mg prot/min

interacting with C, it seems that the expression of its effect is influenced by other components of AC system, mainly G_s.[23] Besides, stimulation by FORSK is additive or more than additive with other activators, including stable GTP analogs.[3,20]

Table 3 shows that AC in CEH membranes exhibits a 16.7-fold activation by 100 μM FORSK within 10 min. Stimulation is linear up to 30 min with no apparent lag phase (not shown) and it is significantly reduced (50%) by the presence of 1 μM SPM. A 55% inhibition of basal activity is observable after 10 min of incubation with SPM alone. The possibility that SPM antagonizes the action of FORSK by favouring the active state of G_i is supported by the fact that only a 10.3-fold increase in AC activity occurs when FORSK is added following a 15 min preincubation with the amine. Table 3 also shows that cardiac AC is

Table 3. In Vitro Effect of SPM on Basal and Stimulated Adenylate Cyclase from Chick Embryo Hearts.

Addition to Preincubation (15 min)	Addition to Incubation (10 min)	Adenylate Cyclase Activity	
		%	%
None (Control)	None	100(32.6)[a]	
	FORSK	1670	100
	FORSK + SPM	831	50
	SPM	45	
SPM (1 μM)	None	41	100
	FORSK	422	1030
GTPγS (1 μM)	None	426	
	FORSK	2018	100
	FORSK + SPM	1372	68

The enzyme was preincubated for 15 min without (control) or with SPM or GTPγS prior to addition of ATP, 100 μM FORSK and/or 1 μM SPM. [a]pmoles cAMP formed/mg protein/min.

activated 4.3-fold by 1 μM GTPγS and that addition of FORSK to GTP γS-prestimulated enzyme leads to an additive stimulatory effect of the two drugs. Again the action of FORSK is antagonized by simultaneous addition of SPM, resulting in a 3.2-fold increase of GTP γS-prestimulateed activity as compared to a 4.7-fold increase in the absence of the amine. Similar results were obtained when 10 μM Gpp(NH)p was substituted for GTP γS or when cultured HC were used as source of AC (not shown). In contrast,SPM alone failed to affect the activity of the enzyme preincubated for 15 min with either GTP γS of Gpp(NH)p, which, in turn, exhibited a lower capacity to stimulate AC preincubated with SPM (not shown). As a whole, these results suggest that the inhibitory action of SPM more likely involves the activation of G_i, rather then the deactivation of G_s.

As shown in various cell types,[1,3,18,20,24] stable GTP analogs not only stimulate basal AC, but also cause inhibition of the enzyme prestimulated by other agents, mainly FORSK. Inhibition occurs through the activation of G_i, which exhibits apparent high affinity for the nucleotides.[18] It follows that, if really SPM inhibits FORSK-stimulated AC via G_i, then the combination of SPM and a GTP analog would produce no greater response than either one alone. Fig.2, clearly shows that this is the case when 100 μM FORSK-stimulated AC from HC was added for 10 min with 10 μM Gpp(NH)p or 1 μM GTP γS in the presence of 1 μM SPM. The activity of FORSK-preincubated AC is strongly reduced by GTP γS, Gpp(NH)p or SPM alone. Again, SPM was uneffective when HC were pretreated with pertussin toxin (not shown). However, the degree of inhibition caused by SPM plus Gpp(NH)p or GTP γS is almost identical to that observed with Gpp(NH)p or GTP γS alone. The study of the time course effect of SPM indicates that the amine causes the acceleration of the inhibitory action of the analogs,which,as reported with other cell types,[19,20] requires a lag phase (not shown).

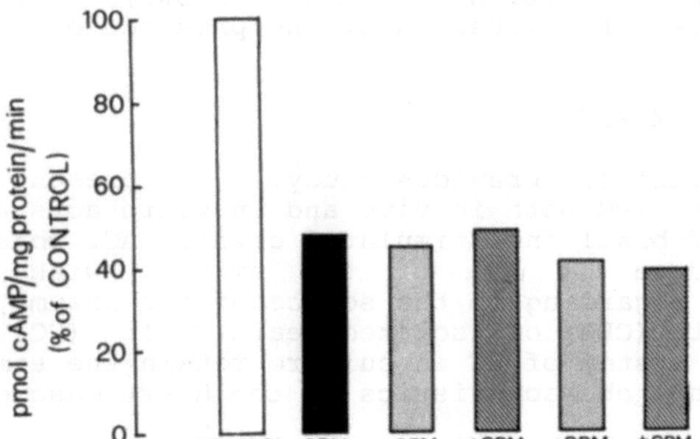

Fig.2. Inhibition of FORSK-stimulated AC from HC by Gpp(NH)p, GTP γS and/or SPM. After a 15 min preincubation with 100 μM FORSK, the assay (10 min) was initiated by adding ATP without (control) or with 10 μM Gpp(NH)p, 1 μM GTP γ S and /or 1 μM SPM. Control activity was 40-fold higher than that of FORSK-unstimulated enzyme.

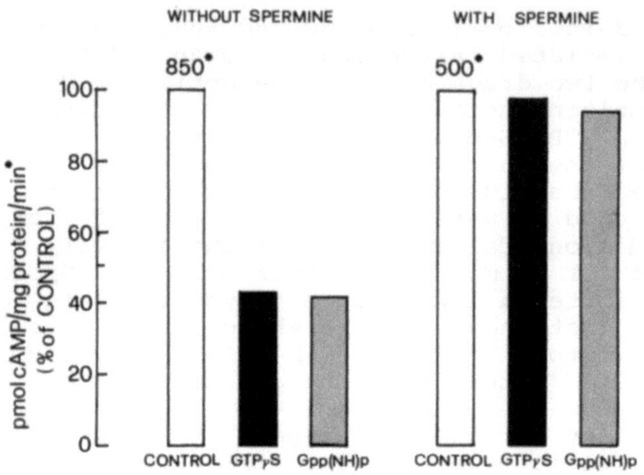

Fig.3. Influence of SPM on stable GTP analogs-induced inhibition of FORSK-stimulated AC from CEH. Experimental conditions were as given in Fig.2, except that SPM was present in the preincubation medium.

The results (Fig.3) of experiments designed to compare the efficiency of stable GTP analogs in inhibiting AC preincubated for 15 min with FORSK alone or in association with SPM, provide further support to the idea that G_i mediates SPM action. In fact,the presence of SPM in the preincubation medium, while reduces (by more than 40%) the stimulation by FORSK of basal activity (24.2 pmol cAMP/mg prot/min), almost completely prevents thereafter the inhibitory effect of 1 μM GTP γS or 10 μM Gpp(NH)p occurring within 10 min. This was oserved either when ATP was added at the end of preincubation (at the time of addition of GTP analogs) as in Fig.3, or when it was included in the preincubation medium (not shown), suggesting that SPM action is not influenced by the presence of the substrate.

CONCLUDING REMARKS

Confirming our previous study,[14] the present data provide evidence that SPM both in vivo and in vitro acts as a negative modulator of basal and stimulated cardiac AC. This occurs at a very low dose (1 μM) of the amine, with no apparent differences regarding to the source of the enzyme, being chick embryo hearts (CEH) or isolated heart cells (HC), indicating that the AC system of HC in culture retain the same structural and functional characteristics of the heart muscle.

As reported with HC exposed to 1 μM SPM,[14] in vitro inhibition of basal AC from CEH is progressive with time, with no apparent lag phase, supporting for an early interaction of SPM with some components of the AC system, as phospholipids or G_i. The possibility that G_i might be a primary site of SPM action is substained by the fact that inhibition by SPM is completely reversed by pretreating HC cultures with pertussin toxin or when AC from CEH was assayed in the presence of 2 mM $MnCl_2$. Besides, the efficiency of SPM appears inversely related to the age of the embryo (not shown) and to the ambient Mg^{2+} concentration.Progressive higher amounts of the ion progressively overcome SPM inhibition of basal activity and of G_s-mediated stimulation of AC by GTPγS or Gpp(NH)p. Assuming that G_i mediates SPM action and that AC activity mirrors the ratio between active G_i and G_s, the speculation may be made that increasing amounts of Mg^{2+} antagonize SPM action by favoring a increaded proportion of active G_s. The exact interplay between Mg^{2+} and SPM in the control of AC activity is still under investigation in our laboratory.

The possibility that SPM action is carried out through activation of the inhibitory imput (G_i) to dually regulated cardiac AC,rather than through deactivation of the stimulatory one (G_s),is further supported by the data obtained with stimulated AC. SPM reduces FORSK stimulation of AC from either HC or CEH, in all the conditions tested, i.e. when added before, after or in association with the diterpene. Furthermore, SPM, while counteracts the stimulatory action of stable GTP analogs within few minutes, i.e. during their lag phase, it was much less or completely ineffective (depending on the concentration of Mg^{2+}) thereafter,when the analogs are likely fully bound to G_s. A reduced capacity of $GTP_\gamma S$ or Gpp(NH)p to stimulate AC is observable following a 15 min preincubation of AC with the amine. The proposal of a G_i mediation of SPM action might account for the non additive inhibitory effect of SPM and $GTP_\gamma S$ or Gpp(NH)p on FORSK-prestimulated AC as well as for the failure of both the analogs to promote inhibition when AC was preincubated with FORSK and SPM. These findings suggest that SPM probably competes with the two analogs for the same site (G_i) on stimulated AC system. Obsiously additional sites of action of SPM on the membrane cannot be rule out. The antagonizing effect of SPM on FORSK is compatible with a possible direct influence of the amine on the catalytic moiety of AC. However the failure of SPM to affect FORSK-induced stimulation of AC from HC treated with pertussin toxin, suggests that G_i plays a major role in mediating SPM action. In this regard, the following hypotheses are currently under investigation in our laboratory: (a) SPM facilitates the dissociation of G_i to α and βγ .The free βγ might in turn mediate SPM inhibition by binding to the α subunit of active G_s, because of the identify of βγ in G_i and G_s,[3,6] (b) SPM favors the binding of GTP to α of G_i, (c) SPM increases the rate at which G_i trasverses the activation/deactivation cycle, (d) SPM favors the coupling of active G_i to C.

ACKNOWLEDGEMENTS

This work was supported by a grant from C.N.R., Rome, Progetto Finalizzato "Oncologia" (n.87.01238.44).

REFERENCES

1. L.E.Limbird, Activation and attenuation of adenylate cyclase, Biochem.J. 195:1 (1981).
2. A.G.Gilam, G proteins and dual control of adenylate cyclase, Cell 36:577 (1984).
3. L.Birnbaumer, J.Codina, R.Mattera, R.A.Cerione, J.D.Hildebrandt, T.Sunyer, F.J.Rojas, M.G.Caron, R.J.Lefkowitz and R.Iyengar,Regulation of hormone receptors and adenylate

cyclases by guanine nucleotide binding N proteins, in: "Recent Progress in Hormone Research", R. O. Greep, ed., Academic Press, New York, 41:41 (1985).

4. A.Levitzki, Regulation of adenylate cyclase by hormones and G-proteins, FEBS Lett. 211:113 (1987).
5. A.P.Dowson, GTP enhances inositol triphosphate-stimulated Ca^{2+} release from rat liver microsomes, FEBS Lett.185:147 (1985).
6. L.Stryer, G proteins: a family of signal transducers, Ann. Rev.Cell.Biol. 2:391 (1986).
7. M.D.Houslay, Ion channels controlled by guanine nucleotide regulatory proteins, Trends Biochem.Sci.12:167 (1987).
8. T.Katada, A.G.Gilman, Y.Watanabe, S.Bauer, and K.H.Jakobs, Protein kinase C phosphorylates the inhibitory guanine nucleotide-binding regulatory component and apparently suppresses its function in hormonal inhibition of adenylate cyclase, Eur.J.Biochem. 151:431 (1985).
9. R.M.O'Brein, M.D.Houslay, G.M.Milligan, and D.Siddle, The insulin receptor tyrosyl kinase phosphorilates holomeric forms of the guanine nucleotide regulatory proteins G_i and G_o, FEBS Lett. 212:281 (1987).
10. T.Higashijima, K.M.Ferguson, P.C.Sternweis, M.D.Smigel, and A.G.Gilman, Effects of Mg^{2+} and the $\beta\gamma$ -submit complex on the interactions of guanine nucleotide with G protein, J. Biol.Chem. 262:762 (1987).
11. J.Codina, J.D.Hildebrandt, L.Birnbaumer, and R.D.Sekura, Effects of guanine nucleotides and Mg on human erythrocyte N_i and N_s , the regulatory components of adenylate cyclase, J. Biol. Chem. 259:11408 (1984).
12. K.H.Jakobs, U.Gehring, B.Gaugler, T.Pfeuffer, and G.Schults, Occurence of an inhibitory guanine nucleotide-binding regulatory component of adenylate cyclase system in cyc variants of S49 lymphoma cells, Eur.J.Biochem.130:605 (1983).
13. D.C.Lehotay, H.K.Lo, and G.S.Levey, Activation of adenylate cyclase: requirement for phospholipids, in: "Cyclic 3':5' Nucleotides: Mechanism of Action", H.Cramer and J.Schultz, eds., John Wiley & Sons, London, pp.1 (1977).
14. C.Clo, C.Pignatti, S.Manfroni, S.Marmiroli, and B.Tantini, Modulation of membrane bound adenylate cyclase by polyamines, in: "Biomedical Studies of Natural Polyamines", C.M. Caldarera, C.Clo and C.Guarnieri, eds., CLUEB, Bologna, pp. 117 (1986).
15. B.Tadolini, L.Cabrini, L.Landi, E.Varani, and P.Pasquali, Polyamine binding to phospholipid vescicles and inhibition of lipid peroxidation, Biochem.Biophys.Res.Commun.122:550 (1984).
16. C.Clo, B.Tantini, M.N.Coccolini, and C.M.Caldarera, Mediation of polyamine-induced decrease of cAMP content by cAMP-phosphodiesterase in chick embryo heart cell cultures, J.Mol.Cell.Cardiol. 13:773 (1981).
17. M.H.Bradford, A rapid and sensitive method for the quantitation of microgram quantities of protein utilizing the principle of protein-dye binding, Anal.Biochem. 72:248 (1976).
18. K.B.Seamon, and J.W.Daly, Guanosine 5'-(β,γ-imido)triphosphate inhibition of forskolin-activated adenylate cyclase is mediated by the putative inhibitory guanine nucleotide regulatory protein, J.Biol.Chem. 257:11591 (1982).
19. K.H.Jakobs, K.Aktories, and G.Schultz, Mechanism of pertussintoxin action on the adenylate cyclase system, Eur. J.Biochem. 140:177 (1984).

20. P.A.Insel, D.Stengel, N.Ferry,and J.Honoune, Regulation of adenylate cyclase of human platelet membranes by forskolin, J.Biol.Chem. 257:7485 (1982).
21. T.H.Hudson and J.N.Fain, Forskolin-activated adenylate cyclase, J.Biol.Chem. 258:9755 (1983).
22. L.R.Forte,D.B.Bylund,and W.L.Zahler,Forskolin does not activate sperm adenylate cyclase,Mol.Pharmacol.24:42 (1983)
23. R.B.Clark, T.J.Goka, D.A.Green, R.Barber, and R.W.Butcher, Differences in the forskolin activation of adenylate cyclase in wild-type and variant lymphoma cells,Mol.Pharmacol. 22:609 (1982).
24. J.D.Hildebrandt and L.Birnbaumer, Inhibitory regulation of adenylyl cyclase in the absence of stimulatory regulation. J.Biol.Chem. 258:13141 (1983).

POLYAMINES IN PLANTS

POLYAMINES AS GROWTH SUBSTANCES IN HIGHER PLANTS

Nello Bagni and Rossella Pistocchi

Department of Biology
Institute of Botany
Via Irnerio 42
40126 Bologna, Italy

GENERAL CONSIDERATIONS

Aliphatic polyamines, recently recognized as a new class of plant growth substances or plant hormones (Bagni et al., 1982), are present in all the plants examined to date and mainly act on cell division processes. However, as other well-known plant hormones (auxins, cytokinins, gibberellins, but also abscisic acid and ethylene), they display a broad spectrum of effects on plant tissues, but very little is understood of the underlying events and processes. In spite of the wide variety of responses, the tendency is to generalize from observations made on a single type of response. We propose, therefore, to examine some of the characteristics of polyamines especially in relation to their role as plant growth substances. The early studies showed that in a plant tissue, temporarily lacking in polyamines, a positive effect on the induction of cell division was demonstrated by supplying exogenous polyamines (Bertossi et al., 1965), while in plants with a normal polyamine level this positive effect could not be recognized. Anyway a direct correlation between endogenous polyamine levels and growth rate was shown in normal and tumour plant tissues (Bagni and Serafini-Fracassini, 1979 and 1985). In the absence of plant mutants deficient in the enzymes responsible for putrescine synthesis, contrary to those isolated in microorganisms (Tabor and Tabor, 1983; Tabor et al., 1983) and animals (Pohjanpelto et al., 1981; Steglich et al., 1983), the approach used until a few years ago was to study the mechanisms of polyamine-induced plant growth by utilizing different competitive and irreversible inhibitors of polyamine synthesis.

The effects on <u>in vivo</u> systems, though not unfavourable (Bagni et al., 1981), were not as decisive as expected due to the difficulty in blocking the different pathways of putrescine synthesis; a complete block can be reached only at higher inhibitors concentrations. The different sensitivity of a plant tissue or organ to applied polyamines, which is assumed to partly depend upon the levels of endogenous polyamines and the different rates of polyamine synthesis, stimulated us to investigate the problems of uptake, transport and translocation of polyamines in order to have some evidence that polyamines could be considered plant hormones. In fact, the lack of evidences on the possibility for polyamines to be transported, their cationic nature and the lack of isolation of a receptor were in favour of their classification as "second messengers" in mediating hormonal effects.

POLYAMINE UPTAKE AND TRANSPORT IN PLANTS AND ISOLATED ORGANS

From our early investigations we noted that by spraying polyamines, at concentrations ranging from 1 to 100 µM, on apple trees at the time of fertilization, fruit-set and yield per tree as well as fruit growth significantly increased, particularly when polyamines were applied during the cell division stage (Costa and Bagni, 1983), suggesting an easy uptake of these substances. In fact, in an isolated apple corymb, fruitlets as well as young leaves absorbed exogenous putrescine and synthesized spermidine and spermine from it. Putrescine, and probably spermidine and spermine, were translocated via the peduncle from leaves to fruitlets and viceversa. This transport does not seem to be polar. The presence of C-label in the peduncle and at the bottom of the spur may suggest that translocation via the sieve tubes had occurred without excluding a transport via parenchyma cells (Bagni et al., 1984). Further research in different seedlings (tomato, maize and pine) demonstrated a rapid uptake and translocation of putrescine from roots to leaves and viceversa. The transport was temperature- and relative humidity-dependent. Polyamines were translocated via the xylem and the sieve tubes plus parenchyma cells. Translocation to the upper part of the plant was partly due to xylem transpiration but, at least in seedlings, it was not interrupted by 100% relative humidity suggesting the involvement of root pressure (Rabiti A.L., Pistocchi R. and Bagni N.; manuscript in preparation). The presence of endogenous polyamines in the xylem and phloem is in favour of their translocation under physiological conditions (Friedman et al., 1986). Contrary to our observations with putrescine, cadaverine in *Nicotiana glauca* moved quickly from roots (where it is mainly synthesized from lysine decarboxylase) to leaves and not viceversa. Cadaverine is a precursor of anabasine, the main alkaloid in *Nicotiana glauca*, whose biosynthesis takes place mainly in the roots (Bagni et al., 1986). For this reason it was not surprising to detect twice as much lysine decarboxylase activity in roots as in stems and leaves.

POLYAMINE UPTAKE IN CULTURED CELLS

The observation made with apple fruitlets (Bagni et al., 1985), seedlings (Rabiti et al.) and *H. tuberosus* explants (Bagni, 1967) that exogenously applied polyamines could be absorbed by plants, induced us to study the characteristics of the transport system involved in polyamine entry into the cells.

Experiments performed with isolated petals of Saintpaulia ionantha (the African violet), floating on a solution containing μM concentrations of labelled polyamines, showed that uptake was linear with time for up to two hours and then decreased (Fig. 1). It was dependent upon the

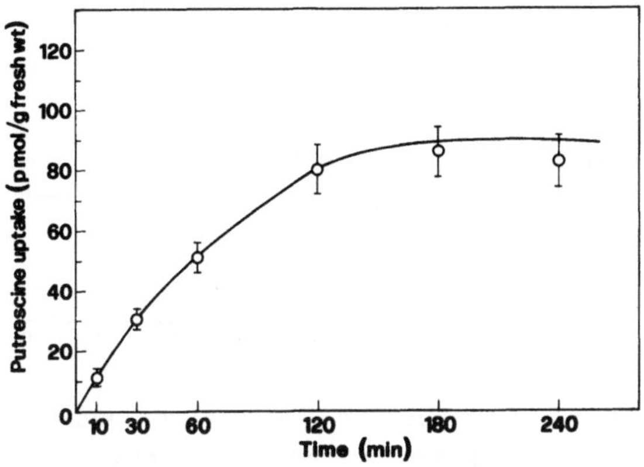

Fig. 1. Time course of putrescine uptake from upper surface of Saintpaulia petals.

external pH and two maxima were observed: one at pH 4-5 and another at pH 8 (Figs. 2 and 3) both at low and high external polyamine

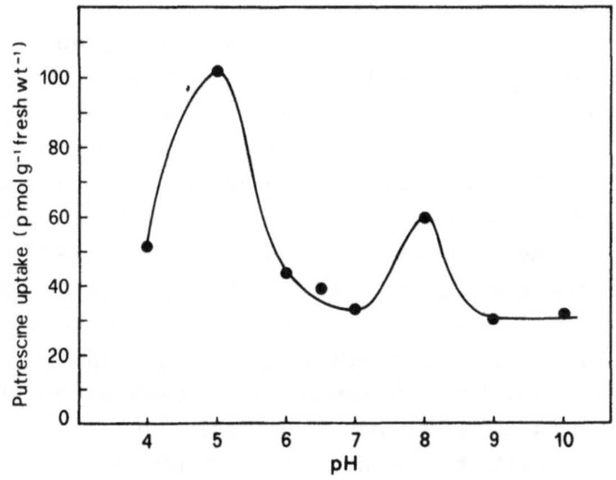

Fig. 2. Effect of pH on putrescine uptake in Saintpaulia petals at low external concentration (0.5 μM).

concentration and with all the polyamines tested (putrescine, spermidine and spermine) (Bagni and Pistocchi, 1985; Pistocchi et al., 1986). A similar dependence on the external pH was observed in the lichen Evernia prunastri (Escribano and Legaz, 1985). The uptake dependence on external concentration showed saturation kinetics and Km values in the

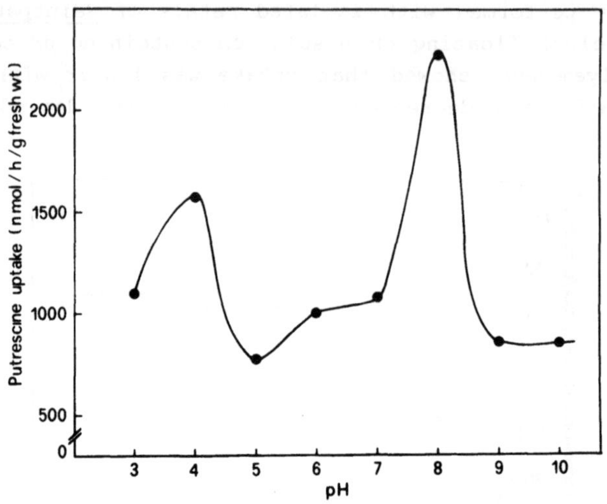

Fig. 3. Effect of pH on putrescine uptake in Saintpaulia petals at high external concentration (50 mM).

mM order of magnitude for the three polyamines tested. At low concentrations (μM) uptake occurred against the concentration gradient but the use of metabolic inhibitors, such as 2,4-dinitrophenol and carbonylcyanide-m-chlorophenylhydrazone at 10 and 100 μM concentration did not affect polyamine transport.

As the cells of a petal are mature and differentiated and probably not very responsive to metabolic events, the study was continued with a more active system, i.e. carrot cells in culture. Cells from friable callus cultures were harvested in the logarithmic phase of growth and transferred to a liquid medium containing the radioactive polyamine. Uptake was stopped by filtering the cells and by washing them with an unlabelled polyamine solution (Pistocchi et al., 1987).

The characteristics of the polyamine transport observed in this system with respect to the previous one, were a more rapid uptake with saturation after 1 - 2 minutes (Fig. 4), and a higher affinity. In this respect the uptake dependence on concentration showed a biphasic pattern. Increasing the external concentration to 5 mM (Fig. 5) gave saturation kinetics with Km values of 41.9 μM and 27.3 μM respectively for putrescine and spermidine (system I). Higher polyamine concentrations of up to 100 mM revealed the existence of a second system which consisted of a saturable component for putrescine (Fig. 6) with a Km value of 29.2 mM and a linear component for spermidine. The choice of short or long periods of time for this type of experiment did not change the pattern in the first system, while in the second system uptake in the short-term experiments was lower. The main point concerning the study on polyamine uptake in carrot cells was the discovery that a substantial amount of polyamines, 25% for putrescine and 77.5% for spermidine, was trapped in the cell walls, so that the presence of this structure interfered considerably with the study on transport across the plasmalemma.

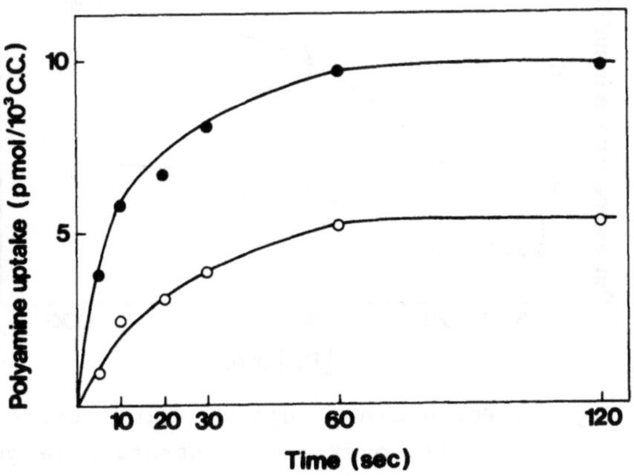

Fig. 4. Time course for putrescine and spermidine uptake in carrot cell cultures. Final concentration: 1 µM. ●, putrescine; o, spermidine. C.C. = cell clusters.

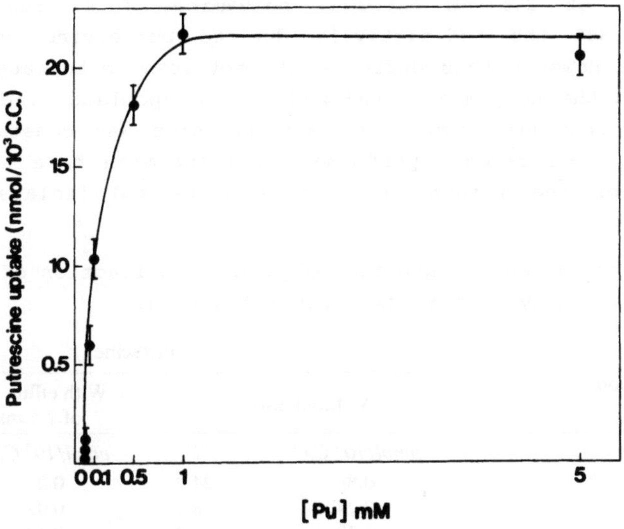

Fig. 5. Putrescine uptake in carrot cells, over a concentration range of 1 µM to 5 mM (system I), as measured in short- and long-term experiments. The experiments were performed over long (30 min) and short (20 sec) time intervals with the same results. C.C. = cell clusters.

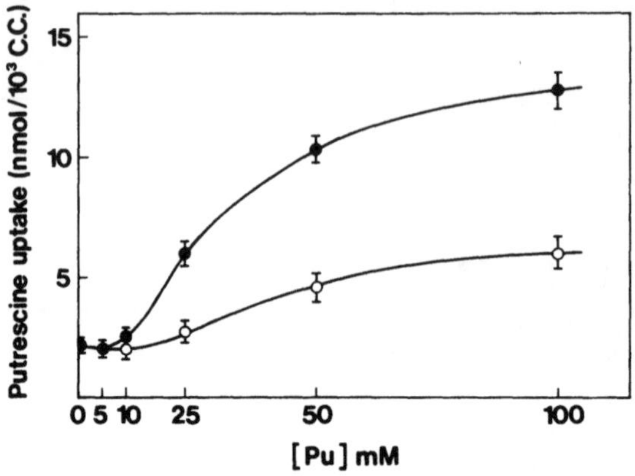

Fig. 6. Putrescine uptake in carrot cells, over a concentration range of 1 to 100 mM (system II), in short- (20 sec, o) and long-term (30 min. •) experiments. C.C. = cell clusters.

Using the fractionation method, putrescine and spermidine were found in the soluble fraction of the cytoplasm (68% and 15% respectively) (Table I). This distribution occurred also after an efflux experiment in the presence of 2.5 mM external polyamine. This treatment reduced by 92% and 84% the radioactivity due to putrescine and spermidine respectively. However this decrease did not seem to be caused only by an exchange with the polyamine localized in the apoplast since the percentage of the polyamine found in each fraction was constant (Table I).

Recently, experiments performed with spermine have shown that the time-course of the uptake into whole cells and isolated cell walls

Table I. Distribution of supplied polyamines in fraction of carrot cells obtained by differential centrifugation.

Fraction	Putrescine			
	Without efflux		With efflux in the presence of 2.5 mM putrescine	
	pmol/10³ C.C.	%	*pmol/10³ C.C.*	%
Cell walls	0.84	24.2	0.07	25.0
Particulate	0.24	6.9	0.02	7.1
Cytoplasmic	2.39	68.9	0.19	67.9
Total pmol uptaken	3.47		0.28	

	Spermidine					
	Without efflux		With efflux in the presence of 2.5 mM spermidine		With efflux in the presence of 2.5 mM Ca^{2+}	
	pmol/10³ C.C.	%	*pmol/10³ C.C.*	%	*pmol/10³ C.C.*	%
Cell walls	11.0	73.0	1.6	64.3	8.2	77.5
Particulate	2.9	19.3	0.5	20.3	1.4	12.9
Cytoplasmic	1.2	7.7	0.4	15.4	1.0	9.6
Total pmol uptaken	15.1		2.5		10.6	

was different: with cell walls there was no increase in spermine uptake with time going from 10 seconds to 10 minutes and the amount of polyamine absorbed was equal to that absorbed by the cells at 10 seconds, showing that polyamine uptake in the short term was not only due to binding to the cell surface except for the shortest time tested (10 sec).

The effect of calcium on polyamine uptake was investigated as calcium has been found to be necessary in several transport studies. In our system we found that putrescine uptake was stimulated at low calcium concentration (10 µM) and inhibited at higher levels (Fig. 7); for spermidine uptake there was an increasing stimulation in the 10 µM to 1 mM calcium concentration range (Table II) and the same was observed for spermine uptake. La^{3+}, an inhibitor of calcium transport, nullified the stimulatory effect of calcium on putrescine uptake (Fig. 7); up to 100 µM concentration, it mimicked the stimulatory effect of calcium on spermidine uptake, but at 1 mM it was highly inhibitory also in the presence of calcium (Table II). These results

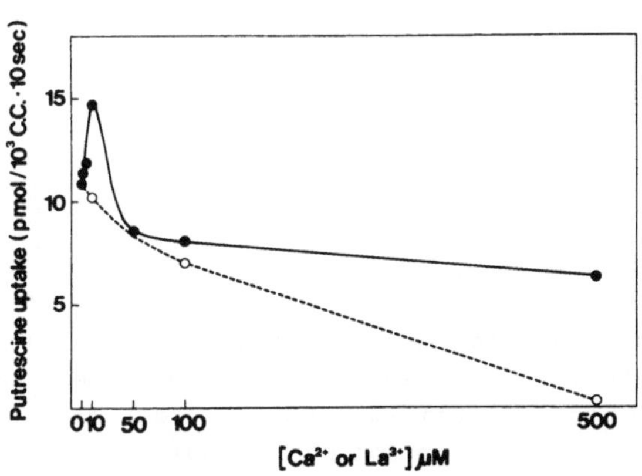

Fig. 7. Effect of increasing concentration of Ca^{2+} and La^{3+} on putrescine uptake in carrot cells. (•) putrescine uptake in the presence of Ca^{2+}; (o) putrescine uptake in the presence of La^{3+}, with or without the same concentration of Ca^{2+}. C.C. = cell clusters.

suggest that spermidine and spermine utilize a different transport system from that of putrescine, one that is activated by calcium, while La^{3+} probably interferes with the sites of polyamine uptake. An apparently higher specificity of the transport system for spermidine was revealed also by competition experiments in which it was observed that only putrescine uptake was affected by the presence of fairly low concentrations of other polyamines, while spermidine uptake was affected only by very high concentrations. Experiments designed to investigate the driving forces for polyamine uptake demonstrated that it was only partially inhibited by metabolic inhibitors.

Table II. Effect of Ca^{2+} and La^{3+} on spermidine uptake in carrot cells.

Cation Concentration	Spermidine Uptake		
	Ca^{2+}	$Ca^{2+}+La^{3+}$	La^{3+}
µM		pmol/10^3 C.C	
0 (control)		5.40 ± 0.54	
10	6.22 ± 0.62	4.72 ± 0.48	5.63 ± 0.20
50	7.46 ± 0.45		
100	8.12 ± 0.52	10.51 ± 0.72	13.32 ± 0.61
500	12.86 ± 0.58		
1000	14.35 ± 0.65	1.83 ± 0.28	1.45 ± 0.33

SPERMIDINE UPTAKE IN ISOLATED PROTOPLASTS AND VACUOLES

Owing to the high degree of binding between polyamines and cell walls, uptake studies were subsequently performed with protoplasts and then extended to isolated vacuoles in order to obtain information on the subcellular compartmentation of polyamines as well.
Protoplasts and vacuoles were isolated from phloem parenchyma of carrot taproots; uptake was measured by adding ^{14}C-labelled spermidine to protoplasts and vacuoles, suspended in their appropriate medium, and stopped by centrifuging them through a layer of silicon oil (Pistocchi et al., 1988).

Spermidine uptake into protoplasts and vacuoles was very rapid reaching a maximum after 1 to 2 min (Fig. 8); the insertion of a washing layer of 100 mM unlabelled spermidine between two silicon oil layers, showed that a significant amount of spermidine binding to the protoplast and vacuole surface occurred (Fig. 8). Spermidine uptake into protoplasts displayed a linear dependence on external concentration (Fig. 9). Uptake into isolated vacuoles, however, displayed a biphasic concentration dependence: a saturable component below 1 mM (Km = 61.8 µM) was followed by a linear component up to the highest concentration tested (50 mM). The pH-dependence of spermidine uptake into protoplasts was shown to increase linearly from pH 5.5 to 7.0; in vacuoles there was a distinct optimum at pH 7.0 (Fig. 10).

By comparing the endogenous polyamine content of protoplasts with that of vacuoles isolated from them, it appeared that 24% and 28% of the intracellular putrescine and spermidine, respectively, were vacuolar in location (Table III). However leakage of polyamines

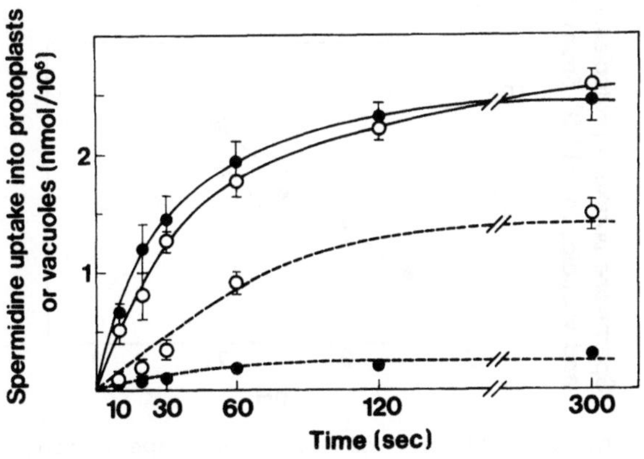

Fig. 8. Time course for spermidine uptake into carrot protoplasts and vacuoles (final concentration: 6.6 µM). (●), protoplasts; (o), vacuoles; (——), no washing layer; (- - -), washing layer with incubation medium plus 100 mM unlabelled spermidine.

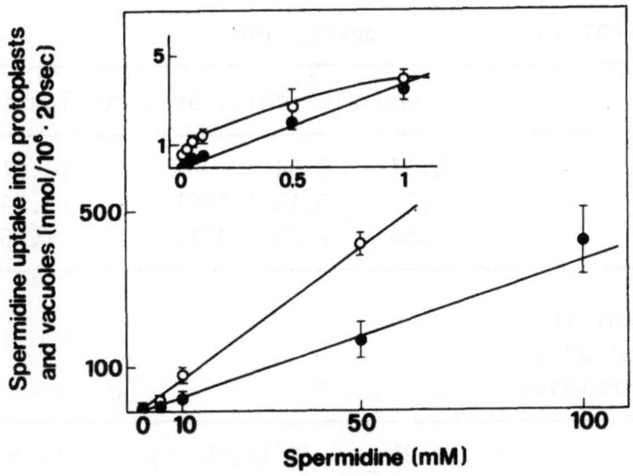

Fig. 9. Spermidine uptake into carrot protoplasts (●) and vacuoles (o) over a concentration range of 6.6 µM to 1 mM and of 1 to 100 mM.

out of the vacuole occurred during the isolation procedure. Preloading the protoplasts with 100 mM cold spermidine and then isolating the vacuoles, caused the total spermidine content to increase 20-fold and that of the vacuoles by a factor of 6 as compared to the endogenous content (Table III). When the compartmentation analysis was performed

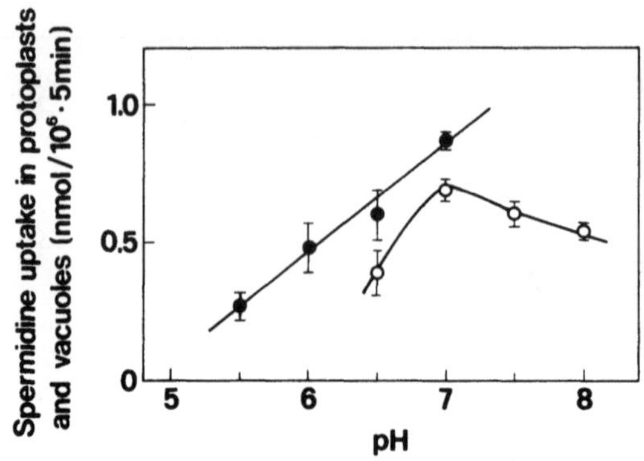

Fig. 10. Effect of pH on spermidine uptake into carrot protoplasts (●) and vacuoles (o).

Table III. Endogenous polyamine content of carrot taproots cells, protoplasts and vacuoles and subcellular localization of exogenously applied spermidine (100 mM cold or 6.6 μM labelled spermidine for 5 min) to protoplasts.

Endogenous content in	putrescine	spermidine
	$nmol/10^6$ cells or protoplasts or vacuoles	
cells	3.57 ± 0.54	10.67 ± 1.44
protoplasts	3.23 ± 0.15 (100%)	3.74 ± 0.38 (100%)
vacuoles	1.36 ± 0.23 (42%)	1.06 ± 0.16 (28%)
Exogenous content in protoplasts preloaded with 100 mM spermidine		
protoplasts	2.98 ± 0.38 (100%)	73.70 ± 2.87 (100%)
vacuoles	1.28 ± 0.33 (43%)	6.04 ± 0.10 (8%)
Exogenous content in protoplasts preloaded with 6.6 μM spermidine		
protoplasts	n.d.[a]	0.41 ± 0.06 (100%)
vacuoles	n.d.	0.11 ± 0.01 (27%)

[a] = not determined

by incubating the protoplasts with 6.6 μM ^{14}C spermidine, the distribution of the polyamine taken up was such that 27% of it was localized in the vacuole.

Thus the vacuole seems to represent a, perhaps temporary, storage site for polyamines with a higher specialization for transport than the plasmalemma, as the results for concentration- and pH- dependent uptake suggest.

CONCLUSIVE REMARKS

Studies aimed at understanding polyamine transport are made difficult by the cationic nature of these compounds which gives them the possibility to interact with many negatively charged sites both outside and inside the cell. We have demonstrated the existence of a high degree of interaction between polyamines and cell walls, plasmalemma and tonoplast; at any rate uptake and transport of polyamines in different plant systems does occur.

In mature plants and seedlings polyamines are taken up by roots and leaves; translocation occurs in both directions and acropetal transport seems to be due partly to xylem transpiration. At the cellular level it has been observed, using Saintpaulia petals and carrot cell cultures, that polyamine uptake is pH-dependent and that it shows saturation kinetics in response to the external concentration. When the interactions with the cell walls were eliminated by using protoplasts, uptake appeared to be linear. For this reason and owing to the impossibility of demonstrating a clear energy-dependence for polyamine transport, at the present time, it can be hypothesized that polyamines enter through the plasmalemma by means of a passive transport, possibly via an exchange with other polyamines and/or cations. The saturation kinetics, sometimes biphasic, observed in the presence of cell walls could signify that this structure partecipates in some way in the transport process, possibly as a reservoir of polyamines with different binding capacities. The transport system located in the tonoplast, which is apparently more specialized, suggests that this compartment intervenes in the regulation of the cytoplasmic level of polyamines.

REFERENCES

Bagni, N., 1967, Absorption of spermine and spermidine and their degradation in explants of Helianthus tuberosus in vitro, Z. Pflanzenphysiol., 57: 22.

Bagni, N., Baraldi, R. and Costa, G., 1984, Translocation and metabolism of aliphatic polyamines in leaves and fruitlets of Malus domestica (cv. "Ruby Spur"), Acta Hortic., 149: 173.

Bagni, N., Creus, J., and Pistocchi, R., 1984, Distribution of cadaverine and lysine decarboxylase activity in Nicotiana glauca plants, J. Plant Physiol., 125: 9.

Bagni, N., and Pistocchi, R., 1985, Putrescine uptake in Saintpaulia petals, Plant Physiol., 77: 398.

Bagni, N., and Serafini-Fracassini, D., 1979, Polyamines and plant tumors, Ital. J. Biochem., 28: 392.

Bagni, N., and Serafini-Fracassini, D., 1985, Involvement of polyamines in the mechanism of break of dormancy in *Helianthus tuberosus*, Bull. Soc. bot. Fr. Actual bot., 132: 119.

Bagni, N., Serafini-Fracassini, D., and Torrigiani, P., 1982, Polyamines and cellular growth processes in higher plants, in: "Plant Growth Substances 1982", P. F. Wareing, ed., Academic Press, London.

Bagni, N., Torrigiani, P., and Barbieri, P., 1981, Effect of various inhibitors of polyamine synthesis on the growth of *Helianthus tuberosus*, Med. Biol., 59: 403.

Bertossi, F., Bagni, N., Moruzzi, G., and Calderara, C.M., 1965, Spermine as a new growth-promoting substance for *Helianthus tuberosus* (Jerusalem artichoke) in vitro, Experientia, 21:80.

Costa, G., and Bagni, N., 1983, Effect of polyamines on fruit set of apple, Hort. Science, 18: 59.

Escribano, M. I., and Legaz, M. E., 1985, Putrescine accumulation does not affect RNA metabolism in the lichen *Evernia prunastri*, Endocyt. C. Res., 2: 239.

Friedman, R., Levin, N., and Altman, A., 1986, Presence and identification of polyamines in xylem and phloem exudates of plants, Plant Physiol., 82: 1154.

Pistocchi, R., Bagni, N., and Creus, J.A., 1986, Polyamine uptake, kinetics and competition among polyamines and between polyamines and inorganic cations, Plant Physiol., 80 : 566.

Pistocchi, R., Bagni, N., and Creus, J. A., 1987, Polyamines uptake in carrot cell cultures, Plant Physiol., 84: 174.

Pistocchi, R., Keller, F., Bagni, N., and Matile, P., 1988, Transport and subcellular localization of polyamines in carrot protoplasts and vacuoles, Plant Physiol., 87: in press.

Pohjanpelto, P., Virtanen, I., and Holtta, E., 1981, Depletion of polyamines results in disappearance of actin filaments and microtubules in a polyamine-auxotrophic Chinese hamster ovary cell line, Nature, 293: 475.

Steglich, C., Choi, J., and Scheffler, E., 1983, Mammalian cell mutants with altered levels of ornithine decarboxylase activity, in: "Advances in Polyamines Research", 4, U. Bachrach, A. Kaye, and R. Chayen, eds., Raven Press, New York.

Tabor, H., and Tabor, C. W., 1983, Polyamines biosynthesis and function in *Escherichia coli*, in: "Advances in Polyamines Research", 4, U. Bachrach, A. Kaye, and R. Chayen, eds., Raven Press, New York.

Tabor, C.W., Tabor, H., and Tyagy, A. K., 1983, Biochemical and genetic studies of polyamines in *Saccharomyces cerevisiae*, in: "Advances in Polyamines Research", 4, U. Bachrach, A. Kaye, and R. Chayen, eds., Raven Press, New York.

POLYAMINES IN GROWTH AND DIFFERENTIATION OF PLANT CELL CULTURES: THE EFFECT OF NITROGEN NUTRITION, SALT STRESS AND EMBRYOGENIC MEDIA

Arie Altman, Nitsa Levin, Pnina Cohen, Madlyn Schneider and Barry Nadel

The Otto Warburg Center for Biotechnology in Agriculture

The Hebrew University of Jerusalem, Faculty of Agriculture

P.O. Box 12, Rehovot, Israel

INTRODUCTION

In vitro culture of plant cells and tissues has an increasingly important role in the advancement of both basic and applied aspects of plant growth and development. This includes the use of plant tissue cultures for the introduction of new traits by cell selection and genetic engineering, clonal micropropagation, pathogen elimination, as well as for elucidation of several molecular and metabolic events. Controlled organogenesis and/or embryogenesis in cell and tissue cultures (i.e. regeneration of new plants) and selection of specific cell lines, are prerequisites for the practical utilization of the aspects mentioned above. Regeneration from tissue cultures is easily achieved in some plant species such as tobacco and carrot. Several agricultural crops and most woody plants are especially recalcitrant. In most cases growth of cell cultures and *in vitro* embryogenesis and organogenesis is manipulated by the use of known plant hormones only, in an empirical manner, and very little is known on the underlying mechanisms, and on the use of additional or alternative means which may regulate regeneration.

Recent investigations confirm that active cell division and treatment with plants hormones result in significant changes in polyamine (PA) content and biosynthesis, in many plant species and with respect to many developmental processes[1,2,3,4,5]. Changes in hormone levels and synthesis in response to PA treatments have been studied only occasionally, primarily in relation to ethylene[6]. The interaction of PA with auxins, gibberellins and ethylene have received more attention than with cytokinins and abscisic acid, and hence more experimental data are available for the former[1]. Likewise, sharp and rapid changes in PA content and biosynthesis have been documented when intact plants, or plant tissues, were exposed to stress conditions such as accelerated senescence, increased medium osmotica and adverse pH, drought and heat conditions[7,8]. Data on the interaction of salinity stress and polyamines are rather scarce, and plant tissue cultures have been used only rarely for such studies[9].

Growth and differentiation of plant tissue cultures is primarily dependent on the composition of the nutrient medium, and especially on appropriate relative concentrations of hormones (auxins and cytokinins, and in some cases also gibberellins, abscisic acid and ethylene). Therefore, the interaction of PA, growth and plant hormones in tissue cultures seem to be more obvious than in other experimental systems, and plant cell cultures offer a unique opportunity to study and elucidate the role of PA in cell growth and differentiation.

Following the pioneering studies which demonstrated that PA play an important role in growth and activation of *Helianthus* tissue *in vitro*[3,10], PA have been studied recently in tissue cultures of other plant species, in relation to growth and morphogenesis[1]. Thus, analysis of PA content of cells and tissues, measurement of ADC, ODC and other enzymes, and the use of specific inhibitors of PA biosynthesis have indicated that PA play a role in induction of embryogenesis in carrot cell suspension cultures[11,12,13,14], callus formation, cell growth, and organogenesis in leaf explants of tobacco[15,16], *Passiflora*[17], *Pinus radiata* cotyledons[18] and *Helianthus* tuberosus[19].
In addition to their effect on growth and morphogenesis of plant tissue cultures, PA content, biosynthesis and metabolism were studied with respect to inhibitor resistance, selection of mutants, and formation of secondary metabolites and conjugates, in suspension cultures of tobacco, carrot and Paul's scarlet rose[20,21,22,23,24]. Few additional studies were directed at elucidating the role of PA in protoplast and cell senescence of *Avena* and *Alnus*[25,26], and the response of tobacco cells to increased media salinization[9].

The above clearly indicate that a better insight into the role of PA can be gained by simultaneous studies of changes in PA and cell development in response to modification of the culture medium. In the following, we shall review some data which were obtained in 3 experimental systems:

1. Changes in polyamine accumulation in tobacco cell cultures in response to nitrogen nutrition and inhibitors of PA biosynthesis.
2. Polyamines in sensitive and NaCl-resistant tobacco cell suspensions.
3. Formation and growth of somatic embryos in embryogenic celery cultures.

I. INTERACTIONS OF INHIBITORS OF POLYAMINE BIOSYNTHESIS AND NITROGEN NUTRITION IN TOBACCO CALLUS AND CELL CULTURES

Unlike their effect in several mammalian tissues and microorganisms, α DFMO and α DFMA do not cause considerable polyamine depletion when applied to intact plants or excised organs and tissues. Neither do they inhibit substantially growth, although inhibition of ODC and ADC was found in many cases[29]. The relatively large pools of inorganic and organic nitrogenous compounds and reserves in plant cells may be one reason for this lack of inhibitor effect on depletion of polyamines, whose synthesis is tightly linked to the glutamine – citrulline – arginine – ornithine cycle[2,28]. Nitrogen content and metabolism of cell cultures can be controlled conveniently and directly by modifying medium composition (total nitrogen, and the ratio of ammonium-N to nitrate-N). Therefore, the effects of nitrogen nutrition, and polyamine inhibitors, alone and in combination, were studied in short-term (1–2 subcultures, every 8–10 days) and long-term (over 10 subcultures, or "generations") of tobacco (*Nicotiana tabacum* x N. gossii) cell suspension cultures.

Tobacco cell cultures exhibit continuous growth during 8–10 days, reaching a stationary phase thereof, as measured by the increase in packed cell volume, fresh and dry weight. The molar ratio of free polyamines is ca. 100 : 10 : 1 (putrescine : spermidine : spermine), and their cellular content decrease in parallel to the decrease in growth rate (Table 1).

Table 1. Growth and free polyamine content of tobacco cell suspension cultures[a].

			polyamines, nmol/g fresh wt		
harvest day	PCV, %	fresh wt., g/flask	Put	Spd	Spm
0	10	1.4	540	25	tr.
4	21	3.0	632	56	tr.
11	63	9.0	432	27	tr.
14	64	8.8	384	18	tr.

[a] Cells were subcultured into fresh standard Murashige & Skoog medium (see also Table 4) at a rate of 10% (v/v), and harvested at the indicated time. Growth (percent packed cell volume, PCV, and fresh weight) and free polyamine content were recorded. tr. : traces.

The presence of inhibitors of polyamine biosynthesis throughout the short-term culture period resulted in considerable changes, specific to the inhibitors, in free PA content of tobacco cells. Thus, both α DFMO and α DFMA inhibited considerably putrescine accumulation, whereas α DFMO, but not α DFMA, brought about an increase of spermidine, and especially spermine, content (Table 2). Cyclohexylamine (CHA), at 2 mM, induced a marked accumulation of free spermidine, as evident also by the total PA content in the presence of CHA. A concentration of 0.5 mM CHA was not as effective. This effect was unexpected, in view of the fact that CHA is regarded as an inhibitor of spermidine synthase[29]. It can be argued that either CHA is not a specific inhibitor in this cell system, or that free spermidine is released from bound or conjugated PA, or else that it affects the overall nitrogen nutrition. MGBG, at 2 mM, caused a considerable increase in putrescine and spermidine content. When DFMO was present continuously in long-term cultures (during more than 10 generations, or subcultures), it did not affect cellular PA content, and the contents of all three PA were very similar to those of control. Moreover, cell lines which grew for several generations in the presence of DFMO evinced resistance to higher concentrations of DFMO when transferred, gradually, to 5 → 7.5 → 10 mM DFMO. This was combined with lower cellular level of putrescine, and elevated spermidine and spermine contents. This "resistance" may indicate selection of cells with modified PA metabolism, possibly overproducing cell lines, but other mechanisms should not be excluded. This aspect is currently being investigated.

Table 2. The effect of inhibitors of polyamine biosynthesis on growth and free polyamine content of tobacco cell suspensions[a].

	PCV,	polyamines, nmol/g fresh wt			
	%	Put	Spd	Spm	total
Control	66	288	13	2	303
α DFMO	54	38	40	10	88
α DFMA	70	70	13	tr.	83
CHA	61	288	560	6	854
MGBG	46	400	40	tr.	540

[a] Cells were cultured for 7 days in a standard medium with the indicated inhibitors (all at 2mM). Growth (% packed cell volume, PCV) and polyamine content were recorded.

The data mentioned above were obtained when cells were cultured in standard Murashige & Skoog Medium (see also Table 4) which contains 2,4–D (1mg/l) as the only hormone. Since growth of çell cultures is dependent on hormone concentrations, and since plant hormones have been shown previously to affect PA content, the effect of 2,4–D concentration was studied (Table 3). Growth was not affected by reducing 2,4–D concentration, or even when it was omitted. This is not unexpected in the first subculture. However, 2,4–D concentration affected considerably both putrescine and spermidine content, though a direct correlation with 2,4–D concentration could not established. This indicated that hormone concentration can affect PA content, independent of their effect on growth.

Table 3. The effect of 2,4–D concentration on growth and polyamine content of tobacco cell suspensions[a].

2,4–D mg/l	fresh wt. g/flask	PCV %	PA content, nmol/g fresh wt. Put	Spd
1	9.8	62	218	32
0.25	12.0	70	120	88
0	10.8	75	173	15

[a] Cells were cultured for 8 days in a standard Murashige & Skoog medium, with different concentrations of the auxin 2,4–D.

The nitrogen levels of the medium, in combination with several inhibitors of PA biosynthesis, were investigated with respect to culture growth and cellular PA content. The data for DFMO (at 5 mM) indicate several significant interactions (Table 4). Growth was only slightly inhibited by DFMO, whereas combinations of DFMO and nitrogen depletion inhibited growth by ca. 50%. Depletion of ammonium nitrate (i.e. both ammonium-N and nitrate-N) resulted in a marked decrease in putrescine content, a small increase in spermidine, with no effect on spermine content. Depletion of potassium nitrate resulted in ca. 50% increase in putrescine content, and a 4 fold increase in spermidine content. Total nitrogen depletion was accompanied by a considerable reduction in the cellular content of all 3 polyamines. The combined effect of DFMO and nitrogen deletion (NH_4NO_3, or KNO_3, or both) are especially noteworthy: while the effect on putrescine content is rather moderate, a dramatic increase in spermidine content (5–7 fold of control) and spermine content (7–14 fold of control) is evident.

Table 4. The effect of nitrogen nutrition and α DFMO, alone and in combination, on growth and free polyamine content of tobacco cell cultures[a].

	DFMO,	PCV	polyamines, nmol/g fresh wt.			
medium	mM	%	Put	Spd	Spm	total
Control	0	69	312	20	3	335
"	5	60	45	22	7	74
$-NH_4NO_3$	0	76	40	22	0	66
"	5	50	117	136	42	295
$-KNO_3$	0	77	448	80	4	532
"	5	40	102	128	39	269
$-NH_4NO_3$ $-KNO_3$	0	49	145	3	0	170
"	5	33	73	108	23	204

[a] Cells were cultured for 8 days in a standard Murashige & Skoog medium[b], or with modified nitrogen regimes[c], without or with α DFMO. Growth (% packed cell volume, PCV) and polyamine content are presented.

[b] Control medium: Murashige & Skoog medium (macronutrients, micronutrients, vitamines, 3% w/v sucrose), containing also 2,4–D (1 mg/l).

[c] Nitrogen content:

control:	NH_4NO_3 (1650 mg/l) and	KNO_3 (1900 mg/l)
$-NH_4NO_3$:	–	KNO_3 (1900 mg/l)
$-KNO_3$:	NH_4NO (1650 mg/l)	–
$-NH_4NO_3$ $-KNO_3$:	–	–

The interactions of nitrogen nutrition and DFMO (and several other inhibitors, as will be reported elsewhere), showing clear additive or synergistic effects, are now being studied. Preliminary data show that the changes in PA content involve also marked changes in ADC and ODC activity, and in the incorporation of several PA precursors. A shift in the ratio of free to "bound" PA is also possible, e.g. release of spermidine and spermine from conjugated compounds, to "compensate" for changes in nitrogen metabolism and cation/anion balance.

2. POLYAMINES IN SENSITIVE AND NaCl-RESISTANCE TOBACCO CELL LINES

Changes in polyamine content and metabolism in plants, and especially accumulation of putrescine and induction of ADC activity, accompany and may be causally related to a variety of stress conditions. This include potassium and magnesium deficiency, ammonium feeding, media acidification, osmotic stress and accelerated senescence[2,7,8]. Salinity stress has not been investigating extensively with respect to PA, or results were not conclusive. Our previous studies indicated that intact plants respond to high NaCl levels by a major increase in ODC activity and incorporation of arginine and ornithine in roots, resulting in translocation and accumulation of putrescine in leaves[28,30]. Changes in PA content, in response to salt stress, were studied also in cell cultures of *Nicotina sylvestris*[9].

A long term (6–10 days) exposure of wild-type, unadapted, cells to NaCl was accompanied by a steady increase in spermidine and spermine contents, whereas cellular putrescine levels fluctuated. A considerable increase in the content of total free amino acids, in response to salt, was noted in both cell lines. Changes in PA in the presence of $CaCl_2$ were smaller than in its absence, indicating that the response to NaCl stress may involve a cation balance. The addition of DFMO in combination with DFMA (each at 1mM) inhibited part of the NaCl-induced increase in putrescine content, whereas DFMO alone and DFMA alone did not modify the response to NaCl. A short exposure of wild type, sensitive, tobacco cells to a 150mM NaCl salt stress resulted in a considerable increase in putrescine content within 24h which, however, reached the control levels at 48h (Fig. 2). The salt-resistant cell line, which have been grown without salt for several subcultures, evinced as well a moderate increase in putrescine content, which continued for 48h (Fig. 2) and up to 10 days (unpublished data).

Selection of cell lines which are resistant to high NaCl concentration in the medium has been achieved by stepwise exposure of cell suspensions and calli of several plant species, including tobacco. The availability of a salt resistance, stable, tobacco cell lines (which grows well in the presence of 250–300 mM NaCl), and of the sensitive wild type[31], was used in our studies with respect to the content of free and bound PA, and the activity of ADC and ODC. Major differences in polyamine content of the lines, when grown in their "standard" medium (i.e. the wild type with no salt, the resistant line with 250mM NaCl) were found. Thus the contents of free putrescine, spermidine and spermine in sensitive, unadapted cells were 2–6 fold higher than in the resistant cell line, throughout the culture period (Fig. 1).

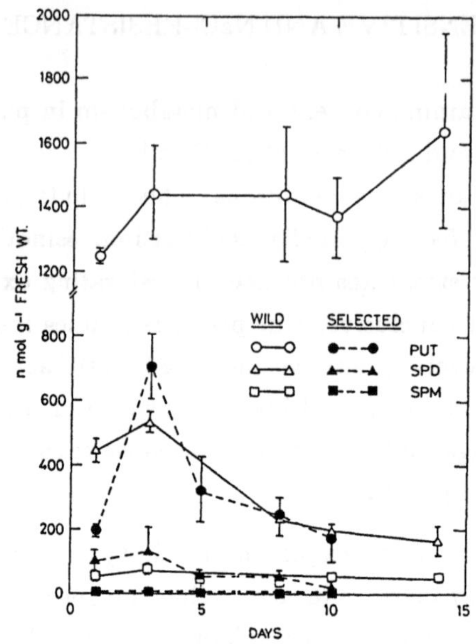

Fig. 1 Free polyamine content in wild type (sensitive) tobacco cells and in a salt resistant cell line.

The sensitive, wild-type, suspension cultures were grown for 14 days in a standard Murashige & Skoog medium; the salt resistant, selected, cell line was grown in the same basal medium, but with 250 mM NaCl.

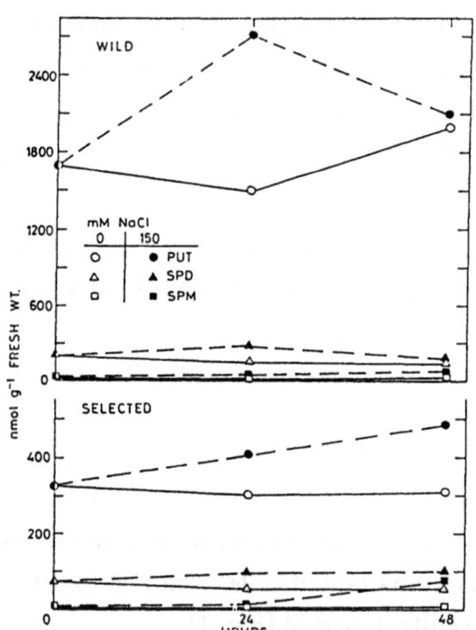

Fig. 2 Changes in free polyamine content upon short exposure of tobacco cells to NaCl. Wild type, sensitive, cells were subcultured in a standard medium. Salt-resistant, selected, cells was subcultured routinely with 250 mM, and were gradually transfered to a medium without NaCl prior to the experiment. Both cell lines were then exposed to 150 mM for 24 and 48 h, and the contents of polyamine were recorded.

3. POLYAMINES AND SOMATIC EMBRYOGENESIS IN CELERY

Asexual (vegetative) reproduction in plants, i.e. regeneration from somatic cells, is achieved *in vitro* by two basic morphogenetic patterns: organogenesis (regeneration and proliferation of shoots and/or roots), and embryogenesis (bipolar organization, leading to simultaneous root and shoot formation, resembling sexual embryogenesis). Somatic embryos usually arise from single cells which undergo two distinctive differentiation stages: (a) induction and initiation of embryogenic cell clumps, in the presence of auxins (mainly 2,4–D) and (b) embryo development and maturation in the absence of auxins. It has been established that several other factors are important: reduced nitrogen source in the medium, competent primary explants, and establishment of permissive condition for embryo maturation, The manipulation of somatic embryogenesis, however, is rather empirical, and the metabolic and molecular control mechanisms are only little understood[32].

Recent studies indicate that putrescine biosynthesis and PA play a role in cellular differentiation during somatic embryogenesis in carrot. Thus, a requirement for polyamines and ADC activity, in interaction with 2,4–D, was established by the use of DFMA, MGBG and CHA[12,13]. It was also noted that inhibition of putrescine biosynthesis via the ODC pathway, by using DFMO, induced somatic embryogenesis in carrot even in the presence of otherwise inhibitory concentrations of the auxin 2,4–D[14]. Somatic embryogenesis in celery (*Apium graveolens* L.) is, in principal, similar to that reported previously for carrot, but unlike carrot it is highly efficient, it can be synchronized, and a clear distinction can be made between the various morphological structures: cells, embryogenic clumps, aggregated embryos, and free-floating single embryos[33]. The role of PA levels and biosynthesis in growth and differentiation of celery embryogenic cell suspensions, is the subject of the present study.

Embryogenic cell suspensions were established from petiole and leaf callus, and subcultured routinely in a "maintenance medium" (with 3.2μM 2,4–D and 0.88 μM benzyl-aminopurine). Development of somatic embryos occurred on a "regeneration medium" (with 2.3 μM kinetin and 4% mannitol). Synchronized cultures were produced by fractionation through a series of screens. We found that induction of somatic embryos in celery, as well as the differentiation and development of somatic embryos, are accompanied by changes in the content of free PA and the activity of ADC and ODC. Moreover, several inhibitors of PA biosynthesis, and addition of certain PA to the medium, affected growth and embryo differentiation. Growth of embryogenic cultures (increase in cell volume, i.e. percent "packed" volume) continued for at least 24 days, with a lag period between the 8th and 12th day. (Fig. 3). Growth of the 200μ–800μ fraction (consisting mainly of cell clumps and globular embryos) was inhibited by 0.2 mM MGBG, and more so by 2 mM, from the 4th day on. Spermidine at 2mM did not relieve the MGBG-induced inhibition, and improved growth in the presence of 2mM MGBG only slightly. Growth of the larger size fraction (800μ–1400μ, consisting of differentiated embryos and clumps of embryos) was completely inhibited by 2mM

MGBG, but the inhibition by 0.2 mM MGBG was relieved once the embryos became more mature, after 20 days in culture. This may indicate that maturation and later stages of embryo development are not dependent on S-adenosyl methionine decarboxylase activity, but the direct participation of PA in this process has yet to be demonstrated. The importance of PA for differentiation of celery embryos, as distinct from growth, was demonstrated in another experimental system. Plating of embryogenic clumps (removed from suspension cultures) on agar-solidified medium permits one to follow the advancement of differentiation, and to record the formation of complete plantlets, as distinct from only shoot or only root formation. MGBG was found to inhibit both growth and normal differentiation, i.e. plantlet formation. Root formation was not inhibited by MGBG, however, suggesting that shoot formation may be more sensitive to inhibition of SAM-decarboxylase activity. The direct involvement of PA in this morphogenetic process is indicated by the fact that addition of spermidine to the medium dramatically promoted and enhanced plantlet formation (Fig. 4). Thus, the majority of the embryogenic clumps formed only shoots, and only a small number of fully-differentiated plantlets were observed in the control medium. In the presence of 0.1mM spermidine, however, almost 80% of the embryogenic clumps formed complete plantlets. In addition, the number of plantlets per clump was much higher in spermidine-enriched medium than in the control.

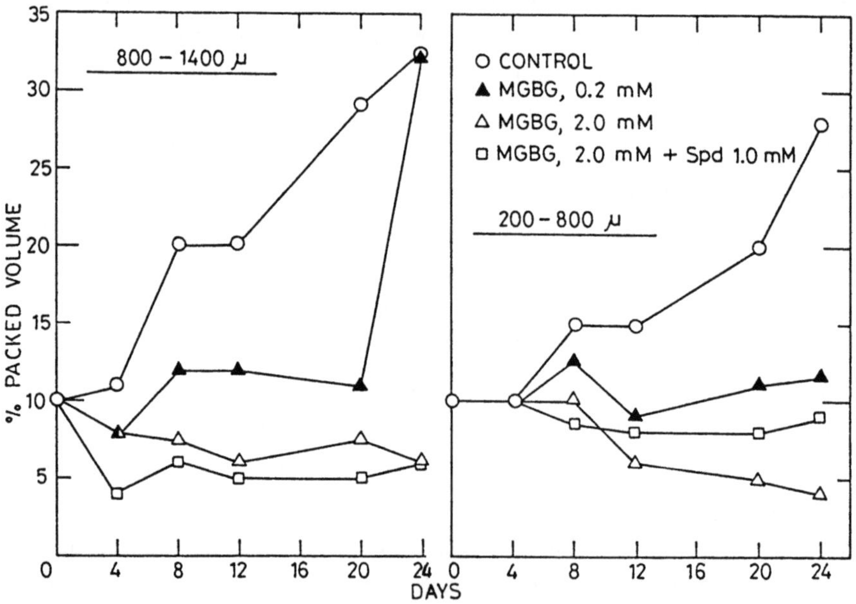

Fig. 3 The effect of MGBG on growth of celery embryogenic cultures.
Embryogenic cell suspensions of celery were sieved through appropriate screens, yielding 2 size fraction: 200–800 μ (mainly cell clumps), and 800–1400 μ (mainly clumps of embryos and larger embryos). MGBG and spermidine were added at the start of the experiments, and growth (measured as packed volume) was recorded at 4 day intervals.

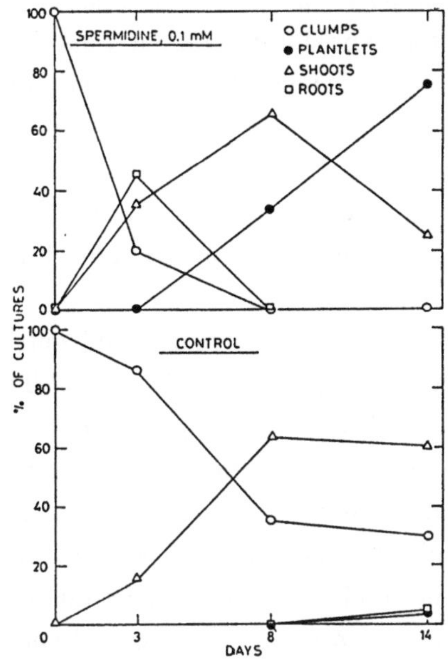

Fig. 4 Morphogenic patterns of embryogenic cell cultures of celery as affected by spermidine.
Embryogenic clumps, separated from celery cell suspensions by sieving, were placed individualy on agar-solidified regeneration medium (control), or with spermidine (0.1 mM). The various developmental stages were scored for two weeks.

CONCLUDING REMARKS

In addition to direct evidence for changes in PA content of plants in response to hormone application, major fluctuation in PA also occur during developmental processes which are normally regulated by plant hormones. Notable, among these, are growth and morphogenesis of tissue cultures, growth of some organs, and certain aspects of senescence. The data mentioned above, and other studies, provide evidence that PA content and biosynthesis play a role in the control of plant tissue culture growth and morphogenesis in a few selected cases. As cell and tissue cultures are dependent on the presence of plant hormones in the medium, PA and plant hormones probably interact in some manner. In most cases, however, the evidence is only circumstancial, and direct evidence to a causal interaction between PA metabolism and plant hormones is lacking. In fact, it can be argued that PA change simultaneously with growth and differentiation of cell cultures, but that these processes are independent of each other. Somatic embryogenesis and cell activation may be notable exceptions.

Further elucidation of the role of polyamines in growth and differentiation of plant cell cultures should be aimed at the following issues:

1. **Timing** – do changes in polyamines biosynthesis, i.e. enzyme activation and synthesis, preceed cell mitosis and differentiation? what are the mechanisms? The use of appropriate inhibitors and molecular hybridization should greatly aid answering this issue.
2. **Localization** – can specific polyamines and enzymes of PA biosynthesis be traced to specific cells, or group of cells, or tissues, during cell division and differentiation? Histochemical localization and tissue *in situ* hybridization seem to be the techniques of choice.
3. **Culture media** – how do growth hormones and nutritional constituents of the culture medium affect PA biosynthesis and content? The study of nitrogen metabolism with respect to PA is of special importance.
4. **"Seconday" metabolism and binding** – how are arginine, ornithine and putrescine diverted into alternative pathways, other than tri-and tetra-amines? Diversion into "secondary" plant metabolites, (e.g. alkaloids, cinnamoyl putrescine) and binding to insoluble cell fraction may mask the interactions between a given growth response and PA content, and should not be overlooked.

ACKNOWLEDGEMENTS

The assistance and collaboration of A. Viterbo, and M. Dekel is gratfully acknowledged.

This research was supported by a grant from the United States-Israel Binational Agricultural Research and Development Fund (BARD) to A.A., and by a grant from Land Baden-Württemberg, University of Hohenheim, FRG.

We thank Dr. P.P. McCann of the Merrell-Dow Research Institute for the gift of DFMO and DFMA.

REFERENCES

1. A. Altman, Polyamines and plant hormones, *in*: "The Physiology of Polyamines", U. Bachrach and Y.M. Heimer, eds., CRC Press, Boca Raton (1988).
2. T.A. Smith, Polyamines, *Ann. Rev. Plant Physiol.* 36:117 (1985).
3. N. Bagni, D. Serafini-Fracassini, and P. Torrigiani, Polyamines and cellular growth processes in higher plants, *in*: "Plant Growth Substances 1982", P.F. Wareing, ed., Academic Press, London (1982).
4. R.D. Slocum, R. Kaur-Sawhney, and A.W. Galston, The physiology and biochemistry of polyamines in plants, *Arch. Biochem. Biophys.*, 235:283 (1984).
5. M. Schwartz, A. Altman, Y. Cohen, and T. Arzee, Localization of ornithine decarboxylase and changes in polyamine content in root meristems of *Zea mays*, *Physiol. Plant.* 67:485 (1986).

6. A. Apelbaum, A. Goldlust, and I. Icekson, Control by ethylene of arginine decarboxylase activity in pea seedlings and its implication for hormonal regulation of plant growth, *Plant Physiol.* 79:635 (1985).
7. H.E. Flores, N.D. Young, and A.W. Galston, Polyamine metabolism and plant stress, *in*: "Cellular and Molecular Biology of Plant Stress", UCLA Symposia on Molecular and Cellular Biology, New Series, Vol. 22, J.L. Key, and T. Kosuge, eds., (1985).
8. A. Altman, R. Friedman, D. Amir, and N. Levin, Polyamine effects and metabolism in plants under stress conditions, *in*: "Plant Growth Substances 1982", P.F. Wareing, ed., Academic Press, London, (1982).
9. N.I. Shevyakova, B.P. Strogonov, and I.G. Kiryan, Metabolism of polyamines in NaCl-resistant cell lines from Nicotina Sylvestris, Plant Growth Reg. 3:365 (1985).
10. N. Bagni, Aliphatic amines and a growth factor of coconut milk as stimulating cellular proliferation of *Helianthus tuberosus* (Jerusalem artichoke) *in vitro*, *Experienta*, 22:732 (1966).
11. M.J. Montague, J.W. Koppenbrink, and E.G. Jaworski, Polyamine metabolism in embryogenic cells of *Daucus carote.* I. Changes in intracellular content and rates of synthesis, *Plant Physiol.* 62:430 (1978).
12. R.P. Feirer, G. Mignon, and J.D. Litvay, Arginine decarboxylase and polyamines required for embryogenesis in the wild carrot, *Science*, 223:1433 (1984).
13. A.A. Fienberg, J.H. Choi, W.P. Lubich, and Z.R. Sung, Developmental regulation of polyamine metabolism in growth and differentiation of carrot cultures, *Planta*, 162:532 (1984).
14. S.C. Minocha, and C. Robie, The role of 2,4–D and polyamines in somatic embryogenesis in carrot cell cultures, *12th Inter Conf. Plant Growth Substances, Abstracts*, 116 (1985).
15. J. Martin-Tanguy, The occurence and possible function of hydroxycinnamoyl acid amides in plants, *Plant Growth Regul.* 3:381 (1985).
16. P. Torrigiani, M.M. Altamura, G. Pasqua, B. Monacelli, D. Serafini-Fracassini, and N. Bagni, Free and conjugated polyamines during de novo floral and vegetative bud formation in thin layers of tobacco, *Physiol. Plant.* 70:453 (1987).
17. H.V. Desai, and A.R. Metha, Changes in polyamine levels during shoot formation, root formation, and callus induction in cultured *Passiflora* leaf discs, *J. Plant Physiol.* 119:45 (1985).
18. S. Biondi, P. Torrigiani, A.A. Sansovini, and N. Bagni, Inhibition of polyamine biosynthesis by dicyclohexylamine in cultured cotyledon of *Pinus radiata*, *Physiol Plant.* 72:471 (1988).
19. P. Torrigiani, D. Serafini-Fracassini, and N. Bagni, Polyamine biosynthesis and effect of dicyclohexylamine during the cell cycle of *Helianthus tuberosus* tuber, *Plant Phsyiol.* 84:148 (1987).
20. J. Berlin, Formation of putrescine and cinnamoyl putrescines in tobacco cell cultures, *Phytochemistry*, 20:53 (1981).

21. T.A. Smith. G.R. Best, A.J. Abott, and E.D. Clements, Polyamines in Paul's scarlet rose suspension cultures, *Planta* 144:63 (1978).
22. A.F. Tiburcio, R. Kaur-Sawhney, R.B. Ingersoll, and A.W. Galston, Correlation between polyamines and pyrrolidine alkaloids in developing tobacco callus, *Plant Physiol.* 78:323 (1985).
23. R.L. Malmberg, Biochemistry, cellular and developmental characterization of a temperature-sensitive mutant of *Nicotiana tabacum* and its second site revertant, *Cell* 22:603 (1980).
24. A.C. Hiatt, J. McIndoo, and R.L. Malmberg, Regulation of polyamine biosynthesis in tobacco, *J. Biol. Chem.* 261:1293 (1986).
25. A. Altman, R. Kaur-Sawhney, and A.W. Galston, Stabilization of oat leaf protoplasts through polyamine-mediated inhibition of senescence, *Plant Physiol.* 60:570 (1977).
26. O. Huhtinen, J. Honkanen, and K. Simola, Orthinine– and putrescine-supported divisions and cell colony formation in leaf protoplasts of alders (*Alnus glutinosa* and *A. incana*), *Plant Sci. Lett.* 28:2 (1982).
27. R.D. Slocum, and A.W. Galston, Inhibition of polyamine biosynthesis in plants and plant pathogenic fungi, *in:* "Inhibition Polyamine Metabolism", P.P. McCann, A.E. Pegg, and A. Sjoerdsma, eds., Academic Press, N.Y. (1987).
28. A. Altman, R. Friedman, and N. Levin, Alternative metabolic pathways for polyamine biosynthesis in plant development, *in:* "Advances in Polyamine Research", Vol. 4, U. Bachrach, A. Kaye, and R. Chayen, eds., Raven Press, New York, (1983).
29. C.W. Tabor, and H. Tabor, Polyamines, *Ann. Rev. Biochem.* 53:749 (1984).
30. R. Friedman. N. Levin, and A. Altman, Presence and identification of polyamines in xylem and phloem exudates of plants, *Plant Physiol.* 82:1154 (1986).
31. A.E.A. Watad, L. Reinhold, and H.R. Lerner, Comparison between a stable NaCl-selected *Nicotiana* cell line and the wild type: K^+, Na^+, and proline pools as a function of salinity, *Plant Physiol.* 73:624 (1983).
32. K. Nomura, and A. Komamine, Molecular mechanisms of somatic embryogenesis, *Oxford Surveys of Plant Molec. and Cell Biol.* 3:456 (1986).
33. B. Nadel, A. Altman, and M. Ziv, Regulaton of somatic embryogenesis in celery cell suspensions I. Promotive effects of mannitol on somatic embryo development, submitted.

THE DI- AND POLYAMINE OXIDASES OF PLANTS

Terence A. Smith and Jacqueline H.A. Barker

Research Station
Long Ashton
Bristol, BS18 9AF
U.K.

INTRODUCTION

Although the di- and polyamine oxidases of plants are apparently limited in their distribution, in some species they are remarkably active. Earlier work on these enzymes has been summarized by Smith (1985a), Morgan (1985), Rinaldi et al., (1986), and Mondovi and Avigliano (1987).

Since these enzymes produce peroxide, they are easily detected and estimated by methods linking peroxidase with the oxidation of a phenolic substrate, though much pioneering work of high quality was accomplished by measuring oxygen uptake using Warburg manometry (Hill and Mann, 1964). Recent research has utilised the oxygen electrode, which has better sensitivity and is more easily adaptable for studies of enzyme kinetics. Activity measurements have also depended on the estimation of the oxidation residue of the amine. For example pyrroline, a product of putrescine oxidation by diamine oxidase, may be measured by formation of its yellow adduct with 2-aminobenzaldehyde (Holmstedt et al., 1961).

Plant amine oxidases can now be obtained by simple procedures in large amounts, and with high specific activity, and they are usually quite stable on storage. These enzymes can be used as a means of estimating amines, or in measuring the activities of enzymes which form amines, like arginine decarboxylase (Smith, 1978; 1979; 1983a) and spermidine synthase (Suzuki et al., 1981; Sindhu and Cohen, 1983). The present review includes an assessment of an improved procedure for the spectrophotometric estimation of arginine decarboxylase based on this principle.

DIAMINE OXIDASE

This enzyme, which is found principally in the Leguminosae, has a broad specificity and oxidizes putrescine and spermidine by the mechanism shown in Fig. 1.

Characterization and purification

The diamine oxidase (DAO) from pea seedlings, which has been studied for many years, is now well characterized. McGowan and Muir (1971), and Yanagisawa et al., (1981) used a multi-step procedure for purification of the epicotyl DAO, but recently DAOs from Pisum, Lens, and Cicer (Angelini et al., 1985; Federico et al., 1985a) have been purified in a single step on an immunoaffinity column. The DAO was eluted in pH 11.5 buffer containing glycerol (1:1). Recovery was about 70%, with purification in excess of 1000-fold to apparent homogeneity.

Each DAO appears to be formed from a dimer of similar subunits, each dimer in the legume DAO having 2 atoms of copper (Mondovi and Avigliano, 1987) which can be removed by chelating agents. The copper does not undergo a valency change during the oxidation cycle. The DAO is colored pink, even when the copper is removed, but it becomes yellow on adding substrate in the absence of oxygen (Rinaldi et al., 1984). DAO from Euphorbia latex or lentil seedlings showed a free-radical type of ESR spectrum on reaction with substrate in the presence of cyanide (Finazzi-Agro et al., 1984).

Lentil DAO which appeared to be homogeneous by ultracentrifugation showed two components by gel HPLC and PAGE, but at least 6-7 isoenzymes were found by electro-focussing. These forms were not interconvertible and the separation pattern could be modified on incubation with glycohydrolases, suggesting that the different forms of DAO may depend, at least in part, on the composition of the carbohydrate side chains (Oratore et al., 1987a).

In recent work, DAO has been purified to homogeneity (135-fold) from leaves of clover (Trifolium subterraneum), and three isoenzymes have been demonstrated (Delhaize and Webb, 1987). The Mr was 150 000, and the copper content was 1.5 atoms/dimer. It appears that some copper may have been displaced during purification. The partially purified DAO from soybean (Glycine max) (Mr 225 000) was characterized by a high activity at pH 9, and an almost negligible rate at pH 7 (Suzuki and Matsumoto, 1985). The DAO from Euphorbia characias latex has been purified to homogeneity (Mr 144 000) and showed two identical subunits with two Cu^{2+} atoms/dimer (Rinaldi et al., 1982). Roots of Nicotiana rustica contain a DAO (Mr 105 000) which is probably concerned in the formation of nicotine from N-methylputrescine (Haslam and Young, 1987).

A DAO from rice seedlings which was purified to homogeneity has a Mr of 123 000 with two apparently identical subunits. This enzyme was unusual since it contained four molecules of FAD per dimer and copper was apparently absent. The substrate specificity was broad, but the enzyme was more active with diamines than with polyamines (Chaudhuri and Ghosh, 1984). An uncharacterized DAO has been detected in extracts of wheat leaves (100 pkat/g fr. wt) using a method in which the pyrroline formed from ^{14}C-labelled putrescine is selectively extracted into toluene for counting (Nadeau and Chouinard, 1987). A similar method has been used to demonstrate DAO activity in Helianthus tuberosus (Torrigiani et al., 1987).

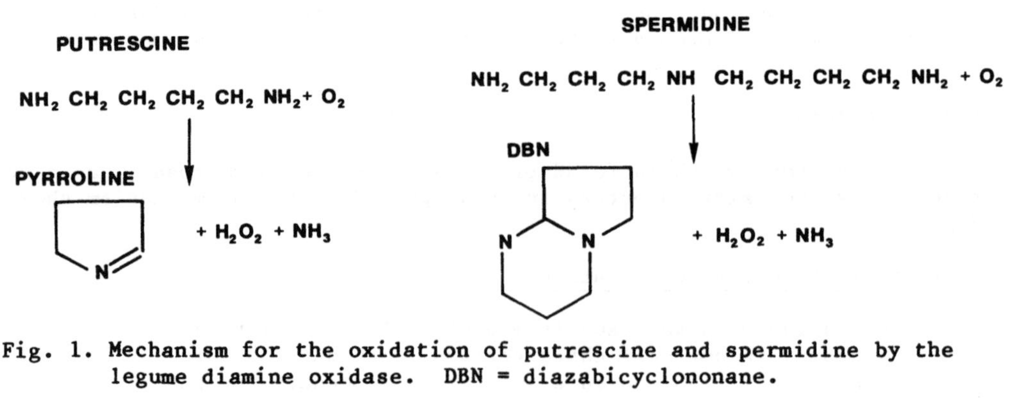

Fig. 1. Mechanism for the oxidation of putrescine and spermidine by the legume diamine oxidase. DBN = diazabicyclononane.

Although it has been known for many years that DAO contains active carbonyl groups, it is only recently that this factor has been tentatively identified as the pink colored pyrroloquinoline quinone (PQQ) (Fig. 2) (Dooley et al., 1987; Glatz et al., 1987), or a similar molecule (Suzuki, 1987). In the Euphorbia DAO two essential -SH groups per mole of dimeric enzyme are exposed on denaturation, though these are unreactive in the native state (Floris et al., 1983).

The spectrum of the pea seedling DAO up to 2 hr after the addition of amine at -15°C is independent of the substrate and identical with that obtained by anaerobic addition of substrate at room temperature. The amine is converted to the corresponding aldehyde in an anaerobic stage, leaving the enzyme in an intermediate, "reduced" form, which is subsequently reconverted to the native state on reaction with O_2, eliminating H_2O_2 and NH_3 (Kluetz et al., 1980).

Physiology

Activity of the pea seedling cotyledon DAO was induced on soaking the seed in 0.1% putrescine, spermidine or ornithine (Srivastava et al., 1977). The embryo DAO was unaffected by these treatments. DAO in 3-day old lentil seedling epicotyls grown in darkness was greatly reduced on subsequent irradiation for four days with continuous red or blue light by comparison with the dark-grown controls. Far red light gave less inhibition. The same effects were reproduced by 10 min daily exposure to these treatments. On irradiation with red light for 10 min, followed by far red for 15 min, it was shown that the inhibitory effect of red light could be reversed (Angelini et al., 1988).

A positive correlation was found between DAO activity and putrescine and cadaverine concentration in the epicotyl of pea and lentil seedlings. By contrast the levels of spermidine and spermine were in general inversely related to DAO content (Federico and Angelini, 1988).

Subcellular localization

The pea seedling DAO occurs at a very high concentration in the epicotyls, where it represents about 3% of the total protein (Yanagisawa et al., 1981). Federico and Angelini (1986) infiltrated the cell walls of lentil and pea epicotyls under vacuum with 10 mM potassium phosphate buffer and collected the apoplastic fluid by centrifugation. Disc gel electrophoresis of this extra-cellular fluid showed only one main band on staining for protein or for DAO activity. The simultaneous presence of DAO and peroxidase in cell walls suggests that the peroxide generated on oxidation of the amines could be utilized by the peroxidase (Angelini and Federico, 1986). It is intriguing to find that most enzymes which contain PQQ are extracellular (cf. serum amine oxidase) (Ferguson, 1984).

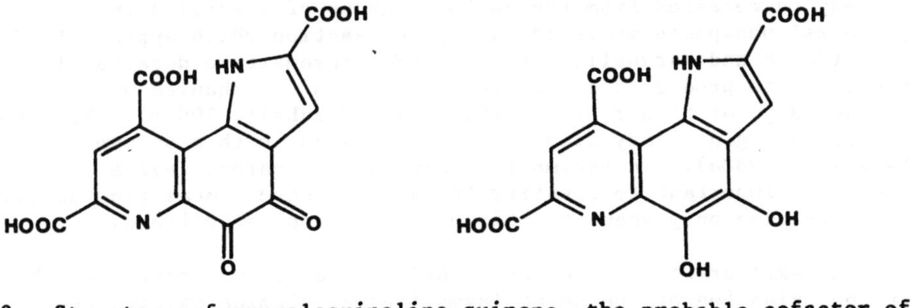

Fig. 2. Structure of pyrroloquinoline quinone, the probable cofactor of diamine oxidase. The reduced form is shown on the right.

Immunological studies

In an investigation of the reaction between lentil DAO and its antibody, the copper-free enzyme and the enzyme modified with phenylhydrazine showed a more efficient lattice formation than the native enzyme. The presence of the copper appears to confer a changed configuration which reduces the interaction with antibody. This was confirmed by the effect of molar guanidine: the copper-free enzyme was readily denatured, while the native enzyme showed much greater stability (Oratore et al., 1987b).

In a radioimmunoassay for DAO, antibodies were incubated with ^{125}I-labelled enzyme and with either unlabelled DAO or with plant extract. The complexes were isolated, and the radioactivity recovered was a measure of the competition attributable to the non-radioactive DAO in the plant extract. The sensitivity was about 5 ng of enzyme protein. Using this assay, it was shown that activity is not determined by the concentration of an inhibitor and is related to changes in absolute amounts of enzyme protein (Federico et al., 1985b). DAO was localized in lentil seedlings by an immunohistochemical method, based on fluorescein-labelled antibodies raised against lentil DAO. DAO was present mainly in association with membranes and cell walls of the leaves. In roots, the enzyme showed a strong association with the xylem in the differentiated zone, and with cell walls and nuclei in differentiating tissue. In the cotyledons, the DAO was found in the plasma membranes, on the membranes of the protein bodies and on the vacuolar surface (Federico et al., 1985c).

On adding cupric ions to copper deficient clover plants, DAO activity was restored to normal levels only in leaves which had not emerged when the copper was added. This suggests that active DAO can only be synthesized in young leaves (Delhaize et al., 1985). This was confirmed by a pulse chase experiment with $^{14}CO_2$, followed by immunoprecipitation of DAO, which again showed that only leaf primordia and very young leaves could synthesize the enzyme. Apo-DAO was absent from copper deficient leaves, indicating that synthesis of DAO is controlled by the copper concentration (Delhaize et al., 1986).

ESTIMATION OF ARGININE DECARBOXYLASE USING DIAMINE OXIDASE

Introduction

In animals, putrescine is formed almost exclusively from ornithine by direct decarboxylation, but in higher plants and in some micro-organisms putrescine is formed from arginine, via agmatine and N-carbamoylputrescine (Smith, 1985b). Arginine decarboxylase (ADC), the enzyme forming agmatine, is often measured by formation of $^{14}CO_2$ from [U-^{14}C]- or [1-^{14}C]-arginine. However, $^{14}CO_2$ may be released from the guanidino group by the combined action of arginase (Wright et al., 1981) and urease (Davies and Shih, 1984), enzymes which are widespread in higher plants. In addition, CO_2 may be released from the carboxyl group of several amino acids by a pyridoxal phosphate-mediated oxidative reaction which appears to depend on Mn^{2+} (Smith and Marshall, 1988a). A DFMO-insensitive decarboxylation of ornithine to produce 4-aminobutanamide by this mechanism was demonstrated in a wide range of higher plants (Smith and Marshall, 1988b). Many amino acids including arginine may also be decarboxylated in this system (Smith and Marshall, 1988c). In estimating amino acid decarboxylase activity it is therefore important to identify the amine product, measuring activity by $^{14}CO_2$ release only when stoichiometry has been established.

In earlier work, a method of ADC estimation was developed which depends on the oxidation of the amine (agmatine) by diamine oxidase (DAO) to produce peroxide, which was then used by peroxidase to form a colored product from guaiacol (Smith, 1978; 1979; 1983a). The main chromogen in this reaction is tetraguaiacol (ε = 26 600/M/cm) which is formed with consumption of 4 moles of hydrogen peroxide (i.e. ε = 6650/M H_2O_2/cm) (Chance and Maehly, 1955). However guaiacol is oxidized to form several other products, some of which are not colored (Mäkinen and Tenovuo, 1982). The experimentally derived value for ε calculated from Smith (1978) was 4370/M/cm (i.e./M H_2O_2/cm).

Peroxidase substrates with much greater absorbancies have been introduced subsequently, notably aminoantipyrine (AAP)/3,5-dichloro-2-hydroxybenzene sulphonic acid (DCHBS) (Barham and Trinder, 1972; Pietta et al., 1982), 3-(dimethylamino)benzoic acid (DMAB)/3-methyl-2-benzothiazolinone hydrazone (MBTH) (Ngo and Lenhoff, 1980) and 2,2'-azino-di-(3-ethylbenzthiazoline-6-sulphonic acid) (ABTS) (Childs and Bardsley, 1975, Shindler et al., 1976; Szutowicz et al., 1984). In the present work, these substrates have been investigated for their ability to substitute for guaiacol in the DAO/peroxidase linked assay for ADC estimation. In addition, the ADC was estimated in these extracts by the direct determination of agmatine by ion pair reverse phase HPLC using post column derivatization and fluorescence detection.

Experimental

Chemicals. AAP, DCHBS (Na salt), and DMAB were from Aldrich; *o*-phthalaldehyde, peroxidase (type II) and ABTS from Sigma; heptane sulphonic acid (Na salt) from BDH; and MBTH from Koch-Light.

Plants. Pea seedlings (*Pisum sativum* L. cv Alaska) were grown in darkness at 25°C for 10 days in vermiculite watered with a nutrient medium containing NH_4NO_3 (8mM); $CaCl_2$ (4mM); $MgSO_4$(1 mM); $NH_4H_2PO_4$ (1 mM); KCl (6 mM); FeEDTA (0.1mM) and micronutrients.

Oat seedlings (*Avena sativa* L. cv Avalanche) were grown in a mixture of equal parts of sand and perlite at 23°C in a 16 hr day with 250 μmol/sec/m^2 light and watered daily with a nutrient medium, as used for the pea seedlings but omitting the KCl.

Enzymes. Diamine oxidase was prepared from the pea seedling epicotyls by the method of McGowan and Muir, (1971) and Smith (1978) up to and including the stage of ethanol fractionation. ADC was prepared by blending 10g of the shoots of 24-day-old potassium deficient oat seedlings in 30 ml of pH 7, 0.1 M sodium phosphate buffer. The extract was filtered through nylon cloth, centrifuged at 4000 *g* for 15 min, and dialysed for 40 hr against the extraction buffer (two changes). The extract was then recentrifuged at 23 000 *g* for 15 min and 2 ml samples were incubated with 0.5 ml 25 mM arginine dihydrochloride or 0.5 ml water in duplicate for 3 hr at 30°C. Aliquots (0.5 ml) were withdrawn for HPLC. The solutions were then boiled for 5 min, and centrifuged. For the spectrophotometric assay, 0.1 ml samples were added to cuvettes which contained peroxidase, AAP, DCHBS, and pH 7 buffer to 2.5 ml, (see caption for Table 1 for details) prior to adding 10 μl DAO (10 nkat). The change in absorbance was determined after completion of the oxidation, usually within 60 min.

HPLC. Trichloroacetic acid (50 μl of 50%) was added to samples of the incubates (0.5 ml), and the protein removed by centrifuging (10 000 *g* for 15 min). Samples (50 μl) of the supernatant were injected without removal of the TCA after passage through a 0.45 μm filter. The HPLC column was 25 cm x 5 mm i.d., Partisil 5 ODS3 (Hichrom), fitted with pre- and guard columns. Injection was by a 200 μl Rheodyne loop.

Eluting buffers. A NaH_2PO_4 (50 mM) : acetonitrile (98:2) containing 8 x 10^{-3}M heptane sulphonic acid (Na salt), and adjusted to pH 2.3 with 3 M H_3PO_4. B NaH_2PO_4 (0.2 M) : acetonitrile (37:13) containing 8 x 10^{-3}M heptane sulphonic acid (Na salt) and adjusted to pH 3.1 with 3 M H_3PO_4. The buffers were degassed initially with ultrasonics under vacuum and with helium during use.

o-Phthalaldehyde reagent. Stock buffer: Boric acid (0.4 M) containing 0.1% Brij-35 and adjusted to pH 10.6 with 10 M KOH. Stock mercaptoethanol: Ethanol (20 ml of 95%) containing 1.2 ml of 2-mercaptoethanol. Detection reagent: o-Phthalaldehyde (100 mg) was dissolved in 4 ml of the ethanolic mercaptoethanol, added to 120 ml of the borate buffer and filtered through a no.1 sintered glass. The solution was purged with nitrogen before and during use, and kept in darkness.

The column was eluted isocratically at 1 ml/min and the o-phthalaldehyde reagent was pumped at 0.5 ml/min into the column effluent. The fluorescence was measured in a Spectro-Glo fluorometer (Gilson) with excitation at 360 nm and emission at 455 nm, and the sensitivity at x 20. The waste from the fluorometer was collected in a 1 litre bottle containing Na hypochlorite solution to destroy the mercaptoethanol. Standards of agmatine (50 µl of 100 µM, 5 nmol) were used for calibration. For HPLC separation after arginine decarboxylase assay, the ratio of buffer A:B was 3:2, giving Rt arginine 5 min and agmatine 8 min (Fig. 3). A ratio of A:B of 17:3 gave Rt ornithine 7 min and putrescine 12.5 min, and buffer A only, gave citrulline 4 min and N-carbamoylputrescine 8 min.

Results

On comparing the three peroxidase substrates (Table 1) ABTS was the most sensitive, although it was particularly subject to interference by the plant extract. The DMAB/MBTH also reacted non-specifically with the plant extract to produce a brown colour and was therefore considered to be

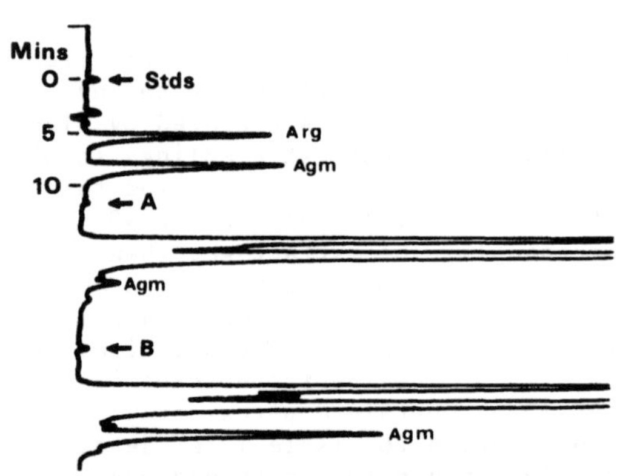

Fig. 3. HPLC separation of incubates of oat leaf extract with arginine, sampled initially, (A) and after 2 hr at 30°C, (B). Separation of a standard mixture of arginine (Arg) and agmatine (Agm) (5 nmol each) is also shown. For details see Experimental section.

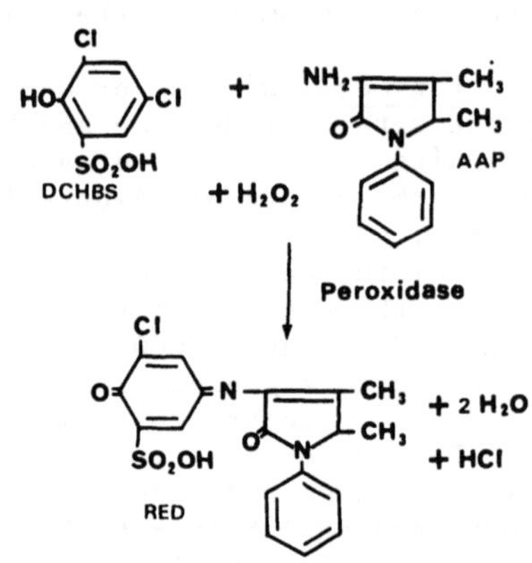

Fig. 4. Catalytic oxidation of aminoantipyrine (AAP) and 3,5-dichloro-2-hydroxybenzenesulphonic acid (DCHBS) effected by peroxidase in the presence of hydrogen peroxide.

Table 1. Comparison of peroxidase substrates

Substrate (final concentration)	DCHBS(1 mM) AAP(0.1 mM)	DMAB(1 mM) MBTH(0.02 mM)	ABTS(1 mM)
λ max	515 nm	590 nm	412 nm
ε (Absorbance/M/cm) (published)	22 000a	47 600b	36 000c 32 400d 24 600e
ε (this work)	26 000	37 800	49 400
ε in presence of 0.1ml oat extract	24 200	38 600	30 200
Relative rate of increase in absorbance	100	140	130

In addition to the chromogens, the cuvettes contained 0.1 ml of peroxidase (1 mg/ml), DAO (10 nkat) and sodium phosphate buffer (pH 7, 0.1 M) to a total volume of 2.5 ml. The reaction was initiated with 5 μl of 2.5 mM putrescine. For substrate abbreviations, see text.
Refs. - a Barham and Trinder, 1972; b Ngo and Lenhoff, 1980; c Childs and Bardsley, 1975; d Shindler et al., 1976; e Szutowicz et al., 1984.

Table 2. Comparison of the amounts of agmatine formed by oat leaf ADC estimated by the HPLC method and by the spectrophotometric method using AAP and DCHBS as peroxidase substrates. The samples were incubated in duplicate with and without arginine for 3 hr at 30°C.

	HPLC method nmol/2.5 ml	Spectrophotometric method nmol/2.5 ml
Without arginine	106, 90	78, 99
With arginine	2790, 2560	2440, 2420
Difference of means	2580	2340

unsuitable. The red colour formed on peroxidative oxidation of the AAP/DCHBS (Fig. 4) was stable and showed little interference on adding the oat extract. Even though it had the smallest extinction of the three systems tested, it was at least 6 times more sensitive than the guaiacol reagent used previously (Smith, 1978; 1979). This system was therefore selected for further study.

On estimating the agmatine formed by the oat leaf arginine decarboxylase with HPLC and with the peroxidase/AAP/DCHBS methods, the mean of the values agreed within 10%. (Table 2). The arginine decarboxylase activity calculated from these results was 22 nkat/g fr. wt.

Discussion

The peroxidase substrates presented here are well suited to the estimation of amine oxidases, having much greater sensitivity than guaiacol. Although these methods could in principle be modified to detect other amino acid decarboxylases, addition of thiols and pyridoxal phosphate could cause interference, and the HPLC method might then be more appropriate. The speed and specificity of these assays may prove useful in establishing stoichiometry. 4-Aminobutanamide formed by the oxidative decarboxylation of ornithine is not a substrate for DAO, and the corresponding amide formed from arginine is unlikely to be attacked. A mixture of peroxidase, AAP and DCHBS appears to be stable indefinitely at -15°C.

Spectrophotometric methods for amino acid decarboxylase assay rely on a source of DAO uncontaminated by the amino acid decarboxylase. We have shown that the DAO which is present at a high specific activity in the apoplast of pea seedling epicotyl (Angelini and Federico, 1986; Federico and Angelini, 1986), may be a convenient source of high purity DAO for this purpose.

Other systems for amine oxidase estimation

A method has been developed like the one adopted here, for estimating polyamines with barley seedling PAO in which the peroxidase-catalysed oxidation product of *N*-ethyl-*N*-3-methyl-phenyl-*N*-acetylethylenediamine is condensed with aminoantipyrine to form a chromogen (ε = 12 100/M/cm) (Suzuki et al., 1982).

Other methods based on peroxidase catalysed reactions could be suitable for measurement of amine oxidase activity or amine concentrations. The peroxidative coupling of 3-methyl-2-benzothiazolinone hydrazone with its formaldehyde azine is particularly sensitive (ε = 69 000/M/cm) (Capaldi and Taylor, 1983). The formation of a fluorescent product on peroxidative oxidation of homovanillic acid and similar substrates is also quite sensitive (Guilbault et al., 1968; Sindhu and Cohen, 1983), though the natural occurrence of UV absorbing compounds in plant extracts can cause considerable interference. Bachrach and Plesser (1986) showed that peroxide generated by pea seedling DAO may be estimated by a very sensitive chemiluminescent technique dependent on luminol and peroxidase.

Some methods for the estimation of amine oxidases depend on the determination of the oxidized diamine product. A method based on the measurement of the yellow 2-aminobenzaldehyde adduct which is formed spontaneously with pyrroline or piperideine, is widely used but has a relatively low sensitivity (ε= 1860/M/cm) (Holmstedt et al., 1961). This adduct may be stabilized for HPLC (Biondi et al., 1984) or gas chromatography (Smith et al., 1986) on oxidation with CrO_3. Another method for pyrroline estimation is based on the colour generated on heating with ninhydrin in strongly acidic medium (ε = 18 400/M/cm) (Naik et al., 1981; Peč et al., 1985).

Oxidation of 1,4-diamino-2-butene (dehydroputrescine) by DAO yields pyrrole which can be measured by reaction with 4-dimethylaminobenzaldehyde with great sensitivity (ε = 63 000/M/cm at 563 nm). However, although dehydroputrescine is oxidized 40% faster than putrescine by hog kidney DAO, activity with the pea seedling enzyme is about 14% of that with putrescine (Macholán et al., 1975).

POLYAMINE OXIDASE

The polyamine oxidase (PAO) found only in the monocotyledonous plants, mainly in the Gramineae, is quite distinct from DAO. Although it is particularly active in oats (Smith, 1983b), it has been characterized in barley (Smith, 1983b), maize (Suzuki and Yanagisawa, 1980, Federico et al., 1987) and millet (Hirasawa et al., 1986) and it has also been found in the water hyacinth (*Eichhornia stricta* monocot., Pontederiaceae) (Yanagisawa et al., 1987). The enzyme is quite specific for spermidine and spermine, and it has FAD as the cofactor, though this is not always easy to demonstrate (Smith, 1983b). The properties of these enzymes are summarized in Table 3.

Table 3. Some properties of plant polyamine oxidases

	Oat	Barley	Maize	Millet	Water hyacinth
Mr (gel filtration)	85 000	85 000	53 000	80 000	87 000
Mr (SDS PAGE)	ND	ND	53 000	ND	60 000
Cofactors	ND	ND	FAD	FAD	FAD
Activity nkat/mg protein	6020	ND	700	185	600
Purification factor	885	ND	4.8	168	449
Rate spermine/spermidine	1	14	0.7	ND	0.7
Km μM spermine	2	30	18	0.5	20
spermidine	8	20	22	60	28
Optimum pH spermine	6.5	4.8	6.5	6.5	6.5
spermidine	6.5	8.0	6.5	6.5	6.5

References -

Oats: Smith, 1976; 1977; 1983b; 1985c.
Barley: Smith, 1972; 1974; 1976; Smith and Bickley, 1974.
Maize: Hirasawa and Suzuki, 1975; Suzuki and Hirasawa, 1973; Suzuki and Yangisawa, 1980; Federico et al., 1988.
Millet: Hirasawa et al., 1986.
Water hyacinth: Yanagisawa et al., 1987. ND = not determined

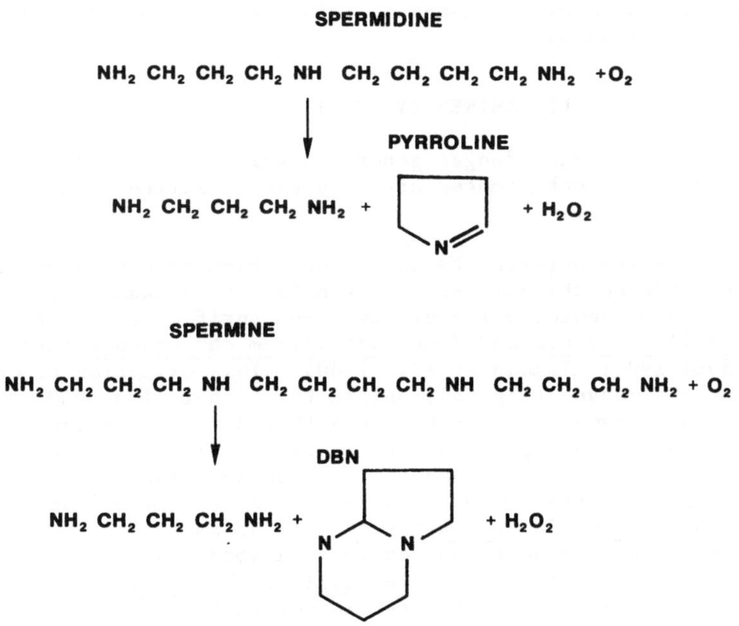

Fig. 5. Mechanism for the oxidation of spermidine and spermine by the cereal polyamine oxidase. DBN = diazabicyclononane.

Like the DAO, PAO occurs in the cell walls (Kaur-Sawhney et al., 1981) and it can be eluted at a high specific activity (Federico et al., 1987). Only two chromatographic steps were needed to purify the eluted enzyme from maize seedlings to homogeneity.

The enzyme oxidizes the polyamines at the secondary amino group giving cyclic products derived from aldehydes, together with diaminopropane and peroxide (Fig. 5). The product of spermine oxidation occurs mainly as the bicyclic form, diazabicyclononane (DBN) (Brändange et al., 1984; Smith et al., 1986), together with pyrroline, in the leaves of various cereals (Smith et al., 1986).

The oat leaf enzyme is stimulated by molar KCl and has a pH optimum of 6.5 for both spermidine and spermine. The fungicide guazatine is a powerful competitive inhibitor (_Ki_ _ca_ 8nM) (Smith, 1985c).

FUNCTIONS OF THE DI- AND POLYAMINE OXIDASES IN HIGHER PLANTS

The amine oxidation products, pyrroline or DBN, could show biological activity, but their properties are at present unknown. However, since both DAO and PAO occur in the cell walls, it might be anticipated that these enzymes should serve an important function at this site. There are several possible reasons for this location.

1) The peroxide formed in the oxidative process may be utilised in the peroxidative lignification process (Angelini and Federico, 1986; Federico and Angelini, 1987). However, this would not explain the high activity also present in the cotyledons of legumes.
2) The peroxide may prevent attack by pathogens.
3) The enzymes may prevent the transfer of amines between cells. This could be important in morphogenesis.

The restricted occurrence of amine oxidases in plants raises questions regarding amine degradation in those species lacking these enzymes. Such plants may have alternative means of amine catabolism which are at present obscure (Smith, 1985a).

OXIDATION OF DI- AND POLYAMINES BY FUNGI

A DAO found in many fungal genera on being grown in diamine supplemented media shows much greater activity with agmatine than with putrescine (Isobe et al., 1982).

Polyamines are oxidized by many fungi which have been grown on media containing these as the sole source of nitrogen (Yamada et al., 1980). FAD-containing polyamine oxidases have been purified to crystallization from _Aspergillus terreus_ and from _Penicillium chrysogenum_ (Isobe et al., 1980a; 1980b; 1985; Yamada et al., 1980). This oxidation produces 3-aminopropionaldehyde (Fig. 6), unlike the higher plant PAO, but like certain animal enzymes. On culturing with 1,3-diaminopropane as an inducer, a _Penicillium_ sp., produced an extracellular FAD-dependent PAO with the same mechanism (Kobayashi and Horikoshi, 1982) which also attacked acetylpolyamines (Kobayashi et al., 1983). A peroxisomal spermidine oxidase found in _Candida boidinii_ with a pH optimum of 10, degraded spermidine and acetylspermidine to putrescine (Haywood and Large, 1984; Gillyon et al., 1987). Since the PAO of _Penicillium chrysogenum_ oxidized spermine but showed little activity with spermidine, while the _Aspergillus terreus_ PAO oxidized both spermine and spermidine, these enzymes were used as the basis of a method for the differential estimation of these polyamines (Isobe et al., 1980c; 1981; 1985).

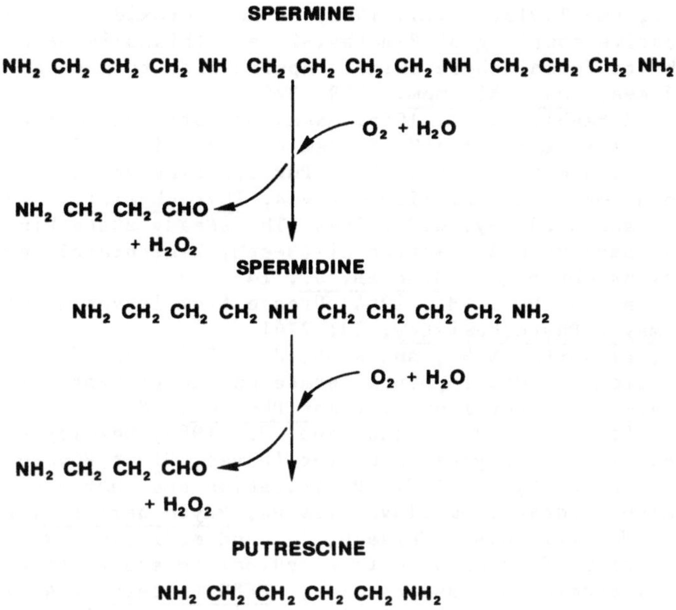

Fig. 6. Oxidation of spermine and spermidine by fungal polyamine oxidase.

ACKNOWLEDGEMENTS

The authors are very grateful to Shell Research Ltd, Sittingbourne, Kent, U.K. for financing this work. Special thanks are due to Mr. E.G. Cotterill for his help with the HPLC.

REFERENCES

Angelini, R., and Federico, R., 1986, Occurrence of diamine oxidase in the apoplast of Leguminosae seedlings, in "Biochemical Studies of Natural Polyamines", C.M. Caldarera, C. Clô, and C. Guarnieri eds., CLUEB, Bologna, p 183.

Angelini, R., Di Lisi, F., and Federico, R., 1985, Immunoaffinity purification and characterization of diamine oxidase from Cicer, Phytochemistry, 24: 2511.

Angelini, R., Federico, R., and Mancinelli, A., 1988, Phytochrome-mediated control of diamine oxidase level in etiolated lentil seedlings, Plant Physiol., (in press).

Bachrach, U., and Plesser, Y.M., 1986, A sensitive, rapid, chemiluminescence-based method for the determination of diamines and polyamines, Anal. Biochem., 152: 423.

Barham D., and Trinder, P., 1972, An improved colour reagent for the determination of blood glucose by the oxidase system., Analyst, 97: 142.

Biondi, P.A., Simonic, T., Secchi, C., Ronchi, S., and Manzocchi, A., 1984, Sensitive assay for diamine oxidase activity using high-performance liquid chromatography, J. Chrom., 309: 151.

Brandänge, S., Eriksson, L.-H., and Rodriguez, B., 1984, Ring-chain tautomerism of an N,N-acetal formed by enzymatic oxidation of spermine or spermidine, Acta Chem. Scand., B38: 526.

Capaldi, D.J., and Taylor, K.E., 1983, A new peroxidase color reaction: oxidative coupling of 3-methyl-2-benzothiazolinone hydrazone (MBTH) with its formaldehyde azine. Application to glucose and choline oxidases, Anal. Biochem., 129: 329.

Chance, B., and Maehly, A.C., 1955, Assay of catalases and peroxidases, in: "Methods in Enzymology", Academic Press, II: 772.

Chaudhuri, M.M., and Ghosh, B., 1984, Purification and characterization of diamine oxidase from rice embryos, Phytochemistry, 23: 241.

Childs, R.E., and Bardsley, W.G., 1975, The steady state kinetics of peroxidase with 2,2'-azino-di-(3-ethylbenzthiazoline-6-sulphonic acid) as chromogen, Biochem. J., 145: 93.

Davies, H.M., and Shih, L.-M., 1984, Urease from leaves of Glycine max and Zea mays, Phytochemistry, 23: 2741.

Delhaize, E., Dilworth, M.S., and Webb, J., 1986, The effects of copper nutrition and developmental state on the biosynthesis of diamine oxidase in clover leaves, Plant Physiol., 82: 1126.

Delhaize, E., Loneragan, J.F., and Webb, J., 1985, Development of three copper metalloenzymes in clover leaves, Plant Physiol., 78: 4.

Delhaize, E., and Webb, J., 1987, Purification and characterization of diamine oxidase from clover leaves, Phytochemistry, 26: 641.

Dooley, D.M., McGuirl, M.A., Peisach, J., and McCracken, J., 1987, The generation of an organic free radical in substrate-reduced pig kidney diamine oxidase-cyanide, FEBS Letters, 214: 274.

Federico, R., and Angelini, R., 1986, Occurrence of diamine oxidase in the apoplast of pea epicotyls, Planta, 167: 300.

Federico, R., and Angelini, R., 1987, On the physiological roles of diamine and polyamine oxidases in the cell wall, Phytochemical Society of Europe Symposium on 'Amines in Plants', University College, London, Abstract.

Federico, R., and Angelini, R., 1988, Distribution of polyamines and their related catabolic enzyme in etiolated and light grown Leguminosae seedlings, Planta, 173: 317.

Federico, R., Di Lisi, F., and Angelini, R., 1985a, Purification of diamine oxidase from Lens culinaris by affinity chromatography, Plant Science, 38: 9.

Federico, R., Angelini, R., Cesta, A., and Pini, C., 1985b, Determination of diamine oxidase in lentil seedlings by enzyme activity and immuno reactivity, Plant Physiol., 79: 62.

Federico, R., Angelini, R., Argento Ceru, M.P., and Manes, F., 1985c, Immunohistochemical demonstration of lentil diamine oxidase, Cell Mol. Biol., 31: 171.

Federico, R., Alisi, C., and Forlani, F., 1988, Properties of the polyamine oxidase from the cell wall of maize seedlings, Phytochemistry, (in press).

Ferguson, S.J., 1984, A novel bacterial dehydrogenase cofactor turns up in a serum amine oxidase, Trends Biochem. Sci., 9: 367.

Finazzi-Agro, A., Rinaldi, A., Floris, G., and Rotilio, G., 1984, A free-radical intermediate in the reduction of plant Cu-amine oxidases, FEBS Letters, 176: 378.

Floris, G., Giartosio, A., and Rinaldi, A., 1983, Essential sulfhydryl groups in diamine oxidase from Euphorbia characias latex, Arch. Biochem. Biophys., 220: 623.

Gillyon, C., Haywood, G.W., Large, P.J., Nellen, B., and Robertson, A., 1987, Putrescine breakdown in the yeast Candida boidinii: subcellular location of some of the enzymes involved and properties of two acetamidoaldehyde dehydrogenases, J. Gen. Microbiol., 133: 2477.

Glatz, Z., Kovár, J., Macholán, L., and Peč, P., 1987, Pea (Pisum sativum) diamine oxidase contains pyrroloquinoline quinone as a cofactor, Biochem. J., 242: 603.
Guilbault, G.G., Brignac, P.J., and Juneau, M., 1968, New substrates for the fluorometric determination of oxidative enzymes, Anal. Chem., 40: 1256.
Haslam, S.C., and Young, T.W., 1987, Partial purification of a diamine oxidase from Nicotiana rustica, Phytochemical Society of Europe Symposium on 'Amines in Plants', University College, London, Abstract.
Haywood, G.W., and Large, P.J., 1984, Partial purification of a peroxisomal polyamine oxidase from Candida boidinii and its role in growth on spermidine as sole nitrogen source, J. Gen. Microbiol., 130: 1123.
Hill, J.M., and Mann, P.J.G., 1964, Further properties of the diamine oxidase of pea seedlings, Biochem. J., 91: 171.
Hirasawa, E., and Suzuki, Y., 1975, Purification and properties of a polyamine oxidase from Zea mays, Phytochemistry, 14: 99.
Hirasawa, E., Watanabe, H., and Suzuki, Y., 1986, Polyamine oxidase of millet shoots, Phytochemistry, 25: 1739.
Holmstedt, B., Larsson, L., and Tham, R., 1961, Further studies of a spectrophotometric method for the determination of diamine oxidase activity, Biochim. Biophys. Acta, 48: 182.
Isobe, K., Tani, Y., and Yamada, H., 1980a, Crystallization and characterization of polyamine oxidase from Penicillium chrysogenum, Agric. Biol. Chem., 44: 2651.
Isobe, K., Tani, Y., and Yamada, H., 1980b, Crystallization and characterization of polyamine oxidase from Aspergillus terreus, Agric. Biol.Chem., 44: 2749.
Isobe, K., Tani, Y., and Yamada, H., 1980c, Kinetic properties of fungal polyamine oxidases and their application to differential determination of spermine and spermidine, Agric. Biol. Chem., 44: 2955.
Isobe, K., Tani, Y., and Yamada, H., 1981, Differential determination procedure for putrescine, spermidine and spermine with polyamine oxidase from fungi and putrescine oxidase, Agric. Biol. Chem., 45: 727.
Isobe, K., Tani, Y., and Yamada, H., 1982, A new enzyme, agmatine oxidase, from fungi, Agric. Biol. Chem., 46: 1345.
Isobe, K., Tani, Y., and Yamada, H., 1985, Formation of polyamine oxidases by fungi and application to differential determination of polyamines, in: "Polyamines: Basic and Clinical Aspects", K. Imahori, F. Suzuki, O. Suzuki and U. Bachrach, eds., VNU Science Press, Utrecht, The Netherlands, p.333.
Kaur-Sawhney, R., Flores, H.E., and Galston, A.W., 1981, Polyamine oxidase in oat leaves: a cell wall-localized enzyme, Plant Physiol., 68: 494.
Kluetz, M.D., Adamsons, K., and Flynn, Jr., J.E., 1980, Cryoenzymology and spectrophotometry of pea seedling diamine oxidase, Biochemistry, 19: 1617.
Kobayashi, Y., Higashi, T., Machida, H., Iwasaki, S., and Horikoshi, K., 1983, Oxidation of acetylpolyamines by extracellular polyamine oxidase produced by Penicillium sp.No. PO-1, Biochim. Biophys. Acta, 743: 431.
Kobayashi, Y., and Horikoshi, K., 1982, Purification and characterization of extracellular polyamine oxidase produced by Penicillium sp.No. PO-1, Biochim. Biophys. Acta, 705: 133.
Macholán, L., Hubálek, F., and Subová, H., 1975, Oxidation of 1,4-diamino-2-butene to pyrrol, a sensitive test of diamine oxidase activity, Coll. Czech. Chem. Commun., 40: 1247.

Mäkinen, K.K., and Tenovuo, J., 1982, Observations on the use of guaiacol and 2,2'-azino-di(3-ethylbenzthiazoline-6-sulfonic acid) as peroxidase substrates, Anal. Biochem., 126: 100.

McGowan, R.E., and Muir, R.M., 1971, Purification and properties of amine oxidase from epicotyls of Pisum sativum, Plant Physiol., 47: 644.

Mondovi, B., and Avigliano, L., 1987, Amine oxidases, Life Chem. Rep., 5(1-4): 187. (Chem.Abs., 108: 90270).

Morgan, D.M.L., 1985, Polyamine oxidases, Biochem. Soc. Trans., 13: 322.

Nadeau, P., Delaney, S., and Chouinard, L., 1987, Effects of cold hardening on the regulation of polyamine levels in wheat (Triticum aestivum L.) and alfalfa (Medicago sativa L.), Plant Physiol., 84: 73.

Naik, B.I., Goswami, R.G., and Srivastava, S.K., 1981, A rapid and sensitive colorimetric assay of amine oxidase, Anal Biochem., 111: 146.

Ngo T.T., and Lenhoff, H.M., 1980, A sensitive and versatile chromogenic assay for peroxidase and peroxidase-coupled reactions, Anal. Biochem., 105: 389.

Oratore, A., D'Andrea, G., Floris, G., Rinaldi, A., and Finazzi-Agro, A., 1987a, Microheterogeneity of lentil seedling (Lens culinaris) amine oxidase, Ital. J. Biochem., 36: 92.

Oratore, A., D'Andrea, G., Strom, R., and Finazzi-Agro, A., 1987b, Relationship between immune lattice formation and compactness of holo- and apo-forms of amine oxidase from Lens culinaris, Biochem. Int., 14: 277.

Peč, P., and Pavliková, M., 1985, New colorimetric method for the determination of diamine oxidase activity by ninhydrine, Biologia, 40: 1217.

Pietta, P., Calatroni, A., and Colombo R., 1982, Determination of diamine oxidase activity by high performance liquid chromatography, J.Chrom., 243: 123.

Rinaldi, A., Floris, G., and Finazzi-Agro, A., 1982, Purification and properties of diamine oxidase from Euphorbia latex, Eur. J. Biochem., 127: 417.

Rinaldi, A., Giartosio, A., Floris, G., Medda, R., and Finazzi-Agro, A., 1984, Lentil seedlings amine oxidase: preparation and properties of the copper-free enzyme, Biochem. Biophys. Res. Communs., 120: 242.

Rinaldi, A., Floris, G., and Giartosio, A., 1986, Plant amine oxidases, in: "Structure and Functions of Amine Oxidases", B. Mondovi, ed., CRC Press Inc., Boca Raton, Florida, U.S.A., p.51.

Shindler, J.S., Childs, R.E., and Bardsley, W.G., 1976, Peroxidase from human cervical mucus, Eur. J. Biochem., 65: 325.

Sindhu, R.K., and Cohen, S.S., 1983, Some properties of the spermidine synthase of Chinese cabbage, in: "Advances in Polyamine Research", U. Bachrach, A. Kaye, and R. Chayen, eds., Raven Press, New York, 4: 371.

Smith, T.A., 1972, Purification and properties of the polyamine oxidase of barley plants, Phytochemistry, 11: 899.

Smith, T.A., 1974, Polyamine oxidation by enzymes from Hordeum vulgare and Pisum sativum seedlings, Phytochemistry, 13: 1075.

Smith, T.A., 1976, Polyamine oxidase from barley and oats, Phytochemistry, 13: 633.

Smith, T.A., 1977, Further properties of the polyamine oxidase from oat seedlings, Phytochemistry, 16: 1647.

Smith, T.A., 1978, Spectrophotometric method for the estimation of arginine decarboxylase, Anal. Biochem., 92: 331.

Smith, T.A., 1979, Arginine decarboxylase of oat seedlings, Phytochemistry, 18: 1447.

Smith, T.A., 1983a, Arginine decarboxylase (oat seedlings) in: "Methods in Enzymology", 94: 176.

Smith, T.A., 1983b, Polyamine oxidase (oat seedlings), in: "Methods in Enzymology", 94: 311.

Smith, T.A., 1985a, The di- and polyamine oxidases of higher plants, Biochem. Soc. Trans., 13: 319.

Smith, T.A., 1985b, Polyamines, Ann. Rev. Plant Physiol., 36: 117.

Smith, T.A., 1985c, The inhibition and activation of polyamine oxidase from oat seedlings, Plant Growth Regulation, 3: 269.

Smith, T.A., and Bickley, D.A., 1974, Further properties of the polyamine oxidase of barley leaves, Phytochemistry, 13: 2437.

Smith, T.A., and Marshall, J.H.A., 1988a, The oxidative decarboxylation of amino acids by plant extracts, Biochem.Soc.Trans., 16: (in press).

Smith, T.A., and Marshall, J.H.A., 1988b, The oxidative decarboxylation of ornithine by extracts of higher plants, Phytochemistry, 27: 703.

Smith, T.A., and Marshall J.H.A., 1988c, Oxidation of amino acids by manganous ions and pyridoxal phosphate, Phytochemistry, 27: 1611.

Smith, T.A., Croker, S.J., and Loeffler, R.S.T., 1986, Occurrence in higher plants of 1-(3-aminopropyl)pyrrolinium and pyrroline: products of polyamine oxidation, Phytochemistry, 25: 683.

Srivastava, S.K., Prakash, V., and Naik, B.I., 1977, Regulation of diamine oxidase activity in germinating pea seeds, Phytochemistry, 16: 185.

Suzuki, O., and Matsumoto, T., 1985, Partial purification and properties of soybean seedling amine oxidase, in: "Polyamines: Basic and Clinical Aspects", K. Imahori, F. Suzuki, O. Suzuki, and U. Bachrach, eds., VNU Science Press, Utrecht, The Netherlands, p 89.

Suzuki, O., Matsumoto, T., Oya, M., Katsumata, Y., and Samejima, K., 1981, A new fluorometric assay for spermidine synthase, Anal. Biochem., 115: 72.

Suzuki, O., Matsumoto, T., Katsumata, Y., and Oya, M., 1982, A new enzymatic method for the quantitation of spermine in human semen, Z. Rechtsmed., 88: 67.

Suzuki, Y., 1987, The target of carbonyl reagents in the pea diamine oxidase activity, Phytochemical Society of Europe Symposium on 'Amines in Plants', University College, London, Abstract.

Suzuki, Y., and Hirasawa, E., 1973, Polyamine oxidase from Zea mays shoots, Phytochemistry, 12: 2863.

Suzuki, Y., and Yanagisawa, H., 1980, Purification and properties of maize polyamine oxidase: a flavoprotein, Plant Cell Physiol., 21: 1085.

Szutowicz, A., Kobes, R.D., and Orsulak, P.J., 1984, Colorimetric assay for monoamine oxidase in tissues using peroxidase and 2,2'-azino(3-ethylbenzthiazoline-6-sulfonic acid) as chromogen, Anal. Biochem., 138: 86.

Torrigiani, P., Serafini-Fracassini, D., and Fara, A., 1987, Evidence for the presence of a diamine oxidase activity in different physiological stages of Jerusalem artichoke tuber, Phytochemical Society of Europe Symposium on 'Amines in Plants', University College, London, Abstract.

Wright, L.C., Brady, C.J., and Hinde, R.W., 1981, Purification and properties of the arginase from Jerusalem artichoke tubers, Phytochemistry, 20: 2641.

Yamada, H., Isobe, K., and Tani, Y., 1980, Oxidation of polyamines by fungal enzymes, Agric. Biol.Chem., 44: 2469.

Yanagisawa, H., Hirasawa, E., and Suzuki, Y., 1981, Purification and properties of diamine oxidase from pea epicotyls, Phytochemistry, 20: 2105.

Yanagisawa, H., Kato, A., Hoshiai, S., Kamiya, A., and Torii, N., 1987, Polyamine oxidase from water hyacinth, Plant Physiol., 85: 906.

Dedicated to the memory of Dr. Keith Scorer, who died in Mexico, May 1988.

CONTROL OF PHYTOPATHOGENS BY INHIBITORS OF POLYAMINE BIOSYNTHESIS

Arthur W. Galston and Leonard H. Weinstein

Yale University, New Haven, CT and
Boyce Thompson Institute, Ithaca, N.Y., USA

INTRODUCTION

It is now clear that polyamines (PAs) are required for the normal growth and development of microorganisms as well as cells of higher animals and plants [1,2,3]. In all organisms, putrescine (Put) is formed from ornithine *via* ornithine decarboxylase (ODC); in some microorganisms and in higher plants, Put may also be formed from arginine *via* arginine decarboxylase (ADC). Put is then transformed to the triamine spermidine (Spd) and the tetramine spermine (Spm) by successive enzyme-mediated aminopropyl transfers from S-adenosylmethionine (SAM).

The work of C.W. Tabor [4] with *Saccharomyces cerevisiae*, an Ascomycete, has shown that in this organism Put can be made only *via* ODC and that mutational blockage of ODC leads to a virtually complete cessation of growth. If the mutation is at the aminopropyltransferase locus, so that the organism can make Put but not Spd or Spm, then growth proceeds at about 1/6 the normal rate. Furthermore, Spd and/or Spm are absolutely required for sporulation and for survival of a double-stranded RNA "killer plasmid." PAs are therefore absolutely required by yeast and specific PAs seem to have specific regulatory functions in the development of this organism.

Since many plant pathogens are Ascomycetes, it occurred to us that selective blockage of ODC could lead to disease control. If we assume (a) that all organisms need PAs, (b) that fungi can make PAs only *via* ODC, and (c) that higher plants can make PAs via both ODC and ADC, then selective blockage of ODC should permit ADC to function unhindered. Indeed, blockage of ODC by the specific suicide inhibitor a-difluoromethylornithine (DFMO, [5]) has been shown not to inhibit ADC activity in oat leaves, but actually to promote it by about 20% [6].

EXPERIMENTS ON FUNGAL GROWTH *IN VITRO*

We began by studying the effects of DFMO and DFMA, (a-difluoromethylarginine, the analogous suicide inhibitor of ADC [7]) on the growth of selected fungi on defined medium [8]. Cultures of *Botrytis* sp., *Botrytis cinerea*, *Rhizoctonia solani* and *Monilinia fructicola* were maintained on Czapek's agar medium [9] were transferred several times to obtain stable growth rates. Then filter-sterilized DFMO or DFMA was added to the medium cooled to 46-48° C so as to obtain final concentrations of 0.1, 0.5 and 1.0 mM concentrations of inhibitor. Solutions were then dispensed aseptically into 20x100 mm sterile plastic petri dishes (20 ml per dish) and allowed to solidify. For experiments involving reversal of inhibitions, we used 0.5 mM DFMO to yield ~50% inhibition of mycelial growth, and concentrations of Put and/or Spd of 0.1 and 1.0 mM to reverse the inhibitions. To initiate growth, plugs of mycelium 7 mm in diameter were cut with a sterile cork borer from the advancing edge of each culture and inverted in the center of control and experimental culture plates. The three replicate plates used for each concentration were incubated at 24°C in the dark and colony diameters in mm were measured at 2, 4 and 6 days after inoculation along two diameters at right angles to one another. Cell length and diameter measurements were also made on the 7th day with the help of a calibrated ocular micrometer.

Both DFMO and DFMA produced successively greater inhibitions of mycelial growth in all fungi as the concentrations were raised from 0.1 to 0.5 to 1.0 mM. Figure 1 shows the effect of PAs and inhibitors on colony diameter of _Botrytis_ sp., the most sensitive species, after 2, 3 and 6 days of incubation in the dark at 24°. It is clear that both DFMO and DFMA are effective inhibitors. The latter is paradoxically somewhat more effective than the former, presumably due to an _in vivo_ arginase-mediated conversion of DFMA to DFMO [10]. In all cases, Put and/or Spd can reverse the inhibition fully; furthermore, it occasionally appeared that cultures containing inhibitor +PA showed >100% reversal, creating an "overshoot" of growth.

Mycelial morphology and cell size were also dramatically altered in inhibitor-containing cultures of _Botrytis_ sp. and _Monilinia fructicola_. Both inhibitors depressed growth sharply, and there were consequently few scattered hyphae as compared with the control, which produced a thick, fluffy mat of aerial mycelium. In control cultures, the mycelia were distinctly white, gradually turning to light brown, whereas in inhibitor-containing cultures, the mycelia were characteristically brown, changing to darker brown as the cultures aged. Cultures with PAs alone or in combination with inhibitors showed normal mycelial morphology.

In inhibitor-containing cultures, cell length was much reduced, cell diameters were increased, and sporangia were absent. Mycelia also contained sharply distinct septae and crinkled branches as compared to normal. In PA-containing cultures, both cell length and cell diameters were increased. In the case of plates containing both PAs and inhibitors, no inhibitions were seen and in fact, a considerable increase in cell length and diameters were recorded in many treatments. The data on cell size measurements of _Botrytis_ sp. are summarized in Table 1.

EXPERIMENTS ON PATHOGENESIS IN PLANTS

Many important pathogens, such as rusts, smuts and mildews, are biotrophic parasites, and cannot be grown on synthetic media. Accordingly, we undertook experiments [12] with DFMO and DFMA in which we sought to affect the growth of selected fungi on host plants. In the first instance, to permit precise quantification of the results, we investigated the bean rust fungus (_Uromyces phaseoli_) growing on Pinto beans. Infection by a single uredospore of _Uromyces_ leads to a discrete and bounded foliar lesion, which can be quantitated with respect to numbers and size.

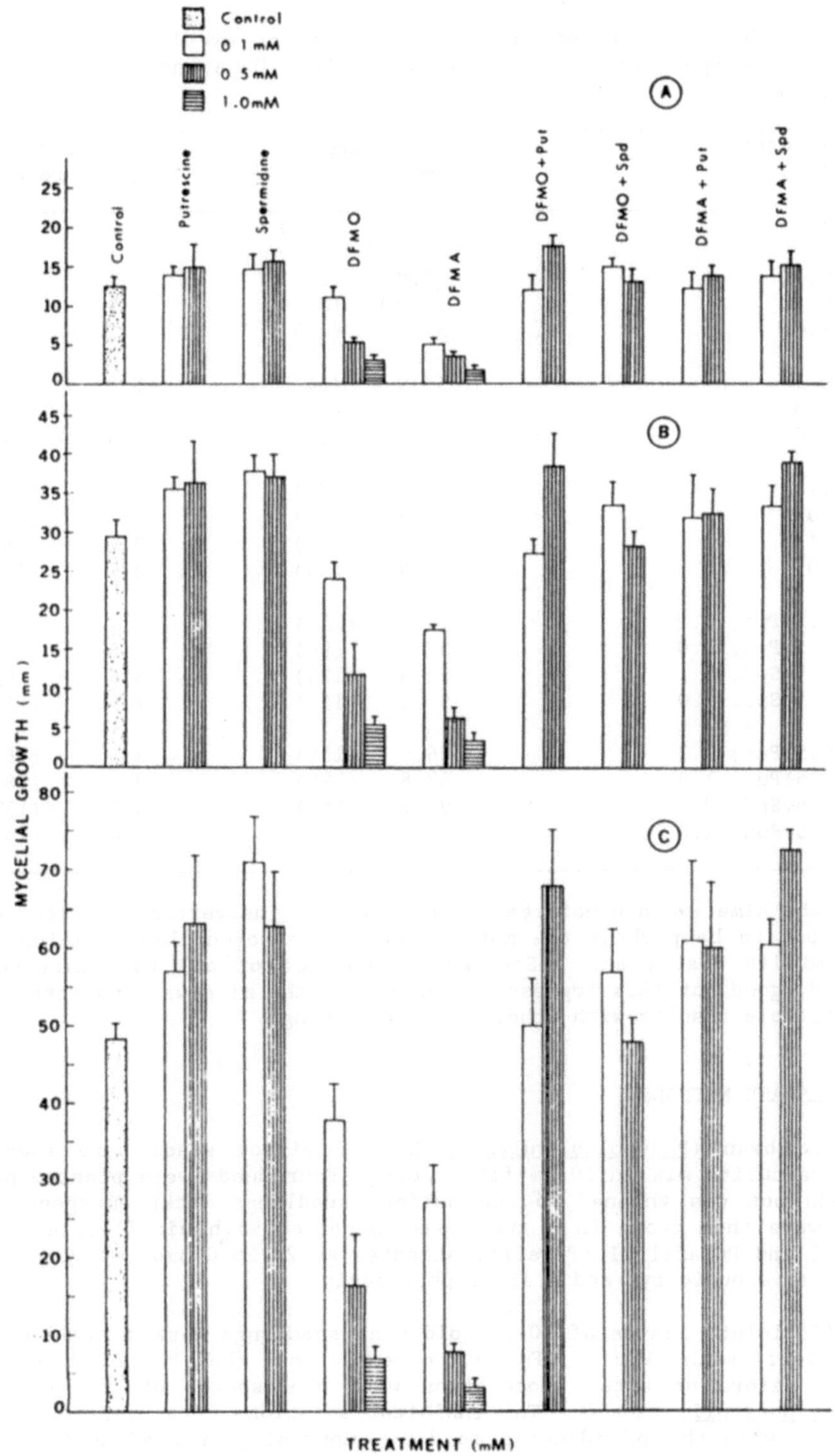

Figure 1. The effect of DFMO, DFMA and PAs, alone and in combination on the colony diameters of Botrytis sp. on the second (A), fourth (B) and sixth (C) days after inoculation. Bars represent mean ± s.d. for colony diameters in millimeters.

Table 1. Cell length and diameters (in mM) of mycelia in Botrytis sp. cultured on Czapek agar supplemented with DFMO, DFMA and PAs, alone and in combination.

Treatment (mM)	Length mM	Length relative	Diameter mM	Diameter relative
Control	63.9	(100)	5.1	(100)
DFMO, 0.1	40.8	(63)	7.6	(149)
DFMO, 0.5	51.9	(81)	7.1	(139)
DFMO, 1.0	31.9	(49)	7.2	(141)
DFMA, 0.1	53.9	(84)	5.8	(113)
DFMA, 0.5	51.5	(80)	7.6	(149)
DFMA, 1.0	41.5	(64)	8.9	(174)
Put, 0.1	78.4	(122)	6.8	(133)
Put, 1.0	81.9	(128)	7.7	(150)
Spd, 0.1	62.5	(97)	9.6	(188)
Spd, 1.0	53.9	(84)	8.7	(170)
DFMO, 0.5+Put, 0.1	74.1	(115)	7.2	(141)
DFMO, 0.5+Put, 1.0	57.8	(90)	9.6	(188)
DFMO, 0.5+Spd, 0.1	82.8	(129)	8.6	(168)
DFMO, 0.5+Spd, 1.0	97.2	(152)	7.7	(150)
DFMA, 0.5+Put, 0.1	86.5	(135)	8.3	(162)
DFMA, 0.5+Put, 1.0	82.8	(129)	9.4	(184)
DFMA, 0.5+Spd, 0.1	99.7	(156)	6.9	(135)
DFMA, 0.5+Spd, 1.0	88.3	(138)	7.6	(149)

At the time, we did not realize that we had inadvertantly selected what turned out to be perhaps the most sensitive pathogen/plant combination, a rust and its host plant. Spectacular control of all rust diseases was obtained, good but less impressive control of the mildews, and often as yet unpredictable results with other classes of fungi.

MATERIALS AND METHODS

Pinto bean (*Phaseolus vulgaris* L. cv. Pinto) seeds were sown in a peat/vermiculite mix in 10-cm fiber pots. Four seeds were planted per pot and each pot was thinned to two uniform seedlings after emergence. The plants were then grown in a greenhouse supplied with air filtered through charcoal and Purafil II (Purafil, Atlanta) at 23-25°C ambient temperature, 70% relative humidity, and a 16-hr photoperiod.

Unifoliolate leaves of 10-day-old bean seedlings were sprayed (*ca*. 400 ml per leaf) with DFMO or DFMA at concentrations that ranged from 0.01 to 5.0 mM, before or after inoculation with uredospores of the bean rust, *Uromyces phaseoli*, race 0. The inhibitor solutions were prepared in 0.01% Tween 20, with the pH adjusted to 7.0. Control plants without inhibitor were sprayed similarly and were allowed to dry before inoculation with the pathogen. After inoculation with rust uredospores (25 mg/100 ml of 0.01% Tween 20) all plants were placed in dew chambers (100% relative humidity) for 16 hr at 19°C in total darkness. After exposures to inhibitors and uredospores, all plants were returned to the greenhouse and arranged randomly. Disease severity was evaluated 7 days after inoculation by counting local foliar lesions. In control plants and those treated with low concentrations of DFMO, uredospores were collected randomly from each treatment to determine percent germination and pathogenicity. For determination of germination, spores were dusted onto petri plates containing 10 mM 2-(N-morpholino) ethanesulfonic acid (Mes) at pH 7.0, 3 mM $CaCl_2$, 2mM $MgSO_4$ and 1% purified agar, and incubated for 3 hr at 19°C in the light. For determination of pathogenicity, collected spores were sprayed onto bean leaves and incubated as described above. Each exposure consisted of 12 plants in six replicate pots. All experiments were repeated at least once, with similar results.

Foliar lesions characteristic of infection by Uromyces phaseoli uredospores were apparent on control plants within 4-5 days after inoculation and attained their maximal size after 10-12 days. Abundant lesions occurred in the control plants, while plants with pre- or postinoculation exposures to increasing concentrations of DFMO developed progressively fewer lesions. No symptoms of the bean rust disease could be detected on plants treated with DFMO at 0.5 mM or higher. At the lower concentrations, inhibition of uredial development was more pronounced on plants given postinoculation exposures to the inhibitors. In such cases, uredial appearance was delayed for 2-6 days in DFMO-treated plants, and the extent of delay was dose-dependent. Increasing the concentration of DFMO from 0.01 to 0.25 mM resulted in a dose-dependent decrease in lesion number and disease severity. The ID_{50} was found to be at 0.05 mM and 0.025 mM for pre- and postinoculation exposure to DFMO, respectively. In contrast, DFMA was ineffective in reducing infection, even at 5.0 mM. The average numbers of lesions per square centimeter produced by pre- and postinoculation exposures to different concentrations of both the inhibitors are summarized in Table 2. Uredospores collected from plants treated with low concentrations of DFMO and DFMA show no decrease in germination or pathogenicity.

The protective effects of DFMO are not limited to the region of application, indicating translocation of either DFMO or some substance formed as a result of its application. Thus, when the petiolar or apical half of an attached unifoliolate leaf was treated with 1 mM DFMO and

Table 2. Mean number of lesions induced by uredospores of Uromyces phaseoli on unifoliolate leaves of bean plants exposed to DFMO or DFMA before or after inoculation.

Treatment: Inhibitor	Treatment: Conc., mM	Lesions per cm^2: Preinoculation Exposure	Lesions per cm^2: Postinoculation Exposure
None		59±4	61±2
DFMO	0.01	47±6	40±5*
	0.025	34±1**	28±2**
	0.05	29±4**	17±2**
	0.075	19±2**	12±1**
	0.10	14±2**	5±1**
	0.25	2±1**	2±1**
	0.50	0**	0**
	1.0	0**	0**
	5.0	0**	0**
DFMA	1.0	51±4	56±3
	5.0	46±4	50±2*

inoculated with uredospores, the other half was also protected against infection. Transfer of the protective effect was better from leaf base to leaf apex than in the reverse direction. In the half-leaf treatment experiments, postinoculation treatment was once again somewhat more effective than preinoculation exposures to DFMO. Similar translocated protective effects were noted when a longitudinal half or the unifoliolate leaf up to the midvein was treated with DFMO, (Table 3).

In further experiments, plants that had been treated with DFMO and pathogen were reinoculated with pathogen after 1 week, at a time when disease symptoms were apparent on the unifoliolate leaves. The number of lesions on the newly emerged trifoliolate leaves was considerably reduced in plants previously treated at 0.05, 0.1, 0.5, and 1.0 mM DFMO, while 5.0 mM afforded complete protection.

Plants sprayed with DFMO at all concentrations appeared healthy, exhibiting no malformation or reduction in growth rate. By contrast, unprotected, infected plants showed a marked reduction in height. Thus it appears that the use of DFMO as a practical antifungal agent is at least feasible, since the small quantities applied appeared to become systemic.

In later studies [13], we probed the kinetics of the escape from control of pathogenesis by DFMO, to answer the practical question of how long after a known infection by uredospores DFMO may be applied with benefit. For this purpose, test plants were exposed to 1.0 mM DFMO 1, 2, 3, 4, or 5 days after inoculation with uredospores of the pathogen. Unifoliolate leaves were inoculated with uredospores and then, at daily intervals, a different group of plants was sprayed with 0.01% Tween 20 lacking or containing 1.0 mM DFMO. Complete protection against the pathogen was found when DFMO was first applied up to 3 days after inoculation. Disease symptoms occurred when the interval between inoculation and DFMO application was 4 or 5 days, although the number of lesions was significantly reduced below control (unprotected) values (Table 4).

To determine whether the effect of DFMO was related to inhibition of PA biosynthesis, we conducted experiments to determine the effects of PA application alone or in combination with 0.05 mM DFMO (which yielded ~ 50% inhibition of uredial formation), on the severity of the disease. Application of a solution of 1 mM Put reduced the number of lesions by about 30%. The number of lesions was reduced by 70% in plants exposed to 0.05 mM DFMO alone; when Spd was supplied 1 hour after DFMO, the inhibition conferred by DFMO was substantially reduced.

Table 3. Evidence for translocation of a protective effect against bean rust disease from unifoliolate leaves of pinto bean partially treated with 1 mM DFMO.

Treatment	Lesions per cm^2			
	Preinoculation Exposure		Postinoculation Exposure	
	Treated side	Untreated side	Treated side	Untreated side
Control	80±12	75±13	71±7	68±8
1 mM DFMO	0**	22±4	0**	0**

Table 4. Kinetics of inhibition of pathogenesis resulting from postinoculation application of DFMO to unifoliolate leaves of bean plants after different intervals of time.

Time between spore inoculation and DFMO spray, days	DFMO Concentration mM	Lesions per cm^2 [a]
1	0	74±6
	1.0	0
2	0	69±6
	1.0	0
3	0	64±4
	1.0	0
4	0	65±6
	1.0	41±5[b]
5	0	69±3
	1.0	53±4[b]

[a] Each value is Mean + SE, based on six replicates (one leaf per plant).
[b] Significant difference at 5% level.

The effect of DFMO and DFMA sprays (0.01, 0.1, and 1.0 mM) on PA titers was also examined. Unifoliolate leaves were sprayed in the usual way and examined 1 and 3 days later, while the first trifoliolate leaves, unexpanded at the time of spray and themselves unsprayed, were examined 8 days later. There were no significant reductions of PA levels in any of the treated leaves; in fact, Put and Spd titers were increased by the highest concentration of DFMO. This appears to be due to the paradoxical promotion of ADC activity by DFMO [1].

Thus in the absence of any depression in the PA titer of the host plant, we reason that the protective effect of DFMO against the fungus results in part from its persistence on the surface of the leaf, where fungal spores germinate and initiate growth. Since rust uredosporelings contain progressively higher Spd as their growth and differentiation progresses [14], it is reasonable to believe that PA deprivation caused by DFMO would inhibit both processes. The translocatability of the protective effect of DFMO [12] indicates that this substance may also initiate internal protective metabolic changes that are not reflected in the PA titer of the host plant.

RANGE OF DISEASES CONTROLLABLE BY DFMO and DFMA

The encouraging results with Uromyces led us to explore related diseases which might be controlled by inhibitors of polyamine biosynthesis.

We soon learned by studies with additional members of the Urediniales that we had selected an order of fungi that was generally sensitive to DFMO treatment. For example, the ID_{50} values for post-inoculation exposures of wheat stem rust (Puccinia graminis f. sp. tritici), wheat leaf rust (P. recondita), common rust of corn (P. sorghi), and oat stem rust (P. graminis f. sp. avenae) (Table 5) were 0.03 mM, 0.008 mM, 0.31 mM, and 0.024 mM, respectively. Somewhat less sensitive to DFMO were the Erysiphales, with ID_{50} values of 0.7 mM for wheat powdery mildew (Erysiphe graminis) (Table 6), and 3.25 mM for apple powdery mildew (Podosphaera leucotricha). Control of bean powdery mildew (Erysiphe polygoni) was poor, and the ID_{50} value was about 8.18 mM [15].

Several other species of fungi have been tested with varying degrees of control. With southern corn leaf blight (Helminthosporium maydis), the ID_{50} for DFMO was 0.23 mM but for DFMA it was lower (0.14 mM) (Table 7; Weinstein, unpublished) a paradoxical result possibly related to conversion to DFMO. Treatment of verticillium wilt (Verticillium dahliae) of tomato with DFMO gave the high ID_{50} value of 7.00 mM [16], but also provided a significant amount of control.

There were at least five notable responses evident from these experiments. Firstly, it was demonstrated that application of DFMO to foliage could produce a positive response in the control of the soil-borne verticillium wilt of tomato (Verticillium dahliae), a fungus disease refractory to chemical control. Secondly, DFMA was often effective in controlling some pathogens and, in a few cases, it was as effective or more effective than DFMO. This was evident in experiments with the biotrophic parasite, oat stem rust (Puccinia graminis f. sp. avenae), and the facultative parasites, southern corn leaf blight (Helminthosporium maydis) and verticillium wilt of tomato, for example. In the latter case, the inhibition was reversed by putrescine [16]. Thirdly, in a number of cases, concentrations of DFMO (or DFMA) as low as 0.01 mM provided 40 to 50% control of a disease, and increasing concentrations did not show an expected dose-response relationship. This was the case with several facultative parasites, including southern corn leaf blight and verticillium wilt of tomato, and biotrophs, including oat and wheat stem rusts. Ample evidence has been presented to show that DFMA can be converted to DFMO by arginase [10,11,16], and kinetic studies suggest that the rate of conversion of arginine and DFMA by the tobacco enzyme is similar (Slocum, Bitonti, McCann, Feirer, personal communication). This does not explain why DFMA is more effective than DFMO in controlling certain pathogens. A possible explanation is that a mechanism to convert DFMO to DFMA exists in bean plants and in certain fungi that was found to be more active than the arginase-mediated conversion of DFMA to DFMO (Slocum and Weinstein, unpublished). This is not understood by us at the present time because DFMO does not appear to serve as a substrate for l-ornithine carbamoyltransferase in plants, the first step in the citrulline cycle. Fourthly, the relatively effective control of certain pathogens at concentrations of inhibitor as low as 0.01 mM points either to a great sensitivity of the pathogen to DFMO (or DFMA), or to a stimulation in the production of naturally-occurring fungal inhibitors such as phytoalexins [17] by the plant. Fifthly, some facultative pathogens, such as cucumber anthracnose (Colletotrichum lagenarium), are barely responsive to the presence of up to 10 mM DFMO, DFMA, to a combination of the two inhibitors, or to other inhibitors of the polyamine pathways when cultured on Czapek-Dox medium or on the cucumber plant. Since neither inhibitor affects germination or growth of the mycelium on a defined medium, it might be concluded that penetration or transport of the inhibitor does not occur in these systems. However, recent studies on the uptake kinetics of ^3H-DFMO and ^3H-DFMA by Colletotrichum lagenarium and the DFMO-sensitive Helminthosporium maydis (Slocum and Weinstein, unpublished), indicated that uptake of these inhibitors was about the same in both organisms, and that 10 mM DFMO had little effect on ^3H-DFMO uptake in either organism.

Table 5. Control of oat stem rust (*Puccinia graminis* f. sp. *avenae*) by DFMO or DFMA sprayed before or after inoculation.

Inhibitor	Conc., mM	Pre-inoculation	Post-inoculation
		% Control of infection	% Control of infection
DFMO	0.00	-	-
	0.01	29.2	44.6
	0.025	30.3	53.9
	0.065	30.3	59.9
	0.10	36.9	64.3
	0.13	35.2	65.4
	0.16	46.2	73.1
	0.20	50.0	79.7
	1.00	71.5	88.5
DFMA	0.00	-	-
	0.01	-	31.9
	0.025	7.2	29.7
	0.065	21.5	37.4
	0.10	23.7	49.0
	0.13	23.7	54.4
	0.16	40.2	54.4
	0.20	46.2	61.6
	1.00	44.6	79.6

Table 6. Control of wheat powdery mildew (*Erysiphe graminis*) on winter wheat sprayed by DFMO before or after inoculation.

Concentration of DFMO, mM	Pre-inoculation	Post-iooculation
	% Control of infection	% Control of infection
0.00	-	-
0.01	41	19
0.025	35	30
0.065	46	46
0.10	51	68
0.20	46	68
1.00	66	76
5.00	95	89

Table 7. Control of southern corn leaf blight caused by _Helminthosporium maydis_ on plants sprayed with DFMO or DFMA before or after inoculation.

	Pre-inoculation		Post-inoculation
Inhibitor	Conc., mM	% Control of infection	% Control of infection
DFMO	0.00	-	-
	0.01	20	41
	0.025	18	49
	0.065	16	49
	0.10	8	59
	0.13	24	55
	0.16	18	57
	0.20	20	57
	1.00	26	69
DFMA	0.00	-	-
	0.01	2.0	55
	0.025	17.7	55
	0.065	-	57
	0.10	11.8	55
	0.13	-	59
	0.16	15.7	57
	0.20	9.9	55
	1.00	23.6	71

This paper reports only on joint research activities at our respective institutions. Some of the recent unpublished information was also conducted by Robert Slocum at Williams College in Williamstown, Massachusetts. Research on polyamine pathways and inhibition in _Helminthosporium maydis_ cultured _in vitro_ has been reported by Birecka _et al_.[18]. Walters [19] has reported in detail on DFMO inhibition of a rust on broad beans, _Uromyces vicia-fabae_. He and his colleagues have conducted extensive studies on barley powdery mildew (_Erysiphe graminis_ f. sp. _hordei_) and its control on barley plants by DFMO and by a number of other inhibitors of polyamine biosynthesis (unpublished).

ACKNOWLEDGEMENTS

This work has been supported by grants from the National Science Foundation, National Institutes of Health and BARD (US-Israel Binational Agricultural Research and Development Fund) to A.W. Galston, and from the Boyce Thompson Institute for Plant Research to L.H. Weinstein.

REFERENCES

1. C. W. Tabor and H. Tabor. Polyamines in microorganisms. _Microbiol. Rev._ 49:81 (1985).
2. R. D. Slocum, R. K. Sawhney, and A. W. Galston. The physiology and biochemistry of polyamines plants. _Arch. Biochem. Biophys._ 235:28 (1984).
3. O. Heby. Role of polyamines in the control of cell proliferation and differentiation. _Differentiation_ 19:1 (1981).

4. C. W. Tabor. Mutants of Saccharomyces cerevisiae deficient in polyamine biosynthesis: studies on the regulation of ornithine decarboxylase. Med. Biol. 59:272 (1981).
5. B. W. Metcalf, P. Bey, C. Danzin, M. J. Jung, P. Casara, and J. P. Vevert. Catalytic irreversible inhibition of mammalian ornithine decarboxylase (E.C. 4.1.1.17) by substrate and product analogues. Jour. Amer. Chem. Soc. 100:2551 (1978).
6. H. E. Flores and A. W. Galston. Polyamines and plant stress: Activation of putrescine biosynthesis by osmotic shock. Science 217:1259 (1982).
7. A. Kallio, P. P. McCann, and P. Bey. DL-a-(difluoromethyl)-arginine: A potent enzyme-activated irreversible inhibitor of bacterial arginine decarboxylases. Biochem. 20:3163 (1981).
8. M. V. Rajam and A. W. Galston. The effects of some polyamine biosynthetic inhibitors on growth and morphology of phytopathogenic fungi. Plant and Cell Physiol. 26:683 (1985).
9. Anonymous. Difco Manual of dehydrated culture media and reagents for microbiological and clinical laboratory procedures. (9th Edition) Difco Labs, Inc., Detroit, Michigan. 350 pp, 1953.
10. R. D. Slocum and A. W. Galston. Arginase-mediated hydrolysis of DFMA to DFMO in vivo. Pl. Physiol. 77 (Suppl): 45 (1985).
11. R. D. Slocum and A. W. Galston. Inhibition of polyamine biosynthesis in plants and plant pathogenic fungi, in: "Inhibition of Polyamine Metabolism," P.M. McCann. A. E. Pegg, and A. Sjoerdsma, eds., Academic Press, New York (1987).
12. M. V. Rajam, L. H. Weinstein, and A. W. Galston. Prevention of a plant disease by specific inhibition of fungal polyamine biosynthesis. Proc. Natl. Acad. Sci. (USA) 82:6874 (1985).
13. M. V. Rajam, L. H. Weinstein, and A. W. Galston. Kinetic studies on the control of the bean rust fungus (Uromyces phaseoli L.) by an inhibitor of polyamine biosynthesis. Pl. Physiol. 82:485 (1986).
14. W. K. Kim. Folate and polyamine content of undifferentiated and differentiated wheat stem rust uredosporelings. Can. Jour. Bot. 49:119 (1971)
15. L. H. Weinstein, J. F. Osmeloski, S. H. Wettlaufer, and A. W. Galston. Protection of wheat against leaf and stem rust and powdery mildew diseases by inhibition of polyamine metabolism. Plant Science 51:311 (1987).
16. H. Mussell, J. Osmeloski, S. H. Wettlaufer, and L. Weinstein. Suppression of verticillium wilt of tomato by difluoromethylornithine, a suicidal inhibitor of polyamine biosynthesis. Plant Disease 71:313 (1987).
17. I. A. M. Cruickshank. Phytoalexins. Ann. Rev. Phytopath . 1:351 (1963).
18. H. Birecka, M. O. Garraway, R. J. Baumann, and P. P. McCann. Inhibition of ornithine decarboxylase and growth of the fungus Helminthosporium maydis. Plant Physiol. 80:798 (1986).
19. D. R. Walters. The effects of a polyamine biosynthesis inhibitor on infection of Vicia faba L. by the rust fungus, Uromyces vicia-fabae (Pers.) Schroet. New Phytol. 104:613 (1986).

RELATIONSHIP BETWEEN POLYAMINE AND ETHYLENE BIOSYNTHESIS IN PLANTS AND ITS SIGNIFICANCE FOR MORPHOGENESIS IN CELL CULTURES[1]

Subhash C. Minocha

Department of Botany and Plant Pathology
University of New Hampshire
Durham, New Hampshire 03824, USA

The regulation of somatic embryogenesis in plant cell cultures is a topic of considerable interest, not only because of its potential applications for micropropagation and genetic engineering in plants but also as a prime example of the process of controlled development and differentiation. Therefore, it is not surprising that a vast amount of literature has accumulated discussing the morphogenetic and biochemical events underlying embryogenic development in cell cultures. The complexity of interaction of various chemical factors within the plant tissue and those that are externally applied, has been stressed in a number of reviews (Steward, 1976; Street, 1977; Ammirato, 1983,a,b, 1984; Sung et al., 1984; Raghavan, 1986).

Among the variety of plants that have been used for the study of organogenesis and somatic embryogenesis, carrot cell culture system is perhaps the best choice. Although it does not fulfill all the requirements of an "ideal" system, it certainly meets most of the important criteria needed for detailed biochemical analysis of the events associated with induced morphogenesis.

Carrot cells can be grown in large quantities under well defined conditions and can be easily induced to differentiate into somatic embryos by the removal of auxin from the growth medium. Embryogenic response is quick and predictable. Detailed information on the timecourse of embryogenesis and a number of quantitative aspects of morphological variations in the somatic embryos have recently become available (Ammirato, 1983b, 1984; Schiavone and Cooke, 1985). Although, the lack of synchrony is still a major drawback of the system, recent progress in partial synchronization of the differentiation process (Fujimura and Komamine, 1979a; Sala et al., 1983; Bradley et al., 1984) and the techniques of concentrating specific developmental stages of differentiating embryos by differential centrifugation and filtration (Warren and Fowler, 1977; Giuliano et al., 1983; LoSchiavo, 1984) can yield sufficient material for most biochemical studies. A fair amount of background information on histochemical and biochemical aspects of embryogenesis in this system is already available (Gregor et al., 1974; Masuda et al., 1984; Sala et al., 1983; Terzi et al., 1985).

---

1. Scientific contribution No. 1582 from the New Hampshire Agricultural Experiment Station. The research was partially supported by National Science Foundation Grant No. DCB-8615945.

Auxin and Somatic Embryogenesis

The earlier studies of Steward et al. (1958) and Reinert (1959) indicated the need for complex nutrient mixtures like coconut milk and several amino acids for embryogenesis in carrot. However, it was soon found that extensive embryogenesis could be obtained in this tissue simply by the removal of auxin from the growth medium (Halperin and Wetherell, 1964). Since then, a variety of physical and chemical factors have been shown to affect embryogenesis in carrot cell cultures, particularly, the source of nitrogen (Halperin and Wetherell, 1965; Reinert, 1973; Wetherell and Dougall, 1976) and K^+ ions (Brown et al., 1976). Although cytokinins and gibberellins can alter the quantitative aspects of embryogenesis (Sung et al., 1984; Fujimura and Komamine, 1975, 1980), auxin remains the most critical controlling factor.

Auxin plays a dual role in the growth and differentiation of carrot cell cultures. First, it is used to induce callus formation and callus growth; second, it suppresses embryogenesis in callus cells. Differentiation is initiated upon the removal of auxin from the growth medium (Kato and Takeuchi, 1966; Reinert, 1959, 1973; Halperin and Wetherell, 1964). Whereas, 2.4-dichlorophenoxyacetic acid (2,4-D) is the most commonly used auxin in carrot cultures, the involvement of endogenous auxin has been demonstrated in a number of cases (Newcomb and Wethrell, 1970; Smith and Street, 1974; Fujimura and Komamine, 1979b).

It has been postulated that 2,4-D in fact may be essential for imparting the embryogenic potential to carrot cells or even initiating embryogenesis while suppressing the growth of embryos (Halperin and Wetherell, 1964; Reinert, 1973; Kato and Takeuchi, 1966). The production of globular embryos seen in low density carrot cultures has been used to support this view (Borkird et al. 1986). Although a number of carrot cell strains (mutants?) have been selected which lack embryogenic potential on removal of auxin (Breton and Sung, 1982), no strains are known that develop normal embryos with high frequency in the presence of exogenous auxin, thus confirming an absolute inhibitory role of auxin in carrot embryogenesis. The question as to the biochemical basis of this inhibitory role still remains unanswered.

Polyamines and Embryogenesis

More recently, it has been suggested that polyamines (putrescine, spermidine and spermine) may play an important regulatory role in the process of somatic embryogenesis. Polyamines have been implicated in a variety of growth and developmental processes in animals as well as plants (Campbell et al., 1978; Tabor and Tabor, 1983, 1984; Galston, 1983; Slocum et al., 1984; Smith, 1985a; Pegg, 1986). Most of the work, however, involves animal cells.

The physiological importance of polyamines in the growth and development of plants is apparent from the following types of observations: (1) increases in polyamine levels precede the induction of cell division in dormant tissues (Bagni et al., 1981; Seraffini-Fracassini et al., 1980, 1984; Huhtinen et al., 1982/1983); (2) polyamine levels increase in association with the germination of seeds, sprouting of tubers, growth of fruits from pollinated ovaries, and the promotion of growth by red light (Bagni et al, 1980, Cohen et al., 1982a,b; Goren et al., 1982; Slocum and Galston, 1985); (3) the senescence of excised leaves is accompanied by a decline in polyamine biosynthesis and the process can be arrested by the application of exogenous polyamines (Kaur-Sawhney and Galston, 1979; Altman and Bachrach, 1981; Kaur-Sawhney et al., 1982; Roberts et al., 1984); (4) inhibition of polyamine biosynthesis affects DNA, RNA, and protein synthesis and also the integrity of the plasma membrane (Tabor and Tabor, 1984; Pegg, 1986); (5) polyamines stabilize protoplasts, thylakoid membranes of chloroplasts and retard senescence in protoplasts (Altman et al., 1977; Galston et al., 1978; Cheng and Kao, 1983; Kaur-Sawhney, et al. 1985); (6) cellular levels of polyamines increase sharply in response to various kinds of physical and chemical stress and it has been suggested that they may help reduce stress-induced damage to the cell (Priebe et al., 1978; Flores and Galston, 1982, 1984a,b; Young and Galston, 1983; Turner and Stewart, 1986); (7) the rates of synthesis of polyamines and the cellular polyamine levels increase significantly prior to the induction of organogenesis and embryogenesis in cell cultures (Montague et al., 1978, 1979; Feirer et al., 1984; Fienberg et al., 1984; Desai and Mehta, 1985, Minocha and Robie, 1985); (8) cellular levels of polyamines are modulated by a variety of plant growth regulators (Suresh et al., 1978; Dai et al., 1982; Kyriakidis, 1983; Cho, 1983; Palavan et al., 1984; Apelbaum et al., 1985); and, (9) polyamine synthesis is regulated by a complex feedback control mechanism involving a number of key biosynthetic enzymes with extremely short half-lives (Tabor and Tabor, 1984; Slocum et al. 1984; Pegg, 1986).

Biosynthetic pathways for the three major polyamines are well established (Fig. 1). In animals, the key enzyme in the pathway is ornithine decarboxylase (ODC), whereas in plants two major enzymes seem to be involved at the first steps (Tabor and Tabor, 1984; Slocum et al., 1984; Smith, 1985a; Morgan, 1987). These are ODC and arginine decarboxylase (ADC).

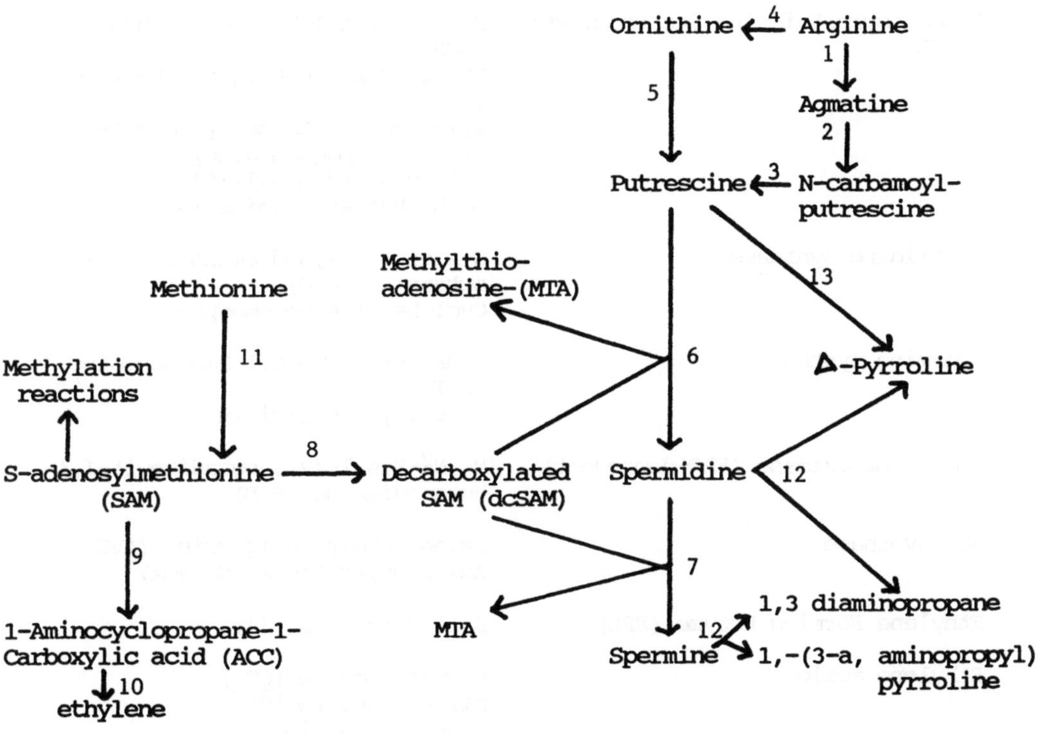

Fig. 1. Combined pathways for the biosynthesis of polyamines and ethylene in plants. The enzymes involved are: 1. Arginine decarboxylase (ADC); 2. Agmatine iminohydrolase; 3. N-carbamoylputrescine amidohydrolase; 4. Arginase; 5. Ornithine decarboxylase (ODC); 6. Spermidine synthase; 7. Spermine synthase; 8. SAM decarboxylase (SAMDC); 9. ACC synthase; 10. Ethylene Forming Enzyme; 11. S-Adenosyl-methionine synthase; 12. Polyamine oxidase (PAO); 13. Diamine oxidase (DAO) (modified from Slocum et al. 1984; Pegg, 1986; Morgan, 1987).

Another important enzyme of the biosynthetic pathway is S-adenosylmethionine decarboxylase (SAMDC), which is found in both plants and animals. Significant increases in the cellular levels of different polyamines are always correlated with increase in the activities of these enzymes. Each of the three major enzymes has a strikingly short turnover time (half-lives are ODC - 10 min, ADC - 60-120 min, SAMDC - 30-60 min) indicating the cell's need for rapid modulation of polyamine levels in response to a variety of stimuli. Specific chemical inhibitors of all these enzymes are also known (Table I).

TABLE 1. Inhibitors of enzymes involved in biosynthesis and metabolism of polyamines and ethylene (modified from Slocum et al. 1984; Pegg, 1986).

Enzyme	Inhibitor
L-Ornithine decarboxylase (ODC)	DL-α-Difluoromethylornithine (DFMO) α-Fluoromethyl-(E)-dehydro-ornithine α -monofluoromethylornithine (MFMO)
L-Arginine decarboxylase (ADC)	Difluoromethylarginine (DFMA)
S-Adenosylmethionine decarboxylase (SAMDC)	Methylglyoxal bis-(guanylhydrazone) (MGBG) Ethylglyoxal bis-(guanylhydrazone) (EGBG) 1,1-[(methylethanediylidene)-dinitrilo] bis-(3-aminoguanidine) (MBAG) S-S'-deoxyadenosyl-(5')-2-methylthioethylamine (SMT)
Spermidine synthase	S-Adenosyl-1,8-diamino-3-thio-octane(AdoDATO) Cyclohexylamine phosphate (CHAP)
Spermine synthase	S-Methyl-5'-methylthioadenosine (MMTA) 5'-methyltubericidin
Polyamine oxidase (FAD-dependent)	N^1,N^2-Bis-1-(2,3-butadienyl)-1,4-butanediamine (BBB)
ACC Synthase	Amino-ethoxyvinylglycine (AVG) Amino-oxyacetic acid (AOA)
Ethylene Forming Enzyme (EFE)	Salicylic acid (SA)
Ethylene action	carbon dioxide (CO_2) silver ions (Ag^{+2}) 2,5-norbornadine

Other enzymes involved in the metabolism of polyamines in plants are: (1) Arginase, which converts arginine to ornithine, which can be either used for putrescine synthesis by ODC or converted back to arginine through ornithine cycle; (2) agmatine iminohydrolase, which converts agmatine (the product of ADC) to carbamoylputrescine, which is then converted to putrescine by a multifunctional enzyme called putrescine synthase (putrescine transcarbamoylase). The synthesis of putrescine by the ADC pathway seems to be primarily regulated at the first step, i.e. ADC. In some plants a citrulline decarboxylase catalyzing the formation of N-carbamoylputrescine has also been suggested as another source of putrescine but this enzyme is not widespread in the plant kingdom. A number of diamine oxidases (DAOs) and polyamine oxidases (PAOs) have been reported in plants which catalyze the degradation of polyamines (Smith, 1985b). PAOs have been reported only from grasses. It should also be mentioned that there are a number of other 'minor' polyamines found in algae and higher plants (Smith, 1981; Hamana and Matsuzaki, 1982), which are interconvertible and may contribute to the cellular pools of the three major polyamines discussed above. None of them have, however, been shown to be a significant source of putrescine, spermidine or spermine.

Increased polyamine biosynthesis has been observed during organogenesis in cell cultures of a number of plants (Slocum et al. 1984; Kaur-Sawhney et al. 1985, Tiburcio et al. 1987; Torrigiani et al. 1987; Litz and Schaffer, 1987). The role of polyamines in embryogenesis in carrot cell cultures has been examined in detail by several workers. Montague et al. (1978, 1979) observed that the induction of embryogenesis by the removal of 2,4-D was accompanied by increased levels of the three major polyamines, Later, Fienberg et al. (1984) and Feirer et al. (1984) confirmed these observations and also showed that enzyme activities for putrescine and spermidine biosynthesis generally increased under embryogenic conditions. Inhibition of ADC by DFMA or of SAMDC by MGBG caused an inhibition of embryogenesis and reduction in the level of cellular polyamines. The effects of these inhibitors were reversible by addition of polyamines to the medium. Bradley et al. (1984), however, reported that exogenous putrescine or arginine actually blocked the growth of somatic embryos beyond the globular stage. A similar effect of spermidine and spermine has been observed in our laboratory (unpublished data). DFMO, an inhibitor of ODC, had no inhibitory effect on somatic embryogenesis, leading to the conclusion that carrot tissue employs the ADC pathway for polyamine synthesis. Other than alluding to an "important role" in embryogenesis, no explanations have been advanced regarding the significance of the correlation between the changes in polyamine levels and embryogenesis in carrot tissue.

While confirming that polyamines do play an important role in somatic embryogenesis in carrot, studies from our laboratory provide further insight into the possible roles of ODC and ADC in the process of differentiation. The results show that while ODC levels do remain low during the first 10-12 days of somatic embryogenesis (development up to the stage of almost normal embryo), a sharp rise in ODC activity is seen after that period, coinciding with the increased growth of embryos into whole plants (Fig. 2). Also, older carrot plants (2-3 month old) growing in vitro show significantly high levels of ODC indicating that ODC activity may be restricted to 'maturing' tissues. ADC, which is the predominant enzyme during the first two weeks in culture, shows reduced activity in older tissues. Even more interesting than the above observation is the morphogenetic effect of DFMO, which is a known specific inhibitor of ODC in most plant and all animal tissues (Table 1). This compound when used at 1 or 10 mM concentration, while having no apparent effect on the extractable ODC activity from either embryogenic or nonembryogenic cells, causes a significant increase in the activity of ADC in the cells (Fig. 2). A concomitant increase in all three major polyamines is seen in the presence of DFMO (Table 2). Moreover, when differentiating embryos from 2,4-D-free medium are transferred to 2,4-D containing medium in the presence of DFMO, normal embryogenic growth continues. Control cultures (minus DFMO, plus 2,4-D) all revert to callus within 2-4 days. Once again, increases in ADC activity and polyamines are seen in the presence of DFMO. DFMO when added to 2,4-D-free medium significantly promotes the development of embryos. Thus, through a manipulation of the polyamine biosynthesis, for the first time it has been possible to obtain somatic embryogenesis in carrot cell cultures even in the presence of well established inhibitory concentrations of 2,4-D. Further studies also show that cultures that are differentiating either in the 2,4-D-free medium or in 2,4-D+DFMO medium produce significantly less ethylene than those grown in the presence of 2,4-D and are not differentiating (data not presented here).

TABLE 2. The effect of 10mM DFMO on the cellular polyamine levels in carrot cell cultures grown in the presence of 0.2 mg/l 2,4-D for 4 and 6 days. Polyamines were extracted in 5% perchloric acid, dansylated, and analyzed by HPLC using a gradient system of acetonitrile and heptanesulphonate. Polyamine concentrations are presented as nmol/g fr. wt. of cells. Data are Mean±S.E. of three replicates. Difference from the control (***) $p \leq .001$; (**) $.001 < p \leq .005$ (T-Test).

Days	Polyamine	minus DFMO	plus DFMO
4	Putrescine	30.6 ± 2.3	80.1 ± 3.0***
	Spermidine	75.1 ± 2.9	137.8 ± 4.8***
	Spermine	30.1 ± 3.8	58.6 ± 3.1**
6	Putrescine	12.9 ± 1.2	38.2 ± 2.8***
	Spermidine	128.8 ± 6.0	184.3 ± 4.0**
	Spermine	51.5 ± 2.7	80.7 ± 3.4**

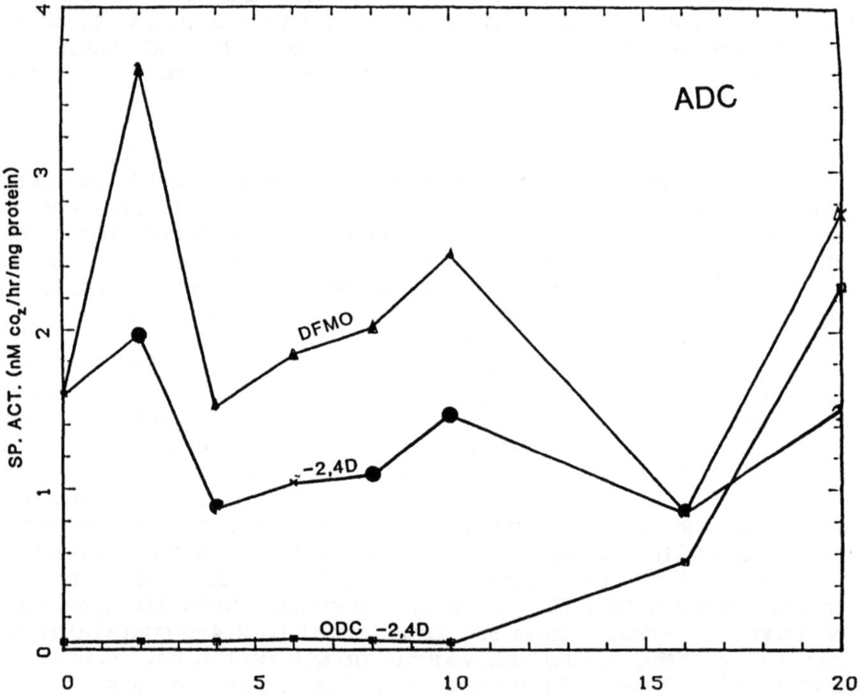

FIG. 2. Changes in the activity of ADC and ODC in carrot cells grown in the presence or absence of DFMO for different times. Each point represents an average of 3 replicates.

Methylglyoxal bis-(guanylhydrazone) (MGBG), an inhibitor of SAMDC, inhibits growth as well as embryogenesis (500 uM) or only embryogenesis (50 uM) when added to the embryogenic media. Likewise, cyclohexylammonium phosphate (CHAP), an inhibitor of spermidine synthase is also a strong inhibitor of embryogenesis in the absence of 2,4-D. Both MGBG and CHAP significantly inhibit spermidine and spermine biosynthesis within 24h of application, while allowing putrescine to accumulate. The effects of MGBG and CHAP are reversed by 1-2 mM concentrations of spermidine or spermine. By itself, exogenously supplied spermidine or spermine also suppresses embryogenesis at concentrations higher than 2 mM.

Auxin, Ethylene, and Polyamines

Whereas the exact role of exogenous auxin in the suppression of embryogenesis in carrot (and of many other tissues) is not known, a possible suggestion comes from the studies of Wochok and Wetherell (1972), Tisserat and Murashige (1977), and Verma and Tarka, (1985). These authors reported that ethylene is a strong inhibitor of embryogenesis in carrot cell cultures. Ethylene is a gaseous plant hormone which is known to affect a number of growth and developmental processes in plants such as fruit ripening, senescence of leaves and flower petals, and abscission (Abeles, 1973; Lieberman, 1979; Yang and Hoffman, 1984). Ethylene biosynthesis is regulated both through auto-catalytic induction as well as through a variety of chemical and environmental stimuli. It is well established that plant tissues grown in vitro produce significant amounts of ethylene and also that ethylene inhibits growth and differentiation in vitro (Everett, 1982; Mele et al., 1982; Balague and Pech, 1985; Verma and Tarka, 1985; Woltering, 1986; Cassells and Tamma, 1986; Sauerbrey et al. 1987, 1988). Furthermore, it is also known that auxins stimulate ethylene formation in various tissues (Burg and Burg, 1966; McBride and Evans, 1977; Sembdner et al., 1980; Yang and Hoffman, 1984). Auxin must be continuously present for ethylene to be produced (Fowler and Morgan, 1972; Abeles, 1973). It is thus possible that the auxin effect on somatic embryogenesis may actually be due to the production of ethylene.

As shown in Fig. 1, the biosynthesis of ethylene involves first the conversion of S-adenosylmethionine to 1-aminocyclopropane-1-carboxylic acid (ACC) by ACC synthase which is then converted to ethylene by a multifunctional membrane-bound ethylene forming enzyme (EFE) (Lurssen et al. 1979; Yang and Hoffman, 1984). Both these enzymes can perform regulatory functions but ACC synthase is believed to be the major regulatory step. ACC synthesis can be inhibited by amino-ethoxyvinylglycine (AVG) and amino-oxyacetic acid (AOA) (Yu and Yang, 1979; Yang and Hoffman, 1984), whereas, salicylic acid has been shown to inhibit EFE activity.

A number of physiological effects of ethylene in plants are known to be antagonized by treatment with polyamines, leading to the suggestion that the two might interact with each other in affecting the growth and development of plants (Apelbaum et al., 1975, 1981; Roberts et al., 1984; Galston and Kaur-Sawhney, 1987). This becomes even more important due to the fact that both ethylene and polyamines share and probably compete for the same intermediate, S-adenosylmethionine (SAM). It has been shown that inhibition of ACC synthesis stimulates the incorporation of radioactivity from methionine into spermidine and promotes the synthesis of spermine (Even-Chen et al., 1982). Also, on inhibition of polyamine synthesis by D-arginine, DFMA, and MGBG, ethylene production was significantly promoted (Roberts et al. 1984). An increased synthesis of spermidine or externally supplied polyamines reduced the endogenous concentrations of ACC and also inhibited the conversion of ACC to ethylene (Apelbaum et al., 1981; Suttle, 1981; Fuhrer et al.,1982).

In summary, it is known that: (i) auxin suppresses embryogenesis; (ii) auxin promotes ethylene biosynthesis; (iii) ethylene inhibits embryogenesis; (iv) promotion of polyamine biosynthesis by DFMO allows embryogenesis to occur even in the presence of auxin; (v) increased polyamine biosynthesis, especially spermidine and spermine, accompanies embryogenesis; (vi) externally supplied polyamines do not replace the need for increased rate of biosynthesis of polyamines in the cells; and (vii) the pathway for spermidine and spermine biosynthesis shares a common precursor (S-adenosylmethionine) with ethylene biosynthesis. From the results obtained so far it has become apparent that although the inhibition of polyamine biosynthesis inhibits embryogenesis, and the promotion of polyamine biosynthesis promotes embryogenesis, the continuous synthesis of polyamines at high rates seems to be more important than the mere presence of elevated levels of polyamines in the cells. One possible significance of this conclusion is that continuous high rates of spermidine and spermine synthesis may reduce the cellular levels of S-adenosylmethionine, thus causing a reduction in ethylene biosynthesis.

The hypothesis to explain the auxin effects on somatic embryogenesis in carrot and its reversal by DFMO has three postulates: (1) ethylene is a major suppressor of embryogenesis and its production is promoted by auxin; (2) continued polyamine biosynthesis is required for cell division and embryogenesis to proceed; and (3) the increased biosynthesis of polyamines, especially spermidine and spermine induced by DFMO probably "steals away" S-adenosylmethionine from ACC synthase, thus causing a reduction in ethylene biosynthesis.

While the hypothesis is quite compatible with most of the published work, further evidence is needed to strengthen this hypothesis and thus gain further knowledge about the regulation of somatic embryogenesis in carrot. This can come from: (1) detailed analysis of the effects of various inhibitors of polyamine and ethylene biosynthesis on embryogenesis and on cellular levels of the enzymes and the metabolites of the two pathways; (2) the analysis of polyamine biosynthesis and the potential to undergo embryogenesis in mutant cell lines which are either overproducers or auxotrophic for specific polyamine(s); and (3) the modulation of the rates of polyamine biosynthesis through regulated action of foreign genes for polyamine biosynthetic enzymes introduced into the test plant. Although a variety of "specific" inhibitors for each step in the biosynthesis of polyamines and ethylene are available (Table 1), and their use in a number of plant and animal systems has provided a wealth of information about the importance of polyamines for a number of growth and developmental phenomena, a critical evaluation of the role of specific polyamine(s) in a developmental process suffers from the following drawbacks: (i) pleiotropic effects of most of the compounds have been poorly evaluated thus far. For some inhibitors (e.g., AVG) which affect enzyme activity through interaction with a co-factor, such as pyridoxal phosphate (Apelbaum et al., 1981; Privalle and Graham, 1987), the effects on other enzymes utilizing such a co-factor are bound to be significant; (ii) most of these inhibitors have been synthesized only in minute quantities and thus are not available for large scale experimental use; (iii) variable rates of uptake, accumulation and metabolism of inhibitors by cells can lead to erroneous interpretation of results; and (iv) the inhibitors, in general, are useful only in reducing the biosynthesis of a metabolite and not in promoting a specific step.

The availability of cell lines which either overproduce a specific polyamine or have reduced levels of an enzyme for a specific step can be extremely valuable in analyzing the importance of that particular step in the developmental process (Tabor and Tabor, 1984; Malmberg and McIndoo, 1983; Malmberg, personal communication; Pegg, 1986). The selection and characterization of such mutant cell lines, however, can be a formidable task in higher plants. Also, such cell lines only provide a constitutive regulation of the step, thus limiting our ability to experimentally manipulate the rates of biosynthesis of individual metabolites.

The recently developed techniques for isolation and cloning of specific genes and for their transfer into heterologous plant systems for expression in a constitutive or an inducible manner (Herrerra-Estrella et al. 1984, 1985; Fluhr et al. 1986; Simpson et al. 1983; Timko et al. 1985) should provide unique means to modulate polyamine biosynthesis in the cells. This should then allow us to regulate the rate of biosynthesis of any one or more metabolite and to study its effects on a morphogenic process.

Our laboratory is currently developing recombinant genes for each of the enzymes involved in the polyamine and ethylene biosynthetic pathway in plants. These genes will be recombined with appropriate promoters so that on transfer to carrot cells, their expression can be induced by stimuli which otherwise have little effect on the rate of growth or embryogenesis. These genes, combined with the use of inhibitors and certain mutant cell lines, should allow us to elucidate the interaction of the two important metabolic pathways (i.e. polyamine and ethylene biosynthesis) and their role in embryogenesis.

REFERENCES

Abeles, F. B. 1973. Ethylene in Plant Biology. Acad. Press, New York.

Altman, A. and U. Bachrach. 1981. Involvement of polyamines in plant growth and senescence. In: Adv. Polyamine Res. vol. 3:365-375.

Altman, A., R. Kaur-Sawhney, and A. W. Galston. 1977. Stabilization of oat leaf protoplasts through polyamine-mediated inhibition of senescence. Plant Physiol. 60:570-574.

Ammirato, P. V. 1983a. Embryogenesis. In: D. A. Evans, W. R. Sharp, P. V. Ammirato, and Y. Yamada (eds.). Handbook of Plant Cell Culture pp. 82-123. Macmillan, N.Y.

Ammirato, P. V. 1983b. The regulation of somatic embryo development in plant cell cultures: suspension culture techniques and hormone requirements. Bio/technology March 1983: 68-74.

Ammirato, P. V. 1984. Induction, maintenance, and manipulation of development in embryogenic cell suspension cultures. In: I. K. Vasil (ed.), Cell Culture and Somatic Cell Genetics of Plants. Vol. 1 p. 139-151. Acad. Press, N. Y.

Apelbaum, A., A. C. Burgoon, J. D. Anderson, M. Lieberman, R. Ben-Arie, and A. K. Mattoo. 1981. Polyamines inhibit biosynthesis of ethylene in higher plant tissue and fruit protoplasts. Plant Physiol. 68:453-456.

Apelbaum, A., A. Goldlust, and I. Icekson. 1985. Control by ethylene of arginine decarboxylase activity in pea seedlings and its implication for hormonal regulation of plant growth. Plant Physiol. 79:635-640.

Bagni, N., B. Malucelli, and P. Torrigani. 1980. Polyamines, storage substances and abscisic acid-like inhibitors during dormancy and very early activation of <u>Helianthus tuberosus</u> tuber tissues. Physiol. Plant. 49:341-345.

Bagni, N., P. Torrigiani, and P. Barbieri. 1981. Effect of various inhibitors of polyamine synthesis on the growth of Helianthus tuberosus. Med. Biol. 59:403-409.

Balague, C. and J. C. Pech. 1985. Relationship between the activity of ethylene-forming enzyme and the level of intracellular 2,4-dichlorophenoxyacetic acid in pear cell cultures in vitro. J. Plant Growth Reg. 4:81-89.

Borkird, C., J. H. Choi, and Z. R. Sung. 1986. Effect of 2,4-dichlorophenoxyacetic acid on the expression of embryonic program in carrot. Plant Physiol. 81:1143-1146.

Bradley, P. M., F. El-Fike, and K. L. Giles. 1984. Polyamines and arginine affect somatic embryogenesis of Daucus carota. Plant Sci. Lett. 34:397-401.

Breton, A. M. and Z. R. Sung. 1982. Temperature-sensitive carrot variants impaired in somatic embryogenesis. Develop. Biol. 90:58-66.

Brown, S., D. F. Wetherell, and D. K. Dougall. 1976. The potassium requirement for growth and embryogenesis in wild carrot suspension cultures. Physiol. Plant. 37:73.

Burg, S. P. and A. E. Burg. 1966. The interaction between auxin and ethylene and its role in plant growth. Proc. Nat. Acad. Sci. USA 55: 262-269.

Campbell, R. A., D. R. Morris, D. Bartos, G. D. Davies, and F. Bartos (eds.). 1978. Advances in Polyamine Research Vols. 1 and 2, Raven Press, N. Y.

Cassells, A. C. and L. Tamma. 1986. Ethylene and ethane release during tobacco protoplast isolation and protoplast survival potential in vitro. Physiol. Plant 66: 303-308.

Cheng, S. H. and C. H. Kao. 1983. Localized effect of polyamines on chlorophyll loss. Plant and Cell Environ. 24:1463-1467.

Cho, S. C. 1983. Effect of cytokinin and several inorganic cations on the polyamine content of lettuce cotyledons. Plant Cell Physiol. 24:27-32.

Cohen, E., S. Arad, Y. M. Heimer, and Y. Mizrahi. 1982a. Participation of ornithine decarboxylase in early stages of tomato fruit development. Plant Physiol. 70:540-543.

Cohen, E., Y. M. Heimer, Y. Mizrahi. 1982b. Ornithine decarboxylase and arginine decarboxylase activities in meristematic tissues of tomato and potato plants. Plant Physiol. 70:544-546.

Dai, Y. R., R. Kaur-Sawhney, and A. W. Galston. 1982. Promotion by gibberellic acid of polyamine biosynthesis in internodes of light-grown dwarf peas. Plant Physiol. 69:103-105.

Desai, H. V. and A. R. Mehta. 1985. Changes in polyamine levels during shoot formation, root formation, and callus induction in cultured Passiflora leaf discs. J. Plant Physiol. 119:45-53.

Even-Chen, Z., A. K. Mattoo, and R. Goren. 1982. Inhibition of ethylene biosynthesis by aminoethoxyvinylglycine and by polyamines shunts label from 3,4-[^{14}C]-methionine into spermidine in aged orange peel discs. Plant Physiol. 69:385-388.

Everette, N. 1982. The determination phase of differentiation. In: Proc. Intl. Cong. Plant Tissue and Cell Culture, Tokyo, A. Fujiwara (ed.), pp. 93-94.

Feirer, R. P., G. Mignon, and J. D. Litvay. 1984. Arginine decarboxylase and polyamines required for embryogenesis in the wild carrot. Science 223:1433-1435.

Fienberg, A. A., J. H. Choi, W. P. Lubich, and Z. R. Sung. 1984. Developmental regulation of polyamine metabolism in growth and differentiation of carrot culture. Planta 162:532-539.

Flores, H. E. and A. W. Galston. 1982. Polyamines and plant stress: Activation of putrescine biosynthesis by osmotic shock. Science 217:1259-1261.

Flores, H. E. and A. W. Galston, 1984a. Osmotic stress-induced polyamine accumulation in cereal leaves. I. Physiological parameters of the response. Plant Physiol. 75:102-109.

Flores, H. E. and A. W. Galston. 1984b. Osmotic stress-induced polyamine accumulation in cereal leaves. II. Relation to amino acid pools. Plant Physiol. 75:110-113.

Fluhr, R., C. Kuhlmeier, F. Nagy and N.-H. Chua. 1986. Organ-specific and light-induced expression of plant genes. Science 232:1106-1112.

Fowler, J. L. and P. W. Morgan. 1972. The relationship of the peroxidative indoleacetic acid oxidase system to in vivo ethylene synthesis in cotton. Plant Physiol. 49: 555-559.

Fuhrer, J., R. Kaur-Sawhney, L. M. Shih and A. W. Galston. 1982. Effect of exogenous 1,3-diaminopropane and spermidine on senescence of oat leaves. Plant Physiol. 70: 1597-1600.

Fujimura, T., A. Komamine. 1975. Effects of various growth regulators on the embryogenesis in a carrot cell suspension culture. Plant Sci. Lett. 5: 359-364.

Fujimura, T. and A. Komamine. 1979a. Synchronization of somatic embryogenesis in a carrot cell suspension culture. Plant Physiol. 64:162-164.

Fujimura, T. and A. Komamine. 1979b. Involvement of endogenous auxin in somatic embryogenesis in a carrot suspension culture. Z. Pflanzenphysiol. 95: 13-19.

Fujimura, T. and A. Komamine. 1980. Mode of action of 2,4-D zeatin on somatic embryogenesis in a carrot cell suspension culture. Z. Pflanzenphysiol. 99:1-8.

Galston, A. W. 1983. Polyamines as modulators of plant development. BioScience 33:382-388.

Galston, A. W., A. Altman, R. Kaur-Sawhney. 1978. Polyamines, ribonuclease and the improvement of oat leaf protoplasts. Plant Sci. Lett. 11:69-79.

Galston, A. W. and R. Kaur-Sawhney. 1987. In: Plant hormones and their role in plant growth and development. P. J. Davies (ed.) pp. 280-295.

Giuliano, G., D. Rosellini and M. Terzi. 1983. A new method for the purification of the different stages of carrot embryoids. Plant cell Rep. 2:216-218.

Goren, R., N. Palavan, H. E. Flores, and A. W. Galston. 1982. Changes in polyamine titer in etiolated pea seedlings following red light treatment. Plant Cell Physiol. 23:19-26.

Gregor, D., J. Reinert and H. Matsumoto. 1974. Changes in chromosomal proteins from embryo induced carrot cells. Plant Cell Physiol. 15:875-881.

Halperin, W. and D. F. Wetherall. 1964. Adventive embryony in tissue cultures of wild carrot Daucus carota. Amer. J. Bot. 51:274-283.

Halperin, W. and D. F. Wetherall. 1965. Ammonium requirement for embryogenesis in vitro. Nature 205:519-520.

Hamana, K. and S. Matsuzaki. 1982. Distribution of polyamines in prokaryotes, algae, plants and fungi. In: Polyamines: Basic and Clinical Aspects. Imahori et al. (eds.) pp. 105-112. VNU Science Press.

Herrera-Estrella, L., G. Van den Broeck, J. Schell, M. Van Montague, M. P. Timko, A. R. Cashmore, and A. P. Mausch. 1985 Use of Chimeric genes to study light-inducible gene expression and chloroplast import of the small subunit of Ribulore-1,5-bisphosphate carboxylase. In: Molecular Biology of the Photosynthetic Apparatus. Cold Spr. Harbor Lab. N. Y. pp. 397-405.

Herrera-Estrella, L., G. Van den Broeck, R. Maenhaut, M. Van Montagu, J. Schell, M. Timko and A. Cashmore. 1984. Light-inducible and chloroplast-associated expression of a Chimaeric gene introduced into Nicotiana tabacum using a Ti plasmid vector. Nature 310:115-120.

Huhtinen, O., J. Honkanen, and L. K. Simola. 1982/83. Ornithine-and putrescine-supported divisions and cell colony formation in leaf protoplasts of alders (Alnus glutinosa and A. incana). Pl. Sci. Lett. 28:3-9.

Kato, H. and M. Takeuchi, 1966. Embryogenesis from the epidermal cells of carrot hypocotyl. Sci. Pap. Coll. Gen. Edu. Univ. Tokyo 16:245-254.

Kaur-Sawhney, R. and A. W. Galston. 1979. Interaction of polyamines and light on biochemical processes involved in leaf sequence. Plant Cell Environ. 2:189-196.

Kaur-Sawhney, R., L. M. Shih, H. E. Flores, and A. W. Galston. 1982. Relation of polyamine synthesis and titer aging and senescence in oat leaves. Plant Physiol. 69:405-410.

Kaur-Sawhney, R., N. S. Shekhawat, and A. W. Galston. 1985. Polyamine levels as related to growth, differentiation and senescence in protoplast-derived cultures of Vigna aconitifolia and Avena sativa. Plant Growth Regulation 3:329-337.

Kyriakidis, D. A. 1983. Effect of plant growth hormones and polyamines on ornithine decarboxylase activity during the germination of barley seeds. Physiol. Plant. 57:499-504.

Lieberman, M. 1979. Biosynthesis and action of ethylene. Annu. Rev. Plant Physiol. 30:533-591.

Litz, R. E. and B. Schafer. 1987. Polyamines in adventitious and somatic embryogenesis in mango (Mangifera indica L.). J. Plant Physiol. 128:251-258.

LoSchiavo, F. 1984. A critical review of the procedure for embryo purification. Plant Mol. Biol. Rep. 2:15-18.

Lurssen, K., K. Naumann, R. Schroder. 1979. 1-Amino-cyclopropane-1-carboxylic acid-an intermediate of the ethylene biosynthesis in higher plants. Z. Pflanzenphysiol. 92:285-294.

Malmberg, R. L. and J. McIndoo. 1983. Abnormal floral development of a tobacco mutant with elevated polyamine levels. Nature 305:623-625.

Masuda, K. Y. Kikuta, and Y. Okazawa. 1984. Embryogenesis and ribosomal DNA in carrot cell suspensions cultured in vitro. Plant Sci. Lett. 33:23-29.

McBride, R. and M.L. Evans. 1977. Auxin inhibition of acid- and fusicoccin-induced elongation in lentil roots. Planta 136:97-102.

Mele, E., J. Messegher, and P. Camprubi. 1982. Effect of ethylene on carnation plants grown in sealed vessels. In: Proc. Intl. Cong. Plant Tissue and Cell Culture. Tokyo, A. Fujiwara (ed.), pp 69-70.

Minocha, S.C. and C. Robie. 1985. The role of 2,4-D and polyamines in somatic embryogenesis in carrot cell cultures (Abst.) 12th Intern. Conf. Pl. Gr. Subs., Aug 26-31, 1985, Heidelberg, W. Germany.

Montague, M. J., T. A. Armstrong, and E. G. Jaworski. 1979. Polyamine metabolism in embryogenic cells of Daucus carota. I. Changes in intracellular content and rates of synthesis. Plant Physiol. 63:341-345.

Montague, M.J., J.W. Koppenbrink, and E. G. Jaworski. 1978.Polyamine metabolism in embryogenic cells of Daucus carota. II. Changes in arginine decarboxylase activity. Plant Physiol. 62:430-433.

Morgan, D. M. L. 1987. Polyamines. In:Essays in Biochemistry. R. D. Marshall and K. F. Tipton (eds.). Vol. 23, pp. 82-115. Academic Press, N.Y.

Newcomb, W. and D. F. Wethrell. 1970. The effects of 2,4,6-trichlorophenoxyacetic acid on embryogenesis in wild carrot tissue cultures. Bot. Gaz. 131:242-245.

Palavan, N., R. Goren, and A. W. Galston. 1984. Effects of some growth regulators on polyamine biosynthetic enzymes in etiolated pea seedlings. Plant Cell Physiol. 25:541-546.

Pegg, A.E. 1986. Recent advances in the biochemistry of polyamines in eukaryotes. Biochem. J. 234:249-262.

Priebe, A., H. Klein, and J. Jager. 1978. Role of polyamines in SO_2 polluted pea plants. J. Exp. Bot. 29:1045-1050.

Privalle, L. S. and J. S. Graham. 1987. Radiolabeling of a wound-inducible pyridoxal phosphate-utilizing enzyme: Evidence for its identification as ACC synthase. Arch. Biochem. Biophys. 253:333-340.

Raghavan, V. 1986. Somatic embryogenesis. Embryogenesis in Angiosperms.

Cambridge Univ. press, Cambridge, U. K. pp. 115-151.

Reinert, J. 1959. Uber die Kontrolle der Morphogenese und die Induktion von Adventivembrynen an Gewebekulturen aus Karotten. Planta 53:318-333.

Reinert, J. 1973. Aspect of organization-organogenesis and embryogenesis. In: H.E. Street (ed.) Plant Tissue and Cell Culture. Blackwell Sci. Publ. Oxford, U. K. pp. 338-355.

Roberts, D. R., M. A. Walker, J. E. Thompson, and E. B. Dumbroff. 1984. The effects of inhibitors of polyamine and ethylene biosynthesis on senescence ethylene production and polyamine levels in cut carnation flowers. Plant Cell Physiol. 25:315-322.

Sala, F., M. G. Galli, E. Nielson, E. Magnien, M. Devreux, G. P. Noy and S. Spadari. 1983. Synchronization of nuclear DNA synthesis in cultured Daucus carota L. cells by aphidicolin. FEBS Lett. 153:204-208.

Saurbrey, E., K. Grossman, and J. Jung. 1987. Is ethylene involved in the regulation of growth of sunflower cell suspension cultures? J. Plant Physiol. 127:471-479.

Saurbrey, E., K. Grossman, and J. Jung. 1988. Ethylene production by sunflower cell suspensions. Plant Physiol. 87:510-513.

Sembdner, G., D. Gross, H. W. Liebisch and G. Schneider. 1980. Biosynthesis and metabolism of plant hormones. In: Hormonal Regulation of Development I. J. MacMillan (ed.) Encycl. Plant Physiology New Series vol. 9 pp. 281-444. Springer-Verlag, Berlin.

Seraffini-Fracassini, D., N. Bagni, and P. G. Cionini, and A. Bennici. 1980. Polyamines and nucleic acids during the first cell cycle of Helianthus tuberosus tissue after the dormancy break. Planta 148:332-337.

Seraffini-Fracassini, D., P. Torrigiani, and C. Branca. 1984. Polyamines bound to nucleic acids during dormancy and activation of tuber cells of Helianthus tuberosus. Physiol. Plant 60:351-357.

Schiavone, F. M., and T. J. Cooke. 1985. A geometric analysis of somatic embryo formation in carrot cell cultures. Can. J. Bot. 63:1573-1578.

Simpson, R. B., M. Lillis, L. Margossian and T. D. McKnight. 1983. DNA transfer to plant cells using the Ti plasmid vector. In: Plant Molecular Biology. R. B. Goldberg (ed.), Alan R. Liss, Inc. N. Y. pp. 23-34.

Slocum, R. D. and A. W. Galston. 1985. Changes in polyamine biosynthesis associated with post-fertilization growth and development in tobacco ovary tissues. Plant Physiol. 79:336-343.

Slocum, R. D. R. Kaur-Sawhney, and A. W. Galston. 1984. The Physiology and biochemistry of polyamines in plants. Arch. Biochem. Biophys. 235:283-303.

Smith, S. M. and H. E. Street. 1974. The decline of embryogenic potential as callus and suspension cultures of carrot (Daucus carota L.) are serially subcultured. Ann. Bot. 38: 233-241.

Smith, T. A. 1981. Amines. In: The Biochemistry to Plants. P. K. Stumpf and E. E. Conn (eds.) vol. 7, pp. 249-268. Acad. Press, N.Y.

Smith, T. A. 1985a. Polyamines. Ann. Rev. Pl. Physiol. 36: 117-143.

Smith, T. A. 1985b. The di- and poly-amine oxidases in higher plants. Biochem Soc. Trans. 610 meeting 13:319-322.

Steward, F. C. 1976. Multiple interactions between factors that control cells and development. In: N. Sutherland (ed.), Perspectives in Experimental Biology. 2: 9-23. Pergamon Press, Oxford.

Steward, F. C., M. O. Mapes, and K. Mears. 1958. Growth and organized development of cultured cells. II. Organization in cultures grown from freely suspended cells. Amer. J. Bot. 45: 705-708.

Street, H. E. 1977. Embryogenesis and chemically induced organogenesis. In: W. R. Sharp, P. O. Larsen, E. F. Paddock, and V. Raghavan (eds.), Plant Cell and Tissue Culture Principles and Applications. pp. 123-153. Ohio St. U. Press, Columbus.

Sung, Z. R., A. Fienberg, R. Chorneau, C. Borkird, I. Furner, J. Smith. 1984. Developmental biology of embryogenesis from carrot culture. Plant Mol. Biol. Rep. 2: 3-14.

Suresh, M. R., S. Ramakrishna, and P. R. Adiga. 1978. Regulation of arginine decarboxylase and putrescine levels in Cucumis sativus cotyledons. Phytochemistry 17:57-63.

Suttle, J.C. 1981. Effect of polyamines on ethylene production. Phytochemistry 20:1477-1488.

Tabor, H. and C. W. Tabor (eds.). 1983. Methods in Enzymology. vol. 94, pp. 1-497 Acad. Press, New York.

Tabor, C. W. and H. Tabor. 1984. Polyamines. Ann. Rev. Biochem. 53:749-790.

Terzi, M., L. Pitto, Z. R. Sung. 1985. Somatic Embryogenesis. IPRA, Rome. pp. 1-202.

Tibercio, A. F., R. Kaur-Sawhney, and A. W. Galston. 1987. Effect of polyamine biosynthetic inhibitors on alkaloids and organogenesis in tobacco callus cultures. Plant Cell, Tissue and Organ Culture 9:111-120.

Timko, M. P., A. P. Kausch, C. Castresana, J. Fassler, L. Herrera-Estrella, G. Van den Broeck, M. Van Montagu, J. Schell and A. R. Cashmore. 1985. Light regulation of plant gene expression by an upstream enhancer-like element. Nature 318:579-582.

Tisserat, B. and T. Murashige. 1977. Effects of ethephon, ethylene, and 2,4-dichlorophenoxyacetic acid on asexual embryogenesis in vitro. Plant Physiol. 60:437-439.

Torrigiani, P., M. M. Altamura, G. Pasqua, B. Monacelli, D. Serafini-fracassini, and N. Bagni. 1987. Free and conjugated polyamines during de novo floral and vegetative bud formation in thin cell layers of tobacco. Physiol. Plant. 70:453-460.

Turner, L. B. and G. R. Stewart. 1986. The effect of water stress upon polyamine levels in barley (Hordeum vulgare L.) leaves. J. Exp. Bot. 37:170-177.

Verma, D. C. and T. Tarka, 1985. Influence of 1-aminocyclopropanecarboxylic acid (ACC) and consequent ethylene biosynthesis on growth and somatic embryogenesis in wild carrot (Daucus carota L.) cell suspension. In: Somatic Embryogenesis. M. Terzi, L. Pitto and Z. R. Sung (eds.), IPRA, Rome, pp 151-158.

Warren, G. S. and M. W. Fowler. 1977. A physical method for the separation of various stages in the embryogenesis of carrot cell culture. Plant Sci. Lett. 9:71-76.

Wetherell, D. F. and D. K. Dougall. 1976. Sources of nitrogen supporting growth and embryogenesis in cultured wild carrot tissue. Physiol. Plant. 37:97-103.

Wochok, Z. S. and D. F. Wetherell. 1972. Restoration of declining morphogenetic capacity in long-term cultures of Daucus carota by kinetin. Experientia 28:104-105.

Woltering, E. J. 1986. Ethylene and carbon dioxide accumulation within various tissue culture systems. Acta Bot. Neerl. 35:50.

Yang, S. F. and N. E. Hoffman, 1984. Ethylene biosynthesis and its regulation in higher plants. Annu. Rev. Plant Physiol. 35:155-189.

Young, N. D. and A. W. Galston. 1983. Putrescine and acid stress: Induction of arginine decarboxylase activity and putrescine acumulation by low pH. Plant Physiol. 71:767-771.

Yu, Y. B. and S. F. Yang. 1979. Auxin-induced ethylene production and its inhibition by aminoethoxyvinylglycine and cobalt ions. Plant Physiol. 64:1074-1077.

PURIFICATION OF POLYAMINE OXIDASE FROM MAIZE SEEDLINGS BY IMMUNOADSORBENT COLUMN

Rodolfo Federico, Chiara Alisi, Alessandra Cona and Riccardo Angelini

Dipartimento di Biologia Vegetale
Università degli Studi di Roma "La Sapienza"
P.le Aldo Moro 5, I-00185 Rome, Italy

INTRODUCTION

Plant polyamine oxidases (PAOs) are typical of the monocotyledones and are responsible for the terminal catabolism of polyamines, producing H_2O_2, diaminopropane and Δ'-pyrroline (from spermidine) or diazabicyclononane (from spermine)(Smith, 1985a; Smith et al., 1986). These enzymes are quite distinct from diamine oxidases (DAOs) which occur in the dicotyledones (Smith, 1985a; Rinaldi et al., 1986) the latter containing pyrroloquinolinequinone (Glatz et al., 1987; Suzuki 1987) as the cofactor while PAOs are flavoproteins (Smith, 1983). In general DAOs have a broader substrate specificity oxidizing diamines but also spermine and spermidine (Chaudhuri and Gosh, 1984).

PAOs have been detected mainly in the Gramineae (Smith, 1985b) and recently in the water hyacinth (Yanagisawa et al., 1987). These enzymes have been well characterized in barley (Smith, 1983), maize (Suzuki and Yanagisawa 1980; Federico et al., 1988), millet (Hirasawa et al., 1987) and water hyacinth (Yanagisawa et al., 1987).

Amine oxidases (AOs) have been shown to be mainly associated with the cell wall (Kaur-Sawhney et al., 1981; Federico and Angelini, 1986). It is intriguing that most of the functions of their physiological substrates are intracellular (Bagni et al 1982; Smith, 1985b). Actually polyamines (PAs) are involved in several aspects of cellular metabolism (Smith, 1985b). In the last few years evidence has been accumulated for a possible localization of PAs in the cell wall (Goldberg and Perdrizet, 1984). The extracellular distribution of AOs has been considered as a clue for the understanding of their physiological roles (Federico and Angelini, 1986). In particular these enzymes have been thought to be involved in the biosynthesis of hydrogen peroxide needed in the peroxidative processes occurring in the apoplast namely lignification and wall stiffening (Goldberg et al., 1985; Angelini and Federico, 1986). In addition it has been suggested that AOs may prevent transfer of PAs between cells (Galston and Kaur-Sawhney, 1982).

Immunohistochemical evidence obtained in lentil suggests for both an intra- and extra-cellular localization of DAO in very young seedlings (Federico et al., 1985). There is the possibility of different cellular distribution and function of AOs, these being related to the state of the cellular differentiation (Torrigiani et al., 1987).

The methods usually employed to purify polyamine oxidases are time-consuming, involve several steps and often require organic solvents. These procedures produce low amounts of PAO and can modify the native structure of the enzyme. In this paper we describe a simple and rapid procedure for the purification of PAO from maize seedlings by affinity chromatography using specific anti-PAO antibodies covalently linked to Sepharose 4B.

EXPERIMENTAL

Chemicals. Heat-killed formalinized cells of Staphylococcus aureus Cowan I(SACI) strain were purchased as Pansorbin from Behring Corp.(San Diego, U.S.A.). CNBr-activated Sepharose 4B was from Pharmacia Fine Chemicals (Uppsala, Sweden). 4-chloro-1-naphtol was obtained from Bio-Rad (Richmond, U.S.A.).

Plants. Maize (Zea mays L.), oat (Avena sativa L.) and barley (Hordeum vulgare L.) seeds were soaked for 12 hr in aerated tap water and then grown on moistened vermiculite for 10 days at 25 C (14 hr day; 300 umol/sec·m^2).

Enzyme purification and assays. PAO was prepared from cell walls of maize seedlings as previously reported (Federico et al., 1988). PAO activity was estimated polarographically in a Hansatech oxigraph (Hansatech Ltd, Norfolk, UK) equipped with a Clark electrode according to Rinaldi et al. (1982). One kat of enzyme is the amount of protein catalyzing the oxidation of 1 mol of substrate/sec under the assay conditions.

Homogenates. The crude homogenates were obtained by grinding the plant tissues in a Waring blender at 4 C with 2 vols of 0.4 M KP_i, pH 6.5, for 4 min. The suspension was strained through cheesecloth and centrifuged at 20000g for 30 min at 4 C. The pellet was discarded and the supernatant used for the immunodiffusion and immunoprecipitation procedure and for the immunoaffinity purification method described below.

Anti-PAO antibodies preparation and purification. Two New Zealand rabbits were immunized as previously described (Federico et al., 1985). Animals were bled twice a month withdrawing 40 ml of blood from each one, and were bled to death a month after the last antigen injection. Sera were tested for anti-PAO antibody content by double diffusion in agar gel using an antigen concentration of 0.5 mg/ml.

The PAO-Sepharose 4B column was prepared by coupling 10 mg of pure enzyme to 4 ml of swollen gel, according to the manufacturer. The coupling efficiency was about 70%. Pooled sera were applied to the column equilibrated with 0.15 M KP_i, pH 6.5 and the specific antibodies eluted with 3.5 M $MgCl_2$ and then dialyzed against 15 mM KP_i/150 mM NaCl, pH 7.2.

Purification of PAO. An immunoadsorbent was prepared by coupling affinity purified anti-PAO antibodies to CNBr-activated Sepharose 4B (3 mg/ml of swollen gel) according to the manufacturer. A high coupling efficiency was obtained (90% antibodies linked to the gel). The crude extract from maize with 0.5 M KP_i pH 6.5 at a flow rate of 30 ml/h. The column was washed with the same buffer until absorbance at 280 nm had fallen to the background level. Elution of the enzyme was obtained with HCl/glycine 0.1 M, pH 3.

Immunoprecipitation. The assay was carried out in 1 ml Eppendorf assay tubes at room temperature. The incubation mixture (0.2 ml) contained: 100 µl of plant extract (17 nkat/sample), 15 mM KP_i pH 7.2, 150 mM NaCl and the appropriate amount of antiserum. After incubation for 60 min at room temperature with shaking, PAO-antibody complexes were removed by addition of 40 µl of a 10% (v/v) suspension of SACI. Samples were subsequently shaken for 60 min to perform immunoadsorption of rabbit antibodies and then centrifuged for 2 min at 10000g to pellet the protein-antibody complex. PAO activity was determined in the supernatants. Determination of the cross reactivity of antiserum was made by comparing the amount of antibody precipitated by maize enzyme to the amount precipitated by oat and barley enzyme.

Analytical methods. Proteins were estimated by the method of Bradford (1976) using bovine γ-globuline as reference standard. SDS electrophoresis was carried out according to Weber and Osborn (1969). Native electrophoresis (pH 4.3) was performed according to Gabriel (1971). Absorption spectra were recorded by a PU 8820 (Philips) spectrophotometer.

RESULTS

Characteristics of the immunoadsorbent. The binding capacity of the anti-PAO immnoadsorbent, investigated applying the purified enzyme to 4 ml of gel, was about 0.7 mg (500 nkat) per ml of swollen gel. About 75% of the adsorbed enzyme was recovered from the column without any loss of specific activity.

One-step enzyme purification. The elution pattern of crude maize seedlings extract from the anti-PAO antibody column is reported in fig.1. The purification of the enzyme was carried out as described in Experimental. About 400 ml of crude sample, corresponding to 1500 nkat, were applied to the immunoadsorbent column (4 ml) preequilibrated with KP_i 0.5 M, pH 6.5 and the column was washed with the same buffer until absorbance at 280 was lower than 0.01 A. The enzyme was then eluted with HCl/glycine 0.1 M, pH 3, and the active fractions were immediatedly neutralized with KH_2PO_4 1 M and collected. All the 1500 nkat applied were fixed to the immunoadsorbent and 1000 nkat were recovered after elution, giving a yield of about 70% . The chromatographic procedures were carried out at room temperature. The mild elution conditions and the stability of the immunoadsorbent permit the column to be reused several times (>10) without any loss of binding capacity.

Enzyme characteristics. The homogeneity of the purified enzyme was evaluated by analytical polyacrylamide gel electrophoresis in native (pH 4.3) and denaturing (SDS) conditions. Native electrophoresis display the same band pattern when stained both for protein and PAO activity and SDS electrophoresis showed a single band after staining with Coomassie blue (fig.2). Maize PAO obtained by this purification technique exhibit properties similar to those observed for the native enzyme obtained by conventional procedures with respect to pH optima and K_ms for Spm and Spd, UV and visible spectra and sugar content (table 1).

Table 1. Some properties of maize PAO.

pH optimum		
	spermine	6.5
	spermidine	6.5
k_m		
	spermine	18 µM
	spermidine	22 µM
A_{max}		278, 380, 456 nm
Sugar content		2.5%

Ouchterlony double diffusion in agar analysis revealed that the specific antibodies for the maize PAO cross-react with oat enzyme with reaction of partial identity, while barley enzyme does not react in the same experimental conditions. These results indicates that oat and maize enzymes share some, but not all, antigenic determinants. Immunoprecipitation tests were then performed to quantitate the extent of cross-reaction. The results of this analysis (fig.3) show that the oat enzyme has a very high immunological homology with the maize enzyme, while the barley PAO share only few antigenic determinants with the latter.

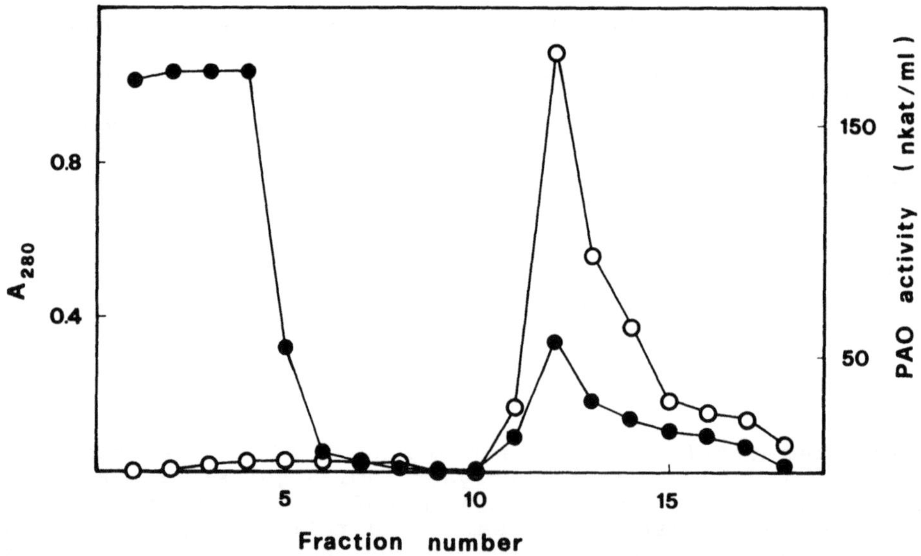

Fig.1. Elution pattern of crude maize seedlings extract on anti-PAO immunoadsorbent column. Empty symbols, enzyme activity; solids symbols, absorbance at 280 nm. Fraction volumes are 10 ml from 1 to 9 and 2.5 ml from 10 to 18.

Fig.2. SDS polyacrylamide gel electrophoresis of PAO purified by immunoadsorbent column.

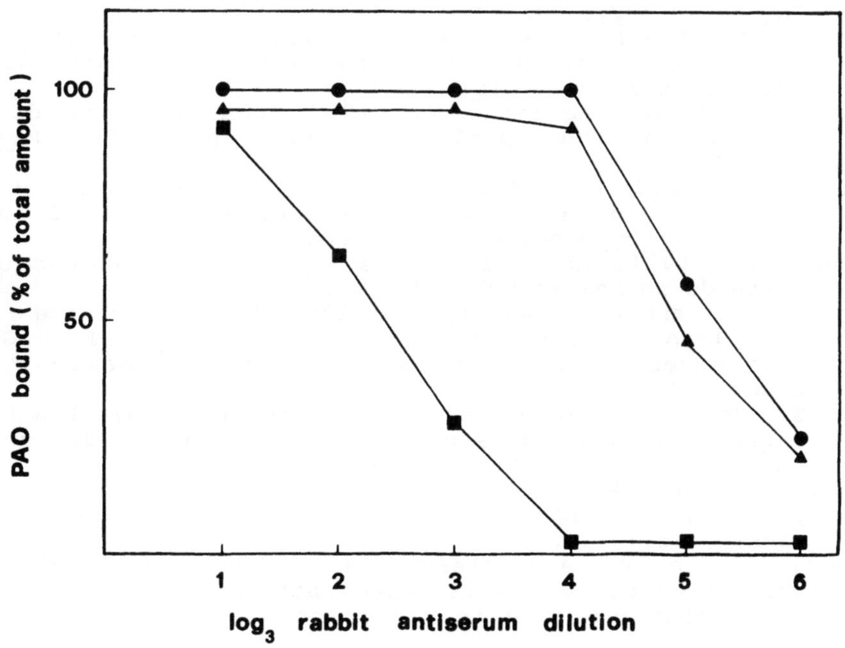

Fig.3. Immunoprecipitation test performed using anti-maize PAO immune serum as described in Experimental.(●) maize; (▲) oat; (■) barley.

CONCLUSION

The immunoaffinity method hereby described for the one-step purification of PAO from maize seedlings proved to be well suited to obtain a fair quantity of homogeneous enzyme with biochemical characteristics similar to those exhibited by conventional pure preparations.

Some immunological characteristics of oat, barley and maize enzymes have been presented. Partial immunological homology has been observed between PAOs from these monocots . This fact is of primary importance in view of performing an immunoaffinity procedure for the purification of PAOs from other species using the immunoadsorbent described above.

The easier availability of pure PAOs from different species will make possible a comparative study and perhaps an evolutionary analysis of this enzyme family.

REFERENCES

Angelini, R., and Federico, R., 1986, Occurrence of diamine oxidase in the apoplast of Leguminosae seedlings, in "Biochemical Studies of Natural Polyamines", C.M. Caldarera, C. Clo, and C. Guarnieri eds., CLUEB, Bologna, p.183.

Bagni, N., Serafini-Fracassini, D., Torrigiani, P., 1982, Polyamines and cellular growth processes in higher plants, in "Plant Growth Substances 1982", P.F. Wareing, ed, Academic Press, London, New York, p.473.

Chaudhuri, M.M., and Ghosh, B., 1984, Purification and characterization of diamine oxidase from rice embryos, Phytochemistry, 23: 241.

Federico, R., Angelini, R., Argento-Ceru, M.P., and Manes, F., 1985, Immunohistochemical demonstration of lentil diamine oxidase, Cell Mol. Biol., 31: 171.

Federico, R., and Angelini, R., 1986 Occurrence of diamine oxidase in the apoplast of pea epicotyls, Planta, 167: 300.

Federico, R.,Alisi, C. and Forlani, F., 1988, Properties of the polyamine oxidase from the cell wall of maize seedlings,Phytochemistry, in press.

Gabriel, O., 1971, Analytical disc gel electrophoresis, in "Methods in Enzymolgy", 22: 565.

Galston, A.W., and Kaur-Sawhney, R., 1982, Polyamines: are they a new class of plant growth regulators?, in "Plant Growth Substances 1982", P.F. Wareing, ed, Academic Press, London, New York, p.451.

Glatz, Z., Kovar, J., Macholan, L., and Pec, P., 1987, Pea (Pisum sativum) diamine oxidase contains pyrroloquinoline quinone as a cofactor, Biochem. J., 242: 603.

Goldberg, R., and Perdrizet, E., 1984, Ratio of free to bound polyamines during maturation in mung-bean hypocotyl cells, Planta, 161: 531.

Goldberg, R., Le, T., and Catesson, A.M., 1985, Localization and properties of cell wall enzyme activities related to the final stages of lignin biosynthesis, J. Exp. Bot., 36: 503.

Kaur-Sawhney, R., Flores, H.E., and Galston, A.W., 1981, Polyamine oxidase in oat leaves: a cell-wall localized enzyme, Plant Physiol., 68: 494.

Rinaldi, A., Floris, G., and Giartosio, A., 1986, Plant Amine Oxidases, in: "Structure and Function of Amine Oxidases", B. Mondovi, ed., CRC Press Inc., Boca Raton, Florida, U.S.A., p.51.

Smith, T.A., 1983, Polyamine oxidase (oat seedlings), in: "Methods in enzymolgy", 94: 311.

Smith, T.A., 1985a, The di- and polyamine oxidases of higher plants, Biochem. Soc. Trans., 13: 319.

Smith, T.A., 1985b, Polyamines, Ann. Rev. Plant Physiol.

Smith, T.A, Croker, S.J., and Loeffler, R.S.T., 1986, Occurrence in higher plants of 1-(3-aminopropyl)pyrrolinium and pyrroline: products of polyamine oxidation, Phytochemistry, 25: 683.

Suzuki, Y., and Yanagisawa, H., 1980, Purification and properties of maize polyamine oxidase: a flavoprotein, Plant Cell Physiol., 21: 1085.

Suzuki, Y., 1987, The target of carbonyl reagents in the pea diamine oxidase activity, Phytochemical Society of Europe Symposium on 'Amines in Plants', University College, London, Abstract.

Torrigiani, P., Serafini-Fracassini, D., and Fara, A., 1987, Evidence for the presence of a diamine oxidase activity in different stages of Jerusalem artichoke tubers, Phytochemical Society of Europe Symposium on 'Amines in Plants, University College, London, Abstract.

Weber, K., and Osborn, M., 1969, The reliability of molecular weight determination by dodecyl sulphate-polyacrylamide gel electrophoresis, J. Biol. Chem., 244: 4406.

Yanagisawa, H., Kato, A., Hoshiai, S., Kamiya, A., and Torii, N., 1987, Polyamine oxidase from water hyacinth, Plant Physiol., 85: 906.

POLYAMINES IN MICROORGANISMS

POLYAMINE POOLS AND THE CONTROL OF ORNITHINE DECARBOXYLASE ACTIVITY

Rowland H. Davis, Glenn R. Barnett and Janet L. Ristow

Department of Molecular Biology and Biochemistry
University of California, Irvine
Irvine, CA 92717, U. S. A.

INTRODUCTION

The metabolism of the polyamines, putrescine, spermidine and spermine is represented in a general way in Fig. 1. During growth, the polyamines are formed from ornithine (and in many species, arginine) and S-adenosylmethionine. The resulting pools of polyamines, in addition to fulfilling the needs of an expanding cell mass, are potentially excreted, sequestered by cell constituents and interconverted or catabolized. The regulation of polyamine biosynthesis is exerted in part by the polyamines themselves and also by signals related to the onset or continuation of the cell cycle and differentiation. In relation to the former signal, polyamines rarely, if ever, exert significant feedback inhibition in the strict sense on their own synthesis. Alternative mechanisms, however, have come to light, among which are the modification, antizyme inhibition and elimination of the ornithine decarboxylase (ODC) protein in response to excess polyamines (Pegg and McCann, 1982; Tabor and Tabor, 1984, 1985).

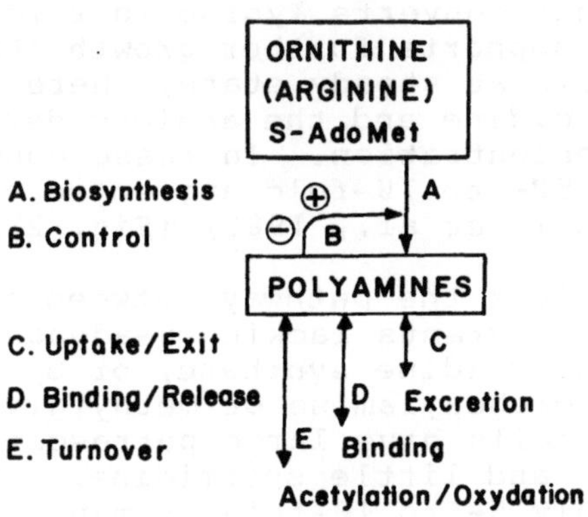

Fig. 1. General scheme of polyamine metabolism, emphasizing factors governing polyamine pool size.

A matter that has clouded our knowledge of the biosynthetic control of the pathway has been the sequestration or compartmentation of polyamines, owing to their polycationic nature and their consequent non-covalent binding to nucleic acids, membranes, and other polyanionic cell constituents (Paulus et al., 1983; Kashiwagi et al., 1986; Frydman et al, 1984). An important question that has rarely been explored is whether the two classes of control signal - polyamine pools and growth-related stimuli - are in fact manifestations of the same thing. That is, do growth stimuli lead to changes in the diffusible fraction of polyamines, which then acts as the proximal regulator of biosynthetic enzymes? This question will not be answered here, but only legitimized by work from our laboratory on a simple eucaryotic organism, the fungus *Neurospora crassa*.

Several features of the organism suit it to study of polyamine flux. Little intercoversion of polyamines takes place in normal circumstances of exponential growth; little spermine is made (Paulus et al., 1982); and the ODC-antizyme mechanism of control appears to be absent (DiGangi and Davis, 1987; Barnett et al., 1988). If ODC modification occurs, it does not lead to enzyme inactivation. Finally, in exponential growth, putrescine is a limiting factor in spermidine synthesis (Davis and Ristow, unpublished), and this is correlated with a small putrescine pool (0.5 - 1.0 nmol per mg, dry weight) and a large spermidine pool (15 - 18 nmol per mg, dry weight), the latter being the major polyamine of this organism (Davis et al., 1985).

CONTROL OF ODC

Our recent studies have revealed the somewhat distinct roles of putrescine and spermidine in the negative control of ODC (Davis et al, 1985; Barnett et al., 1988). An arginaseless strain of *N. crassa* allows us to starve the organism of ornithine at will: when arginine is added to cultures growing in minimal medium, the *de novo* synthesis of ornithine is feedback-inhibited, and without arginase, the cells lack the normal alternate source of ornithine (Davis et al., 1970). Cells continue to grow, though slowly, owing to a highly derepressed ODC that converts lysine inefficiently to cadaverine, which supports further growth (Paulus et al., 1982). In such cells at steady state, there is little or no putrescine or spermidine and the analogs derived from lysine are very low in concentration. In these conditions, ODC activity is about 50- to 70-fold augmented over growth in minimal medium (Davis et al., 1985) (Fig. 2).

We may also block the pathway between putrescine and spermidine by using mutants lacking S-adenosylmethionine decarboxylase or spermidine synthase, or by adding inhibitors of these steps (cyclohexylamine or methylglyoxal *bis*guanylhydrazone). Such cells have large putrescine pools, some of which is excreted, and little spermidine. In these conditions, ODC activity is only 5- to 10-fold augmented over growth in minimal medium (Davis et al., 1985). The difference in this result compared to the effect of ornithine starvation suggested that spermidine deprivation induced an increase in enzyme level that was limited by the accumulation of putrescine.

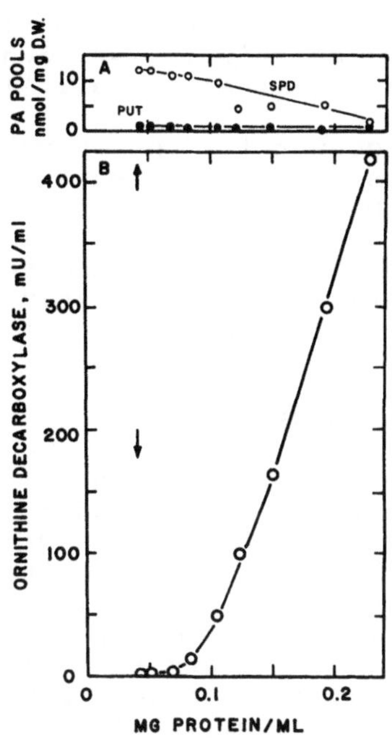

Fig. 2. Behavior of polyamine pools (A) and the differential rate of increase of ornithine decarboxylase activity (B) after the addition of arginine (arrow) to an exponential culture of N. crassa). The slope of the curve in (B) is equivalent to the specific activity of the enzyme in new cell mass, and changes over 70-fold during the experiment. Abbreviations: D.W., dry weight, PA, polyamine.

Finally, growth of prototrophic strains or strains (spe-1) with catalytically impaired ODC protein in the presence of 1 mM spermidine yields mycelia that have a large spermidine pool and little or no putrescine. In these conditions, ODC protein levels are 2- to 4-fold lower than in minimal-growth cultures (Barnett et al., 1988).

We may correlate the three metabolic situations - no polyamine synthesis, high putrescine and low spermidine, and low putrescine and high spermidine - with the activity, synthesis and turnover of ODC. We find, first, that ODC activity is determined by the amount of immunologically detectable ODC; there appear to be no changes in the specific activity of the enzyme protein (DiGangi and Davis, 1987). Second, when either putrescine or spermidine is synthesized normally or present in high amounts, radioactive methionine and immunoprecipitation techniques show the half-life of the enzyme to be short: approximately 50 - 55 min. When both are lacking, the enzyme is 10-fold more stable, with a half-life of about eight hours (Barnett et al., 1988).

Third, when spermidine synthesis is impaired, whether or

not putrescine is absent or present in high amounts, the synthetic rate of ODC protein is 5-fold higher than it is in minimal-grown cultures (Barnett et al., 1988)

These results account for the enzyme activities in steady-state cultures: 5-fold increased in cultures with high putrescine and low spermidine; 50- to 70-fold increased in cultures with neither polyamine. Because only spermidine represses the enzyme, and both putrescine and spermidine promote its inactivation, it is unlikely that control is exerted by a single polyamine-binding entity.

SEQUESTRATION OF POLYAMINE POOLS

Given the clear evidence of control exerted by polyamines, we may extend our analysis to the onset of enzyme augmentation in response to ornithine deprivation (Fig. 2). The striking fact is that the cellular polyamine content hardly changes as the differential rate of ODC synthesis rises by 70 fold. We interpret this to mean that normally only a very small proportion of the putrescine and spermidine pools are diffusible and thus active in regulation. When their synthesis is impaired by ornithine deprivation, the diffusible pools of polyamine are immediately sequestered as a normal consequence of continued growth (Davis et al., 1985).

What direct evidence of polyamine compartmentation do we have? Early studies in our laboratory showed that most of the putrescine and spermidine in the cell were not on the main line of biosynthesis (Paulus et al., 1983). By adding radioactively labelled ornithine to growing cells, we found by isotope dilution analysis that only about 20 percent of the endogenous polyamines (a maximal estimate) mixed with the newly synthesized radioactive polyamines derived from ornithine as they went through the pathway to spermine. Only one-third of the sequestered fraction could be accounted for by vacuolar pools, bound to stored polyphosphate. We believe the rest is bound to ribosomes, DNA and membranes in a relatively slowly exchangeable form. This belief is supported by studies in other systems (Kashiwagi et al, 1986; Igarishi et al., 1982).

More recent, though circumstantial, evidence of compartmentation of spermidine has been obtained in studies of internal spermidine pools in relation to spermidine in the medium. In short term studies of uptake rate versus polyamine concentration, the uptake of spermidine in the growth medium appears to be largely by diffusion, with only a small saturable component of uptake (Davis and Ristow, unpublished). To study this rigorously in long-term tests of steady-state cultures, we grew an ornithine decarboxylase-less mutant, which can make no endogenous polyamines, in medium containing various concentrations of spermidine, ranging from 0.5 to 5.0 mM. Correlating with the largely "diffusional" character of polyamine entry is the fact that the slope of increase of internal spermidine in cell water versus polyamine in the medium is about 1 during expontential growth. However, there is a strikingly high y-intercept of this function, implying that at low concentrations of spermidine, substantial binding or ion-trapping takes place before the equilibration of external and internal spermidine can take place. The value of the y-intercept is approximately 14 to 18 nmol per mg, dry weight. This approximates the normal spermidine pool in cells that can make their own polyamines. This implies that in normal cells, most of the spermidine is bound, and that this binding capacity must be saturated before equilibration of external and the internal, diffusible pool takes place. It also accounts for the rapid and complete loss of negative control of ODC upon brief starvation for polyamines brought about by ornithine limitation.

A last observation is that upon permeabilization with butanol, cells lose about 80% of their putrescine and 40% of their spermidine to the medium (Davis and Ristow, unpublished). This implies binding of spermidine to cell constituents *in vivo*, but owing to the change in ionic strength upon permeabilization, no hard conclusions can be drawn. However, when the cellular polyamines are radioactively labelled, they can be shown to be almost wholly displaced by cold polyamines after cell permeabilization. Whatever the trapping mechanism, it is subject to exchange in these conditions.

CONCLUSIONS

The relationship between endogenous polyamines and control of their biosynthesis suggests that only polyamines in the diffusible state exert control. We feel the evidence of extensive intracellular polyamine sequestration is good, and our data indicate that the binding in vivo is reversible, if slowly. Granting the implications of these data, we may return to the questions posed at the outset: do all growth stimuli that evoke regulatory responses of ODC and other polyamine biosynthetic enzymes do so by way of changes in the delicately poised concentration of free polyamines? Or do growth stimuli have a direct input into the mechanisms of transcription, translation and enzyme turnover? Perhaps the best means of answering this legitimate question is by studying mutations that derange control in response to various stimuli. We hope eventually to have such mutations in *N. crassa*.

REFERENCES

Barnett, G. R., Seyfzadeh, M., and Davis, R. H., Putrescine and spermidine control degradation and synthesis of ornithine decarboxylase in *Neurospora crassa*. *J. Biol. Chem.*, in press.

Davis, R. H., Krasner, G. N. DiGangi, J. J., and Ristow, J. L., 1985, Distinct roles of putrescine and spermidine in the regulation of ornithine decarboxylase in *Neurospora crassa*. *Proc. Natl. Acad. Sci. U. S.*, 82:4105.

Davis, R. H., Lawless, M. B., and Port, L. A., 1970, Arginaseless *Neurospora*: genetics, physiology and polyamine synthesis. *J. Bacteriol.*, 102:299.

DiGangi, J. J., and Davis R. H., 1987, Ornithine decarboxylase from *Neurospora crassa*. Purification, characterization and regulation by inactivation. *J. Biol. Chem.*, 262:7889.

Frydman, B., Frydman, R. B., de los Santos, C., Garrido, D. A., Goldemberg, S. H., and Algranati, I. D., 1984, Putrescine distribution in *Escherichia coli* studied in vivo by ^{13}C nuclear magnetic resonance. *Biochim. Biophys. Acta*, 8Ø5: 337.

Igarishi, K., Sakamoto, I., Goto, N., Kashiwagi, K, Honma, R., and Hirose, S., 1982, Interaction between polyamines and nucleic acids or phospholipids. *Arch. Biochem. Biophys.* 219:438.

Kashewagi, K., Kobayashi, H., and Igarishi, K., 1986, Apparently unidirectional polyamine transport by proton motive force in polyamine-deficient *Escherichia coli*. *J. Bacteriol.*, 165:972.

Paulus, T. J., Cramer, C. L., and R. H. Davis, 1983, Compartmentation of spermidine in *Neurospora crassa*. *J. Biol. Chem.*, 258:86Ø8.

Paulus, T. J., Kiyono, P., and Davis, R. H., 1982, Polyamine-deficient *Neurospora crassa* mutants and synthesis of cadaverine. *J. Bacteriol.*, 152:291.

Pegg, A. E. and McCann, P. P., 1982, Polyamine metabolism and function. *Am. J. Physiol.* 243: C212.

Tabor, C. W., and Tabor, H., 1984, Polyamines. *Annu. Rev. Biochem.* 53: 749.

Tabor, C. W., and Tabor, H., 1985, Polyamines in microorganisms. *Microbiol. Rev.* 49:81.

BIOCHEMICAL PROPERTIES OF UNUSUAL POLYAMINES FOUND IN AN EXTREME THERMOPHILE, *Thermus thermophilus*

Tairo Oshima, Nobuko Hamasaki and Taketoshi Uzawa

Department of Life Science, Tokyo Institute of Technology, Nagatsuta, Yokohama 227, Japan

INTRODUCTION

An extreme thermophile, *Thermus thermophilus* was isolated from a Japanese hot spring (Oshima and Imahori, 1974) which can grow in a temperature range of 50 °C to 85 °C. This unique microorganism produces a variety of polyamines. So far thirteen polyamines and two diamines have been detected; their trivial names and chemical structures are listed in Table 1. Most of these polyamines have been extracted from the thermophile, purified and identified by NMR, mass, and IR spectra. Their identification was finally confirmed by chemical synthesis (Oshima, 1983A). Unusually long polyamines and branched polyamines produced by the thermophile may play important roles in the thermophile cells and may exhibit unique activities in biochemical reactions *in vitro*.

We investigated the effects of such unusual polyamines on protein biosynthesis and conformational stability of nucleic acids. The main findings described in this review are: 1) A polyamine containing a quaternary ammonium nitrogen was most effective in stimulation of protein synthesis catalyzed by the thermophile extract at high temperatures. 2) Longer polyamines are more effective in stabilizing DNA and tRNA.

UNUSUALLY LONG POLYAMINES

T. thermophilus cells produce many polyamines longer than spermine (Table 1). The longest polyamine so far detected is 1, 20-diamino-4,8,12,16-tetraazaeicosane (Oshima and Senshu, 1985). Most of these polyamines were discovered for the first time from the thermophilic bacterium.

These polyamines can be easily synthesized from shorter polyamines by reaction with N-(3-bromopropyl)phthalimide or N-(4-bromobutyl)phthalimide. Both reagents are commercially available. The reaction product is then hydrolyzed to remove the phthalic acid moiety. Thus an aminopropyl or aminobutyl group can be added to the amino terminal(s) of a shorter polyamine. In

Table 1. Polyamines in the cells of an extreme thermophile, *T. thermophilus*.[1)]

Trivial name	Systematicame	Chemical structure
Caldine[2)]	1,7-Diamino-4-azaheptane	$NH_2(CH_2)_3NH(CH_2)_3NH_2$
Spermidine	1,8-Diamino-4-azaoctane	$NH_2(CH_2)_3NH(CH_2)_4NH_2$
sym-Homospermidine	1,9-Diamino-5-azanonane	$NH_2(CH_2)_4NH(CH_2)_4NH_2$
Thermine[3)]	1,11-Diamino-4,8-diazaundecane	$NH_2(CH_2)_3NH(CH_2)_3NH(CH_2)_3NH_2$
Spermine	1,12-Diamino-4,9-diazadodecane	$NH_2(CH_2)_3NH(CH_2)_4NH(CH_2)_3NH_2$
Thermospermine	1,12-Diamino-4,8-diazadodecane	$NH_2(CH_2)_3NH(CH_2)_3NH(CH_2)_4NH_2$
Homospermine	1,13-Diamino-4,9-diazatridecane	$NH_2(CH_2)_3NH(CH_2)_4NH(CH_2)_4NH_2$
Caldopentamine	1,15-Diamino-4,8,12-triazapentadecane	$NH_2(CH_2)_3NH(CH_2)_3NH(CH_2)_3NH(CH_2)_3NH_2$
Homocaldopentamine	1,16-Diamino-4,8,12-triazahexadecane	$NH_2(CH_2)_3NH(CH_2)_3NH(CH_2)_3NH(CH_2)_4NH_2$
Caldohexamine	1,19-Diamino-4,8,12,16-tetraazanonadecane	$NH_2(CH_2)_3NH(CH_2)_3NH(CH_2)_3NH(CH_2)_3NH(CH_2)_3NH_2$
Homocaldohexamine	1,20-Diamino-4,8,12,16-tetraazaeicosane	$NH_2(CH_2)_3NH(CH_2)_3NH(CH_2)_3NH(CH_2)_3NH(CH_2)_4NH_2$
	Tris(3-aminopropyl)amine	$(NH_2(CH_2)_3)_3N$
	Tetrakis(3-aminopropyl)ammonium	$(NH_2(CH_2)_3)_4N^+$

1)The thermophile contains two diamines, 1,3-diaminopropane and putrescine.
2) =Norspermidine 3) =Norspermine

some cases, the use of an iodinate derivative is better than bromoalkylphthalimide. NMR and mass spectra data of these polyamines of *T. thermophilus* are summarized elsewhere (Oshima, 1983B; Oshima, in press).

In *T. thermophilus*, longer polyamines seem to be synthesized by aminopropyltransfer from decarboxylated S-adenosylmethionine. Aminopropyltransferase activity was observed in a cell-free extract of *T. thermophilus* when a variety of polyamines was incubated in the presence of decarboxylated S-adenosylmethionine (the activity was followed by measuring the formation of methylthioadenosine, Pegg, 1983).

Biosynthetic pathways of most of longer polyamines listed in Table 1 could be explained by the aminopropyltransferase activity of the extract (unpublished data); polyamines without butyl group can be synthesized by successive aminopropyltransfer from 1,3-diaminopropane (Paulin et al., 1983), and those containing a butyl group can be synthesized by successive aminopropyltransfer to the amino nitrogen of the terminal aminopropyl group from spermidine. At present, it is premature to speculate on the biosynthetic path for polyamines containing two butyl groups such as *sym*-homospermidine and homospermine.

THE PRESENCE OF N-BRANCHED POLYAMINES

Among the unusual polyamines found in *T. thermophilus*, tetrakis(3-aminopropyl)ammonium is the first branched polyamine found in nature (Oshima et al., 1987). This polyamine containing a quaternary ammonium nitrogen was chemically synthesized by a reaction of tris(3-aminopropyl)amine with N-(iodoaminoproryl)phthalimide in dioxane, after the terminal amino groups had been protected. Tris(3-aminopropyl)amine was synthesized by reduction of either tris(2-cyanoethyl)amine, which was synthesized by a reaction of ammonia with an excess amount of acrylonitrile, or commercially available 3,3',3"-nitrilotrispropionamide with lithium aluminum hydride in tetrahydrofuran.

The identification of tetrakis(3-aminopropyl)ammonium suggested us the presence of tris(3-aminopropyl)amine in cells of *T. thermophilus* as a precursor in the biosynthetic pathway of the quaternary ammonium polyamine. When tris(3-aminopropyl)-amine was added to the thermophile cell suspension at 75°C, the increment of tetrakis(3-aminopropyl)ammonium content in the cell was observed (Hamasaki, unpublished). Aminopropyltransferase activity in a cell-free extract of *T. thermophilus* in the presence of decarboxylated S-adenosylmethionine was observed when tris(3-aminopropyl)amine was used as a substrate, though the reaction product in this experiment has not yet been identified. These experiments suggest that the tetrakis(3-aminopropyl)ammonium is synthesized by aminopropyltransfer to the central nitrogen atom of tris(3-aminopropyl)amine in the thermophile.

Unfortunately conventional analytical HPLC using a cation exchange resin column could not separate tris(3-aminopropyl)-amine from thermine under various elution conditions. Recently, in cooperation with Dr.Gamou in Kouchi University, we found an analytical method for this tertiary amine by the use of HPLC:

tris-(3-aminopropyl)amine can be separated from thermine and determined on column chromatography using a Shim-pack CLC-ODS eluted with a sodium perchlorate buffer containing sodium 1-hexanesulfonate and methanol (details will be published elsewhere). Thus we confirmed the presence of tris(3-amino propyl)amine in cells of *T. thermophilus*. The content of this branched polyamine is about one-tenth of that of thermine in the cells grown at 75°C and harvested at the late log phase (roughly 1.8 μmol of thermine is present per g of wet cells).

IN VITRO PROTEIN SYNTHESIS

Spermine or at least one of the other tetraamines is essential for *in vitro* protein synthesis at high temperature catalyzed by a cell-free extract of *T. thermophilus* (Ohno-Iwashita et al., 1976). The presence of polyamine is required for formation of active ternary complex between the messenger, ribosomes, and aminoacyl-tRNA. No activity was observed in the absence of polyamine, if the reaction was carried out at 60 °C or higher temperature. Aminoacylation and chain elongation steps were also stimulated by the addition of polyamine, but the addition was not essential for these steps (Ohno-Iwashita et al., 1975).

Newly discovered branched polyamines were tested in our laboratory for their stimulating activities of protein synthesis at high temperatures. The results are shown in Fig. 1, compared to the reaction in the presence of spermine. Low concentration of tetrakis(3-aminopropyl)ammonium, a polyamine containing a quaternary nitrogen atom, enhanced most effectively the protein

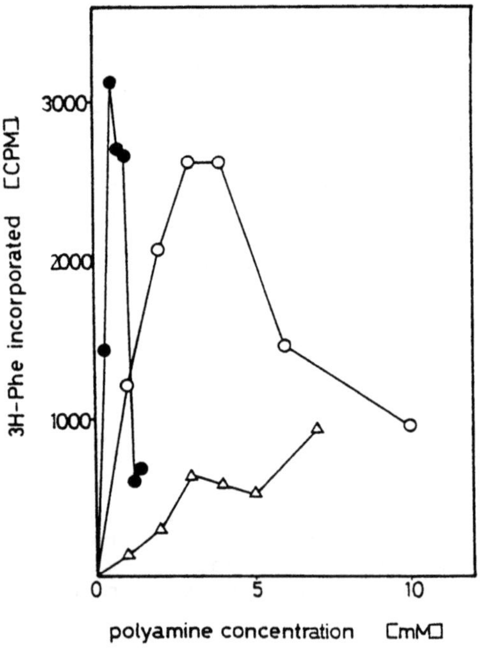

Fig. 1. Effects of spermine (open circle), tetrakis (3-aminopropyl) ammonium (filled circle), and tris(3-aminopropyl) amine (triangle) on poly (Phe) synthesis catalyzed by an extract of *T. thermophilus* at 65°C for 10 min.

synthesis. The optimum concentration of this unique polyamine was almost one-fifth of those of other larger polyamines. This result suggests that tetrakis(3-aminopropyl)ammonium plays an important role in regulation of in vivo protein synthesis even though this polyamine is a minor component and only a small amount is present in the cells of T. thermophilus. The optimum concentration for protein synthesis seems to be close to the cellular concentration (0.3 μmol/g wet cell). It is also noteworthy that at higher concentrations, the quaternary amine inhibited the protein synthesis. The other branched polyamine, tris(3-aminopropyl)amine was the least active among tetraamines tested in stimulating the in vitro protein synthesis. We reconfirmed the previous result that spermidine is much less active than spermine.

Though only a trace amount of spermine is present in the cells of T. thermophilus, spermine restored the protein synthesis at high temperature. In the presence of thermine, one of the major polyamines in the thermophile cell, the rate of protein synthesis at 65°C was about two-third of that in the presence of spermine. Our previous study showed that thermospermine, the other major polyamine, supports the protein synthesis at 65°C as much as spermine does (Oshima, 1978).

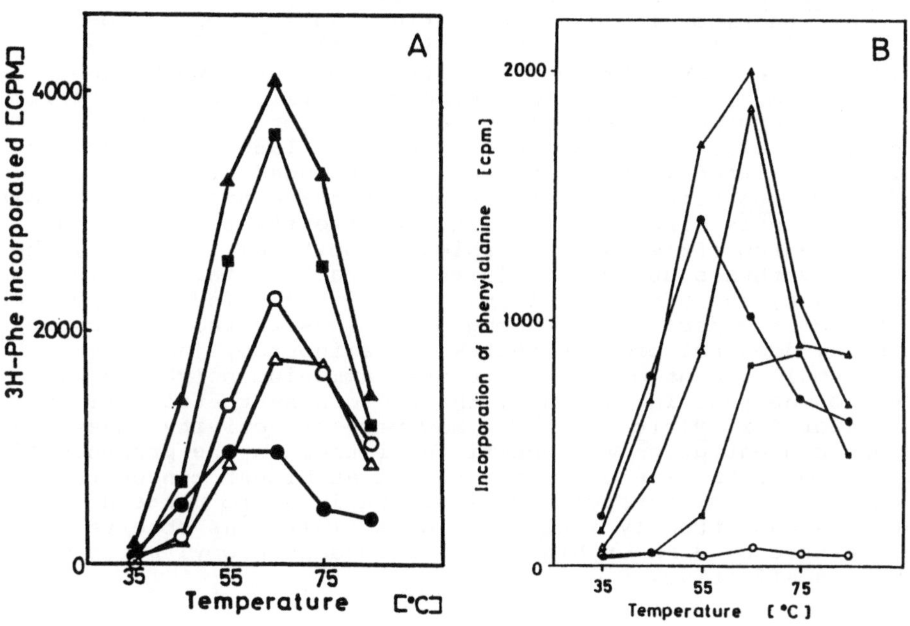

Fig. 2. Effects of temperature on poly(Phe) synthesis catalyzed by an extract of T. thermophilus. The reactions were carried out for 10 min. (A) The addition of 3 mM spermine (square), 3 mM thermine (open circle), 3 mM tris (3-aminopropyl)amine (filled circle), 3 mM caldopentamine (open triangle), and 0.8 mM tetrakis(3-aminopropyl) ammonium (filled circle). (B) No addition (open circle), and the addition of 1 mM (filled circle), 2 mM (filled triangle), 3 mM (open triangle), and 7 mM (square) spermine.

Polyamine composition of the thermophile varies according to the growing temperature (Oshima,1985):at a high temperature (80°C) cells contained more thermine, pentaamines, and hexaamines than at a lower temperature (60°). On the other hand, thermospermine and triamines (caldine, spermidine, and _sym_-homospermidine) were rich in the cells grown at 60°C.

We have investigated the effect of temperature on the stimulation of unusual polyamines on the cell-free protein synthesis catalyzed by _T. thermophilus_ extract. Some of the results are shown in Fig. 2. The highest activity was observed in the presence of 0.8 mM tetrakis(3-aminopropyl)ammonium at 65°C, as shown in Fig. 2A. Even at 80°, the quaternary ammonium polyamine is the best among polyamines tested for stimulating the reaction. The optimum temperature for the _in vitro_ protein synthesis in the presence of 3mM thermine was 65°C, and the activity at 80°C was much less than that at 65°C, though thermine in the thermophile cell was more abundant at 80°C than at 65°C. The optimum temperature of the reaction in the presence of caldopentamine was slightly higher than those in the presence of other polyamines. In the presence of 3 mM tris(3-aminopropyl) amine, the optimum temperature was lowered to 60°. These findings are only partly consistent with the polyamine compositions at different growth temperatures. The optimum temperature increases upon increasing the concentration of spermine(Fig. 2B). A similar result was obtained for tetrakis(3-aminopropyl)ammonium.

In vitro protein synthesis catalyzed by an extract of an extremely thermophilic archaebacterium, _Sulfolobus acidocaldarius_ (or closely related species), also requires the addition of a polyamine in the reaction mixture (Cammarano et al. 1985; Friedman, 1985). This thermoacidophilic archaebacterium produces thermine and spermidine as the major components (Kneifel et al., 1986; Cacciapuoti et al., 1986; Friedman and Oshima, submitted). Spermine was not produced by some strain; _S. acidocaldarius_ strain 7 contains only a trace amount of spermine which can be incorporated from the yeast extract used in the culture medium (Friedman and Oshima, submitted). However, the protein synthesis using a cell-free extract of this strain was supported by the addition of spermine (3 mM). Indeed, spermine was the most effective one among polyamines tested. This archaebacterium does not produce tetrakis(3-aminopropyl)ammonium. When a small amount of the quaternary ammonium polyamine was added to the reaction mixture, the protein synthetic activity was restored _in vitro_ at 65°C. However the activity under the optimum concentration was one-fifth of that in the presence of optimum concentration of spermine, and at higher concentrations of tetrakis(3-aminopropyl)ammonium, the _in vitro_ protein synthesis was strongly inhibited; no activity was observed in the presence of 3 mM or higher. Tris(3-aminopropyl)amine was much less effective.

STABILIZING EFFECT OF POLYAMINES ON DNA

It has been known that polyamines bind with nucleic acids, induce or stabilize specific conformations of nucleic acids, and protect thermal denaturation of DNA and RNA. (Tabor,1962; Mahler et al., 1961; Behe and Felsenfeld, 1981; Thomas and Bloomfield, 1984; Vertino et al., 1987). To investigate the possible

Table 2. Melting temperatures of calf thymus DNA (A_{258}=1.0) in the presence of various polyamines. Tm was measured in 1 mM sodium cacodylate buffer containing 50 mM sodium chloride, pH7.0.

Polyamine	Melting temperature [^{o}C]
None	78.5
Spermine	79.5
Thermine	79.0
Tris(3-aminopropyl)amine	78.5
Caldopentamine	79.5
Tetrakis(3-aminopropyl)ammonium	79.5
Caldohexamine	80.0
Caldoheptamine	82.5

physiological functions of unusual polyamines, change of melting temperature(Tm) of DNA by the addition of polyamine was measured. As shown in Table 2, higher Tm was observed when longer polyamine was added. Our study supported the general conclusions of previous investigations (Vertino et al., 1987) that, firstly the effects of polyamines are proportional to the number of protonated nitrogen atoms in the molecule, and, secondly heptaamine and hexaamine were more effective in stabilizing calf thymus DNA than pentaamines or tetraamines.

Concerning two branched polyamines, the stabilizing effects on DNA were different from each other; the quaternary ammonium polyamine showed a similar stabilizing effect to the corresponding straight polyamines, whereas tris(3-aminopropyl) amine was least active in stabilizing DNA. For instance, increment of Tm in the presence of 5 μM of tetrakis(3-aminopropyl)ammonium was roughly same as that of the presence of caldopentamine (both polyamines consist of four propyl groups and five amino or aza nitrogen atoms and their chemical formula in a neutral solution are same $C_{12}H_{36}N_5^+$). In contrast, tris(3-aminopropyl)amine which contains four nitrogen atoms was much less active in stabilizing DNA than the other tetraamines. Perhaps the distance between the terminal nitrogen atoms is important for stabilizing DNA. In this context, it is interesting to note that spermine was more effective than thermine in stabilizing DNA. The same result has been reported by Basu and Marton(1987).

STABILIZING EFFECT OF POLYAMINES ON $tRNA_f^{Met}$

Native tRNA contains at least two tightly bound polyamine molecules (Sakai et al., 1975; Sakai and Cohen, 1976) and their binding sites have been elucidated by X-ray crystallography (Quigly et al., 1978; Holbrook et al., 1978). Purified tRNA prepared by alcohol precipitation is free from polyamine and Tm of purified tRNA is raised by the addition of polyamines. In

Table 3. Melting temperatures of *E. coli* $tRNA_f^{Met}$ (A_{258}=0.5) in the presence of 5 μM of various polyamines. Tm was measured in a 1 mM sodium cacodylate buffer containing 50 mM sodium chloride, pH 7.0. Magnesium was absent in the buffer. Concentration of the tRNA was adjusted as A_{258}=0.5.

Polyamine	Melting temperature [°C]
None	58.0
Spermine	61.5
Thermine	64.5
Tris(3-aminopropyl)amine	----[1)]
Caldopentamine	61.0
Tetrakis(3-aminopropyl)ammonium	62.0
Caldohexamine	73.5
Caldoheptamine	78.5

[1)] No sharp melting profile was observed.

our preliminary study, $tRNA_f^{Met}$ from *Escherichia coli* was used and dissolved in a 1 mM sodium cacodylate buffer containing 50 mM sodium chloride in the absence of Mg^{2+}, pH 7.0 (Table 3). As in the case of DNA stabilization, longer polyamines are more effective than shorter ones such as spermine. In the presence of 5 μM of caldoheptamine, Tm of $tRNA_f^{Met}$ was raised by 20°C whereas an increase of only 3-6.5°C was observed by the addition of pentaamines (including tetrakis(3-aminopropyl)ammonium) or tetraamines, as shown in Table 3. The exact number of binding sites and the dissociation constants are being determined in our laboratory.

CONCLUSIONS

Though our study is still preliminary and future investigations are necessary to draw final conclusions, they suggest an important role for the unusual polyamines in stabilizing cellular components and biochemical reactions in the thermophilic bacteria. DNA and tRNA are protected from thermal denaturations by addition of longer polyamines. *In vitro* protein synthesis was supported by the addition of low concentrations of tetrakis(3-aminopropyl)ammonium. Though hexaamines and tetrakis(3-aminopropyl)ammonium are minor components in *T. thermophilus* and their cellular concentrations are low, our study suggested that they may be important in thermophily of this unique microorganism because these unusual polyamines stabilize nucleic acid or ribosomal complex at relatively low concentrations. For instance, the cellular concentration of tetrakis(3-aminopropyl)ammonium is close to the optimum concentration of this polyamine in stimulating the *in vitro* protein synthesis at 75°C.

Our study also suggests new applications for polyamines. For instance, tetrakis(3-aminopropyl)ammonium strongly inhibited the protein synthesis at moderate concentrations. This branched polyamine may exhibit unique pharmacological activities for cell growth, though in our preliminary experiments, multiplication of *E. coli* was not inhibited by the addition of this polyamine in the medium. Since it has been known that polyamines also protect DNA from digestions by restriction nucleases(Shishido, 1985), longer polyamines may be used when partial hydrolysis of DNA is desired in experiments of genetic manipulation. Effects of unusual polyamines on restriction endonucleases are currently investigated in our laboratory.

References

Basu, H. S., and Marton, L. J., 1987, The interaction of spermine and pentaamines with DNA, *Biochem. J.*, 244:243.

Behe, M., and Felsenfeld, G., 1981, Effects of methylation on a synthetic polynucleotide: The B-Z transition in poly(dG-m^5dC)·poly(dG-m^5dC), *Proc. Natl. Acad. Sci. U. S. A.*, 78:1619.

Cacciapuoti, G., Porcelli, M., Carteni-Farina, M., Gambacorta, A., and Zappia, V., 1986, Purification and characterization of propylamine transferase from *Sulfolobus solfataricus*, an extreme thermophilic archaebacterium, *Eur. J. Biochem.*, 161:263.

Cammarano, P., Teichner, A., Londei, P., Acca, M., Nicolaus, B., Sang, J. L., and Amils, R., 1985, Insensitivity of archaebacterial ribosomes to protein synthesis inhibitors, *EMBO J.*, 4:811.

Friedman, S. M., 1985, Protein synthesis in cell-free extracts from a thermophilic archaebacterium, *System. Appl. Microbiol.*, 6:1.

Holbrook, S. R., Sussman, J. L., Warant, R. W., and Kim, S. H., 1978, Crystal structure of yeast phenylalanine transfer RNA II. Structural features and functional implications, *J. Mol. Biol.*, 123:631.

Kneifel, H., Stetter, K. O., Andressen, J. R., Wiegel, J., König, H., and Schöberth, S. M., 1986, Distribution of polyamines in representative species of archaebacteria, *System. Appl. Microbiol.*, 7:241.

Mahler, H. R., Mehrotra, B. D., and Sharp, C. W., 1961, Effects of diamines on the thermal transition of DNA, *Biochem. Biophys. Res. Commun.*, 4:79.

Ohno-Iwashita, Y., Oshima, T., and Imahori, K., 1975, *In vitro* protein synthesis at elevated temperature by an extract of an extreme thermophile, *Arch. Biochem. Biophys.*, 171:490.

Oshima, T., 1978, Novel polyamines of extremely thermophilic bacteria, *in*: "Biochemistry of Thermophily," S. M. Friedman, *ed.*, Academic Press, New York.

Oshima, T., 1983A, Novel polyamines in *Thermus thermophilus*: Isolation, identification and chemical synthesis, *Methods Enzymol.*, 94:401.

Oshima, T., 1983B, Unusual polyamines in an extreme thermophile, *Thermus thermophilus*, *Adv. Polyamine Res.*, 4:479.

Oshima, T., in press, Polyamines in thermophiles, *in*: "The Physiology of Polyamines," U. Bachrach and Y. M. Heimer, eds., CRC Press, Boca Raton.

Oshima, T., and Imahori, K., 1974, Description of *Thermus thermophilus* (Yoshida and Oshima) comb. nov., a nonsporulating thermophilic bacterium from a Japanese thermal spa, *Intern. J. Syst. Bacteriol.*, 24:102.
Oshima, T., and Senshu, M., 1985, Unusual long polyamines in a thermophile, *in*: "Polyamines: Basic and Clinical Aspects," K. Imahori, F. Suzuki, O. Suzuki, and U. Bachrach, eds., VNU Science Press, Utrecht.
Oshima, T., Hamasaki, N., Senshu, M., Kakinuma, K., and Kuwajima, I., 1987, A new naturally occurring polyamine containing a quaternary ammonium nitrogen, *J. Biol. Chem.*, 262:11979.
Paulin, L., Ruohola, H., Nykanen, I., and Pösö, H., 1983, The incorpolation of 1,3-diaminopropane into thermine by an extreme thermophile: A novel route for the biosynthesis of polyamines, *FEMS Microbiol. Lett.*, 19:299.
Pegg, A. E., 1983, Assay of aminopropyltransferases, *Methods Enzymol.*, 94:260.
Quigley, G. Y., Teeter, M. T., and Rich, A., 1978, Structural analysis of spermine and magnesium ion binding to yeast phenylalanine transfer RNA, *Proc. Natl. Acad. Sci. U. S. A.* , 75:64
Sakai, T. T., and Cohen, S. S., 1976, Effects of polyamines on the structure and reactivity of tRNA, *Prog. Nucleic. Acids Res. Mol. Biol.*, 1:15
Sakai, T. T., Forget, R., Jacqueline, I., Freda, C. E., and Cohen, S. S., 1975, The binding of polyamines and of ethidium bromide to tRNA, *Nucleic Acids Res.*, 2:1005
Shishido, K., 1985, Effect of spermine on cleavage of plasmid DNA by nuclease S_1 and Bal31, *Biochim. Biophys. Acta*, 826:147
Tabor, H., 1962, The protective effect of spermine and other polyamines against heat denaturation of deoxyribonucleic acid, *Biochemistry*, 1:496
Thomas, T. J., and Bloomfield, V. A., 1984, Troidal condition of Z DNA and identification of an intermediate in the B to Z transition of poly(dG-m^{5}dC)·poly(dG-m^{5}dC), *Biochemistry*, 24:713
Vertino, P. M., Bergenon, R. J., Cavanaugh, Jr., P. F., and Porter, C. W., 1987, Structural determinations of spermidine-DNA interactions, *Biopolymers*, 26:691

POLYAMINES IN THE CELL CYCLE OF THE MALARIAL PARASITE

PLASMODIUM FALCIPARUM

U. Bachrach, L. Abu-Elheiga, Y.G. Assaraf, J. Golenser* and D.T. Spira*

Department of Molecular Biology and Parsitology*, Hebrew University-Hadassah Medical School, Jerusalem, Israel

INTRODUCTION

Malaria is a major health problem in many developing nations. More than 800 million people suffer from the disease with 1.5 milion deaths each year[1], Plasmodium falciparum causes the majority of the cases and is responsible for the fatal disease in man. The parasites undergo asexual developmental cycles in the erythrocyte and produce up to 32 merozites.

Despite the public health importance of malaria in the various tropical and subtropical regions, the physiology of the parasites and the host-parasite relationship are only poorly understood. Any biochemical or physiological difference between the host and parasite may have theoretical and practical implications. Recent studies have demonstrated that human erythrocytes lack ornithine decarboxylase (ODC) activity and contain only small amounts of polyamines[2]. A remarkable increase in polyamine content and in the activity of ODC was noted upon infecting the erythrocytes with the malarial parasite[2]. These changes were stage-dependent and were maximal during the advanced stage (schizont) of the cell cycle[2]. Polyamine synthesis appears to be essential for the multiplication of the parasite in cultured erythrocytes. Inhibitors of its biosynthesis, such as α-difluoromethylornithine (DFMO) blocked the completion of the cell cycle of the parasite, inhibited its growth, and facilitated the synchronization of the culture[3]. Addition of putrescine to the DFMO-treated cultures resulted in the resumption of growth[4]. This suggests that a polyamine-dependent reaction is required for schizogony, as the development of the parasites in polyamine-depleted systems is arrested at the early trophozoite stage.

This study deals with the identification of one of the polyamine-dependent steps in the malarial cell cycle and with the characterization of one of the enzymes involved.

MATERIALS AND METHODS

Plasmodium falciparum was cultured according to the candle jar method[5]. Proteins were labeled either with [^{3}H]-leucine (38.3 Ci/mmol, 1μCi/well or with [^{35}S]-methionine (40 μCi/ml) and separated by electrophoresis on polyacrylamide gels as described[2]. Nucleic acids were labeled by the addition of [^{3}H]-hypoxanthine (10 Ci/mmol, 1μCi/well) and RNA digested with 0.3 N NaOH. DNA was then precipitated with trichloroacetic acid (TCA), collected on glass-fibre filters and the radioactivity was counted[6].

The activity of DNA polymerase was determined according to a published method[7] as follows: the reaction mixture contained 20 mM phosphate buffer, pH 7.2, 0.5% sucrose, 25 μg/ml bovine serum albumin, 0.1 mM EDTA, 0.5 mM dithiothreitol, 8 mM $MgCl_2$, 6 μg of activated calf thymus DNA (Sigma), 100 μM dGTP, dCTP, dATP and 36 μM dTTP (Sigma) and 3 uCi [methyl-^{3}H] dTTP (125-150 cpm/pmole, Amersham) in a final volume of 25 μl. Reaction was started by the addition of 2 μl of enzyme (0.1 - 2 units). After incubation at 37° for 30 min, TCA-insoluble material was collected on glass-fibre filters and counted as above.

The activity of DNA polymerase was also determined in polyacrylamide gels following electrophoresis[8]. Enzyme extract (1-3 units) in 65 mM Tris-HCl buffer, pH 7.4, was incubated at 37° for 3 min with 1% SDS and 10% glycerol. Samples were electrophoresed at 4° and 60V in 7.5% polyacrylamide gels containing 1% SDS and 150 μg/ml activated calf thymus DNA. After 16 h, the gel was washed for 90 min with 65 mM Tris buffer pH 7.4 and then kept at 4° for 3h for renaturation, followed by incubation at 37° for 17 h in 70 mM Tris-HCl 7.5, 7mM $MgCl_2$, 10 mM 2-mercaptoethanol, 50 μCi [^{32}P]dTTP (2000 Ci/mmol, Amersham). Unincorporated radioactive phosphate was washed out with cold 5% TCA and 1% sodium pyrophosphate. Gel was finally dried and autoradiographed.

Neutralization assays of DNA polymerase by specific antibodies was carried out as described by Tanaka et al[9]. Monoclonal antihuman SjK 287-38 antibodies (0.5-3.0 g) was added to 0.1 units enzyme preparation and sedimented by centrifugation after adding protein A Sepharose. The supernatant was assayed for DNA polymerase activity by precipitation with TCA as above.

The molecular weight of DNA polymerase was determined by filtration through a Bio gel 0.5A column (88 x 1.2 cm) equilebrated with 10mM Tris pH 6.0; 1mM EDTA; 150 mM KCl (flow rate 14ml per hour).

RESULTS

DNA synthesis in malaria-infected erythrocytes

Recent studies from our laboratory demonstrated that DFMO inhibited the incorporation of [^{3}H]-leucine into malarial proteins[6]. Further studies showed that DFMO inhibited the synthesis of specific polypeptides and that putrescine stimulated the synthesis of some of them: polyacrylamide gel electrophoresis of [^{35}S]-methionine-labeled proteins indicated that DFMO arrested the synthesis of various proteins including two of molecular weights of 170 and 103 KD. Putrescine, which reversed the inhibitory effect elicited by DFMO, stimulated the synthesis of those two proteins[6].

We speculated that one of the proteins could be an enzyme involved in the replication of the malarial DNA[6]. This speculation was supported by the findings that the synthesis of malarial DNA was blocked by DFMO and reversed by putrescine[6]. Preliminary studies demonstrated that the synthesis of DNA in malaria-infected erythrocytes is a highly synchronized reaction. Usually DNA synthesis starts approximately 30 hours after erythrocyte infection, and reaches a peak 10-15 hours later. The kinetics of synthesis of malarial DNA correlated well with the activity of the DNA polymerase assayed *in vitro* (Fig. 1)

DNA synthesis is a vital event in the cell cycle of any organism. In eukaryotic cells various types of DNA polymerases have been described[10]. In parasites, on the other hand, mainly the α type of DNA polymerase has been detected[11-13]. These enzymes can be inhibited by aphidicolin[14] unlike other DNA biosynthetic enzymes found in many mammalian cells[10].

It appears that the malarial parasites contain an α-like DNA polymerase, as aphidicolin at a concentration of 2 μg/ml caused a 90% inhibition of DNA polymerase activity. N-ethylmaleimide (20mM) another inhibitor of DNA polymerase α, also inhibited the enzyme by 85%.

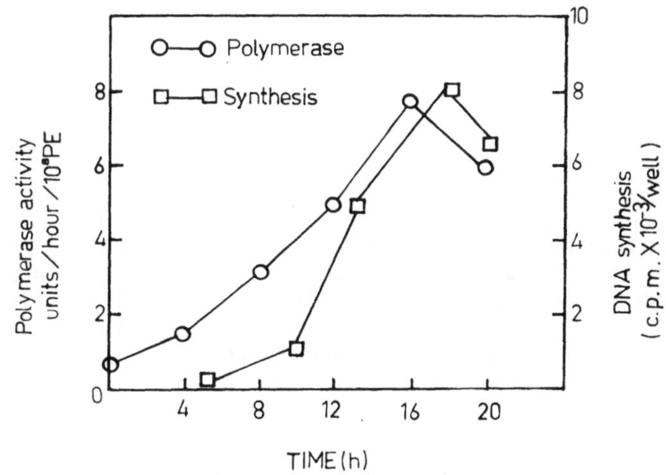

Fig. 1. DNA synthesis and DNA polymerase activity in malaria-infected erythrocytes. Cultured erythrocytes were infected with Plasmodium falciparum. When cultures reached the mature trophozoite stage, 10mM DFMO was added and incubation continued for 42 hours. Putrescine (250 μM) was then added (zero time) and the activity of DNA polymerase (O) and DNA synthesis (□) were determined at various times.

Monoclonal antibodies against human DNA polymerase-α neutralized approximately 80% of DNA polymerase activity (0.1U) of NIH 3T3 cells transfected with ras oncogene. On the other hand similar concentrations of the monoclonal antibody (2-3 μg/assay) had no effect on the activity of 0.1 U of malarial DNA polymerase.

Molecular weight of malarial DNA polymerase

The molecular weight of malarial DNA polymerase was determined by gel filtration and by detecting the activity of a protein band separated by electrophoresis on SDS polyacrylamide gels.

It may be seen (Fig. 2) that the enzyme, which was purified approximately 1000-fold had a molecular weight of approximately 100 KD when determined by gel filtration.

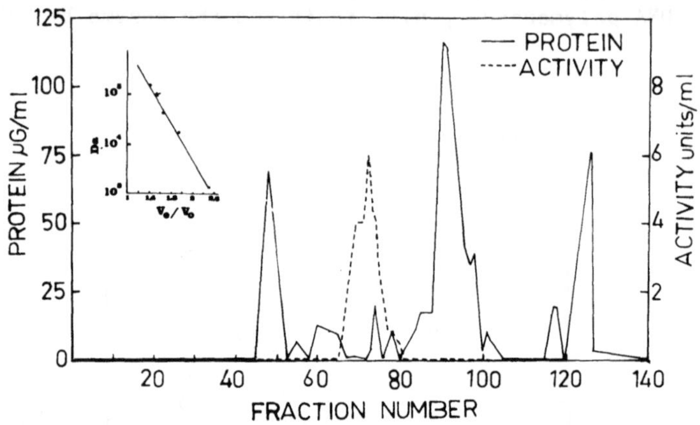

Fig. 2. Determination of the molecular weight of malarial DNA polymerase by gel filtration. Crude extracts from free parasites (3.0 mg, 250 U in 2.0 ml 50 mM Tris pH 7.4 and 10% sucrose) were applied onto a Bio Gel 0.5A colum. Fractions of 2.25 ml each were collected and assayed for DNA polymerase activity (- - -) and protein (——).

The second method for the determination of the molecular weight of malarial DNA polymerase was based on the separation of proteins by electrophoresis on polyacrylamide gels and detecting the activity of the DNA polymerase band.

The radioautogram (Fig. 3) clearly shows that a protein of a molecular weight of 103 KD had DNA polymerase activity. Other bands of lower molecular weight also exhibited some DNA polymerase activity.

Effect of polyamines on malarial DNA synthesis

The above described findings imply that the malarial DNA polymerase is an α-like enzyme with a molecular weight of approximately 100 KD. It should also be noted that a protein of a similar molecular weight (103 KD) is synthesized only in small amounts in the presence of DFMO when polyamine synthesis is blocked[6]. The addition of putrescine to the DFMO-treated cultures resulted in the accumulation of the 103 KD protein[6]. To test the possibility that this protein has DNA polymerase activity, proteins were extracted from DFMO-treated cultures after the addition of putrescine and separated by electrophoresis on polyacrylamide gels. It may be seen that the addition of putrescine to DFMO treated cultures led to the increase in the activity of the 103 KD molecular weight DNA polymerase (Fig. 3 lanes e, f, g).

These results thus suggest that putrescine stimualtes the synthesis of malarial DNA polymerase although a small amount of enzyme is formed even in polyamine-depleted cells.

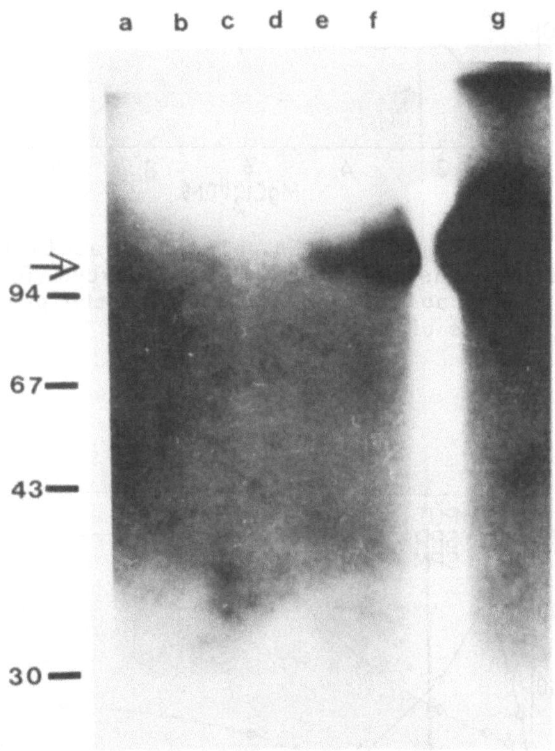

Fig. 3. Determination of molecular weight and activity of malarial DNA polymerase by polyacrylamide gel electrophoresis. Crude extracts from 10^8 infected erythrocytes were separated by electrophoresis on SDS polyacrylamide gels and activity determined as described under Methods. DFMO was added to the cultures and released by putrescine as described in the legend to Fig. 1 above. Samples were taken 42 hours after the addition of DFMO - a. at 48 hours - b. at 54 hours - c. at 60 hours - d. Putrescine was added after 42 hours and samples analyzed after 6 hours - e. after 12 hours - f. and after 18 hours - g.

In addition to their effect on DNA polymerase synthesis, polyamines could also affect the activity of the enzyme. Fig. 4 shows that malarial DNA polymerase requires Mg^{++} for optimal activity. Polyamines could replace Mg^{++} and in the presence of 2mM $MgCl_2$, 0.2mM spermine and 1-2mM spermidine caused a 5-fold stimulation of enzyme activity (Fig. 5).

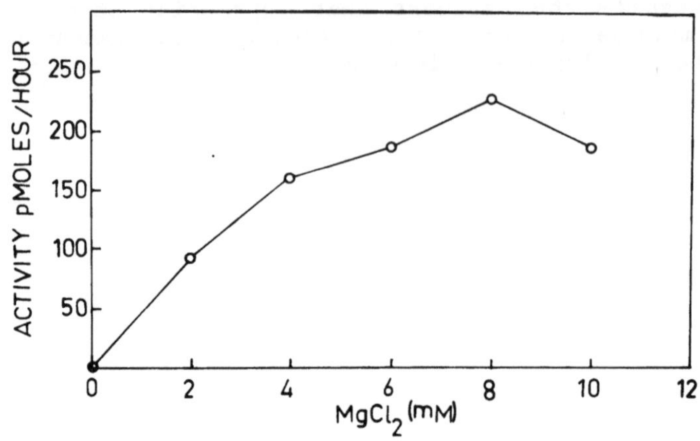

Fig. 4. Effect of $MgCl_2$ on DNA polymerase activity. Crude extracts from parasite-infected erythrocytes (0.3 U) were assayed for DNA polymerase activity in the presence of increasing amounts of $MgCl_2$.

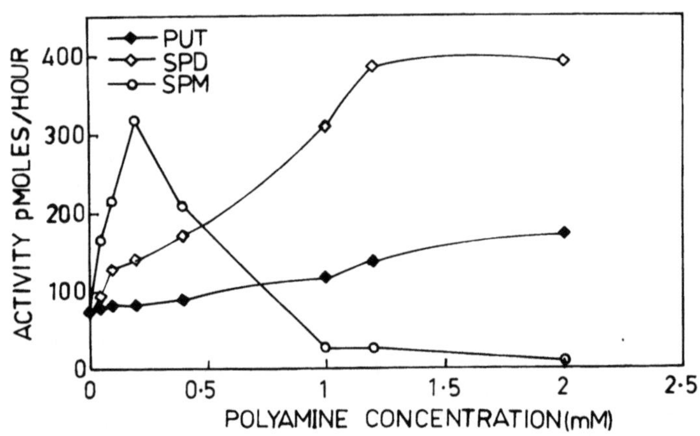

Fig. 5. Effect of polyamines on the activity of malarial DNA polymerase. Crude extracts from parasite-infected erythrocytes (0.3 U) were assayed for DNA polymerase activity in the presence of 2mM $MgCl_2$ and increasing amounts of putrescine (◆), spermidine (◇) and spermine (○).

DISCUSSION

It has been the goal of molecular biologists and parasitologists to detect differences in the properties of key enzymes of parasites and hosts. Cohen[15] postulated that new chemotherapeutic strategies might be developed following the identification of metabolic and enzymological differences between host and parasite. Polyamine biosynthetic enzymes, their products or metabolites may be the target for such an approach. It has been shown that ODC from parasites, including Trypanosoma brucei and Plasmodium falciparum differ from the host enzyme in their antigenic properties[16] and in the case of T. brucei also in the structure of the gene[17]. The present study shows that the α-like DNA polymerase from P. falciparum was not neutralized by monoclonal antibodies which interacted with the mammalian enzyme. This suggests that the two enzymes differ structurally.

Polyamine synthesis occurs in parasite-infected cells during the early stages of the parasite cell cycle. To study the role of polyamines in the development of the malarial parasites, inhibitors were added to the infected erythrocytes. It became apparent that polyamines are required for the synthesis and the activity of an α-like DNA polymerase (Fig. 3). The accumulation of polyamines at a certain stage of the cell cycle, triggered the replication of malarial DNA and apparently also stimulated the synthesis of "late proteins". In this respect the malarial parasite resembles the bacteria-phage system, in which DNA synthesis is the transition stage between the synthesis of "early" and "late" proteins. The possibility that the newly synthesized DNA differs from the DNA in the mature parasite should be considered. Polyamines also stimulate DNA synthesis in lectin-stimulated lymphocytes. According to Igarashi, MGBG which inhibits polyamine synthesis, interfered with the synthesis of cellular DNA[18].

Our findings that the synthesis of malarial DNA is catalyzed by a polyamine dependent α-like DNA polymerase may open new avenues in the development of new antimalarial drugs.

REFERENCES

1. W.P. Rogers, Advances in parasitology: 1886-1986. Inter. J. Parasitol. 17: 1-13 (1987).

2. Y.G. Assaraf, J. Golenser, D.T. Spira, and U. Bachrach, Polyamine levels and the activity of their biosynthetic enzymes in human erythrocytes infected with the malarial parasite, Plasmodium falciparum, Biochem. J. 222: 815-819 (1984).

3. Y.G. Assaraf, J. Golenser, D.T. Spira, and U. Bachrach. Plasmodium falciparum: synchronization of cultures with DL- - difluoromethylornithine an inhibitor of polyamine biosynthesis, Exp. Parasitol. 61: 229-235 (1986).

4. Y. G. Assaraf, J. Golenser, D.T. Spira, G. Messer, and U. Bachrach, Cytostatic effect of DL-α-difluoromethylornithine against Plasmodium falciparum and its reversal by diamines and spermidine. Parsoitol. Res. 73: 313-318 (1987).

5. W. Trager and J.B. Jensen, Human malaria parasites in continuous culture, Science, 193: 673-675 (1976).

6. Y.G. Assaraf, L. Abu-Elheiga, D.T. Spira, H. Desser, and U. Bachrach, Effect of polyamine depletion on macromolecular synthesis of the malarial parasite, Plasmodium falciparum, cultured in human erythrocytes. Biochem. J. 242: 221-226 (1987).

7. N. Khan, G. Wright, L. Dudycz, and N. Brown, Butylphenyl dGTP a selective and potent inhibitor of DNA polymerase alpha, Nucleic Acid Res. 12: 3695-3706 (1984).

8. A. Spanos, S.G. Sedwick, G.T. Yarranton, U. Hübscher, and G.R. Banks, Detection of the catalytic activities of DNA polymerases and their associated exonucleases following SDS-polyacrylamide gel electrophoresis. Nucleic Acid Res. 9: 1825-1839 (1981).

9. S. Tanaka, S-Z. Hu, T. S-F. Wang, and D. Korn, Preparation and preliminary characaterization of monoclonal antibodies against human DNA polymerase α. J. Biol. Chem. 257: 8386-8390 (1982).

10. L.A. Loeb, P.K. Liu, and M. Fry, DNA polymerase α: Enzymology, function, fidelity and mutagenesis. Prog. Nucleic Acid Res. Mol Biol. 33: 57-110. (1986).

11. A. M. Holmes, E. Cheriathunam, A. Kalineki, and L.M.S. Cheng, Isolation and partial characterization of DNA polymerases from Crithidia fasiculata, Mol. Biochem. Parasitol. 10: 195-205 (1984).

12. T.W. North, and D.J. Wyler, DNA synthesis in promastigotes of Leishmania major and L. donovanis, Mol. Biochem. Parasitol. 22: 215-221 (1987).

13. L.M.S. Chang, E. Cheriathunam, E.M. Mahoney, and A. Cerami, DNA polymerases in parasitic protozoa differ from host enzymes, Science, 208: 510-511 (1980).

14. S. Ikegami, T. Taguchi, M. Ohashi, M. Oguro, H. Nagano, and Y. Mano, Aphidicolin prevents mitotic cell division by interfering with the activity of DNA polymerase-α, Nature 275: 458-460 (1978).

15. S.S. Cohen, Comparative biochemistry and drug design for infectious disease, Science 205: 964-971 (1979).

16. Y.G. Assaraf, The role of polyamines in the development of malaria parasite. Ph.D. thesis, Hebrew University Jerusalem, 1986.

17. M.A. Phillips, P. Coffino, and C.C. Wang, Cloning and sequencing of ornithine decarboxylase gene from Trypanosoma brucei. J. Biol. Chem. 262: 8721-8727 (1987).

18. K. Igarashi, C.W. Porter, and D.R. Morris, Comparison of specificity of inhibition of polyamine synthesis in bovine lymphocytes by ethylglyoxal bis(guanylhydrozone) and methylglyoxal bis(guanylhydrazone), Cancer Res. 44: 5326-5331 (1984).

THE ROLE OF POLYAMINES IN THE GROWTH AND TRANSFORMATION OF THE AFRICAN TRYPANOSOME

Bruce F. Giffin

Department of Biology
University of Dayton
Dayton, Ohio U.S.A.

INTRODUCTION

The transformation from bloodstream to procyclic trypomastigote is a crucial step in the life cycle of the African trypanosome. The alteration in cellular architecture is accompanied by biochemical changes which permit survival of the bloodstream form trypanosome in the midgut of the insect vector. While a specific inducer of transformation has not yet been identified, recent studies have implicated the polyamines as having some role to play in this process.

THE LIFE CYCLE OF THE AFRICAN TRYPANOSOME

The *Trypanosoma brucei* subspecies is transmitted by the tsetse fly of the genus *Glossina*. In the course of taking a blood meal from an infected mammal, the fly ingests bloodstream form trypanosomes (Figure 1) which undergo significant developmental changes in the midgut (Vickerman, 1965). Here the transformation to procyclic trypomastigotes occurs, a step involving dramatic biochemical and morphological alterations (Bowman and Flynn, 1976; Bienen et al., 1983). The procyclic trypomastigotes lose the variant surface glycoprotein coat that is characteristic of an individual trypanosome clone (Cross, 1975), becoming non-infective for the vertebrate host (Brown et al., 1973). In addition to the morphological changes, the bloodstream form trypomastigote switches from an extramitochondrial respiration to mitochondrial respiration which permits it to survive in the midgut of the vector. After three weeks of proliferation and continued development in the midgut of the fly as procyclic trypomastigotes, the parasites move to the salivary gland. Here they develop into metacyclic trypanosomes which are morphologically similar to the bloodstream forms, and have re-acquired both their surface coat and infectivity (Vickerman, 1969; LeRay et al., 1978). When an infected fly takes a blood meal, the metacyclic trypomastigotes are injected into the bloodstream of the host and transform to long slender bloodstream forms within 24 hours (Brun et al., 1984). During this transformation the trypanosome switches from mitochondrial respiration back to an exclusive dependency on glycolysis for ATP production (Opperdoes and Borst, 1977). The long slender bloodstream forms undergo rapid proliferation in the bloodstream of the host and are capable of antigenic variation. During the course of a parasitemia, pleomorphic strains transform from

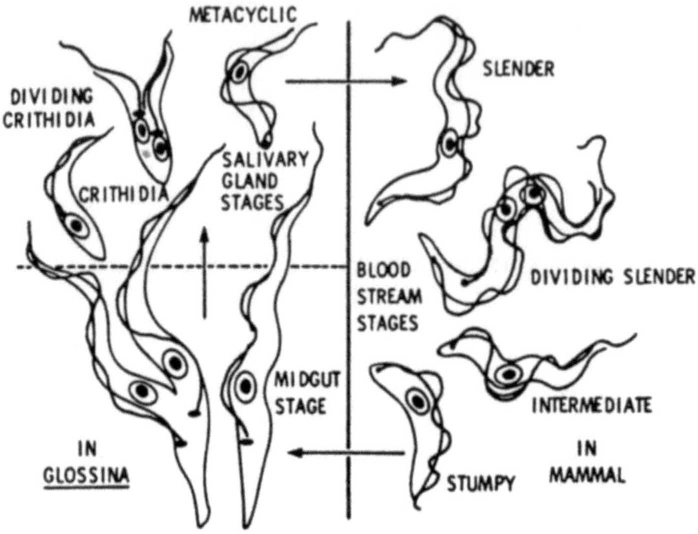

Figure 1. Changes in the morphology of the Trypanosoma brucei subspecies during the course of its life cycle (Redrawn after Vickerman, 1965).

long slender bloodstream forms to short stumpy forms via an intermediate form (Vickerman, 1965). The short stumpy forms are generally considered to be biochemically preadapted for life in the tsetse fly midgut (Hill, 1976). They are, however, terminal forms in the mammal in that they do not undergo binary fission; accumulate large amounts of plasma lipoprotein (Venkatesan and Ormerod, 1976); oxidize proline and, unless taken up by the insect vector, will be removed from the blood by the host's immune system. When the tsetse fly takes a blood meal the short stumpy form transforms to the procyclic trypomastigote in the midgut, thus completing the life cycle of the trypanosome (Wijers and Willett, 1960).

Currently there is some controversy as to which of the bloodstream forms is, in fact, transmissible to the vector (Giffin, et al., 1986a).

POLYAMINES AND THE AFRICAN TRYPANOSOME

Polyamines have long been implicated in cellular growth processes, particulary in cells undergoing rapid proliferation. Intracellular levels of ornithine decarboxylase and polyamines increase in dividing cells. The increased intracellular polyamine content, which can be up to four times higher than in non-dividing cells, precedes the increase in amount of RNA, DNA or protein (Janne, et al., 1978; Pegg and McCann, 1982; Tabor and Tabor, 1984). A role for polyamines in regulating the differentiation of mammalian cells has recently been reviewed (Pegg and McCann, 1982). In some cell systems polyamine depletion prevents differentiation (Bethell and Pegg, 1981; Erwin et al., 1983), while in others (Schindler et al., 1983 and 1985; Heby et al., 1983; Chen et al., 1983; Kapyako and Janne, 1983) differentiation is induced.

The role of polyamines in the African trypanosome is not clearly understood but is currently under investigation. When trypanosomes are exposed to difluoromethylornithine (DFMO), an irreversible inhibitor of ornithine decarboxylase, which is the first and rate-limiting enzyme in the polyamine biosynthetic pathway, there is an inhibition of cell division and a change in cellular morphology (Bacchi et al., 1983, deGee et al., 1984; Giffin et al., 1986a). These effects can be reversed by coadministration of putrescine, spermidine or spermine (Nathan et al., 1981). The use of DFMO in the treatment of acute infections of Trypanosoma brucei brucei (Bacchi et al., 1980), Trypanosoma brucei rhodesiense (McCann et al., 1981a) and Trypanosoma brucei gambiense (McCann et al., 1981b) has been demonstrated. The implications of polyamine biosynthesis inhibition as a chemotherapeutic strategy for treating several protozoal infections has recently been reviewed (McCann et al., 1983; Sjoerdsma and Schechter, 1984; Bacchi and McCann, 1987).

Polyamines and Growth Rates

Studies both in vivo and in vitro have demonstrated a correlation between the intracellular content of the polyamines and trypanosomal doubling times. When long slender bloodstream forms of Trypanosoma brucei brucei (EATRO 110) are cultured with a monolayer of embryonic bovine trachea fibroblasts in a modified Eagle's Minimal Essential Medium (Brun et al., 1981), they continue to proliferate as long slender bloodstream forms and can be subpassaged indefinitely. If, however, the culture medium contains 100 uM DFMO, trypanosomal growth continues normally only for the first 24 hours, after which there is a cessation of cellular replication (Figure 2). The inhibition of ornithine decarboxylase by DFMO has a dramatic effect on the intracellular polyamines of the trypanosome within the first 24 hours: putrescine is depleted and spermidine is decreased to greater than 50% of control levels. Only putrescine and spermidine are found in significant concentrations in the bloodstream form trypanosome (Bacchi et al., 1977). However, trypanosomes treated with DFMO begin to accumulate a significant amount of spermine. Since this polyamine is also found in the procyclic trypomastigote, its accumulation may be related to the concomitant morphological and biochemical changes that follow the decrease in intracellular putrescine and spermidine.

When a 250 g rat is given an inoculum of 10^7 trypanosomes, the growth curve for the trypanosome population has a characteristic lag phase which lasts for approximately 24 hours (Figure 3). This essentially represents the time prior to which trypanosomes are detectable in the peripheral blood of the animal. The log phase represents that period of exponential increase in cell number following the lag phase and is associated with a characteristic doubling time of 4-5 hours. In monomorphic strains of trypanosomes there is no transformation to the short stumpy form. Consequently the trypanosome population continues to divide and the level of parasitemia reaches in excess of 10^8 cells/ml of blood by 72 hours post-inoculation resulting in death of the animal. Toward the end of the log phase, the doubling time exhibits a greater than 2-fold increase. Presumably this is due to stresses on the population such as lack of glucose availability, elevated levels of pyruvate in the blood, and reduced blood oxygen tension (unpublished data). In contrast, pleomorphic strains of trypanosomes undergo transformation during the log phase. Consequently there is a dramatic increase in the population doubling time as a plateau phase is reached and cell division ceases. These bloodstream forms, consisting of a large number of non-dividing short stumpy forms, do not kill the host and are soon eliminated by the host's immune system (Ormerod, 1979; deGee et al., 1983). Prior to the production of circulating antibodies, a small number of trypanosomes have changed their glycoprotein antigenic cell coat and thus are able to effect a recrudescence of the parasitemia in several days (Vickerman, 1965; Roelants and Pinder, 1984). This process, known as antigenic variation, enables the trypanosome to maintain a successful habitation of the host's bloodstream and produce the characteristic cyclic parasitemia.

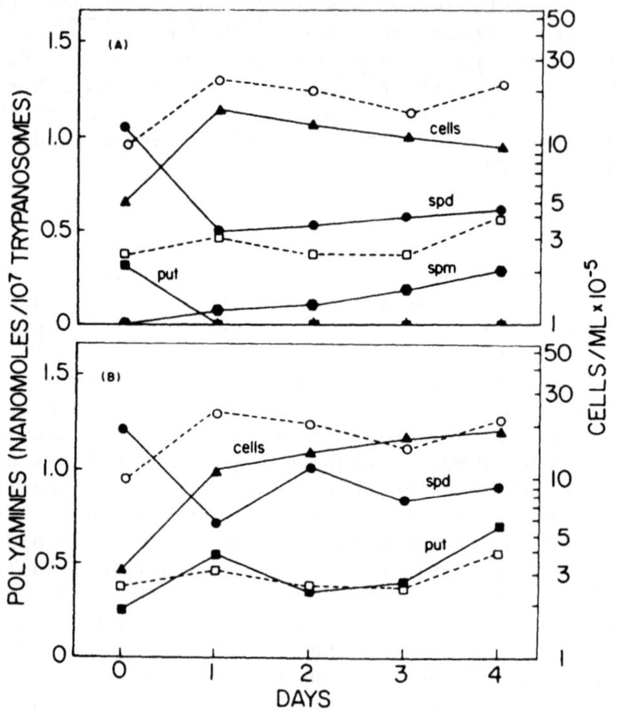

Figure 2. (A) Cell numbers and intracellular polyamine levels in bloodstream form trypanosomes during *in vitro* culture with 100 uM DFMO and (B) 100 uM DFMO and 100 uM putrescine. The broken lines represent putrescine (put) and spermidine (spd) levels in untreated controls. Spermine (spm) is detectable in the growth-inhibited cells only. (From Giffin et al., 1986a.)

We have studied polyamine profiles as they relate to doubling times in monomorphic and pleomorphic trypanosome populations both *in vivo* and *in vitro*. An inverse correlation exists between intracellular polyamine levels and doubling times of trypanosomes cultured *in vitro* and those taken from infected rodents (Figure 4). These data clearly support the role of polyamines in cells that are undergoing rapid proliferation. The efficacy of DFMO as a chemotherapeutic agent for trypanosomiasis may in fact be related to its ability to effect a cessation of cell division subsequent to polyamine depletion followed by the induction of transformation (Giffin et al., 1986a). How trypanosomiasis can be cured by a rapid, simultaneous transformation of the long slender bloodstream forms to short stumpy forms will be the topic of the next section.

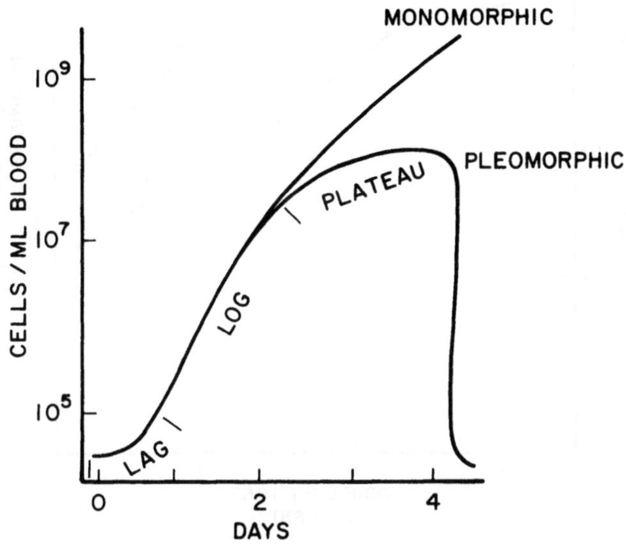

Figure 3. Diagram of a generalized growth curve for a pleomorphic and monomorphic strain of African trypanosome.

Polyamines and the Induction of Transformation

Morphological changes. Following the decreases in the polyamines and the cessation of cell division which result from the treatment of trypanosomes with DFMO either *in vitro* or *in vivo*, there is the induction of a morphological and biochemical transformation (Bacchi et al., 1983; deGee et al., 1984; Giffin et al., 1986a). The morphological alterations were determined to be essentially the same as those which occur during the course of a pleomorphic parasitemia in nature by the following criteria: (a) The intermediate forms measure between 20-25 um in length and are approximately 2.75 um in width with a less well-developed undulating membrane and a shorter free flagellum as compared to the long slender forms. The kinetoplast is found closer to the posterior end of the cell body in this cell type. (b) The short stumpy forms measure 13-22 um in length with a width of approximately 3.5 um. The undulating membrane is poorly developed with a greatly diminished or completely absent free flagellum. The kinetoplast assumes a terminal position at the posterior end of the cell body. Trypanosomes exposed to DFMO during *in vitro* culture are depleted of putrescine within 24 hours and become intermediate in their morphology by 48 hours. These intermediate forms then transform to short stumpy forms by 96 hours of culture (Figure 5). That this transformation is specifically

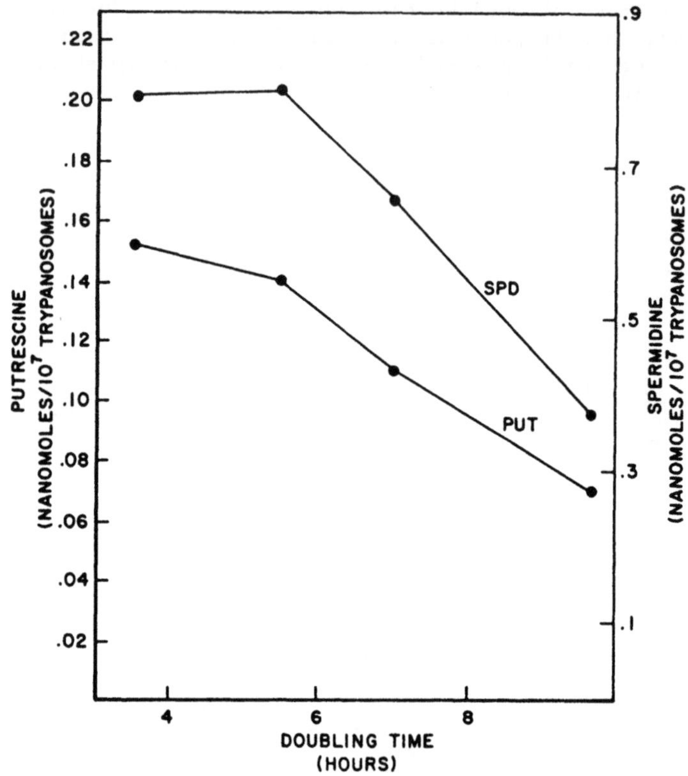

Figure 4. The correlation between intracellular putrescine (put) and spermidine (spd) levels and doubling times in a monomorphic strain of *Trypanosoma brucei* (EATRO 110) measured *in vivo*.

induced by the decrease in polyamines was demonstrated by the fact that the coadministration of exogenous putrescine abrogated the effects of DFMO alone, i.e., intracellular polyamine levels remain similar to control levels, the inhibition of cell proliferation was prevented and there is no alteration in the morphology of the long slender bloodstream forms (Figure 2 and Figure 5B).

When DFMO is administered to rodents infected with acute infections of *Trypanosoma brucei brucei* (EATRO 110) a similar sequence of events transpires with a slightly different chronology. Putrescine is depleted within 24 hours and the intermediate bloodstream forms are present in the peripheral blood of the animal by 36 hours. These then transform into short stumpy forms by approximately 60 hours following the initiation of DFMO chemotherapy. And, as is the case in the *in vitro* culture experiments, the transformation can be abrogated *in vivo* by a single injection of putrescine (300 mg/kg body weight) 18 hours after the initiation of DFMO treatment to the rodent. Under these circumstances putrescine and spermidine levels in the trypanosome remain at control levels, the doubling time is not increased, transformation does not occur, and the long slender trypanosome population quickly kills the host.

Biochemical changes. The long slender bloodstream form trypanosome depends exclusively on glycolysis for ATP production. There is no func-

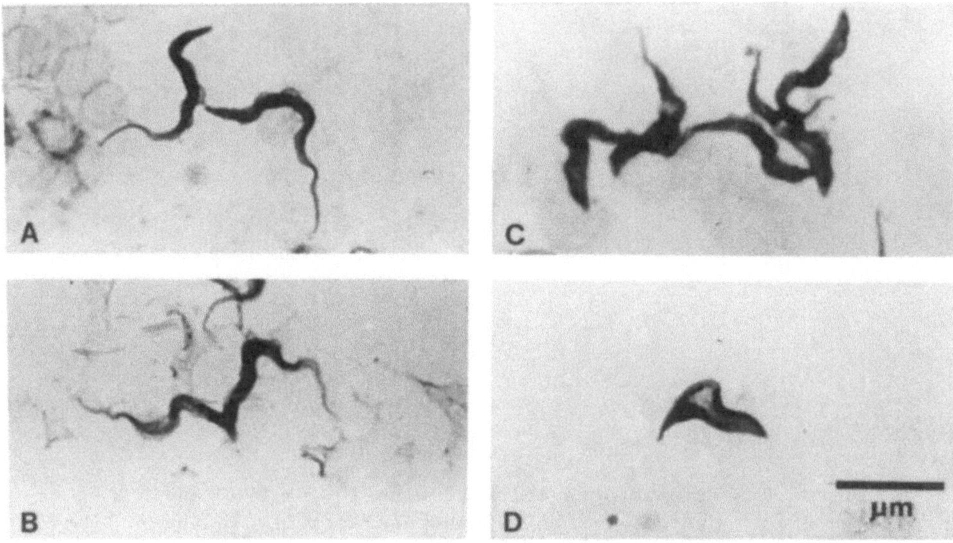

Figure 5. Morphological alterations in long slender bloodstream form trypanosomes after exposure to 100 uM DFMO in culture. (A) Control long slender forms. (B) Long slender forms after 4 days of culture with both 100 uM DFMO and 100 uM putrescine. The effects of DFMO are abrogated and the cells retain their long slender morphology. (C) Long slender forms become intermediate in their morphology after 48 hours of culture. (D) Short stumpy forms are present after 4 days of culture. Note the progressive increase in cell volume and the regression of the flagellum.

tional Krebs cycle or cytochrome system in the single, peripheral, tubular mitochondrion which characteristically contains very few cristae. The morphological transformation to the procyclic trypomastigote is accompanied by mitochondrial proliferation and the synthesis of tricarboxylic acid cycle enzymes and cytochromes (Bowman and Flynn, 1976; Bienen et al., 1983). These biochemical changes permit the survival of the trypanosome in the midgut of the tsetse fly where glucose and oxygen levels are considerably lower than in the bloodstream of the vertebrate host. Biochemical activation of the mitochondrion can be detected by electron microscopy and cytochemical staining techniques. Both the DFMO-induced intermediate and short stumpy bloodstream form trypanosomes stain positive for NADH-tetrazolium reductase activity (Figure 6) and demonstrate mitochondrial proliferation and cristae development when examined by electron microscopy.

That these biochemical changes are in themselves sufficient to permit adaptation of the trypansome to the vector has been demonstrated by introducing the trypanosomes into SDM-79 medium (Brun and Schonenberger, 1979). This medium, which simulates tsetse fly hemolymph, permits the transformation of biochemically transformed bloodstream form trypanosomes into procyclic trypomastigotes. When DFMO-induced intermediate and short stumpy trypanosomes were introduced into SDM-79 medium (Figure 7), only the inter-

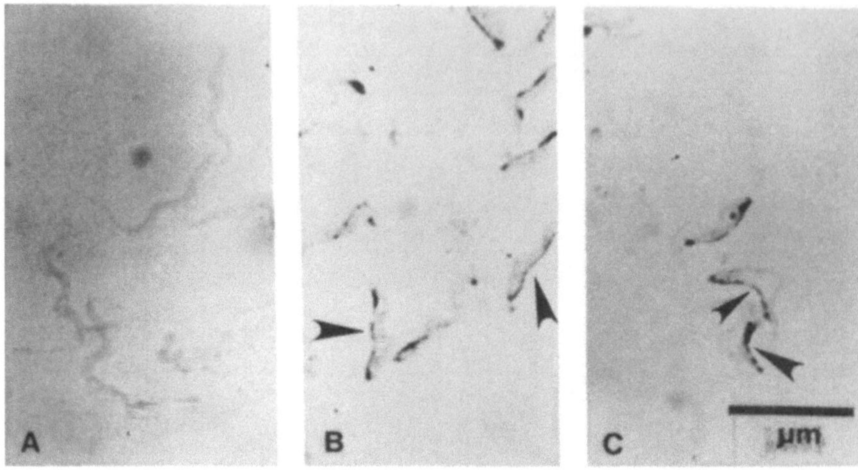

Figure 6. Trypanosomes cultured with 100 uM DFMO and stained for NAD diaphorase activity in the mitochondrion. (A) Control long slender forms are negative for NAD diaphorase activity. (B) Intermediate forms after 48 hours and (C) short stumpy forms after 96 hours of culture with 100 uM DFMO are positive for NAD diaphorase activity. Foci of formazan deposition along the single tubular mitochondrion indicate activation of this organelle (arrows).

mediate forms successfully transformed. This data suggests that the intermediate form may be the cell type that is transmissible to the vector and not the short stumpy form, which has generally been considered the vector-infective bloodstream form. This would support the previous speculation by Ormerod (1979), that the short stumpy is a degenerate cell type that is incapable of either binary fission or further transformation into the procyclic trypomastigote. In addition it is possible to stimulate the transformation of long slender bloodstream form trypanosomes into procyclic trypomastigotes with citric acid cycle intermediates (Brun and Shonenberger, 1981). Under these conditions the long slender trypanosomes become intermediate in their morphology prior to the cellular alterations characteristic of the procyclic trypomastigote, and they by-pass the short stumpy stage altogether. Furthermore, intermediate bloodstream form trypanosomes from a pleomorphic strain (STIB 366A) were found to have the capacity to transform to procyclic trypomastigotes without the transition to the short stumpy form (Ghiotto et al., 1979).

Some additional support for the transmissibility of the intermediate form and the degenerate nature of the short stumpy form comes from preliminary experiments in which DFMO-induced intermediate and short stumpy forms were used to infect tsetse flies (unpublished data). Only tsetse flies fed intermediate forms were found to have midgut infections and, although not statistically significant since only 3% of the flies were positive, it is significant that the trypanosome strain used in these studies (STIB 355-A) is notorious for its inability to infect the vector. Additional studies

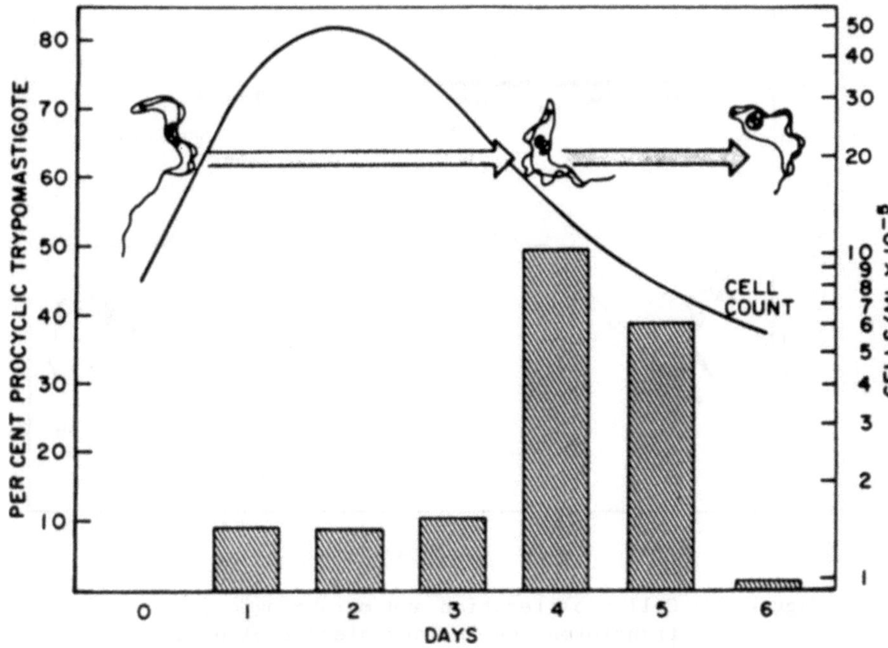

Figure 7. Long slender trypanosomes cultured with 50 uM DFMO for 6 days. Each day trypanosomes were introduced into SDM-79 medium. Bars indicate the percent of procyclic trypomastigotes present after three days as determined by a differential cell count using a Neubauer hemacytometer. The percentage of transformation is highest at days 4 and 5 when the cultured trypanosomes had an intermediate bloodstream form morphology. (From Giffin, et al., 1986a.)

indicated that there is only a weak mitochondrial activation in the DFMO-induced intermediate and short stumpy forms from this strain, and neither form had the capacity to transform to procyclic trypomastigotes when cultured in SDM-79 medium. Further work on the EATRO 110 isolate and other strains that are more easily transmitted to the vector is needed to support the speculation that the intermediate, and not the short stumpy form, is the bloodstream form that is vector-infective.

Polyamines and Transformation to Procyclic Trypomastigote

Brun and Schonenberger (1981) have shown that transformation of the long slender bloodstream form trypanosomes can be initiated by citric acid cycle intermediates. The use of this technique and the application of recent methods of *in vitro* culture of bloodstream (Brun et al., 1981) and procyclic trypomastigotes (Brun and Schoenberger, 1979) enables the synchronous transformation of monomorphic strains of trypanosomes (Overath et al., 1983). We used this technique to determine whether or not polyamines were required for transformation to the procyclic trypomastigote.

The transformation protocol (Figure 8) consisted of culturing long slender bloodstream trypanosomes in the presence of citric acid cycle

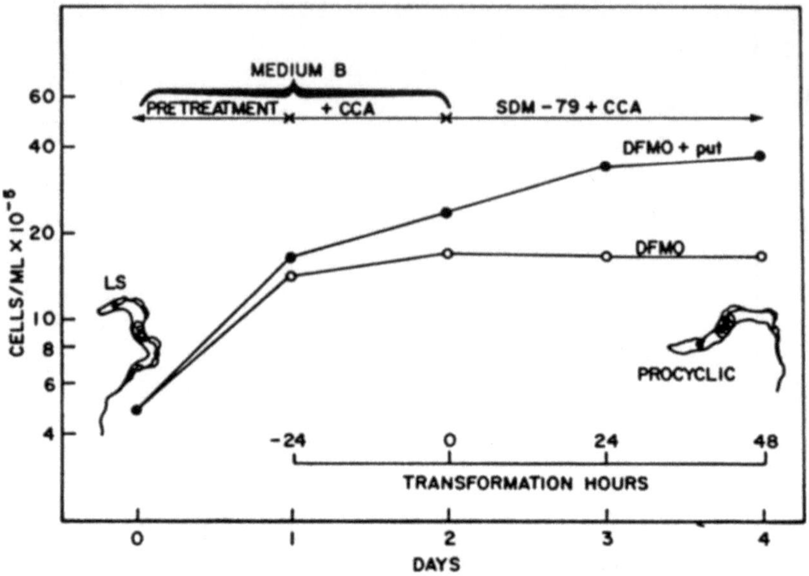

Figure 8. Cell proliferation and morphological transformation in long slender bloodstream trypanosomes depleted of polyamines prior to the initiation of transformation to procyclic trypomastigotes using citric acid cycle (CCA) intermediates. Pretreatment for 24 hours depleted the long slender forms of putrescine and reduced spermidine by greater than 50%. The transformation was stimulated by exposing the trypanosomes to citric acid cycle intermediates as described in the text. Transformation to procyclic trypomastigote was accompanied by cell proliferation in cells exposed to 100 uM DFMO and 100 uM putrescine (put) whereas cells exposed to DFMO alone successfully transformed but did not proliferate.

intermediates (100 mM citric acid, 100 mM cis-aconitate, 200 mM magnesium acetate) with DFMO or DFMO and putrescine added to the medium to give a final concentration of 100 uM. After 24 hours the trypanosomes were removed from the medium, resuspended in SDM-79 medium and incubated for 48 hours at 28° C. The data indicate that the process of transformation to procyclic trypomastigotes is independent of cellular proliferation and does not have a requirement for putrescine. The proliferative component of the transformation process is dependent on the trypanosome replenishing its intracellular content of putrescine. The coadministration of putrescine prevented the cytostatic effect of DFMO alone, since polyamine-depleted cells have an active transport system that can take up exogenous polyamines when intracellular levels are depleted (Pegg and McCann, 1982). These observations are also supported by Blahushova et al. (1984) who reported that their pleomorphic strain of *Trypanosoma brucei brucei* (STIB 247) successfully completed transformation to procyclic trypomastigotes when DNA replication and cell division were blocked by hydroxyurea.

CONCLUSIONS

Insights into the Life Cycle of the Trypanosome

The results of studies presented in this paper support and disprove various aspects of the life cycle of Trypanosoma brucei as previously described. The sequence of transformation in both in vivo and in vitro systems using DFMO is long slender bloodstream form transforming to a short stumpy form via an intermediate form. However, in contrast to the "classical hypothesis" (Robertson, 1912 and 1913; Wijers, 1959; Wijers and Willet, 1960), the data support the existence of a transitional intermediate form that can transform to a short stumpy form or revert back to the long slender bloodstream form (Figure 9). We conclude that the short stumpy form is not infective for the fly, at least in terms of its ability to successfully transform when introduced into SDM-79 medium. Ongoing tsetse infectivity studies will ultimately result in a definitive conclusion. The degenerate nature of the short stumpy form as originally stated by Cunningham et al. (1963) and Venkatesen and Ormerod (1976) is confirmed by these studies, in that the DFMO-induced short stumpy form not only has lost infectivity for the vertebrate host, but also fails to survive in a variety of culture conditions that support the growth of the long slender and intermediate bloodstream forms.

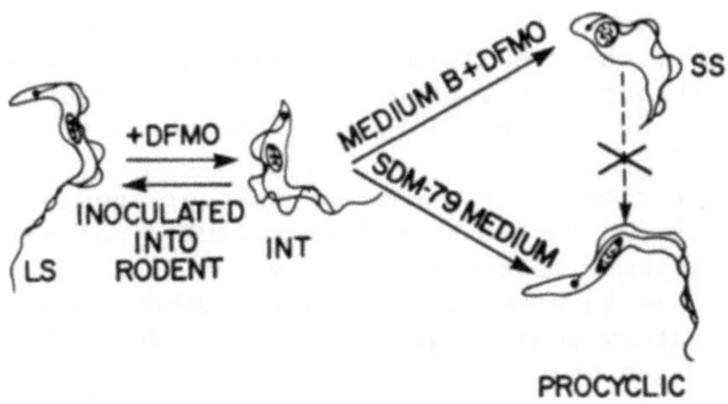

Figure 9. Summary of the DFMO-induced alterations in Trypanosoma brucei brucei. (From Giffin, et al., 1986a).

Both the intermediate and short stumpy forms have activated mitochondria. However, the short stumpy form is degenerate, and unable to transform into a procyclic trypomastigote. This suggests that the transformation from long slender form to short stumpy form may not only be a developmental pathway to produce the vector-infective stage, but may also serve some other function in the vertebrate. The production of a non-viable, degenerate cell type would appear to be a wasteful developmental event. However, if one considers that pleomorphism could be a regulatory mechanism evolved by the trypanosome to insure long-term survival of the host, then it is possible for the production of a degenerate cell type to make good biological sense. When pleomorphic strains are repeatedly subpassaged in the laboratory, there is selection against pleomorphism and these strains become monomorphic.

Under these circumstances the only bloodstream form present throughout the parasitemia is the long slender form trypanosome. Characteristic of these monomorphic strains is a virulent infection that results in the death of the host in a matter of days. This is in contrast to the cyclical remission and recrudescence of parasitemia seen in pleomorphic strains.

From the parasites' point of view, a reasonable strategy is to not kill the host quickly. The phenomenon of antigenic variation, which has permitted the trypanosome to successfully evade the host's immune system, is the mechanism by which the parasite is guaranteed long-term survival in the host. However, in monomorphic strains the trypanosomes proliferate so rapidly that the host is killed before it can mount an immune response to a specific variant antigenic type. One can consider that in pleomorphic strains the production of a degenerate, non-dividing cell type is an adaptation to regulate its population density, conserve the limited glucose stores present in the host's bloodstream, and give the host's immune system time to eliminate the major antigenic variant. Thus, the remission and recrudescence of parasitemia in pleomorphic strains keeps the host alive and permits sufficient time for transmission of the intermediate form to the vector.

The fact that a pertrubation in polyamines following DFMO treatment is capable of inducing a monomorphic strain to become pleomorphic is of interest. Monomorphic strains in the laboratory may have lost their ability to respond quickly to natural signal(s) for transformation, whether produced by the host or the trypanosome population. Perhaps a decreased sensitivity to the signal(s) causes the long slender bloodstream forms to proliferate unchecked and kill the host. However, DFMO, because of its cytostatic effect, regulates the extent of the parsitemia and the host lives longer. Under these circumstances the long slender forms have adequate time to become sensitized to the transformation-stimulating signal(s) and subsequently progress through the sequence of transformation. The degenerate short stumpy forms can then be eliminated by the host's immune system.

At the present time it is not clear whether the increase in doubling time and the associated decrease in intracellular polyamine content which precedes the transformation event is the actual inducer of transformation, or occurs in response to some other signal. Other studies (unpublished data) in our laboratory would suggest that any event that stresses the trypanosome population can result in transformation. It needs to be established whether or not the modulation of intracellular polyamine content is the common mechanism for all transformation-inducing agents or events.

Synchronous Transformation as a Strategy for a Cure

The cure of trypanosomiasis by DFMO may be the result of many alterations in the cell biology of the parasite, which might include a decreased rate of oxygen utilization (Giffin et al., 1986b), alterations in trypanothione metabolism (Fairlamb et al., 1985), or a decreased rate of variant surface glycoprotein synthesis (Bitonti et al., 1988). However, the rapid synchronous induction of transformation is in itself appealing. Both monomorphic and pleomorphic strains are shifted to the non-dividing, degenerate short stumpy form following the depletion of intracellular putrescine. The rapidity of this event is such that the small percentage of long slender forms which would normally have undergone antigenic switching do not, and the entire trypanosome population becomes short stumpy. These non-dividing forms, all of which have the same glycoprotein coat, are then eliminated by the host's immune system.

Effective and economical immunological control of African trypanosomiasis has been investigated for many years but has not been a viable possibility because of antigenic variation. Rational approaches to controlling parasitic diseases depend on our ability to elucidate the significant differences between the biology of the parasite and the host. The transformation from the bloodstream form to procyclic trypomastigote is a crucial step in the life cycle of this parasite about which little is known in terms of its regulation at the molecular level. Consequently continued work is needed for a complete understanding of the biology of transformation. Only then will we be able to take advantage of the novel therapeutic strategy of inhibiting or initiating the transformation process.

REFERENCES

Bacchi, C.J., and McCann, P.P., 1987, Parasitic protozoa and polyamines, in: Inhibition of Polyamine Metabolism, Academic Press, New York.

Bacchi, C.J., Lipschik, G.Y., and Nathan, H.C., 1977, Polyamines in trypanosomatids, J. Bacteriol., 131:657.

Bacchi, C.J., Nathan, H.C., Hutner, S.H., McCann, P.P., and Sjoerdsma, A., 1980, Polyamine metabolism: A potential therapeutic target in trypanosomes, Science, 210:332.

Bacchi, C.J., Garafalo, J., Mochenhaupt, D., McCann, P.P., Diekma, K.A., Pegg, A.E., Nathan, H.C., Mullaney, E.A., Chunosoff, L., Sjoerdsma, A., and Hutner, S.H., 1983, In vivo effects of difluoromethylornithine on the metabolism and morphology of Trypanosoma brucei brucei, Mol. Biochem. Parasitol., 7:209.

Bethell, D.R.H., and Pegg, A.E., 1981, Polyamines are needed for the differentiation of 3T3-L1 fibroblasts into adipose cells, Biochem. Biophys. Res. Comm., 102:272.

Bienen, E.J., Hill, G.C., and Shin, K.-O., 1983, Elaboration of mitochondrial function during Trypanosoma brucei differentiation, Mol. Biochem. Parsitol., 7:75.

Bitonti, A.J., Cross-Doerson, D.E., and McCann, P.P., 1988, Effects of difluoromethylornithine on protein synthesis of the variant-specific glycoprotein (VSG) in Trypanosoma brucei brucei, Biochem. J., 250:295.

Bowman, I.B.R., and Flynn, I.W., 1976, Oxidative metabolism of trypanosomes, in: "Biology of the Kinetoplastidia," Vol. 1, Academic Press, New York.

Brown, R.C., Evans, D.A., and Vickerman, K., 1973, Int. J. Parasitol., 3:691.

Brun, R., and Schonenberger, M., 1979, Cultivation and in vitro cloning of procyclic culture forms of Trypanosoma brucei in a semi-defined medium, Acta Trop., 36:289.

Brun, R., Jenni, L., Schonenberger, M., and Schell, K.F., 1981, In vitro cultivation of bloodstream forms of Trypanosoma brucei, T. rhodesiense and T. gambiense, J. Protozool., 28:470.

Brun, R., Hecker, H., Jenni, L., and Moloo, S.K., 1984, A quantitative ultrastructural study on the transformation of Trypanosoma brucei brucei metacyclic to bloodstream forms in vitro, Acta Tropica, 41:117.

Chen, L., Nav, D., and Liu, A., 1983, Effects of inhibitors of ornithine decarboxylase on the differentiaton of mouse neuroblastoma cells, Cancer Res., 43:2812.

Cross, G.A.M., 1975, Identification, purification and properties of clone-specific glycoprotein antigens constituting the surface coat of Trypanosoma brucei, Parasitol., 71:393.

Cunningham, M.P., Van Hoeve, K., and Lumsden, W.H.R., 1963, Variable infectivity of organisms of the T. brucei subgroup during acute relapsing infections in rats, related to parasitemia, morphology and antibody response. Annual Report of the East African Trypanosomiasis Research Organization, 1962-1963, p. 21.

deGee, A.L.W., Carstens, P.H.B., McCann, P.P., and Mansfield, J.M., 1984, Role of antibody in the elimination of trypanosomes after difluoromethylornithine chemotherapy, J. Parasitol., 69:818.

Erwin, B., Enton, D., Florini, J., and Pegg, A., 1983, Polyamine depletion inhibits the differentiation of L6 myoblast cells, Biochem. Biophys. Res. Comm., 114:944.

Ghiotto, V., Brun, R., Jenni, L., and Hecker, H., 1979, Trypanosoma brucei: morphometric changes and loss of infectivity during transformation of bloodstream forms to procyclic culture forms in vitro, Exp. Parasitol., 48:447.

Giffin, B.F., McCann, P.P., Bitonti, A.J., and Bacchi, C.J., 1986a, Polyamine depletion following exposure to difluormethylornithine both in vivo and in vitro initiates morphological alterations and mitochondrial activation in a monomorphic strain of Trypanosoma brucei brucei, J. Protozool., 33:238.

Giffin, B.F., McCann, P.P., and Bacchi, C.J., 1986a, Effect of putrescine on the respiration of Trypanosoma brucei brucei, Mol. Biochem. Parasitol., 20:165.

Heby, O., Oredsson, S., Olsson, I., and Marton, L., 1983, A role for the polyamines in mouse embryonal carcinoma (F9 and PCC3) cell differentiabut not in human promyelocytic leukemia (HL-60) cell differentiation, Adv. Polyamine Res., 4:727.

Hill, G.C., 1976, Electron transport systems in kinetoplastida, Biochem. Biophys. Acta, 456:149.

Janne, J., Poso, H., and Raina, A., 1978, Polyamines in rapid growth and cancer, Biochem. Biophys. Acta, 473:241.

Kapyako, K., and Janne,J., 1983, Stimulation of melanotic expression in murine melanoma cells exposed to polyamine antimetabolites, Biochem. Biophys. Res. Comm., 113:18.

LeRay, D., Bary, J.D., and Vickerman, K., 1978, Antigenic heterogeneity of metacyclic forms of Trypanosoma brucei, Nature, 273:300.

McCann, P.P., Bacchi, C.J., Clarkson, Jr., A.B., Seed, J.R., Nathan, H.C., Amole, B.O., Hutner, S.H., and Sjoerdsma, A., 1981a, Further studies on difluoromethylornithine in African trypanosomes, Med. Biol., 59-434.

McCann, P.P., Bacchi, C.J., Hanson, W.L., Cain, G.D., Nathan, H.C., Hutner, S.H., and Sjoerdsma, A., 1981b, Effect on parasitic protozoa of difluoromethylornithine - an inhbitor of ornithine decarboxylase, in: "Advances in Polyamine Research," Vol. 3, Raven Press, New York.

McCann, P.P., Bacchi, C.J., Nathan, H.C., and Sjoerdsma, A., 1983, Difluoromethylornithine and the rational development of polyamine antagonists for the cure of protozoan infections, in: "Mechanism of Drug Action," Academic Press, New York.

Nathan, H.C., Bacchi, C.J., Hutner, S.H., Rescigno, D., McCann, P.P., and Sjoerdsma, A., 1981, Antagonism by polyamines of the curative effects of difluoromethylornithine in Trypanosoma brucei brucei infections, Biochem. Pharmacol., 30:3010.

Opperdoes, F.R., and Borst, P., 1977, Localization of nine glycolytic enzymes in a microbody-like organelle in Trypanosoma brucei: the glycosome, FEBS Letters, 80:360.

Ormerod, W.E., 1979, Development of Trypanosoma brucei in the mammalian host, in: "Biology of the Kinetoplastida," Academic Press, New York.

Overath, P., Czichos, J., Stock, U., and Nonnengaesser, C., 1983, Repression of glycoprotein synthesis and release of surface coat during transformation of Trypanosoma brucei, EMBO J., 2:1721.

Pegg, A.E., and McCann, P.P., 1982, Polyamine metabolism and function: A brief review, Am. J. Physiol., 243:C212.

Robertson, M., 1912, Notes on the life history of Trypanosoma gambiense, Proc. Roy. Soc, Series B, 85:66.

Robertson, M., 1913, Notes on the behavior of a polymorphic trypanosome in the bloodstream of the mammalian host, Report of the Sleeping Sickness Commission of the Royal Society, 13:111.

Roelants, G.E., and Pinder, M., 1984, Immunobiology of African trypanosomiasis, Cont. Topics Immunol., 12:225.

Schindler, J., Kelly, M., and McCann, P.P., 1983, Inhibition of ornithine decarboxylase induces embryonal carcinoma cell differentiation, Biochem. Biophys. Res. Comm., 114-410.

Schindler, J., Kelly, M., and McCann, P.P., 1985, The response of several murine embryonal carcinoma cell lines to stimulation of differentiation by difuloromethylornithine. J. Cell Physiol. 122:1.

Sjoerdsma, A., and Scheckter, P.J., 1984, Chemotherapeutic implications of polyamine biosynthesis inhibition, Clin. Pharmacol. Ther., 35-287

Tabor, C.W., and Tabor, H., 1984, Polyamines, Ann. Rev. Biochem. 53:749.

Venkatesan, S., and Ormerod, W.E., 1976, Lipid content of the slender and stumpy forms of Trypanosoma brucei rhodesiense: a comparative study, Comp. Biochem. Physiol., 538:481.

Vickerman, K., 1965, Polymorphism and mitochondrial activity in sleeping sickness trypanosomes, Nature, 20:762.

Wijers, D.J.B., 1959, Polymorphism in Trypanosoma gambiense and Trypanosoma rhodesiense, and the significance of the intermediate forms, Ann. Trop. Med. Parasitol., 53:59.

Wijers, D.J.B., and Willett, K., 1960, Factors that may influence the infection rate of Glossina palpalis with Trypanosoma gambiense II. The number and the morphology of the trypanosomes present in the blood of the host at the time of the infected feed, Ann. Trop. Med. Parasitol., 54:341.

THE ROLE OF GLUTATHIONYLSPERMIDINE AND TRYPANOTHIONE IN REGULATION OF INTRACELLULAR SPERMIDINE LEVELS DURING GROWTH OF *CRITHIDIA FASCICULATA*

Alan H. Fairlamb

Department of Medical Protozoology
London School of Hygiene and Tropical Medicine
London, U.K.

INTRODUCTION

Trypanosomatids are unusual in that a considerable proportion of their intracellular spermidine and glutathione (GSH) is found as the conjugate N^1-glutathionylspermidine (GSH-SPD) and N^1-,N^8-bis-(glutathionyl)spermidine ($T[SH]_2$)[1]. The latter compound is unique to the parasitic and free-living Trypanosomatidae and has been assigned the trivial name trypanothione[2].

These unicellular flagellated organisms contain neither glutathione reductase nor glutathione peroxidase[3], but possess an analogous trypanothione reductase and trypanothione peroxidase system for the maintenance of intracellular thiols in the reduced state and for the removal of hydrogen peroxide[4,5]. Thus, trypanothione plays a central role in the trypanosomatid's defence mechanisms against oxidative damage [Figure 1].

The majority of the intracellular trypanothione is maintained as the dithiol ($T[SH]_2$) by means of the $NADP^+$-dependent enzyme trypanothione reductase (Figure 1, reaction 6). The purified enzyme has a number of physical and chemical properties similar to that of glutathione reductase and therefore is a member of the family of flavoprotein disulphide oxidoreductases that includes glutathione reductase, lipoamide reductase and mercury reductase[6,7]. Other intracellular disulphides such as GSSG can be reduced non-enzymatically by thiol-disulphide exchange with $T[SH]_2$ (Figure 1, reaction 5). Although GSSG is not a substrate for the enzyme, it should be noted that glutathionylspermidine disulphide and the mixed disulphide of glutathionylspermidine and glutathione can serve as alternative substrates to trypanothione disulphide ($T[S]_2$)[8] so that these organisms may not be entirely dependent on $T[SH]_2$ for maintenance of intracellular redox.

The same cannot be said for the removal of hydrogen peroxide and other alkyl peroxides. This reaction is catalyzed by a $T[SH]_2$-dependent enzyme, where GSH or GSH-SPD will not replace the requirement for $T[SH]_2$ (Figure 1, reaction 4)[4]. The enzyme ('trypanothione peroxidase') would appear to be involved in the removal of H_2O_2 produced as a by-product of aerobic metabolism in these organisms. This is of particular interest for chemotherapy, because certain nitrofurans are thought to exert their trypanocidal action through 'futile redox-cycling'[9]. The overall sequence is boxed in Figure 1, in which nitrofurans (RNO_2) undergo enzyme-catalysed 1-electron reduction by NADH or NADPH (reaction 1) to form the RNO^-_2 radical species. This compound is non-enzymatically oxidized (reaction 2) by molecular oxygen to reform RNO_2 and superoxide anion, O_2^-. Superoxide dismutes into H_2O_2 either spontaneously or catalysed by the parasite's Fe-containing superoxide dismutase[10]. Thus, if the rate of reactions 1 and 2 exceeds the organism's capacity to remove O_2^- and H_2O_2 by superoxide dismutase and trypanothione peroxidase, then the highly reactive hydroxyl radical (or a related species) can be formed by the metal catalysed Haber-Weiss reaction, resulting in damage to DNA and membrane lipids.

By modifying the side chain of the nitrofuran drug, nifurtimox, it has proved possible to design 'subversive substrates ' of trypanothione reductase.[11] These 'tailor-made' molecules have a double-edged effect in not only inhibiting trypanothione reductase (Figure 1, reaction 6), but also acting as pseudosubstrates for the enzyme producing O_2^- and H_2O_2 (Figure 1, reactions 1-3). Other trypanocidal agents also have interactions with the overall metabolic scheme in Figure 1. Difluoromethylornithine (DFMO, reaction 9)[12-14], buthionine sulfoximine (BSO, reaction 10)[5,16] and pentamidine and berenil (reaction 8)[17] inhibit key enzymes for the biosynthesis of spermidine and glutathione, the precursors of trypanothione. Trivalent arsenicals are taken up by *T.brucei* and sequester $T[SH]_2$ as the melaminophenyl arsenical adduct, MEL T. Arsenicals and MEL T are also effective inhibitors of trypanothione reductase.[18] It should be emphasised that these are probably not the only actions of these drugs, but

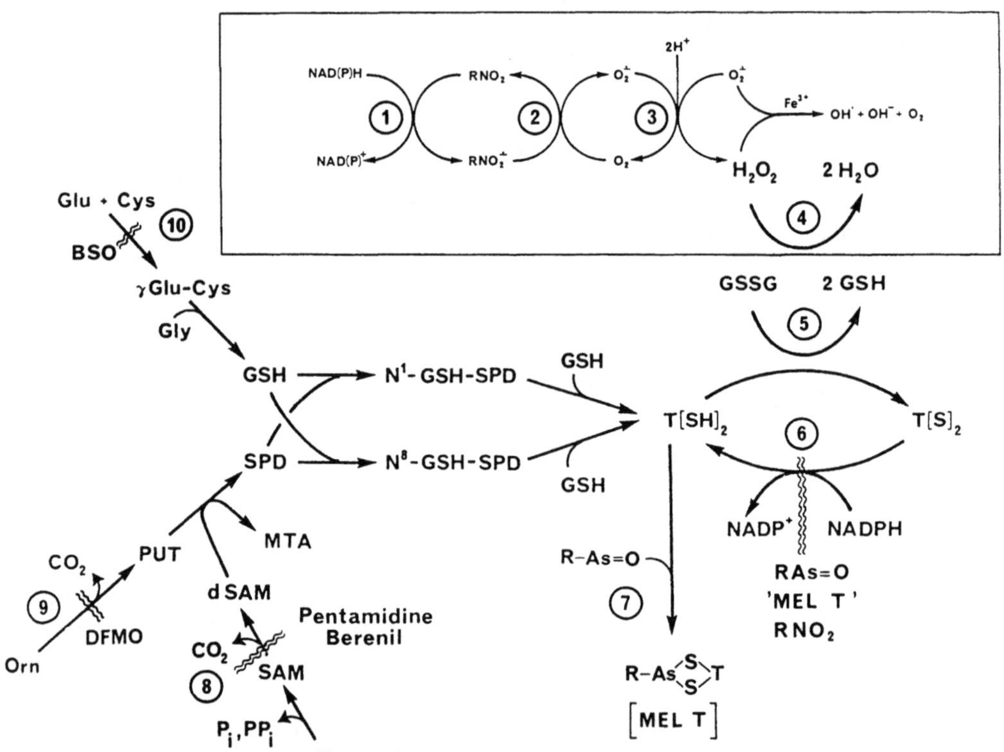

Figure 1. Metabolism and functions of trypanothione, showing the sites of action of some trypanocidal compounds. Futile redox-cycling of nitroheterocyclic compounds is enclosed within a box. See text for the abbreviations used.

Table 1. Levels of glutathione and glutathionylspermidine conjugates in trypanosomatids in exponential phase of growth.

Organism	Total GSH[a] nmol/10^8 cells	%-age GSH recovered as: GSH	GSH-SPD	$T[SH]_2$	Refs
T.brucei bloodstream forms	5.56	24	5	71	13
T.brucei procyclic forms	11.35	27	4	69	14
C.fasciculata	51.1	12	20	68	23
L.b.guyanensis promastigotes	13.7	14	4	82	24
L.donovani promastigotes	15.6	14	4	82	b)

a) Total GSH = [GSH] + [GSH-SPD] + $2[T(SH)_2]$ b) *Howard & Fairlamb,* unpubld

merely offer an hypothetical explanation for the synergistic effect of DFMO with some of these compounds,[19] as well as suggesting new drug combinations that could increase therapeutic efficacy and decrease undesirable host toxicity.

Another potential target for chemotherapeutic attack is the unique biosynthetic pathway from GSH and spermidine to trypanothione. Biosynthesis of trypanothione proceeds via ATP-dependent reactions with GSH-SPD as obligatory intermediate.[1] Purification of the enzymes involved is under investigation. GSH-SPD has previously been identified in *Escherichia coli* under stationary phase and anaerobic conditions and it has been suggested that spermidine or glutathione (or both) could be sequestered as GSH-SPD as a means of regulating cell division.[20,21] When stationary phase *E.coli* are diluted into fresh medium, essentially all of the GSH-SPD is rapidly converted to spermidine and GSH.[20,21] In contrast, *C.fasciculata* contains different amounts of GSH-SPD in all phases of growth.[1,22] In order to understand the function of this metabolite in trypanosomatids, we have examined the changes in thiols and polyamines in response to various growth conditions.

MATERIALS AND METHODS

A cloned isolate of *C.fasciculata* (WTC-HS6) was cultured as described previously.[22] For analysis of polyamines, 5 x 10^7 cells were pelleted by centrifugation, extracted with trichloroacetic acid and the supernatants hydrolyzed with propionic acid/HCl.[23] Components were separated by reversed phase HPLC and detected by post-column derivatization with fluorescamine.[13] Thiols were measured after pre-derivatization with the fluorescent reagent monobromobimane and separation by HPLC.[13]

RESULTS

The levels of glutathione and glutathionylspermidine conjugates from a selection of free-living and parasitic trypanosomatids in exponential phase of growth is given in Table 1. Although the total amount of GSH (free or in peptide linkage to spermidine) varies over a ten-fold range, approximately 70-80 % of the total GSH is found as $T[SH]_2$ in logarithmically

Table 2. Thiol and polyamine levels in *C.fasciculata*

Metabolite Content (nmol per 10^8 cells)	Growth phase		Net Change
	Exponential	Stationary	(S - E)
A. Total putrescine	74.7	35.6	-39.1
B. Total spermidine	57.4	49.0	- 8.4
C. GSH-SPD	10.4	26.2	+15.8
D. $T[SH]_2$	17.4	4.3	-13.1
Total spermidine peptide (C+D)	27.7	30.5	+ 2.8
Free spermidine [B-(C+D)]	29.7	18.5	-11.2

growing cells. In contrast, exponentially growing and stationary phase *C.fasciculata* show marked differences in thiol levels (Table 2). Note that $T[SH]_2$ is the principal intracellular thiol in exponentially growing cells, whereas GSH-SPD is the major thiol in stationary phase. Both total putrescine and total spermidine levels are decreased in non-dividing cells and the amount of spermidine covalently linked to GSH increases from 48 % to 62 % of the total spermidine on cessation of growth. From these data it can be calculated that the level of unconjugated or "free" spermidine decreases by 38 % (11.2 nmol/10^8 cells) in stationary phase.

Examination of the thiol content throughout the growth curve (Figure 2) shows that although the total GSH decreases progressively from early to late exponential phase (51 to 40 nmol/10^8 cells), the percentage distribution of glutathione as GSH, GSH-SPD and $T[SH]_2$ remains essentially constant. The transition from $T[SH]_2$ to GSH-SPD as principal thiol (Figure 2, left panel) coincides with i) entry into stationary phase and ii) a decrease in the pH of the medium to below 5 (Figure 2, right panel).

The effect of restoring favourable growth conditions was examined by diluting stationary phase organisms into fresh medium (Figure 3). Reversal of the percentage distribution of GSH in GSH-SPD and $T[SH]_2$ can be detected

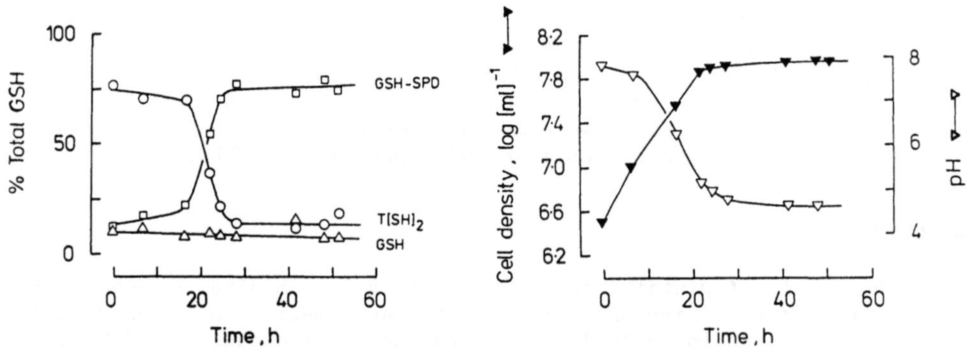

Figure 2. Percentage distribution of glutathione and glutathionylspermidine conjugates during growth of *C.fasciculata*. Reproduced from [22].

within 5 minutes of dilution and is essentially complete within 1 hour (Figure 3a). Since the total GSH and spermidine is unchanged during this interval (Figure 3b), it can be concluded this effect precedes *de novo* synthesis of these metabolites and must therefore result from a redistribution of existing GSH-SPD and $T[SH]_2$. Since the amount of free GSH does not change, then the following stoichiometry must apply:

$$2\ \text{GSH-SPD} \longrightarrow T[SH]_2 + \text{spermidine}$$

In this experiment, the levels of peptide-bound spermidine at zero time and after 60 min were 31 and 18 nmol/10^8 cells, respectively, indicating a net production of 13 nmol/10^8 cells of free spermidine, in good agreement with the data presented in Table 2 (11.2 nmol/10^8 cells).

The role of pH in regulation of GSH-SPD and $T[SH]_2$ was examined by diluting stationary phase cells into medium of different pH (Table 3). Fresh medium adjusted to pH 5 results in full recovery of $T[SH]_2$ levels as found for fresh medium at pH 8. Thus pH *per se* is not the regulatory trigger. Nonetheless, when spent medium is adjusted to pH 8, an intermediate pattern between that found for exponential phase and stationary phase is observed. Supplementing spent medium with glucose (10 mM) and one or all of the amino acid precursors (glutamate, cysteine, methionine and ornithine 0.1 mM) failed to reverse the GSH-SPD:$T[SH]_2$ ratio (not shown).

DISCUSSION

From these results, it can be concluded that the level of free spermidine in *C.fasciculata* is decreased by one third in stationary phase cells by conversion to GSH-SPD. This finding is similar to that found previously for *E.coli*[20,21], although it should be emphasised that $T[SH]_2$ is not present in these bacteria. The significance of the pronounced decrease in $T[SH]_2$ levels in stationary phase cells is not clear at present. It could simply serve as a source of GSH for formation of GSH-SPD, but it is also possible that the decrease in $T[SH]_2$ also acts as a regulatory signal in some other unknown manner. Whether GSH-SPD accumulation is the signal for these cells to cease growth or is merely a secondary effect is not understood. However, it is clear that, on restoration of growth, GSH-SPD is converted to $T[SH]_2$, rapidly elevating free spermidine prior to *de novo* synthesis of polyamines. Possibly this mechanism allows macromolecular

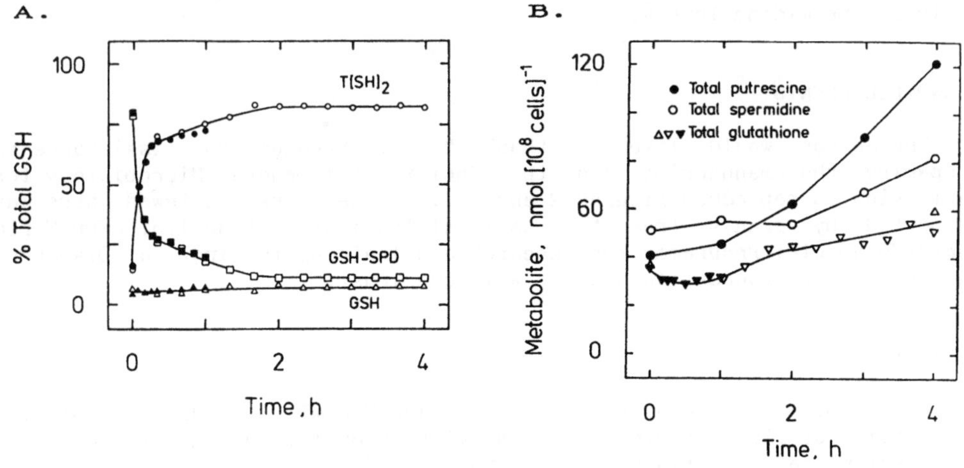

Figure 3. Effect of diluting stationary phase cells into fresh medium. Reproduced from [22].

Table 3. Effect of pH of medium on thiol levels in *C.fasciculata*.

Treatment[a]	pH	Total GSH (nmol/10^8 cells)	Percentage GSH recovered as: GSH	GSH-SPD	$T[SH]_2$
Spent medium	5	36	5	83	12
Spent medium[b]	8	42	8	34	58
Fresh medium	8	68	10	22	68
Fresh medium[c]	5	67	8	23	69

[a] Stationary phase cultures were diluted 10-fold and incubated for 4 h before analysis.
[b] Spent medium adjusted to pH 8 with 5M NaOH.
[c] Fresh medium adjusted to pH 5 with 5M HCl.

biosynthesis to be initiated immediately, rather than awaiting induction of ornithine decarboxylase and *de novo* synthesis of spermidine. This may be relevant to the significant observation that ornithine decarboxylase of a related trypanosomatid *T.brucei* differs strikingly from its mammalian counterpart in that it is not subject to regulation by rapid turnover of the enzyme.[25] The factors responsible for regulation of growth in these organisms remain unclear, although simple deficiency of the amino acid precursors of trypanothione or limiting energy supply can be ruled out. Whether other unicellular prokaryotic and eukaryotic organisms use similar mechanisms for sequestration of spermidine in stationary phase deserves investigation.

CONCLUSIONS

Although studies in the role of glutathionylspermidine and trypanothione are still in their infancy, our curent knowledge points to a pivotal role for these peptides in three key areas: regulation of intracellular thiol redox; defence against oxidative stress; and regulation of intracellular spermidine levels.

ACKNOWLEDGEMENTS

The author would like to thank Ms C. Strobos for assistance in preparing the manuscript and the Society for General Microbiology for permission to reproduce Figures 2 and 3.. The work reviewed here was supported by grants from the National Institutes of Health, UNDP/World Bank/WHO Special Programme for Research and Training in Tropical Diseases, The Rockefeller Foundation and the Wellcome Trust.

REFERENCES

1. Fairlamb, A. H., Henderson, G. B., and Cerami, A., 1986, The biosynthesis of trypanothione and N^1-glutathionylspermidine in Crithidia fasciculata, Mol. Biochem. Parasitol., 21:247-257.
2. Fairlamb, A. H., Blackburn, P., Ulrich, P., Chait, B. T., and Cerami, A., 1985, Trypanothione: A novel bis(glutathionyl)spermidine cofactor

for glutathione reductase in trypanosomatids, Science, 227:1485-1487.
3. Fairlamb, A. H., and Cerami, A., 1985, Identification of a novel, thiol-containing co-factor essential for glutathione reductase enzyme activity in trypanosomtids, Mol. Biochem. Parasitol., 14:187-198.
4. Henderson, G. B., Fairlamb, A. H., and Cerami, A., 1987, Trypanothione dependent peroxide metabolism in Crithidia fasciculata and Trypanosoma brucei, Mol. Biochem. Parasitol., 24:39-45.
5. Fairlamb, A. H., and Henderson, G. B., 1987, Metabolism of trypanothione and glutathionylspermidine in trypanosomatids, in: "Host-Parasite Molecular Recognition and Interaction in Protozoal Infections," K.-P. Chang and D. Snary, eds., pp. 29-40, NATO-ASI Series, Heidelberg.
6. Shames, S. L., Fairlamb, A. H., Cerami, A., and Walsh, C. T., 1986, Purification and characterization of trypanothione reductase from Crithidia fasciculata, a newly discovered member of the family of disulfide-containing flavoprotein reductases, Biochemistry, 25:3519-3526.
7. Krauth-Siegel, R. L., Enders, B., Henderson, G. B., Fairlamb, A. H., and Schirmer, R. H., 1987, Trypanothione reductase from Trypanosoma cruzi: Purification and characterization of the crystalline enzyme, Eur. J. Biochem., 164:123-128.
8. Henderson, G. B., Fairlamb, A. H., Ulrich, P., and Cerami, A., 1987, Substrate specificity of the flavoprotein trypanothione disulphide reductase from Crithidia fasciculata, Biochemistry, 26:3023-3027.
9. Docampo, R., and Moreno, S. N. J., 1986, Free radical metabolism of antiparasitic agents, Fed. Proc., 45:2471-2476.
10. Le Trang, N., Meshnick, S. R., Kitchener, K., Eaton, J. W., and Cerami, A., 1983, Iron-containing superoxide dismutase from Crithidia fasciculata, J. Biol. Chem., 258:125-130.
11. Henderson, G.B., Ulrich, P.U., Fairlamb, A.H., Rosenberg, I., Pereira, M., Sela, M. and Cerami, A. (1988) "Subversive substrates" for the enzyme trypanothione disulfide reductase: A new approach to chemotherapy of Chagas' disease. Proc. Natl. Acad. Sci., USA, in press.
12. Bacchi, C.J., Nathan, H.C. and Hutner, S.H., 1980, Polyamine metabolism: A potential therapeutic target in trypanosomes, Science, 210: 332-334.
13. Fairlamb, A.H., Henderson, G.B., Bacchi, C.J., and Cerami, A., 1987, In vivo effects of difluoromethylornithine on trypanothione and polyamine levels in bloodstream forms of Trypanosoma brucei, Mol. Biochem. Parasitol., 24:185-191.
14. Bellofatto, V., Fairlamb, A.H., Henderson, G.B., and Cross, G.A.M., 1987, Biochemical changes associated with alpha-difluoromethylornithine uptake and resistance in Trypanosoma brucei, Mol. Biochem. Parasitol., 25:227-238.
15. Griffith, O.W. and Meister, A., 1979, Potent and specific inhibition of glutathione synthesis by buthionine sulphoximine (S-n-butyl homocysteine sulphoximine), J.Biol.Chem., 254:7558-7560.
16. Arrick, B.A., Griffith, O.W. and Cerami, A., 1981, Inhibition of glutathione synthesis as a chemotherapeutic strategy for trypanosomiasis, J. Exp. Med., 153:720-725.
17. Bitonti, A.J., Dumont, J.A. and McCann, P.P., 1986, Characterization of Trypanosoma brucei brucei S-adenosyl-L-methionine decarboxylase and its inhibition by Berenil, pentamidine and methylglyoxal bis(guanylhydrazone), Biochem. J., 237:685-689.
18. Fairlamb, A.H., Henderson, G.B. and Cerami, A., 1988, Complex-formation of dihydrotrypanothione with the trivalent arsenical drug, melarsen oxide, and its effect as an inhibitor of trypanothione reductase in Trypanosoma brucei, submitted for publication.
19. McCann, P.P., Bacchi, C.J., Nathan, H.C. and Sjoerdsma, A., 1983, Difluoromethylornithine and the rational development of polyamine antagonists for the cure of protozoan infections in: "Mechanisms of

Drug Action," Singer, T.P. and Ondarza, R.N., eds., Academic Press, New York.

20. Tabor, H., and Tabor, C. E., 1975, Isolation, characterization, and turnover of glutathionylspermidine from Escherichia coli, J. Biol. Chem., 250:2648-2654.
21. Tabor, C.E. and Tabor, H., 1970, The complete conversion of spermidine to a peptide derivative in Escherichia coli. Biochem. Biophys. Res. Comm., 41:232-238.
22. Shim, H., and Fairlamb, A. H., 1988, Levels of polyamines, glutathione and glutathione-spermidine conjugates during growth of the insect trypanosomatid Crithidia fasciculata, J. Gen. Microbiol., 134:807-817.
23. Westhall, F. and Hesser, H., 1974, Fifteen-minute acid hydrolysis of peptides. Anal. Biochem., 61:610-613.
24. Keithly, J.S. and Fairlamb, A.H., 1988, Inhibition of Leishmania species by α-difluoromethylornithine in: "NATO-ASI Series: The First Centenary (1885-1985) New Strategies for Control," Hart, D.T., ed., in press.
25. Phillips, M.A., Coffino, P. and Wang, C.C., 1988, Cloning and sequencing of the ornithine decarboxylase gene from Trypanosoma brucei: Implications for enzyme turnover and selective difluoromethylornithine inhibition, J.Biol.Chem., 262:8721-8727

INHIBITORS OF POLYAMINE
BIOSYNTHESIS IN THERAPY

REGULATION OF POLYAMINE BIOSYNTHETIC ACTIVITY BY SPERMIDINE AND SPERMINE ANALOGS --- A NOVEL ANTIPROLIFERATIVE STRATEGY

Carl W. Porter and Raymond J. Bergeron

Grace Cancer Drug Center
Roswell Park Memorial Institute
Buffalo, New York 14263 (CWP)
and
Department of Medicinal Chemistry
J. Hillis Miller Health Center
University of Florida
Gainesville, Florida 32610 (RJB)

Introduction

Our current rationale for targeting polyamines as a mechanism-based drug discovery initiative is based on several diverse yet interrelated considerations (Table 1) some of which derive from findings to be reviewed here. Of these, the most critical are (a) the strict dependence of proliferating cells on polyamine availability (b) the increased rate of polyamine biosynthesis and transport associated with neoplastic tissues relative to their normal counterparts and (c) the novelty of polyamine biosynthesis and/or function as a departure from more cytotoxic DNA-directed approaches. While admittedly, an absolute basis for selectivity has not yet been defined, recognition must be given to the fact that no chemotherapeutic strategy described to date is wholly selective for tumor tissue and yet total curability of a number of malignant diseases can be achieved. We believe this justifies the evaluation of new and novel approaches even in the absence of absolute metabolic rationale for selective tumor cell kill. Although still confined to animal model systems, encouraging antitumor data with polyamine analogs lend credence to this belief (Bergeron et al., 1988).

Although inviting as a target for chemotherapeutic intervention (Pegg, 1988), the polyamine biosynthetic pathway is inherently resistant to the use of single enzyme inhibitors by virtue of several unique properties. All enzymes, and particularly ornithine decarboxylase (ODC) and S-adenosylmethionine decarboxylase (AdoMetDC), turn over very rapidly so that even irreversible inhibitors require a sustained presence in order to register a significant impact on the pathway and, subsequently, on cell proliferation. In addition and perhaps more importantly, the key rate limiting enzymes, ODC and AdoMetDC, are highly regulated by intracellular polyamine pools. Through mechanisms involving selective protein translation (Holtta et al., 1986; Kahana and Nathans, 1985; Kameji and Pegg, 1987; Persson et al., 1988) and degradation (Murakami and Hayashi, 1986), enzyme activity is maintained in a

Table 1. Rationale for Targeting Polyamine Biosynthesis and/or Function as an Anticancer Strategy

Specific inhibitors of polyamine biosynthesis inhibit neoplastic cell growth. (Mamont et al., 1978; Tang et al., 1980; Pegg et al., 1988; Kramer et al., 1988).
Polyamines and polyamine biosynthetic activity are elevated in neoplastic tissues over surrounding normal tissues even when the latter are, themselves, rapidly proliferating (Kingsnorth et al., 1984; Luk et al., 1984; Porter et al., 1987b; Rhozin et al., 1984).
Polyamine uptake, and presumably polyamine analog uptake, is greater in proliferating than in non-proliferating tissues (Pohanpelto, 1976; Porter and Sufrin, 1986).
Polyamine analogs and inhibitors are not overtly cytotoxic to most normal and neoplastic cell types but rather induce a gradual slowing of cell growth (Fig. 1). Host toxicity, therefore, may be minimized or controlled.
Certain tumor types (i.e. human lung carcinomas and melanoma) are uniquely sensitive to polyamine analogs and inhibitors and exhibit a distinctly cytotoxic response to them (Casero et al., 1987a,b).
Polyamine biosynthetic enzymes are unique to that pathway and there is no alternative route for polyamine biosynthesis in mammalian cells (Pegg, 1986).
Associations between sensitivity to polyamine inhibitors and analogs and oncogene expression have been identified (Chang et al., 1988; Celano et al., 1988).
At least one enzyme of polyamine metabolism (5'-deoxy-S-methylthioadenosine phosphorylase) is absent in a significant proportion of certain malignant cell types (Fitchen et al., 1987; Traweek et al., 1988; Yamanaka et al., 1987).
Polyamine analogs which suppress polyamine biosynthesis demonstrate significant _in vitro_ and _in vivo_ antitumor activity (Bergeron et al., 1988).

depressed state by intracellular polyamine pools. Thus, perturbations which decrease those pools, such as enzyme inhibition, result in substantial increases in one or both enzyme activities -- ODC inhibitors increase AdoMetDC activity (Mamont et al., 1981; Poso and Pegg, 1981); AdoMetDC inhibitors increase ODC activity (Pegg et al., 1988; Kramer et al., 1988) and synthase inhibitors increase either or both enzyme activities (Pegg, 1986; Pegg et al., 1987). While under cell culture conditions, sufficient amounts of inhibitor can be used to sustain a pathway blockade, it is likely, in whole animal systems, that the effectiveness of such inhibitors will be markedly compromised by these regulatory responses. One way by which these problems could be effectively countered is by the concomitant use of two enzyme inhibitors (Pegg et al., 1988; Kramer et al., 1988) or by the use of a single inhibitor with dual specificity towards both ODC and AdoMetDC.

Table 2. Polyamine Analogs with Potentially Useful Enzyme Regulatory Properties

Acronym*	Structure	IC_{50}** (uM)
BEP	$CH_3CH_2NH(CH_2)_4NHCH_2CH_3$	<1000
BES	$CH_3CH_2NH(CH_2)_3NH(CH_2)_4NHCH_2CH_3$	50-100
BESm	$CH_3CH_2NH(CH_2)_3NH(CH_2)_4NH(CH_2)_3NHCH_2CH_3$	5-10
BEHSm	$CH_3CH_2NH(CH_2)_4NH(CH_2)_4NH(CH_2)_4NHCH_2CH_3$	1-5
6-Spermyne	$NH_2(CH_2)_3NHCH_2C{\equiv}CCH_2NH(CH_2)_3NH_2$	<300

*Abbreviations include BEP, N^1,N^4,bis(ethyl)putrescine; BES, N^1,N^8-bis(ethyl)spermidine; BESm, N^1,N^{12}-bis(ethyl)spermine and BEHSm, N^1,N^{14}-bis(ethyl)homospermine

**Against cultured L1210 cells.

In recognition of these problems and in search of an alternative approach to the use of enzyme inhibitors, we have examined the usefulness of polyamine analogs. Although such compounds have obvious potential for affecting polyamine function at sites distal to the pathway itself, our findings to date indicate that a significant, if not total proportion of the cellular impact of one class of analogs involves interference with polyamine biosynthesis. The remainder of this review will focus on these effects and the potential of polyamine analogs as an antiproliferative strategy. A more detailed review of the polyamine regulatory responses as they relate to this approach can be found elsewhere (Porter and Bergeron, 1988).

Regulatory Approach

The regulatory mechanisms which control ODC and AdoMetDC activities in cells are apparently designed to maintain steady-state polyamine pool sizes since, in addition to responding to decreases in polyamine pools, they also respond to increases in polyamine pools. Thus, as has been recognized for some time (Kay and Lindsay, 1973; Clark and Fuller, 1975), exposure of cells to exogenous polyamines rapidly and dramatically lowers ODC and AdoMetDC activities (Porter et al., 1988). Despite the loss in biosynthetic activity, cells continue to proliferate by using the exogenously provided polyamines in place of those which they would otherwise synthesize.

We have sought to exploit the regulatory mechanisms inherent in the polyamine pathway to a chemotherapeutic advantage by identifying polyamine analogs which can substitute for the natural polyamines in negatively regulating key biosynthetic enzyme activities but which are incapable of substituting for cellular functions required for cell proliferation. Thus, by suppressing polyamine biosynthesis, the analogs should eventually bring about polyamine depletion and concomitant inhibition of cell growth. To our knowledge this approach is novel among antimetabolite strategies and is made feasible by unique aspects of the polyamine pathway such as (a) the nature of the regulatory controls (b) the rapid turnover of ODC and AdoMetDC, (c) the segregation of determinants for enzyme regulation and growth-dependent functions on the polyamine and (d) the direct dependence of cell growth on polyamine biosynthetic activity.

SPD Analogs as Regulators of Enzyme Activity.

From a relatively large series of systematically modified N-acyl and N-alkyl SPD (spermidine) analogs, detailed structure-function relationships were delineated which gave rise to a strategy for carrying out our objectives (Porter et al., 1985). Specifically, it was determined that structural determinants for growth-related functions were partially segregated on the SPD molecule from those involved in enzyme regulation. The central amine appears to be more critical to enzyme regulation than the terminal amines while the reciprocal relationship obtains for uptake and growth-related functions (Porter and Sufrin, 1986). From the perspective of our interests, this was a fortuitous finding indeed, since if the enzyme regulation and growth functions were similarly dependent on the same molecular determinants, the regulatory strategy would not have been feasible. It should be emphasized, however, that these relationships are relative so that modification of a given amine(s) does not fully abolish cellular functions related to that amine and, conversely, is not without consequence to functions related to other amines.

Consistent with the above findings, N^1,N^8-bis(ethyl)spermidine (BES, Table 2) emerged as the prototypic compound for the regulatory strategy (Porter et al., 1985). At its approximate IC_{50} concentration (~50-100 µM), it suppressed ODC activity, depleted cells of putrescine (PUT), SPD and SPM (spermine) and was only partially effective in supporting cell growth in the presence of the ODC inhibitor, α-difluoromethylornithine (DFMO; Mamont et al., 1978) -- an assay which was developed for evaluating the ability of an analog to substitute for SPD in functions related to cell growth (Porter and Bergeron, 1983). Further detailed biochemical studies comparing the cellular effects of BES, a regulator of ODC, with DFMO, an inhibitor of the enzyme, revealed close similarities in the kinetics of growth inhibition and polyamine pool changes, cell cycle perturbations, precursor incorporation effects and so on (Porter et al., 1986) indicating the that, as with DFMO, growth inhibition by BES was probably related to SPD depletion secondary to drug interference with ODC. A more definitive demonstration of this relationship was not possible since, in either product prevention or reversal studies, exogenous SPD would undoubtedly diminish the intracellular accumulation of BES and render such experiments uninterpretable. While sharing certain cellular effects with DFMO, BES offered several advantages over the inhibitor. As a polyamine analog, it penetrated cells more effectively than DFMO and inhibited cell growth at significantly lower IC_{50} concentrations (i.e. 50-100 uM versus 1-5 mM for DFMO). In addition, it did not elicit a typical compensatory increase in AdoMetDC activity which typically accompanies polyamine depletion. As a possible consequence of this, BES accomplishes a decrease in SPM pools as opposed to an increase caused by DFMO. Finally, the depletion in polyamine pools was not accompanied by an increase in polyamine uptake as occurs with DFMO (Alhonen-Hongisto

Table 3. Relative Effects of DFMO and Various Polyamines and Polyamine Analogs on L1210 Cell Polyamine Enzymes and Pools. (Data expresssed as percent of control).

Treatment	Decarboxylase Activity		Polyamines			SSAT Activity
(48 hr)	ODC	AdoMetDC	PUT	SPD	SPM	(at 24 hr)
1 mM DFMO	10	310	nd	nd	135	70
1 mM BEP	65	118	nd	60	92	90
100 μM BES	2	87	nd	nd	56	1300
10 μM BESm	1	11	nd	nd	27	1500
1 μM BEHSm	3	10	nd	nd	61	300
10 μM SPM	9	12	nd	12	210	50
10 μM 6-Spermyne	95	10	130	26	11	200

et al., 1980) -- a phenomenon which may be of little consequence under cell culture conditions but which may significantly compromise drug effectiveness in animal systems.

SPM Analogs as Regulators of Enzyme Activity

The near concordance of BES-induced polyamine effects with those proposed hypothetically for an ideal regulatory analog led to the synthesis and study of comparable PUT and SPM analogs and homologs (Table 2, Porter et al., 1987c; Bergeron et al., 1988). In a comparative study of the cellular effects of BES with those the SPM analog, N^1,N^{12}-bis (ethyl)spermine (BESm), it was found that, while the two shared certain regulatory mechanisms, they differed distinctly in their enzyme specificity. Whereas BES lowered ODC activity and, at the same time, prevented AdoMetDC activity from increasing as a result of polyamine depletion, BESm dramatically lowered both ODC and AdoMetDC activities (Table 3, Porter et al., 1987c). In doing so, SPM pools were decreased by approximately 70% bringing about the most total depletion of all polyamines achieved thus far by any single agent. By contrast, the bis(ethyl) derivative of PUT was completely ineffective in regulating either enzyme (Table 3).

Polyamine Depletion and Growth Inhibition

To further evaluate the cellular effects of BESm and the relationship of polyamine depletion to growth inhibition, we have recently completed a detailed study of the cellular effects of the combination of DFMO, with S-(5'-deoxy-5'-adenosyl)-methylthioethylhydroxylamine (AMA), an irreversible inhibitor of AdoMet decarboxylase synthesized by the Khomutovs (Khomutov et al., 1983; Artamonova et al., 1986). Interestingly, it was determined (Kramer et al., 1988) that, while either agent can inhibit polyamine biosynthesis, DFMO elicited a more cytotoxic response in L1210 cells than AMA, and that the combination of the the inhibitors resulted in additive growth inhibition. A further analysis of this effect indicated that the extent of growth inhibition was closely dependent on the amount of intracellular polyamines remaining after pathway blockade and that, although only SPD could fully support growth at control levels, both PUT and SPM were capable of contributing to the process.

Thus, maximum growth inhibition is achieved by maximum polyamine depletion, and in L1210 cells at least, this is best accomplished by the combined inhibition of both ODC and AdoMetDC. In order to evaluate the relationship of polyamine depletion by BESm with growth inhibition, we compared the kinetics of the latter with the combination of DFMO and AMA (Fig. 1) at concentrations which gave an approximate 50% growth inhibition at 48 hr. In extended incubations (i.e. 120 hr) growth inhibition by 10 µM BESm was more similar to that by DFMO alone than by the combination even though polyamine depletion was comparable to that achieved by the latter. Although the reason(s) for this apparent inconsistency is still unclear, we are inclined to attribute it to the ability of BESm to partially substitute for the natural polyamines supporting cell growth. Accordingly, the extent of this substitution may be similar to that achieved by the residual SPM pool in DFMO-treated cells.

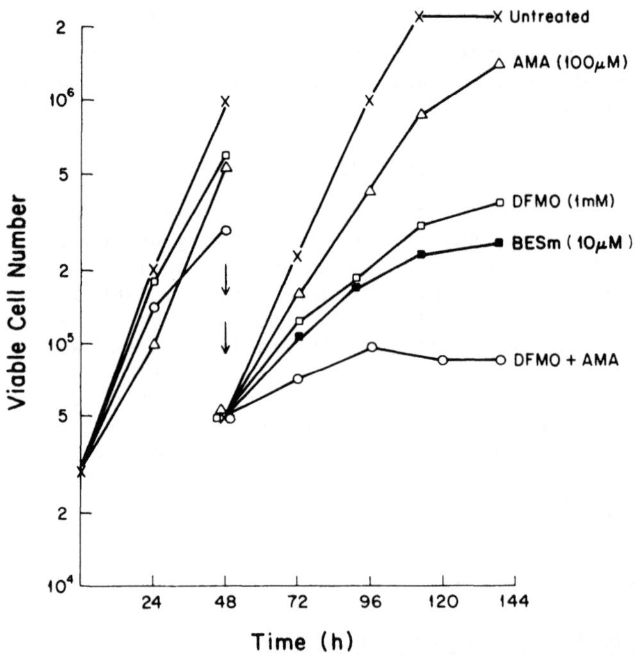

Fig. 1. A comparison of long term growth inhibition by BESm, DFMO, AMA and the combination of DFMO and AMA. Individual drugs were used at concentrations which inhibited cell growth by approximately 50% at 48 hr after which, cells were reseeded and treatment continued for an additional 96 hr.

It is not evident how the ability of regulatory analogs to substitute for natural polyamines in growth-related functions might be further reduced. As indicated by polyamine prevention studies with DFMO and AMA, SPD is apparently the obligatory polyamine for growth of L1210 cells. Thus, SPM analogs, such as BESm, should be less effective in supporting growth under conditions of polyamine depletion than SPD analogs, such as BES. This might partially contribute to the greater effectiveness of the former over the latter in inhibiting cell growth. For the same purposes, we have also determined that the size of the N-alkyl group is optimized at 2 carbons since bis(methyl) and bis-(propyl) analogs of SPD or SPM are both less effective (Porter et al., 1985; Bergeron et al., 1988). Preliminary indications are that bis-(methyl) analogs substitute more readily for the polyamines than bis-(ethyl) analogs and bis(propyl) analogs enter cells less effectively than bis(ethyl) analogs. One possibility for decreasing polyamine substitution activities that is currently under study involves altering in aliphatic chain lengths separating the amines.

Mechanism(s) of Enzyme Regulation by Analogs

To date, the preponderance of evidence indicates that regulation of decarboxylase activities by exogenous natural polyamines occurs at the level of translation (Kahana and Nathans, 1985; Dirks et al., 1986; Sertich and Pegg, 1987; Holtta et al., 1986; Persson et al., 1986; Katz and Kahana, 1987) and/or, in the case of ODC, by post-translational mechanisms leading to enhanced degradation (Heller et al., 1976; McCann et al., 1979; Seely and Pegg, 1983; Murakami and Hayashi, 1986). From the outset it was presumed that both BES and BESm lowered ODC or ODC and AdoMetDC activities by the same mechanisms utilized by the natural polyamines. To the extent that their effects have now been characterized, this now seems to be the case (Porter et al., 1987a). As with the natural polyamines, the analogs exert no direct inhibitory effects at concentrations up to 1 mM. At 3 μM, the kinetics of ODC suppression by SPD or BES are nearly identical and reach a maximum by less than 2 hr. As an aside, it is relevant to note that enzyme suppression is rapidly reversible so that upon removal of BESm, ODC activity fully recovers in 2 hrs (Porter et al., 1987a) and AdoMetDC, in 12 hr (unpublished data). Thus, another way by which analog antiproliferative activity might be improved is by increasing their binding affinity for sites involved in regulating the enzymes.

The rapid decline in enzyme activity is accompanied by a concomitant decrease in ODC protein but not in ODC mRNA (Porter et al., 1987a) indicating control at the levels of translation or post-translational modification. In unpublished studies with BESm, identical conclusions were also reached regarding control of AdoMetDC activity. Perhaps the best evidence for regulation at the level of translation has been obtained with *in vitro* translation systems using filtered reticulocyte lysates. Kameji and Pegg, (1987) and Persson *et al*. (1988) have convincingly demonstrated that the natural polyamines can selectively interfere with the translation of ODC and AdoMetDC protein. Recently, Pegg et al., (1988) have shown that this is also the case for BES and BESm. Interestingly, although BES can suppress translation of both ODC and AdoMetDC, much higher analog concentrations are required for the latter suggesting that the apparent selective effects of the analog on ODC in intact cells may not be strictly structure-dependent but also dose dependent--a phenomenon which, in turn, could be controlled by the cellular uptake of the analog.

Further indication that ODC and AdoMetDC might be differentially regulated by structural determinants of the polyamine or polyamine analog molecules was recently obtained with an unsaturated analog of SPM, 6-spermyne (Table 2; Porter et al., 1988). Whereas treatment of L1210 cells with either SPM or BESm for 48 hrs suppressed both ODC and AdoMetDC activities, identical treatment with 6-spermyne decreased only AdoMetDC. The effect was accompanied by a shift in polyamine pools which were consistent with the changes in enzyme activity (Table 3) -- PUT increased slightly while SPD and SPM decreased in a manner similar to that seen in cells treated with an AdoMetDC inhibitor (Pegg et al., 1988; Kramer et al., 1988). Since there was no evidence for direct inhibition of either ODC or AdoMetDC by 6-spermyne, it was concluded that enzyme suppression via the same regulatory events utilized by SPM itself was involved.

An analysis of the temporal development of the enzyme effect of 6-spermyne on decarboxylase activities unexpectedly revealed that, in similarity to SPM, there was an initial decline in ODC activity at 1-2 hr which, with time, eventually returned to control levels. Since AdoMetDC activity decreased in parallel with that in cells treated with SPM and remained similarly suppressed over the entire 48 hr, degradation of 6-spermyne was excluded as the reason for the ODC recovery. This was further supported by chromatographic evidence. Rather, on the basis of polyamine pool analysis, it was concluded that the initial decrease in ODC activity was due to a displacement by the analog of intracellular polyamines from natural binding sites. These, in turn, regulated ODC activity so long as they were present in sufficient amounts within the cell. With time, however, the effects of increasing amounts of 6-spermyne on AdoMet decarboxylase produced a steady decline in SPD and SPM pools allowed for recovery of ODC activity. Thus, evidence seems to indicate that, by virtue of the triple bond in its PUT moiety, 6-spermyne is rendered incapable of regulating ODC activity and hence selective in this regard for AdoMetDC activity.

At a minimum, our recent results with the analog provide strong indication that ODC and AdoMetDC are regulated by two distinct and functionally separable sites. This interesting structure-function relationship may be exploited in designing polyamine analogs with anti-proliferative potential. 6-Spermyne was synthesized for other purposes and is not growth inhibitory (Table 2). Although it markedly depletes SPD and SPM pools, it is probably very capable of substituting for them in growth-related functions. From previous studies (Porter et al., 1985; Porter et al., 1987), it would seem that bis N-alkylation of 6-spermyne would interfere with this activity and render it growth inhibitory. Beyond this possibility, the analog itself should find usefulness in dissecting out mechanisms involved in the regulation of ODC and AdoMetDC. It would be interesting, for example, to determine whether 6-spermyne interferes with the translation of AdoMetDC but not ODC in the reticulocyte lysate translation system (Kameji and Pegg, 1987; Persson et al., 1988 and Pegg et al., 1988) thus increasing the likelihood that these acellular findings are actually representative of cellular events.

Other Polyamine Effects by Analogs

Erwin and Pegg (1986) reported that in L6 cells depleted of PUT and SPD by DFMO, the SPD analog, BES, produces a very substantial rise in SPD/SPM acetyltransferase (SSAT) activity. The enzyme catalyzes the transfer of an acetyl group from acetyl coenzyme A to the aminopropyl primary amine of SPD and SPM and is probably best known for its role in the metabolic interconversion of polyamines (Bolkenius and Seiler, 1981; Pegg et al., 1981; Mamont et al., 1981; Seiler, 1987). A second enzyme of the pathway, polyamine oxidase, eliminates the acetylated aminopropyl moiety thus catalyzing the formation of SPD from SPM and PUT from SPD. Although the physiological significance of this pathway is uncertain since its inhibition is of negligible consequence to cell growth (Claverie et al., 1987), there is increasing indication that it may play a role in the transport of polyamines from cells (Wallace, 1987; Seiler, 1987). Since, in the case of the polyamine analogs such an effect could contribute to their lowering of polyamine pools, we have taken an interest in examining the relationship of SSAT induction and polyamine depletion. Dr. Libby of our laboratory has recently compared the ability of various polyamine analogs and inhibitors to induce SSAT activity in unperturbed L1210 cells (Libby et al., 1988). Over a 24 hr period, BES and BESm increased SSAT activity by 13- and 16-fold, respectively, making them the most potent inducers of the enzyme thus far identified. The increase was found to be due to elevated protein synthesis predominantly at the level of translation and to an apparent prolongation of enzyme half-life related to enzyme stabilization.

Interestingly, and consistent with the known substrate specificity of the enzyme for aminopropyl moieties (Erwin and Pegg, 1986; Seiler, 1987), the SPM analog, N^1,N^{14}-bis(ethyl)homospermine (BEHSm, Table 2) produced only a 2-3 fold rise in SSAT activity. Because BEHSm has the same regulatory effects on ODC and AdoMetDC as BESm, but does not induce SSAT activity, the analog afforded a convenient opportunity to examine the contribution of the latter to polyamine depletion. In L1210 cells treated with equimolar concentrations of BESm or BEHSm for 48 hr (Table 3), SPM pools were decreased to a significantly greater extent in BESm-treated cells (30% of control) than in BEHSm-treated cells (60% of control). Although alternative interpretations exist, the results seem to suggest that induction of SSAT activity may be involved in polyamine depletion by BES and BESm. We have not yet definitely determined whether the greater depletion of SPM achieved with BESm over BEHSm is due to increased excretion of acetylated polyamines (Wallace, 1987) or to the increased conversion of the higher polyamines to PUT which then can be further catabolized to γ-aminobutyric acid (Seiler, 1981).

Another recently elucidated activity of BESm involves an apparent anti-mitochondrial effect which, thus far, has been documented as a disappearance of mitochondrial DNA detectable by Southern blot analysis of treated L1210 cells (Bergeron et al., 1988). It is not evident whether this is due solely to a direct effect of the analog on the DNA or the organelle itself or whether it is secondarily related to polyamine depletion. However, since in unpublished studies, we have observed that nuclear DNA is not similarly affected, it is tempting to speculate that, because mitochondrial DNA lacks histones, it may be uniquely sensitive to polyamine depletion and/or to the direct effects of certain polyamine analogs. The contribution of this effect to the antiproliferative activity of the analogs is currently under study. Its role is expected to be much different from the antimitochondrial effects of the anticancer agent, methylglyoxal-bis(guanylhydrazone) (MGBG), since the latter elicits a distinctly cytotoxic response in cultured cells (Mikles-Robertson et al., 1978) whereas the analogs do not.

Antiproliferative Potential of Regulatory Analogs

Typically in most cell lines thus far examined, the nature of growth inhibitory effects elicited by the polyamine analogs, BES and BESm, can be classified as cytostatic as opposed to cytotoxic or cytolytic. The response of L1210 cells to these analogs is representative (Fig. 1). Cell growth is slowed continuously and, only after several days, is it significantly halted. During incubations of less than 96 hrs, cell growth continues but at a much reduced rate. Further, this growth inhibition requires the continued presence of the drug since, after a delay, cell cultures exhibit an ability to recover control growth characteristics following drug removal. Overall, these responses are nearly identical to those registered with specific polyamine inhibitors such as DFMO, AMA (Fig. 1) and the inhibitor of SPD synthase, AdoDATO (Tang et al., 1980).

Two exceptions to this generality have been noted. Fortuitously, both are representative of neoplastic diseases which are regarded as highly chemo-refractile. In the first of these, Casero et al., (1987a) have found that NCI H157 undifferentiated human large cell lung carcinoma line is highly sensitive to the SPD analog, BES. Furthermore the sensitivity is associated with an obvious cytotoxic response which is apparent by a decline in cell number within the culture at relatively low drug concentrations. This sensitivity has since been confirmed with BESm and other related analogs and has been shown to be accompanied by enzyme regulatory events and polyamine depletion typical of that seen in L1210 cells (Casero, personal communciation). In contrast to large cell lung carcinoma, small cell lung carcinoma cell line, NCI H82, is insensitive to BESm and is only minimally affected by relatively high concentrations of the drug (Casero et al., 1987b). Interestingly, these same cell lines are inversely sensitive to DFMO -- small cell lung carcinoma is cytotoxically affected while large cell lung carcinoma is relatively refractile to the drug (Luk et al., 1981; 1982; 1986)

Another disease uniquely sensitive to the polyamine analogs appears to be human melanoma. We have observed that MALME-3 melanoma exhibits high sensitivity and a cytostatic response characterized by cell loss in response to BESm. Preliminary results, however, suggest that cytotoxicity does not seem to correlate with analog-induced polyamine depletion suggesting that an alternative mode of action could be involved with these particular cells. Cellular differentiation accompanied by increased melanogenesis represents one possibility. Unlike with human lung tumor lines where responsiveness to the polyamine analogs is opposite to that for DFMO (Casero et al., 1987b; Luk et al., 1982), melanoma seems to be uniquely sensitive to the analogs and DFMO (Kapyaho and Janne, 1983; Kapyaho et al., 1985; Sunkara et al., 1985).

As indication of the generality of the above observations, it is pertinent to report that, when BESm was tested in the NCI in vitro tumor screening panel, 6 of 9 non small cell lung carcinoma lines and 4 of 5 melanoma cell lines exhibited IC_{50} values which were at least 1 log concentration lower than that of the mean for all 48 cell lines. Interestingly, 5 of 5 small cell lung carcinomas had IC_{50} values which were significantly higher than the mean. We believe, therefore, that the large cell lung carcinoma and melanoma represent diseases to be targeted during the continued preclinical development of these analogs.

Differential Sensitivity with Oncogene Transfection

Because the polyamine analogs are dependent on regulatory mechanisms in altering biosynthetic enzyme activities, and since oncogene expression is known to affect such processes, we have endeavored to determine the influence of oncogene transfection on the responsiveness of rat fibroblasts to BESm and certain polyamine inhibitors (Chang et al., 1988). Rat 1 cell lines transfected with the EJ2-ras or n-myc oncogene were obtained from the Robert Weinberg (Whitehead Institute, Cambridge MA). Oncogene-transfected lines grew at rates comparable to the parent Rat-1 line with doubling lines of 14 to 18 hr but differed from the latter by continuing to grow past confluency. During 96 hr incubations, the 3 cell lines were similarly sensitive to the specific enzyme inhibitors, DFMO and AdoDATO -- but differentially sensitive, however, to BESm. Cells transfected with n-myc had an IC_{50} of 10 μM compared to >200 μM for ras transfected cells and the parent Rat-1 cells. Polyamine pool sizes and decarboxylase activities were markedly affected in the n-myc cells but relatively unperturbed in the ras-transfected and parent lines. The basis for this sensitivity was determined to be due to increased intracellular accumulation of BESm by the n-myc transfected cells. A similar differential sensitivity to MGBG, which utilizes the polyamine carrier mechanism confirmed that increased transport was responsible. We are currently examining the generality of this phenomenon among other cell types. It should be noted, however, that Rat-1 cells may be uniquely suited for demonstrating an association between n-myc transfection and activation of polyamine transport since the parent cells appear to be unusually impermeable to the polyamine analogs, BESm (IC_{50}, >200 μM). The effect, none-the-less opens intriguing possibilities for inquiry into the regulation of polyamine uptake which here-to-fore has only been observed to occur following polyamine depletion with inhibitors such as DFMO (Alhonen-Hongisto et al., 1980; Porter et al., 1982). If found to be a generality, the association may have implications for the projected clinical use of BESm or other agents transported by the polyamine carrier.

Antitumor Activity of Regulatory Analogs

Thus far, the *in vivo* effectiveness of these agents has only been evaluated against murine L1210 leukemia. In studies recently published (Bergeron et al., 1988), it was found that under appropriate scheduling, BESm was curative in a significant proportion of animals. Perhaps because BESm is rapidly excreted, dosing schedule was found to be highly critical. At the optimal schedule (15 mg/kg q 8 hr x 6 i.p.), BESm produced a 376% increase in life span in mice inoculated i.p. with 10^5 L1210 cells and more than 50% long term survivors. Future studies will examine the *in vivo* effectiveness of BESm against human melanoma and large cell lung carcinoma xenographs in athymic mice.

Summary

Interference with polyamine biosynthesis by analog-mediated regulatory mechanisms represents a viable alternative to the use of specific enzyme inhibitors as an antiproliferative strategy. The approach is unique among antimetabolite approaches and is made possible by unusual characteristics inherent to the polyamines and their biosynthetic pathway. Current antitumor data obtained with these analogs provides indication of their potential usefulness as antitumor agents but, at the same time, demonstrates the need for improvement. This latter might be attained by the rational design of analogs which (a) bind more tightly at enzyme regulatory sites, (b) which are less able to substitute for natural polyamines in growth related functions and (c) which are eliminated less rapidly from tumor-bearing animals. At the same time, the continued preclinical development of available analogs might proceed most productively by targeting large cell lung carcinoma and melanoma and by examining the generality of the relationship between oncogene expression and the accompanying sensitivity to regulatory analogs.

Acknowledgements

This research was supported by grants CA-37606, CA-22153 and CA-24538 from the National Institutes of Health, Bethesda, MD.

References

Alhonen-Hongisto, L., 1980, Regulation of S-adenosylmethionine decarboxylase by polyamines in Ehrlich ascites-carcinoma cells grown in culture, Biochem. J., 190:747-754.

Artamonova, E.Y., Zavalova, L.L., Khomutov, R.M., and Khomutov, A.R., 1986, Irreversible inhibition of S-adenosylmethionine decarboxylase by hydroxylamine-containing analogues of decarboxylated S-adenosylmethionine, Biorg. Khim., 12:206-212.

Bergeron, R.J., Neims, A.J., McManis, J.S., Hawthorne, T.R., Vinson, J.R.T., Bortell, R., and Ingeno, M.J., 1988, Synthetic polyamine analogues as antineoplastics, J. Med. Chem., in press.

Casero, R.A., Jr., Go, B., Theiss, H.W., Smith, J., Baylin, S.B., and Luk, G.D., 1987a, Cytotoxic response of the relatively difluoromethylornithine-resistant human lung tumor cell line NCI H157 to the polyamine analogue N^1,N^8-bis(ethyl)spermidine, Cancer Res., 47:3964-3967.

Casero, R.A., Luk, G.D., Celano, P., and Baylin, S.F., 1987b, Differentiation sensitivity to the spermidine analog, N^1,N^8-bis-(ethyl)spermidine (BES), in human lung cancer cell lines, Proc. Am. Assoc. Cancer Res., 28: 318.

Celano, P., Baylin, S.B., Giardiello, F.M., Nelkin, B.D., and Casero, R.A., Jr., 1988, Effect of polyamine depletion on c-myc expression in human colon carcinoma cells. J. Biol. Chem. 263:5491-5494, 1988.

Chang, B.K., Porter, C.W., and Bergeron, R.J., 1988, Response to polyamine analogs and inhibitors in pancreatic cancer cell lines, submitted for publication.

Claverie, N., Wagner, J., Knodgen, B., and Seiler, N., 1987, Inhibition of polyamine oxidase improves the antitumoral effect of ornithine decarboxylase inhibitors, Anticancer Res., 7:765-772.

Dirks, L. Grens, A., Slezyneger, T.C., and Scheffler, I.E., 1986, Posttranscriptional regulation of ornithine decarboxylase activity, J. Cell Physiol., 126:371-378, 1986.

Erwin, B.G. and Pegg, A.E., 1986, Regulation of spermidine/spermine N^1-acetyltransferase in L6 cells by polyamines and related compounds, Biochem. J., 238:581-587.

Heller, J.S., Fong, W.F., and Canellakis, E.S., 1976, Induction of a protein inhibitor of ornithine decarboxylase by the end products of its reaction, Proc. Natl. Acad. Sci. U.S.A., 73:1858-1862.

Holtta, E., 1986, Control of ornithine decarboxylase in Chinese hamster ovary cells by polyamines, J. Biol. Chem., 261:9502-9508.

Kahana, C. and Nathans, D., 1985, Translational regulation of mammalian ornithine decarboxylase by polyamines, J. Biol. Chem., 260:-15390-15393.

Kameji, T. and Pegg, A.E., 1987, Effect of putrescine on the synthesis of S-adenosylmethionine decarboxylase, Biochem. J., 243:285-288.

Kapyako, K., Sinervirta, R., and Janne, J., 1985, Effects of inhibitors of polyamine biosynthesis on the growth and melanogenesis of murine melanoma cells, Cancer Res., 45:1444-1448.

Kapyaho, K. and Janne, J., 1983, Stimulation of melanotic expression in murine melanoma cells exposed to polyamine antimetabolites, Biochim. Biophys. Res. Commun., 113:18-23.

Katz, A. and Kahana, C., 1987, Transcriptional activation of mammalian ornithine decarboxylase during stimulated growth, Mol. Cell. Biol., 7:2641-2643.

Khomutov, R.M., Zavalova, L.L., Shrku, V.I., Artamonova, E.Y., and Khomutov, A.R., 1983, Effective inhibitor of S-adenosylmethionine decarboxylase, Bioorganic Chem. (Russ.), 9:130-131.

Kingsnorth, A.N., Lumdsen, A.B., and Wallace, H.M., 1984, Polyamines in colorectal cancer, Br. J. Surg. 71:791-794.

Kramer, D.L., Khumotov, R.M., Bukin, Y.V., and Porter, C.W., 1988, The relative abilities of individual polyamines to sustain growth and viability of L1210 cells treated with specific irreversible inhibitors of ornithine and S-adenosylmethionine decarboxylases, submitted for publication.

Libby, P.R., Bergeron, R.J., and Porter, C.W., 1988, Structure-function correlations of polyamine analog induced increases in spermidine/ spermine acetyltransferase activity, submitted for publication.

Luk, G.D., Abeloff, M.D., McCann, P.P., Sjoerdsma, A., and Baylin, S.B., 1986, Long-term maintenance therapy of established human small cell variant lung carcinoma implants in athymic mice wih a cyclic regimen of difluoromethylornithine, Cancer Res., 46:1849-1853.

Luk, G.D. and Baylin, S.B., 1984, Ornithine decarboxylase as a biologic marker in familial colonic polyposis, New Engl. J. Med. 331:80-83.

Luk, G.D., Goodwin, G., Gazdar, A.F., and Baylin, S.B., 1982, Growth inhibitory effects of DL-α-difluoromethylornithine in the spectrum of human lung carcinoma cells in culture, Cancer Res., 42:3070-3073.

Luk, G.D., Goodwin, G., Marton, L.J., and Baylin, S.B., 1981, Polyamines are necessary for the survival of human small-cell lung carcinoma in culture, Proc. Natl. Acad. Sci. U.S.A., 78:2355-2358.

Mamont, P.S., Duchesne, M-C., Grove, J., and Bey, P., 1978, Antiproliferative properties of DL-α-difluoromethylornithine in cultured cells. A consequence of irreversible inhibition of ornithine decarboxylase, Biochem. Biophys. Res. Commun., 81:58-66.

McCann, P.P., Tardif, C., Hornsperger, J.M., and Bohlen, P., 1979, Two distinct mechanisms for ornithine decarboxylase regulation by polyamines in rat hepatoma cells, J. Cell. Physiol., 99:183-190.

Mikles-Robertson, F., Feuerstein, B., Dave, C., and Porter, C.W., 1979, The generality of methylglyoxal-bis(guanylhydrazone) (MGBG)-induced mitochondrial damage and the dependence of this effect on proliferative activity. Cancer Res. 39:1919-1926, 1979.

Murakami, Y. and Hayashi, S-I., 1985, Role of antizyme in degradation of ornithine decarboxylase in HTC cells., Biochem. J., 226:893-896.

Pegg, A.E., 1986, Recent advances in the biochemistry of polyamines in eukaryotes, Biochem. J., 234:249-262.

Pegg, A.E., 1988, Polyamine metabolism and its importance in neoplastic growth and as a target for chemotherapy, Cancer Res., 48:759-774.

Pegg, A.E., Jones, D.B., and Secrist, J.A., 1988, Effect of inhibitors of S-adenosylmethionine decarboxylase, Biochem. J., 263:285-288.

Pegg, A.E., Madhubala, R., Kameji, T., and Bergeron, R.J., 1988, Control of ornithine decarboxylase activity in difluoromethylornithine-resistant L1210 cells by polyamines and synthetic analogues, J. Biol. Chem., in press.

Pegg, A.E. and McCann, P.P., 1982, Polyamine metabolism and function, Am. J. Cell. Physiol., 243:C212-C221.

Pegg, A.E., Tang, K-C., and Coward, J.K., 1982, Effects of S-adenosyl-1,-8-diamino-3-thiooctane on polyamine metabolism, Biochem., 21:5082-5089.

Pegg, A.E., Wechter, R., and Pajunen, A., 1987, Increase in S-adenosylmethionine decarboxylase in SV-3T3 cells treated with S-methyl5'-methylthioadenosine, Biochem. J., 244:49-54.

Persson, L., Holm, I., and Heby, O., 1986, Translational regulation of ornithine decarboxylase by polyamines, Fed. Eur. Biochem. Soc. Letters, 205:175-178.

Persson, L., Holm, I., and Heby, O., 1988, Regulation of ornithine decarboxylase mRNA translation by polyamines, J. Biol. Chem., 263:3528-3533.

Pohjanpelto, P., 1976, Putrescine transport is greatly increased in human fibroblasts initiated to proliferate, J. Cell. Biol., 68:512-520.

Porter, C.W., Berger, F.G., Pegg, A.E., Ganis, B., and Bergeron, R.J., 1987a, Regulation of ornithine decarboxylase activity by spermidine or the spermidine analog, N^1,N^8-bis(ethyl)spermidine (BES), Biochem. J., 242:433-440.

Porter, C.W. and Bergeron, R.J., 1983, Spermidine requirement for cell proliferation in eukaryotic cells: structural specificity and quantitation, Science, 219: 1083-1085.

Porter, C.W. and Bergeron, R.J., 1988, Enzyme regulation as an approach to interference with polyamine biosynthesis -- an alternative to enzyme inhibition. in: "Advances in Enzyme Regulation," G. Weber, ed., Pergamon Press, New York and Oxford, Vol. 27, pp. 57-79.

Porter, C.W., Bergeron, R.J., and Stolowich, N.J., 1982, Biological properties of N^4-spermidine derivatives and their potential in anticancer chemotherapy, Cancer Res., 42: 4072-4078, 1982.

Porter, C.W., Cavanaugh, P.F., Jr., Stolowich, N., Ganis, B., Kelly, E., and Bergeron, R.J., 1985, Biological properties of N^4 and N^1,N^8-spermidine derivatives in cultured L1210 leukemia cells, Cancer Res., 45:2050-2057.

Porter, C.W., Ganis, B., Vinson, T., Marton, L.J., Kramer, D.L., and Bergeron, R.J., 1986, A comparison and characterization of growth inhibition by α-difluoromethylornithine (DFMO), an inhibitor of ornithine decarboxylase and N^1,N^8-bis(ethyl)spermidine (BES), an apparent regulator of the enzyme, Cancer Res., 46:6279-6285.

Porter, C.W., Herrera-Ornelas, L., Pera, P., Petrelli, N.F., and Mittelman, A., 1987b, Polyamine biosynthetic activity in normal and neoplastic human colorectal tissues, Cancer 60:1275-1281.

Porter, C.W., McManis, J., Casero, R.A., and Bergeron, R.J, 1987c, Relative abilities of bis(ethyl) derivatives of putrescine, spermidine, and spermine to regulate polyamine biosynthesis and inhibit L1210 leukemia cell growth, Cancer Res., 47:2821-2825.

Porter, C.W. and Sufrin, J.R., 1986, Interference with polyamine biosynthesis and/or function by analogs of polyamines or methionine as a potential anticancer chemotherapeutic strategy, Anticancer Res., 6:525-542.

Rozhin, J., Wilson, P.S., Bull, A.W., and Nigro, N.D., 1984, Ornithine decarboxylase activity in the rat and human colon, Cancer Res. 44:3226-3230.

Seely, J.E. and Pegg, A.E., 1983, Effect of 1,3-diaminopropane on ornithine decarboxylase enzyme protein in thioacetamide-treated rat liver, Biochem. J., 216: 701-707.

Seiler, N., 1987, Functions of polyamine acetylation, Can. J. Physiol. Pharmacol., 65:2024-2035, 1987.

Seiler, N., Turnover of polyamines, 1981, in "Polyamines in Biology and Medicine", D.R. Morris and L.J. Marton, eds., Marcel Dekker, New York and Basel, pp. 169-180.

Sunkara, P.S., Chang, C.C., Prakash, N.J., and Lackmonn, P.J., 1985, Effect of inhibition of polyamine biosynthesis by DL-difluoro methylornithine on the growth and melanogenesis of B16 melanoma in vitro and in vivo, Cancer Res, 45:4067-4070.

Tang, K.C., Pegg, A.E., and Coward, J.K, 1980, Specific and potent inhibition of spermidine synthase by the transition state analog, S-adenosyl-Lthio-1,8-diaminoctane, Biochem. Biophys. Res. Commun., 96: 1371-1377.

Traweek, S.T., Riscoe, M.K., Ferro, A.J., Braziel, R.M., Magenis, R.E., and Fitchen, J.H., 1988, Methylthioadenosine phosphorylase deficiency in acute leukemia: Pathologic, cytogenetic and clinical features. Blood, in press.

Wallace, H.M., 1987, Polyamine catabolism in mammalian cells: excretion and acetylation, Med. Sci. Res., 15:1437-1440, 1987.

FLUORINE-CONTAINING POLYAMINES : BIOCHEMISTRY AND POTENTIAL APPLICATIONS

Pierre S. Mamont, Nicole Claverie and Fritz Gerhart

Merrell Dow Research Institute, 16, rue d'Ankara
67084 Strasbourg Cedex, France

INTRODUCTION

Polyamine metabolism and its significance in diseases characterized by rapid cell proliferation have been reviewed recently.[1,2] These reviews emphasize the importance of the polyamine biosynthetic pathway as a promising target for the design of new therapeutic agents and summarize the progress made in the field.

At present, among all the experimental approaches which have been used for manipulating polyamine metabolism - direct or indirect inhibition of L-ornithine (ODC) and of S-adenosyl-L-methionine (SAM-DC) decarboxylases, of the aminopropyl transferases[1,2] or of the polyamine interconversion pathway[3] - only direct inhibition of ODC by specific irreversible inhibitors has proved to exert beneficial effects against animal models of proliferative diseases.[4-7] However, despite the striking efficacy of difluoromethylornithine (DFMO ; eflornithine) against both the early and late stage of T.b. gambiense trypanosomiasis in humans, the antineoplastic effects of this ODC inhibitor in clinical studies have been disappointing.[8] As discussed,[1,8] the apparent discrepancies between the marked antitumor effects of the ODC inhibitors in animal models and their ineffectiveness in the clinical situation might result from multiple factors including differences in antitumor protocols, insufficient polyamine deficiency due to both incomplete blockade of ODC activity in vivo and to availability of endogenous polyamines which can be mobilized from other tissues via the interconversion pathway and be taken up by the tumor tissues.

There is some experimental evidence that even in cultured cells DFMO does not inhibit entirely ODC activity.[9] As hypothesized,[9] residual ODC activity allows enough spermine synthesis for maintenance of residual growth and cell viability. This notion has been further supported by the fact that (2R,5R)-6-heptyne-2,5-diamine (MAP), a more potent irreversible inhibitor of ODC than DFMO,[10] depletes all polyamines including spermine, causes a complete cessation of rat hepatoma (HTC) cell replication and a time-dependent loss of cell viability.[11] However, in animal models, long-term administration of MAP or of the esters of (E)-2-(fluoromethyl)dehydroornithine, another potent inhibitor of ODC,[12] scarcely affects spermine concentrations.[6,13] As a consequence, these compounds may not offer possibly any advantage over DFMO for studying the pharmacological consequences of polyamine deficiency <u>in vivo</u> or in cancer chemotherapy when used as monotherapy. It becomes more and more apparent that approaches other than inhibition of ODC should be used in the future, i.e., inhibition of the other biochemical reactions of polyamine metabolism with particular attention to specific and irreversible inhibition of SAM-DC. Promising candidates for this enzyme inhibition have been recently synthesized and have proved to be transiently effective in cultured cells.[14] Their potential usefulness in <u>in vivo</u> studies remains however to be determined.

In the late 70's, another approach based on the repression of ODC expression by polyamines,[15,16] and called "gratuitous" repression of polyamine biosynthesis, has been proposed.[17] Early experimental attempts using this concept have been unsuccessful because of lack of specificity of action of the amines utilized and many other reasons that have been already fully discussed.[18] Reinforced by the demonstration that SAM-DC expression like ODC can also be negatively regulated by spermidine, spermine and closely-related analogues,[19] the approach has been recently validated with the synthesis of N-alkyl derivatives of polyamines which share with the natural biogenic amines the properties to repress the decarboxylase enzymes but not the functions for growth of cultured cells.[20,21]

Following the same strategy, we now report on some of the biochemical and biological properties of newly synthesized fluorinated spermidine analogues and discuss their potential applications.

Table 1. Structural Formulae of the Fluorinated Spermidine Analogues.

Spermidine	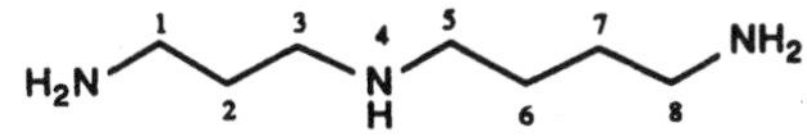
6-monofluorospermidine	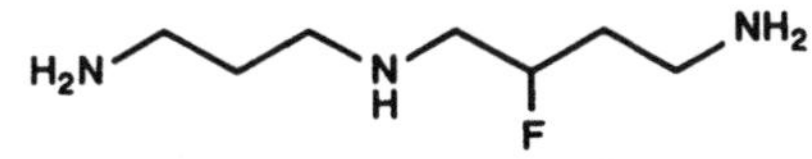
7-monofluorospermidine	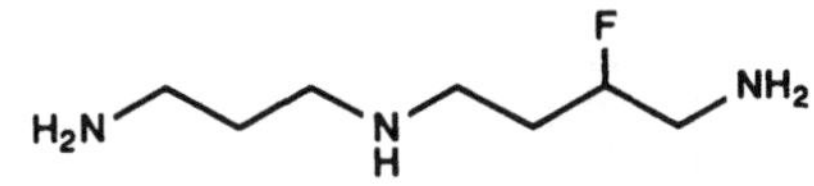
6,6-difluorospermidine	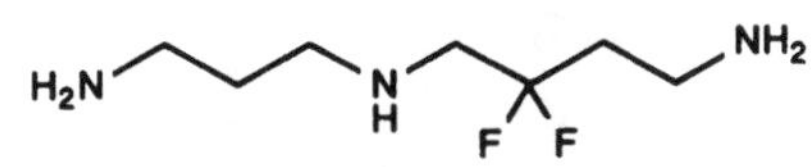
7,7-difluorospermidine	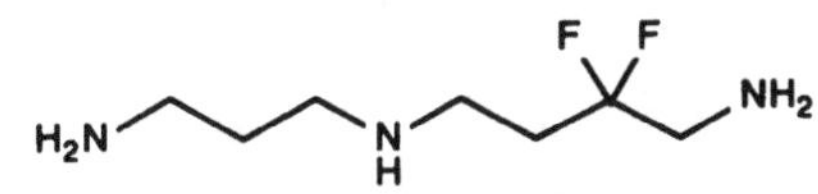

BIOCHEMICAL PROPERTIES OF THE FLUORINATED ANALOGUES OF SPERMIDINE

A series of monofluorinated and geminal difluorinated analogues of spermidine (4-azaoctane-1,8-diamine) listed in Table 1 have been tested <u>in vitro</u> as potential substrates of partially purified HTC cell spermine synthase.[22] All compounds are substrates of this enzyme, the product of the reaction 6-monofluorospermine and 6,6-difluorospermidine, when 6- or 7-monofluorospermidine and 6,6- or 7,7-difluorospermidine are used in the reaction mixture respectively, being identified by chromatographic comparison with authentic samples. However, substrate capability of these fluorinated analogues of spermidine varies markedly as a function of the number of fluorine substituents and of their position relative to the amine functions. As seen in Table 2, di- but not monofluorosubstitution of the hydrogen atoms of the methylene group at position 6 causes a 5-fold increase of the affinity constant for the enzyme relative to spermidine. On the contrary, mono- and difluoro-substitution at position 7 decreases 8- and 105-fold respectively the Km values.

It is known that substitution by fluorine atoms of the methylene hydrogens in β-position of an amino group results in a marked decrease of the basicity of the amine function.[23] This is verified for the fluorinated analogues of spermidine.[22] As determined by potentiometric

Table 2. Comparative Biochemical and Biological Activities of Mono- and Difluorinated Analogues of Spermidine in HTC Cells

	SPM Synthase	Inhibition of Decarboxylases		Polyamines					Growth (b)	
							Fluorinated			
	Km (μM)	ODC	SAM-DC	PUT	SPD	SPM	SPD(c)	SPM(c)	- DFMO	+ DFMO
		(IC_{50} : μM)		(% of control)						
SPD	88	0.1	1.2							↑
6-monofluoro SPD	93	0.4	1.5	14	10	25	145	118	0	↑
7-monofluoro SPD	11	0.15	2.4	1	24	13	154	166	0	↑
6,6-difluoro SPD	458	1.0	(a)	88	99	111	76	17	0	0
7,7-difluoro SPD	0.5	4.0	5.6	81	175	17	88	146	↗	↓

a : 13 % decrease at 10 μM concentration
b : 10 μM concentration of the amines were used.
0 : no effect ; ↗ : partial effect ; ↑ : enhanced growth ; ↓ : reduced growth
c : % relative to control SPD and SPM values.
ODC, SAM-DC activities were determined 2 and 6 h respectively after addition of the amines to the culture medium. Cell extracts were prepared from 1-day culture for ODC and from 2-day culture in the presence of 5 mM DFMO for SAM-DC. Polyamines were measured 4 days after addition of 10 μM of the amines. Control values for ODC and SAM-DC activities were 1.10 and 14.6 nmol CO_2 h^{-1} mg $protein^{-1}$ respectively and 0.42, 2.90 and 3.30 nmol/10^6 cells for PUT, SPD and SPM respectively.

titration the pKa value of the amine function at position 4 of 6-mono and 6,6-difluorospermidine is decreased by 1.22 and 2.70 units respectively relative to spermidine (pKa value = 8.40 at pH = 7.4 and 25°C). Similarly, the pKa value of the terminal butylamino group (N^8) of 7-mono and 7,7-difluorospermidine is decreased from 10.81 to 7.80 and 6.64 respectively. These findings allow the establishment of a relationship between the sites of protonation and the capability for the spermidine analogues to serve as acceptor in the spermine synthase reaction.[22] Protonation of N^1 and N^4 seems essential for efficient aminopropylation whereas decrease of protonation of the terminal butylamino group appears to correlate inversely with the increase of binding affinity.

COMPARATIVE BIOLOGICAL EFFECTS OF THE FLUORINATED ANALOGUES OF SPERMIDINE IN CULTURED HTC CELLS

ODC and SAM-DC REGULATION

Spermidine exerts regulatory control of both ODC and SAM-DC mainly at the translational and post-translational levels.[24-26] 7-mono- or 6-monofluorospermidine added to the culture medium of growing HTC cells decreased ODC activity with IC_{50} values of the same order of magnitude as spermidine (Table 2). Similarly, these compounds reduced DFMO-induced SAM-DC activity at concentrations equivalent to spermidine. Disubstitution by fluorine atoms either at position 6 or at position 7 considerably diminished the repressive effects on the decarboxylases, 6,6-, in contrast to 7,7-difluorospermidine, being almost completely ineffective against SAM-DC when used at 10 µM concentration (Table 2).

Effects on Polyamine Metabolism and Growth

Monofluorinated spermidine. As expected, based on their capabilities to repress the decarboxylases and to be substrates of spermine synthase, 6- or 7-monofluorospermidine, when added at 10 µM concentration to the culture medium of growing HTC cells, depleted these cells from their polyamine contents within 4 days (Table 2). Both fluorinated spermidine analogues accumulated intracellularly to reach a concentration value exceeding that of the natural endogenous spermidine, 7-monofluorospermidine being converted into 6-monofluorospermine to a greater extent than the 6-fluorinated spermidine analogue. Despite the marked induced-depletion of the natural endogenous amines, growth continued unabated. These findings suggest that the polyamine analogues can substitute spermidine and spermine in their function associated with cell replication. In fact, when medium of DFMO-inhibited HTC cells was supplemented with 10 µM 6- or 7-monofluorospermidine, growth resumed following a lag period of one generation time.

Difluorinated spermidine. 10 µM 6,6-difluorospermidine affected neither the intracellular polyamine concentrations nor the growth of HTC cells and was scarcely converted into 6,6-difluorospermine (Table 2). These results are consistent with the biochemical and biological properties of this spermidine analogue since it is a relatively poor "repressor" of the decarboxylases and a poor substrate of spermine synthase. Therefore, 6,6-difluorospermidine did not inhibit HTC cell spermine synthesis *in situ* as measured by the rate of conversion of radiolabelled spermidine into spermine (Table 3). In similar experimental conditions, 7,7-difluorospermidine reduced by almost 90 % the transformation of radioactive spermidine into spermine (Table 3) and was efficiently metabolized into 6,6-difluorospermine (Table 2). The spermine content of HTC cells was decreased by 83 % within 4 days upon exogenous addition of 10 µM 7,7-difluorospermidine whereas spermidine levels increased 1.75-fold relative to spermidine control values (Table 2). Growth was not affected for 3 days and ceased thereafter. Cell viability as determined by measurement of cloning efficiency was decreased by 33 % (Table 4).

Inhibition of polyamine biosynthesis by MAP causes putrescine and spermidine deficiency in HTC cells within 1 day after its addition to the culture medium, and a time-dependent decrease of the spermine content of the cells. This leads ultimately to growth arrest and to a progressive loss of cell viability.[11] When combined with 7,7-difluorospermidine a much earlier decrease of the spermine content of the cells was observed (46 % at day 1) whereas spermidine concentration was maintained at 70 % of control basal values (Table 5). In those conditions, MAP induced-increase of SAM-DC activity was blocked. 7,7-difluorospermidine and 6,6-difluorospermine, its metabolite, accumulated at concentrations 2.3- and 1.5-fold higher than those of control spermidine and spermine respectively. Growth ceased after one generation time and a time-dependent decrease of cell viability occurred, 44 % cell death being already observed at day 1 of cell

Table 3. Effects of 6,6-Difluorospermidine or 7,7-Difluorospermidine on Spermine Synthesis in HTC Cells

	Intracellular Polyamines					Radiolabelled Polyamines			
	PUT	SPD	SPM	DFSPD	DFSPM	SPD		SPM	
	(nmol/10^6 cells)					d.p.m./cell	d.p.m./pmol	d.p.m./cell	pmol formed/10^6 cells
Treatment									
None	0.86	2.87	1.90	0	0	12.60	4390	0.29	66.0
10 µM 6,6-DFSPD	0.92	3.25	2.67	1.22	0.15	12.20	3754	0.25	66.6
10 µM 7,7-DFSPD	0.37	2.88	1.00	2.19	1.12	15.10	5240	0.05	9.5

HTC cell cultures were pre-incubated for 24 h in the presence or absence of 6,6-DFSPD or 7,7-DFSPD. Cells were washed and resuspended in fluorinated polyamine-free medium. They were then pulse-labelled for 1 h with 5 µCi/ml (terminal methylenes-^{3}H) SPD (23 Ci/mmol). Intracellular polyamine contents, SPD specific radioactivities and newly-formed radiolabelled SPM were determined after separation by HPLC as previously described.[11]

culture (Table 4). At day 4, the combined treatment caused almost one log of cell kill.

Table 4. Effects of MAP Used Alone or in Combination with 7,7-Difluorospermidine on Cloning Efficiency.

Addition	Absolute cloning efficiency (%)			
	Day 1	Day 2	Day 3	Day 4
None	45 ± 3	48 ± 3	43 ± 3	42 ± 1
100 µM MAP	45 ± 3	37 ± 2	42 ± 1	20 ± 1
7,7-DFSPD	38 ± 4	ND	ND	28 ± 2
100 µM MAP + 10 µM 7,7-DFSPD	25 ± 1	15 ± 2	10 ± 1	5 ± 1

HTC cell cultures (1×10^5 cells/ml) were incubated for 4 days in the presence or absence of the drugs. Cloning efficiency was determined at the indicated time as previously described.[11]
ND : not determined.

Altogether these findings suggest that elevation of spermidine concentration over control values, due to inhibition of spermine synthesis by 7,7-difluorospermidine used alone, may compensate for the deficiency of spermine. Similar conclusions have been drawn from experiments of inhibition of spermine synthase by S-methyl-5'-methylthioadenosine in cultured SV-3T3 cells.[27] Thereafter growth arrest may possibly result from the progressive accumulation of the fluorinated spermidine analogue and of its metabolite reaching concentrations sufficient to antagonize the spermidine and spermine growth functions. Combined-treatment with MAP or with DFMO, which reduces both spermidine and spermine synthesis but also induces a further accumulation of the fluorinated polyamine analogues, precipitates and accentuates this cytotoxicity.

Table 5. Effects of MAP Alone or in Combination with 7,7-Difluorospermidine on HTC Cell Growth and Polyamine Content

Addition	Time (days)	Cell Number 1×10^{-5}/ml	Polyamines or MAP (nmol/10^6 cells) PUT	SPD	SPM	DFSPD	DFSPM	MAP
None	0	0.11	0.12	1.46	2.36			
	1	0.22	0.24	1.82	2.44			
	2	0.44	0.38	2.38	2.90			
	3	0.63	0.34	3.08	3.46			
100 μM MAP	1	0.22	ND	0.17	2.79			1.43
	2	0.27	ND	ND	2.32			2.21
	3	0.32	ND	ND	1.91			2.40
100 μM MAP + 10 μM 7,7-DFSPD	1	0.20	ND	0.94	1.28	4.44	1.60	1.49
	2	0.22	0.02	0.97	1.03	6.40	3.64	2.26
	3	0.20	0.04	1.05	1.02	7.12	5.08	2.68

Growth and polyamine analysis were determined as previously described.[11]
ND : not detected. $A

ANTITUMORAL EFFECTS OF 7,7-DIFLUOROSPERMIDINE IN MORRIS HEPATOMA-BEARING RATS

The synergistic antiproliferative and cytotoxic effects of 7,7-difluorospermidine in combination with MAP observed in cultured cells prompt us to perform such studies in vivo.

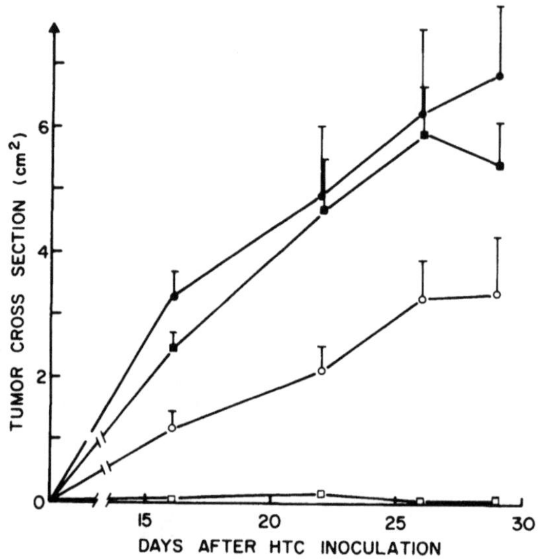

Fig. 1. Effects of MAP or of 7,7-difluorospermidine used alone or in combination on the growth of Morris hepatoma in rats.

Treatment with MAP (2 g/l in drinking water) and with 7,7-difluorospermidine (200 mg/kg/day infused via minipumps) started immediately after i.m. inoculation of HTC cells to Buffalo rats. Tumor size was determined as previously described.[6] ●, control (n = 5) ; ■, 7,7-DFSPD (n = 4) , o, MAP (n = 5) ; □, MAP + 7,7-DFSPD (n = 5).

Effects at Early Stage of Tumor Development

As illustrated in Figure 1, 7,7-difluorospermidine (200 mg/kg bodyweight/day infused via minipumps) administered 4 h after tumor cell inoculation was ineffective. MAP (0.2 % in drinking water ; drug intake : 150 mg/kg bodyweight/day) reduced the tumor cross section by 40 % in confirmation of previously published results.[6] The combined treatment blocked completely the development of the tumor. This has never been achieved so far in vivo by manipulation of polyamine metabolism.

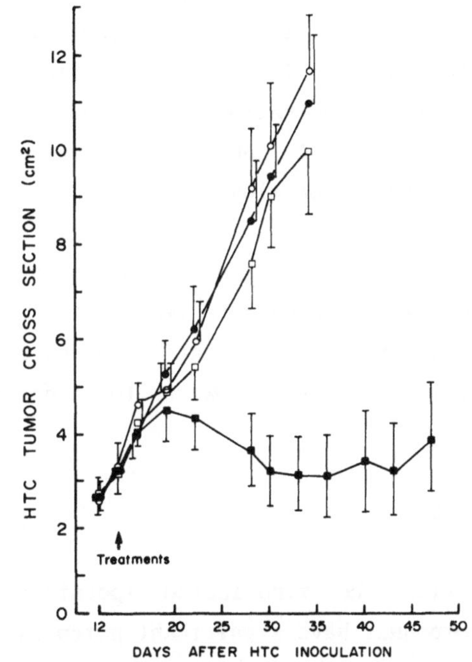

Fig. 2. Effects of MAP or of 7,7-difluorospermidine used alone or in combination on the growth of preimplanted Morris hepatoma in rats.

Experimental protocol was as described in the legend of Fig. 1 except that treatments started at day 14 after tumor inoculation. ●, control (n = 7) ; o, 7,7-DFSPD (n = 7) ; □, MAP (n = 7) ; ■, 7,7-DFSPD + MAP (n = 8).

Effects at Late Stage of Tumor Development

When administered to rats at day 14 after tumor cell inoculation, 7,7-difluorospermidine (200 mg/kg bodyweight/day/infused) or MAP (0.2 % in drinking water) did not affect the ongoing development of the tumor (Figure 2). When administered in combination tumor growth continued unaltered for 5 days. Thereafter, stabilization of tumor size and tumor regression was observed for 5 and 3 of the 8 animals of the group respectively. On average, the tumor size was reduced by 75 % at day 34 and the mean tumor weight at sacrifice (day 47) was decreased by 62 % relative to control values. 7,7-difluorospermidine administered alone reduced by 55 % the tumor spermine concentration but did not affect those of putrescine and spermidine. In combination with MAP, putrescine concentration was decreased by 71 % and deficiency in spermine was further accentuated (75 % relative to control values). Spermidine concentration remained at control values. The fluorinated spermidine analogue accumulated at concentration equivalent to spermidine and 6,6-difluorospermine built up at concentration value 2.7-fold higher than that of the normal tumor spermine.

It is the first time that such spermine deficiency can be achieved in *in vivo* tumor studies and even more important, that regression of preimplanted tumor is observed. The potential values of such combination in other animal tumor models remains however to be determined.

SUMMARY AND CONCLUSIONS

Investigations with the fluorinated spermidine analogues show clearly that these compounds have significant potential for studying the metabolism and functions of the polyamines. However, the biochemical and biological properties of these analogues are dissimilar. This is due to the influence of the fluorine substituent(s) on the basicity of the amine function proximal to the fluoromethylene group, this effect being amplified by geminal disubstitution.

The monofluorinated spermidine analogues compare well with the natural amine in their ability to regulate the expression of the decarboxylase enzymes, to be substrates of spermine synthase and to support growth of polyamine-deficient cells. It is also likely that 6-monofluorospermine, formed biochemically *in situ*, shares with spermine similar functions. These findings raise the possibility of using these spermidine analogues to study the metabolism and pharmacology of polyamines *in vivo* but also to provide more insight into the regulatory role of spermidine in ODC and SAM-DC expression. Another potential application may be the use of these analogues as probes in tumor imaging and therapy control. This indication has been inferred by studies in tumor-bearing animals, using ^{19}F-NMR spectroscopy determination of tissue fluorospermidine and fluorospermine, formed biochemically from the precursors 2-fluoro or 2,2-difluoroputrescine,[28,29] and which demonstrate preferential accumulation in tumor versus normal tissue.[30] Finally, these monofluorinated spermidine analogues may exert beneficial effects in pathological states associated with polyamine deficiency. These diseases remain however to be identified.

Among the difluorinated spermidine analogues, 7,7-difluorospermidine possesses the most interesting properties. This spermidine analogue still possesses ODC and SAM-DC repressing activities although at much higher concentration than spermidine. More importantly it is a potent inhibitor of spermine synthesis both in cultured cells and *in vivo* due to its efficient competition with spermidine in the spermine synthase reaction. This compound not only depletes tumor cell of its spermine content but, in addition, appears to exert by itself and/or via 6,6-difluorospermine, the product of its metabolism, polyamine antagonist effects. Combined with MAP but also with DFMO, two potent irreversible inhibitors of ODC which block the synthesis of the natural endogenous polyamines, 7,7-difluorospermidine causes an immediate decrease of viability in cultured HTC cells and promotes tumor regression and stabilization in hepatoma-bearing rats. These findings are of sufficient interest for pursuing research aims at the design of new polyamine analogues with the underlying ideas that they should fulfill some but not all the functions of the natural amines. However, rational design of such analogues is entirely dependent on a better understanding at the molecular levels of the steric and charge distribution requirements of the polyamines for binding to their active sites and a better knowledge of the structure of the key polyamine macromolecular acceptors which are involved in the control of polyamine synthesis and of cellular replication and differentiation.

ACKNOWLEDGMENTS

The authors wish to thank Mrs. M. Siat, Miss A. Bernhardt and Mrs. I. Fornieles for expert technical assistance.

REFERENCES

1. A.E. Pegg, Polyamine metabolism and its importance in neoplastic growth and as a target for chemotherapy, Cancer Res. 48:759 (1988).
2. P.P. McCann, A.E. Pegg, and A. Sjoerdsma,"Inhibition of Polyamine Metabolism, Biological Significance and Basis for New Therapies", Academic Press Inc., Orlando (1987).
3. N. Seiler, Functions of polyamine acetylation, Can. J. Physiol. Pharmacol. 65:2024 (1987).
4. C.J. Bacchi, and P.P. McCann, Parasitic protozoa and polyamines, in : "Inhibition of Polyamine Metabolism, Biological Significance and Basis for New Therapies", P.P. McCann, A.E. Pegg, and A. Sjoerdsma, eds., Academic Press Inc., Orlando, pp. 317-344 (1987).
5. P.S. Sunkara, and S.B. Baylin, Inhibitors of polyamine biosynthesis : cellular and in vivo effects on tumor proliferation, in : "Inhibition of Polyamine Metabolism, Biological Significance and Basis of New Therapies", P.P. McCann, A.E. Pegg, and A. Sjoerdsma, eds., Academic Press Inc., Orlando, pp. 121-140 (1987).
6. J. Bartholeyns, P. Mamont, and P. Casara, Antitumor properties of (2R,5R)-6-heptyne-2.5-diamine, a new potent enzyme-activated irreversible inhibitor of ornithine decarboxylase in rodents, Cancer Res. 44:4972 (1984).
7. N. Claverie, J.-L. Pasquali, P.S. Mamont, C. Danzin, M. Weil-Bousson, and M. Siat, Immunosuppressive effects of (2R,5R)-6-heptyne-2,5-diamine, and inhibitor of polyamine synthesis. II. Beneficial effects on the development of a lupus-like disease in MRL-lpr/lpr mice, Clin. Exp. Immunol. 72:293 (1988).
8. P.J. Schechter, J.L.R. Barlow, and A. Sjoerdsma, Clinical aspects of inhibition of ornithine decarboxylase with emphasis on therapeutic trials of eflornithine (DFMO) in cancer and protozoan diseases, in : "Inhibition of Polyamine Metabolism, Biological Significance and Basis of New Therapies", P.P. McCann, A.E. Pegg, and A. Sjoerdsma, eds., Academic Press Inc., Orlando, pp. 345-364 (1987).
9. P.S. Mamont, M.-C. Duchesne, A.-M. Joder-Ohlenbusch, and J. Grove, Effects of ornithine decarboxylase inhibitors on cultured cells, in :"Enzyme-Activated Irreversible Inhibitors", N. Seiler, M.J. Jung, and J. Koch-Weser eds., Elsevier/North Holland Biochemical Press, Amsterdam, pp. 43-54 (1978).

10. C. Danzin, P. Casara, N. Claverie, B.W. Metcalf, and M.J. Jung, (2R,5R)-6-heptyne-2,5-diamine, an extremely potent inhibitor of mammalian ornithine decarboxylase, Biochem. Biophys. Res. Commun. 116:237 (1983).

11. P.S. Mamont, M. Siat, A.-M. Joder-Ohlenbusch, A. Bernhardt, and P. Casara, Effects of [2R,5R]-6-heptyne-2,5-diamine, a potent inhibitor of L-ornithine decarboxylase, on rat hepatoma cells cultured in vitro, Eur. J. Biochem. 142:457 (1984).

12. P. Bey, F. Gerhart, V. Van Dorsselaer, and C. Danzin, α-(fluoromethyl) dehydroornithine and α-(fluoromethyl) dehydroputrescine analogues as irreversible inhibitors of ornithine decarboxylase, J. Med. Chem. 26:1551 (1983).

13. P.S. Mamont, C. Danzin, M. Kolb, F. Gerhart, P. Bey, and A. Sjoerdsma, Marked and prolonged inhibition of mammalian ornithine decarboxylase in vivo by esters of (E)-2-(fluoromethyl) dehydroornithine, Biochem. Pharmacol. 35:159 (1986).

14. A.E. Pegg and D.B. Jones, J.A. Secrits III, Effects of inhibitors of S-adenosylmethionine decarboxylase on polyamine content and growth of L1210 cells, Biochem. 27:1408 (1988).

15. D.M. Pett, and H.S. Ginsberg, Metabolism of polyamines in KB cells, Fed. Proc. 27:615 (1968).

16. J.E. Kay, and V.J. Lindsay, Control of ornithine decarboxylase activity in stimulated human lymphocytes by putrescine and spermidine, Biochem. J. 132:791 (1973).

17. J. Jänne, H. Pösö, and A. Raina, Polyamines in rapid growth and cancer, Biochem. Biophys. Acta 473:241 (1978).

18. O. Heby, and J. Jänne, Polyamine antimetabolites : Biochemistry, specificity, and biological effects of inhibitors of polyamine synthesis, in : "Polyamines in Biology and Medicine", D.R. Morris, and L.J. Marton, eds., Marcel Dekker, New York, pp. 243-310, (1981).

19. P.S. Mamont, A.-M. Joder-Ohlenbusch, M. Nussli, and J. Grove, Indirect evidence for a strict negative control of S-adenosyl-L-methionine decarboxylase by spermidine in rat hepatoma cells, Biochem. J. 196:411 (1981).

20. C.W. Porter, B. Ganis, T. Vinson, L.J. Marton, D.L. Kramer, and R. Bergeron, Comparison and characterization of growth inhibition in L1210 cells by α-difluoromethylornithine, an inhibitor of ornithine decarboxylase, and N^1,N^8-Bis(ethyl)spermidine, an apparent regulator of the enzyme, Cancer Res. 46:6279 (1986).

21. C.W. Porter, J. McManis, R.A. Casero, and R. Bergeron, Relative abilities of bis (ethyl) derivatives of putrescine, spermidine and spermine to regulate polyamine biosynthesis and inhibit L1210 leukemia cell growth, Cancer Res. 47:2821 (1987).

22. J.C. Baillon, P.S. Mamont, J. Wagner, F. Gerhart, and P. Lux, Fluorinated analogues of spermidine as substrates of spermine synthase, Eur. J. Biochem., in press.

23. R.D. Chambers, The influence of fluorine or fluorocarbon groups on some reaction centres, in : "Fluorine Organic Chemistry", R.D. Chambers, ed., John Wiley and Sons, New York,pp. 64-96, (1973).

24. A.E. Pegg, Recent advances in the biochemistry of polyamines in eukaryotes, Biochem. J. 234:249 (1986).

25. L. Persson, I. Holm, and O. Heby, Regulation of ornithine decarboxylase mRNA translation by polyamines, J. Biol. Chem. 263:3528 (1988).

26. T. Kameji, and A.E. Pegg, Inhibition of translation of mRNAs for ornithine decarboxylase and S-adenosyl methionine decarboxylase by polyamines, J. Biol. Chem. 262:2427 (1987).

27. A.E. Pegg, and J.K. Coward, Growth of mammalian cells in the absence of the accumulation of spermine. Biochem. Biophys. Res. Commun. 133:82 (1985).

28. S. Sarhan, B. Knödgen, F. Gerhart, and N. Seiler, Chain-fluorinated polyamines as tumor markers. I. In vivo transformation of 2,2-difluoroputrescine into 6,6-difluorospermidine and 6,6-difluorospermine, Int. J. Biochem. 19:843 (1987).

29. F. Dezeure, S. Sarhan, and N. Seiler, Chain-fluorinated polyamines as tumor markers. IV. Comparison of 2-fluoroputrescine and 2,2-difluoroputrescine as substrates of spermine synthase in vitro and in vivo, Int. J. Biochem., submitted for publication.

30. W.E. Hull, W. Kuntz, R.E. Port, and N. Seiler, Chain-fluorinated polyamines as tumor markers. III. Determination of geminal difluoropolyamines and their precursor 2,2-difluoroputrescine in normal tissues and experimental tumors by in vitro and in vivo ^{19}F-NMR spectroscopy, NMR in Biomed., in press.

MECHANISM OF ANTITUMOR ACTIVITY OF NORSPERMIDINE, A STRUCTURAL HOMOLOGUE OF SPERMIDINE

Prasad S. Sunkara, John H. Zwolshen, Nellikunja J. Prakash and Terry L. Bowlin

Merrell Dow Research Institute
2110 E. Galbraith Road
Cincinnati, Ohio 45215 USA

INTRODUCTION

The availability of α-difluoromethylornithine (DFMO), a specific and irreversible inhibitor of ornithine decarboxylase (ODC) made it possible to demonstrate an essential role for polyamines in cell proliferation and differentiation.[1,2,3] Polyamine depletion induced by DFMO treatment in a number of *in vitro* and *in vivo* model systems resulted in growth retardation[4,5,6] and cell death.[7,8] Further, the remarkable low toxicity of DFMO and its significant antitumor and antimetastatic activity in experimental tumor models[7,10,11] suggested that approaches aimed at perturbing polyamine metabolism should be explored in identifying novel antitumor agents.

The stringent structural specificity of the natural polyamines for mammalian cell growth suggests that appropriate structural analogues of these polycations could serve as potential antineoplastic agents through polyamine antagonism. Methylglyoxal bis(guanylhydrazone) (MGBG), a proven antineoplastic agent,[1,2] is a structural analogue of spermidine (Spd) with which it competes for entry into cells via an active transport system[13]. Further, the antitumor activity of MGBG can be potentiated following Spd depletion by DFMO.[11,14] Porter and associates[15,16] have studied a series of structural analogues of the natural polyamines and have provided a reasonable basis for exploring the therapeutic potential of such an approach for the control of neoplastic growth. The above mentioned considerations led us to investigate the antitumor activity of Norspermidine (Nspd) (Fig. 1), a close structural analogue of Spd, alone or in combination with DFMO.[17] The objective of the present investigation is to understand the mode of antiproliferative activity of Nspd in cell cultures.

$$H_2N\text{-}CH_2\text{-}CH_2\text{-}CH_2\text{-}NH\text{-}CH_2\text{-}CH_2\text{-}CH_2\text{-}CH_2\text{-}NH_2$$

SPERMIDINE

$$H_2N\text{-}CH_2\text{-}CH_2\text{-}CH_2\text{-}NH\text{-}CH_2\text{-}CH_2\text{-}CH_2\text{-}NH_2$$

NORSPERMIDINE

Figure 1. Structural Formulae of Spermidine and Norspermidine

MATERIALS AND METHODS

Chemicals. DL[1-^{14}C]ornithine (57 mCi/mmol) and S-adenosyl-L-[1-^{14}C]methionine (59 mCi/mmol) [^{3}H]thymidine (2 Ci/nmol)·[^{3}H]leucine (300 mCi/mmol) and [^{3}H]uridine (50 Ci/mmol) were purchased from New England Nuclear (Boston, MA). Nspd was obtained as the free base from the Aldrich Chemical Company (Milwaukee, WI). The base was neutralized to pH 7.0 with HCl prior to use. DFMO was synthesized in our laboratories according to previously published procedures.[1] All other chemicals were of the reagent grade obtained from commercial sources.

Cells. HeLa cells were cultured in Eagle's minimal essential medium supplemented with 10% heat inactivated serum, non-essential amino acids, sodium pyruvate, glutamine and penicillin-streptomycin mixture in an atmosphere containing 5% CO_2 at 37°C.

Determination of Antiproliferative Activity of Norspermidine. HeLa cells were plated at a density of 2.5 x 10^{5}/100 mm dish and after overnight incubation, the medium was replaced with medium containing different concentrations of Nspd. At different time periods, the cells were collected by trypsinization and counted by Coulter counter.

Animals. Male BDF1 mice were obtained from Charles River Breeding Laboratories (Wilmington, MA). Male C57BL/6 and DBA/2 mice were purchased from Jackson Laboratories (Bar Harbor, ME). The animals were housed in plastic cages under standard laboratory conditions, with free access to food and water.

L1210 Leukemia Tumor Propagation and Therapeutic Evaluation. Tumors were maintained by weekly i.p transplantation of L1210 cells into DBA/2 mice. Cells were collected by peritoneal lavage with ice-cold PBS and enumerated by counting in a Coulter counter (Coulter Electronics Inc., Hialeah, FL). For evaluation of the antitumor effects of test compounds, 10^{5} tumor cells were inoculated intraperitoneally into groups of BDF1 mice. Antitumor effect was evaluated by comparing the mean survival times of treated animals to those of untreated control animals.

3LL Subcutaneous Lewis Lung Carcinoma. Solid tumors were maintained in C57BL/6 mice by subcutaneous inoculation of 10^{6} viable dissociated 3LL carcinoma cells into the interscapular region of the mice. Antitumor activity was evaluated by comparing the tumor weights, observed at the time of autopsy on day 18 after inoculation, to that of control tumor-bearing animals. The primary tumor in this model metastasizes to the lungs within 2 weeks after subcutaneous transplantation and consequently allows the evaluation of effects on metastatic spread as well. Metastases were quantified by counting the lung foci at the time of autopsy as previously described.[11]

EL4 Lymphoma. Solid tumors were propagated in C57BL/6 mice by the procedure similar to that described for 3LL solid carcinoma. Therapeutic evaluation was carried out employing tumor weight data, according to the procedure described for 3LL tumor earlier, and by comparing mean survival times of treated and control animals.

Biochemical Assays

Determination of Macromolecular Synthesis. The effect of norspermidine on DNA, RNA and protein synthesis was determined by the ability of cells to incorporate [^{3}H]thymidine, [^{3}H]uridine and [^{3}H]leucine respectively, as described earlier.[18]

Determination of Mitotic Index. Both the supernatant medium and attached HeLa cells from control and Nspd treated were collected, counted and cytocentrifuge preparations were made. The cells were fixed in

Carnoy's fixative and stained with aceto-orcein. The percentage of mitotic cells in at least three different microscopic fields were determined.

Determination of Ornithine Decarboxylase (ODC) and S-adenosylmethionine Decarboxylase (AdoMetDC). HeLa cells from control and norspermidine treated cultures were collected. Cells were then disrupted by two cycles of freeze-thawing and sonication for 2 minutes. The resulting suspension was centrifuged at 10,000 x g for 20 min. The supernatant was employed for ODC and AdoMetDC activities by procedures described earlier.[9] Protein was quantified by the Bradford method.[19]

Determination of Polyamines. Hela cells from control and norspermidine treated cultures were extracted with 0.5 to 1 ml of 0.4 N $HClO_4$. The resulting suspension was spun at 5000 x g for 10 minutes, the supernatant employed for polyamine determinations by RP-HPLC as described earlier.[20]

In Vitro Polymerization of Bovine Brain Tubulin. Tubulin was isolated from bovine brain according to the method described by Williams and Lee.[21] This purification procedure involved cycles of assembly and disassembly followed by chromatography on phosphocellulose. Polymerization of purified tubulin into microtubules was initiated by the addition of GTP as originally described by Gaskin et al.[22] with modifications of Williams and Lee.[21] Briefly, the reaction mixture contained in a total volume of 1.5 ml, 2.25 mg of tubulin, 100 mM PIPES, 2 mM EGTA, 1 mM $MgSO_4$ pH 6.9. The assembly was initiated at 37°C by the addition of 1 mM GTP and was followed turbidimetrically at 350 nm with a Perkin-Elmer spectrophotometer. Norspermidine and other polyamines were tested at different concentrations for inhibition of tubulin assembly.

RESULTS

The results of our studies presented in Fig. 2 indicate a dose-dependent inhibition of HeLa cell growth by Nspd with an IC_{50} of 0.05 mM at the end of 96 h. Nspd treatment also resulted in an unexpected accumulation of cells in mitosis. Maximum mitotic arrest was observed on day 3 with 0.05 mM Nspd treatment.

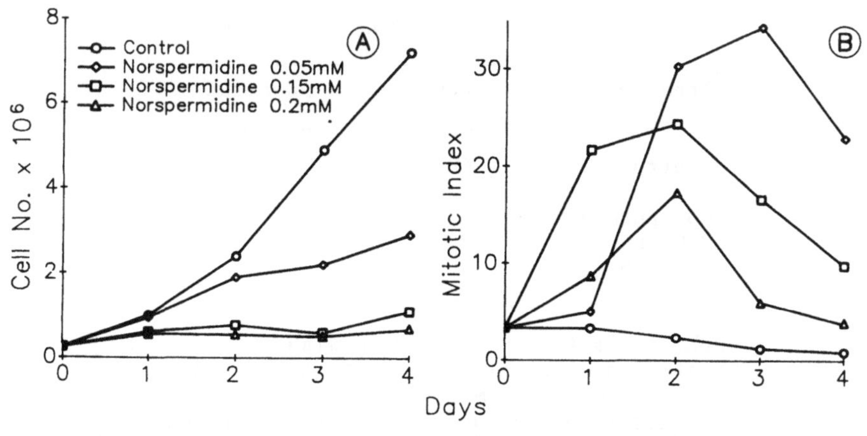

Figure 2. Effect of Norspermidine on the Growth and Mitotic Accumulation of HeLa Cells.

A: Growth Inhibition **B:** Mitotic Accumulation

Table 1. Effect of Norspermidine on the Intracellular Polyamine Levels in HeLa Cells in Culture.

Treatment	Polyamine concentration ($p.moles/10^6$ cells, Mean; n=2)			
	Put	Spd	Spm	Nspd
Control	1105	3973	1370	-
Norspermidine				
0.05 mM	62	942	1269	4828
0.15 mM	59	445	777	6128
0.20 mM	50	516	919	2370

Cells were obtained as described in Figure 2 and polyamine analysis was performed by HPLC.

These biological effects were accompanied with accumulation of Nspd and depletion of putrescine, spermidine and spermine by 95,77 and 46 percent, respectively, compared to control cells (Table 1). Nspd also showed inhibition of ODC activity by and 100% in 0.05 and 0.15 mM cell-free extracts obtained from treated cells while no inhibition of AdoMetDC by Nspd was observed (Table 2). However, Nspd failed to show inhibition of ODC activity when incubated with cell free extracts containing the enzyme, suggesting that the effect observed is an indirect effect of Nspd on the enzyme activity.

Table 2. Effect of Norspermidine on ODC and AdoMetDC Activities of HeLa Cells in Culture.

Treatment	Enzyme activity (p.moles/mg/h; Mean; n=2)	
	ODC	AdoMetDC
Control	68.9	129.3
Norspermidine		
0.05 mM	27.5	145.7
0.15 mM	0	158.4

HeLa cells (2.5×10^5/100 mm dish) were plated and incubated overnight at 37°C in a CO_2 incubator. Medium containing different concentrations of Nspd was added and incubated for 96 h with a change of medium at 48 h. Cells were collected, washed and the enzyme activity was determined.

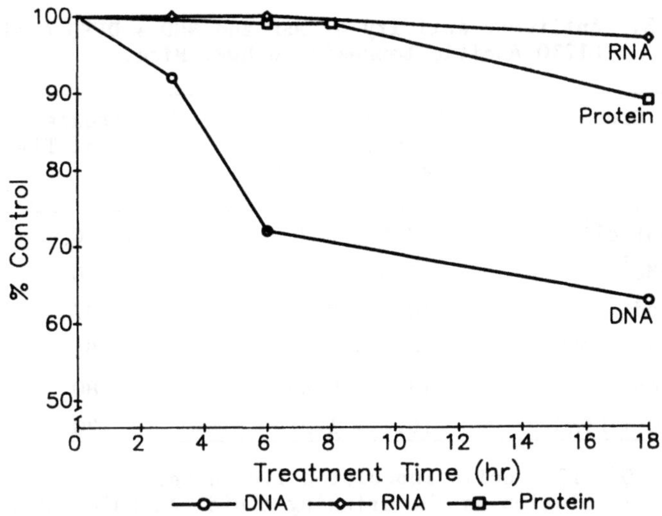

Figure 3. Effect of Norspermidine (0.15 mM) on Macromolecular Synthesis in HeLa Cells.

To understand the mechanism of antiproliferative activity of Nspd, the effect of Nspd on the macromolecular synthesis of HeLa cells was investigated. The results presented in Fig. 3 indicate that over a period of 18 h Nspd inhibited DNA synthesis by 40% without any significant effect on RNA or protein synthesis. We could not reverse the antiproliferative activity of Nspd by exogenous putrescine, spermidine or spermine in cell cultures (data not presented).

To further understand the effect of Nspd on the accumulation of cells in mitosis, we have evaluated the effect of Nspd on the <u>in vitro</u> polymerization of bovine brain tubulin. The data presented in Fig. 4 indicate a dose-dependent inhibition of tubulin polymerization of 30, 70 and 95 percent inhibition at 2, 4 and 10 mM Nspd concentrations, respectively.

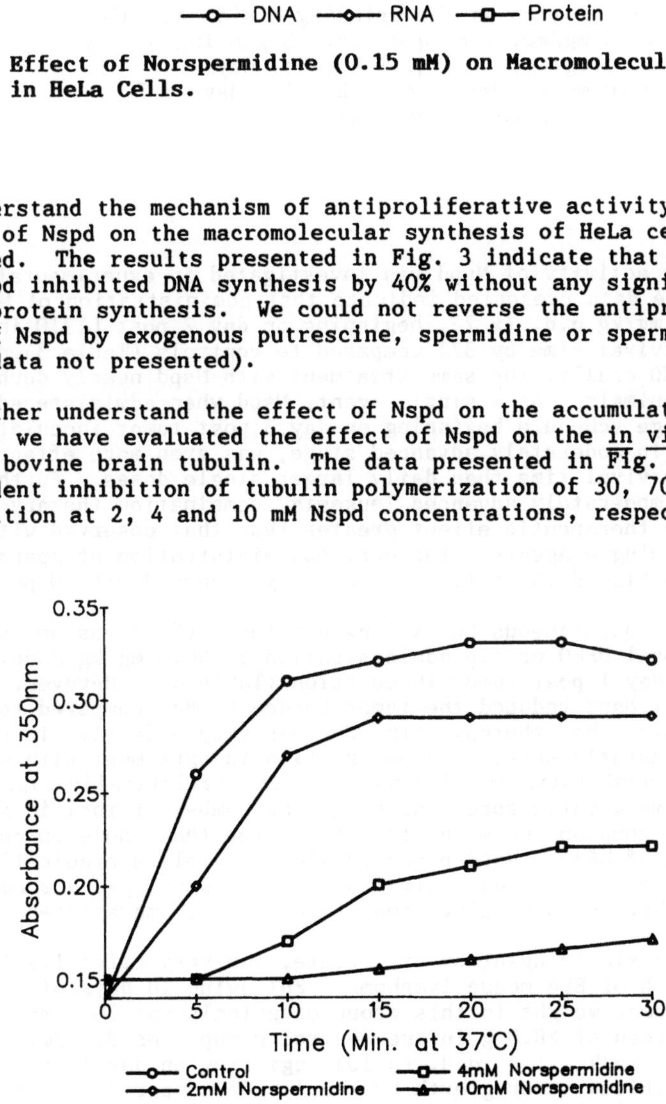

Figure 4. Effect of Norspermidine on <u>In Vitro</u> Tubulin Polymerization.

Table 3. **Antitumor Activity of NSD and NSD + DFMO Against L1210 Ascitic Leukemia in BDFI Mice.**

Group	Survival Time in Days (Mean ± SE, N=6)	% Increase in Survival Time
Control[1]	7.6 ± 0.2	0
DFMO[2]	7.6 ± 0.2	0
Nspd[3]	10.4 ± 0.2*	37
Nspd + DFMO	14.8 ± 0.1*	95
Nspd[4]	13.7 ± 0.4*	80
NSd + DFMO[5]	16.7 ± 1.2*	120

1: 10^5 L1210 cells inoculated i.p on day 0
2: 2% solution as the drinking-fluid beginning on day 0
3: 75.0 mg/kg, i.p, q.d. (x7) beginning on day 2
4: 20.0 mg/kg, i.p, q. 3 h (x4), days 3-7
5: 20.0 mg/kg, Nspd, q. 3 h (x4), days 3-7
* $p<0.01$ compared to controls

Antitumor activity of Nspd was investigated in experimental animal tumors.[17] The data presented indicate that administration of Nspd, at a dose of 75.0 mg/kg q.d. (x 7), beginning on day 2 post L1210 inoculation increased survival time by 37% compared to controls (Table 3). In animals receiving DFMO orally, the same treatment with Nspd nearly doubled the life span of the animals. As a single agent, Nspd when administered on a frequent dosage schedule beginning on day 3 post tumor inoculation, when the tumor is in moderately advanced stage, was even more effective in prolonging survival time than daily larger single doses. Furthermore, in animals with moderately advanced leukemia, combination therapy with DFMO + Nspd provided therapeutic effect greater than that observed with these two compounds as single agents. Further, coadministration of spermidine almost completely abolished the antitumor activity observed with Nspd + DFMO.[17]

Growth of subcutaneous Lewis lung carcinoma (3LL) was not significantly effected by oral DFMO or i.p administration of 50.0 mg/kg Nspd, q.d. (x 14) beginning on day 1 post tumor inoculation (Table 4). However, a combination of DFMO + Nspd reduced the tumor burden by 83% compared to untreated control animals. Monotherapy with DFMO was very effective in reducing 3LL pulmonary metastatic spread, an observation in agreement with our previous work.[11] Nspd administration did not produce statistically significant reduction in metastatic spread although the number of foci in animals receiving the compound were consistently lower than those in controls. A combination of DFMO + Nspd produced almost complete eradication of metastatic spread. Only one animal out of six developed a single pulmonary focus, the other animals being free of any visible metastases.

DFMO, as a single agent by oral route, was very effective in suppressing the growth of EL4 mouse lymphoma. Following 18 days of DFMO therapy, the average tumor weight in this group of animals was 70.0 mg compared to a tumor burden of >8.0 g in the control group. Nspd, administered at a dose of 50.0 mg/kg, i.p, q.d. (x 10) beginning on day 1 post-tumor inoculum, inhibited tumor growth by nearly 80% compared to untreated controls (Table 5). A combination of DFMO + Nspd adminis growth arrest as evidenced by the absence of any visible tumor at the site of inoculation.

Table 4. Antitumor, Antimetastatic, and Biochemical Effects of Nspd and Nspd + DFMO Against Subcutaneous 3LL Solid Tumors in Mice.

Group	Tumor Weight (g) (Mean ± SE, N=5)	% Inhibition	Number of Lung Foci (Mean ± SE, n=5)	% Inhibition
Control[1]	5.16 ± 0.51	-	31.0 ± 10.0	-
Nspd[2]	3.46 ± 0.81	33	10.2 ± 4.8	67
DFMO[3]	4.93 ± 0.79	4	4.7 ± 1.3	85
DFMO + Nspd	0.86 ± 0.16*	83	0.2 ± 0.2*	99

1: 10^6 viable 3LL cells were inoculated subcutaneously into the interscapular region on day 0; animals were autopsied on day 18, tumors were weighed and subsequently analyzed for polyamines.
2: 50.0 mg/kg, i.p, q.d. (x14) beginning on day 1
3: 2.0% in drinking water beginning on day 1 until day 18
* p<0.01 compared to controls

In a separate experiment, animals were observed for mortality. The results again demonstrated improved antitumor activity with Nspd + DFMO compared to the results obtained when these two agents are administered individually. In the group receiving DFMO + Nspd, one animal survived for more than 96 days from time of tumor inoculation.

DISCUSSION

The data presented in this article indicate that antiproliferative activity of norspermidine is due to its pertubation of polyamine metabolism. Norspermidine treatment resulted in a inhibition of ODC activity without any significant effect on AdoMetDC activity.

Table 5. Antitumor Activity of Nspd Alone or in Combination with DFMO Against EL4 Solid Tumors in Mice.

Group	Survival time in days, (Mean ± SE, N = 10)	% Increase in survival	Mean tumor weight (g) (Mean ± SE, N = 10)	% Tumor inhibition
Control[1]	21.8 ± 1.1		8.36 ± 0.82	
DFMO[2]	50.9 ± 4.1**	133	0.07 ± 0.04**	99.2
Nspd[3]	26.2 ± 1.5*	20	1.87 ± 0.51**	77.6
Nspd + DFMO	69.2 ± 7.3**◆	217	Undetectable	100.0

1: 10^6 dissociated viable tumor cells were inoculated subcutaneously into the interscapular region of C57BL/6 mice on day 0; animals were autopsied on day 18.
2: 2% in drinking water beginning on day 1
3: Administered 50.0 mg/kg, i.p, q.d. (X10) i.p. beginning on day 1.

* p<0.05 compared to controls
** p<0.001 compared to controls
◆ One animal in this group survived for >96 days

Nspd, when incubated with ODC in vitro, did not inhibit the enzyme. Polyamines are known to inhibit ODC through induction of the inhibitory protein "antizyme."[23] Recently it has been reported that spermidine and spermine inhibit ODC and AdoMetDC activities by interfering with the translation of the mRNAs for these enzymes.[24] We are currently investigating whether Nspd, being a close structural analogue of spermidine, could inhibit ODC at mRNA translation level or by the induction of antizyme.

The depletion of intracellular polyamines and accumulation of Nspd in HeLa cells not only resulted in antiproliferative activity, but also an unexpected accumulation of cells in mitosis. To further understand the mitotic arrest, we have examined the effect of Nspd on the in vitro polymerization of bovine brain tubulin. The data clearly indicated a dose-dependent inhibition of tubulin polymerization into microtubules at mM concentrations that can easily be achieved intracellularly, based on the high levels of accumulation of Nspd seen in HeLa cells.

Nspd also showed significant in vivo antitumor activity, alone or in combination with DFMO, against L1210 leukemia, 3LL carcinoma and EL4 lymphoma in mice. However, our data[17] suggest that observed antineoplastic activity might be, at least in part, due to competition with Spd for uptake and subsequent binding to putative biologically functional sites. Further, in L1210 leukemia model, the antitumor activity of DFMO + Nspd combination was reversed by Spd but not Put. However, in cell cultures we could not reverse the effects of Nspd by any natural polyamine (data not presented). Testing high concentrations of polyamines in vitro is not feasible because of the toxicity to cells. However, in animals when equivalent doses of Nspd and Spd were administered (50 mg/kg, q.d. [x7]) reversal was observed. DFMO treatment resulted in a nearly 2-fold increase in the tumoral accumulation of Nspd and significant decreases in Put and Spd concentrations. Increased cellular accumulations following Spd depletion by DFMO has been reported for another antineoplastic agent MGBG, also a structural analogue of Spd.[26] Again, co-administration of Spd but not Put abolished the DFMO-induced increases in the tumor concentration of Nspd in L1210 cells. Based on the data presented here, the antiproliferative and antitumor activity observed with norspermidine may be due to its 1) effect on the polyamine metabolism and 2) direct effect on the microtubule assembly of tumor cells.

Extensive studies with a variety of in vitro and in vivo systems have clearly established the requirement for the natural polyamines in cell growth.[2,3] Although the exact biochemical mechanism by which these organic polycations exert their effects remains elusive, the structural specificity for growth and binding to macromolecules indicates that polyamines are involved in specific events pertaining to cellular proliferation. Such a specific role for Spd has been reported recently[27] in the modulation of the activity of translation initiation factor eIF-4D. In the past a number of structural analogues of the polyamines have been synthesized and demonstrated to exhibit antitumor activity in vivo against murine tumor models.[28] The cytotoxicity of these polyamine analogues was attributed to the toxic metabolites produced via the action of plasma amine oxidase on the compounds. Subsequent work by other investigators[26,14] with a series of structural analogues of the natural polyamines has demonstrated the relatively strict structural specificity of these compounds for entry into cells and for promoting cell growth.[29,30] Weinstock et al.[31] synthesized and tested a number of acetylated derivatives of Spd and a few of these compounds did show activity against B16 melanoma in vivo. These experimental observations and results presented herein warrant further studies aimed at developing polyamine analogues as potential antineoplastic agents.

ACKNOWLEDGEMENTS

The authors would like to thank Mr. Phillip J. Lachmann and Mr. Gregory F. Davis for their technical assistance, and Ms. Debbra Wagner and Mrs. Rita Hoeper for their secretarial help.

REFERENCES

1. B. W. Metcalf, C. Danzin, M.T. Jung, P. Casara and J.P. Vevert, Catalytic Irreversible Inhibition of Mammalian Ornithine Decarboxylase by Substrate and Product Analogues. J. Am. Chem. Soc. 100:2551, 1978.
2. A. E. Pegg, and P.P. McCann, Polyamine Metabolism and Function: A Review. Am. J. Physiol. 243:212, 1982.
3. P. S. Sunkara, and N.J. Prakash, Inhibitors of polyamine biosynthesis as antitumor and antimetastatic agents, in: Novel Approaches to Cancer Chemotherapy. P.S. Sunkara, ed., pp 93, Academic Press, New York (1982).
4. P. S. Mamont, M.C. Duchesne, J. Grove, and P. Bey, Antiproliferative Properties of D,L-α-difluoromethylornithine in Cultured Cells. A Consequence of Irreversible Inhibition of Ornithine Decarboxylase. Biochem. Biophys. Res. Commun. 81:58 (1978).
5. J. Seidenfeld, J.W. Gray, and L.J. Marton, Depletion of 9L Rat Brain Tumor Cell Polyamine Content by Treatment with D.L-α-difluoromethyl-ornithine Inhibits Proliferation and the G_1 to s transition. Exp. Cell Res, 131:209 (1981).
6. P. S. Sunkara, S.K. Fowler, K. Nishioka, and P.N. Rao, Inhibition of Polyamine Biosynthesis by α-Difluoromethylornithine Potentiates the Cytotoxic Effects of Arabinosyl Cytosine in HeLa Cells. Biochem. Biophys. Res. Commun.95:423 (1980).
7. G. D. Luk, G. Goodwin, L.J. Marton, and S.B. Baylin, Polyamines are Necessary for the Survival of Human Small-Cell Lung Carcinoma in Culture. Proc. Natl. Acad. Sci. USA, 78:2355 (1981).
8. P. S. Sunkara, C.C. Chang, N.J. Prakash, and P.J. Lachmann, Effect of Inhibition of Polyamine Biosynthesis by DL-α-difluoromethyl-ornithine on the Growth and Melanogenesis of B16 Melanoma in vitro and in vivo. Cancer Res. 45:4067 (1985).
9. N. J. Prakash, P.J. Schechter, P.S. Mamont, J. Grove, J. Koch-Weser, and A. Sjoerdsma, Inhibition of EMT6 Tumor Growth by Interference with Polyamine Biosynthesis: Effects of α-difluoromethylornithine, an Irreversible Inhibitor of Ornithine Decarboxylase. Life Sci. 26:181 (1980).
10. P. S. Sunkara, N.J. Prakash, G.D. Mayer, and A. Sjoerdsma, Tumor Suppression with a Combination of α-difluoromethylornithine and Interferon. Science Wash.DC) 219:851 (1983).
11. P. S. Sunkara, N.J. Prakash, and A.L. Rosenberger, An Essential Role for Polyamines in Tumor Metastases. FEBS Lett. 150:397 (1982).
12. W. A. Knight, R.B. Livingston, C. Fabin, and J. Costanzi, Methylglyoxalbisguanylhydrazone (Methyl GAG, MGBG) in Advanced Human Malignancy. Proc. Am. Soc. Clin. Oncol. 20:319 (1979).
13. C. Dave, and L. Caballeri, Studies on the Uptake of Methylglyoxal Bis(guanylhydrazone) (CH_3-G) and Spermidine (SPd) in Mouse Leukemia L1210 Sensitive and Resistant to CH_3-G. Fed. Proc. Fed. Am. Soc. Exp. Biol. 32:763 (1973).
14. J. Janne, E. Holtta, A. Kallio, and K. Kapyaho, Role of Polyamines And Their Antimetabolites in Clinical Medicine. Special topics in: Endocrinology and Metabolism 5:227 (1983).
15. C. W. Porter, and R.J. Bergeron, Spermidine Requirement for Cell Proliferation in Eukaryotic Cells. Structural Specificity and Quantitation. Science 219:1083 (1983).

16. C. W. Porter, J. McManis, R.A. Casero, and R.J. Bergeron, Relative Abilities of Bis(ethyl) Derivatives of Putrescine, Spermidine and Spermine to Regulate Polyamine Biosynthesis and Inhibit L1210 Leukemia Cell Growth. Cancer Res. 47:2821 (1987).
17. N. J. Prakash, T.L. Bowlin, G.F. Davis, P.S. Sunkara, and A. Sjoerdsma, Antitumor Activity of Norspermidine, a Structural Analogue of the Natural Polyamine, Spermidine. Anticancer Res. (In press).
18. R. J. Mans, and G.D. Novelli, Measurement of the Incorporation of Radioactive Aminoacids into Protein by a Filter-paper Disk Method. Arch. Biochem. Biophys. 94:48 (1961).
19. M. M. Bradford, A Rapid and Sensitive Method for the Quantitation of Microgram Quantities of Proteins Utilizing the Principle of Protein-dye Binding. Anal. Biochem. 72:248 (1976).
20. T. L. Bowlin, B.J. McKown, and P.S. Sunkara, Ornithine Decarboxylase Induction and Polyamine Biosynthesis are Required for the Growth of Interleukin-2 and Interleukin-3 Dependent Cell Lines. Cellular Immunol. 98:341 (1986).
21. R. C. Williams, and J.C. Lee, Preparation of Tubulin From Brain. Methods in Enzy. 85:376 (1982).
22. F. Gaskin, C.R. Cantor, and M.L. Shelanski, Turbidimetic Studies of the In Vitro Assembly and Disassembly of Porcine Neurotubules. J. Mol. Biol. 89:737 (1974).
23. E. S. Canellakis, J.S. Heller, and D.A. Kyriakidis, The interaction of ornithine decarboxylase with its antizyme. in: Adv. Polyamine Res. 3:1 (1981).
24. T. Kameji, and A.E. Pegg, Inhibition of Translation of mRNAs for Ornithine Decarboxylase and S-adenosylmethionine Decarboxylase by Polyamines. J. Biol. Chem. 262:2427 (1987).
25. L. Alhonen-Hongisto, P. Seppanen, and J. Janne, Intracellular Putrescine and Spermidine Deprivation Induces Increased Uptake of the Natural Polyamines and Methylglyoxal Bis(guanylhydrazone). Biochem. J. 192:941 (1980).
26. P. S. Sunkara, N.J. Prakash, C.C. Chang, and A. Sjoerdsma, Cytotoxicity of Methylglyoxal Bis(guanyl hydrazone) in Combination with α-difluoromethylornithine Against HeLa Cells and Mouse L1210 Leukemia. J. Natl. Cancer Inst. 40:505 (1983).
27. M. H. Park, S.I. Chung, H.L. Colper, and J.E. Folk, The Mammalian Hypusine-containing Protein, Eukaryotic Initiation Factor 4D. J. Biol. Chem.259:4563 (1984).
28. M. Israel, E.C. Zoll, N. Muhammad, and E.J. Modest, Synthesis and Antitumor Evaluation of Presumed Cytotoxic Metabolites of Spermine and N, N^1-bis(3-aminopropyl) Nonane-1,9-diamine. J. Med. Chem. 16:1 (1974).
29. D. R. Morris, In Vivo Studies of the Roles of Putrescine and Spermidine in Escherichia coli. in: Polyamines in Biology and Medicine, D.R. Morris and L.J. Marton, eds, Marcel-Dekker, New York (1981).
30. C. W. Porter, J. McManis, R.A. Casero, and R.J. Bergeron, Relative Abilities of Bis(ethyl) Derivatives of Putrescine, Spermidine and Spermine to Regulate Polyamine Biosynthesis and Inhibit L1210 Leukemia Cell Growth. Cancer Res. 47:2821 (1987).
31. L. T. Weinstock, W.J. Post, and C.C. Cheng, Synthesis of New Polyamine Derivatives for Cancer Chemotherapeutic Studies. J. Pharm. Sci. 70:956 (1981).

THE EFFECTS OF POLYAMINE ANALOGUES ON MALARIA PARASITES IN VITRO AND IN VIVO

Alan J. Bitonti, Peter P. McCann and Albert Sjoerdsma

Merrell Dow Research Institute
Cincinnati, Ohio

INTRODUCTION

Malaria remains one of the most important human infectious diseases in the world. Resistance of Plasmodium falciparum to chloroquine and other antimalarial drugs has created an urgent need for new drugs which are effective against resistant strains (1). Radical departure from the chemical classes of compounds effective in the past is needed to circumvent the problems of cross-resistance between drugs. We have studied a new approach to the chemotherapy of malaria which involves the disruption of the biosynthesis and/or intracellular function of the polyamines, putrescine, spermidine and spermine.

The polyamines are important regulators of growth and differentiation in a wide variety of cell types including parasitic protozoa (2). α-Difluoromethylornithine (DFMO; eflornithine), an irreversible inhibitor of the first enzyme in polyamine biosynthesis, ornithine decarboxylase, inhibits the proliferation of a number of human-infective parasites including P. falciparum (3,4). We have recently shown that DFMO also inhibits the growth of P. berghei in mice but does not cure these infections (5).

Another approach to interference with polyamine biosynthesis and function, which has been applied successfully in cancer cells (6), takes advantage of the ability of the natural polyamines (7,8) and non-physiological polyamine analogs (9,10) to suppress ornithine decarboxylase activity in the cell and also possibly to block the intracellular function of the natural polyamines by displacing them from critical binding sites such as appear to be present on DNA (11). In this approach, it is preferred that the polyamine analogs not subserve the normal physiological roles of putrescine, spermidine and spermine, since then they would simply replace the natural polyamines and growth inhibition would not occur. We have found that a series of bis(benzyl) polyamine analogs have marked antimalarial activity against both chloroquine-sensitive and -resistant Plasmodium falciparum in vitro and when administered in combination with DFMO, cure murine malaria.

EXPERIMENTAL

Culture of Plasmodium falciparum

P. falciparum was maintained in continuous culture in human erythro-

6% suspension of human (type O+) erythrocytes in RPMI 1640 medium supplemented with 10% human serum, 25 mM HEPES and 0.2% $NaHCO_3$ and were incubated at 37°C in an atmosphere of 90% N_2, 5% O_2 and 5% CO_2.

Microtiter Cultures for Drug Testing

Drug testing *in vitro* was carried out using standard procedures (13). In 96-well microtiter cultures, *P. falciparum* was at a starting parasitemia of 0.5% in a 1% suspension of erythrocytes in a total volume of 0.2 ml. [^{3}H]Hypoxanthine (1 µCi/well) was added 24 h after initiation of cultures and the incorporation of [^{3}H]hypoxanthine into plasmodial RNA and DNA was measured by harvesting the organisms onto glass fiber filters 18 h later using an automatic cell harvester and subsequent liquid scintillation counting of the filters.

Protein and Nucleic Acid Synthesis

Microtiter cultures were used to compare the acute effects of MDL 27695 on [^{3}H]isoleucine (to measure protein synthesis) and [^{3}H]hypoxanthine (to measure RNA and DNA synthesis) incorporation. Starting parasitemia in these experiments was 4% in a 1% suspension of erythrocytes. [^{3}H]Isoleucine (1 µCi/well) and [^{3}H]hypoxanthine (1 µCi/well) were each added to their respective cultures at the same time as MDL 27695 and the incorporation of the tritium-labelled compounds was determined as above.

To determine if the effects of MDL 27695 were specifically on the synthesis of either DNA or RNA, or both, parasitized erythrocytes were labelled with [^{3}H]hypoxanthine, washed with cold culture medium and lysed with a solution of 50 mM Tris·HCl (pH 7.0) containing 10 mM EDTA, 2% Triton X-100 (v/v) and 10 µg calf thymus DNA/ml. Total nucleic acids were extracted with trichloroacetic acid as described previously (14). RNA was digested by incubating the pellet of total nucleic acids with 0.3 N NaOH for 18 h at 37°C. The remaining DNA was again precipitated with trichloroacetic acid and the radioactivity incorporated into DNA and total nucleic acids was measured by collecting the precipitates on glass fiber filters followed by scintillation counting. Radioactive incorporation into RNA was determined by subtracting the radioactivity present in DNA from that present in the total nucleic acids.

Thermal Denaturation of Calf Thymus DNA

Calf thymus DNA (15 µg/ml) in a solution of 30 mM NaCl and 1 mM potassium dimethylglutarate (pH 6.2), was incubated in a spectrophotometer equipped with temperature controlled cuvette holders. Either spermine or polyamine analogs were added to the cuvettes and the temperature was raised in intervals of 5°C. The DNA solution was held at each temperature for 10 min before the absorbance at 260 nm was recorded. Melting temperatures (T_m) were determined to be the temperature at which the change in absorbance at 260 nm was one-half of the maximum change.

Plasmodium berghei Infections

The *P. berghei* infection model used to study the blood schizonticidal activity of the polyamine analogs was that of Peters et al. (16). Outbred CD-1 mice weighing 20-22 g, were infected by intravenous injection of 10^7 parasitized erythrocytes taken from a mouse with a rising parasitemia of 20-30%. The NK65 strain of *P. berghei* was used for all studies.

The polyamine analog, MDL 27695, was administered intraperitoneally in 0.2 ml of water once (within 4-6 hr after infection) on the day of infection (Day 0) and three times per day (8:30 AM, 12:30 PM and 4:30 PM) for 3 days (Days 1-3). DFMO was provided ad libitum as a 2% solution in the drinking water for 4 consecutive days beginning on Day 0 at the same time as the first injection of MDL 27695.

Degrees of parasitemia were determined using thin smears of venous blood stained with Giemsa. Results are expressed as the percentage of erythrocytes infected with at least one parasite.

Chemicals

[8-^{3}H]Hypoxanthine (10 Ci/mmol) and L-[4,5-^{3}H(N)]isoleucine (127 Ci/mmol) were purchased from ICN Radiochemicals and New England Nuclear, respectively. The polyamine analogs and α-difluoromethylornithine (DFMO, eflornithine) were synthesized at Merrell Dow Research Institute, Cincinnati, Ohio.

RESULTS

Growth Inhibition of P. falciparum by Polyamine Analogs

A continuous culture of *P. falciparum* (clone D6) was used to determine the antimalarial activity of a series of polyamine analogs[3]. Growth inhibition was assessed by measuring the decreased uptake of [^{3}H]hypoxanthine by infected erythrocytes during a 48 h exposure to the analogs. All the compounds tested (Table 1) are spermine analogs which have elongated central methylene chains and either benzyl-substituted or free terminal amines.

Table 1. Growth Inhibition of *Plasmodium falciparum in vitro* by polyamine analogs.

	Y	Compound Number	Antimalarial Activity IC$_{50}$ µM
Spermine:			
$H_2N(CH_2)_3NH(CH_2)_YNH(CH_2)_3NH_2$	4		
Bis-Benzyl Analogs:			
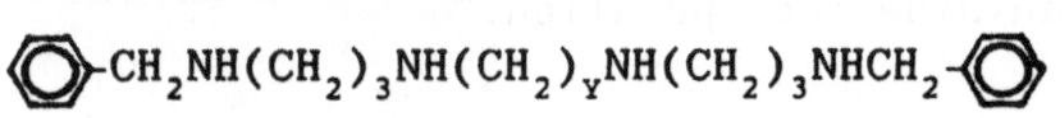	4	MDL 27847	14
	5	MDL 27957	6
	6	MDL 27693	5.3
	7	MDL 27695	3.0
	8	MDL 27391	0.83
	9	MDL 27701	0.69
	10	MDL 27700	0.48
	12	MDL 27994	0.21
Free Amine Analogs:			
$H_2N(CH_2)_3NH(CH_2)_YNH(CH_2)_3NH_2$	7	MDL 26752	4,300
	8	MDL 26547	1,800

The antimalarial activity of the polyamine analogs was measured by following the incorporation of [^{3}H]hypoxanthine into *P. falciparum* clone D6 as described in the Experimental section. IC$_{50}$ represents the concentrations of drugs which inhibit [^{3}H]hypoxanthine incorporation by 50% when compared to untreated controls after approximately 48 h of drug exposure.

The bis(benzyl) analogs were all potent inhibitors of P. falciparum growth, having IC_{50} values ranging from 0.2 - 14 µM; potency increasing as the length of the central methylene chain was increased from 4 to 12. The most potent of the polyamine analogs, MDL 27994, was approximately equipotent with quinine, a standard antimalarial drug, which had an IC_{50} = 0.19 µM when tested against clone D6. In marked contrast to the bis(benzyl) polyamines, the free-amine analogs were only weak inhibitors of P. falciparum growth in vitro with IC_{50} values 1500-2000 times higher than their bis(benzyl) counterparts.

From the series of analogs shown in Table 1, MDL 27695 was selected as a prototype with which to carry out further work because initial tests in mice showed that this compound had a better therapeutic index than the other compounds of this series. MDL 27695 was tested against a chloroquine-resistant P. falciparum strain (FCR3) and the analog was found to be equally effective against FCR3 or the chloroquine-sensitive clone D6 even though the two strains differ 15-fold in their sensitivities to chloroquine (Fig. 1).

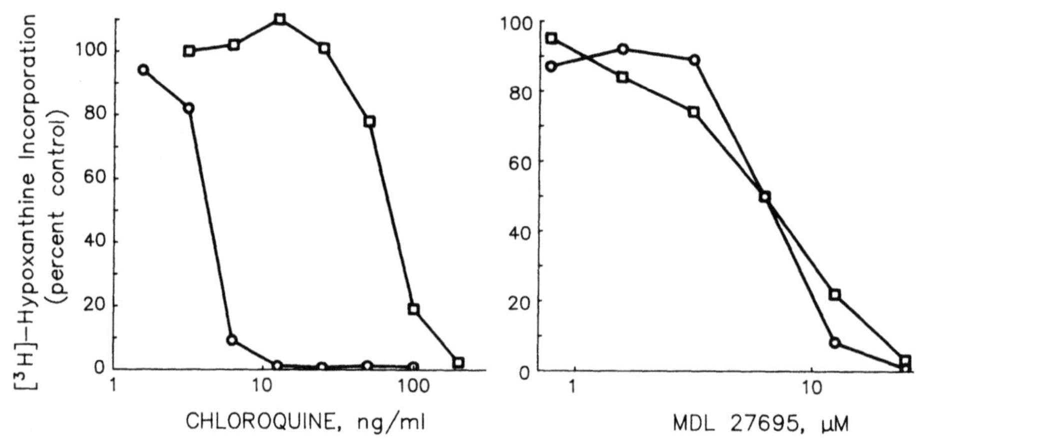

Figure 1. Response of Chloroquine-Sensitive and Chloroquine-Resistant P. falciparum to Chloroquine and MDL 27695

Chloroquine-sensitive (clone D6,O) and chloroquine-resistant (FCR-3 strain,□) strains of P. falciparum were cultured as described in the Experimental section. Parasite growth was measured by [^{3}H]hypoxanthine incorporation.

Effects of Polyamine Analogs on Protein and Nucleic Acid Synthesis in P. falciparum

The early effects of MDL 27695 on macromolecular biosynthesis were studied to elucidate potential mechanisms of action for the polyamine analogs. The incorporation of [^{3}H]hypoxanthine and [^{3}H]isoleucine was measured as an indicator of DNA/RNA synthesis and protein synthesis, respectively. The incorporation of [^{3}H]hypoxanthine is inhibited after 6 h in the presence of either 20 µM or 100 µM MDL 27695 (Fig. 2) and this inhibition increases with time of exposure to the drug. The incorporation of [^{3}H]isoleucine (Fig. 2), by contrast, is not inhibited by either 20 µM or 100 µM MDL 27695 until 14 h, after which time 100 µM MDL 27695 was markedly inhibitory. A similar effect on [^{3}H]hypoxanthine incorporation was noted with 10 µM and 50 µM MDL 27391 (data not shown), suggesting that this phenomenon was common to the entire series of polyamine analogs. The effect of MDL 27695 on the incorporation of [^{3}H]hypoxanthine into P. falciparum nucleic acids was apparently due to an inhibition of both DNA and RNA synthesis (Table 2).

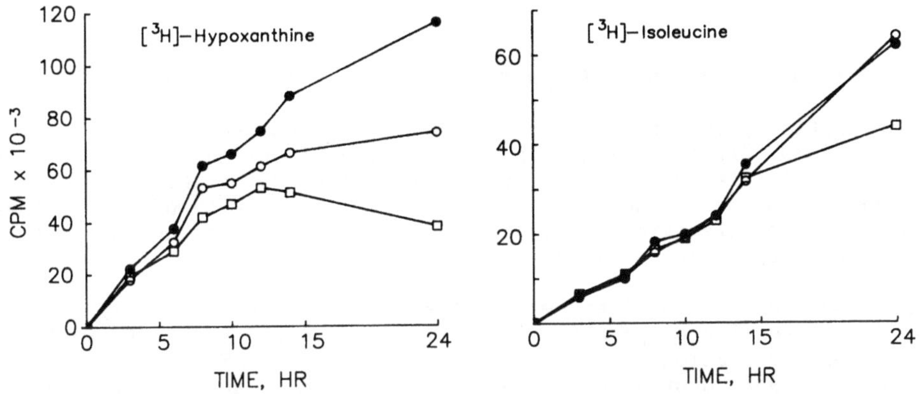

Figure 2. Inhibitory Effects of MDL 27695 on Nucleic Acid ([^{3}H]Hypoxanthine Incorporation) and Protein ([^{3}H]Isoleucine Incorporation) Synthesis in P. falciparum

P. falciparum (clone D6) was incubated in microtiter plates without treatment (●) or in the presence of 20 μM (o) or 100 μM (□) MDL 27695. Parasites were harvested at the indicated times as described in the Experimental section.

TABLE 2. Inhibitory effects of MDL 27695 on DNA and RNA synthesis in P. falciparum.

Incubation Time	Additions	[^{3}H]Hypoxanthine Incorporation: Total Nucleic Acids	DNA	RNA
		cpm/well (% control)		
3 hr	None	21,723 (100)	732 (100)	20,990 (100)
	50 μM MDL 27695	18,959 (87)	741 (101)	18,218 (87)
	250 μM MDL 27695	13,062 (60)	555 (76)	12,507 (60)
6 hr	None	50,907 (100)	1,993 (100)	48,914 (100)
	50 μM MDL 27695	42,957 (84)	1,136 (57)	41,821 (85)
	250 μM MDL 27695	14,305 (28)	715 (36)	13,590 (28)

P. falciparum (clone D6) was incubated with [^{3}H]hypoxanthine and incorporation of radioactivity into total nucleic acids, DNA and RNA was measured as described in the Experimental section.

Stabilization of DNA Against Thermal Denaturation

MDL 27695, MDL 26752 and spermine stabilized calf thymus DNA against thermal denaturation. The melting temperature (T_m) for DNA in the absence of polyamines was 78°C. Addition of 5 μM of either spermine, MDL 27695 or MDL 26752 raised the T_m to 87°C, 83°C and 86°C (Fig. 3), respectively, suggesting that all three polyamines bound directly to DNA.

Effects of MDL 27695 plus DFMO on P. berghei Infections in Mice

The initial promising results with *P. falciparum* led us to test the efficacy of MDL 27695 *in vivo*. *In vivo* testing of antimalarial compounds is generally initiated using the rodent-infective, *P. berghei*. The use of *P. berghei* is necessary because of the lack of infectivity of *P. falciparum* and other human-infective malaria parasites for laboratory animals other than a few rare species of monkeys.

As shown in Table 3, treatment with MDL 27695 alone (Group 2) resulted in a 70% suppression of parasitemia by Day 4 and a 21% cure rate (3/14), whereas the combination of 15 mg/kg MDL 27695 and 2% DFMO (Group 3) resulted in a 95% suppression of parasitemia on Day 4 and a 100% cure rate (14/14). Overall 86/96 (90%) mice were cured in several separate experiments with 10 or 15 mg/kg MDL 27695 plus 2% DFMO. Mice cured in these experiments were rechallenged with *P. berghei* 4 months after the initial infection, and although parasitemia developed, it was transient, suggesting that immunity had developed.

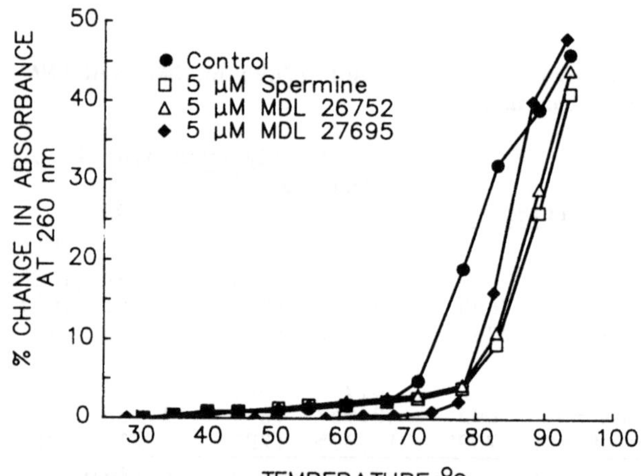

Figure 3. Stabilization of Calf Thymus DNA Against Heat Denaturation by Spermine and Polyamine Analogues

Calf thymus DNA was heat-denatured as descrbed in the Experimental Section. Incubations included (●), no additions; (□), 5 μM spermine; (Δ), 5μM MDL 26752; or (◆), 5 μM MDL 27695.

TABLE 3. Effects of MDL 27695 plus DFMO on *P. berghei* in mice.

Group #	Treatment MDL 27695	Treatment DFMO	Percent Parasitemia Day 4	Average Day of Death	Number of Animals Cured/Total
1	0	0	12.5 ± 3.5	16.7 ± 4.7 (n=14)	0/14
2	15 mg/kg	0	3.7 ± 1.2	23.2 ± 5.1 (n=10)	3/14
3	15 mg/kg	2%	0.13 ± 0.07	---	14/14

Mice were infected and treated following the standard protocol outlined in the Experimental section. Mice were considered cured when they were aparasitemic on periodic examination and lived for longer than 4 months after infection. MDL 27695 was administered intraperitoneally while DFMO was given in drinking water. The average day of death is calculated based only on the animals which died.

It was next found that MDL 27695 plus DFMO was also effective against severe "established" infections of *P. berghei* in mice (Table 4). Mice were allowed to develop a high parasitemia over a 7 day period prior to beginning drug treatments. All mice appeared moribund when drug treatments commenced. Subsequent treatment with MDL 27695 plus DFMO over a two week period resulted in cures for 15/15 animals which finished the full course of chemotherapy. Five mice in the drug-treated group, all of which had parasitemias in excess of 40% at the time drug treatment began, died during the first two days of treatment of infections.

TABLE 4. Effects of MDL 27695 plus DFMO on "established" *P. berghei* infections in mice.

Treatment MDL 27695	Treatment DFMO	Percent Parasitemia Day 7	Percent Parasitemia Day 11	Average Day of Death	Number of Animals Cured/Total
0	0	37.9 ± 13.1	N.D.[a]	14.7 ± 2.6	0/20
15 mg/kg, i.p.	2%	27.5 ± 12.3	7.0 ± 4.5	---	15/20[b]

[a] N.D. = not determined

[b] 5/20 treated mice died during the first two days of treatment due to overwhelming malaria infections.

Mice were infected as described in the Experimental section and the infections were allowed to develop for 7 days prior to drug treatments. Beginning on Day 7, MDL 27695 was given by intraperitoneal injection, 3 times per day for 4 days along with a 2% DFMO solution in place of drinking water, followed by an additional 5 days of treatment beginning on Day 14 and continuing for 5 days (until Day 18). Treated mice became aparasitemic by Day 15. Criteria for cure were as noted in Table 3.

DISCUSSION

The bis(benzyl) polyamine analogs we have described herein were markedly cytotoxic to malaria parasites in vitro and in vivo. The growth inhibitory activity of the analogs against P. falciparum in vitro increased as the length of the central methylene chain (designated by Y in Table 1) was lengthened from 4-12. The increasing potency appears to be related to the analogs' lipid solubilities and, therefore, possibly to the uptake of the compound into the host erythrocytes. In line with this hypothesis is the finding that free-amine analogs, which are much more water soluble, were at least 1000 times less potent against P. falciparum than their bis(benzyl) counterparts. In preliminary studies conducted in our laboratories, the bis(benzyl) analogs were accumulated by human erythrocytes to a much greater extent than were the free-amine analogs. Ready access to the intraerythrocytic compartment may, in part, explain why the polyamine analog, MDL 27695, was also effective against P. berghei in mice.

The first biochemical effect of MDL 27695 which we have noted in P. falciparum was inhibition of RNA and DNA synthesis within 3-6 h after addition of the analog. The depression in DNA/RNA synthesis preceded by several hours the inhibition of [^{3}H]isoleucine incorporation into proteins. These data lead us to propose that one mechanism for destruction of malaria parasites by polyamine analogs is relatively rapid inhibition of macromolecular synthesis leading to cell death. The analogs may bind to DNA in a manner similar to the natural polyamines (17) or the compounds may act as bifunctional intercalating agents such as the diacridine polyamine analogs described by Canellakis et al. (18). The mechanism of action of the diacridine polyamines was found to be specific inhibition of RNA synthesis, purportedly through direct interaction with DNA (19,20).

Binding of the polyamine analogs to DNA is also an attractive explanation for the apparent potentiation by DFMO of the effects of MDL 27695 in vivo. It was shown previously that depletion of polyamines by DFMO, sensitized tumor cells to the cytocidal effects of drugs thought to act through binding to DNA (21). It was also shown that the addition of DFMO to P. falciparum in vitro resulted in blockade of the cell cycle at the late trophozoite stage (3,22), a stage in parasite development which may possibly be particularly sensitive to the effects of polyamine analogs.

We have examined the effects of DFMO and MDL 27695 against asynchronous cultures of P. falciparum (clone D6) and found that the drugs are merely additive rather than synergistic. Future studies may be directed at more clearly defining the biochemical effects of DFMO and MDL 27695 in synchronized cultures.

These polyamine analogs may additionally repress polyamine biosynthesis as do other polyamine analogs (10). Depletion of polyamines by DFMO inhibits erythrocytic schizogony of P. falciparum in vitro (3,4) and P. berghei in vivo (5), exoerythrocytic schizogony of P. berghei (23,24) and the sporogonous cycle in the mosquito vector (25). Therefore, repression of polyamine biosynthesis by our polyamine analogs might be a potential mechanism for slowing growth of the malaria parasite. However, exogenous putrescine, spermidine or spermine did not reverse the growth inhibition caused by polyamine analogs either in vitro or in vivo (unpublished observations). This lack of reversal by the natural polyamines would be expected if the polyamine analogs block the intracellular binding of the polyamines or displace intracellular polyamines from their binding sites. Furthermore, inhibition of polyamine biosynthesis by DFMO generally results only in cytostasis rather than cell death (2), but the bis(benzyl) polyamine analogs seem to be cytotoxic. Thus, while polyamine biosynthesis may be repressed by our polyamine analogs, we do not believe that this inhibition is the major toxic event in malaria parasites. Regardless of mechanism, the polyamine analogs represent a new class of antimalarial drugs which may be useful for the chemotherapy of *P. falciparum* infections resistant to standard agents.

REFERENCES

1. W. Peters, The Problem of Drug Resistance in Malaria. *Parasitology* 90, 705-715 (1985).
2. A. E. Pegg, P. P. McCann, Polyamine Metabolism and Function. *Am. J. Physiol.* 234, C212-C221 (1982).
3. P. P. McCann, C. J. Bacchi, W. L. Hanson, G. D. Cain, H. C. Nathan, S. H. Hutner, and A. Sjoerdsma, Effect on Parasitic Protozoa of α-Difluoromethylornithine - An Inhibitor of Ornithine Decarboxylase. *Adv. Polyamine Res.* 3, 97-110 (1981).
4. J. M. Whaun and N. D. Brown, Ornithine Decarboxylase Inhibition and the Malaria-Infected Red Cell: A Model for Polyamine Metabolism and Growth. *J. Pharm. Exp. Ther.* 233, 507-511 (1985).
5. A. J. Bitonti, P. P. McCann and A. Sjoerdsma, *Plasmodium falciparum* and *Plasmodium berghei*: Effects of Ornithine Decarboxylase Inhibitors on Erythrocytic Schizogony. *Exp. Parasitol.* 64, 237-243 (1987).
6. C. W. Porter and J. R. Sufrin, Interference with Polyamine Biosynthesis and/or Function by Analogs of Polyamines or Methionine as a Potential Anticancer Chemotherapeutic Strategy.
7. J. E. Kay and V. J. Lindsay, Control of Ornithine Decarboxylase Activity in Stimulated Human Lymphocytes by Putrescine and Spermidine. *Biochem. J.* 132, 791-796 (1973).
8. P. P. McCann, C. Tardif, J.-M. Hornsperger and P. Bohlen, Two Distinct Mechanisms for Ornithine Decarboxylase Regulation by Polyamines in Rat Hepatoma Cells. *J. Cell Physiol.* 99, 183-190 (1979).
9. P. P. McCann, C. Tardif, A. E. Pegg and K. Diekema, The Dual Action of the Non-Physiological Diamines 1,3 Diaminopropane and Cadaverine on Ornithine Decarboxylase of HTC Cells. *Life Sci.* 26, 2003-2010 (1980).
10. C. W. Porter, F. G. Berger, A. E. Pegg, B. Ganis and R. J. Bergeron, Regulation of Ornithine Decarboxylase Activity by Spermidine and the Spermidine Analogue N^1N^8-Bis(Ethyl)Spermidine. *Biochem. J.* 242, 433-440 (1987).
11. L. J. Marton and B. G. Feuerstein, Polyamine-DNA Interactions: Possible Site of New Cancer Chemotherapeutic Intervention. *Pharm. Res.* 3, 311-317 (1986).
12. W. Trager and J. B. Jensen, Human Malaria Parasites in Continuous Culture. *Science* 193, 673-675 (1976).
13. R. E. Desjardins, C. J. Canfield, J. D. Haynes and J. D. Chulay, Quantitiative Assessment of Antimalarial Activity In Vitro by a Semiautomated Microdilution Technique. *Antimicrob. Agents Chemother.* 16, 710-718 (1979).
14. J. Inselburg and H. S. Banyal, Synthesis of DNA During the Asexual Cycle of *Plasmodium falciparum* in Culture. *Mol. Biochem. Parasitol.* 10, 79-87 (1984).
15. H. Tabor, The Protective Effect of Spermine and Other Polyamines Against Heat Denaturation of Deoxyribonucleic Acid. *Biochemistry* 1, 496-501 (1962).
16. W. Peters, J. H. Portus and B. L. Robinson, The Value of Drug-Resistant Strains of *P. berghei* in Screening for Blood Schizontocidal Activity. *Ann. Trop. Med. Parasitol.* 69, 155-171 (1975).
17. B. G. Feurerstein, N. Pattabiraman and L. J. Marton, Spermine-DNA Interactions: A Theoretical Study. *Proc. Nat. Acad. Sci. USA* 83, 5948-5952 (1986).
18. E. S. Canellakis, Y. H. Shaw, W. E. Hanners and R. A. Schwartz, Chemistry, Physical Chemistry and Growth Inhibitory Properties. *Biochem. Biophys. Acta* 418, 277-289 (1976).

19. E. S. Canellakis and R. A. Ballantone, The Biological Effects of Putrescine, Spermidine and Spermine Diacridines on HeLa Cells and on the L-1210 and P-388 Leukemia Cells. Biochem. Biophys. Acta 418, 290-299 (1976).
20. E. S. Canellakis, V. Boni, R. A. Bellantone, J. S. Krakow, R. M. Fico and R. A. Schulz, Definition of the General Site of Action. Biochem. Biophys. Acta 418, 300-314 (1976).
21. P. J. Tofilon, S. M. Oredsson, D. F. Deen and L. J. Marton, Polyamine Depletion Influences Drug-Induced Chromosomal Damage. Science 217, 1044-1046 (1982).
22. Y. G. Assaraf, J. Golenser, D. T. Spira and U. Bachrach, Synchronization of *P. falciparum* Cultures by DL-α-Difluoromethylornithine and Putrescine. J. Protozool. 31, A78-A79 (1984).
23. J. M. Gillet, G. Bone and F. Herman, Inhibitory action of α-Difluoromethlornithine on Rodent Malaria (*Plasmodium berghei*). Trans. Roy. Soc. Trop. Med. Hyg. 76, 776-777 (1982).
24. M. R. Hollingdale, P. P. McCann and A. Sjoerdsma, *Plasmodium berghei*: Inhibitors of Ornithine Decarboxylase Block Exoerythrocytic Schizogony. Exp. Parasitol. 60, 111-117 (1985).
25. J. M. Gillet, J. Charlier, G. Bone and P. L. Mulamba, *Plasmodium berghei*: Inhibition of the Sporogonous Cycle by α-Difluoromethylornithine. Exp. Parasitol. 56, 190-193 (1983).

INHIBITION OF ORNITHINE OR ARGININE DECARBOXYLASE AS AN EXPERIMENTAL APPROACH TO AFRICAN OR AMERICAN TRYPANOSOMIASIS

Peter P. McCann[1], Cyrus J. Bacchi[2], Alan J. Bitonti[1], Felipe Kierszenbaum[3] and Albert Sjoerdsma[1]

[1]Merrell Dow Research Institute, Cincinnati, Ohio; [2]Haskins Laboratories of Pace University, New York, New York; and [3]Michigan State University, East Lansing, Michigan, U.S.A.

INTRODUCTION

One of the more rewarding areas of recent research in polyamines has been the elucidation of their role in protozoal growth (Pegg and McCann, 1982; 1988). The first indication of the impact that parasitic protozoa would have on the polyamine field was the discovery by Bacchi et al. (1980) that α-difluoromethylornithine (DFMO), a specific catalytic inhibitor of ornithine decarboxylase, would totally cure acute infections of the African trypanosome T. b. brucei in mice. This dramatic finding led to the remarkably swift use of DFMO in what would have been fatal cases of drug-resistant late-stage human sleeping sickness in Africa (Schechter et al., 1987).

The initial discovery of the antiproliferative effects of DFMO in trypanosomes led to numerous studies wherein the role of polyamines in trypanosome growth and metabolism was explored (reviewed in Bacchi and McCann, 1987). Among the findings in T. b. brucei were that the trypanosomal enzyme has approximately the same sensitivity to DFMO as does mammalian ODC and that, as in mammalian cells, DFMO enters by passive diffusion and not by facilitated transport (Bitonti et al., 1986a). A further interesting point is that polyamine depletion caused by DFMO causes biochemical and morphological alterations of entire trypanosome populations, changing them from bloodstream forms to nonreplicating "short stumpy" forms, a differentiation event that seems independent of a chronologically earlier block of bloodstream form replication (Giffin et al., 1986).

Another facet of polyamine metabolism in trypanosomes is a novel spermidine-containing cofactor that is obligatory for glutathione reductase activity. The cofactor, termed trypanothione, has been identified as N^1, N^8-bis(L-γ-glutamyl-L-hemicystinyl-glycyl)spermidine and has also been shown to be uniquely present in trypanosomatids (Fairlamb et al., 1985; reviewed in Bacchi and McCann, 1987). Recent studies have demonstrated that DFMO-induced spermidine depletion greatly reduces the amount of trypanothione in T. b. brucei, indicating another likely target for the effects of DFMO on trypanosome infections (Fairlamb et al., 1987).

Recently, several other studies have been completed from which we can postulate further reasons why inhibition of ornithine decarboxylase by DFMO is so effective in restricting growth of African trypanosomes and also why DFMO has no effect on T. cruzi, the cause of American trypanosomiasis or Chagas' Disease.

EFFECT OF DFMO ON TRYPANOSOME VARIANT-SPECIFIC GLYCOPROTEIN

In elucidating factors which play a role in the inhibition of trypanosomal growth by DFMO, it is worth noting that one of the earliest effects ascribed to DFMO in mammalian cells is an inhibition of protein synthesis, as evidenced by a change in leucine incorporation as well as a perturbation in polyribosome formation (Rudkin et al., 1984). These changes parallel the depletion of spermidine caused by DFMO and are readily prevented by the presence of putrescine or spermidine in addition to DFMO. The occurrence of a similar rapid decrease in protein synthesis has not been described in trypanosomes. In fact, it was reported previously (Bacchi et al., 1983) that there is a paradoxical 2-4-fold stimulation of protein synthesis in trypanosomes during DFMO treatment. A negative effect of DFMO on protein synthesis in trypanosomes would be interesting, not only because of the general depression of metabolic activity that could ensue, but also because of potential effects on the synthesis of the variant-specific glycoprotein (VSG) which completely coats the trypanosome's plasma membrane and protects it from immune attack (Vickerman, 1969).

The major immune response to trypanosomes is an antibody response directed towards the VSG (Vickerman, 1978; Boothroyd, 1985). Trypanosomes normally undergo antigenic variation by switching their VSG coat, thus avoiding the host immune response (Vickerman, 1978; Boothroyd, 1985). However, if synthesis of new VSG was inhibited either specifically or concurrent with a general decrease in protein synthesis, then the trypanosome could not evade the immune system and parasites would presumably be eliminated. We showed previously that the immune response to trypanosomes is a necessary component of DFMO-induced cures in rats and mice (Bitonti et al., 1986b; deGee at al., 1983). The latter finding is reasonable, since DFMO is regarded as a cytostatic rather than a cytotoxic agent. Thus it is possible that, by limiting protein synthesis (VSG synthesis), DFMO allows the host's immune response to the trypanosome effectively remove the parasites from the blood (Bitonti et al., 1988).

Table 1 shows that trypanosomes isolated from rats treated with 2% DFMO in drinking water were markedly deficient in their ability to incorporate [^{3}H]leucine into protein (Table 1, parts A and B). After exposure to DFMO for 1 day, [^{3}H]leucine incorporation was decreased by 30-40%, whereas 2 days of exposure to DFMO resulted in a decrease in incorporation of 80-90%. A similar decrease in protein synthesis was noted in trypanosomes which were exposed to 100 μM DFMO for 24 h or 48 h in vitro (Table 1, part C). In the latter case, however, maximal inhibition was attained by 24 h and protein synthesis did not apparently decrease any further during 48 h of DFMO exposure. The decrease in protein synthesis at 48 h could not be reversed by adding putrescine (100 μM) or spermidine (100 μM) to incubations of DFMO-treated trypanosomes plus [^{3}H]leucine (results not shown).

The effects of DFMO on the synthesis of VSG were also examined under the same conditions as for general protein synthesis. The incorporation of [^{3}H]myristic acid into VSG was shown, by SDS/PAGE, to be markedly inhibited in trypanosomes harvested from rats which had been treated for 24 h with DFMO (Fig. 1). Treatment with DFMO for 48 h resulted in a further decline of [^{3}H]myristic acid incorporation of VSG. Similar results were obtained if VSG synthesis in typanosomes (bloodstream trypomastigotes) grown in culture was measured. DFMO (100 μM) exposure decreased VSG synthesis at both 24 h and 48 h (Bitonti et al, 1988).

Table 1. [^{3}H]-Leucine Incorporation in T. brucei Treated In Vivo and In Vitro With DFMO

Treatment Time		[^{3}H]-Leucine Incorporation CPM/4x10^6 Trypanosomes (± S.E.M.)	% of Control
A	0 hr	87,524 ± 3,663	100
	24 hr	54,706 ± 15,724	63
	48 hr	9,798 ± 1,932	11
B	0 hr	75,280 ± 10,105	100
	24 hr	52,591 ± 5,524	70
	48 hr	13,862 ± 293	18
C	24 hr - Control	52,197 ± 4,266	100
	24 hr - DFMO	17,056 ± 1,025	33
	48 hr - Control	52,815 ± 600	100
	48 hr - DFMO	17,134 ± 1,335	32

T. b. brucei was grown in rats (A,B) or in tissue culture (C) as described (Bitonti et al, 1988). In each case, the trypanosomes were washed twice in leucine-free RPMI 1640 and then incubated for 60 min in the presence of 5 µCi [^{3}H]-leucine at a cell density of 2 x 10^7/ml.

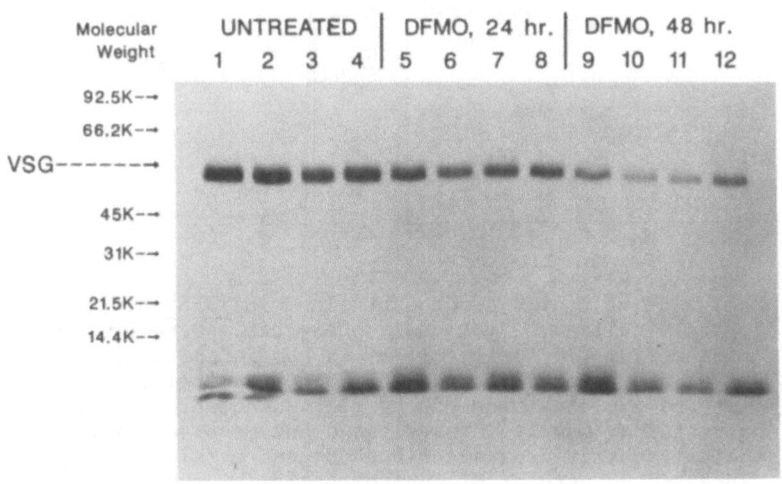

Fig. 1. [^{3}H]Myristic acid incorporation was carried out as described (Bitonti et al, 1988) with trypanosomes obtained from untreated rats (lanes 1-4) or trypanosomes exposed to DFMO in vivo for either 24 h (lanes 5-8) or 48 h (lanes 9-12). Radioactive proteins were separated by SDS/PAGE. Protein markers and their M_r values were: phosphorylase b, 92500; BSA, 66200; ovalbumin, 45000; carbonic anhydrase, 31000; soya-bean trypsin inhibitor, 21500; lysozyme, 14400.

HALF-LIFE OF TRYPANOSOMAL ORNITHINE DECARBOXYLASE

A recent finding that may indicate another reason why DFMO is particularly effective against African trypanosomes is that their ornithine decarboxylase appears to be an appreciably longer-lived molecule than the mammalian enzyme and, as such, does not appear to be rapidly replaced in the cell, even when completely inhibited by DFMO. Initial data came from Phillips et al. (1987) who cloned the ODC gene from *T. b. brucei*, expressed it in *E. coli* and determined the nucleotide sequence. Although there was a 61.5% homology in the amino acid sequences of the mouse and protozoan enzymes, the trypanosome sequence lacked a 36 amino acid C-terminal peptide. This peptide, rich in proline, glutamic acid, aspartic acid, threonine and serine is believed to constitute a "PEST" region which subjects proteins to rapid degredation in mammalian cells (Rogers et al., 1986).

Phillips et al. (1987), thus, found that the ODC from procyclic trypomastigotes was exceedingly stable with no detectable turnover. Confirming these observations, our own current studies (Bacchi et al., 1988) have demonstrated that the $t_{½}$ for the ODC from procyclic forms is >6 h as shown in Figure 2A.

We also measured the half-life of ODC in bloodstream trypomastigotes in mice with a 48 h infection treated with cycloheximide. In Fig. 2B, it can be seen that the apparent $t_{½}$ under these conditions was 5 h.

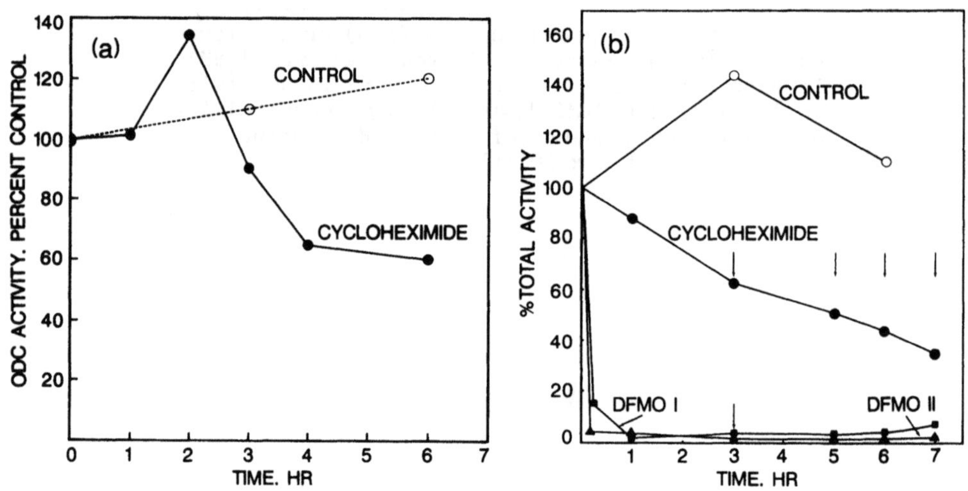

Fig. 2A. Half-life of ornithine decarboxylase in *T. b. brucei* procyclic forms in culture. Stationary phase cells were passaged into fresh medium (10^6 cells/ml) as in a standard enzyme induction curve, and allowed to grow for 16 h. At that time (0 h) sterile-filtered cycloheximide (50 ug/ml) was added. Flasks were harvested at hourly intervals and enzyme preparations and activity obtained as described (Bacchi et al, 1988). Results are expressed as percent of control activity at 0 h (22.3 nmoles/mg protein/h). The experiment was repeated 3 times with similar results.

Fig. 2B. Half life of ODC as measured in *T. b. brucei* bloodstream trypomastigotes; effects of DFMO on enzyme activity *in vivo*. Groups of 5 mice with 48 h infections were given 50 mg/kg cycloheximide (i.v.) at 0,3,5,6 and 7 h. DFMO was administered once, at 0 time (DFMO I) or twice, at 0 and 3 h (DFMO II). Cells were harvested and enzyme preparations made as described (Bacchi et al., 1988). Data is expressed as percent of control activity at 0 h (62.3 nmoles/mg protein/h).

In light of the numerous biochemical effects of DFMO on trypanosomes in vivo (Bacchi and McCann, 1987), we were interested in determining the short-term effect of DFMO on trypanosome ODC activity. As shown in Fig. 2B, after infected animals were given a low dose of DFMO (50 mg/kg) and parasite-infected blood was collected, trypanosome ODC activity decreased to 15% of control levels 15 min after the first injection of DFMO, and remained <10% of control values for the 7 h duration of the experiment. Trypanosome preparations from groups of animals given DFMO doses both at time 0 and 3 h showed a similar reduction of ODC activity.

Recovery of ODC in trypanosomes and host tissues over a 48 h period was also examined after a single 200 mg/kg (i.p.) dose of DFMO, as shown in Fig 3. Although trypanosome ODC activity remained depressed during the first 8 h after treatment (<10% of controls), liver enzyme began to recover after 3 h, and by 24 h both liver and prostate ODC activity had fully recovered. Trypanosome activity, in fact, recovered by only 50% after 48 h.

American Trypanosomiasis - Chagas' Disease

It has remained puzzling why the causative agent of Chagas' Disease, Trypanosoma cruzi, an intracellular-infective cousin of the Leishmania species, was not sensitive either in vitro or in vivo to DFMO, although DFMO did, in fact, impair the normal mammalian macrophage's ability to bind or ingest the parasite (Kierszenbaum et al., 1987a). However, α-difluoromethylarginine (DFMA), a specific irreversible inhibitor of arginine decarboxylase, as well as other catalytic irreversible inhibitors of the same enzyme (Bitonti et al., 1987) markedly inhibited the infection of macrophages by T. cruzi blood forms when the parasites were treated with the inhibitors prior to addition of the parasites to the macrophage cultures (Kierszenbaum et al., 1987b). As shown in Table 2, DFMA, monofluoromethylagmatine (FMAg), and monofluoromethyldehydroarginine (Δ-MFMA) all markedly decreased the capacity of T. cruzi to infect untreated macrophages. The specificity of these inhibitors was demonstrated first by the fact that concentrations of DL-arginine up to 50 mM had no effect on T. cruzi infection of macrophages. Secondly, both agmatine and putrescine completely reversed the effects of DFMA (Table 3) and the other two inhibitors of arginine decarboxylase (not shown).

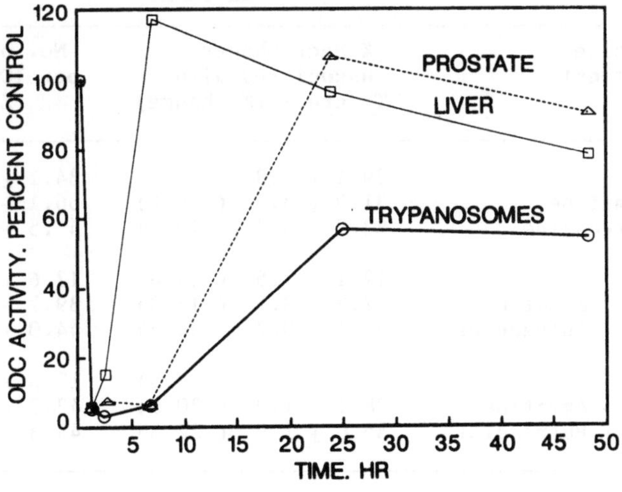

Fig. 3. Recovery of ODC activity after a single dose of DFMO. Male rats (300 g) with a 48 h infection were given a single i.p. injection of DFMO (200 mg/kg); control infected rats were untreated. Trypanosomes, livers and prostates were collected at the time points indicated (3 rats/point) and ODC activity determined in frozen-thawed preparations (trypanosomes) or homogenates (tissues).

Table 2. Effects of Pretreatment of Blood Forms of *T. cruzi* With DFMA, FMAg or Δ-MFMA Significantly Inhibited Its Association With Untreated Macrophages *In Vitro*.

Reagent tested	Concentration (mM)	% Macrophages with parasites (% change)	No. of parasites per 100 macrophages (% change)
Control		35.7 ± 1.6	48.5 ± 3.8
DFMA	3	34.5 ± 2.4 (-3.4)	52.0 ± 5.6 (7.2)
	12	31.4 ± 0.6 (-12.0)	36.4 ± 0.8 (-24.9)
	50	18.9 ± 4.0 (-47.1)	23.0 ± 5.2 (-52.6)
FMAg	3	23.7 ± 1.3 (-33.6)	27.4 ± 1.8 (-43.5)
	12	11.3 ± 3.2 (-68.3)	13.1 ± 4.4 (-73.0)
	50	5.1 ± 0.8 (-85.7)	5.1 ± 0.8 (-89.5)
Δ-MFMA	3	26.8 ± 2.4 (-24.9)	33.2 ± 5.5 (-31.5)
	12	21.1 ± 1.4 (-40.9)	27.0 ± 3.9 (-44.3)
	50	8.6 ± 1.3 (-75.9)	8.7 ± 1.4 (-82.1)

The trypomastigotes were incubated at 37°C for 3 h with medium alone or medium containing the indicated concentrations of DFMA, FMAg, Δ-MFMA, washed, and cocultured with untreated macrophages for 2 hr. All differences between the experimental values obtained with FMAg, Δ-MFMA, or with concentrations of DFMA ≥12 mM and the corresponding control values were statistically significant (P<0.05). Values in parentheses are the percent change. These concentrations of inhibitor did not affect parasite viability or motility.

Table 3. Agmatine or Putrescine Prevent the Inhibitory Effect of DFMA Pretreatment of *T. cruzi*

Parasite treatment	% Macrophages associated with *T. cruzi* (% change)		No. of parasites per 100 macrophages (% change)	
No DFMA	29.1 ± 2.8		34.1 ± 4.8	
0.15mM Agmatine	31.9 ± 3.2	(9.6)	36.1 ± 5.4	(5.9)
0.15mM Putrescine	33.5 ± 3.2	(15.1)	41.5 ± 4.4	(21.7)
30mM DFMA	12.1 ± 1.5	(-58.4)	12.6 ± 1.5	(-63.0)
30mM DFMA + Agmatine	32.9 ± 3.1	(13.0)	39.7 ± 3.0	(16.4)
30mM DFMA + Putrescine	35.4 ± 0.7	(21.6)	44.0 ± 2.0	(29.0)
50mM DFMA	11.8 ± 0.9	(-59.5)	12.5 ± 1.0	(-63.3)
50mM DFMA + Agmatine	38.1 ± 1.0	(30.9)	43.7 ± 2.3	(28.1)
50mM DFMA + Putrescine	35.6 ± 1.9	(22.3)	44.3 ± 3.2	(29.9)

The trypomastigotes were pretreated as indicated (at 37°C for 3 h), washed, and then cocultured with untreated macrophages for 1 h as described (Kierszenbaum et al., 1987b). Only the differences between the values obtained with parasites treated with DFMA alone and the corresponding control values were statistically significant (P<0.05). Values in parentheses are the percent change.

Finally, DFMA significantly interfered with actual intracellular replication of T. cruzi, both in macrophages and in non-phagocytic rat myoblasts (Fig. 4). Exogenous putrescine also prevented this effect (Kierszenbaum et al., 1987b). Although these results are strongly suggestive of the presence of arginine decarboxylase rather than ornithine decarboxylase as a source of putrescine, neither arginine

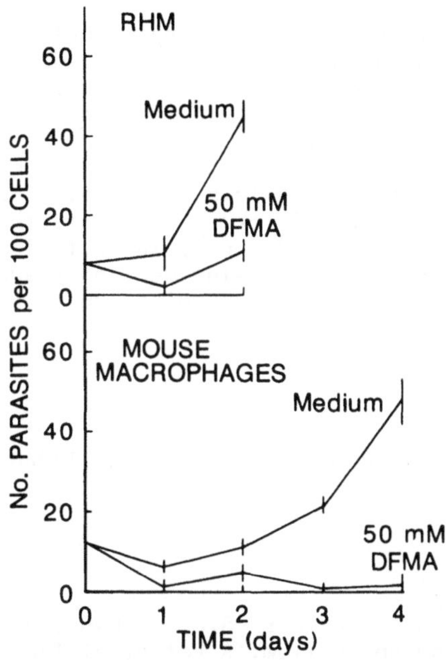

Fig. 4. Effect of DFMA on the capacity of T. cruzi to proliferate intracellularly in rat heart myoblasts (RHM) and mouse macrophages. Untreated trypomastigotes were co-cultured with the host cells at 37°C for 1 h (macrophages) or 2 h (myoblasts), and the free organisms were removed by three washings with medium. Fresh medium alone or containing DFMA was then added to replicated cultures that were further incubated for the indicated periods of time. The fluid phase was replaced with fresh medium without and with DFMA, as appropriate, every 12 h until termination of the experiment.

decarboxylase nor ornithine decarboxylase could be detected in cell extracts containing up to 2.4 x 10^9 trypanosomes. Collectively, however, these data do imply that there may be a specific arginine decarboxylase involved in polyamine metabolism of T. cruzi which might yield a unique therapeutic target for this now incurable disease.

CONCLUSIONS

1. In the African trypanosome, T. b. brucei incorporation of [^{3}H]myristic acid into VSG is inhibited by DFMO. Such inhibition may be related to the overall effect of DFMO on the trypanosome in vivo, inasmuch as it is known that the VSG is the major antigenic determinant of the trypanosome against which host antibodies respond and antibody responses are necessary for DFMO-induced cures of trypanosomiasis. Slowing of VSG synthesis would seemingly allow a better antibody response to be mounted against the trypanosome, thus rendering DFMO more effective in curing a parasitic infection.

2. ODC in African trypanosomes is an appreciably longer-lived molecule than the mammalian enzyme and consequently does not appear to be rapidly replaced in the cell. Lack of ODC regulation and turnover in trypanosomes might be important in in vivo situations where levels of DFMO may vary in the host.

3. The effect of specific inhibitors of arginine decarboxylase against the American trypanosome, T. cruzi, suggests the presence of a unique enzyme target previously found only in bacteria and higher plants. Therefore, the absence of arginine decarboxylase in mammalian cells indicates this enzyme might be a selective target for chemotherapy for T. cruzi infection.

ACKNOWLEDGEMENTS

This research was supported by N.I.H. grant AI 17340 to C.J.B.

REFERENCES

Bacchi, C.J., Nathan, H.D., Hutner, S.H., McCann, P.P. and Sjoerdsma, A., 1980. Polyamine Metabolism: A Potential Therapeutic Target in Trypanosomes, Science, 210:332-334.

Bacchi, C.J., Garofalo, J., Mockenhaupt, D., McCann, P.P., Diekema, K.A., Pegg, A.E., Nathan, H.C., Mullaney, E.A., Chunosoff, L., Sjoerdsma, A., and Hutner, S.H., 1983. In Vivo Effects of α-DL-Difluoromethylornithine on the Metabolism and Morphology of Trypanosoma Brucei Brucei. Molec. Biochem. Parasitol., 7:209-225.

Bacchi, C.J., and McCann, P.P., 1987. Parasitic Protozoa and Polyamines, in Inhibition of Polyamine Metabolism: Biological Significance and Basis for New Therapies (P.P. McCann, A.E. Pegg and A. Sjoerdsma, eds.) Academic Press, Orlando, pp. 317-344.

Bacchi, C.J., Garofalo, J., Santana, A., Hannan, J.C., Bitonti, A.J., and McCann, P.P., 1988. Trypanosoma brucei brucei: Induction of Ornithine Decarboxylase in Procyclic Forms. Submitted.

Bitonti, A.J., Bacchi, C.J., McCann, P.P. and Sjoerdsma, A., 1986a. Uptake of α-Difluoromethylornithine by Trypanosoma brucei brucei. Biochem. Pharmacol., 35:351-354.

Bitonti, A.J., McCann, P.P. and Sjoerdsma, A., 1986b. Necessity of Antibody Response in the Treatment of African Trypanosomiasis with α-Difluoromethylornithine. Biochem. Pharmacol., 35:331-334.

Bitonti, A.J., Casara, P.J., McCann, P.P. and Bey, P., 1987. Catalytic Irreversible Inhibition of Bacterial and Plant Arginine Decarboxylase Activities by Novel Substrate and Product Analogs, Biochem. J., 242:69-74.

Bitonti, A.J., Cross-Doersen, D.E., and McCann, P.P., 1988. Effects of α-Difluoromethylornithine on Protein Synthesis and Synthesis of the Variant Specific Glycoprotein (VSG) in Trypanosoma brucei brucei. Biochem. J., 250:295-298.

Boothroyd, J.C., 1985. Antigenic Variation in African Trypanosomes. Ann. Rev. Microbiol., 39:475-502.
deGee, A.L.W., McCann, P.P. and Mansfield, J.M., 1983. Role of Antibody in the Elimination of Trypanosomes after DL-α-Difluoromethylornithine Chemotherapy in Mice, J. Parasitol., 69:818-822.
Fairlamb, A.H., Blackburn, P., Ulrich, P., Chait, B.T. and Cerami, A., 1985. Trypanothione: a novel bis(glutathionyl)spermidine cofactor for glutathione reductase in trypanosomatids. Science, 227:1485-1487.
Fairlamb, A.H., Henderson, G.B., Bacchi, C.J. and Cerami, A., 1987. In Vivo Effects of Difluoromethylornithine on Trypanothione and Polyamine Levels in Bloodstream Forms of Trypanosoma brucei. Molec. Biochem. Parasitol., 24:185-191.
Giffin, B.F., McCann, P.P., Bitonti, A.J. and Bacchi, C.J., 1986. Polyamine Depletion Following Exposure to DL-α-difluoromethylornithine both In Vivo and In Vitro Initiates Morphological Alterations and Mitochondrial Activation in a Monomorphic Strain of Trypanosoma brucei brucei. J. Protozool., 33:238-243.
Kierszenbaum, F., Wirth, J.J., McCann, P.P., and Sjoerdsma, A., 1987a. Impairment of Macrophage Function by Inhibitors of Ornithine Decarboxylase Activity, Infect. Immun., 55:2461-2464.
Kierszenbaum, F., Wirth, J.J., McCann, P.P., and Sjoerdsma, A., 1987b. Arginine Decarboxylase Inhibitors Reduce the Capacity of Trypanosoma cruzi to Infect and Multiply in Mammalian Host Cells, Proc. Nat. Acad. Sci. U.S.A., 84:4278-4282.
Pegg, A.E. and McCann, P.P. 1982. Polyamine Metabolim and Function: A Review. Am. J. Physiol., 243:C212-C221.
Pegg, A.E. and McCann, P.P. 1988. Polyamine Metabolism and Function in Mammalian Cells and Protozoans, I.S.I. Atlas of Science: Biochemistry, 1:11-18.
Phillips, M.A., Coffino, P., and Wang, C.C., 1987. Cloning and Sequencing of the Ornithine Decarboxylase Gene from Trypanosoma brucei. J. Bio. Chem., 262:8721-8727.
Rogers, S., Wells, R. and Rechsteiner, M., 1986. Amino Acid Sequences Common to Rapidly Degraded Proteins: The PEST Hypothesis. Science, 234:364-368.
Rudkin, B.B., Mamont, P.S. and Seiler, N., 1984. Decreased Protein-Synthetic Activity Is an Early Consequence of Spermidine Depletion in Rat Hepatoma Tissue-Culture Cells. Biochem. J., 217:731-741.
Schechter, P.J., Barlow, J.L.R., and Sjoerdsma, A., 1987. Clinical Aspects of Inhibition of Ornithine Decarboxylase with Emphasis on Therapeutic Trials of Eflornithine (DFMO) in Cancer and Protozoan Diseases, in Inhibition of Polyamine Metabolism: Biological Significance and Basis for New Therapies (P.P. McCann, A.E. Pegg and A. Sjoerdsma, eds.) Academic Press, Orlando, pp. 345-364.
Vickerman, K., 1969. On the Surface Coat and Flagellar Adhesion in Trypanosomes. J. Cell Sci., 5:163-193.
Vickerman, K., 1978. Antigenic Variation In Trypanosomes. Nature, 273:613-617.

THE POLYAMINE METABOLISM OF FILARIAL WORMS AS CHEMOTHERAPEUTIC TARGET

Sylke Müller, R.-M. Wittich, and R.D. Walter

Bernhard-Nocht-Institut für Tropenmedizin
Abteilung für Biochemie
2000 Hamburg 4
F. R. Germany

ABSTRACT

Parasite-specific putrescine-N-acetyltransferase and polyamine oxidase, both involved in the reversed pathway of polyamine metabolism, were demonstrated for Ascaris suum and Onchocerca volvulus. Berenil-treatment was found to be correlated with accumulation of polyamines, especially spermine, obviously due to blockaded polyamine interconversion. Furthermore it was shown that added spermine to the culture medium led to the death of worms. These specificities might be exploited for chemotherapy of filarial infections.

Polyamines are widely distributed in the nature. They are found in higher and lower eucaryotes and in procaryotes as well as in viruses (Tabor and Tabor, 1984). During the last years there have been many approaches to examine the role of polyamines in cell growth and differentiation in vertebrates (Tabor and Tabor, 1984; Pegg, 1986). The polyamine metabolism of parasites also has attracted increasing interest, e.g. in African trypanosomes the initial enzyme of polyamine synthesis - ornithine decarboxylase - has been exploited as a target for chemotherapy by using DFMO (DLα-difluoromethylornithine) (Bacchi et al., 1980; Bacchi et al., 1983; Fairlamb et al., 1985; Giffin et al., 1986).

The polyamine metabolism of filaria and other helminths is still a neglected area of research, although there are reports about distribution pattern of polyamines and some peculiarities of polyamine metabolism in filarial worms (Srivastava et al., 1980; Wittich et al., 1987; Walter, 1988).

DFMO and MGBG (methylglyoxal bis-(guanylhydrazone)), both of which are potent inhibitors of polyamine synthesis in mammals, do not significantly effect the viability of filarial worms (Wittich et al., 1987). If synthesis via ornithine as well as arginine decarboxylase could finally be denied, the absolute dependence of filarial worms on their host for a supply with polyamines would offer some leads for chemotherapy of river blindness and lymphatic filariasis. It should be demonstrated for filaria and other helminths that blockade and disturbance of polyamine synthesis and distribution leads to death of worms. Furthermore the enzymes involved in the interconversion pathway should be identified and their potential as targets for design and developement of enzyme inhibitors to antifilarial drugs should be studied.

A Novel Putrescine N-Acetyltransferase from *Ascaris suum* and *Onchocerca volvulus*

Recently N-acetylputrescine was reported as main excretion product of the polyamine metabolism of filaria (Wittich et al., 1987). The putrescine N-acetyltransferase involved in the reversed pathway of polyamine metabolism shows novel and parasite-specific properties compared to isofunctional enzyme from vertebrates. Since putrescine acetylating activity was found to be just above the limit of detection when assaying the 100,000 g supernatant of *A. suum* and *O. volvulus* a procedure was established to release activity from still unknown inhibitory compounds. Freezing, thawing and intensive dialysis proved to be a suitable method. The purification procedure included DEAE-chromatography and affinity-chromatography on putrescine-Sepharose and resulted in about 1000-and 400-fold purification for *A. suum* and *O. volvulus* enzyme, respectively (Wittich and Walter, unpublished).

Substrate specificity. The enzyme from *A. suum* showed high affinity for several diamines such as putrescine, cadaverine, 1,3-diaminopropane and 1,6-diaminohexane with K_M-values of 0.25 mM, 0.1 mM, 1.25 mM and 0.4 mM, respectively. Spermidine and spermine as well as histones were poor substrates for the N-acetylation (Table 1). The Figures 1a and 1b demonstrate the separation of reaction products from acetylation of various diamines by ion-pair HPLC. Putrescine was found to be the prefered substrate of the nematodic putrescine N-acetyltransferase.

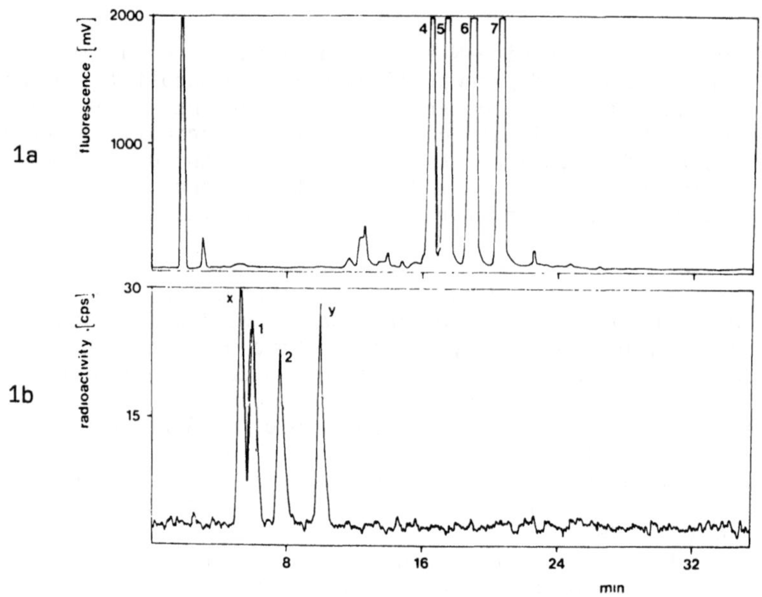

Fig. 1. Mixed samples of individually processed diamines separated by ion-pair HPLC (a) before and (b) after enzyme reaction. (a) Several substrates for N-acetyltransferase 4: 1,3-diaminopropane; 5: putrescine; 6: cadaverine; 7: 1,6-diaminohexane. (b) The resulting respective radioactive acetylated derivatives of the enzyme reaction are 1: N-acetylputrescine; 2: N-acetylcadaverine; x and y are considered as acetylated derivatives of 1,3-diaminopropane and 1,6-diaminohexane, respectively.

Table 1. Substrate Specificity of Putrescine N-Acetyltransferase

Substrate	Reaction rate (%)	
	Onchocerca volvulus	Ascaris suum
1,4-Diaminobutane	100	100
1,3-Diaminopropane	129	116
1,5-Diaminopentane	45	39
1,6-Diaminohexane	74	50
sym-Norspermidine	42	59
sym-Norspermine	8	11
Spermidine	2	3
Spermine	3	4
Histones	1	1

Beside substrate specifity the nematodic enzyme is clearly distinguished from spermine/spermidine N-acetyltransferases of mammals by its molecular weight of about 40,000 as determined by HPLC on a DuPont Zorbax GF-250 gel filtration column (Wittich et al., unpublished).

Effect of Berenil and Spermine on Polyamine Metabolism of Filarial Worms

Berenil was reported to interact with various enzymes of polyamine metabolism and its antifilarial activity was demonstrated (Müller and Walter, 1988; Müller and Walter, in press). It is shown that 100 µM Berenil added to the culture medium killed adult _Brugia pahangi_ within 13 days of treatment. To ensure the effect of Berenil on the polyamine metabolism of filaria, adults of _Dirofilaria immitis_ were treated at non-lethal dosage. The worms were incubated in Franke-Weinstein-medium (Franke and Weinstein, 1983) supplemented with 10 and 100 µM Berenil, respectively. Worms were incubated, polyamines were extracted according to Wittich et al. (1987) and the distribution pattern of polyamines was determined by ion-pair HPLC. As shown in Table 2 and Figure 2 the Berenil-treatment resulted in elevated levels of spermine and spermidine, e.g. addition of 0.1 mM Berenil to maintenance medium led to 20-and 10-fold accumulation of spermine and spermidine, respectively.

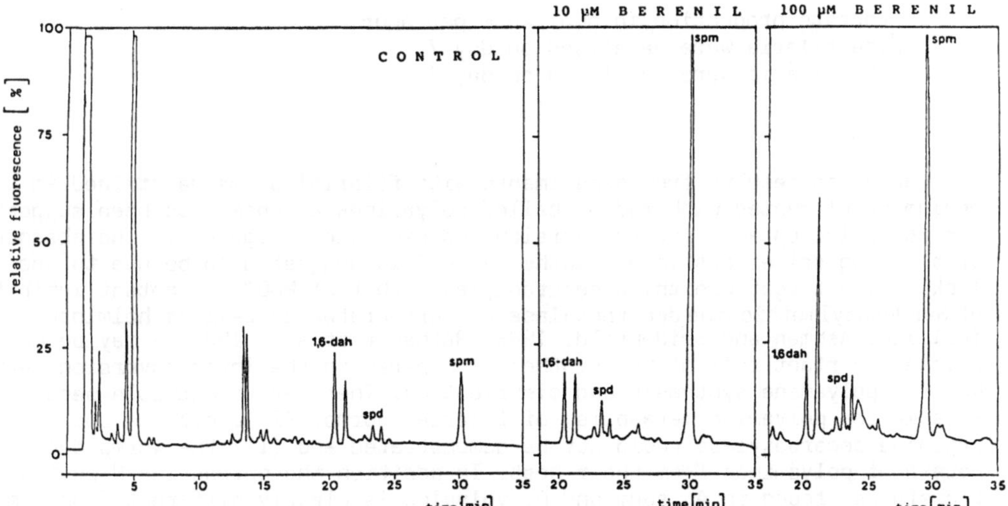

Fig. 2. Separation of 20 µl aliquots of derivatized polyamines from homogenates of _D. immitis_ incubated in Franke-Weinstein-medium without Berenil; with 10 µM Berenil; with 100 µM Berenil. Abbreviations: 1,6-dah, 1,6-diaminohexane; spd, spermidine; spm, spermine.

Table 2. Distribution of Polyamines in Dirofilaria immitis

	Spermidine nmol/mg protein	Spermine nmol/mg protein
Control	0.147	1.688
10 μM Berenil	0.187	12.202
100 μM Berenil	1.182	27.201

Accumulation of polyamines by Berenil-treatment is remarkable because spermine showed a lethal effect on adult B. pahangi in vitro. As shown in Table 3 addition of spermine at concentrations of 0.01 and 0.1 mM to the maintenance medium killed the worms within 7 and 10 days, respectively. As serum was omitted from the medium this lethal effect did not result from aldehydes produced by serum oxidase from added spermine as reported before (Ferrante et al., 1986; Morgan et al., 1986). Addition of serum to the medium reduced the survival period drastically, obviously due to aldehyde production.

Table 3. Lactate Production of Brugia pahangi Maintained in Medium Supplemented with Spermine in Absence and Presence of Serum

	day 2	day 4	day 6	day 8
(a) without serum				
Control	28.25	26.16	21.8	18.16
10 μM Spermine	32.15	32.3	33.5	15.0
100 μM Spermine	15.15	12.2	5.1	- [a]
(b) 5 % fetal calf serum added				
Control	26.0	24.0	28.0	
10 μM Spermine	26.5	2.0	- [b]	
100 μM Spermine	21.3	9.1	- [b]	

Lactate production in nmol/worm per hour
[a] The filaria were paralysed on day 7
[b] The filaria were paralysed on day 3

Based on results from experiments with filarial worms maintained in medium supplemented with radiolabelled polyamines a scheme has been suggested for polyamine uptake, interconversion and excretion (Figure 3). The absence of a biological effect in helminths by DFMO is suggested to be due to the lack of its target ornithine decarboxylase; that of MGBG - a potent inhibitor of S-adenosylmethionine decarboxylase of vertebrates as well as helminths (Williams-Ashman and Seidenfeld, 1986, Rathauer et al., 1988) - may be due to a less important role of the synthetic compared to the interconversion pathway for polyamine synthesis and distribution. There are remarkable peculiarities in the polyamine metabolism of filarial worms: (I) ornithine and arginine decarboxylase could not be demonstrated and (II) the worms depend on uptake of polyamines from their host. In addition the putrescine N-acetyltransferase found in A. suum and O. volvulus is clearly different from mammalian histone, polyamine and putrescine N-acetyltransferases reported so far with respect to substrate specificity and molecular weight (Seiler and Al-Therib, 1974; Libby, 1978; Libby, 1980). The nematodic enzyme shows high conversion rates and correspondingly low K_M-values for diamines with preference for putrescine.

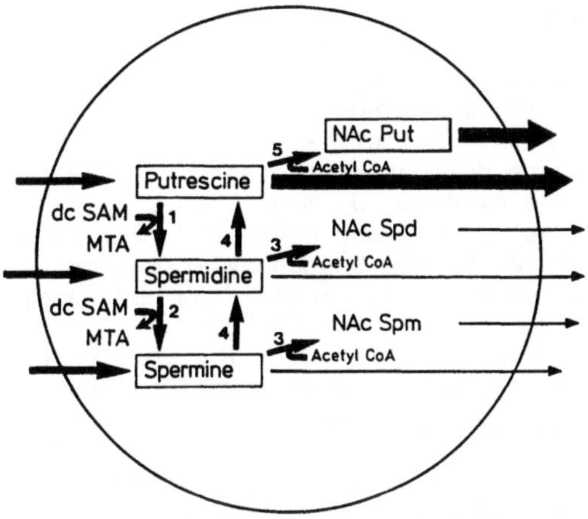

Fig. 3. Proposed scheme for polyamine metabolism in filarial worms.
Enzymes:1, spermidine synthase; 2, spermine synthase; 3, spermine/spermidine N-acetyltransferase; 4, polyamine oxidase; 5, putrescine N-acetyltranferase.
Abbreviations:dcSAM, decarboxylated S-adenosylmethionine; NAc-Put, N-acetylputrescine; NAc-Spd, N-acetylspermidine; NAc-Spm, N-acetylspermine.

This novel putrescine N-acetyltransferase plays obviously a unique role in the degradation of excessive putrescine in parasitic nematodes compared to the mammalian host.
Results presented here on the lethal effect of spermine also provide evidence for the potential of polyamine metabolism of filarial worms for chemotherapy. The accumulation of polyamines caused by Berenil points to the important role of polyamine oxidase for maintaining adequate and crucial levels of spermine and spermidine. Preliminary results, which show the specificity of Ascaris-polyamine oxidase for spermine and spermidine but not for the N-acetylated derivatives, indicate a different function and role of the nematodic enzyme compared with the mammalian polyamine oxidase. Berenil, reported to inhibit S-adenosylmethionine decarboxylase as well as diamine oxidase - both involved in polyamine metabolism - also shows a potent inhibitory effect on Ascaris-polyamine oxidase (McCann et al., 1983; Karvonen et al., 1985; Balana-Fouce, et al., 1986; Bitonti et al., 1986; Rathauer et al., 1988). MDL 72527 a specific inhibitor of mammalian polyamine oxidase had no inhibitory effect on the nematodic enzyme (Bey et al., 1985; Bolkenius and Seiler, 1987; Müller et al., in press).
These findings indicate striking differences of role, function and structure between enzymes of host and parasite which might be exploited for chemotherapeutic attack.

REFERENCES

Bacchi, C. J., Nathan, H. C., Hunter, S. H., 1980, Polyamine metabolism:A potential therapeutic target in trypanosomes, Science, 210:332-334.
Bacchi, C. J., Garofalo, J., Mockenhaupt, D., McCann, P. P., Diekema, K. A., Pegg, A. E., Nathan, H. C., Mullany, E. A., Chunosoff, L., Sjoerdsma, A., Hunter, S. H., 1983, In vivo effects of -D,L-difluoromethylornithine on the metabolism and morphology of trypanosoma brucei brucei, Mol. Biochem.,Parasitol., 7:209-225.

Balana-Fouce, R., Garzon-Pulido, T., Ordonez-Escudero, D., and Garrido-Pertierra, A., 1986, Inhibition of diamine oxidase and S-adenosylmethionine decarboxylase by diminaceneaceturate (berenil), Biochem. Pharmacol., 35 (9):1597-1600.

Bey, P., Bolkenius, F. N., Seiler, N., and Casara, P., 1985, N-2,3-butadienyl-1,4-butanediamine derivatives:Potent irreversible inactivators of mammalian polyamine oxidase, J. Med. Chem., 28 (1):1-2.

Bitonti, A. J., Dumont, J. A., and McCann, P. P., 1986, Characterisation of trypanosoma brucei brucei S-adenosyl-L-methionine decarboxylase and its inhibition by berenil, pentamidine and methylglyoxal bis(guanylhydrazone), Biochem. J., 237:685-689.

Bolkenius, F. N., and Seiler, N., 1987, The role of polyamine reutilization in depletion of cellular stores of polyamines in non-proliferating tissues, Biochem. Biophys. Acta, 923:125-135.

Fairlamb, A. H., Blackburn, P., Ulrich, P., Chait, B. F., and Cerami, A., 1985, Trypanothione. A novel bis(glutathionyl)spermidine cofactor for glutathione reductase in trypanosomatids, Science, 227:1485-1487.

Ferrante,A., Ljungstrom, I., Rzepzyk, C. M., and Morgan, D. M. L., 1986, Differences in sensitivity of Schistosoma mansoni schistosomula, Dirofilaria immitis microfilariae and Nematospiroides dubius third-stage larvae to damage by polyamine oxidase-polyamine system, Infect. Immun., 53 (3):606-610.

Franke, E. D., and Weinstein, P. P., 1983, Dipetalonema vitae (Nematoda: Filarioidea):Culture of third-stage larvae to young adults in vitro, Science, 221:161-163.

Giffin, B. F., McCann, P. P., Bitonti, A. J., and Bacchi, C. J., 1986, Polyamine depletion following exposure to DL -difluoromethylornithine both in vivo and in vitro initiates morphological alterations and mitochondrial activation in a monomorphic strain of trypanosoma brucei brucei, J. Protozool., 33 (2):238-243.

Karvonen, E., Kauppinen, L., Partanen, T., and Pösö, H., 1985, Irreversible inhibition of putrescine-stimulated S-adenosyl-L-methionine decarboxylase by berenil and pentamidine, Biochem. J., 231:165-169.

Libby, P. R., 1978, Calf liver nuclear N-acetyltransferase, J. Biol. Chem., 253:233-237.

Libby, P. R., 1980, Rat liver nuclear N-acetyltransferase:Seperation of two enzymes with both histone and spermidine acetyltransferase activity, Arch. Biochem. Biophys., 203:384-389.

McCann, P. P., Bacchi, C. J., Nathan, H. C., and Sjoerdsma, A., 1983, in "Mechanism of Drug Action", T. P. Singer and R. N. Ordarza, eds., Academic Press, New York.

Morgan, D. M. L., Bachrach, U., Assaraf, Y. G., Harari, E., and Golenser, J., 1986, The effect of purified aminoaldehydes produced by polyamine oxidation on developement in vitro of Plasmodium falciparum in normal and glucose-6-phosphate-dehydrogenase-deficient erythrocytes, Biochem. J., 236:97-101.

Müller, S., and Walter, R. D., 1988, Effect of berenil and spermine on viability and polyamine metabolism of filarial worms in vitro, International Symposium on Polyamines in Biochemistry and Clinical research, Sorrento, Naples, Italy, 13-17 June 1988 (Abstract).

Müller, S., Wittich, R.-M., and Walter, R. D., Characteristics and function of polyamine oxidase in nematodes, Zbl. Bakt. Hyg., in press.

Pegg, A. E., 1986, Recent advances in the biochemistry of polyamines in eukaryotes, Biochem. J., 234:249-262.

Rathauer, S., Wittich, R.-M., and Walter, R. D., 1988, Ascaris suum and Onchocerca volvulus:S-adenosylmethionine decarboxylase, Exp. Parasitol., 65:277-281.

Seiler, N., and Al-Therib, M. J., 1974, Putrescine catabolism in mammalian brain, Biochem. J., 144:29-35.

Srivastava, D. K., Roy, T. K., and shukla, O. P., 1980, Polyamines of helminths, Ind. J. Parasitol., 4 (2):187-189.
Tabor, C. W., and Tabor, H., 1984, Polyamines, Annu. Rev. Biochem., 53: 749-790.
Walter R. D., 1988, Polyamine metabolism in filaria and allied parasites, Parasitol. Today, 4 (4):18-20.
Williams-Ashman, H. G., and Seidenfeld, J., 1986, Aspects of the biochemical pharmacology of methylglyoxal bis(guanylhydrazone), Biochem. Pharmacol., 35 (8):1217-1225.
Wittich, R.-M., Kilian, H.-D., and Walter, R. D., 1987, Polyamine metabolism in filarial worms, Mol. Biochem. Parasitol., 24:155-162.
Wittich, R.-M., and Walter, R. D., 1988, A novel putrescine (diamine) N-acetyltransferase from Onchocerca volvulus and Ascaris suum, International Symposium on Polyamines in Biochemistry and Clinical Research, Sorrento, Naples, Italy, 13-17 June 1988 (Abstract).

LARGE BOWEL TUMOR PROMOTION BY DIAMINE OXIDASE INHIBITION: ANIMAL MODEL AND CLINICAL ASPECTS

Jürgen Kusche [1,3,4], Rudolf Menningen[1], Lothar Leisten[1] and Bernhard Krakamp[2]
[1]IInd Department of Surgery of the University of Cologne and [2]Ist Medical Clinic, Köln-Merheim, FRG

ABSTRACT

In the intestinal mucosa, diamine oxidase probably is involved in a feedback regulation mechanism for the termination of proliferation. This was proven by a considerable promotion of large bowel tumors induced by diamine oxidase inhibition in a rat model. Carcinogenesis was accompanied by characteristic alterations in mucosal diamine oxidase activity.
In large bowel cancer patients similar changes in intestinal diamine oxidase activity had been observed as in the animal model.
Conclusion: Events entailing a reduction of diamine oxidase activity are suspicious for large bowel cancer promotion under experimental and clinical conditions.

INTRODUCTION

The physiological function of diamine oxidase (DAO) (EC 1.4.3.6) is not very appreciated until now, inspite the enzyme deminates putrescine, the key compound of the polyamine metabolism, and inspite it inactivates histamine, a highly active mediator. Concerning the latter substrate histaminosis and shock aggravation (1,2) had been described as a result of intestinal DAO defiency. Concerning the substrate putrescine, intestinal mucosal hyperproliferation and cancer promotion must be regarded as a consequence of the reduction of DAO activity, as it is shown in the present paper. Furthermore, the dysregulation of intestinal DAO activity may be related to an increased risk of large bowel cancer development also under clinical conditions.

MATERIAL AND METHODS

Patients

In 1985 - 1986 specimens of the large bowel mucosa and neoplasms had been studied in 112 patients (58 males, 54 females, age 72 (31 - 86) years (median and range).
30 patients proved to be without disease after colonoscopy. These patients,

---

[3]Address for correspondence: Prof. Dr. J. Kusche, Dr. Madaus GmbH & Co., Ostmerheimer Str. 198
[4]Supported by a grant of the Deutsche Forschungsgemeinschaft Ku 464/2-3

serving as controls, had been examined for the exclusion of tumor disease or transanal bleeding because of hemorrhoids (stage I and II).
20 patients were affected with polyps. In these altogether 50 patients biopsy specimens had been studied.
62 patients suffered from large bowel cancer. In these patients tissue samples had been taken after operation.

Animals

Eight to ten-week-old, male Fisher-344 rats (n = 90) with body weights ranging from 180 - 230 g had been used. They were allowed free access standard pellet feed and tap water. The enviromental condition were regulated to an automatic 12 h light/dark cycle and room temperature held at 18 - 22 ^{o}C.

Experimental tumor induction in rats. The rats were allocated by random digits to the following groups:
(1) Controls: normal saline (100 µl/kg) administered orally with a probe, 3 times/week for 24 weeks (n = 15).
(2) AG (100 mg/kg) administered orally 3 times/week for 24 weeks (n = 15).
(3) AOM (2.5 mg/kg) administered subcutaneously once/week for 10 weeks (n = 2 x 15).
(4) AG + AOM (in doses as mentioned above for groups 2 and 3) (n = 2 x 15).

Sample taking

1) Patients. During colonoscopy biopsy specimens were obtained from the large bowel mucosa, weighed and submerged in 1 ml of sodium-potassium phosphate buffer (0.1 mol/l; pH 7.2). When polyps were removed endoscopially, only a small sample was taken from the head of the polyp along with a biopsy specimen from the non-affected mucosa 2-5 cm distant from the polyp. In operated patients, the resected part of the large bowel was opend lengthwise and samples were taken from the tumor and from the macroscopially healthy mucosa 5-10 cm proximal to the tumor. In samples assigned for DAO determination were weighed and put into buffer, giving a dilution of 1:10 (w/v).

2) Animals. After a duration of the carcinogenesis experiments for 24 weeks (group 3 and 4), or 52 weeks (for all other animals), the rats were exsanguinated under general anesthesia using ether. The large bowel was immediately excized, its lumen opend lengthwise with scissors and feces gently removed. The number of tumors was counted and full thickness samples (approx. 100 mg) were taken from the transverse colon and from tumors. The tissue samples were wrapped with metal foil in liquid nitrogen and stored at 20 ^{o}C. One sample was flattened (without stretching it!) on a sheet of smooth paper and its circumference exactly marked with a pencil. This was necessary for the determination of the bowel wall mass (see below).

Determinations

(1) DAO activity. Full thickness samples of the transverse colon were homogenized with 9 vol of ice-cold sodium-potassium phosphate buffer (0.1 mol/l, pH 7.2). The homogenate was centrifuged for 30 min at 37000 g in a refrigerated (4 ^{o}C) centrifuge. DAO activity of the supernatant was measured using the ^{14}C-putrescine assay. In this test, deamination of putrescine leads to γ-aminobutyraldehyde which is converted to Δ_1-pyrroline or its polymers by a non-enzymatic cyclization. This compound is extracted into toluene while the amine fraction remains in the aqueous phase (for detailed description of the test procedure see (3).

AG: Bis-(1-aminoguanidinum) sulphate (Merck-Schuchardt, Munich, FRG).
AOM: Azoxymethane (Ash Stevens, Detroit, USA).

(2) Planimetric analysis: The mass of the large bowel wall per unit of area was used as an indicator of the thickness of the intestinal mucosa. The circumferences of the tissue samples were scanned with a computerized MOP-AM 02 figure analyser (Contron, Munich, FRG). The area was calculated in cm^2 and divided by the weight of the sample (mg). The resulting quotient (mg/cm^2) has been proven to correlate with the microscopically determined height of the intestinal mucosa ($r = 0.85$).
(3) Statistics: For the descriptive statistic the median ($\tilde{x}$) and range for testing the null-hypothesis, the Wilcoxon test for independent data and the Fisher exact probability test were used.

RESULTS

Tumor incidence

Table 1 shows the incidence of tumors found in the 4 groups of animals. No tumor was found in controls and none in AG-treted rats. In the animals treated with AOM, only one tumor of the small bowel was detected after 24 weeks. After 52 weeks none of the animals treated with AOM only was found to have a tumor. In the AOM + AG group there was no tumor detectable after 24 weeks but altogether 15 tumors in 10/15 animals were detectable after 52 weeks ($p = 0.0003$). The tumors were located mainly in the distal colon and rectum. Four animals simultaneously developed an adenoma and an adenocarcinoma. Overall we counted 11 adenomas, 4 of which exhibited severe dysplasia, and 4 invasive carcinomas, mainly signet-ring cell carcinomas.

Proliferative response of the intestinal mucosa

Enhancement of the mass of the bowel wall was used as an indicator of mucosal hyperproliferation. Twenty four weeks after the start of the experiment, a tendency towards increased mucosal proliferation was noticeable. Proliferation was seen to continue until the 52nd week of the experiment. At this time, enhancement of the mucosal mass of the AOM + AG treated group was found to be statistically significant (Fig. 1).
This was the same group in which all large bowel tumors were seen to occur.

Table 1. Development of large bowel tumors under various conditions

Group of animals	Duration of the experiment (weeks)	Tumor incidence	Number of tumors
Controls (0.9% NaCl)	52	0/15	0
AG	52	0/15	0
AOM	24	1/15	1
AOM + AG	24	0/15	0
AOM	52	0/15	0
AOM + AG	52	10/15*	15

AG = aminoguanidine; AOM = azoxymethane; *p = 0.00003 (Fisher's exact proability test). For further information, especially regarding drug dosage see Materials and Methods.

DAO activity of the large bowel during experimental carcinogenesis

DAO activity of the large bowel wall was studied in AOM treated rats at specific pertinent time points: at the start and at the end of the AOM application period; 6 weeks thereafter; 24 weeks thereafter (half time point of the experiment) and finally at the end of the experiment after 52 weeks

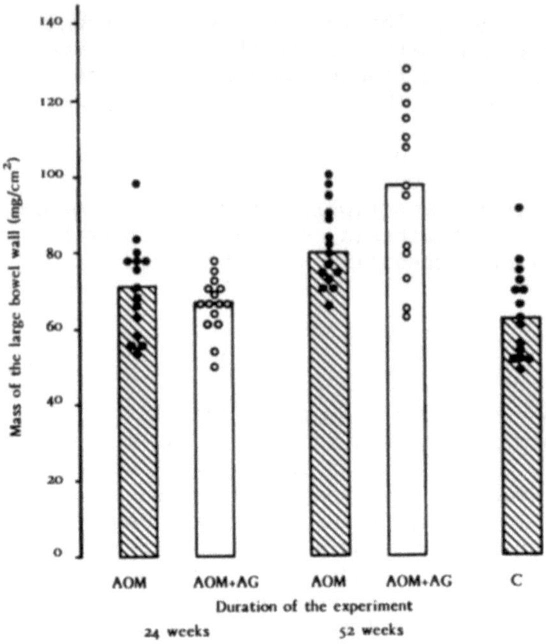

Figure 1. The influence of azoxymethane (AOM) and aminoguanidine (AG) on large bowel wall mass.
C = controls: 0.9% NaCl (100 µl/kg orally) 3 times per week for 24 weeks. AOM = AOM (2.5 mg/kg) once a week for 10 weeks. AOM + AG = AOM as above + AG (100 mg/kg orally) 3 times per week for 24 weeks. The columns represent the median values. Each point or circle represents one animal.
For C vs AOM or AOM + AG (52 weeks duration of the experiment), $p < 0.003$ (Wilcoxon test for independent data).

(Fig. 2). As had previously been found, the enzymic activity of DAO was decreased during the period of AOM administration. Immediately after termination of the carcinogenic injections, however, DAO activity rose to values on average 100 times greater than the initial activity. At the end of the experiment, DAO activity decreased but was still found to be significantly elevated when compared to controls.
At any time until the 24th week during the period of AG administration, DAO activity in group AG animals was almost zero when tested 48 h after the last oral tube feeding dose. These values were therefore not shown in Fig.2.

When we looked at DAO activity in tumor tissue and compared it to that of macroscopically normal mucosa of the transverse colon, a significant decrease of enzymatic activity was found in the tumor (Fig.3). Altogether 15 tumors were found in the AOM + AG group. Of these, 5 tumors were used in toto for histological examination and therefore could not be tested for DAO activity.

DAO activity in human large bowel cancer tissue and in the mucosa distant to the tumors

The DAO activity in the tumor (2.6 (0-45.6) nmol/min·g) was strongly diminished as compared to the surrounding mucosa (17.7 (0-140.6 nmol/min·g). ($p < 0.001$).
We were most interested in comparing the mucosal enzymic activity of cancer patients with that of normal individuals or those with non-malignant disease. However, samples from normal mucosa were obtained only from the left colon, therefore our comparison was limited to this anatomic region (Table 2). Indeed the DAO activity was diminished in the mucosa of cancer patients by

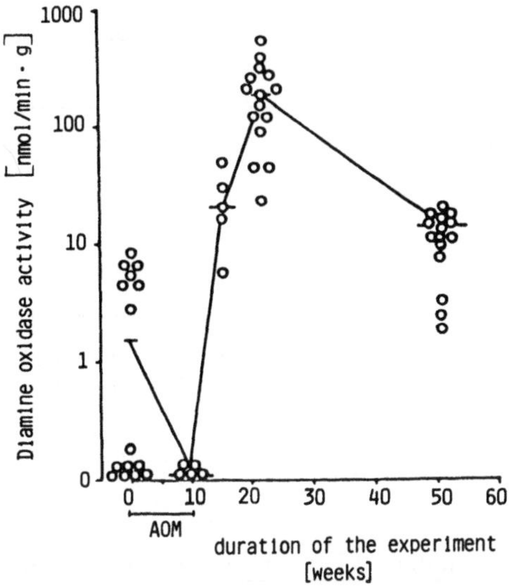

Figure 2. Diamine oxidase activity under the influence of the carcinogen azoxymethane (AOM). The period of AOM application is indicated in the figure. AOM (2.5 mg/kg) was administered s.c. once a week for 10 weeks. Each circle represents a single animal. The values were calculated as the mean from 3 parallel determinations. The line connects the median values. The logarithmic scale on the ordinate should be noted.

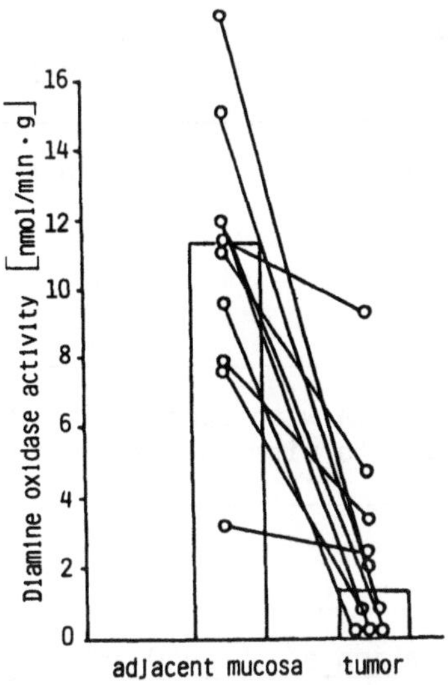

Figure 3. Diamine oxidase activity in large bowel tumors and in the intestinal wall 2 cm proximal to the tumors. Mucosal values and tumor values of one animal are connected by a line. Values were measured in 10 of 15 detected tumors (5 tumors were used completely for histological examination). Tumor DAO activity was significantly reduced ($2\alpha = 0.00006$; Mann-Whitney test for paired data).

22%, but this result was not statistically significant ($p = 0.1$). In cancer tissue, the DAO activity was significantly reduced ($p < 0.001$).

DAO activity in human mucosa and polyps

Studying large bowel polyps, the distribution of DAO activity was simito cancer tissue. The enzymic activity was significantly reduced in polyps as compared with the surrounding mucosa (Table 3). However, in the mucosa distant to the polyps, the measured DAO activity was twice that of the mucosa of cancer patients.

Table 2. Diamine oxidase (DAO) activity in mucosal or cancer tissue of the left colon.

Tissue	n	DAO activity (nmol/min·g)
I Normal mucosa	30	22.8 (3.7-136.5)
II Mucosa of cancer patients	17	17.7 (0.6-47.0)
III Cancer tissue	42	2.6 (0-45.6)

DAO activity: nmol/min·g; $\tilde{x}$ (range)
I/II: $p = 0.1$; II/III: $p = 0.006$. n = number of patients.

Table 3. Diamine oxidase (DAO) activity in human mucosa or tumor tissue (cancer/polyps) of the large bowel.

Subjects	DAO activity (nmol/min g) in:				
	n	Tumors	n	Mucosa 5-10 cm distal to tumors	
Patients with cancer	62	2.6(0-89.3)	61	17.7(0-140.6)	$p<0.001$
Patients with polyps	20	13.4(0-53.0)	14	37.5(5.0-68.0)	$p=0.007$
		n.t.		$p=0.026$	

Evaluation of the DAO activity as an indicator of mucosal hyperproliferation

Hyperproliferative changes of the mucosa were diagnosed histologically if branching and/or hyperplastic glands (Ø>0.1 mm) were found. In 39 patients, surface sections of the mucosa and determinations of DAO activity were performed concommittantly. 51% of these patients showed signs of hyperproliferation distant to the tumor. Mucosal hyperproliferation was associated with a reduced DAO activity (< 20 nmol/min·g) in 15/20 cases. (sensitivity 75%). However, in 11/19 patients the DAO activity was reduced without histological signs of proliferation (specificity 42%).

DISCUSSION

Carcinogenesis, hyperproliferation of the intestinal mucosa and DAO activity

Carcinogenesis by AOM induces an alteration in DNA (cancer initiation) and hyperproliferation of the intestinal mucosa (cancer promotion). Both events are equally important for the large bowel tumor development (4). During the process of AOM induced tumor production at least three important variations of intestinal DAO activity were observed:

1) The DAO activity decreased during the period of AOM injections. This reduction of enzymic activity was probably an expression of regulatory processes in the mucosa, because AOM concentrations as used in the experiment showed no direct inhibition of the enzyme (5,6).

2) The increase in DAO activity detected soon after stopping injections of the carcinogen can be interpreted as an antiproliferative overshoot reaction if our hypothesis of a DAO antiproliferative function is correct. It was observed regularly in a number of experimental models after cessation of a proliferative stimulus (5,7,8), and may parallel the experimental biphasic response of ornithine decarboxylase in response to colon carcinogenesis (9).

3) Concerning the gradual decrease in DAO activity observed at the end of the experiment, one may speculate as follows:

a) 42 weeks after stopping the AMO injections, the mucosa may be returning to a normal state of regulation of mucosa proliferation. If intestinal DAO is involved in a feedback control mechanism for mucosal proliferation(Fig.4);

b) AOM-induced carcinogenesis is associated with inflammatory changes of the bowel wall. The decrease in mucosal DAO activity may therefore be due to the effect of this inflammatory reaction (10).

c) Lastly, the decrease in DAO activity at the end of the experiment may be the result of an overwhelming proliferative stimulus (Fig.4). Indeed, a similar breakdown of the enzymic activity has already been observed in the early phase of adaptive mucosal proliferation (8). Other evidence may be the reduction in DAO activity in intestinal tumors which behave as an autonomous proliferating tissue.

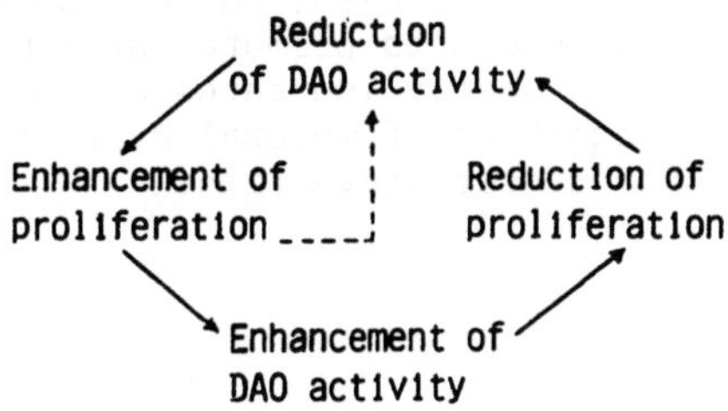

Figure 4. Postulated interrelationship between diamine oxidase (DAO) activity and mucosal proliferation. The continous line demonstrates a feedback control mechanism of mucosal growth regulation. A proliferative stimulus will produce an antiproliferative answer mediated by increased DAO activity. This will reduce mucosal proliferation followed by a decrease in DAO activity, and so on. A pathological (autonomous) or overwhelming proliferation which renders the mucosa out of control will induce a breakdown of the normal enzymatic control mechanism (broken line). All these events have been observed in a number of animal experiments using different models.

Aminoguanidine, a large bowel cancer promotor

A priori, AG seems to be a safe agent since this substance has already been used clinically for antihypertensives(11), for prevention of vascular damage in diabetic patients (12) and for treatment of medullary thyroid cancer (13).

In our experimental study, however, together with carcinogenic initiation of cells, AG acted as a cancer promotor in the large bowel. This effect may be due to a specific blockade of the function of DAO in the polyamine metabolism.

This is the first experimental model of large bowel tumor promotion, based on theoretical considerations being specifically related to a pathobiochemical mechanism like polyamine metabolism.

Changes in mucosal DAO activity of patients

In patients, actually cancer initiation may be impossible to be detected, but there are some methods to diagnose hyperproliferation, e.g. histology, autoradiography, flowcytometry and perhaps alterations in DAO activity. Therefore, mucosal hyperproliferation, being indeed a risk factor of tumor development, includes by its recognition the chance of cancer prevention.

The values of DAO activity found in large bowel cancer and in the surrounding mucosa confirmed our experimental result of a reduction of the enzymic activity in autonomous growing tissue and in impaired mucosa. In a more mild disease, like polyps, the mucosa still seems to be able to regulate proliferation by the activation of the antiproliferative principle DAO (biphasic response of DAO activity).

When we evaluated the DAO activity as an indicator of mucosal hyperproliferation we found a sensitivity of 75% but a specificity of only 42%. The question remains open whether the latter value is due to any, not cancer-related event in the mucosa or whether the DAO is a more sensitive marker of proliferative alterations of the mucosa than the histological picture. This suggestion finds some confirmation in our previous studies concerning appendicitis (10) and gastric intestinal metaplasia (14) where we found alterated DAO activities in tissue which was not yet affected histologically.

CONCLUSION

The DAO was suggested to have a protective and antiproliferative effect on the intestinal mucosa. Any event which reduces mucosal DAO activity should therefore be suspicious to promote large bowel cancer. Mucosal hyperproliferation probably indicates an increased risk of tumor development. However, recognition of mucosal hyperproliferation may initiate therapeutic manoeuvres for cancer prevention.

REFERENCES

1. J. Sattler, and W. Lorenz, Food inducted histaminosis as an epidemiological problem: plasma histamine elevation and haemodynamic alterations after oral histamine administration and blockade of diamine oxidase (DAO). Agents Actions 23: 361-365 (1988).
2. J. Kusche, W. Lorenz, C.-D. Stahlknecht, H. Richter, R. Hesterberg, A. Schmal, E. Hinterlang, D. Weber, and C. Ohmann, Intestinal diamine oxidase and histamine release in rabbit mesenteric ischemia, Gastroenterology 80: 980-987 (1981).
3. J. Kusche and W. Lorenz, Diamine oxidase, in: "Methods of enzymatic analysis", H.U. Bergmeyer ed., Verlag Chemie, Weinheim (1983).
4. A. P. Maskens, Mechanism of colorectal carcinogenesis in animal models: possible implications in cancer prevention, in: Precancerous lesions of the gastrointestinal tract, P. Sherlock, B.C.(1983).
5. J. Kusche, J.R. Izbicki, R. Menningen, A. Curt, and J. V. Parkin, The effect of experimental carcinogenesis on intestinal diamine oxidase, a polyamine deaminating enzyme, Cancer Prev.Detect.9:17-29(1986).
6. J. Kusche, R. Menningen, and L. Leisten, Early alterations of rat intestinal diamine oxidase activity by azoxymethane, an intestinal carcinogen. Agents Actions (in press) (1989).
7. R. Menningen, J. Kusche, L. Leisten, and K. Erpenbach, Diamine oxidase (DAO) activity and intestinal mucosa integrity: influence of suture techniques, Agents Actions 20: 277-280 (1987).
8. R. Menningen, J. Kusche, and K. Erpenbach, Adaptive response of the rat small bowel to 70% resection: is the intestinal diamine oxidase involved in mucosal growth regulation?, Biogenic Amines 5:55-68 (1987).
9. G. D. Luk, S.R. Hamilton, D.O'Ceallaigh, J. Smith, E. Natuzzi, and S. B. Baylin, Azoxymethane (AMO) induces a generalized biphasic increase in intestinal ornithine decarboxylase (ODC) during colon carcinogenesis, Gastroenterology 82: 1121 (1982).
10. R. Menningen, M. Günther, M. Bönninghoff, C.-D. Stahlknecht, I. Kubitza, and J. Kusche, Diamine oxidase as a marker of intestinal integrity in acute appendicitis, Agents Actions 18: 38-40 (1986).
11. L. Vio, and M. G. Mamolo, Potential antihypertensive agents IV-Aminoguanidine and hydrazinopyrinidine derivatives, Il Farmaco 38: 255-264 (1983).
12. M. Brownlec, H. Vlassara, A. Kooney, P. Ulrich, and A. Cerami, Aminoguanidine prevents diabetes induced arterial wall protein crosslinking, Science 232: 1629-1632 (1986).
13. S. B. Baylin, Z. Horakova, and M. A. Beaven, Increased food consumption and growth after treatment with aminoguanidine, Experientia 31: 562-562 (1975).
14. J. Kusche, W. Lorenz, C.-D. Stahlknecht, A. Friedrich, K. Boo, and G. Reichert, Diamine oxidase activity in gastric and duodenal mucosa of man and other mammals with special reference to the pyloric junction, Agents Actions 8: 366-371 (1978).

CLOSING REMARKS :

CHANGING PERSPECTIVES IN POLYAMINE RESEARCH

H.G. Williams-Ashman

Ben May Institute, and Departments of Biochemistry & Molecular Biology and Pharmacological & Physiological Sciences, University of Chicago, Chicago, IL 60637, USA

The extraordinary diversity of new observations and experiments that were presented during this Symposium attest to the vigor and creativity of contemporary research on polyamines. The ever-increasing influx into our field of highly talented young investigators is particularly encouraging. In the restricted time available it would be impossible to comment on most of the outstanding advances announced at this meeting. So I shall consider only very few topics, the choice of which is heavily biased by my personal scientific interests and experiences.

On several occasions over the last four days, I was flooded with recollections of my attendance, almost twenty years ago, at the first international congress ever devoted to the chemical physiology of polyamines. This was a conference organized and published by the New York Academy of Sciences in 1970. That Symposium differed from this present one in several respects. It was held in the Waldorf Astoria Hotel in Manhattan, the environs of which are not exactly as beautiful as Sorrento. At that time, the status of polyamine research was almost primitive in comparison with now. The population of investigators then working on polyamine biochemistry and its applications to medicine and agriculture was relatively tiny. Moreover, in 1970, the entire literature on polyamines was still sufficiently small (perhaps I should add, mercifully) that any fairly industrious and scholarly individual could become acquainted with the majority of it without excessive trouble. (Uriel Bachrach's cyclopedic treatise on Function of Naturally Occurring Polyamines, completed in 1972, had only 1099 reference citations!) Now I put it to you that if one attempted to digest the deluge of reports on polyamines published over the intervening period, it would quickly become evident that this would be a hopeless endeavor, even if one devoted onself to it full time and without ever going to sleep.

In New York in 1970, and likewise in Sorrento this week, I started to wonder how some of the most puzzling phenomena in polyamineology that seemed inexplicable might be comprehended more than a generation later in the future. Fortunately, I was rescued from dwelling too long on such notions by remembering a fantasy once related to me by my scientific mentor, Charles Huggins of the University of Chicago, which went as follows. Suppose you are working away in your office, and becoming more and more miserable as a result of failing to fit some contradictory experimental findings into the framework of any reasonable hypothesis. Suddenly, there is a knock on the door. You get up and open it, and there stands an angel smiling at you, while holding a large envelope. The angel tells you that the envelope contains a twenty-page document that summarizes all that will be understood fifty years later about the matters that bother you. Question : would you yield to the temptation to open the envelope and read what is inside it?. Presuming that the document is couched in terms of concepts accepted half a century later, maybe a wise answer is No, because one probably wouldn't be able to understand any of it. (As stated by Otto Warburg, Annu. Rev. Biochem., 33, 1, 1964) : "A scientist must have the courage to attack the great unsolved problems of his time. Solutions usually have to be forced by carrying out innumerable experiments without much critical hesitation".)

It is unquestionable that recent successful applications of molecular biological techniques (by investigators too numerous to mention) have thrown entirely new light on the structure and expression of genes that code for certain polyamine biosynthetic enzymes, and on the regulation of polyamine turnover in living cells. cDNA's cognate to the mRNA's for ODC and AdoMetDC in several species have now been cloned, as has the genomic DNA for ODC. The primary structures of active ODC and AdoMetDC molecules from diverse species have been deduced from appropriate nucleotide sequence data. It has become feasible to study the intracellular levels of the mRNA's corresponding to these enzymes in relation to variations in the physiological status of cells that contain them; the translation of these mRNA's in cell-free systems; amplification of ODC genes that may occur in cells under certain experimental circumstances; effects of transfecting cells with cDNA's corresponding to ODC and AdoMetDC messages that were attached to suitable plasmid vectors; the origin of the pyruvoyl prosthetic groups of AdoMetDC's from various sources and the nature of the corresponding proenzyme molecules; factors influencing expression of ODC and AdoMetDC genes at both transcriptional and translational levels; and so on and so forth. Furthermore, one can safely predict that similar approaches will, in the forseeable future, permit elucidation of the structure of the genes and mRNA's for all of the other enzymes involved in pathways for both de novo synthesis as well as interconversions of polyamines. This will open the way to experimental analysis of how these genes are expressed under varying physiological circumstances in vivo. One of the great current enigmas is the mechanistic basis of the fact that in higher animal tissues, ODC, AdoMetDC and polyamine N^1-acetyltransferase are in most instances present in such minuscule amounts in resting, non-malignant cells; have such short apparent intracellular half-lives; and can be induced to enormous extents in response to various regenerative, or hormonal and other chemical, stimuli. By contrast, the two polyamine synthases, and also the intracellular flavoprotein polyamine oxidase of the interconversion pathway, are usually present in non-proliferating cells in much higher abundance; have much longer biological half-lives; and are not easily inducible. In the latter contexts, we must face the reality that everything known today about intracellular degradation of all of these enzymes has not progressed beyond pure phenomenology. We still do not understand the mechanisms of turnover of polyamine synthesizing and interconverting enzymes. Even if the amino acid sequences of every one of these enzyme proteins may be uncovered fairly soon, so that one can search for PEST sequences or whatever which might provide clues as to sites of proteolytic attack, and even if much more is found out about potential roles of post-translational modifications of these enzymes in relation to their degration in vivo, it is doubtful that many surprises may not be in store for us as research on these difficult problems progresses.

Another important recent advance has been the synthesis by medicinal chemists of many types of drugs that specifically and potently inhibit various enzymes of polyamine metabolism. Compounds that act as enzyme-activated irreversible ("suicide") inhibitors have proved to be especially valuable as probes into the dynamics of intracellular polyamine turnover, and as potential chemotherapeutic agents. The laboratories of

Phillipe Bey and others at the Merrell-Dow Research Institutes, of James Coward, and of the Southern Research Institute in Birmingham, Alabama have been at the forefront of these spectacular research developments. Application of these enzyme inhibitors has provided many insights into the roles of putrescine, spermidine and spermine in the physiology of a wide variety of cells. And it seems likely that certain polyamine analogs that have been synthesized over the last few years (such as the fluorinated polyamines developed at Merrell Dow, as well as the alkylated polyamine derivatives prepared by Raymond Bergeron) may have comparably useful research applications. A major impetus to production of these drugs has been, of course, the eventuality that they may have therapeutic applications, particularly in cancerous and other proliferative diseases. DFMO and certain other selective inhibitors of polyamine biosynthetic enzymes can profoundly depress the growth of certain experimental animal tumors. As yet, however, DFMO and other suicide inhibitors of ODC have not induced long-lasting remissions, let alone cures, in patients with any types of malignant neoplasms. Although DFMO can effectively deplete various human and animal tumors of putrescine and spermidine, this drug has little effect on spermine levels; this is attributable in part to the fact that DFMO evokes huge increases in AdoMetDC activities and hence of cellular decarboxylated AdoMet contents, so that most of the spermidine present before application of the inhibitor is converted to spermine. Nevertheless, increasing evidence is accruing for selective toxic actions of DFMO that are of therapeutic benefit in other pathologic situations. The truly curative actions of DFMO on pneumonias caused by Pneumocystis carinii in AIDS patients, on African sleeping sickness in cattle and humans, and on T. b. brucei trypanosome infections in mice, are almost miraculous. In this context of selective toxicity, an intruiging report presented at this Symposium by Arthur Galston is worth noting: pathological lesions on leaves of a higher plant induced by certain rust-causing fungi were prevented by administration of DFMO, in situations where the plant host cells could synthesize putrescine via pathways involving the actions of both ODC and arginine decarboxylase, whereas the fungal pathogens utilized only ODC as a means to generate putrescine.

Currently, more attention is being given to polyamines as building blocks for other types of biomolecules besides those derivatives that are the end-products of spermidine and spermine degradation produced via the actions of appropriate amine oxidases. That polyamines can be covalently attached to various chemical substances in cells suggests that such reactions may be of physiological utility, in addition to the numerous regulatory effects of spermidine and spermine that may relate to their non-covalent interactions with nucleic acids, acidic phospholipids, biomembranes and certain enzyme proteins. Alan Fairlamb presented new findings on the turnover of the reduced form of trypanothione [bis(glutathiontyl)spermidine] during the growth cycle of an insect trypanosomatid. Previously, Fairlamb demonstrated that trypanothione is apparently uniquely present in trypanosomatids and serves as a cofactor for the glutathione reductase of these parasites; he also found that spermidine depletion evoked by DFMO resulted in marked decline in the trypanothione content of certain trypanosomatids, which may have chemotherapeutic implications. It is worth remembering in this context that glutathionylspermidine, discovered much earlier by Dubin and the two Tabors in E. coli, is also present in certain trypanosomatids. And indeed, it is well established that polyamines occur in certain plants or microorganisms as conjugates with sugars, steroids, peptides, and phospholipids, and as structural subunits in various antibiotics and plant alkaloids, as well as in the forms of amides of various organic acids in plants. It seems safe to predict that many more polyamine conjugates of fairly low molecular weights will be discovered in due course, and that some of these substances may well turn out to be functionally significant.

A related series of reactions in which polyamines serve as substrates for post-translational modifications of proteins have recently been uncovered. Polyamines are excellent substrates for certain types of transglutaminases, which catalyze the formation of N-mono-γ-glutamyl or N,N-bis(γ-glutamyl) polyamine adducts of selected glutaminyl residues in proteins that are amine acceptor (a.k.a. acyl donor) substrates for this class of enzymes. When putrescine, spermidine or spermine react in transglutaminase-mediated reactions to form N-mono-glutamyl-polyamine adducts of appropriate amine acceptor protein substrates, and the remaining primary amine groups on the latter adducts then further react with a glutaminyl residue on another substrate polypeptide chain to form the N,N-bis(γ-glutamyl)polyamine derivatives, this results in the formation of polyamine-containing cross bridges between protein chains. So far, the only physiological (rather than in vitro model) system in which such polyamine cross-links between proteins has been demonstrated unequivocally is the biologically recondite situation of the formation of copulation plugs formed in the vaginal barrel after coitus in rats, where N,N-bis(γ-glutamyl)-spermidine and -spermine cross bridges clearly contribute alongside N^{ε}-(γ-glutamyl)lysine cross links (all produced by transglutaminase action) collectively contribute to the very insoluble coagulum of seminal plasma proteins. The latter reactions take place extracellularly. However, it is now evident that intracellular transglutaminases can also promote the synthesis of polyamine-containing cross links between proteins. In his lecture during this Symposium, Simone Beninati reported that bis(γ-glutamyl)spermidine is present together with N^{ε}-(γ-glutamyl)lysine in exhaustive proteolytic digests of cell envelopes from the epidermis of normal human skin. Even more interestingly, the relative levels of bis(γ-glutamyl)spermidine cross bridges are much larger than those due to the γ-glutamyl-ε-lysine isodipeptide in epidermal cell envelopes from the flaky skin lesions in patients with psoriasis. This important observation may well spark better insight into the pathogenesis of psoriasis, and might even suggest novel approaches to the therapy of this disease. It also appears from the studies of Beninati, Piacentini and Autuori that N-mono-γ-glutamyl adducts of proteins (especially those containing γ-glutamylputrescine) are formed in certain cells, and one wonders whether these sorts of post-translational modifications may alter the biological properties of certain proteins.

It was studies in John Folk's laboratory on polyamines as transglutaminase substrates that led to the discovery by him and Myung Hee Park that spermidine can also act as a precursor for a quite different sort of post-translational modification of a protein. This relates to the biosynthesis of hypusine in peptide linkage in the eIF-4D initiation factor of eukaryotic protein biosynthesis. In her lecture, Myung Hee Park described how she has now worked out many details of the mechanism of enzymatic formation of hypusine in eIF-4D. This process entails transfer of the butylamine moiety of spermidine to a lysine residue on a precursor protein to form deoxyhypusine in peptide linkage, followed by a charge modification of the protein, and finally hydroxylation of the deoxyhypusine moiety to yield hypusine. Myung Hee Park has gone a long way in characterizing the enzymes and protein substrates of the hypusine biosynthetic pathway. Although the role of the eIF-4D initiation factor in eukaryotic protein biosynthesis remains to be completely clarified, Park's excellent work suggests that spermidine, by virtue as acting as a precursor for the hypusine present in eIF-4D, may contribute to the regulation of protein synthesis by additional mechanisms besides the well known non-covalent interactions of this polyamine with polynucleotides that are involved in gene transcription and mRNA translation.

The major pathway for de novo biosynthesis of spermidine and spermine in animal cells produces 5'-deoxy-5'-methylthioadenosine (MTA) as a by-product. The discovery in 1969 of a specific MTA phosphorylase that splits MTA to adenine and methylthioribose-1-phosphate (MTR-1-P) eventually led several research groups to uncover an unsuspected pathway of general biochemical significance, viz., the conversion of MTR-1-P to methionine by a series of enzymatic steps involving the penultimate formation of 4-S-methylthioketobutyrate, which finally undergoes transamination to yield L-methionine. Taken together with the demonstration that the adenine liberated from MTA by MTA phosphorylase action is effectively converted to ATP, the conversion of MTR-1-P to methionine suggests that, in some cell types at least, there need be no net consumption of ATP and methionine during the extensive production of decarboxylated AdoMet that is required for the de novo synthesis of spermidine and spermine in higher animal tissues. Several presentations at this Symposium indicated the extensive advances in our understanding of the enzymology of MTA phosphorylase that have been accomplished very recently. Of medical interest is the virtually complete lack of MTA phosphorylase in some genetic variants of certain human and animal malignant tumors; Dennis Carson in his lecture discussed how this may be related in the human situation to defects of chromosome 9, on which the gene for MTA phosphorylase is located in our species. Another fascinating finding, presented by Marina Porcelli, is that the xylose analog of MTA accumulates in remarkably large amounts in the nudibranch mollusk Doris verrucosa; the available evidence hints that MTA is readily transformed into xylosyl-MTA in this organism.

Finally, let me again express the deep thanks of all of us to Professor Zappia, Dr. Della Ragione, and their other Italian colleagues on the Organizing Committee for doing so much to make this Symposium so memorable and rewarding.

CONTRIBUTORS

ABBRUZZESE, ALBERTO
Dipartimento di Biochimica e Biofisica
Università di Napoli
Via Costantinopoli, 16
80138 Napoli, ITALY

ABU-ELHEIGA, L.
Department of Molecular Biology
Hebrew University, Hadassah Medical School
P.O. Box 1172
Jerusalem, 91010 ISRAEL

AGOSTINELLI, ENZO
Dipartimento di Scienze Biochimiche
Centro di Biologia Molecolare del CNR
Università di Roma "La Sapienza"
00185 Roma, ITALY

ALHONEN, LEENA
Department of Biochemistry
University of Kuopio
P.O. Box 6
70211 Kuopio, FINLAND

ALISI, CHIARA
Dipartimento di Biologia Vegetale
Università degli Studi di Roma "La Sapienza"
Piazzale Aldo Moro, 5
00185 Roma, ITALY

ALTMAN, ARIE
Otto Warburg Ctr. Biotechnol. in Agriculture
The Hebrew University of Jerusalem
P.O. Box 12
Rehovot, 76-100 ISRAEL

ALVES, FRAUKE
Division of Gastroenterology & Endocrinology
Department of Medicine
Georg-August-University
3400 Göttingen, FRG

ANDERSSON, LEIF C.
Department of Pathology
University of Helsinki
00380 Helsinki
FINLAND

ANDREANA, AUGUSTO
Dipartimento di Terapia Medica
I Facoltà di Medicina
Università di Napoli
80134 Napoli, ITALY

ANGELINI, RICCARDO
Dipartimento di Biologia Vegetale
Università degli Studi di Roma "La Sapienza"
Piazzale Aldo Moro, 5
00185 Roma, ITALY

ANTON, DAVID L.
Central Research and Development
E.I. du Pont de Nemours Co.
Wilmington, DE
19898 USA

ASSARAF, Y.G.
Department of Molecular Biology
Hebrew University, Hadassah Medical School
P.O. Box 1172
Jerusalem, 91010 ISRAEL

BACCHI, CYRUS J.
Haskins Laboratories
Pace University
41 Park Row
New York, NY 10038 USA

BACHRACH, URIEL
Department of Molecular Biology
Hebrew University, Hadassah Medical School
P.O. Box 1172
Jerusalem, 91010 ISRAEL

BAGNI, NELLO
Department of Biology
Institute of Botany
Via Irnerio, 42
40126 Bologna, ITALY

BARBIROLI, BRUNO
Cattedra di Biologia Molecolare
Università di Bologna
Via Irnerio, 48
46126 Bologna, ITALY

BARKER, JACQUELINE H.A.
Long Ashton Research Station
University of Bristol
Bristol, BS18 9AF
ENGLAND

BARNETT, GLENN R.
Department of Molecular Biology
and Biochemistry
University of California
Irvine, CA 92717 USA

BASILION, JAMES P.
Department of Pharmacology
University of Texas
Medical School at Houston
Houston, TX 77225 USA

BASU, HIRAK S.
Brain Tumor Research Center
Department of Neurological Surgery
University of California, San Francisco
San Francisco, CA 94143 USA

BENINATI, SIMONE
Lab. of Cellular Development and Oncology
NIDR, National Institutes of Health
Bldg. 30, Rm. 211
Bethesda, MD 20892 USA

BERGERON, RAYMOND J.
Department of Medicinal Chemistry
J. Hillis Miller Health Center
University of Florida
Gainesville, FL 32610 USA

BITONTI, ALAN J.
Merrell Dow Research Institute
2110 E. Galbraith Road
Cincinnati, OH
45215 USA

BOWLIN, TERRY L.
Merrell Dow Research Institute
2110 E. Galbraith Road
Cincinnati, OH
45215-6300 USA

BROSNAN, MARGARET E.
Department of Biochemistry
Memorial University of Newfoundland
St. Johns, Newfoundland
A1B 3X9 CANADA

CACCIAPUOTI, GIOVANNA
Istituto di Biochimica delle Macromolecole
I Facoltà di Medicina, Università di Napoli
Via Costantinopoli, 16
80138 Napoli, ITALY

CALDARERA, CLAUDIO M.
Dipartimento di Biochimica
Università di Bologna
Via Irnerio, 48
40126 Bologna, ITALY

CANDIANI, ROSSELLA
Istituto di Patologia Generale, Università,
Centro Ricerca in Patologia Cellulare, CNR
Via Mangiagalli, 31
20133 Milano, ITALY

CANELLAKIS, E.S.
Department of Pharmacology
Yale University School of Medicine
New Haven, CT
06510-8066 USA

CANELLAKIS, ZOE NAKOS
Department of Pharmacology
Yale University School of Medicine
P.O. Box 3333
New Haven, CT 06510-8066 USA

CARRERA, CARLOS J.
Department of Basic and Clinical Research
Research Institute of Scripps Clinic
10666 North Torrey Pines Road
La Jolla, CA 92037 USA

CARSON, DENNIS A.
Department of Basic and Clinical Research
Research Institute of Scripps Clinic
10666 North Torrey Pines Road
La Jolla, CA 92037 USA

CARTENI'-FARINA, MARIA
Istituto di Biochimica delle Macromolecole
I Università di Napoli
Via Costantinopoli, 16
80138 Napoli, ITALY

CASTAGNINI, PAOLO
Istituto di Chimica Biologica
Università di Padova
Via F. Marzolo, 3
35131 Padova, ITALY

CHATEL, M.
Groupe Recherche en Thérap.Anticancéreuse
Centre Hospitalier Universitaire
2, Ave. du Professeur Leon-Bernard
35043 Rennes Cedex, FRANCE

CHEN, H.J.
Department of Biological Sciences
and Chemistry
Northern Illinois University
DeKalb, IL 60115-2861 USA

CHIOCCA, ANTONIO
Departments of Pharmacology and Medicine
University of Texas
Medical School at Houston
Houston, TX 77225 USA

CIMINO, GUIDO
Istituto per la Chimica di Molecole di
Interesse Biologico, CNR
Via Toiano, 6
80072 Arco Felice (Napoli), ITALY

CLAVERIE, NICOLE
Merrell Dow Research Institute
16, rue d'Ankara
67084 Strasbourg Cedex
FRANCE

CLEFFMANN, ULRIKE
Division of Gastroenterology & Endocrinology
Department of Medicine
Georg-August-University
3400 Göttingen, FRG

CLO', CARLO
Dipartimento di Biochimica
Facoltà di Medicina e Chirurgia
Via Irnerio, 48
40126 Bologna, ITALY

COHEN, PNINA
Otto Warburg Ctr. Biotechnol. in Agriculture
The Hebrew University of Jerusalem
P.O. Box 12
Rehovot, 76-100 ISRAEL

COLEMAN, CATHERINE S.
Clinical Pharmacology Unit
Department of Medicine & Terapeutics
University of Aberdeen
Aberdeen AB9 2ZD, SCOTLAND, UK

COLOMBATTO, SEBASTIANO
Dipart. di Medicina e Oncologia Sperimentale
Università di Torino
Via M. Buonarroti, 27
10126 Torino, ITALY

CONA, ALESSANDRA
Dipartimento di Biologia Vegetale
Università degli Studi di Roma "La Sapienza"
Piazzale Aldo Moro, 5
00185 Roma, ITALY

COURTHEOUX, P.
Groupe de Recherche en Thérapeutique
Anticancéreuse
Centre Hospitalier Universitaire
Caen, FRANCE

COWARD, JAMES K.
Depts. of Chemistry and Medicinal Chemistry
University of Michigan
Ann Arbor, MI
48109-1055 USA

CREUTZFELDT, WERNER
Division of Gastroenterology & Endocrinology
Department of Medicine
Georg-August-University
3400 Göttingen, FRG

CROZAT, ANNE
The Population Council and
The Rockefeller University
1230 York Avenue
New York, NY 10021 USA

DARCEL, F.
Groupe Recherche en Thérap.Anticancéreuse
Centre Hospitalier Universitaire
2, Ave. du Professeur Leon-Bernard
35043 Rennes Cedex, FRANCE

DAVIES, PETER J.A.
Departments of Pharmacology and Medicine
University of Texas
Medical School at Houston
Houston, TX 77225 USA

DAVIS, ROWLAND H.
Department of Molecular Biology
and Biochemistry
University of California
Irvine, CA 92717 USA

DE ROSA, MARIO
Istituto di Biochimica delle Macromolecole
I Facoltà di Medicina, Università di Napoli
Via Costantinopoli, 16
80138 Napoli, ITALY

DELLA RAGIONE, FULVIO
Istituto di Biochimica delle Macromolecole
I Facoltà di Medicina, Università di Napoli
Via Costantinopoli, 16
80138 Napoli, ITALY

DESIDERIO, ALFONSINA
Istituto di Patologia Generale
Università di Milano
Via Mangiagalli, 31
20133 Milano, ITALY

EISENBERG, LEONARD
The Population Council and
The Rockefeller University
1230 York Avenue
New York, NY 10021 USA

ELORANTA, TERHO
Department of Biochemistry
University of Kuopio
P.O. Box 6
70211 Kuopio, FINLAND

ESPOSITO, CARLA
Dipartimento di Biochimica e Biofisica
I Università di Napoli
Via Costantinopoli, 16
80138 Napoli, ITALY

FAIRLAMB, ALAN H.
Department of Medical Protozoology
London School of Hygiene & Tropical Medicine
Keppel Street
London WC1E 7HT, ENGLAND

FASULO, LUISA
Dipart. di Medicina e Oncologia Sperimentale
Università di Torino
Via M. Buonarroti, 27
10126 Torino, ITALY

FEDERICO, RODOLFO
Dipartimento di Biologia Vegetale
Università degli Studi di Roma "La Sapienza"
Piazzale Aldo Moro, 5
00185 Roma, ITALY

FERIOLI, MARIA ELENA
Istituto di Patologia Generale, Università,
Centro Ricerca in Patologia Cellulare, CNR
Via Mangiagalli, 31
20133 Milano, ITALY

FERRO, ADOLPH J.
Epitope Inc.
15425 S.W. Koll Parkway
Beaverton, OR
97006 USA

FEUERSTEIN, BURT G.
Departments of Laboratory Medicine
and Pediatrics
University of California, San Francisco
San Francisco, CA 94143 USA

FIORETTI, MASSIMO
Istituto di Biochimica delle Macromolecole
I Facoltà di Medicina, Università di Napoli
Via Costantinopoli, 16
80138 Napoli, ITALY

FITCHEN, JOHN H.
Medical Research Service
Veterans Administration Medical Center
P.O. Box 1034
Portland, OR 97207 USA

FLAMIGNI, FLAVIO
Dipartimento di Biochimica
Università di Bologna
Via Irnerio, 48
40126 Bologna, ITALY

FOLK, J.E.
Lab. of Cellular Development and Oncology
NIDR, National Institutes of Health
Bldg. 30, Rm. 211
Bethesda, MD 20892 USA

FÖLSCH, ULRICH R.
Division of Gastroenterology & Endocrinology
Department of Medicine
Georg-August-University
3400 Göttingen, FRG

FULGOSI, BARBARA
Dipart. di Medicina e Oncologia Sperimentale
Università di Torino
Via M. Buonarroti, 27
10126 Torino, ITALY

GALSTON, ARTHUR W.
Department of Biology
Yale University
P.O. Box 666
New Haven, CT 06511 USA

GAMBACORTA, AGATA
Istituto per la Chimica di Molecole di
Interesse Biologico, CNR
Via Toiano, 6
80072 Arco Felice (Napoli), ITALY

GAVAGNIN, MARGHERITA
Istituto per la Chimica di Molecole di
Interesse Biologico, CNR
Via Toiano, 6
80072 Arco Felice (Napoli), ITALY

GERHART, FRITZ
Merrell Dow Research Institute
16, rue d'Ankara
67084 Strasbourg Cedex
FRANCE

GERNER, EUGENE W.
Department of Radiation Oncology
and Biochemistry, University of Arizona
Health Sciences Center
Tuscon, AZ 85724 USA

GIFFIN, BRUCE F.
Department of Biology
University of Dayton
300 College Park
Dayton, OH 45469-0001 USA

GOLENSER, J.
Department of Parasitology
Hebrew University, Hadassah Medical School
P.O. Box 1172
Jerusalem, 91010 ISRAEL

GONZALEZ-BOSCH, CARMEN
Instituto Investigaciones Citológicas de la
Caja de Ahorros de Valencia
Amadeo de Saboya, 4
46010 Valencia, SPAIN

GRAGNANIELLO, VINCENZO
Istituto di Biochimica delle Macromolecole
I Università di Napoli
Via Costantinopoli, 16
80138 Napoli, ITALY

GRILLO, MARIA ANGELICA
Dipart. di Medicina e Oncologia Sperimentale
Università di Torino
Via M. Buonarroti, 27
10126 Torino, ITALY

GRISOLÍA, SANTIAGO
Instituto Investigaciones Citológicas de la
Caja de Ahorros de Valencia
Amadeo de Saboya, 4
46010 Valencia, SPAIN

GUARNIERI, CARLO
Dipartimento di Biochimica
Università di Bologna
Via Irnerio, 48
40126 Bologna, ITALY

GUEGAN, Y.
Groupe Recherche en Thérap.Anticancéreuse
Centre Hospitalier Universitaire
2, Ave. du Professeur Leon-Bernard
35043 Rennes Cedex, FRANCE

GYNTHER, JUKKA
Department of Pharmaceutical Chemistry
University of Kuopio
70211 Kuopio
FINLAND

HAKIM, GABRIELE
Dipartimento di Biochimica
Università di Bologna
Via Irnerio, 48
40126 Bologna, ITALY

HAMASAKI, NOBUKO
Department of Life Science
Tokyo Institute of Technology
Nagatsuta, Midori-Ku
Yokohama 227, JAPAN

HAVOUIS, R.
Groupe Recherche en Thérap.Anticancéreuse
Centre Hospitalier Universitaire
2, Ave. du Professeur Leon-Bernard
35043 Rennes Cedex, FRANCE

HAYASHI, SHIN-ICHI
Department of Nutrition
The Jikei University School of Medicine
Minato-KU
Tokyo 105 JAPAN

HEBY, OLLE
Department of Zoophysiology
University of Lund
Helgonar 3
233 62 Lund, SWEDEN

HERNÁNDEZ-YAGO, JOSÉ
Instituto Investigaciones Citológicas de la
Caja de Ahorros de Valencia
Amadeo de Saboya, 4
46010 Valencia, SPAIN

HICKOK, NOREEN J.
The Population Council and
The Rockefeller University
1230 York Avenue
New York, NY 10021 USA

HICKS, M.F.
Department of Biological Sciences
and Chemistry
Northern Illinois University
DeKalb, IL 60115-2861 USA

HIRVONEN, ARI
Department of Biochemistry
University of Kuopio
P.O. Box 6
70211 Kuopio, FINLAND

HOLM, INGVAR
Department of Zoophysiology
University of Lund
Helgonar 3
233 62 Lund, SWEDEN

HÖLTTÄ, ERKKI
Department of Pathology
University of Helsinki
00380 Helsinki
FINLAND

HSIEH, J.T.
Department of Human Oncology
University of Wisconsin
Clinical Cancer Center
Madison, WI 53792 USA

HU, YU-WAN
Department of Biochemistry
Memorial University of Newfoundland
St. Johns, Newfoundland
A1B 3X9 CANADA

HUANG, S.C.
Department of Pharmacology
Yale University School of Medicine
New Haven, CT
06510-8066 USA

HYVÖNEN, TAPANI
Department of Biochemistry
University of Kuopio
P.O. Box 6
70211 Kuopio, FINLAND

IGARASHI, KAZUEI
Faculty of Pharmaceutical Sciences
Inohana Campus, Chiba University
1-8-1 Inohana
Chiba City, 280 JAPAN

ITO, KIYOSHI
Faculty of Pharmaceutical Sciences
Inohana Campus, Chiba University
1-8-1 Inohana
Chiba City, 280 JAPAN

JÄNNE, JUHANI
Department of Biochemistry
University of Kuopio
P.O. Box 6
70211 Kuopio, FINLAND

JÄNNE, OLLI A.
The Population Council and
The Rockefeller University
1230 York Avenue
New York, NY 10021 USA

JÄRVINEN, TOMI
Department of Pharmaceutical Chemistry
University of Kuopio
70211 Kuopio
FINLAND

JULKUNEN, MERVI
The Population Council and
The Rockefeller University
1230 York Avenue
New York, NY 10021 USA

KAJANDER, E. OLAVI
Department of Biochemistry
University of Kuopio
P.O. Box 6
70211 Kuopio, FINLAND

KALKKINEN, NISSE
Department of Gene Technology
University of Helsinki
Valimotie 7
00380 Helsinki, FINLAND

KALLIO, ARJA
Orion
Genetic Engineering Laboratory
Helsinki
FINLAND

KAMEJI, T.
Department of Nutrition
The Jikei University School of Medicine
Minato-KU
Tokyo 105 JAPAN

KANAMOTO, R.
Department of Nutrition
The Jikei University School of Medicine
Minato-KU
Tokyo 105 JAPAN

KASHIWAGI, KEIKO
Laboratory of Biochemical Pharmacology
NIDDKD, National Institutes of Health
Bldg. 8, Rm. 223
Bethesda, MD 20892 USA

KAUPPINEN, LEILA
Department of Biochemistry
University of Kuopio
P.O. Box 6
70211 Kuopio, FINLAND

KIERSZENBAUM, FELIPE
Cellular Immunology & Immunoparasitology
Department of Microbiology & Public Health
Michigan State University
East Lansing, MI 48824-1101 USA

KRAKAMP, BERNHARD
Ist Medical Clinic
University of Cologne
5000 Köln-Merheim
GERMANY

KUBOTA, MASARU
Department of Basic and Clinical Research
Research Institute of Scripps Clinic
10666 North Torrey Pines Road
La Jolla, CA 92037 USA

KUĆAN, ŽELJKO
Department of Chemistry, Faculty of Science
University of Zagreb
Strossmayerov trg, 14
41000 Zagreb, YUGOSLAVIA

KULOMAA, MARKKU
Department of Biomedical Sciences
University of Tampere
P.O. Box 607
Tampere, 33101 FINLAND

KUSCHE, JÜRGEN
Dr. Madaus GmbH & Co.
Ostmerheimer Str.198
5000 Köln-91
GERMANY

KUTNY, RUSTY
Central Research and Development
E.I. du Pont de Nemours Co.
Wilmington, DE
19898 USA

KYRIAKIDIS, D.A.
Laboratory of Biochemistry
Faculty of Chemistry
University of Thessaloniki
Thessaloniki, GREECE

LAPINJOKI, SEPPO
Department of Pharmaceutical Chemistry
University of Kuopio
70211 Kuopio
FINLAND

LEISTEN, LOTHAR
Department of Surgery
University of Cologne
5000 Köln-Merheim
GERMANY

LEVIN, NITSA
Otto Warburg Ctr. Biotechnol. in Agriculture
The Hebrew University of Jerusalem
P.O. Box 12
Rehovot, 76-100 ISRAEL

LIGUORI, V.
Dipartimento di Biochimica e Biofisica
Università di Napoli
Via Costantinopoli, 16
80138 Napoli, ITALY

LINNALA-KANKKUNEN, ANNIKKA
Department of Biochemistry
University of Kuopio
P.O. Box 6
70211 Kuopio, FINLAND

LORENZINI, ERMA CECILIA
Istituto di Patologia Generale, Università,
Centro Ricerca in Patologia Cellulare, CNR
Via Mangiagalli, 31
20133 Milano, ITALY

LÖSER, CHRISTIAN
Division of Gastroenterology & Endocrinology
Department of Medicine
Georg-August-University
3400 Göttingen, FRG

MALORNI, A.
Istituto per la Chimica di Molecole di
Interesse Biologico, CNR
Via Toiano, 6
80072 Arco Felice (Napoli), ITALY

MAMONT, PIERRE S.
Merrell Dow Research Institute
16, rue d'Ankara
67084 Strasbourg Cedex
FRANCE

MARCOTE, MARIA-JESÚS
Instituto Investigaciones Citológicas de la
Caja de Ahorros de Valencia
Amadeo de Saboya, 4
46010 Valencia, SPAIN

MARINO, GENNARO
Istituto di Chimica Organica e Biologica
I Università di Napoli
Via Mezzocannone, 16
80134 Napoli, ITALY

MARMIROLI, SANDRA
Dipartimento di Biochimica
Università di Bologna
Via Irnerio, 48
40126 Bologna, ITALY

MARTON, LAURENCE J.
Dept. of Laboratory Medicine, and Brain
Tumor Research Ctr, Dept. Neurolog. Surgery
University of California, San Francisco
San Francisco, CA 94143 USA

MATSUFUJI, S.
Department of Nutrition
The Jikei University School of Medicine
Minato-KU
Tokyo 105 JAPAN

McCANN, PETER P.
Merrell Dow Research Institute
2110 E. Galbraith Road
Cincinnati, OH
45215 USA

MEGGIO, FLAVIO
Dipartimento di Chimica Biologica
Università di Padova
Via F. Marzolo, 3
35131 Padova, ITALY

MELANITOU, EVIE
The Population Council and
The Rockefeller University
1230 York Avenue
New York, NY 10021 USA

MENNINGEN, RUDOLF
Department of Surgery
University of Cologne
5000 Köln-Merheim
GERMANY

METAFORA, SALVATORE
International Institute of Genetics
and Biophysics, CNR
Via Marconi, 10
80125 Napoli, ITALY

MEZZETTI, GABRIELE
Istituto di Chimica Biologica
Università di Modena
Via Campi, 287
41100 Modena, ITALY

MILLER, RICHARD L.
Wellcome Research Laboratories
3030 Cornwallis Road
Research Triangle Park
NC 27709 USA

MINOCHA, SUBHASH C.
Department of Botany and Plant Pathology
University of New Hampshire
Durham, NH
03824 USA

MIRALLES, VICENTE J.
Instituto Investigaciones Citológicas de la
Caja de Ahorros de Valencia
Amadeo de Saboya, 4
46010 Valencia, SPAIN

MITCHELL, John L.A.
Department of Biological Sciences
and Chemistry
Northern Illinois University
DeKalb, IL 60115-2861 USA

MONDOVI', BRUNO
Dipartimento di Scienze Biochimiche
Centro di Biologia Molecolare del CNR
Università di Roma "La Sapienza"
00185 Roma, ITALY

MONTI, MARIA GIUSEPPINA
Istituto di Chimica Biologica
Università di Modena
Via Campi, 287
41100 Modena, ITALY

MORRIS, DAVID R.
Department of Biochemistry
University of Washington
J405 Health Sciences Bldg., SJ-70
Seattle, WA 98195 USA

MORUZZI, MARIA STELLA
Istituto di Chimica Biologica
Università di Modena
Via Campi, 287
41100 Modena, ITALY

MOULINOUX, JACQUES-PHILIPPE
Groupe Recherche en Thérap.Anticancéreuse
Centre Hospitalier Universitaire
2, Ave. du Professeur Leon-Bernard
35043 Rennes Cedex, FRANCE

MULLER, SYLKE
Bernhard-Nocht-Institut für Tropenmedizin
Abteilung für Biochemie
Bernhard-Nocht-Strasse 74
2000 Hamburg-4, GERMANY

MURAKAMI, Y.
Department of Nutrition
The Jikei University School of Medicine
Minato-KU
Tokyo 105 JAPAN

MURPHEY, ROBERTA J.
Department of Radiation Oncology
and Biochemistry, University of Arizona
Health Sciences Center
Tuscon, AZ 85724 USA

MUSTELIN, TOMAS
Scripps Clinic and Research Institute
10666 North Torrey Pines Road
La Jolla, CA
92037 USA

NADEL, BARRY
Otto Warburg Ctr. Biotechnol. in Agriculture
The Hebrew University of Jerusalem
P.O. Box 12
Rehovot, 76-100 ISRAEL

NARANDA, TATJANA
Department of Chemistry, Faculty of Science
University of Zagreb
Strossmayerov trg, 14
41000 Zagreb, YUGOSLAVIA

NISHIYAMA, M.
Department of Nutrition
The Jikei University School of Medicine
Minato-KU
Tokyo 105 JAPAN

NOBORI, TSUTOMU
Department of Basic and Clinical Research
Research Institute of Scripps Clinic
10666 North Torrey Pines Road
La Jolla, CA 92037 USA

NÖTHIG-LASLO, VESNA
Rugjer Boskovic Institute
Bijenicka cesta 54
41000 Zagreb
YUGOSLAVIA

NOVELLO, MARIA CRISTINA
Istituto di Chimica Biologica
Università di Padova
Via F. Marzolo, 3
35131 Padova, ITALY

NUTTALL, MARK E.
Clinical Pharmacology Unit
Department of Medicine & Terapeutics
University of Aberdeen
Aberdeen AB9 2ZD, SCOTLAND, UK

OGASAWARA, TOSHIHIRO
Faculty of Pharmaceutical Sciences
Inohana Campus, Chiba University
1-8-1 Inohana
Chiba City, 280 JAPAN

OLIVA, ADRIANA
Istituto di Biochimica delle Macromolecole
I Facoltà di Medicina, Università di Napoli
Via Costantinopoli, 16
80138 Napoli, ITALY

OSHIMA, TAIRO
Department of Life Science
Tokyo Institute of Technology
Nagatsuta, Midori-Ku
Yokohama 227, JAPAN

PAJUNEN, ANTTI
Department of Biochemistry
University of Oulu
Linnanmaa
90570 Oulu, FINLAND

PALUMBO, ROSANNA
Istituto di Biochimica delle Macromolecole
I Facoltà di Medicina, Università di Napoli
Via Costantinopoli, 16
80138 Napoli, ITALY

PANAGIOTIDIS, C.A.
Department of Pharmacology
Yale University School of Medicine
New Haven, CT
06510-8066 USA

PARK, MYUNG HEE
Lab. of Oral Biology and Physiology
NIDR, National Institutes of Health
Bldg 30, Rm 211
Bethesda, MD 20892 USA

PEGG, ANTHONY E.
Department of Physiology
Milton S. Hershey Medical Center
Pensylvania State University
Hershey, PA 17033 USA

PERIN, ANTONIO
Istituto di Patologia Generale
Università di Milano
Via Mangiagalli, 31
20133 Milano, ITALY

PERSSON, LO
Department of Physiology
University of Lund
Helgonar 3
233 62 Lund, SWEDEN

PESSA, TIINA
Department of Pathology
University of Helsinki
00380 Helsinki
FINLAND

PICCININI, GIORGIO
Istituto di Chimica Biologica
Università di Modena
Via Campi, 287
41100 Modena, ITALY

PINNA, LORENZO A.
Dipartimento di Chimica Biologica
Università di Padova
Via F. Marzolo, 3
35131 Padova, ITALY

PISTOCCHI, ROSSELLA
Department of Biology
Institute of Botany
Via Irnerio, 42
40126 Bologna, ITALY

PLOHL, MIROSLAV
Rugjer Boskovic Institute
Bijenicka cesta 54
41000 Zagreb
YUGOSLAVIA

PODDAR, SUSMITA
Department of Pharmacology
University of Texas
Medical School at Houston
Houston, TX 77225 USA

PONG, R.C.
Department of Human Oncology
University of Wisconsin
Clinical Cancer Center
Madison, WI 53792 USA

PONTONI, GABRIELE
Istituto di Biochimica delle Macromolecole
I Università di Napoli
Via Costantinopoli, 16
80138 Napoli, ITALY

PORCELLI, MARINA
Istituto di Biochimica delle Macromolecole
I Facoltà di Medicina, Università di Napoli
Via Costantinopoli, 16
80138 Napoli, ITALY

PORTA, RAFFAELE
Dipartimento di Biochimica e Biofisica
I Università di Napoli
Via Costantinopoli, 16
80138 Napoli, ITALY

PORTER, CARL W.
Grace Cancer Drug Center
Roswell Park Memorial Institute
666 Elm Street
Buffalo, NY 14363 USA

PRAKASH, NELLIKUNJA J.
Merrell Dow Research Institute
2110 E. Galbraith Road
Cincinnati, OH
45215-6300 USA

PUCCI, PIERO
Istituto di Chimica Organica e Biologica
I Università di Napoli
Via Mezzocannone, 16
80134 Napoli, ITALY

QUEMENER, V.
Groupe Recherche en Thérap.Anticancéreuse
Centre Hospitalier Universitaire
2, Ave. du Professeur Leon-Bernard
35043 Rennes Cedex, FRANCE

RICCIO, PIERLUIGI
Dipartimento di Scienze Biochimiche
Centro di Biologia Molecolare del CNR
Università di Roma "La Sapienza"
00185 Roma, ITALY

RISCOE, MICHAEL K.
Medical Research Service
Veterans Administration Medical Center
P.O. Box 1034
Portland, OR 97207 USA

RISTOW, JANET L.
Department of Molecular Biology
and Biochemistry
University of California
Irvine, CA 92717 USA

ROMANO, TERESA
Istituto di Biochimica delle Macromolecole
I Facoltà di Medicina, Università di Napoli
Via Costantinopoli, 16
80138 Napoli, ITALY

RUGGIERO, GIUSEPPE
Dipartimento di Terapia Medica
I Facoltà di Medicina
Università di Napoli
80134 Napoli, ITALY

RUSSO, GIAN LUIGI
Istituto di Biochimica delle Macromolecole
I Facoltà di Medicina, Università di Napoli
Via Costantinopoli, 16
80138 Napoli, ITALY

SACCHI, PIETRO
Dipartimento di Biochimica
Facoltà di Medicina e Chirurgia
Via Irnerio, 48
40126 Bologna, ITALY

SAKAI, YUKO
Faculty of Pharmaceutical Sciences
Inohana Campus, Chiba University
1-8-1 Inohana
Chiba City, 280 JAPAN

SCALABRINO, GIUSEPPE
Istituto di Patologia Generale, Università,
Centro Ricerca in Patologia Cellulare, CNR
Via Mangiagalli, 31
20133 Milano, ITALY

SCHNEIDER, MADLYN
Otto Warburg Ctr. Biotechnol. in Agriculture
The Hebrew University of Jerusalem
P.O. Box 12
Rehovot, 76-100 ISRAEL

SEILER, NIKOLAUS
Merrell Dow Research Institute
16, rue d'Ankara
67084 Strasbourg Cedex
FRANCE

SESSA, ANGELA
Istituto di Patologia Generale
Università di Milano
Via Mangiagalli, 31
20133 Milano, ITALY

SILIPRANDI, DAGMAR
Istituto di Chimica Biologica
Università di Padova
Via F. Marzolo, 3
35131 Padova, ITALY

SILIPRANDI, NORIS
Istituto di Chimica Biologica
Università di Padova
Via F. Marzolo, 3
35131 Padova, ITALY

SINERVIRTA, RIITTA
Department of Biochemistry
University of Kuopio
P.O. Box 6
70211 Kuopio, FINLAND

SJOERDSMA, ALBERT
Merrell Dow Research Institute
2110 E. Galbraith Road
Cincinnati, OH
45215 USA

SMITH, STEVEN S.
Division of Molecular Surgery
City of Hope National Medical Center
Duarte, CA
91010 USA

SMITH, TERENCE A.
Long Ashton Research Station
University of Bristol
Bristol, BS18 9AF
ENGLAND

SODANO, GUIDO
Istituto di Chimica
Università della Basilicata
Potenza
ITALY

SPIRA, D.T.
Department of Parasitology
Hebrew University, Hadassah Medical School
P.O. Box 1172
Jerusalem, 91010 ISRAEL

STANLEY, BRUCE
Department of Pharmacology
Milton S. Hershey Medical Center
Pensylvania State University
Hershey, PA 17033 USA

STEIN, JOSEPH P.
Department of Medicine
University of Texas
Medical School at Houston
Houston, TX 77225 USA

STJERNBORG, LOUISE
Department of Zoophysiology
University of Lund
Helgonar 3
233 62 Lund, SWEDEN

SUNKARA, PRASAD S.
Merrell Dow Research Institute
2110 E. Galbraith Road
Cincinnati, OH
45215-6300 USA

TABOR, CELIA W.
Laboratory of Biochemical Pharmacology
NIDDKD, National Institutes of Health
Bldg. 8, Rm. 223
Bethesda, MD 20892 USA

TABOR, HERBERT
Laboratory of Biochemical Pharmacology
NIDDKD, National Institutes of Health
Bldg. 8, Rm. 223
Bethesda, MD 20892 USA

TADOLINI, BRUNA
Dipartimento di Biochimica
Università di Bologna
Via Irnerio, 48
40126 Bologna, ITALY

TANEJA, Sushil K.
Laboratory of Biochemical Pharmacology
NIDDKD, National Institutes of Health
Bldg. 8, Rm. 223
Bethesda, MD 20892 USA

TANTINI, BENEDETTA
Istituto di Citomorfologia
CNR
Via Irnerio, 48
40126 Bologna, ITALY

THÉRON, J.
Groupe de Recherche en Thérapeutique
Anticancéreuse
Centre Hospitalier Universitaire
Caen, FRANCE

TOME, MARGARET E.
Department of Radiation Oncology
and Biochemistry, University of Arizona
Health Sciences Center
Tuscon, AZ 85724 USA

TONINELLO, ANTONIO
Centro Studio Fisiologia Mitocondriale
Università di Padova
Via F. Marzolo, 3
35131 Padova, ITALY

TOORCHEN, DAVID P.
Wellcome Research Laboratories
3030 Cornwallis Road
Research Triangle Park
NC 27709 USA

TSIRKA, S.A.
Laboratory of Biochemistry
Faculty of Chemistry
University of Thessaloniki
Thessaloniki, GREECE

UTILI, RICCARDO
Dipartimento di Terapia Medica
I Facoltà di Medicina
Università di Napoli
80134 Napoli, ITALY

UZAWA, TAKETOSHI
Department of Life Science
Tokyo Institute of Technology
Nagatsuta, Midori-Ku
Yokohama 227, JAPAN

VERMA, AJIT K.
Department of Human Oncology
University of Wisconsin
Clinical Cancer Center
Madison, WI 53792 USA

WAHLFORS, JARMO
Department of Biochemistry
University of Kuopio
P.O. Box 6
70211 Kuopio, FINLAND

WALLACE, HEATHER M.
Clinical Pharmacology Unit
Department of Medicine & Terapeutics
University of Aberdeen
Aberdeen AB9 2ZD, SCOTLAND, UK

WALTER, ROLF D.
Bernhard-Nocht-Institut für Tropenmedizin
Abteilung für Biochemie
Bernhard-Nocht-Strasse 74
2000 Hamburg-4, GERMANY

WEINSTEIN, LEONARD H.
Boyce Thompson Institute for Plant Research
Cornell University
Tower Road
Ithaca, NY 14853-1801 USA

WEYGAND-DURAŠEVIĆ, IVANA
Department of Chemistry, Faculty of Science
University of Zagreb
Strossmayerov trg, 14
41000 Zagreb, YUGOSLAVIA

WHITE, MICHAEL W.
Department of Biochemistry
University of Washington
J405 Health Sciences Bldg., SJ-70
Seattle, WA 98195 USA

WILLIAMS-ASHMAN, H. GUY
Ben-May Institute and Departments of
Biochemistry & Molecular Biology
University of Chicago
Chicago, IL 60637 USA

WITTICH, R.-M.
Bernhard-Nocht-Institut für Tropenmedizin
Abteilung für Biochemie
Bernhard-Nocht-Strasse 74
2000 Hamburg-4, GERMANY

WOLFF, E.C.
National Institute of Dental Research
National Institutes of Health
Bldg. 30, Rm. 211
Bethesda, MD 20892 USA

XIE, Q.-W.
Laboratory of Biochemical Pharmacology
NIDDKD, National Institutes of Health
Bldg. 8, Rm. 223
Bethesda, MD 20892 USA

YAMANAKA, HISASHI
Department of Basic and Clinical Research
Research Institute of Scripps Clinic
10666 North Torrey Pines Road
La Jolla, CA 92037 USA

ZAPPIA, VINCENZO
Istituto di Biochimica delle Macromolecole
I Facoltà di Medicina, Università di Napoli
Via Costantinopoli, 16
80138 Napoli, ITALY

ZWOLSHEN, JOHN H.
Merrell Dow Research Institute
2110 E. Galbraith Road
Cincinnati, OH
45215-6300 USA

INDEX

MIX
Papier aus verantwortungsvollen Quellen
Paper from responsible sources
FSC® C105338

FSC
www.fsc.org

If you have any concerns about our products,
you can contact us on
ProductSafety@springernature.com

In case Publisher is established outside the EU,
the EU authorized representative is:
Springer Nature Customer Service Center GmbH
Europaplatz 3, 69115 Heidelberg, Germany

Printed by Libri Plureos GmbH
in Hamburg, Germany